Oxford Resources for IB
Diploma Programme

IN COOPERATION WITH

2024 EDITION

ENVIRONMENTAL SYSTEMS AND SOCIETIES

COURSE COMPANION

Jill Rutherford
Gillian Williams

OXFORD
UNIVERSITY PRESS

OXFORD
UNIVERSITY PRESS

Great Clarendon Street, Oxford, OX2 6DP, United Kingdom

Oxford University Press is a department of the University of Oxford. It furthers the University's objective of excellence in research, scholarship, and education by publishing worldwide. Oxford is a registered trade mark of Oxford University Press in the UK and in certain other countries.

British Library Cataloguing in Publication Data
Data available

9781382044011

10 9 8 7 6 5 4 3 2 1

Paper used in the production of this book is a natural, recyclable product made from wood grown in sustainable forests.

The manufacturing process conforms to the environmental regulations of the country of origin.

Printed in China by Shanghai Offset Printing Products Ltd

Authors' Acknowledgements

Writing this Course Companion would not have been possible without a team approach. Many IB Diploma Programme teachers have contributed in varying ways and the authors are most grateful to these busy people, living in different countries and biomes, ecosystems and environments. We believe that this team approach gives the book a truly international flavour, with case studies from many countries and different viewpoints. Thank you to all these teachers. The book would not have been written at all were it not for past students who, with goodwill, enthusiasm and interest, were willing to get hot or cold, wet and muddy for the purpose of collecting data and gaining understanding.
Any errors or omissions are entirely those of the authors and we welcome communication from you to point out where these are and to suggest improvements and updates.

Acknowledgements

The "In cooperation with IB" logo signifies the content in this textbook has been reviewed by the IB to ensure it fully aligns with current IB curriculum and offers high-quality guidance and support for IB teaching and learning.

The publisher wishes to thank the International Baccalaureate Organization for permission to reproduce their intellectual property.

The publisher would like to thank the following for permissions to use copyright material:

Cover: Brett Monroe Garner / Getty Images.

Photos: p2: cuellar / Getty Images; p9(t): Alexandros Michailidis / Shutterstock; p9(tm): Jeffrey Mayer / Alamy Stock Photo; p9(tb): Associated Press / Alamy Stock Photo; p9(b): Adam Seward / Alamy Stock Photo; p10(t): Tim Graham / Alamy Stock Photo; p10(b): Associated Press / Alamy Stock Photo; p11: Vladimir Melnik / Shutterstock; p12: anekoho / Shutterstock; p13(l): Associated Press / Alamy Stock Photo; p13(r): AP Photo / Volodymyr Repik / Alamy Stock Photo; p17(l): Jack Cook, Adam Nieman, Woods Hole Oceanographic Institution; p17(r): Science Photo Library; p20: Aleksander Bolbot / Shutterstock; p22: Manuela Durson / Shutterstock; p25: Tom Brakefield / Digital Vision / Getty Images; p27: Duven Diener / Shutterstock; p29: Janelle Lugge / Shutterstock; p46: MintArt / Shutterstock; p51: Kate Raworth and Christian Guthier (CC-BY-SA 4.0); p61: Russell Millner / Alamy Stock Photo; p70: Marco Bottigelli / Getty Images; p73: Olney Vasan / Stone Sub / Getty Images; p74(t): ssuaphotos / Shutterstock; p74(b): Videohotdogs / Shutterstock; p76(t): EdBockStock / Shutterstock; p76(b): Hugh Adams / Shutterstock; p77: Vereshchagin Dmitry / Shutterstock; p78: Zerbor / Shutterstock; p81(t): Fly Of Swallow Studio / Shutterstock; p81(b): Court of Justice of the European Union; p83(l): Jurgen Vogt / Shutterstock; p83(r): Sarnia / Shutterstock; p84: Itzchaz / Shutterstock; p87: WvdMPhotography / Shutterstock; p88: Mark Schwettmann / Shutterstock; p89(l): crystal51 / Shutterstock; p89(r): everst / Shutterstock; p91: canadastock / Shutterstock; p92: Holger Motzkau / Wikipedia / Wikimedia Commons (CC BY-SA 3.0); p94: MintArt / Shutterstock; p98(t): David Ashley / Shutterstock; p98(bl): mycteria / 123RF; p98(br): Pakhnyushchy / Shutterstock; p99: Andrzej Kubik / Shutterstock; p100: MidoSemsem / Shutterstock; p101(t): age Fotostock / easyFotostock; p101(b): Nhemz / Shutterstock; p102(t): CDC/ James Hicks / Public Health Image Library / Centers for Disease Control and Prevention; p102(b): James Gathany /

Public Health Image Library / Centers for Disease Control and Prevention; p104: Crispin la valiente / Getty Image; p106: Kelsey Green / Shutterstock; p108(tl): Mogens Trolle/Shutterstock; p108(tr): Federico Veronesi / Getty Images; p108(m): Jeff Rzepka / Shutterstock; p108(b): Federico Rostagno / Shutterstock; p110: Tim Graham / Alamy Stock Photo; p111(tl): Shutterstock; p111(tr): SIMON SHIM / Shutterstock; p111(b): Dr. Morley Read / Shutterstock; p112(tl): TTphoto / Shutterstock; p112(tm): oksmit / Shutterstock; p112(tr): Nigel Cattlin / Alamy Stock Photo; p112(bl): Stephen Lew / Shutterstock; p112(bm): Laszlo Csoma / Shutterstock; p112(br): D. Kucharski K. Kucharska / Shutterstock; p115(l): blickwinkel / Alamy Stock Photo; p115(r): imageBROKER.com GmbH & Co. KG / Alamy Stock Photo; p119: pashabo / Shutterstock; p120: NejroN / 123rf; p124(t): This_is_JiHun_Lee / Shutterstock; p124(b): Sabena Jane Blackbird / Alamy Stock Photo; p125(l): blickwinkel / Alamy Stock Photo; p125(r): Paul Hirst / Wikimedia Commons (CC BY-SA 2.5); p132(4i): Adrian Assalve / iStock; p132(4ii): Paul Reeves Photography; p132(4iii): Bildagentur Zoonar GmbH / Shutterstock; p132(4iv): Mark Medcalf / Shutterstock; p133: Mitrofanov Alexander/ Shutterstock; p149(l): tangyan / Shutterstock; p149(r): sirtravelalot / Shutterstock; p159: Harris, N. and Gibbs, D. (2021) / World Resources Institute; p160: Will Iredale / Shutterstock; p162(l): Sopotnicki / Shutterstock; p162(m): kajornyot / Shutterstock; p162(r): Joe Gough / Shutterstock; p164(tl): VASILY BOGOYAVLENSKY / Stringer / Getty Images; p164(tr): Modfos / Shutterstock; p164(b): John Bill / Shutterstock; p166: hddigital / Shutterstock; p168: Worachat Tokaew / Shutterstock; p169: Lano Lan / Shutterstock; p181: leungchopan / Shutterstock; p182: Leonardo Gonzalez / Shutterstock; p183: Greir / Shutterstock; p184: Florin Mihai / Shutterstock; p185: O. Alamany & E. Vicens / Getty Images; p186(t): Norbert Wu/Minden Pictures/Getty Images; p186(b): littlesam/Shutterstock; p199: Walter Bilotta / Shutterstock; p200: Evgeniyqw / Shutterstock; p208: NASA's Earth Observatory; p212: Holly Kuchera/Shutterstock; p214: Bobby Ware/Getty Images; p215: Adam Burton / Alamy Stock Photo; p221: Silvina Alvarez / Shutterstock; p225(t): twomeows / Getty Images; p225(b): BlueRingMedia / Shutterstock; p233: Dani Ber / Shutterstock; p236(t): photong/Shutterstock; p236(b): Mikko Suonio / Alamy Stock Photo; p244(l): kryzhov / Shutterstock; p244(m): KreativKolors / Shutterstock; p244(r): Marko Cermak / Shutterstock; p246(1A): Connormah / Wikimedia Commons (CC BY-SA 2.0); p246(1B): Wirestock Creators / Shutterstock; p246(1C): Jim Harper / Wikimedia Commons (CC BY-SA 2.5); p246(1D): Werner Layer/ mauritius images GmbH / Alamy Stock Photo; p246(3): Oregon State University; p246(4): Oregon State University; p247(5l): DJ Sudermann / Shutterstock; p247(5m): Jurgen Vogt / Shutterstock; p247(5r): Vu Hoang / Alamy Stock Photo; p247(6l): History and Art Collection / Alamy Stock Photo; p250: Paul Souders / Getty Images; p254(t): sekernas / 123RF; p254(b): Arthur van der Kooij / Shutterstock; p257: Alfredo Maiquez / Shutterstock; p262(l): uzuri / Shutterstock; p262(r): Eric Isselee / Shutterstock; p264: cs333 / Shutterstock; p265: Johan W. Elzenga / Shutterstock; p267(t): Steffen Foerster / 123RF Limited; p267(b): Eric Isselee / Shutterstock; p269: Fotos593 / Shutterstock; p275: Joseph Sohm / Shutterstock; p278: zebra0209 / Shutterstock; p279: kakteen / Shutterstock; p280: Toa55 / Shutterstock; p282: Tropper2000 / Shutterstock; p283: Bailey Schwarz / Shutterstock; p284: Norhayati / Shutterstock; p287(t): SJ Travel Photo and Video / Shutterstock; p287(b): ANDRIJA ILIC / Getty Images; p288(t): VVO / Shutterstock; p288(b): Historic Collection / Alamy Stock Photo; p290: David Steele / Shutterstock; p291: Nokuro / Shutterstock; p292: Jearu / Shutterstock; p293: Joyce Mar / Shutterstock; p294: Simon Bratt / Shutterstock; p295: Dave WATTS / Getty Images; p298: Rich Carey / Shutterstock; p299(t): 145 / Ocean/ Corbis; p299(b): Dr Morley Read / Shutterstock; p302: Uwe Bergwitz / Shutterstock; p308: Arvind Balaraman / Shutterstock; p310: MarcAndreLeTourneux / Shutterstock; p312: Khrystyna Hurelych / Shutterstock; p314(l): ravl / Shutterstock; p314(r): J Wheeler / Shutterstock; p316(t): James King-Holmes / Alamy Stock Photo p316(b): Corel Corporation 1994; p321: Landsat Project Science Office at NASA's Goddard Space Flight Center; p324: Berendje Fotografie / Shutterstock; p326: Roman Tiraspolsky / Alamy Stock Photo; p331: MintArt / Shutterstock; p334: - / Staff / Getty Images; p335: Nancie Lee / Shutterstock; p338: Marine Debris Protection / NOAA; p340: Tatsiana Volskaya / Getty Images; p347: Calvin Chan / Shutterstock; p348: Daniel J. Rao / Shutterstock; p350(t): Mikhail Ryazanov / Wikimedia Commons; p350(b): Maslov Dmitry / Shutterstock; p351: gary yim / Shutterstock; p352: Oxford University Press; p368: Taras Vyshnya / Shutterstock; p370: irabel8 / Shutterstock; p371: Hiromi Ito Ame / Shutterstock; p372: Luis Carlos Torres / Shutterstock; p373: photosil / Shutterstock; p374: Lano Lan / Shutterstock; p387: Beat Bieler / Shutterstock; p392: Papilio / Alamy Stock Photo; p394: BlueOrange Studio/Shutterstock; p397: Food and Agriculture Organization of the United Nations (FAO); p398: Iakov Kalinin / Shutterstock; p407: HotFlash / Shutterstock; p409: Tom Wang / Shutterstock; p412: Andrei Armiagov / Shutterstock; 421(l): Corel Corporation 1994; p421(r): Sean Steininger / Shutterstock; p427: Songquan Deng / Shutterstock; p428: NOAA / Woods Hole Sea Grant / Statista; p429: Shane Gross / Shutterstock; p430: Elime / Shutterstock; p431(l): Marti Bug Catcher / Shutterstock; p431(r): Jan Andries Van Franeker / Solent / Shutterstock; p432: Trong Nguyen / Shutterstock; p435: chloe7992 / Shutterstock; p437: Imaginechina Limited / Alamy Stock Photo; p439: EyeMark / Alamy Stock Photo;

continued on page 782

FSC MIX
Paper from responsible sources
FSC® C109093
www.fsc.org

Contents

What is Environmental Systems and Societies?

Welcome to the exciting world of IB Environmental Systems and Societies (ESS). Whether you are taking ESS as a certificate or as part of an IB Diploma programme, it is not easy, but it is rewarding. ESS helps you make sense of the state of the Earth now, and of the world in which we all live in.

ESS is an interdisciplinary course which meets requirements for either Group 3 and Group 4, or both. This course is offered at standard level (SL) and higher level (HL), and this book is designed for both SL and HL students. ESS helps you understand socio-environmental issues of today, the science that we know about the environment, and the social, economic, political and ethical contexts of environmental issues. This includes how decisions are shaped by different perspectives of the world, and the species within it. ESS will also help you to understand how we strive to live more sustainably, knowing that our use of natural resources is greater than the biosphere can currently provide.

ESS aims to empower and equip you to:

1. Develop an understanding of your own environmental impact, in the broader context of the impact of humanity on the Earth and its biosphere.

2. Develop knowledge of diverse perspectives to address issues of sustainability.

3. Engage and evaluate the tensions around environmental issues using critical thinking.

4. Develop a systems approach to provide a holistic lens for the exploration of environmental issues.

5. Be inspired to engage in environmental issues across local and global contexts.

The good news

Yes, there is good news about the environment although it is often overshadowed by less good news.

While anthropogenic carbon emissions are still significantly high, in 2022 they only rose globally by 1%, which is less than predicted and this may be over the peak. The USA and China are developing sources of renewable energy production very quickly. The USA produced 40% of its energy in 2022 from carbon-free resources. Renewable sources of energy are getting cheaper each year.

There are global agreements on many aspects of the environment, e.g. water pollution, air quality, biodiversity. In 2022, the UN Biodiversity Conference (COP 15) agreed a framework to address biodiversity loss, restore ecosystems and protect indigenous rights. The plan includes concrete measures to halt and reverse nature loss, including putting 30% of the planet and 30% of degraded ecosystems under protection by 2030.

Although you are part of a generation facing the often wasteful use of natural resources by generations before you, it is key to know that you have the power to shape the future, and make positive contributions to the environment. We hope that this book gives you ideas on what you can do to make a difference for the better.

What do you need to know before you start?

There is no specific prior learning that you must have to study ESS. You will develop the skills you need as you study the course.

What will you study?

Scan the book and you will see a wide range of topics from ecology, water, and food, to planning the cities of the future and waste management. The book begins with a foundation topic, and it is strongly suggested that you study this first, as it frames three concepts that will run throughout the course: perspectives, systems and sustainability. HL students study these concepts in more depth in Topics 2–8 and this is marked in the book by a red AHL (additional HL) banner down the side of the page. At HL there are also three lenses: environmental law, economics and ethics. These lenses give you perspectives and concepts from these disciplines which will deepen your understanding.

How is the course structured?

This book follows the topic order of the IB Environmental Systems and Societies guide.

Apart from Topic 1 (which should be studied first), the order in which the syllabus is arranged does not prescribe the order in which it is taught. It is up to individual teachers and students to decide on an arrangement that suits your circumstances. Figure 1 show how the inter-related strands of the subject can be integrated.

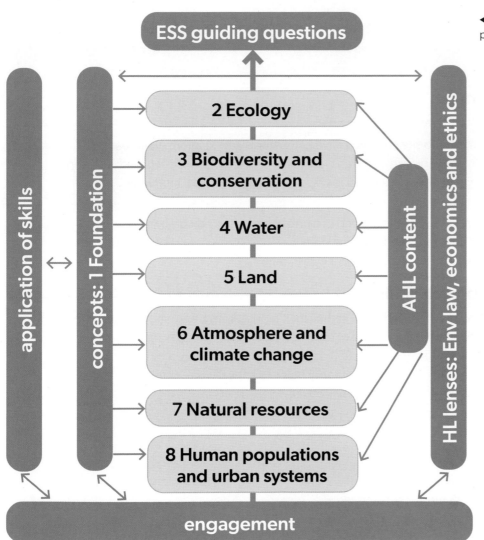

◀ **Figure 1** Relationships between parts of the ESS course

The ESS course content covers concepts which are universal and places these in contexts which vary over time and space and are global or local.

▶ **Figure 2** Relationships between concepts, contexts, content and engagement in ESS

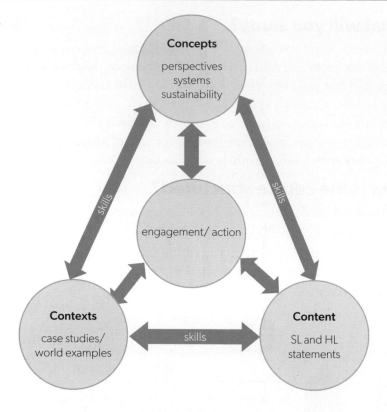

What about assessment?

There are both formative and summative assessments in ESS.

Formative assessments are used in class to see what you have understood and to inform your teacher. These can take many forms, e.g. informal quizzes, written class tests, preparatory work at home, learning practical skills, class debates, reflective journals and feedback, observations in class, interviews, self-assessments or in-class activities.

Summative assessments evaluate your learning at the end of a unit or a course and compare your performance against a standard or criteria and assess your learning. These include end-of-topic or end-of-term tests, the IB marked exams and IB internal assessment investigation.

Practical work and applying skills

This includes activities in the classroom, in the field and in the lab. It ranges from designing and carrying out surveys and questionnaires, online simulations, use of statistical tests, data research and selection to lab experiments and fieldwork. Many aspects of the ESS course can only be covered by going into ecosystems and investigating them. Wherever you are, it really helps to know a local ecosystem well and use it to provide a context for many of the concepts you will cover.

The authors

Jill Rutherford has taught IB Diploma courses for many years. She has administrative and board experience within schools in Hong Kong, the UK and other international schools. She has been a chief examiner for ESS and vice-chair of the IB examining board and led ESS workshops. She was academic director of Ibicus and founding Director of the IB Diploma at Oakham School, England. She holds two degrees from the University of Oxford.

Gillian Williams graduated from Reading University and has taught Environmental Systems, Geography and TOK on the international circuit since 1993. In her international career Gillian has held various leadership positions including Deputy Head, Head of Year and Head of Department. In 2011 she began advising on the IB Environmental Systems and Societies curriculum review.

Course book definition

The IB Diploma Programme course books are resource materials designed to support students throughout their two-year Diploma Programme course of study in a particular subject. They will help students gain an understanding of what is expected from the study of an IB Diploma Programme subject while presenting content in a way that illustrates the purpose and aims of the IB. They reflect the philosophy and approach of the IB and encourage a deep understanding of each subject by making connections to wider issues and providing opportunities for critical thinking.

The books mirror the IB philosophy of viewing the curriculum in terms of a whole-course approach; the use of a wide range of resources, international mindedness, the IB learner profile and the IB Diploma Programme core requirements, theory of knowledge, the extended essay, and creativity, activity, service (CAS).

Each book can be used in conjunction with other materials and, indeed, students of the IB are required and encouraged to draw conclusions from a variety of resources. Suggestions for additional and further reading are given in each book and suggestions for how to extend research are provided.

In addition, the course companions provide advice and guidance on the specific course assessment requirements and on academic honesty protocol. They are distinctive and authoritative without being prescriptive.

IB mission statement

The International Baccalaureate aims to develop inquiring, knowledgeable and caring young people who help to create a better and more peaceful world through intercultural understanding and respect.

To this end, the organization works with schools, governments and international organizations to develop challenging programmes of international education and rigorous assessment.

These programmes encourage students across the world to become active, compassionate and lifelong learners who understand that other people, with their differences, can also be right.

The IB Learner Profile

The aim of all IB programmes to develop internationally minded people who work to create a better and more peaceful world. The aim of the programme is to develop this person through ten learner attributes, as described below.

Inquirers: They develop their natural curiosity. They acquire the skills necessary to conduct inquiry and research and snow independence in learning. They actively enjoy learning and this love of learning will be sustained throughout their lives.

Knowledgeable: They explore concepts, ideas and issues that have local and global significance. In so doing, they acquire in-depth knowledge and develop understanding across a broad and balanced range of disciplines.

Thinkers: They exercise initiative in applying thinking skills critically and creatively to recognize and approach complex problems, and to make reasoned, ethical decisions.

Communicators: They understand and express ideas and information confidently and creatively in more than one language and in a variety of modes of communication. They work effectively and willingly in collaboration with others.

Principled: They act with integrity and honesty, with a strong sense of fairness, justice and respect for the dignity of the individual, groups and communities. They take responsibility for their own action and the consequences that accompany them.

Open-minded: They understand and appreciate their own cultures and personal histories, and are open to the perspectives, values and traditions of other individuals and communities. They are accustomed to seeking and evaluating a range of points of view, and are willing to grow from the experience.

Caring: They show empathy, compassion and respect towards the needs and feelings of others. They have a personal commitment to service, and to act to make a positive difference to the lives of others and to the environment.

Risk-takers: They approach unfamiliar situations and uncertainty with courage and forethought, and have the independence of spirit to explore new roles, ideas and strategies. They are brave and articulate in defending their beliefs.

Balanced: They understand the importance of intellectual, physical and emotional balance to achieve personal well-being for themselves and others.

Reflective: They give thoughtful consideration to their own learning and experience. They are able to assess and understand their strengths and limitations in order to support their learning and personal development.

A note on academic integrity

It is of vital importance to acknowledge and appropriately credit the owners of information when that information is used in your work. After all, owners of ideas (intellectual property) have property rights. To have an authentic piece of work, it must be based on your individual and original ideas with the work of others fully acknowledged. Therefore, all assignments, written or oral, completed for assessment must use your own language and expression. Where sources are used or referred to, whether in the form of direct quotation or paraphrase, such sources must be appropriately acknowledged.

How do I acknowledge the work of others?

The way that you acknowledge that you have used the ideas of other people is through the use of footnotes and bibliographies.

Footnotes (placed at the bottom of a page) or endnotes (placed at the end of a document) are to be provided when you quote or paraphrase from another document or closely summarize the information provided in another document. You do not need to provide a footnote for information that is part of a 'body of knowledge'. That is, definitions do not need to be footnoted as they are part of the assumed knowledge.

Bibliographies should include a formal list of the resources that you used in your work. 'Formal' means that you should use one of the several accepted forms of presentation. This usually involves separating the resources that you use into different categories (e.g. books, magazines, newspaper articles, Internet-based resources, CDs and works of art) and providing full information as to how a reader or viewer of your work can find the same information. A bibliography is compulsory in the Extended Essay.

What constitutes malpractice?

Malpractice is behaviour that results in, or may result in, you or any student gaining an unfair advantage in one or more assessment component. Malpractice includes plagiarism and collusion.

Plagiarism is defined as the representation of the ideas or work of another person as your own. The following are some of the ways to avoid plagiarism:

- words and ideas of another person to support one's arguments must be acknowledged

- passages that are quoted verbatim must be enclosed within quotation marks and acknowledged

- email messages, websites on the internet and any other electronic media must be treated in the same way as books and journals

- the sources of all photographs, maps, illustrations, computer programs, data, graphs, audio-visual and similar material must be acknowledged if they are not your own work

- when referring to works of art, whether music, film dance, theatre arts or visual arts and where the creative use of a part of a work takes place, the original artist must be acknowledged.

Collusion is defined as supporting malpractice by another student. This includes:

- allowing your work to be copied or submitted for assessment by another student

- duplicating work for different assessment components and/or diploma requirements.

Other forms of malpractice include any action that gives you an unfair advantage or affects the results of another student. Examples include, taking unauthorized material into an examination room, misconduct during an examination and falsifying a CAS record.

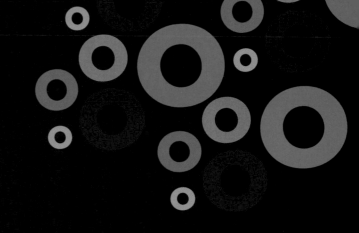

Take learning online with Kerboodle

What is Kerboodle?

Kerboodle is a digital learning platform that works alongside your print textbooks to create a supportive learning environment. Available for UK and international curricula, Kerboodle helps you save time and reinforces student learning with a range of supportive resources.

Use Kerboodle to:

- Enable learning anywhere with online and offline access to digital books
- Enhance student engagement with activities and auto-marked quizzes
- Boost performance and exam confidence with assessment materials
- Support independent learning with easy access across devices
- Deliver responsive teaching underpinned by in-depth reports
- Save time with tools to help you plan, teach, and monitor student progress
- Improve the classroom experience by highlighting specific content
- Get fast access with single sign-on via school Microsoft or Google accounts

Find out more and sign up for a free trial!

For the best teaching and learning experience use Kerboodle with your print resources!

For more information, visit:

www.oxfordsecondary.com/kerboodle

Need help?

Contact your local Educational Consultant: www.oxfordsecondary.com/contact-us

How to use this book

This ESS course companion is underpinned by a concept-based approach. These concepts are put into various contexts in case studies or skills activities, and you will add more examples of local and global interest.

There are various boxes and activities throughout the book.

Developing conceptual understanding

Guiding questions

At the start of every chapter, guiding questions are included to engage you with some of the questions that might arise as they study the material.

Understandings

These are content statements about what you will study. They are from the IB ESS guide.

Key terms

These introduce the definitions of important terminology used to explain environmental systems and societies.

Connections

These boxes link content with other parts of the course.

Theory of knowledge

This is an important part of the IB Diploma course. It focuses on critical thinking and understanding how we arrive at our knowledge of the world. The TOK features in this book are modelled on the TOK Exhibition and pose questions for you that highlight these issues.

AHL Sections marked as additional higher level (AHL) are required for HL students only.

Check your understanding

The statements and questions in this section serve as a checkpoint for you to see if you understood the material covered in the subtopic.

Applying and analysing

ATL Activity

These give you an opportunity to apply your ESS knowledge and approaches to learning (ATL) skills.

Data-based questions

Frequent examples of data-based questions have been included, both embedded within chapters as well as at the end of chapters. Many of these questions come from previous IB exams. Data-based questions teach the skills of data presentation, processing and analysis. In this syllabus, relatively more topic statements focus on presentation and mathematical analysis of data generated from experiments.

Case study

These are real life examples of global politics issues, outlining the context, the stakeholders involved and the impact on the world.

Practical

These can be carried out in the classroom, a laboratory or out in the field. They are an important part of the ESS course.

Exam-style questions

Use these questions at the end of each theme to draw together concepts from that chapter and other parts of the book, and to practise answering exam-style questions. Many of these are past IB biology exam questions.

A note on terminology used to describe countries

In this book, the terms "HIC", "MIC" and "LIC" are used to describe high-, middle- and lower-income countries.

You may have heard different terms in other subject lessons to describe the status of different countries, such as developed and developing countries, less economically developed, more economically developed, newly industrialized, emerging markets, BRIC or BRICS countries.

A note on soil classification terminology

Soil classification terminology depends where you live and which system you adopt. The US Department of Agriculture (USDA) taxonomy is the one used in this book. In it are 12 soil orders based on dominant properties of soils affected by their location. All the names end in -sol and ones mentioned in Topic 5 are mollisols, alfisols, gelisols and oxisols.

The FAO (Food and Agriculture Organization of the UN) has a system which includes podzols and chernozems. Various countries use different classifications often focused on soils in that country. You may also read about soils called laterite (high in iron oxide, acidic, reddish soils in the tropics), and black or brown earths (chernozems), which are high in organic matter and fertility, and very good for growing crops. But these are not terms in the USDA system of classification.

Topic 1

Foundation

> *You cannot get through a single day without having an impact on the world around you. What you do makes a difference, and you have to decide what kind of difference you want to make.*
>
> Jane Goodall (born 1934), *primatologist*

1.1 Perspectives

- How do different perspectives develop?
- How do perspectives affect the decisions we make concerning environmental issues?

Understandings

1. A perspective is how a particular situation is viewed and understood by an individual. It is based on a mix of personal and collective assumptions, values and beliefs.
2. Perspectives are informed and justified by sociocultural norms, scientific understandings, laws, religion, economic conditions, local and global events, and lived experience among other factors.
3. Values are qualities or principles that people feel have worth and importance in life.
4. The values that underpin our perspectives can be seen in our communication and actions with the wider community. The values held by organizations can be seen through advertisements, media, policies and actions.
5. Values surveys can be used to investigate the perspectives shown by a particular social group towards environmental issues.
6. Worldviews are the lenses shared by groups of people through which they perceive, make sense of and act within their environment. They shape people's values and perspectives through culture, philosophy, ideology, religion and politics.
7. An environmental value system is a model that shows the inputs affecting our perspectives and the outputs resulting from our perspectives.
8. Environmental perspectives (worldviews) can be classified into the broad categories of technocentric, anthropocentric and ecocentric.
9. Perspectives and the beliefs that underpin them change over time in all societies. They can be influenced by government or non-governmental organization (NGO) campaigns or through social and demographic change.
10. The development of the environmental movement has been influenced by individuals, literature, the media, major environmental disasters, international agreements, new technologies and scientific discoveries.

What is an environmental value system?

Key terms

A **perspective** is how a particular situation is viewed and understood by an individual.

An **argument** is a statement or statements made to support a personally held perspective or to counter a different one.

Values are qualities or principles that people feel have worth and importance in life. They may be individual or held by a group.

Worldviews are the lenses shared by groups of people through which they perceive, make sense of, and act within their environment.

TOK

Is the decision to shoot Harambe a value-free judgement?

To what extent is it ethical to put greater value on human life than that of animals?

Case study 1

Harambe the gorilla

In 2016, a three-year-old boy climbed into a gorilla enclosure in Cincinnati Zoo, USA. He crawled through bushes and into a shallow moat. Harambe, a lowland gorilla who had lived in zoos for 17 years, picked up the boy from the moat and carried him on to dry land. As onlookers screamed, Harambe became more agitated and a zoo worker, who feared for the boy's life in the hands of a 200 kg gorilla, shot the animal dead. The incident was filmed and is, of course, on YouTube.

The boy had minor injuries. His parents were investigated by police but not charged. His mother was targeted by harassment on social media. Several celebrities criticized the shooting. Jane Goodall, a well-respected primatologist, said that although Harambe appeared to be protecting the child, he was so powerful that even with the best intentions, a swipe of his arm could have killed the boy.

Questions

1. What is your **perspective** on the shooting of Harambe?

2. List the **arguments** for and against the shooting.

3. What **values** do you think the zookeeper who shot Harambe held?

4. What values do you think the people who harassed the mother held?

5. What **worldviews** do you think the zoo employees hold?

DAILY NEWS

Gorilla Killed at Cincinnati Zoo

Cincinnati Zoo's Dangerous Animal Response Team had to fatally shoot a gorilla named Harambe, after a toddler entered the enclosure. The decision has sparked widespread debate among critics and animal experts …

▲ Figure 1 Harambe's story was widely reported and discussed

The spectrum of environmental value systems

- Different societies hold different environmental philosophies and comparing these helps explain why societies make different choices.

- The **environmental value system** (EVS) each individual holds will be influenced by cultural, religious, economic and socio-political contexts.

- The environment or any organism can have its own intrinsic value regardless of its value to humans. How we measure this value is a key to understanding the value we place on our environment.

For much of history, our viewpoint has been that the Earth's resources are unlimited and that we can exploit them with no fear of them running out. And for much of history that has been true. The much smaller human population in the past was just one species among many. The words and phrases we use describe how we have seen the environment: "fighting for survival", "battle against nature", "man or beast", "conquering Everest", "beating the elements". It is only in very recent times that humans have been able to control our environment and even think about terraforming (altering conditions to make them habitable for humans) on Mars.

The Industrial Revolution of the 1800s in Europe and North America brought technological development. This meant humans were driven to explore, conquer and subdue the planet for industrial growth. This ideology has the worldview that economic growth improves the lot of all. But now it is clearer that the Earth's resources are finite because the Earth is finite. Humans may be the first species to change the conditions on Earth and so make it unfit for human life.

Key term

An **environmental value system** (EVS) is a model that shows the inputs affecting our perspectives and the outputs resulting from our perspectives.

TOK

How does our perspective influence the language we use to describe our relationship with the Earth?

To what extent does evaluative language affect our perspective on the environment?

What is your environmental worldview?

You have a view of the world that is formed through your experiences of life—your background, culture, education and the society in which you live. This is your paradigm or worldview.

ATL Activity 1

ATL skills used: thinking, research, self-management

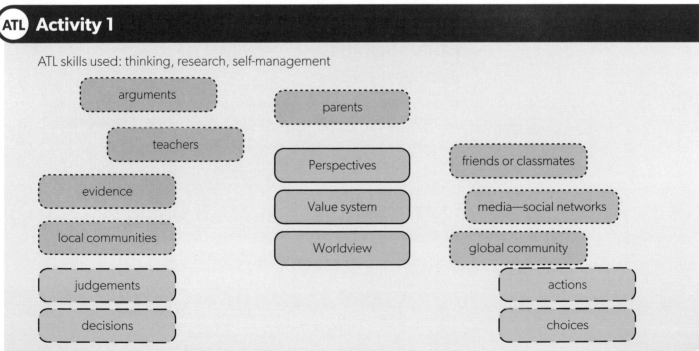

▲ Figure 2 Relationships and influences on your EVS

1. On a large sheet of paper, draw three rectangles and label them "Perspectives", "Value system" and "Worldview".

2. On sticky notes or pieces of card of one colour, write the influences on your environmental worldview. Some possible influences are shown in Figure 2 but you may also have others. Vary the text size to show the weighting and importance of that influence on you.

3. Now think of a local environmental issue that affects you. This could relate to pollution, loss of habitat, climate change or something else. On sticky notes or card of a different colour, write the decisions, judgements, choices and actions you would take on this issue.

4. Stick the sticky notes or cards from (2) and (3) on the sheet of paper and draw arrows to the relevant boxes.

5. Keep this either on the classroom wall or in your file so you can refer to it half-way through the course and at the end to see if any of your cards should change.

(ATL) Activity 2

ATL skills used: thinking, communication, social

Consider these words: environment, natural and nature.

1. Think about what they mean to you. Write down your responses.

2. Now discuss what you wrote with two of your classmates. Do you agree?

3. What have you written that is similar or different?

4. Why do you think your responses may be different?

5. How different do you think the responses of someone from a different century or culture may be? Discuss some examples.

6. Do you consider yourself to be optimistic or pessimistic about the environment (Figure 3)?

▲ Figure 3 Are you optimistic or pessimistic in outlook? Do you see the glass as half full or half empty?

One classification of different environmental philosophies

The ecocentric worldview:

- puts ecology and nature as central to humanity and emphasizes a less materialistic approach to life with greater self-sufficiency of societies

- gives the natural world pre-eminent importance and intrinsic value

- favours small scale, low-technology lifestyles with restraint in the use of all natural resources

- is life-centred—which respects the rights of nature and the dependence of humans on nature so has a holistic view of life which is Earth-centred.

Extreme ecocentrists are called deep ecologists.

The anthropocentric worldview:

- views humankind as being the central, most important element of existence

- splits into a wide variety of views

- believes humans must sustainably manage the global system, including through the use of taxes, environmental regulation and legislation

- is human-centred—in which humans are not dependent on nature but nature is there to benefit humankind.

The technocentric worldview:

- believes that technological developments can provide solutions to environmental problems

- assumes all environmental issues can be resolved through technology

- believes there can be unlimited economic growth.

ATL Activity 3

ATL skills used: thinking, communication, social, self-management

Get into groups of two or three in class.

Working individually, decide if you agree or disagree with these statements and think about your reasons. Then share your views with your group, giving your reasons.

1. I would like to drive a fast, powerful car.

2. I am concerned about how humans are destroying rainforests.

3. I do not care about climate change.

4. I believe that technology will fix any damage humans do to the environment.

5. I only drink bottled water.

Collate your answers and decide if you call yourself a technocentric, anthropocentric or ecocentric person, or perhaps a mix of these.

What does this tell you about (a) your perspectives, (b) your values and (c) your environmental value system?

What, if any, worldviews does your class share?

The environmental movement—a long time coming

Although the modern environmental movement originated in the 1960s, humans noticed their effect on their environment long before:

- The Romans noted problems such as air and water pollution.

- Between the late 14th century and the mid 16th century, pollution was associated with the spread of epidemic disease in Europe.

- Soil conservation was practised in China, India and Peru as early as 2,000 years ago.

Connections: This links with material throughout the course.

The global human population has been estimated at 50 million in 1,000 BCE, 200 million in the year 0 and 1 billion by 1800. Until the last 200 years, there seemed to be an infinite supply of natural resources for humans to use. Now however, there are over 8 billion humans on Earth. To understand modern environmentalism, we need to look at:

- events that have caused concern over their environmental impacts

- the responses to these impacts of individuals, groups of individuals, governments, businesses and the United Nations (UN).

A few influential individuals and independent pressure groups, with their use of media, have catalysed the movement, making it a peoples' or "grass roots" movement. But there has been a continuing divide in philosophy between those who think the reason to conserve nature is to continue to supply goods and services to humankind in a sustainable way (environmental managers) and those who believe conservation should be unconditional and for its spiritual value (deep ecologists). It becomes a simple question: Do we save nature for our sake or for its sake?

Who is involved?

We all are. The development of the environmental movement has been influenced by:

- individuals

- literature

- the media

- major environmental disasters

- international agreements

- new technologies

- scientific discoveries.

Until recently, most people in the world probably did not focus on environmental issues unless such issues were directly brought to their attention. But norms of behaviour (for example, reducing energy use, purchasing choices such as electric vehicles, dolphin-friendly tuna and recycling) and political choices (e.g. the successes of green parties, Extinction Rebellion, Greenpeace) have changed. We no longer think human activity in the environment cannot affect us—it already has.

There is general agreement that the modern environmental movement was catalysed by Rachel Carson's book, *Silent Spring*, published in 1962. She warned of the effects of pesticides on insects—both pests and beneficial insects. She also explained how this was being passed along the food chain to kill other animals including large birds of prey. What really gained people's attention was her belief that pesticides such as DDT (dichlorodiphenyltrichloroethane, a persistent, synthetic insecticide) were finding their way into animals and people, and accumulating in fatty tissues, causing higher risks of cancer. Chemical industries tried to ban the book, but many scientists shared her concerns. When an investigation ordered by US president John F. Kennedy confirmed her fears, DDT was banned.

Climate change is **the** issue of today. There is an increasing understanding of what causes it and what its effects are, but less action on changing our activities to combat it.

The overflowing bathtub analogy of climate change:

- If we run a bath to overflowing, we run in and turn off the taps. We don't stop and discuss what to do with others in the house. But the bath is still full.

- Analogously, we are pumping greenhouse gases into the atmosphere while we know they are changing the global climate. But we keep discussing what to do about it while still pumping out the gases.

Some environmental campaigners

Greta Thunberg (born 2003) is a Swedish environmental activist who challenges world leaders to act now to mitigate climate change. She started the Friday school strike for climate on her own and by 2019 there were over one million students involved in the movement "Fridays for Future". She has spoken at the UN, the UN Climate Change Conference (COP), the World Economic Forum in Davos, Glastonbury Festival in the UK and many other high-profile events. She created *The Climate Book* published in 2022. She has strong support and strong criticism but there is no doubt that the "Greta effect" has galvanized her peers to act, and perhaps also business and political leaders.

▲ Figure 4 Greta Thunberg at Glastonbury 2022

Wanjiku "Wawa" Gatheru (born 1999) is an environmental justice advocate helping to inspire a generation of environmentalists and tell the stories of those most adversely impacted by climate change. When she was 15, she stumbled into an environmental science class where she realized that climate change was an issue of justice. She is the founder of Black Girl Environmentalist, a Rhodes Scholar, and a climate communicator.

▲ Figure 5 Wawa Gatheru

Tokata Iron Eyes (born 2003/4) is Native American and a member of the Standing Rock Sioux tribe. Iron Eyes was environmentally active from age 9, opposing a uranium mine and in 2016 the Dakota Access Pipeline. This 1,886 km underground pipeline takes shale oil from the Bakken site across four states in the USA from North Dakota to Illinois. The route goes under Lake Oahe and across sacred sites of the Standing Rock Sioux tribe. Although environmental impact assessments (EIAs) (subtopic 7.1) were carried out and the pipeline route changed slightly, there was and is concern about freshwater pollution from pipeline rupture or leakage.

▲ Figure 6 Iron Eyes at a climate forum at the Lakota People's Law Project in 2019

Sir David Attenborough (born 1926) has had a very long career producing and narrating many TV series and many books on natural history and the natural world. During his early life, the Earth's biocapacity (the amount of ecological resources the Earth is able to generate that year) was far greater than humanity's ecological footprint (humanity's demand for that year). But since early in the 1970s, humanity has been using more natural resources annually than the Earth can produce. His recent programmes have focused on biodiversity loss, deforestation, plastics in the oceans and climate change and have had a large impact on educating people.

▲ Figure 7 Sir David Attenborough

▲ Figure 8 Dame Ellen MacArthur

▲ Figure 9 James Hansen

Dame Ellen MacArthur (born 1976) was the fastest solo round-the-world sailor in 2005. She took everything she needed on her trimaran and completed the voyage in 71 days. She realized that the world was no different from her boat. It has a finite supply of the resources we use and no more. We use and dispose of materials in a linear economy instead of reusing materials and eliminating waste in a circular economy (see subtopic 1.3). She founded the charity "The Ellen MacArthur Foundation" in 2010 to create a circular economy and this is having an impact on industry and global trade.

Hans Cosmas Ngoteya is a Tanzanian conservationist who promotes peaceful co-existence between wildlife and humans. He is co-founder of the Landscape and Conservation Mentors Organization. This organization focuses on promoting, supporting and improving community livelihoods through sustainable environmental practices.

He also created "Vijana na Mazingira" (VIMA) a youth-focused project that provides conservation education and alternative livelihood options in an effort to reduce pressure on natural resources.

James Hansen (born 1941) worked at NASA (National Aeronautics and Space Administration, USA) and for a long time studied the Earth's atmosphere, evaluating how anthropogenic (human-released) gases affect the Earth's climate. His global climate models help in understanding climate and how humans affect it. He is famous for his research work in climatology, and his 1988 testimony at the USA Congressional Committee on climate change. He calculated that global temperatures had risen by between 0.5°C and 0.7°C in the past century and are 0.8°C warmer than in the previous century.

Other influences on the environmental movement

Independent pressure groups often use awareness campaigns to influence the public and thus influence government and corporate business organizations. These independent groups are called non-governmental organizations (NGOs). There are many, both international and national. Examples are:

- Friends of the Earth (FoE)

- World Wide Fund for Nature (WWF)

- Global Footprint Network (GFN)

- Greenpeace

- Earthjustice.

They raise awareness of issues and run specific campaigns. Examples of NGO campaigns are:

- WWF campaigned for years to achieve the ban on the international commercial trade in ivory in 1989. This trade threatened the African elephant due to poaching. By 2017, China, the world's biggest ivory market, announced that all ivory sales within the country would be banned.

(ATL) Activity 4

ATL skills used: communication, research, self-management

Research three environmental campaigners where you are living and write brief notes on their impact.

- Earthjustice is suing the US government over the decision to go ahead with the Willow Project in 2023. This is in Western Arctic, Alaska where drilling for oil on public lands was given the go ahead. Five million people signed petitions to stop it. What has happened since then?

Corporate businesses are often involved because, in supplying consumer demand, they are using resources and creating environmental impact (for example, mining for minerals or burning fossil fuels). Many of these businesses are multinationals and transnational corporations (TNCs). Recent campaigns on reducing one-use plastics and increasing sustainability have influenced their activities.

Governments make policy decisions including environmental ones (for example, planning permission for land use). They also apply legislation (laws) to manage the country (for example, emissions controls over factories). They meet with other governments to consider international agreements (for example, the United Nations Environment Programme, UNEP). Different countries are at different stages of environmental awareness, as are different individuals. Legislating about emissions is important, but so is making sure there is enough food for the population. While different countries may put environmental awareness at different levels of priority, all are aware of the issues facing the Earth and all must be involved in finding solutions.

Inter-governmental bodies such as the UN have become highly influential by holding Earth Summits to bring together governments, NGOs and corporations to consider global environmental and world development issues.

▲ Figure 10 Rusty fuel and chemical drums on the Arctic coast

ATL Activity 5

ATL skills used: thinking, research

Books that have influenced environmental action include:

- Rachel Carson *Silent Spring* (1962)

- James Lovelock's books on the Gaia hypothesis

- Bill McKibben *The End of Nature* (1989)

- David Wallace-Wells *The Uninhabitable Earth* (2019).

Films include:

- Al Gore's documentary *An Inconvenient Truth* (2006)

- *No Impact Man* (2009)

- *Breaking Boundaries* (2021).

1. Research **one** of these books and **one** of these films or **one** other environmental book and film of local or global relevance, and write a short paragraph on each. Include:

   - what they covered and when

   - their main messages

   - what environmental worldview you think they hold

   - your perspective on their messages.

2. Research the impact and results of an environmental campaign run by an NGO in your local area.

TOK

Using a global environmental issue of your choice, evaluate how one of the ways of knowing influences our EVS approach.

To what extent have your emotions affected your response to this issue?

Case study 2

Oil company advertising

In 2007 Royal Dutch Shell, a giant oil company, launched an ad campaign with the slogan "Don't throw anything away. There is no away." The slogan is very true but the accompanying picture is of chimneys of an oil refinery emitting flowers. The text in some forms of the ad conveyed that if we had a magic bin, then we could make our rubbish disappear. The advert also suggests that we can actively find ways to recycle. For example, greenhouses use carbon dioxide to help us grow flowers, and our waste sulfur to make concrete. These are examples of energy solutions.

The body that oversees false advertising judged that the image was acceptable as it was conceptual and fanciful. But it also found that the advert was misleading because the wording implied a significant amount of the emissions from Shell oil refineries were recycled. Shell disputed the judgement. The advert was never reused.

▲ Figure 11 An oil refinery

Questions

1. Do you agree with the premise of the advert that there is no "away"?

2. Is it true that waste carbon dioxide can grow flowers?

3. Is it true that waste sulfur can make superstrong concrete?

4. Do you think Shell was mispresenting the environmental impact of its activities?

5. Do you think this advertisement could be used today?

This is an example of "greenwashing" or "green sheen" where marketing or advertising aims to persuade the public that the company does not harm the environment in making its products.

6. Look up earth.org/greenwashing-companies-corporations/. Do you use any of the products mentioned?

7. Does it matter to you that the companies are accused of greenwashing?

8. Find examples of the companies mentioned which are disputing the accusations.

9. Investigate earth.org a little more. It is a charity NGO run from a Hong Kong office. What values do you think it holds?

10. Do you think its information is accurate and factual? What is your evidence?

11. Find a response from a company (for example, BP, Nestlé) that it mentions as greenwashing. Does this response convince you?

BP is one of the oil giants, and was rebranded in 2002 from British Petroleum to Beyond Petroleum. The pledge then was to hold its carbon emissions constant and to be a planetary steward.

In 2006 a BP pipeline fractured off Alaska, causing one of the largest oil spills there.

In 2010 the Deepwater Horizon oil rig had an explosion that caused the largest marine oil spill ever.

The Beyond Petroleum branding was quietly dropped.

Now BP has these aims:

- By 2030 reduce its oil and gas production by 40%

- Increase investment in renewable energy via LightsourceBP to 50 gigawatts of energy produced by 2030

- Increase electric car charging points

- Undertake no new oil exploration in countries where it does not currently work

- Become net zero by 2050 and reduce emissions by 50% in 2030.

12. What does "net zero" mean for BP?

13. Has the perspective of the company changed?

14. What do you think may have changed it?

15. Find recent oil company adverts and compare these with older ones. How have the adverts changed?

In 2022–23, the energy crisis caused by the Ukraine–Russia conflict increased the price of gas and oil. The major global oil and gas companies then made very large profits because prices had risen. Some governments imposed a windfall tax (extra levy imposed by a government on unexpected profit) and used this to subsidize consumer energy bills.

16. Do you think imposing a windfall tax in this way was the right thing to do? Give reasons for your argument.

17. All oil and gas companies know their product is finite. Research what they are doing to diversify and write a short review of your findings.

Case study 3

Some major environmental disasters

Minamata disaster (1956)

Mercury is a heavy metal and is poisonous to animals. It affects the nervous system causing loss of vision, hearing and speech; and lack of coordination in arms and legs. Severe poisoning causes insanity or death. Mercury was used in the hat-making industry into the 20th century. Hat makers were known to often suffer mental illnesses although the source of such illnesses was unknown. This is the basis of the name of the "Mad Hatter" character in Lewis Carroll's *Alice in Wonderland* and the phrase "mad as a hatter".

▲ Figure 12 A woman holds a Minamata disease victim

The Chisso Corporation built a chemicals factory in Minamata, Japan and was very successful. One product was acetaldehyde, which requires the use of a mercury compound as a catalyst. Wastewater containing methylmercury from this process was released into Minamata Bay and bioaccumulated (see subtopic 2.2) in the food chain. People of Minamata traditionally ate a lot of shellfish and were poisoned by mercury. It took over 30 years to recognize the cause of their illnesses and compensation is still being given by the Chisso Corporation although the mercury release stopped in 1968.

Bhopal disaster (1984)

Bhopal is a city in the state of Madhya Pradesh in India. In the early hours of the morning of 3 December 1984, in the centre of the city, a Union Carbide pesticide plant released 40 tonnes of methyl isocyanate (MIC) gas. Half a million people were exposed to the gas and nearly 3,000 people died immediately. This is considered to be the world's worst industrial disaster.

Chernobyl (1986)

▲ Figure 13 The Chernobyl nuclear reactor plant after the explosion in 1986

In 1986, at Chernobyl, the worst nuclear disaster ever occurred. Chernobyl is a few miles north of Kiev, the capital of Ukraine (then part of the USSR). An explosion followed by a fire resulted in a level 7 event (the highest) in reactor number 4. The reactor vessel containing the uranium radioactive material split and exposed the graphite moderator to air, which caused it to catch fire. The reactor went into uncontrollable meltdown and a cloud of highly radioactive material drifted over much of the USSR and Europe, as far west as Wales and Scotland. Fission products from the radioactive cloud (for example, isotopes of caesium, strontium and iodine) have a long half-life and were accumulated in food chains. In 2009, there were still restrictions on selling sheep in some Welsh farms due to their levels of radiation. There is much debate about how many people have been affected by the radiation as long-term effects, such as cancers and deformities at birth, are difficult to link to one event. Thirty-one workers died of radiation sickness because they were exposed to high levels of radiation while trying to shut down the reactor. Some had a lethal dose of radiation within one minute of exposure. Estimates of later deaths vary but some sources state that about 1,000 extra cases of thyroid cancer and 4,000 other cancers were caused by the fall-out cloud. Other estimates state that one million people will have died as a result of the disaster.

The authorities of the day did not "announce" the disaster. The world was alerted to the disaster when monitoring in Sweden picked up fall-out on clothing of workers at one of their nuclear plants.

Even today, reactor 4 is still dangerous. It was encased in a concrete shell, but the other reactors continued to run until 2000. In 2017, a metal structure was built to surround reactor 4 to completely seal it.

An exclusion zone of 30 km was established around the reactor site and is still in force. There is still some contamination of the area but, with the exclusion of humans, the ecosystem has recovered, and biodiversity has increased.

Fukushima Daiichi (2011)

In 2011, there was another nuclear accident at the Fukushima Daiichi nuclear plant in Japan. An earthquake set off a tsunami which caused damage resulting in the meltdown of three reactors in the plant. The water flooding these became radioactive and will take many years to remove. Although the radiation leak was about 30% that of Chernobyl and radiation levels in the air were low, a third of a million people were evacuated from the densely populated area. Later reports showed the accident was caused by human error. The plant was not built to withstand a tsunami even though it was close to the sea in an earthquake zone. The plant is still not secured.

After the disaster, there were anti-nuclear demonstrations in other countries and Germany announced it was closing older reactors and phasing out nuclear power generation. France, Belgium and Switzerland all had public votes to reduce or stop nuclear power plants. In other countries, plans for nuclear plants were abandoned or reduced.

All these and other environmental disasters cause suffering or deaths and are reported in news stories. But each year there are many more deaths worldwide due to air pollution caused by burning fossil fuels or wood.

Questions

1. Two of these disasters are chemical releases and two are radioactivity releases. What do they have in common?

2. Chernobyl has become synonymous with the dangers of nuclear power. The green political lobby argued that all nuclear power generation should stop. But nuclear reactor accidents are very rare and safety levels ever higher as new plants are developed.

   The views of some people about the rights and wrongs of using nuclear power are not based on evidence but on emotions. The rare accidents in nuclear power plants (Five Mile Island, Chernobyl, Fukushima) have resulted in some countries banning nuclear power generation. But our need for more and more energy may mean it has to be used.

   a. To what extent do you think the arguments about nuclear power are based on emotion rather than reason?

   b. Do you agree or disagree with producing energy from nuclear power? Give your reasons.

   c. Look up cold fusion nuclear energy. Do you think this is a viable way of meeting human energy needs?

3. The *Exxon Valdez* oil tanker spilled 11 million gallons of oil off Alaska in 1989. The Deepwater Horizon oil well spill released 210 million gallons in the Gulf of Mexico in 2010.

   a. What were the short-term and longer-term environmental impacts of both spills?

   b. Who paid financial compensation and to whom?

   c. Are oil spills less likely now given what has been learned?

4. Research and find data about the number of human deaths per year caused by air pollution due to fossil fuel burning.

5. Discuss in class your views on strategies for energy generation.

(ATL) Activity 6

ATL skills used: thinking, research

Environmental disasters cause loss of life, illness and major damage to ecosystems but there is also good news. Societies can work together to reduce the likelihood of these accidents, improve regulations and agree treaties and international laws as well as advancing technology to benefit the environment and humanity.

Select **one** item from each of the three categories below to research. For each item you select:

- note its aims and achievements

- evaluate its results and successfulness.

Category 1: International agreements	Category 2: Technological developments	Category 3: Scientific discoveries
Rio Earth Summit of 1992	The Green Revolution	Toxicity of pesticides
Rio+20 of 2012	Genetically modified organisms (GMOs) for crops and livestock	Biodiversity discovery and loss
United Nations Framework Convention on Climate Change (UNFCCC)	Artificial fertilizers	Habitat degradation, e.g. deforestation, trawling
COP27 in 2022	Meat analogues from plants or fungi	
A more recent COP	Renewable energy generation	A local discovery which benefits ecosystems and societies
A local or global agreement on environmental protection	A technology from an indigenous culture that benefits ecosystems and societies	

▲ Table 1 Categories for research

Check your understanding

In this subtopic, you have covered:

- the difference between perspectives, values, worldviews and value systems

- what these terms mean

- different environmental value systems

- the development of the environmental movement with examples.

How do different perspectives develop?

1. Discuss what influences your own perspectives.

2. Suggest which are the stronger influences.

How do perspectives affect the decisions we make concerning environmental issues?

3. If your perspectives lead you to a more technocentric or anthropocentric EVS, explain how you would change your lifestyle and in what way.

4. If you are led to a more ecocentric EVS, explain how you would change your lifestyle and in what way.

❯❯ Taking it further

- Debate or discuss your own perspectives and how they might influence your behaviour.

- Check how others in your class differ.

- Design appropriately persuasive materials to support an environmental or social cause that you care about.

- Be an example to show how personal actions can create change towards a more sustainable society.

- Investigate the role of politics, intergovernmental organizations (IGOs), non-governmental organizations (NGOs) and individuals (through social media) in solving an environmental issue.

- Take part in a Model United Nations (MUN) group on environmental issues.

1.2 Systems

- How can the systems approach be used to model environmental issues at different levels of complexity and scale?

Understandings

1. Systems are sets of interacting or interdependent components.
2. A systems approach is a holistic way of visualizing a complex set of interactions, and it can be applied to ecological or societal situations.
3. In system diagrams, storages are usually represented as rectangular boxes and flows as arrows, with the direction of each arrow indicating the direction of each flow.
4. Flows are processes that may be either transfers or transformations.
5. Systems can be open or closed.
6. The Earth is a single integrated system encompassing the biosphere, the hydrosphere, the cryosphere, the geosphere, the atmosphere and the anthroposphere.
7. The concept of a system can be applied at a range of scales.
8. Negative feedback loops occur when the output of a process inhibits or reverses the operation of the same process in such a way as to reduce change. They are stabilizing as they counteract deviation.
9. As an open system, an ecosystem will normally exist in a stable equilibrium, either in a steady-state equilibrium or in one developing over time (for example, succession), and will be maintained by stabilizing negative feedback loops.
10. Positive feedback loops occur when a disturbance leads to an amplification of that disturbance, destabilizing the system and driving it away from its equilibrium.
11. Positive feedback loops will tend to drive the system towards a tipping point.
12. Tipping points can exist within a system where a small alteration in one component can produce large overall changes, resulting in a shift in equilibrium.
13. A model is a simplified representation of reality; it can be used to understand how a system works and to predict how it will respond to change.
14. Simplification of a model involves approximation and, therefore, loss of accuracy.
15. Interactions between components in systems can generate emergent properties.
16. The resilience of a system, ecological or social, refers to its tendency to avoid tipping points and maintain stability.
17. Diversity and the size of storages within systems can contribute to their resilience and affect their speed of response to change (time lags).
18. Humans can affect the resilience of systems through reducing these storages and diversity.

What is a system?

There are a lot of key concepts involved in understanding **systems**, but it is important to understand them. After all, this course is called Environmental Systems and Societies.

You are made up of systems (e.g. digestive, nervous, skeletal and blood systems). You are also part of various systems (e.g. family, school, city, nation, the Earth) as well as your ecological, socio-economic, political and cultural systems.

Decisions about the environment are rarely based only on politics or science or economics. We may want to save the tigers, but we will be constrained by economic, societal and political systems which may alter our value systems.

Key term

A **system** is a set of inter-related parts working together to make a functioning whole.

The human place in the biosphere

The biosphere is all the parts of the Earth where life exists—all ecosystems. It includes living organisms, soil, water and air. Humans and all other organisms live within this thin layer, yet we know little about how it is regulated or self-regulates, or about the effects the human species is having on it.

The anthroposphere is the part of the Earth that is made or modified by humans —this includes cities, towns, roads, all our machines, mines, ports, energy networks, and land cultivated for crops and livestock. It reduces the biosphere because humans degrade land and damage habitats.

The other spheres are:

- air (atmosphere)

- rocks and soil (geosphere, which is composed of the lithosphere (rocks) and the pedosphere (soil))

- water (hydrosphere) and frozen water (cryosphere).

Now look at Figures 1 and 2. They show the comparative volumes of the Earth's water and air. Humans and all living things need both but there is not as much as you might think.

TOK

To what extent do these images change the way you interpret the world?

▲ Figure 2 All the air on Earth (atmosphere) compressed into a sphere. The atmosphere forms a thin layer around the Earth. Half the air is within 5 km of the surface and becomes less dense with increasing altitude

◀ Figure 1 The Earth's water (hydrosphere) compressed into a series of spheres in comparison to the size of the Earth. The largest sphere represents all the water (in oceans, ice caps, lakes, rivers, groundwater, atmospheric water and living things). The medium-sized sphere is all the Earth's liquid freshwater. The smallest sphere is water in rivers and lakes

Systems diagrams

- Inputs and outputs to a system may be matter or energy.

- Storages (stores) may be living or non-living.

- Flows are movements of matter or energy.

- Sizes of storage boxes represent the relative sizes of the storages.

Connections: This links to material covered in subtopics 2.4, 5.1 and 8.2.

ATL Activity 7

ATL skills used: thinking, communication

Using the model and key in Figures 3 and 4, draw your own systems diagram for the following systems:

- a candle
- a mobile phone
- a green plant
- a pond
- you
- your school.

Label the inputs, outputs, storages and flows.

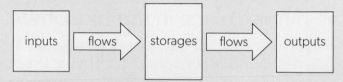

▲ Figure 3 A general system diagram

All systems have:	Represented by:		
Storages or stores of matter or energy	a box or circle		
Flows into, through and out of the system	arrows	⇒ or	→
Inputs	arrows in	⇒ or	→
Outputs	arrows out	⇐ or	←
Boundaries	lines	———	
Processes which transfer or transform energy or matter from storage to storage	e.g. respiration, precipitation, diffusion		

▲ Figure 4 Key to the general system diagram

ATL Activity 8

ATL skills used: thinking, research

Study Figure 5, a systems diagram of a deciduous forest ecosystem.

Herbivores are animals that eat plants; carnivores are animals that eat other animals. Decomposers and detritivores break down dead organic matter such as leaves (detritus) or dead animals.

1. List the storages.

2. List the flows.

3. Give three examples of plants that may be growing here.

4. Give two examples of herbivores.

5. Give two examples of carnivores.

6. Which input is brought in from outside the system?

7. How may it be brought in?

8. What flows out of the system?

9. What is the process by which it flows out?

10. What else leaves the system?

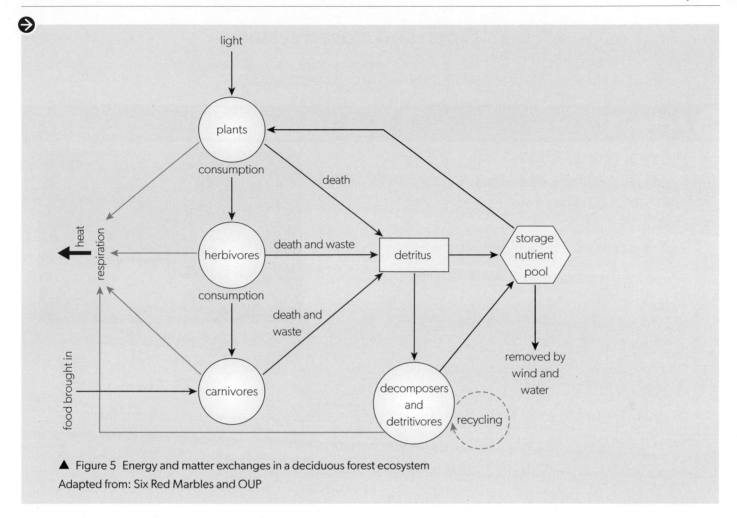

▲ Figure 5 Energy and matter exchanges in a deciduous forest ecosystem
Adapted from: Six Red Marbles and OUP

Transfers and transformations

Both matter (or material) and energy move or flow through ecosystems as transfers or transformations.

In transfers, the matter or energy moves from one place to another but does not change its state or chemical nature. Examples include:

- water moving from a river to the sea

- chemical energy in the form of sugars moving from a herbivore to a carnivore when a carnivore eats the herbivore

- energy in the form of heat moving in ocean currents.

In transformations, the matter or energy changes state (such as liquid to gas; light to chemical energy). Examples include:

- matter to matter (soluble glucose converted to insoluble starch in plants)

- energy to energy (light converted to heat by radiating surfaces)

- matter to energy (burning wood)

- energy to matter (photosynthesis).

Both types of flow (transfers and transformations) require energy. Transfers are simpler—they require less energy and are more efficient than transformations.

Connections: This links to material covered in subtopics 2.3, 4.1 and 5.1.

Open and closed systems

Most systems are **open systems**. All ecosystems are open systems exchanging matter and energy with their environment.

ATL Activity 9

ATL skills used: thinking, self-management

In the forest ecosystem, which of the following processes are transfers and which are transformations?

▲ Figure 6 A deciduous forest (deciduous trees shed their leaves each year)

1. Plants fix (capture) energy from light entering the system and use it in photosynthesis to make sugars from carbon dioxide and water.

2. Soil bacteria fix nitrogen from the air into nitrates.

3. Herbivores that live within the forest may graze in adjacent ecosystems such as a grassland but when they return, they enrich the soil with faeces.

4. Forest fires expose the topsoil, which may be removed by wind and rain.

5. Mineral nutrients are leached (washed) out of the soil and transported in groundwater to streams and rivers.

6. Water is lost through evaporation and transpiration from plants.

7. Heat is exchanged with the surrounding environment across the boundaries of the forest.

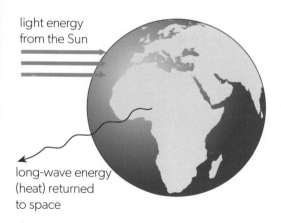

light energy from the Sun

long-wave energy (heat) returned to space

▲ Figure 7 A closed system—the Earth

Open system models can even be applied to the remotest oceanic island—energy and matter are exchanged with the atmosphere, surrounding oceans and even migratory birds.

A **closed system** exchanges energy but not matter with its environment.

System	Energy exchanged	Matter exchanged
Open	Yes	Yes
Closed	Yes	No

▲ Table 1 The difference between open and closed systems

Closed systems are extremely rare in nature. No natural closed systems exist on Earth but the Earth itself can be thought of as an "almost" closed system.

Light energy in large amounts enters the Earth's system and some is eventually returned to space as long-wave radiation (heat) (see Figure 7). Because a small amount of matter is exchanged between the Earth and space, it is not truly a closed system. **What types of matter can you think of that enter the Earth's atmosphere and what types leave it?**

Most examples of closed systems are artificial and are constructed for experimental purposes. An aquarium or terrarium may be sealed so that only energy in the form of light and heat but not matter can be exchanged. Bottle gardens or sealed terraria usually do not survive for long because the system becomes unbalanced (not in equilibrium). For example, there may not be enough oxygen or carbon dioxide for plants and animals, and the organisms die.

An example of a closed system that went wrong is Biosphere 2 (see Case study 4). An example of a closed system that is in equilibrium is a bottle garden that is sealed for decades with only light entering.

Equilibrium

Open systems tend to exist in a state of balance or **stable** equilibrium which avoids sudden changes. This does not mean that systems do not change, but that if change exists, it tends to exist between limits.

Examples of a steady-state equilibrium

- If a water tank fills at the same rate that it empties, there is no net change but the water flows in and out. It is in a steady state.

- In economics, a market may be stable but there are flows of capital in and out of the market.

- In ecology, a population of ants or any organism may stay the same size but individual organisms are born and die. If these birth and death rates are equal, there is no net change in population size.

- A mature climax ecosystem, like a forest, is in **steady-state** equilibrium with no long-term changes. It usually looks much the same for long periods of time, although all the trees and other organisms are growing, dying and being replaced by younger ones. There are flows in and out of the system— light inputs from the Sun; energy outputs as heat lost through respiration; matter inputs in rainwater and gases; matter outputs in salts lost in leaching and rain washing away the soil. Over years, the inputs and outputs balance.

- The maintenance of a constant body temperature is another example. We sweat to cool ourselves and shiver to warm up, but our core body temperature is about 37°C.

Maintenance of a steady-state equilibrium is achieved through negative feedback mechanisms (see Figure 8).

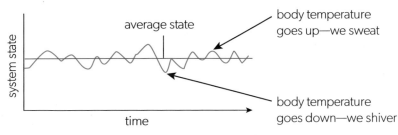

▲ Figure 8 **Steady-state equilibrium**

Unstable and stable equilibria

Systems can be stable or unstable.

In a stable equilibrium the system tends to return to the same equilibrium after a disturbance. In an unstable equilibrium the system returns to a new equilibrium after disturbance. The differences are shown in Figure 9.

Is the Earth's climate in an unstable equilibrium now?

(a)

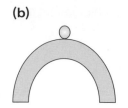

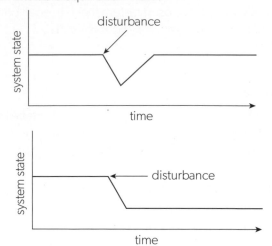

(b)

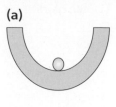

▶ Figure 9 Diagrams of (a) stable and (b) unstable equilibria

Case study 4

Biosphere 2

Biosphere 2 was a prototype space city, a human attempt to create a habitable closed system on Earth. Built in Arizona at the end of the 1980s, Biosphere 2 was a greenhouse covering three hectares and intended to explore the use of closed biospheres in space colonization. Two major "missions" were conducted but both ran into problems. The biosphere never managed to produce enough food to sustain the participants adequately and, at times, oxygen levels became dangerously low and needed augmenting. The inhabitants opened the windows and doors thus making it an open system.

▲ Figure 10 Biosphere 2

Inside Biosphere 2 were various ecosystems: a rainforest, coral reef, mangroves, savanna, desert, an agricultural area and living quarters. Electricity was generated from natural gas and the whole building was sealed off from the outside world.

For two years, eight people lived in Biosphere 2 in a first trial. But oxygen levels dropped from 21% to 14% and of the 25 small animal species put in, 19 became extinct, while ants, cockroaches and katydids thrived. Bananas grew well but there was not enough food to keep the eight people from being hungry. Oxygen levels gradually fell, and it is thought that soil microbes respired much of this. Carbon dioxide levels fluctuated widely. A second trial started in 1994 but closed after a month when two of the team vandalized the project by opening up windows and doors to the outside. Cooling the massive greenhouses was an issue, using three units of energy from air conditioners to cool the air for the input of every one unit of solar energy. So, there were social, biological and technological problems with the project as the team split into factions and questions were asked as to whether this was a scientific, business or artistic venture.

The result was to show how difficult it is to make a sustainable closed system when the complexities of the component ecosystems are not fully understood.

Search for Jane Poynter's TED talk on Biosphere 2 online and watch it.

Questions

1. Why do you think this project was called Biosphere 2?

2. Biosphere 2 has been described as a "closed system". What does this mean?

3. Was it truly a closed system?

4. What type of equilibrium was reached in Biosphere 2?

5. Discuss what was learned from the Biosphere 2 experiments.

6. What does this suggest about humans' ability to restore ecosystems once they have been destroyed?

Feedback loops

Systems are continually affected by information from outside and inside the system. Here are two simple examples.

1. If you start to feel cold, you can either put on more clothes or turn the heating up. The sense of cold is the information, putting on clothes or turning up the heating is the reaction.

2. If you feel hungry, you have a choice of reactions as a result of processing this information: eat food or do not eat and feel more hungry.

Examples of negative feedback

1. Your body temperature starts to rise above 37°C because you are walking in the tropical sun and the air temperature is 45°C. The sensors in your skin detect that your surface temperature is rising so you start to sweat and go red as blood flow in the capillaries under your skin increases. Your body attempts to lose heat.

2. A thermostat in a central heating system is a device that can sense the temperature. It switches a heating system on when the temperature decreases to a predetermined level, and off when it rises to another warmer temperature. So a room, a body or a factory can be maintained within narrow limits of temperature.

3. Global temperature rises causing ice caps to melt. More water in the atmosphere means more clouds. More solar radiation is reflected by the clouds so global temperatures fall. (This is shown in Figure 11, but compare with Figure 13, which interprets the situation differently.)

> **Connections:** This links to material covered in subtopics 2.1, 2.5, 3.3, 4.4, 5.2, 6.2 and 7.3.

Key terms

A **feedback loop** is when information starts a reaction that may input more information which may start another reaction.

Negative feedback loops occur when the output of a process inhibits or reverses the operation of the same process in such a way as to reduce change. They stabilize the system and counteract deviation.

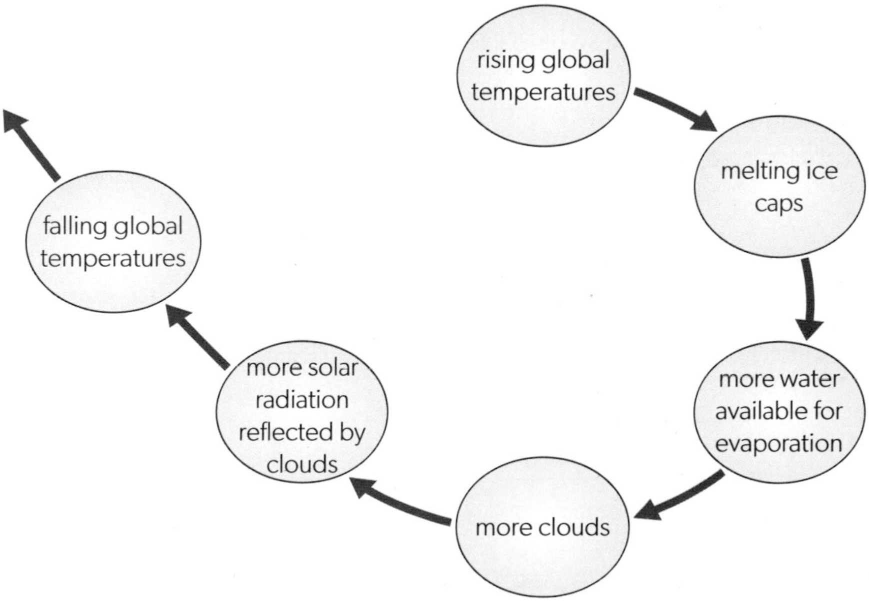

rising global temperatures → melting ice caps → more water available for evaporation → more clouds → more solar radiation reflected by clouds → falling global temperatures

▲ Figure 11 Negative feedback dampening change

4. Predator–prey interactions. The Lotka–Volterra model (proposed in 1925 and 1926) is also known as the predator–prey model and shows the effect of changing numbers of prey on predator numbers. When prey populations (e.g. mice) increase, there is more food for the predator (e.g. owls). The predators eat more and breed more, resulting in more predators which eat

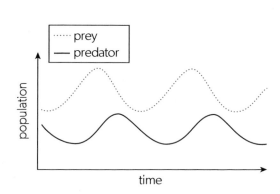

▲ Figure 12 Cycles of predator and prey in the Lotka–Volterra model

more prey so the prey numbers decrease. If there are fewer prey, there is less food, and the predator numbers decrease. The change in number of predators lags behind the change in prey numbers (Figure 12).

The snowshoe hare and Canadian lynx is a well-documented example of this (see Data-based questions).

5. Some organisms have internal feedback systems, physiological changes occurring that prevent breeding when population densities are high, promoting breeding when they are low. It is negative feedback loops such as these that maintain "the balance of nature".

Examples of positive feedback

1. You are lost on a high snowy mountain. When your body senses that it is cooling below 37°C, various mechanisms such as shivering attempt to raise your core body temperature again. But if these are insufficient to restore normal body temperature, your metabolic processes start to slow down, because the enzymes that control them do not work so well at lower temperatures. As a result, you become lethargic and sleepy and move around less and less, so your body cools even further. Unless you are rescued at this point, your body will reach a new equilibrium: you will die of hypothermia.

2. In some low-income countries (LICs), poverty causes illness and contributes to poor standards of education. In the absence of knowledge of family planning methods and hygiene, this contributes to population growth and illnesses, adding further to the causes of poverty. This is sometimes referred to as "a vicious circle of poverty".

3. Global temperature rises causing ice caps to melt. Dark soil is exposed so more solar radiation is absorbed. This reduces the **albedo** (reflecting ability of a surface) of the Earth so global temperature rises. Compare Figure 13 with Figure 11 and you can see that the same change can result in positive or negative feedback. This is one reason why predicting climate change is so difficult.

Key terms

Positive feedback loops occur when a disturbance leads to an amplification of that disturbance which destabilizes the system and drives it away from its equilibrium towards a tipping point.

A tipping point is the minimum change in a system that destabilizes it and shifts the regime to a new equilibrium or stable state.

Albedo is a measure of how much a surface reflects. The more reflective, the higher the albedo, measured from 0 to 1 or as a percentage.

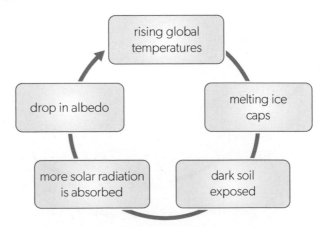

▲ Figure 13 **Positive feedback in global warming**

Both natural and human systems are regulated by feedback mechanisms. Generally, we wish to preserve the environment in its present state, so negative feedback is usually helpful and positive feedback is usually undesirable. However, there are situations where change is needed and positive feedback is advantageous. For example, if students enjoy their Environmental Systems and

Societies lessons, they want to learn more so they attend classes regularly and complete assignments. Consequently, they move to a new equilibrium of being better educated about the environment.

We shall come back to feedback loops in various sections of this book, particularly in those relating to climate change and sustainable development.

Data-based questions

Predator–prey interactions and negative feedback

The Hudson Bay Trading Company in Northern Canada kept very careful records of pelts (skins) brought in and sold by hunters over almost a century (Figure 15). This is a classic set of data and shows this relationship because the hare is the only prey of the lynx and the lynx is the only predator of the hare. Usually, things are more complicated.

It is assumed that the numbers of animals trapped was small compared to the total populations and that the numbers trapped were roughly proportional to total population numbers. Also, that the prey always has enough food so does not starve. Given that, the cycles are remarkably constant, with the lynx populations always smaller than and lagging behind the hare ones.

▲ Figure 14 Canadian lynx chasing snowshoe hare

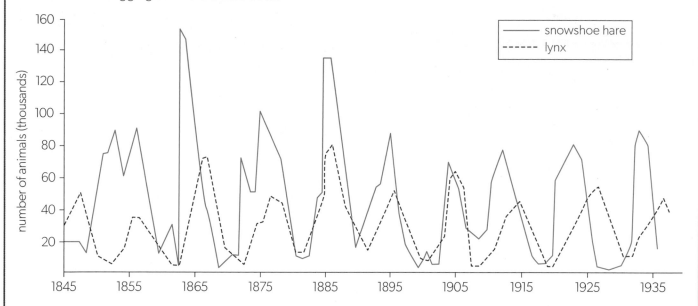

▲ Figure 15 Snowshoe hare and Canadian lynx population numbers from 1845 to 1940

1. On average, what was the cycle length of the lynx population?

2. On average, what was the cycle length of the hare population?

3. Why do lynx numbers lag behind hare numbers?

4. Why are lynx numbers smaller than hare numbers?

5. Things are never as straightforward in ecology as we expect, though. In regions where lynx died out, hare populations continued to fluctuate. Why do you think this was?

(ATL) Activity 10

ATL skills used: thinking, research

Examples of feedback

Below are seven examples of positive and negative feedback mechanisms that might operate in the physical environment. No one can be sure which of these effects is likely to be most influential, and consequently we cannot know if the Earth will manage to regulate its temperature despite human interference with many natural processes.

- Label each of the seven examples as either positive or negative feedback.

- From the same seven examples select one example of positive feedback and one of negative feedback. Draw diagrams to show how the feedback in your selections affects the system. Include feedback loops on your diagrams.

1. As carbon dioxide levels in the atmosphere rise:

   - the temperature of the Earth rises

   - as the Earth warms, the rate of photosynthesis in plants increases so more carbon dioxide is removed from the atmosphere by plants, reducing the greenhouse effect and reducing global temperatures.

2. As the Earth warms:

   - ice cover melts, exposing soil or water

   - albedo decreases (albedo is the fraction of light that is reflected by a body or surface)

   - more energy is absorbed by the Earth's surface

   - global temperature rises

   - more ice melts.

3. As the Earth warms:

   - upper layers of permafrost melt, producing waterlogged soil above frozen ground

   - methane gas is released in an anoxic environment

   - the greenhouse effect is enhanced

   - the Earth warms more, melting more permafrost.

4. As the Earth warms:

   - increased evaporation of water produces more clouds

   - clouds increase albedo, reflecting more light away from the Earth

   - temperature falls

   - rates of evaporation fall.

5. As the Earth warms:

   - organic matter in soil is decomposed faster so more carbon dioxide is released

   - enhanced greenhouse effect occurs

   - the Earth warms further

   - rates of decomposition increase.

6. As the Earth warms:

   - evaporation increases

   - snowfall at high latitudes increases

   - ice caps enlarge

   - more energy is reflected by increased albedo of ice cover

   - the Earth cools

   - rates of evaporation fall.

7. As the Earth warms:

   - polar ice caps melt releasing large numbers of icebergs into oceans

   - warm ocean currents such as the Gulf stream are disrupted by additional freshwater input into oceans

   - reduced transfer of energy to the poles reduces temperature at high latitudes

   - ice sheets reform and icebergs retreat

   - warm currents are re-established.

The Gaia hypothesis—a model of the Earth

In the 1970s, James Lovelock and Lynn Margulis put forward the Gaia hypothesis. They argued that the Earth and its biological systems act as a single entity which has self-regulating negative feedback loops to keep conditions on Earth within a range favourable to life. The name "Gaia" was used because it is the name of an Ancient Greek Earth goddess.

The argument is based on these facts:

- The temperature at the Earth's surface is constant even though the Sun is giving out 30% more energy than when the Earth was formed.

- The composition of the atmosphere is constant with 79% nitrogen, 21% oxygen and 0.03% carbon dioxide. Oxygen is a reactive gas and should be reacting, but it does not.

- The oceans' salinity is constant at about 3.4% but rivers washing salts into the seas should increase this.

- Despite volcanic eruptions and meteor impacts the Earth remains habitable.

After much criticism, James Lovelock and Andrew Watson developed a DaisyWorld as a mathematical simulation to show that feedback mechanisms can evolve from the activities of self-interested organisms—black and white daisies in this case.

In Lovelock's book *The Revenge of Gaia*, he makes a strong case for the Earth being an "older lady" now, more than halfway through her existence as a planet and so not being able to bounce back from changes as well as she used to. He suggests that we may be entering a phase of positive feedback when the previously stable equilibrium will become unstable—passing a tipping point—and we will shift to a new, hotter, equilibrium state. Controversially, he suggests that the human population would survive but with a 90% reduction in numbers.

Tipping points

Small changes occur in systems and may not make a huge difference. But when these changes tip the equilibrium over a threshold, known as a **tipping point** (Figure 17), the system may transform into a very different one. Then positive feedback loops drive the system to a new steady state.

Characteristics of tipping points

- They involve positive feedback which makes the change self-perpetuating (for example, deforestation reduces regional rainfall, which increases fire-risk, which causes forest dieback).

- There is a threshold beyond which a fast shift of ecological states occurs.

- The threshold point cannot be precisely predicted.

- The changes are long-lasting.

- The changes are hard to reverse.

- There is a significant time lag between the pressures driving the change and the appearance of impacts, creating great difficulties in ecological management.

ATL **Activity 11**

ATL skills used: research

Search on the web for a DaisyWorld or DaisyBall computer simulation and have a go at the game.

Key term

An ecological **tipping point** is reached when an ecosystem experiences a shift to a new state in which there are significant changes to its biodiversity and the services it provides.

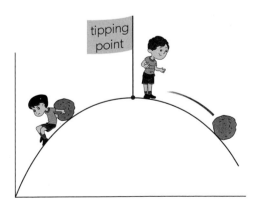

▲ Figure 17 **Illustrating a tipping point**

Keystone species are species which have a large influence on an ecosystem relative to their population size. They are the "glue" of their ecosystem and their loss alters the ecosystem greatly. They may be animals, plants or microorganisms. Without them, the ecosystem may not exist or may exist in a very different form. Examples include elephants, sea stars, sea otters and grey wolves.

TOK

Does the use of the metaphor "tipping point" influence our perspective on this knowledge?

Examples of tipping points

1. **Lake eutrophication**—if nutrients are added to a lake ecosystem, it may not change much until enough nutrients are added to shift the lake to a new state—then plants grow excessively, light is blocked by decomposing plant material, oxygen levels fall, and animals die. The lake is described as eutrophic. It takes a great effort to restore such a lake to its previous state.

2. **Extinction of a keystone species**—if a **keystone species** is lost from an ecosystem, the system may be transformed to a new state which cannot be reversed.

3. **Coral reef death**—if ocean acidity levels rise enough, reef coral dies and cannot regenerate.

Tipping points are well-known in local or regional ecosystems but there is debate about whether or not we are reaching a global tipping point. Some people say that climate change caused by human activities will force the Earth to a new, much warmer state—as much as 8°C warmer than today. But evidence is that we see warming in one region and cooling in others, wetter in some and drier in others. The global system is extremely complex and every ecosystem responds differently.

If there are global tipping points, there are major implications for decision-makers. Some may think that below this point not much would change but once it is reached, all is lost as society could not respond fast enough. That could lead to inaction or despair: a point of view summed up as: "What's the point? There is nothing we can do now."

The best approach may be the precautionary one. This means that although we do not know what will happen, we can take steps to modify what we do and so minimize risk. Such risk management is the responsible route to take.

Complexity and stability

Most ecosystems are very complex. There are many feedback links, flows and storages. It is likely that a high level of complexity makes for a more stable system that can withstand stress and change better than a simple one. This may be because another pathway can take over if one is removed.

Imagine a road system where one road is blocked by a broken-down truck; vehicles can find an alternative route on other roads. If a community has a number of predators and one is wiped out by disease, the other predators will increase in number because there is more prey for them to eat. So prey numbers will not increase in these circumstances.

Tundra ecosystems are fairly simple and populations in them may fluctuate widely (e.g. lemming population numbers).

Monocultures (farming systems in which there is only one major crop) are vulnerable to the sudden spread of a pest or disease through a large area with devastating effect. The spread of potato blight through Ireland in 1845–48 provides an example. Potato was the major crop grown over large areas of the island, and the cultural, biological, economic and political consequences were severe.

Resilience of systems

The resilience of a system measures how it responds to a disturbance. Resilience is the ability of a system to return to its initial state after a disturbance. If it has low resilience, it will enter a new state (see Figure 18).

Resilience can be modelled as a ball in a bowl. If the ball is pushed upwards, it returns to the bottom of the bowl—its initial state. But if it is pushed enough, it will leave the bowl and settle elsewhere—in an additional state. The higher the walls of the bowl, the more resilience the system has because the more energy you need to push it out of the bowl. See Figure 17 as well.

Resilience is generally considered a good thing, whether in a society, individual or ecosystem, as it maintains stability of the system.

In eucalypt forests of Australia, fire is seen as a major hazard. But eucalypts have evolved to survive forest fires. Their oil is highly flammable and the trees produce a lot of litter which also burns easily. But the trees regenerate quickly after a fire because they have buds within their trunks and plants that would have competed with them are destroyed. The eucalypts are resilient. But when the indigenous eucalypts are replaced by tree species that cannot withstand fire, it can be devastating.

In managed systems, such as agriculture, we want stability so we can predict that the amount of food grown is about the same each year. If this does not happen, there can be disastrous consequences (for example, the 1840s Irish potato famine, and the 2022 East Africa drought and famine).

But resilience is not always good. For example, a pathogenic bacterium causing a fatal disease could be very resilient to antibiotics. This would mean that it would kill many people. In this case, resilience of the bacterium to antibiotics is not good for us.

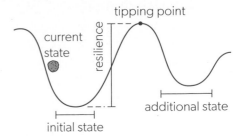

▲ Figure 18 The more resilient a system, the more disturbance it can deal with

▲ Figure 19 Fire in a eucalypt forest in Australia

Factors affecting ecosystem resilience

- The more diverse and complex an ecosystem, the more resilient it tends to be as there are more interactions between different species.

- The greater the species biodiversity of the ecosystem, the greater the likelihood that there is a species that can replace another if it dies out and so maintain the equilibrium.

- The greater the genetic diversity within a species, the greater the resilience. A monoculture of wheat or rice can be wiped out by a disease if none of the plants have resistance which is more likely in a diverse gene pool.

- Species that can shift their geographical ranges are more resilient.

- The larger the ecosystem, the more resilience as animals can find each other more easily and there is less edge-effect. For example, a lake is more resilient than a puddle.

- The climate affects resilience—in the Arctic, regeneration of plants is very slow as the low temperatures slow down photosynthesis and so growth. In the tropical rainforests, growth rates are fast as light, temperature and water are not limiting.

- A species that reproduces at a faster rate can recover more quickly. Those with a fast reproductive rate can recolonize the ecosystem faster.

- Humans can remove or mitigate the threat to the system (e.g. remove pollutants, reduce invasive species) and this will result in faster recovery.

- Humans can decrease resilience by reducing biodiversity, e.g. monoculture crops in the North American prairies; reducing size, e.g. making roads through ecosystems and splitting them into smaller areas; and reducing genetic diversity, in cloning e.g. bananas (see subtopic 3.1).

Emergent properties

An emergent property is a property of a system but not of the individual parts of the system.

Examples of emergent properties

1. Your face: your eyes, ears, nose and mouth are all sensory organs but only together do they make your face what it is.

2. Chlorine is a toxic gas and sodium is a very reactive metal. Bonded together, they make the compound sodium chloride (salt).

3. Honey bees: a honey bee colony shows division of labour in which the queen lays eggs, workers collect pollen and nectar and care for the eggs, and drones mate with the queen. Individually a bee is not a functioning colony, collectively bees are.

4. Predator–prey oscillations (see Figure 12).

5. Trophic cascades (see subtopic 2.5).

Models of systems

A model is a simplified version of the real thing. Humans use models to help us understand how a system works and to predict what happens if something changes. Systems work in predictable ways, following rules. But as humans, we do not always know what these rules are. A model can take many forms. It could be:

Connections: This links to material covered in subtopic 1.1.

- a physical model (e.g. a wind tunnel or river, a globe or model of the solar system, an aquarium or a terrarium)

- a software model (e.g. of climate change or evolution such as DaisyWorld)

- mathematical equations

- data flow diagrams.

Models have limitations as well as strengths. They may omit some of the complexities of the real system (through lack of knowledge or for simplicity), but they enable us to predict the effects of change to the inputs of the system.

The strengths of models are that they:

- are easier to work with than complex reality

- can be used to predict the effect of a change of input

- can be applied to other similar situations

- can help us see patterns

- can be used to visualize really small things (atoms) and really large things (the solar system).

The weaknesses of models are:

- accuracy is lost because the model is simplified

- if our assumptions are wrong, the model will be wrong

- predictions may be inaccurate.

Check your understanding

In this subtopic, you have covered:

- systems being: living or non-living (e.g. biological, social, economic); open (exchanging energy and matter) or closed (exchanging matter only); at any scale—small or large (a cell is a system; so are you, a bicycle, a car, a home, a pond, an ocean, a smart phone, a farm, the Earth)

- systems as sets of interacting or interdependent components which are organized to make a functioning whole

- systems having storages and flows between storages

- flows providing the inputs and outputs of energy and matter to a system and sometimes inputs are stored in storage or stock

- material and energy undergo transfers and transformations in flowing from one storage to the next

- feedback loops may be negative (reduce change and stabilize the equilibrium) or positive (increase change and destabilize equilibrium leading to a tipping point)

- models have their limitations but can be useful in helping us to understand systems

- emergent properties are those properties of a system which are not held by its individual components

- resilience of a system is its ability to resist change or adapt to disturbance; it is reduced in lower diversity or smaller systems, and by humans who fragment ecosystems or reduce their biodiversity.

How can the systems approach be used to model environmental issues at different levels of complexity and scale?

1. Define a system.

2. Draw an open theoretical system diagram showing storages, flows, inputs and outputs.

3. List four ecosystems in increasing order of size.

4. Explain how

   a. negative and

   b. positive

   feedback loops affect a system in terms of the equilibrium reached.

5. Describe an example of a tipping point.

6. Explain what a model is and what its advantages and disadvantages are.

7. Explain an emergent property using a named example.

8. Describe resilience of a system and explain how scale and complexity affect it.

►► Taking it further

- Build a bottle ecosystem, or other lab-based ecosystem (i.e. a mesocosm, terrarium, or aquarium) and use it to construct a systems diagram.

- Use your skills of system analysis to help solve a whole-school problem.

- Educate your peers about the importance of tipping points.

- What is sustainability and how can it be measured?
- To what extent are challenges of sustainable development also ones of environmental justice?

Understandings

1. Sustainability is a measure of the extent to which practices allow for the long-term viability of a system. It is generally used to refer to the responsible maintenance of socio-ecological systems such that there is no diminishment of conditions for future generations.
2. Sustainability is comprised of environmental, social and economic pillars.
3. Environmental sustainability is the use and management of natural resources that allows replacement of the resources, and recovery and regeneration of ecosystems.
4. Social sustainability focuses on creating the structures and systems, such as health, education, equity, community, that support human well-being.
5. Economic sustainability focuses on creating the economic structures and systems to support production and consumption of goods and services that will support human needs into the future.
6. Sustainable development meets the needs of the present without compromising the ability of future generations to meet their own needs. Sustainable development applies the concept of sustainability to our social and economic development.
7. Unsustainable use of natural resources can lead to ecosystem collapse.
8. Common indicators of economic development, such as gross domestic product (GDP), neglect the value of natural systems and may lead to unsustainable development.
9. Environmental justice refers to the right of all people to live in a pollution-free environment, and to have equitable access to natural resources, regardless of issues such as race, gender, socio-economic status, nationality.
10. Inequalities in income, race, gender and cultural identity within and between different societies lead to disparities in access to water, food and energy.
11. Sustainability and environmental justice can be applied from at the individual to the global scale.
12. Sustainability indicators include quantitative measures of biodiversity, pollution, human population, climate change, material and carbon footprints, and others. These indicators can be applied on a range of scales, from local to global.
13. The concept of ecological footprints can be used to measure sustainability. If these footprints are greater than the area or resources available to the population, this indicates unsustainability.
14. The carbon footprint measures the amount of greenhouse gases (GHGs) produced, measured in carbon dioxide equivalents (in tonnes). The water footprint measures water use (in cubic metres per year).

15. Biocapacity is the capacity of a given biologically productive area to generate an ongoing supply of renewable resources and to absorb its resulting wastes.
16. Citizen science plays a role in monitoring Earth systems and whether resources are being used sustainably.
17. There is a range of frameworks and models that support our understanding of sustainability, each with uses and limitations.
18. The UN Sustainable Development Goals (SDGs) are a set of social and environmental goals and targets to guide action on sustainability and environmental justice.
19. The planetary boundaries model describes the nine processes and systems that have regulated the stability and resilience of the Earth system in the Holocene epoch. The model also identifies the limits of human disturbance to those systems, and proposes that crossing those limits increases the risk of abrupt and irreversible changes to Earth systems.
20. The doughnut economics model is a framework for creating a regenerative and distributive economy in order to meet the needs of all people within the means of the planet.
21. The circular economy is a model that promotes decoupling economic activity from the consumption of finite resources. It has three principles: eliminating waste and pollution, circulating products and materials, and regenerating nature.

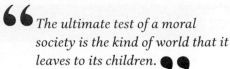

The ultimate test of a moral society is the kind of world that it leaves to its children.

Dietrich Bonhoeffer (1906–1945), German theologian

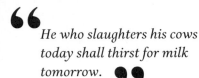

He who slaughters his cows today shall thirst for milk tomorrow.

Muslim proverb

Sustainability

Being **sustainable** means living within the means of nature, on the interest or sustainable **natural income** generated by **natural capital**. This means, for example, not overfishing. Humans consume natural resources all the time. You use electricity to power machines, heat or cool homes, you eat and drink, wear clothes, use transport. But sustainability is a word that may mean different things to different people and sustainability may be **environmental**, **social** or **economic**.

The word "sustainable" is often used as an adjective in front of words such as resource, development and population.

Connections: This links to material covered in subtopics 1.2 and 7.2.

Key terms

Sustainability is a measure of the extent to which human activities allow for the long-term viability of a system. It is generally used to refer to the responsible maintenance of socio-ecological systems such that there is no diminishment of conditions for future generations.

Natural income is the yield or harvest from natural resources.

Natural capital is the stock of natural resources on Earth. This includes rocks, soil, water, air and all living things. It also includes the services that support life such as photosynthesis and the water cycle.

Environmental sustainability is the use and management of natural resources that allows replacement of the resources, and the recovery and regeneration of ecosystems.

Social sustainability focuses on creating the structures and systems that support human well-being, including health, education, equity, community and culture such as belief systems and language.

Economic sustainability focuses on creating the economic structures and systems to support production and consumption of goods and services that will support human needs into the future.

(ATL) Activity 12

ATL skills used: thinking, research

Compare the two models in Figures 1 and 2.

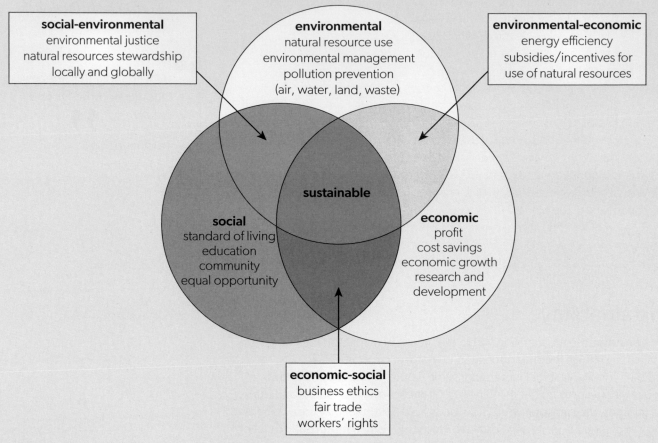

social-environmental
environmental justice
natural resources stewardship
locally and globally

environmental
natural resource use
environmental management
pollution prevention
(air, water, land, waste)

environmental-economic
energy efficiency
subsidies/incentives for
use of natural resources

sustainable

social
standard of living
education
community
equal opportunity

economic
profit
cost savings
economic growth
research and
development

economic-social
business ethics
fair trade
workers' rights

▲ Figure 1 The spheres of a sustainable model. Only when all three overlap is there sustainability

1. Why do the social and economic circles of the Venn diagram overlap the environmental circle?

2. Is culture relevant to these models of sustainability? Where would you draw it in?

3. Does the choice of model change how we treat our environment?

4. Evaluate these models. (Consider their strengths and weaknesses.)

▶ Figure 2 An alternative model of sustainability representing all other systems within the environmental system

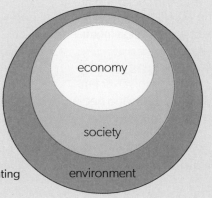

economy

society

environment

Why sustainability?

Economists may have a different view from environmentalists on what "sustainable development" means. You make judgements and decisions all the time about the actions you take: what you eat, what you do, what you use, how you behave to others. A sustainable society tries to ensure that it is a fair and just one, a healthy one and one that leaves a healthy environment with enough natural income to provide for future generations.

Any society that supports itself in part by depleting essential forms of natural capital is unsustainable. There is a finite amount of materials on Earth and we are using much of it unsustainably. In economic terms, we are living on the capital as well as the interest. Our societies and economies cannot grow or make progress outside environmental limits. Sensible use of **renewable natural capital** is sustainable. Mining **non-renewable natural capital** (e.g. coal, metal ores) to supply resources is not sustainable and the capital will eventually run out.

Look back at the key terms:

- social sustainability

- economic sustainability

- environmental sustainability.

Only when these three pillars are in balance can true sustainability be achieved (Figure 1).

Environmental sustainability is managing resources to conserve biodiversity, reduce or remove pollution and avoid depleting resources. Enhancing resilience of systems enhances sustainability.

But environmental improvement is not sustainability unless it includes economic and social factors.

Investing in social and human sustainability improves the human capital of society. This means better health programmes, better nutrition, skills and knowledge. It also means preserving the rights of future generations to these.

Economic sustainability is about efficient production of goods and services and maintaining this over time. Indicators of economic development usually measure monetary value and neglect the value of natural systems.

A common indicator of economic development is gross domestic product (GDP)—the total value of goods and services produced by a country over a period of time. But the challenge is to add the monetary value of natural goods and services to GDP for a true measure of sustainability.

Green GDP is an indicator of economic growth with environmental factors taken into account alongside the standard GDP of a country. Green GDP factors include biodiversity losses and costs attributed to climate change.

Is it possible to balance these three pillars of sustainability? That is what society is trying to do now.

Key terms

Renewable natural capital can be generated or replaced as fast as it is being used. This includes all life and ecosystems as well as non-living systems such as the ozone layer or groundwater.

Non-renewable natural capital is either irreplaceable or can only be replaced over geological timescales (e.g. fossil fuels, soil, water in aquifers and minerals).

Connections: This links to material covered in subtopic 1.2.

ATL Activity 13

ATL skills used: communication, social, research, self-management

There are some hard facts here. Is it all bad? No; it is up to you to make a difference.

Circles of sustainable positivity

Work in groups of three or four. Each person in the group should choose a sustainable area (from social, economic and environmental sustainability). Examples are:

- **social**: good nutrition, good healthcare, good education, good housing, aesthetics, cultural opportunities

- **economic**: increased wealth in a society, increased equity, decent employment opportunities, public infrastructure improvements, renewable energy supplies

- **environmental**: conserving biodiversity, habitat restoration, cleaner air, cleaner water, improving soils.

Each person should draw a large circle on a piece of paper. Do some research and fill your circle with positive news, facts and images (nothing negative). Think creatively.

Pass your circle to the next person in your group so they can add to it. Then repeat so each person has added something positive to each circle. Finally, share your ideas with the class.

The term **sustainable development** was introduced in the Brundtland Report *Our Common Future* (1987). It was defined as "development that meets the needs of the present without compromising the ability of future generations to meet their own needs".

It took three years for the commission to investigate and then write the report. The report stated that "critical global environmental problems were primarily the result of the enormous poverty of the South and the non-sustainable patterns of consumption and production in the North".

The Brundtland Report was a crucial publication because it recognized that human resource development (socio-economic development) **and** environmental conservation were both necessary:

- in reducing poverty

- in promoting equity of wealth, gender and justice.

It also recognized limits to growth in industrialized countries.

ATL Activity 14

ATL skills used: communication, research

Following on from the Brundtland Report came the Earth Summit of 1992 which led to Agenda 21, the Rio Declaration and documents on forest conservation, biodiversity and climate change.

1. Look up and make brief notes on the Brundtland Report.

2. Who was Brundtland?

3. What are the three principles of sustainability outlined in the report?

4. Were any results of subsequent actions legally binding on countries?

5. How many countries signed the Rio Declaration?

6. How relevant is the Brundtland Report today?

Ecological overshoot

The World Wide Fund for Nature (WWF), an NGO, notes that "humanity's annual demand on the natural world has exceeded what the Earth can renew in a year since the 1970s". This "ecological overshoot" has continued to grow over the years, reaching a 50% deficit in 2008 and a 75% deficit in 2018. This means that it takes 1.75 years for the Earth to regenerate the renewable resources that people use and absorb the carbon dioxide waste they produce, in a year.

How can this be possible when there is only one Earth? Renewable resources can be harvested faster than they can be re-grown. In the same way, it is possible to withdraw money from a bank account faster than waiting for any interest this money generates. And, just like overdrawing from a bank account, using a renewable resource faster than it can regenerate leads to depletion of the resource.

At present, people are often able to shift their sourcing when this happens. However, at current consumption rates, the new sources will also run out of resources. And some ecosystems will collapse even before the resource is completely gone. This overshoot is due to the level of overall consumption and per capita consumption. It is more in some parts of the world and cannot continue indefinitely.

Connections: This links to material covered in subtopics 1.2 and 4.3.

Earth Overshoot Day

Earth Overshoot Day marks the date when humanity's demand for ecological resources and services in a given year exceeds what the Earth can regenerate in that year. We—as a global population—at present consume resources as if we had 1.75 planets. Some use more than others. See Figures 3 and 4.

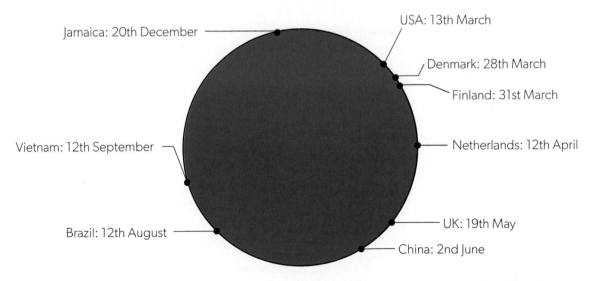

Earth Overshoot Day 2022

Jamaica: 20th December

USA: 13th March

Denmark: 28th March

Finland: 31st March

Netherlands: 12th April

Vietnam: 12th September

Brazil: 12th August

UK: 19th May

China: 2nd June

▲ Figure 3 Earth Overshoot Day 2022 for selected countries (from Global Footprint Network (GFN) data)

Adapted from: National Footprint and Biocapacity Accounts, 2022 (CC BY-SA 4.0)

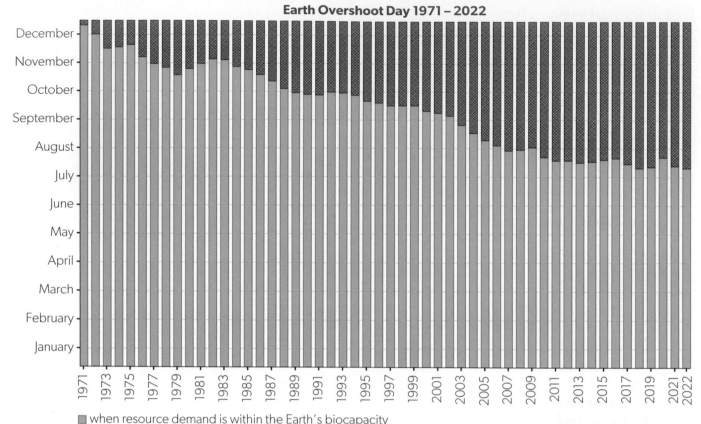

Earth Overshoot Day 1971 – 2022

■ when resource demand is within the Earth's biocapacity
▨ when resource demand overshoots (exceeds) the Earth's biocapacity

▲ Figure 4 Earth Overshoot Day up to 2022

Data from: National Footprint and Biocapacity Accounts, 2023 (CC BY-SA 4.0)

The Earth Overshoot Day is the estimated date each year when the GFN calculates that all the natural income for the year has been used. In 2022 it was on 28 July. Look up the GFN site and find out what the Earth Overshoot Day is now.

While Earth Overshoot Day is an estimate, we know that globally humans are not living within sustainable limits. Some of us are, but some are using far, far more than is sustainable. There is huge inequality globally in use of natural resources. What to do about it is challenging all humans.

Key term

Environmental justice is the right of all people regardless of race, gender, socio-economic status or national origin to:

- live in a pollution-free environment free from hazardous waste

- have equitable access to natural resources

- have fair treatment through laws and regulations.

Environmental justice

Environmental justice is also known as "distributive justice". This means that the risks and benefits of environmental exploitation are equally distributed. The global environmental justice movement developed in the 1980s, following many cases in which local peoples were impacted by multinational corporations extracting resources or creating industrial complexes that caused harm.

Sadly, there are many examples of environmental injustice. They can be classified as examples of:

- hazardous waste dumping
- insufficient safety measures
- resource extraction
- land appropriation
- climate justice issues.

You have already read about the Bhopal, Minamata and Deepwater Horizon disasters. These all have environmental justice implications. Other examples are:

- landfill or incinerator sites for domestic or industrial waste that are situated near communities of low income (e.g. Ghazipur in Delhi, India)

- waste plastic disposal from high-income countries (HICs) to low-income countries (LICs) and associated pollution of water and air (e.g. to Malaysia, Thailand, Vietnam)

- mining tailings dams (e.g. in Chile)

- Maasai land rights in Tanzania and Kenya (see Case study 5)

- indigenous land rights in Australia and North America

- indigenous property rights in the Amazon

- disparity in energy, electricity or water supplies for different groups in a society.

> Connections: This links to material covered in subtopics 1.1, 3.3, 4.2, 4.3, 5.2, 5.3, 6.3, 7.2, 8.3 and HL.a.

Case study 5

Maasai land rights

From websites.umich.edu/~snre492/Jones/maasai.htm

There are over one million indigenous Maasai people who live in the African Great Lakes region of southern Kenya and northern Tanzania (Figure 5). They are traditionally transhumance nomadic pastoralists who live by herding their livestock—mostly cattle—over vast areas of land in search of grazing and water. Transhumance is the practice of moving livestock seasonally from one grazing area to another. Maasai people mostly live on the blood and milk of their animals and rarely kill them for meat.

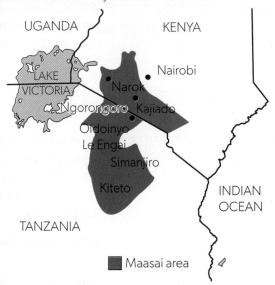

▲ Figure 5 Maasai area

There was a view during colonization by the British empire that the Maasai way of life was not compatible with wildlife in the area and national parks were created with no consideration for local peoples. It is easier for authorities to regulate and know about a sedentary population who stay in one place (for example, farmers) than nomads. Also, land was privatized and rancher colonists moved in to displace the Maasai.

Parks and animal reserves have proliferated as governments encourage tourism, development and sedentary lifestyles in towns, and Maasai traditional lands have been taken from them.

The British and, after independence, the Kenyan governments valued wildlife for the income from tourism more than the indigenous Maasai. Various land exchanges moved the Maasai from their traditional grazing areas to smaller land areas which were not compatible with pastoralism. The belief was that the Maasai herds would overgraze to the detriment of wildlife.

The Serengeti with its iconic wildebeest herds was an area mostly lost to the Maasai and in the 1950s they were moved to the Ngorongoro Conservation Area (NCA) in Tanzania. Here is the highest concentration of wildlife on Earth when the annual migration of more than one million wildebeest, half a million gazelles and quarter of a million zebras come over from the Serengeti. It contains a high population of lions and cheetahs and is a refuge for elephants and black rhinos. While management plans were supposed to protect the Maasai, their lives became worse not better as the plans prioritized wild animals, and thus tourism income, not the Maasai.

The Maasai have practised pastoralism for at least 200 years and adjust their herd numbers in order to live sustainably. But various threats are damaging them and driving them to more poverty, not less (Figure 6). Recently, they have begun to organize their struggle for land rights. This has not happened earlier because the nature of their way of life means they have no single leader; nor is there any international advocate for their case.

The issues are:

- scale of tourism industry

- lack of income from tourism to Maasai communities

- cultural differences between various groups

- lack of inclusion of Maasai in decision-making

- lack of communication between groups

- land grabs by various groups

- climate change causing droughts

- increasing populations.

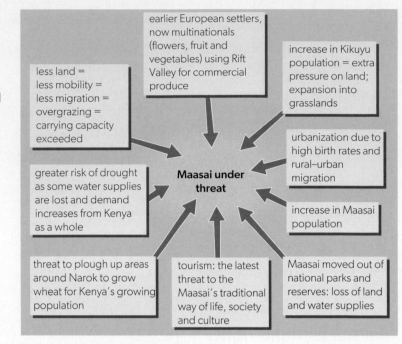

▲ Figure 6 Threats to the Maasai

Activity 15

ATL skills used: communication, research, self-management

Read Case study 5 on Maasai land rights. Research **one** local and **one** global example of an environmental justice issue. Make a display of your choice to show:

- the issue

- the stakeholders—such as companies, peoples, governments, local authorities, NGOs and pressure groups

- the impacts of the activity

- the actions taken by all parties

- the results.

Sustainability indicators

How we measure sustainability is crucial, and there are many ecological and socio-economic indices we can use. These could be anything from air quality, environmental vulnerability and water poverty to GDP per capita (per head of population), life expectancy or gender equity.

We can also measure sustainability on scales from local to global. The smaller the scale, the more accurate it can be, but we need a global measurement to get the whole picture.

Both environmental justice and sustainability can be viewed at operating scales ranging from the individual to intergovernmental organizations (Figure 7).

	Your individual decision on what to buy and use and how to dispose of it, where to live, how to behave
increasing scale	A business or company on its aims and mission, how and where it operates, what resources it uses, what it produces, who invests in it
	A cultural, religious or political community on how it sets and regulates its rules and norms (e.g. peaceful protest to reduce traffic or development)
	A city and a country on setting policies and laws and socio-economic systems
	A global organization such as the UN on agreeing intergovernmental treaties and laws

▲ Figure 7 Scales of sustainability and environmental justice

Sustainability indicators include:

- biodiversity

- pollution levels

- human population size, health and income

- energy production and consumption

- climate change

- ecological and carbon footprints.

Ecological footprints

An **ecological footprint (EF)** is a model used to estimate the demands that human populations place on the environment. The measure takes into account the area required to provide all the resources needed by the population, plus the assimilation of all wastes. When the EF is greater than the area available to the population, the situation is unsustainable: the population exceeds the **carrying capacity** (see subtopic 8.4) of the environment.

EFs may vary significantly from country to country and person to person and include aspects such as lifestyle choices (EVS), productivity of food production systems, land use and industry.

The ecological footprint indicator of sustainability can be applied at an individual to global scale, and it is visual so can be interpreted easily.

There are other measures of ecological footprint. A carbon footprint is the amount of greenhouse gases produced, measured in carbon dioxide equivalents. A water footprint measures water use.

Carrying capacity is the inverse of EF. Carrying capacity is the number of a population that a unit of land can support. EF is the area of land that is needed to support that population.

Biocapacity and ecological footprints are methods of measuring human impact on the environment.

If an area's EF is greater than its biocapacity, the situation is unsustainable.

Key terms

An **ecological footprint (EF)** is the area of land and water required to sustainably provide all the resources required at the rate of consumption and to assimilate all wastes at the rate of production by a given population.

Carrying capacity is the maximum number of individuals of a species that the environment can sustainably support.

Biocapacity is the capacity of a biologically productive area to generate a supply of renewable resources and to absorb its waste.

ATL Activity 16

ATL skills used: thinking, communication, social, research

1. Measure your EF at www.footprintcalculator.org/home

2. In what ways does your EF differ from those of your classmates and why?

3. Plot your class EFs graphically. Comment on the graph.

4. How might you reduce your EF?

5. How might the EF of other socio-economic groups differ?

6. Evaluate the accuracy of using an EF calculator.

Practical

Skills:

- Tool 1: Experimental techniques
- Tool 2: Technology
- Inquiry 1: Exploring and designing

Produce a questionnaire of your own to calculate the EF of a person. You can use the ideas in Figure 8 to help you.

carbon
represents the amount of forest land that could sequester CO_2 emissions from the burning of fossil fuels, excluding the fraction absorbed by the oceans which leads to acidification

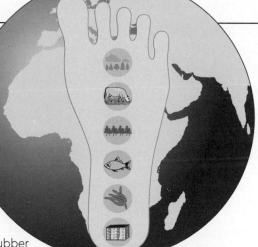

cropland
represents the amount of cropland used to grow crops for food and fibre for human consumption as well as for animal feed, oil crops and rubber

grazing land
represents the amount of grazing land used to raise livestock for meat, dairy, hide and wool products

forest
represents the amount of forest required to supply timber products, pulp and fuel wood

built-up land
represents the amount of land covered by human infrastructure, including transportation, housing, industrial structures and reservoirs for hydropower

fishing grounds
calculated from the estimated primary production required to support the fish and seafood caught, based on catch data for marine and freshwater species

▲ Figure 8 Ecological footprint calculator

In 1961, it was calculated that the EF of all people on Earth was 0.72 global hectares (gha) per person. Then the biocapacity of the Earth was within sustainable limits.

In 2018, the EF was equivalent to 1.75 Earths or 2.8 gha per person. It would take 21 months to regenerate one year's worth of resources that humans use. Since the 1970s, we have been in ecological overshoot: our annual demand on the natural world exceeds what it can supply.

Connections: This links to material covered in subtopics 4.2, 6.2, 6.3 and 8.3.

Data-based questions

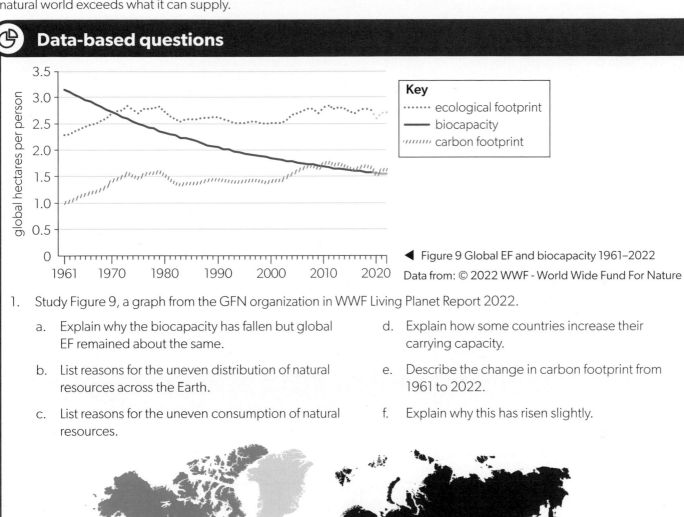

◀ Figure 9 Global EF and biocapacity 1961–2022
Data from: © 2022 WWF - World Wide Fund For Nature

1. Study Figure 9, a graph from the GFN organization in WWF Living Planet Report 2022.

   a. Explain why the biocapacity has fallen but global EF remained about the same.

   b. List reasons for the uneven distribution of natural resources across the Earth.

   c. List reasons for the uneven consumption of natural resources.

   d. Explain how some countries increase their carrying capacity.

   e. Describe the change in carbon footprint from 1961 to 2022.

   f. Explain why this has risen slightly.

Key
- < 1.7 gha/person
- 1.7–3.4 gha/person
- 3.4–5.1 gha/person
- 5.1–6.7 gha/person
- >6.7 gha/person
- insufficient data

▲ Figure 10 Global EF per person

WWF (2022) Living Planet Report 2022 - Building a nature-positive society. Almond, R.E.A., Grooten, M., Juffe Bignoli, D. & Petersen, T. (Eds). WWF, Gland, Switzerland.

2. Study Figure 10. The Earth's biocapacity was 1.6 gha per person in 2022.

   a. List the regions on Earth in which the EF per person is lower than global biocapacity.

   b. Describe possible reasons for high EFs in North America, Australia and Mongolia.

   c. Where else in the world is EF very high? Explain why.

   d. State possible reasons for insufficient data from Greenland and Iceland.

   e. State possible reasons for insufficient data from the country at the Horn of Africa.

The ecological footprint of a country depends on several factors: its population size and consumption per capita—how many people and how much land each one uses. It includes the cropland and other land that is needed to grow food, grow biofuels, graze animals for meat, produce wood and dig up minerals, and the area of land needed to absorb wastes—not just solid waste but wastewater, sewage and carbon dioxide.

Sustainability models

Sustainability models, like all models, are simplified versions of reality and therefore have both uses and limitations. Here we look at four sustainability frameworks and models:

- Sustainable Development Goals (SDGs)

- planetary boundaries model

- doughnut economics model

- circular economy.

ATL Activity 17

ATL skills used: social, research, self-management

The **Millennium Ecosystem Assessment (MEA)** is funded by the UN and started in 2001. It is a research programme that focuses on how ecosystems have changed over the last decades and predicts changes that will happen. In 2005, it released the results of its four-year study of the Earth's natural resources. It was not happy reading. The report said that natural resources (food, freshwater, fisheries, timber, air) are being used in ways that degrade them so as to make them unsustainable in the longer term.

Key facts reported in 2005 were:

- 60% of world ecosystems have been degraded

- about 25% of the Earth's land surface is now cultivated

- we use between 40% and 50% of all available freshwater running off the land and water withdrawals have doubled over the past 40 years

- over 25% of all fish stocks are overharvested

- since 1980, about 35% of mangroves have been destroyed

- about 20% of corals have been lost in 20 years and another 20% degraded

- nutrient pollution has led to eutrophication of waters and dead coastal zones

- species extinction rates are now between 100 and 1,000 times above the background rate

- we have had more effect on the ecosystems of the Earth in the last 50 years than ever before.

Some recommendations were to:

- remove subsidies to agriculture, fisheries and energy sources that harm the environment

- encourage landowners to manage property in ways that enhance the supply of ecosystem services, such as carbon storage, and the generation of freshwater

- protect more areas from development, especially in the oceans.

1. Research statistics for the key indicators listed and attempt to complete Table 1.

2. Add other data to the key indicators that you think is relevant.

3. Compare your data with others in the class.

	2005	Current year
Degraded ecosystems percentage	60%	
Land surface area cultivated	25%	
Freshwater use	40–50%	
Fish stocks overharvested	25%	
Mangroves destroyed	35%	
Corals lost	20%	
Species extinction rates	100–1,000× background levels	

▲ Table 1 Table of key indicators

Sustainable Development Goals

Sustainable Development Goals (SDGs) are 17 goals formulated by the UN in 2015 which run until 2030 or beyond. They are assessed each year.

They followed on from the Millennium Development Goals (MDGs), which ended in 2015.

The MDGs were eight goals to be achieved by 2015 that responded to the world's main development challenges. They were designed for low-income countries (LICs) and middle-income countries (MICs) only. The MDGs were:

Goal 1: Eradicate extreme poverty and hunger

Goal 2: Achieve universal primary education

Goal 3: Promote gender equality and empower women

Goal 4: Reduce child mortality

Goal 5: Improve maternal health

Goal 6: Combat HIV/AIDS, malaria and other diseases

Goal 7: Ensure environmental sustainability

Goal 8: Develop a Global Partnership for Development

TOK

To what extent are you optimistic about the results of the impact of humans on Earth? Use examples to support your ideas.

To what extent do governments have the responsibility to take steps to safeguard the environment and protect humans from suffering? Support your ideas with examples.

Connections: This links to material covered in subtopics 4.2, 5.2, 7.2 and 8.3.

There were successes of the MDGs but the results were mixed:

- up to 471 million people lifted out of poverty—halving the total number

- gender disparity reduced

- more clean drinking water worldwide

- 21 million lives saved.

But there was greater progress among LICs than MICs. The global economic crisis of 2008 reduced funding. Environmental progress was low and, above all, the goals were created top-down. This meant planning did not involve the people the goals were meant to help.

Critics of the MDGs said that reduction in poverty would have happened anyway due to economic growth.

The SDGs have followed on from the MDGs. The SDGs aim to build on the work already done by the MDGs. But there is quite a lot of disagreement on whether this is the best way to go about it.

▲ Figure 11 Sustainable Development Goals

Each goal shown in Figure 11 has typically 8 to 12 targets. Each target has one to four indicators in three tiers to measure progress. The targets are either outcomes or means of implementation. Some targets have internationally recognized methodologies for assessing whether or not they are reached. Some do not. This complex arrangement was reviewed in 2020 with some changes.

There is no question that these are admirable and necessary goals for the world. And progress is being made. But the challenges are huge. Compare the goals in Figure 11 with the comments below.

1. Extreme poverty means living on less than USD 2.15 per day (2017 prices) and in 2021 this included 9% of the world's population—700 million people. Most were in South Asia or Sub-Saharan Africa.

2. Around the world, one in nine people are undernourished, mostly in LICs.

3. Healthcare is improving and life expectancy mostly increasing.

4. Major progress has been made in primary education and reducing discrimination in education.

5. Gender equality, reduction in female genital mutilation (FGM) and empowering women are mostly happening.

6. Water stress (see subtopic 4.2) limits access to clean water for 2.2 billion people.

7. Progress on clean energy has happened in India, Bangladesh and Kenya but 840 million people do not have access to electricity.

8. Economic growth and decent work are variable with trafficking and exploitation of youth.

9. Decoupling of carbon emissions from economic growth is improving but only 53% of the world's population use the internet.

10. Reducing inequality and discrimination against women, minorities and those with disabilities has had mixed progress.

11. About 3.9 billion people globally live in cities and about 25% of these live in slums.

12. Reducing plastic waste, increasing recycling and removing fossil fuel subsidies are all happening.

13. A just transition to combating climate change and reducing risks is the goal but wildfires, droughts, floods and hurricanes appear to be more frequent and those in vulnerable conditions suffer more.

14. Reducing marine plastic pollution and illegal fishing are key but three billion people rely on marine life for their livelihood.

15. Deforestation and desertification increase biodiversity losses; poaching, illegal logging and soil acidification add to the issues. Species extinctions are increasing.

16. Reducing violence, abuse, exploitation, crime and corruption are clear goals.

17. International cooperation is essential for any of these goals to be met— cooperation not competition.

According to the World Bank in 2019, 659 million people on Earth lived below the poverty line of USD2.15 per day. This was a reduction but may have been due to economic growth which takes people out of poverty.

Many people still live in conditions that do not meet their fundamental daily needs, and that is unacceptable.

Uses of the SDGs

- A collective agenda with common ground—a global agenda for sustainable development at the highest level means that governments can use political pressure on challengers or each other.

- Universal goals for all countries, and not only LICs—all countries are working towards the same goals.

- A galvanizing agenda—while the goals are not legally binding, there is more freedom for countries to address inequalities within and between countries.

- The goal of ending poverty is tangible and the MDGs made progress towards it, which the SDGs continue.

Limitations of the SDGs

- The goals do not go far enough or fast enough. The targets move slowly up to 2030 but unanticipated things (conflict, famine, floods) happen each year to deflect the course.

- Top-down and bureaucratic—there are some 169 targets and much reporting and gathering of necessary evidence, but on the ground progress may be small.

- Ignoring local contexts—while there must be a balance between respecting local context and working at the international level, it is hard to find this balance.

- Lack of reliable data—the wider the data collection, the less reliable it may be, yet data is crucial to know if progress is being made.

- The goals are not binding—therefore there are no penalties for not acting and no accountability. It is up to individual governments to create the systems.

- Ignoring underlying inequalities—economic theory is based on financial values of companies and GDP of countries. If mainstream economics is fundamentally the wrong measure for all types of sustainability, that has to change.

You may wonder why this continues if we all know it to be so. It is perhaps due to many factors, including the following:

- Inertia—this is when changing what we do seems too difficult to achieve.

- The result of the "tragedy of the commons" (see subtopics 3.2 and 6.3)— this is when so many individuals act in their own interest to harvest a resource that they destroy the long-term future of that resource. Hence, there is none for anyone. It may be obvious that this will happen, but because each individual benefits from taking the resource they continue to do so. For example, you may know that hunting an endangered species will result in its extinction. But if your family are starving and it is the only source of food, you will probably hunt it to eat it.

Some people think that the real worth of natural capital is about the same as the value of the gross world product (total global output)—about USD 65 trillion per year. But we are only just beginning to give economic value to soil, water and clean air, and to measure the cost of loss of biodiversity.

At the same time, the average person alive now is in a better situation than anyone in history.

The planetary boundaries model

Connections: This links to material covered in subtopics 2.3, 3.2, 4.2, 5.2, 6.2, 6.4 and 8.1.

In 2009, Johan Rockström and 28 internationally renowned scientists identified nine key processes that regulate the stability and resilience of our planet. They proposed quantitative "planetary boundaries" within which humanity can continue to develop and thrive for generations to come. Crossing these boundaries increases the risk of generating large-scale abrupt or irreversible environmental changes.

The planetary boundaries model attempts to answer the question: **up to what limits will the Earth system be able to absorb human activities without compromising the living conditions of species?**

These models from 2009 (Figure 12) and 2022 (Figure 13) describe the nine processes and systems that regulate the stability and resilience of the Earth in the Holocene Epoch. The traffic light system shows how human activity has moved the process or system to a danger zone in which there is risk of fast and irreversible change. Green is within the boundary, yellow is the zone of uncertainty and orange is high risk.

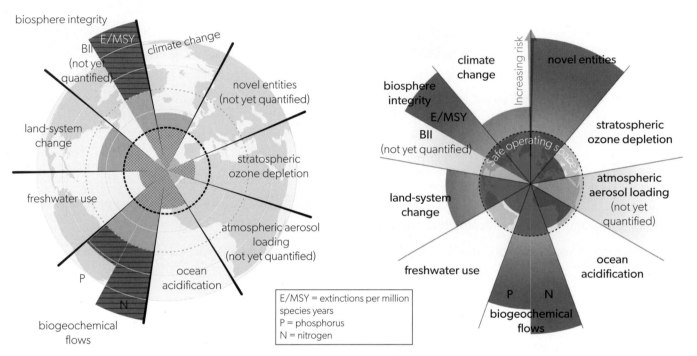

▲ Figure 12 The nine planetary boundaries model in 2009
Adapted from: J. Lokrantz/Azote based on Steffen et al. 2015 (CC BY 4.0)

▲ Figure 13 The nine planetary boundaries model in January 2022
Adapted from: Azote for Stockholm Resilience Centre, based on analysis in Persson et al. 2022 and Steffen et al. 2015 (CC BY 4.0)

E/MSY = extinctions per million species years
P = phosphorus
N = nitrogen

Uses of the planetary boundaries model

- Identifies science-based limits to human disturbance of Earth systems.

- Focuses on more than just climate change (which dominates discussion).

- Alerts the public and policymakers to the urgent need for action.

Limitations of the planetary boundaries model

- Focuses only on ecological systems and does not consider the human dimension necessary to take action for environmental justice.

- Is a work in progress—assessments of boundaries are changing as new data becomes available.

- The focus on global boundaries may not be a useful guide for local and country-level action.

Note: Novel entities = chemical pollution

- There are about 350,000 different manufactured chemicals on the market today.

- Plastic production increased by 79% between 2000 and 2015.

- The total mass of plastics on the planet is now more than twice the mass of all living mammals.

- About 80% of all plastics ever produced remain in the environment.

- Plastics contain more than 10,000 other chemicals, so their environmental degradation creates new combinations of materials and unprecedented environmental risks.

(ATL) Activity 18

ATL skills used: thinking, research

1. State which planetary boundaries in this model were crossed in 2009.

2. Describe the differences between the 2009 and 2022 models.

3. In May 2022 it was reported that the boundary for freshwater use had been crossed. Research the state of the boundaries model now and redraw the model to show this.

4. Use the paragraphs below as a guide to explain crossed planetary boundaries.

Loss of biosphere integrity (biodiversity loss and extinctions)

BII stands for Biodiversity Intactness Index. Similar to the extinction rate, this measures how well the biodiversity of a given area has been maintained. A well-functioning and resilient ecosystem has a BII of over 90%. At-risk ecosystems have a BII of less than 30%.

The Millennium Ecosystem Assessment of 2005 concluded that changes to ecosystems due to human activities were more rapid in the past 50 years than at any time in human history, increasing the risks of abrupt and irreversible changes. The main drivers of change are the demand for food, water and natural resources, causing severe biodiversity loss and leading to changes in ecosystem services. These drivers are either steady, showing no evidence of declining over time, or are increasing in intensity. The current high rates of ecosystem damage and extinction can be slowed by efforts to protect the integrity of living systems (the biosphere), enhancing habitat, and improving connectivity between ecosystems while maintaining the high agricultural productivity that humanity needs.

Biogeochemical flows: Nitrogen and phosphorus flows to the biosphere and oceans

The biogeochemical cycles of nitrogen and phosphorus have been radically changed by humans as a result of many industrial and agricultural processes. Nitrogen and phosphorus are both essential elements for plant growth, and production and application of fertilizers is the main concern.

Human activities now convert more atmospheric nitrogen into reactive forms than all of the Earth's natural processes combined. Much of this new reactive nitrogen is emitted to the atmosphere in various forms rather than taken up by crops. When it rains, it pollutes waterways and coastal zones or accumulates in the terrestrial biosphere.

A relatively small proportion of phosphorus fertilizers applied to food production systems is taken up by plants; much of the phosphorus mobilized by humans also ends up in aquatic systems. These can become depleted of oxygen when bacteria consume algal blooms that grow in response to the high nutrient supply.

A significant fraction of the applied nitrogen and phosphorus makes its way to the sea and can push marine and aquatic systems across ecological thresholds of their own.

The doughnut economics model

The doughnut economics model is a framework for creating a **regenerative** and **distributive** economy that meets the needs of all people within the means of the planet.

Search on the web for Kate Raworth's 2018 TED talk on doughnut economics. Then examine Figure 14, the doughnut of social and planetary boundaries.

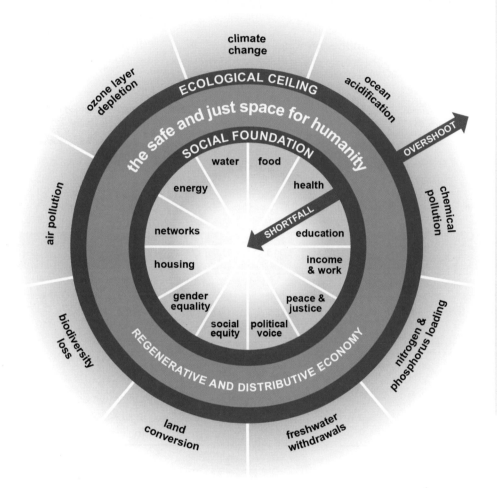

▲ Figure 14 **Model of doughnut economics**

There are two concentric rings:

- a social foundation to ensure that no one is left falling short on life's essentials

- an ecological ceiling to ensure that humanity does not collectively overshoot the planetary boundaries that protect Earth's life-support systems.

According to the Doughnut Economics Action Lab website, between them is a "doughnut-shape space that is both ecologically safe and socially just: a space in which humanity can thrive".

The social foundation (inner boundary of the doughnut) is based on the social SDGs. The ecological ceiling (outer boundary of the doughnut) is based on planetary boundaries science. Together the social foundation and the ecological ceiling represent the minimum conditions for an economy that is ecologically safe and socially just.

Key terms

Regenerative design is a principle in which products or services contribute to systems that renew or replenish themselves. This means the materials and energy that go into a product or process can be reintroduced into the same process or system, requiring little to no inputs to maintain it. At the heart of regenerative design is a strong connection to the place in which a product or process is extracted, produced, used and disposed of at end of life.

Distributive design is about designing our activities in such a way that they share the value from the start, instead of redistributing it afterwards. This value could be money, but also land, companies and the ability to create money. Distributive economics is an economic paradigm which promotes the equitable distribution of wealth through a combination of open design, flexible fabrication, and open business models, towards replicability.

Today, billions of people still fall short of the social foundation, while humanity has collectively overshot most of the planetary boundaries.

The goal is to move into the doughnut and create an economy that enables humanity to thrive in balance with the rest of the living world.

This will only be achieved if economies become regenerative and distributive by design. A regenerative economy works with and within the cycles and limits of the living world. A distributive economy shares value and opportunity far more equitably among all stakeholders.

Uses of the doughnut economics model

This model has both ecological and social elements, so it:

- supports the concept of environmental justice

- has reached popular awareness

- is being used at different scales (e.g. countries, cities, neighbourhoods, businesses)

- supports action on sustainability.

ATL Activity 19

ATL skills used: thinking, research, self-management

Visit the Doughnut Economics Action Lab (DEAL) website or find a copy of the book *Doughnut Economics* by Kate Raworth.

1. State the five principles of doughnut economics in practice.

2. Describe the mindset suggested in the website or book for a 21st century economist.

3. Figure 15 combines the planetary boundaries model and the doughnut economics model.

    a. Several of the shortfalls are large. Choose three and explain why they are so large, giving an example for each one.

    b. Evaluate the benefits and disadvantages of combining the two models.

4. Look at doughnuteconomics.org/tools/173 which is a visualization of the doughnut model for different countries.

    a. List the data sets used to create this model.

    b. Select four countries from the list and comment on the overshoot of planetary boundaries and shortfall of life essentials of these countries.

    c. Explain reasons for these differences.

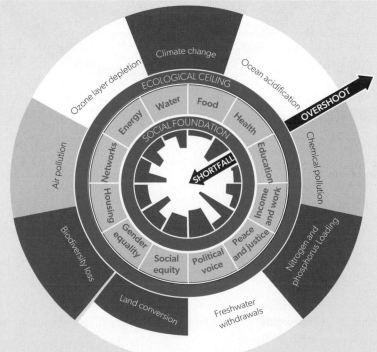

▲ Figure 15 **Combining the planetary boundaries and doughnut models**
Adapted from: Steffen et al. 2015 in Folke, C. et al. (2021). Our future in the anthropocene biosphere. Ambio, 50(4), 834–869.
https://doi.org/10.1007/s13280-021-01544-8 (CC BY 4.0)

Limitations of the doughnut economics model

- Is a work in progress.

- Will not work as long as individuals are focused on their comparative wealth and income levels.

- Advocates broad principles of regenerative and distributive practice

- Does not propose specific policies.

The circular economy model

After her solo voyage, Ellen MacArthur realized how the Earth is like a ship. It has all the resources on board but no more. And we cannot get more. She asked why we have a linear economy of take, make, use and dispose instead of a circular one of take, make and use but then reuse, repair, remake, remove waste and regenerate natural systems. This is the circular economy. The circular economy is based on three principles, driven by design:

1. Eliminate waste and pollution—in our linear economy, raw materials are extracted, processed into products, used and mostly then become waste. Waste is recycled, incinerated or goes to landfill or the oceans. As natural resources are finite, this is not sustainable.

   If product design was such that materials re-entered the economy at the end of their use, waste could be eliminated or very much reduced. For example, a crisp packet is used once and may be made of various materials which cannot be recycled. If it were designed to enter the circular economy, it could be remade into another crisp packet or other product.

2. Circulate products and materials (at their highest value)—this can be in a technical or a biological cycle (see Figure 16— the **butterfly model** systems diagram).

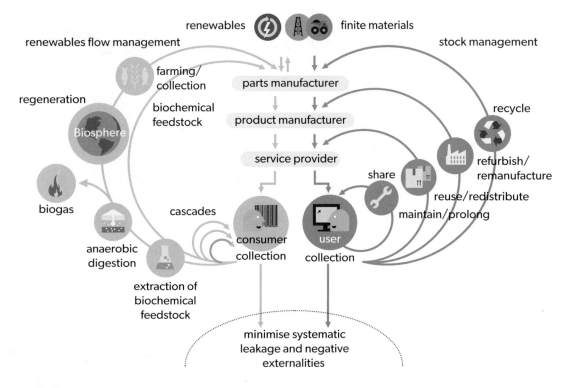

◀ Figure 16 The butterfly model of the circular economy

Adapted from: Ellen Macarthur Foundation / drawing based on Braungart and Mcdonough Cradle to Cradle (C2C)

In the **technical cycle**, it is most efficient to maintain and reuse entire products. For example, a phone is useful as a phone, not as a pile of components. Sharing instead of owning is efficient, while resale or donation of unwanted goods is excellent. Eventually the product can be deconstructed into its raw materials which are used, not wasted.

In the **biological cycle**, food by-products which are biodegradable can be composted or anaerobically digested, with the nutrients captured and used to regenerate soil or grow more food.

3. Regenerate nature—shift the focus from extraction to regeneration and build natural capital. There is no waste in nature, so we emulate that. For example, the food industry works regeneratively, restoring soil fertility and resilience and biodiversity. Agroforestry, conservation ecology and organic farming, for example, all have a place here.

Uses of the circular economy model

*   Regeneration of natural systems.

*   Reduction of greenhouse gas emissions.

*   Improvement of local food networks and supporting local communities.

*   Reduction of waste by extending product life cycle.

*   Changed consumer habits.

Limitations of the circular economy model

*   Lack of environmental awareness by consumers and companies.

*   Lack of regulations enforcing recycling of products.

*   Some waste not recyclable and technical limitations.

*   Lack of finance.

Check your understanding

In this subtopic, you have covered:

*   what sustainability is and different uses of the term

*   what sustainable development means

*   measuring sustainability and economic development

*   environmental justice issues

*   ecological footprints

*   models used to measure sustainability—SDGs, planetary boundaries, doughnut economics and circular economy.

What is sustainability and how can it be measured?

1. List and define the three types of sustainability.

2. Explain why sustainability is so important.

3. List five sustainability indicators.

4. Explain the concept of an ecological footprint.

To what extent are challenges of sustainable development also ones of environmental justice?

5. Define sustainable development.

6. Define environmental justice.

7. Describe one example of environmental injustice.

8. List and evaluate SDGs which involve environmental and social justice.

⟫ Taking it further

- Present research on examples of environmental injustice and inequalities leading to problems of access to resources.

- Promote the doughnut economic model or circular economy strategies for the school community.

- Use an SDG to advocate for a particular issue.

- Investigate the whole-school carbon footprint and produce a plan to reduce the school's carbon emissions.

- Design a sustainability walk to highlight sustainable options locally.

1.4 Practical work: Questionnaires and surveys

Within the ESS is a list of skills you need to acquire. You will need to master these skills **before** you start on your individual investigation.

Practical work is expected to take up 30 hours of your lesson time. This is in addition to the 10 hours you need to complete your individual investigation (see subtopic 2.6).

Most of the skills are explained in the relevant topics but there are two extra subtopics in this book: 1.4 and 2.6. These contain skills and experimental methodologies that you need to know and practise until they become effortless. Subtopics 1.4 and 2.6 apply to both SL and HL students.

In this subtopic (1.4), you will study how to develop:

- questionnaires and surveys

- behaviour over time graphs.

1

Questionnaires and surveys are suggested at various places in the ESS guide

- 1.1.5 Design and carry out questionnaires/surveys/interviews, using online collaborative survey tools, to correlate perspectives with attitudes towards particular environmental or sustainability issues. Select a suitable statistical tool to analyse this data.

- 3.3.12 Use questionnaires to assess the impact of ecotourism or the values that it promotes.

- 5.2.17 Create a survey to investigate food preferences and the worldviews of various groups.

- 6.3.4 Create surveys to investigate attitudes to a proposed solution in the school or community to mitigate climate change.

- 7.1.7 Create a survey to investigate the value that members of the school community place on different ecosystem services.

Questionnaires and surveys

These are useful tools to collect information about topics including:

- correlating worldviews and perspectives with attitudes towards particular environmental or sustainability issues

- ecological footprints and their link with age, gender, income, educational level

- animal conservation status and level of development of a country.

Questionnaires allow for the collection of up-to-date data that is specific to the purpose of the study. Well-designed questionnaires allow for graphical displays and statistical analysis. In some cases you may need to weight the response for really useful data.

Questionnaire design

Here are some basic rules for designing a questionnaire.

1. Keep it short, no more than five to seven questions. People may be in a hurry, so a short simple questionnaire allows them to continue with their daily routine without too much interruption.

2. Keep it simple. You might know everything there is to know about what you are investigating but your target audience may not. Avoid technical terms.

3. Stick to closed questions; they are easier to analyse and display graphically. Closed questions are where you give a limited number of options to a particular question (Figure 1).

4. Question order

   a. Start with a screening question to establish whether or not the person should complete the questionnaire. You may only want people who are resident in the country of study or you may only want tourists.

   b. Move on to simple questions. You may need to know age group or education level. These are often the independent variable aspect of your survey.

   c. Then ask the harder questions. These are probably the ones that involve the substance of your study (maybe the dependent variable).

5. Do not use biased questions that will push the respondent in a particular direction.

6. For sensitive questions (for example, about age, income or education level) make sure you use closed questions with categories for responses. See Figure 2 for an example.

7. Run a pilot with friends and family to make sure the questionnaire gives you the information you need.

8. Make any necessary adjustments based on the pilot.

9. Make arrangements for the respondent to complete the questionnaire anonymously. In face-to-face interviews give the respondent the questionnaire and let them place it in an envelope.

10. Always ask yourself, "does this question relate clearly and directly to my investigation?" If the answer is "no", remove the question and think again.

Advantages of questionnaires

1. Standardized questions and responses are reasonably objective.

2. Data collection is quick.

3. You can collect a lot of information in a short period of time, especially if you are doing face-to-face data collection.

Key term

A **questionnaire** is a series of questions with a limited set of responses designed to obtain information about a particular topic.

How do you get to school?
Tick the appropriate response and move quickly on to the next question.

Walk	
Cycle	
School bus	
Public transport	
Car	
Other	

▲ Figure 1 Example of a closed question

How old are you?

Under 21	
21–40	
41–60	
Over 60	

▲ Figure 2 Example of how to tackle sensitive questions

Disadvantages of questionnaires

1. Data collection may be quick but good questionnaires take time to design.

2. You are sometimes relying on the memory of the respondents—not always very accurate.

3. If you are not conducting the questionnaire face to face you cannot explain the questions if they are not clear to the respondent.

4. Open-ended questions are difficult and time-consuming to analyse effectively, so do not use too many.

5. If the questionnaire is too long there is a risk that responses will be superficial, so keep it short.

6. People may not answer honestly if they feel embarrassed or if they think the response will damage them in some way.

Distribution of the questionnaire

Having designed and piloted the questionnaire you now have to collect the information. There are a number of options for this.

1. Face-to-face interviews

This involves you going out and conducting the survey face to face with the target population. This method can be time-consuming but it does tend to produce plenty of data. Where you go to conduct the survey will depend on the topic but generally speaking you can:

*   go to local centres of population (towns, villages, cities)

*   go to tourist sites, beaches and parks

*   visit local shopping centres

*   use your school population—students, teachers and parents.

You may need to seek permission for some of these options, so check first.

2. Email or online

If you are using your school population, you may be able to get access to the school's list of email addresses for students and staff and send your survey out by email. Many websites now have a questionnaire facility so you could send the survey out as an online survey.

3. Mail—very traditional but rather slow. Not recommended.

Sampling methods

It is usually impossible to complete your questionnaire for a whole population. Therefore, it is necessary to ask a representative **sample** of the population. However, the sample must be unbiased, representative of the whole population and include at least 30 people. The more people the better and more reliable, but also the more time-consuming.

Key term

Sampling is a statistical technique that allows you to obtain representative data from a small portion of the whole population.

There are four techniques that can be used to achieve an unbiased sample for questionnaires:

- systematic
- random
- stratified
- convenience.

Systematic sampling

This is sometimes called "nth" sampling—you simply ask people at a given interval, for example, every fifth person. The interval will depend on where you are and whether or not the area is busy. This method is very straightforward and the most useful method if you are dealing with an unknown population—for example, asking people on the street or in a mall. Here's what to do.

1. Select an interval (e.g. ask every fifth person that passes you).

2. Select a strategic sampling position.

3. Stop every fifth person that passes you and ask them if they mind completing your questionnaire.

4. Conduct your questionnaire with that person.

5. Repeat this process until you have a minimum of 30 completed questionnaires.

Stratified sampling

This is the method to use if you are targeting particular groups for the questionnaire. For example, you may want to ask just men, just women or a particular age group. This opens up two possibilities.

- Do you want an even number of respondents in each of your groups?

- Do you want a number of respondents in each category that is representative of the population as a whole? This may be hard to achieve.

The choice is yours.

Using the earlier example:

- You may want an equal number of people from each of four age groups: under 21, 21–40, 40–60 and over 60.

- In this case, you can stop every person that passes you and ask which age group they fit into. Only continue with the questionnaire if they are in the age group(s) that interests you.

- Repeat until you have a minimum of 30 completed questionnaires in total.

Use random.org to generate a random number

1. Go to www.random.org.

2. Go to "True random number generator" and set the minimum number to 1 and the maximum number to the last number on your list.

3. Click "generate"—this will give you a number.

4. Repeat until you have the required number of participants.

Random sampling

In this method, every member of the population has an equal chance of being selected. The method involves the use of random number tables or generators. With questionnaires, this can only be done if you have access to a list of the whole population. For example, you may be conducting the questionnaire for the whole school, a particular year group or all parents. Here's what to do.

1. Take a list of the entire population and allocate a number to each member of the population.

2. Generate a random number.

3. Find that number on your list—this person will be one of your respondents.

4. Repeat this until you have a minimum of 30 names to conduct the questionnaire with.

Convenience sampling

In this case, the selection of respondents is based on the fact that they are easy for the researcher to access. They may be in a good geographic location or available at the right time or just happy to participate. There are problems here of selection bias and sampling bias. This must be noted in analysis and evaluation.

Possible uses of a questionnaire

Questionnaires are a tool which could form part of your investigation. They are designed to generate data. They do not form a full individual investigation.

Example 1: Investigate the relationship between age and environmental attitudes

Research question: Is there a relationship between age and environmental attitudes?

The independent variable could be: age group—for example, over 25 or under 25.

This could be used to practise the following internal assessment (IA) skills.

- Planning: if you use this as an example then plan your own investigation

- Results, analysis and conclusion

- Discussion and evaluation

- Application

- Communication

How to tackle this research question:

1. Select a number of environmental issues that are covered in the ESS syllabus.

2. Design a questionnaire so that each question has three responses:

    a. ecocentric

    b. anthropocentric

    c. technocentric.

Collecting the data: the questionnaire

The responses on the questionnaire in Figure 3 are given a letter (e, a or t) to make it easier for you to understand. Do not identify answers in this way on the questionnaire you use.

*   Figure 3 shows some possible questions on a few of the topics you study— the aim is to give you some ideas. Make your own questionnaire to suit your area and your target audience.

*   Do not ask too many questions. The ideal number is five to seven.

*   Mix up the responses so there is no clear pattern for respondents to follow.

*   Adjust the responses to suit your audience, especially if you are going to ask the general public.

*   Pictures are not essential, but relevant pictures give people an idea of what you are talking about.

*   Other possible issues that you could use include loss of biodiversity; eutrophication; soil degradation; ozone depletion; tropospheric ozone or photochemical smog; resource depletion; solid domestic waste.

Reminder: There are three main environmental value systems (EVSs):

*   The **ecocentric** worldview puts ecology and nature as central to humanity and emphasizes a less materialistic approach to life with greater self-sufficiency of societies. It is life-centred— respects the rights of nature and the dependence of humans on nature so has a holistic view of life which is Earth-centred. Extreme ecocentrists are deep ecologists.

*   The **anthropocentric** worldview believes humans must sustainably manage the global system. This might be through the use of taxes, environmental regulation and legislation. It is human-centred —humans are not dependent on nature but nature is there to benefit humankind.

*   The **technocentric** worldview believes that technological developments can provide solutions to environmental problems. Environmental managers are technocentrists. Extreme technocentrists are cornucopians.

	1. The photograph shows one of the impacts of global climate change. This is considered by some to be a major environmental issue. Which one of the following solutions do you think is most suitable?
	We must educate people to encourage the reduction of greenhouse gases (GHGs)— such as use public transport, reduce electricity consumption, change diets. (e)
	We must regulate the production of GHGs through legislation and taxes. (a)
	We must look to technology for solutions—such as renewable energy, scrubbers, hybrid cars. (t)

2. Human population growth is significant. Most people agree that this will cause problems. What is the solution?	
Further scientific research is needed to ensure we can increase space, food production, water supply and resources. (t)	
It does not matter if people become less materialistic and more self-sufficient. (e)	
Policies such as China's "One child policy" should be employed to bring population growth under control. (a)	

3. Acid deposition is a serious problem in some areas of the world. How should we deal with this issue?	
We must educate people to encourage them to reduce the combustion of fossil fuels that cause acid deposition. (e)	
We must use legislation and impose taxes in order to reduce the production of the gases that cause acid deposition. (a)	
We must look to technology for solutions—such as renewable energy, scrubbers, hybrid cars. (t)	
4. The Great Pacific Garbage Patch (GPGP) is a mass of plastic in the middle of the Pacific Ocean. How can we avoid adding more plastic to it?	
Clean up the GPGP. (a)	
Reduce, reuse and recycle. (t)	
Raise awareness (through education) of the concept of biorights and the need for humans to self-regulate consumption of plastics. (e)	

▲ Figure 3 A section of a possible questionnaire on the subject of environmental attitudes

Once you have designed the questionnaire, select a sampling technique then go out and gather your data.

Collating the data

1. The first step is simple—separate the questionnaires into piles according to the independent variable. In this case, that is age.

2. Use a blank questionnaire and the five-bar tally system to record how many people responded to each question in each of the response categories. An example is shown in Figure 4.

The photograph shows one of the impacts of global climate change. This is considered by some to be a major environmental issue. Which one of the following solutions do you think is most suitable?	
We must educate people to encourage the reduction of GHGs— such as use public transport, reduce electricity consumption, change diets.	//// //// //// //// ////
We must regulate the production of GHGs through legislation and taxes.	////
Look to technology for solutions—such as renewable energy, scrubbers, hybrid cars.	/

▲ Figure 4 Example of collation of answers by under 25s for question 1 (Figure 3). This uses the five-bar tally system, which makes questionnaire collation very easy

3. Full collation will give you a table that summarizes the data as in Table 1.

	Ecocentric		Anthropocentric		Technocentric	
	Over 25s	Under 25s	Over 25s	Under 25s	Over 25s	Under 25s
Global climate change	5	6	3	5	7	5
Human population growth	4	9	4	4	8	4
Acid deposition	3	9	3	5	9	3
GPGP	2	5	2	6	5	2

▲ Table 1 Collation of results of our fictitious example. Sample size 118 of which 55 over 25 and 63 under 25. (Note—these are purely fictitious numbers.)

Presenting the data

The numbers from your table can then be used to produce a suitable presentation form, such as the bar chart in Figure 5.

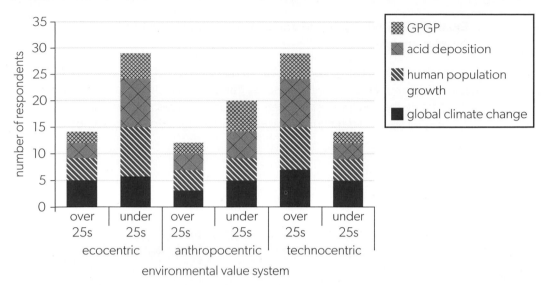

▲ Figure 5 Bar chart showing distribution of environmental attitudes according to age

Adapted from: Six Red Marbles and OUP

Example 2: Environmental attitude survey

Table 2 shows a range of quotes with which participants are invited to agree or disagree.

Statement	Agree	Disagree
"Let us a little permit Nature to take her own way; she better understands her own affairs than we." Michel de Montaigne		
"For the first time in the history of the world, every human being is now subjected to contact with dangerous chemicals, from the moment of conception until death." Rachel Carson		
"There are no passengers on Spaceship Earth. We are all crew." Marshall McLuhan		
"We shall require a substantially new manner of thinking if mankind is to survive." Albert Einstein		
"Every creature is better alive than dead, men and moose and pine-trees, and he who understands it aright will rather preserve its life than destroy it" H.D. Thoreau		
"We do not inherit the earth from our ancestors, we borrow it from our children." Attributed to Chief Seattle		
"The system of nature, of which man is a part, tends to be self-balancing, self-adjusting, self-cleansing. Not so with technology". E.F. Schumacher		
"Your grandchildren will likely find it incredible—or even sinful—that you burned up a gallon of gasoline to fetch a pack of cigarettes!" Paul MacCready Jr		
"Don't blow it—good planets are hard to find." Quote in *Time* magazine		
"A thing is right when it tends to preserve the integrity, stability and beauty of the biotic community. It is wrong when it tends otherwise." Aldo Leopold		

▲ Table 2 Section of a questionnaire on environmental attitudes

You do not have to use these quotes; you can use any others that you prefer. Go online and select your own quotes. You can find some that show anthropocentric or technocentric attitudes.

This could be used to practise the following IA skills.

- Planning: if you use this as an example then plan your own investigation

- Results, analysis and conclusion

Example 3: Environmental value systems and natural resources—a bipolar analysis

A bipolar questionnaire poses a series of questions to which the respondent rates their feelings or ideas on a sliding scale. You could use this type of questionnaire to compare the differences between, for example, ages, genders, income groups. Or you could gather the overall data to check opinions in a population as a whole.

Collecting the data

Table 3 shows the form of a bipolar questionnaire.

Worldviews can be considered on a sliding scale from most ecocentric to most technocentric. Using a scale of 1–5 (where 1 is the most ecocentric viewpoint and 5 is the most technocentric), rate your response to each of the natural resource issues below.	ECOCENTRIC People can be educated to see the environment holistically. People must exercise self-restraint and be: • less materialistic • self-sufficient.		ANTHROPOCENTRIC People can manage the environment sustainably through: • taxes, environmental regulation and legislation • debate to reach a consensual, pragmatic approach to solving environmental problems.		TECHNOCENTRIC Technological developments can provide solutions to environmental problems and improve the lot of humanity. This can be done through scientific research to form policies and understand how systems can be controlled, manipulated or exchanged to solve resource depletion.
Fossil fuel depletion	1	2	3	4	5
Water shortages	1	2	3	4	5
Soil depletion and desertification	1	2	3	4	5
Precious metals or gemstone depletion	1	2	3	4	5
Biodiversity loss	1	2	3	4	5
Landscape degradation	1	2	3	4	5
Deforestation	1	2	3	4	5

▲ Table 3 Part of a bipolar questionnaire on distribution of environmental attitudes

Again, you could support these ideas with pictures to help your respondents visualize the problems associated with the loss or excess use of these natural resources.

Once you have designed the questionnaire, select a sampling technique then go out and gather your data.

To save paper, you could have a single questionnaire and fill in the responses on a digital version.

Collating the data

1. Again, the first step for the paper version is to collate the questionnaire data. If you are gathering general data for the population you do not need to separate into piles (if you have an independent variable you are interested in then you do have to).

2. a. Use a blank questionnaire and the five-bar tally system to record how many people responded to each question in each of the response categories.

   b. If you used a digital version to record responses, you will already have everything in your digital version. Warning: this method requires a lot of concentration—it is easy to get distracted and forget.

Presenting the data

This data can be presented as a bar graph, as in the previous examples and Figure 6, or it can be shown in other ways. The most appropriate way to present your data will be a way that assists analysis.

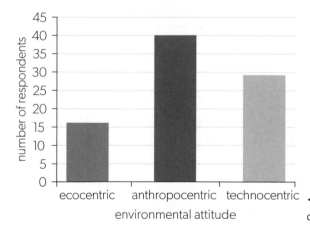

◄ Figure 6 Bar graph (again from completely fictitious data) to show total number of responses in each category

Behaviour over time graphs

Behaviour over time graphs show how a variable changes over time. Time is plotted along the horizontal (x) axis. The "behaviour" of the variable that changes over time is plotted on the vertical (y) axis. Often behaviour over time graphs are used to make projections of possible trends. They focus on patterns of change not on a single event. By plotting time into the future with different scenarios, behaviour over time graphs help us to focus on underlying causes.

The first section of the graph in Figure 7 (solid line) is behaviour observed up to the present. The second section (dotted lines) represents predicted alternative scenarios.

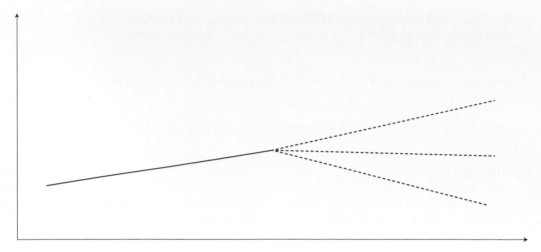

▲ Figure 7 The form of a behaviour over time graph

Questions to ask in behaviour over time graphs include the following:

- What is changing?

- How is it changing?

- Why is it changing?

- What are the relationships of the system?

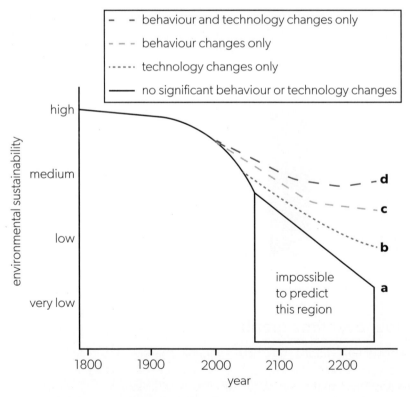

▲ Figure 8 Behaviour over time graph showing possible declines in environmental sustainability. The box represents a region where environmental sustainability levels are unknown; perhaps when a tipping point is reached

Adapted from Dennis C Hopkins

ATL Activity 20

ATL skills used: thinking

Look at the behaviour over time graph in Figure 8.

1. List at least three more variables (not listed below) which could be on the
    y-axis.

2. Describe two possible changes in behaviour that would cause the line to
    decrease more slowly.

3. Describe two possible changes in technology that would cause the line
    to decrease more slowly.

4. Describe two different changes in both behaviour and in technology that
    may cause sustainability to fall.

Behaviour over time graphs and UN population predictions

Figure 9 shows UN predictions about world population through the 21st century.

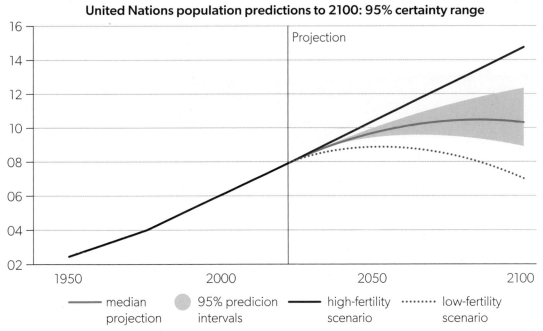

▲ Figure 9 Behaviour over time graph from the UN concerning world population predictions

Adapted from Population Matters. Data from: United Nations, 2022.

The y-axis could be used to show predictions about any variable that changes over time.
Examples include:

- biodiversity levels
- sea level rise
- global temperature changes
- carbon dioxide levels
- meat-eating in a group
- replacement of indigenous traditional lifestyles
- smoking or vaping in an age group
- littering.

Exam-style questions

1. To what extent can the different environmental value systems improve the sustainability of food production? [9]

2. In addressing environmental issues, mitigation strategies may be seen as primarily ecocentric and adaptation strategies as primarily technocentric.

   To what extent is this view valid in the context of named strategies for addressing the issue of global warming? [9]

3. Discuss the role of feedback mechanisms in maintaining the stability and promoting the restoration of plant communities threatened by human impacts. [9]

4. a. i. Identify **one** transfer and **one** transformation process shown in **Figure 1**. [2]

      ii. Outline how urbanization might impact **two** of the storages in **Figure 1**. [2]

   b. Run-off from agricultural land can result in excess nutrients entering water bodies. Outline **one** indirect measure of organic pollution. [3]

5. Outline how a positive feedback loop can impact an ecosystem. [4]

6. To what extent would different environmental value systems be successful in reducing a society's ecological footprint? [9]

7. To what extent do the approaches and strategies of different environmental value systems improve access to freshwater? [9]

8. a. Outline how the concept of sustainability can be applied to managing natural capital. [4]

   b. Explain how environmental indicators are used to assess sustainability. [7]

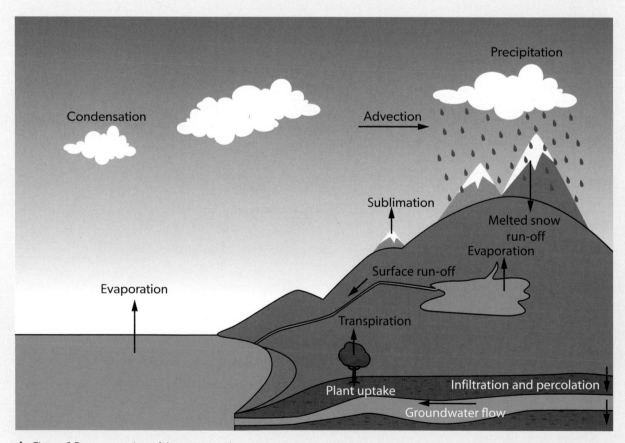

▲ Figure 1 Representation of the water cycle

9. a. With reference to processes occurring within the atmospheric system identify **two** transformations of matter. [2]

   b. With reference to processes occurring within the atmospheric system identify **two** transfers of energy. [2]

10. Outline the factors that lead to different environmental value systems in contrasting cultures. [4]

11. Outline how feedback loops are involved in alternate stable states and the tipping points between them. [4]

12. Outline how the model shown in **Figure 2** demonstrates positive feedback. [2]

13. Outline **two** historical influences on the development of the modern environmental movement. [4]

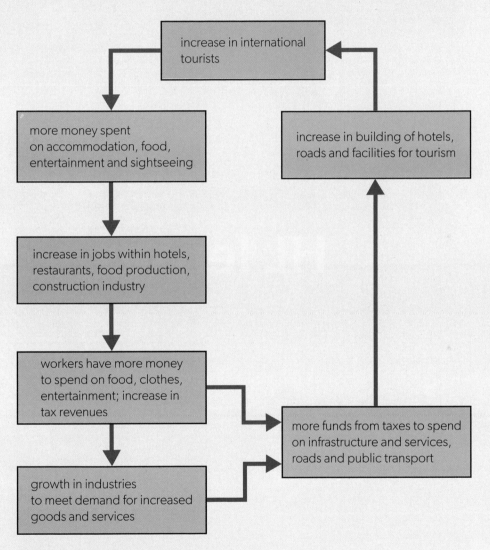

▲ Figure 2 Tourism multiplier effect

HL lenses

Introduction to the HL lenses

You should have already worked through Foundation (Topic 1), which introduces the three concepts—perspectives, systems and sustainability. As you work through Topics 2–8, these concepts will become clearer as we refer to them in each topic.

In Topics 2–8, there is additional HL (AHL) content which gives you greater depth and breadth.

In this section are the three HL lenses:

- HL.a—environmental law

- HL.b—environmental economics

- HL.c—environmental ethics.

These lenses give you viewpoints from different disciplines and should help you to make more connections between areas of the course. The aim is to increase your knowledge and understanding of the complexities of environmental issues, solutions and management. Figure 1 shows how the parts and concepts of the ESS course are linked.

We suggest that as an HL student you are aware of the content of these lenses early on in your study of the course and then refer to them when considering other topics. Of course, SL students are welcome to have a look as well but HL students are expected to synthesize and analyse with greater depth and breadth, and to use these lenses where appropriate.

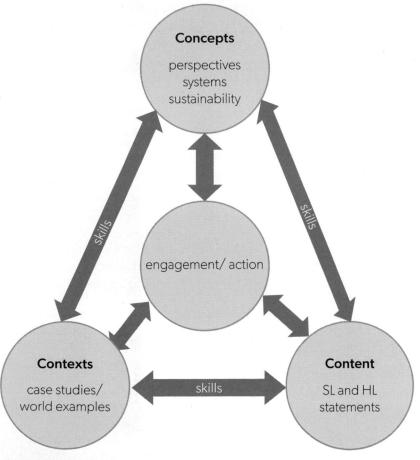

▲ Figure 1 HL relationships between concepts, contexts, content and HL lenses

ATL Activity 1

ATL skills used: thinking, communication, self-management

1. Read the understandings for the three HL lenses.

2. For each HL lens, create a table with three columns and as many rows as there are understandings.

3. Fill in the understandings in the first column.

4. In the second column, add relevant examples for as many understandings as you can.

   You will not be able to do this all at once but keep looking back to your tables and add examples as they arise in the course.

5. In the third column, link these understandings and the concepts to issues that you cover in the course.

HL.a Environmental law

Guiding question

- How can environmental law help to ensure the sustainable management of Earth systems?

Understandings

1. Laws are rules that govern human behaviour and are enforced by social or governmental authority.

2. Environmental law refers specifically to the rules about how human beings use and impact natural resources, with the aim of improving social and ecological sustainability.

3. Environmental laws can have an important role in addressing and supporting environmental justice, but they can be difficult to approve due to lobbying.

4. Environmental law is built into existing legal frameworks, but its success can vary from country to country.

5. Environmental constitutionalism refers to the introduction of environmental rights and obligations into the constitution.

6. Environmental laws can be drafted at the local, national or international level.

7. International law provides an essential framework for addressing transboundary issues of pollution and resource management.

8. UN conferences produce international conventions (agreements) that are legally binding, and protocols that may become legally binding, to all signatories.

9. International agreements can generate institutions or organizations to aid their implementation.

10. The application of international environmental law has been examined within international courts and tribunals.

11. There is an increasing number of laws granting legal personhood to natural entities in order to strengthen environmental protection.

12. Both legal and economic strategies can play a role in maintaining sustainable use of the environment.

▲ Figure 1 Example of a symbol representing a law—a "no entry" road sign

What are laws?

Laws are rules that govern human behaviour and are made and enforced by social or governmental authority. They are binding on the conduct of a community, aim to protect against abuses, and ensure rights as citizens.

Why have laws?

- Laws defend us from people who may wish to do us harm for no good reason. The International Criminal Court seeks to do this by trying individuals for genocide, war crimes, crimes against humanity and aggression.

- Laws promote the common good. If we all pursue self-interest only, all eventually suffer. The tragedy of the commons demonstrates this and there are examples throughout the book.

- Laws help in resolving disputes over limited resources. The Antarctic Treaty ensured that Antarctica was used for peaceful purposes with no disputes over sovereign territory.

- Laws help people to do the right thing. This could be in the form of moral laws or laws about doing no harm to others. Doing no harm to minorities is part of this.

What is environmental law?

Environmental laws are the legal frameworks that protect the natural environment.

Environmental law refers specifically to the rules about how human beings use and impact natural resources, with the aim of improving social and ecological sustainability.

The most important areas of environmental law relate to:

- pollution management—of air, soil and water

- wildlife and habitat conservation

- conservation of resources—environmental impact assessments (EIAs) on projects such as mining, development, industrial complexes, housing growth to address sustainability of forests, minerals, fisheries

- climate change.

Why have environmental laws?

Our environment is the total of all living and non-living things that influence human life. Clearly if laws only applied to actions between humans or to property belonging to humans, much of the environment would not be covered by the rule of law and common resources would be open to exploitation by a minority. Environmental laws attempt to prevent overexploitation and degradation of natural resources for the short-term interests of a minority, above the long-term interests of the common good. Laws can support or ensure ethical behaviour when economic systems incentivize environmental and social harm.

Levels of drafting of environmental laws

Environmental laws can be drafted at the local, national or international level. Generally speaking, laws made at higher national or international levels supersede those made at lower or local levels, depending on the scope of power of the authority involved.

International

International laws or bilateral agreements may be created and applied to transboundary environmental issues related to pollution and resource management. International environmental agreements exist regarding, for example, fisheries, air pollution and trade in endangered species.

Examples include the Montreal Protocol, Kyoto Protocol, CITES and IUCN.

National

Environmental laws around the world vary in scope, but typically share a common purpose: they use the legal system to protect the natural world and promote sustainability and environmental protection. This may involve forming an environment agency, or another body which oversees and enforces the laws and regulations, and ensures remedial work is done.

The USA Environmental Protection Agency (EPA) was founded in 1970 to establish and enforce environmental laws. The mission of the EPA is to protect human health and the environment.

TOK

Laws necessarily limit freedoms and therefore do harm to some as an unintended side-effect.

How true is it that laws must limit freedom and do have unintended side-effects?

▲ Figure 2 Environmental laws are designed to protect the environment

Connections: This links to material covered in subtopics 1.3, 3.3, 4.3, 7.1 and 7.2.

Connections: This links to material covered in subtopics 3.2, 3.3, 6.3 and 6.4.

The EPA works to ensure that:

- Americans have clean air, land and water

- national efforts to reduce environmental risks are based on the best available scientific information

- federal laws protecting human health and the environment are administered and enforced fairly, effectively and as Congress intended

- environmental stewardship is integral to USA policies concerning natural resources, human health, economic growth, energy, transportation, agriculture, industry and international trade, and these factors are similarly considered in establishing environmental policy

- all parts of society—communities, individuals, businesses, and state, local and tribal governments—have access to accurate information sufficient to effectively participate in managing human health and environmental risks

- contaminated lands and toxic sites are cleaned up by potentially responsible parties and revitalized

- chemicals in the marketplace are reviewed for safety.

Other national laws cover specific areas of the environment.

1. Air quality

Environmental law seeks to reduce air pollution levels and carbon emissions, particularly where they are harmful to human health. It does this through outlawing or limiting specific chemicals and pollutants, mandating air quality targets and enforcing regulations. About 97% of the world's population live in areas where air quality is usually worse than World Health Organization guidelines.

Air quality indices (AQI) measure pollutants such as ozone, sulfur dioxide, nitrogen dioxide and particulate matter, but exactly what is measured depends on the country. There are websites giving global, national and local data on AQI in real time so that authorities can either issue warnings or stop outdoor activities.

China's New Air Law of 2015 has reduced air pollution nearly as much in seven years as the USA did in three decades, helping to bring down average global smog levels. Cities are required to meet national air pollution targets including greenhouse gases (GHGs) by controlling coal use and transport emissions. However, it should be remembered that not all bad air quality is caused by anthropogenic sources. Beijing suffers the effects of sandstorms from the Gobi Desert, in which case the AQI can go above 1,000.

2. Water quality

Water pollution laws govern what can, and cannot, be discharged into water sources. They also control sewage treatment and disposal, plastics waste reduction in water sources, wastewater and animal waste management in industrial and agricultural sectors, and surface water run-off in urban areas and construction sites.

▲ Figure 3 Air pollution from car combustion engines

▲ Figure 4 Water pollution

In the USA, the Clean Water Act (CWA) established the basic structure for regulating discharges of pollutants into the waters of the USA and regulating quality standards for surface waters. This Act was created in 1948 and significantly reorganized and expanded in 1972. It was a most important modern environmental law.

3. Soil quality

Soil quality laws mandate what types of chemical discharges are allowed on soil and sediment, how soil health should be measured and monitored, and how pollutants should be cleaned up in the event of spills or other contaminant emergencies.

The EU Soil Health Law of 2023 sets a strategy for all soils to be in healthy condition by 2050 and to make protecting, restoring and sustainably using soils the norm. This is a comprehensive legal framework for soil protection.

4. Sustainability

These are environmental laws that protect biodiversity, natural habitats and existing ecosystems. Most governments have national laws that govern resource exploitation, both terrestrial and aquatic. These cover hunting and fishing, mineral resources, water resources and forestry.

In India, the Biological Diversity Act of 2002 was in response to obligations under the UN Convention on Biodiversity (CBD). It aims to protect biodiversity, facilitate sustainable management of biological resources, and allow for equitable sharing of benefits.

To encourage renewable energy investment, environmental laws help to regulate and support industries such as wind, solar, biomass and hydroelectric. Nations may introduce subsidies or incentives to use renewable energy sources or reduce energy use.

On recycling of products, Germany has the highest recycling rate in the world at over 66%. This is due to strong government rulings.

- Under the Packaging Ordinance (Verpackungsverordnung – VerpackV) of 1991, manufacturers and distributors of goods are responsible for ensuring packaging is either returned or recycled. This applies to commercial and domestic goods. Quotas are set for the recovery of materials, which manufacturers and distributors must abide by.

- Under the Closed Substance Cycle and Waste Management Act of 1996, producers of goods are responsible for the avoidance, recycling, reuse, and correct disposal of any waste. This is the circular economy model (see subtopic 1.3).

- The green dot on packaging means it is accepted for recycling and that the manufacturer has contributed to the cost of recycling it. Companies pay a licence fee to join the scheme, if they do not have to collect recyclable packaging themselves.

In addition, the culture of recycling is strong in the community and made easy for consumers with a national system of bins for different wastes. There is also a "refund on return" scheme for bottles and cans.

▲ Figure 5 Recycling bin with glass, plastic and metal

Local

At the local level, councils can have laws about recycling and waste disposal. In the UK, local councils collect waste and there are 338 local authorities in England. Each council has slightly different rules, so there is considerable variation in "household waste" recycling rates (from 18–64% in 2020–21). This is confusing for consumers and many give up trying to "do the right thing" if littering is not punished in some way.

Other laws and regulations at local level may be about air pollution caused by traffic congestion, noise levels, littering control, planning regulations, recycling requirements and so on.

Many cities around the world are working on schemes to reduce air pollution. Some schemes offer incentives for compliance, others apply penalties for non-compliance.

(ATL) Activity 2

ATL skills used: thinking, communication, research

Find an example of a local environmental law or regulation where you live—for example, relating to air, water or soil pollution.

1. Describe what it says.

2. Explain why it is in force.

3. Explain what it aims to achieve.

4. Evaluate its effect.

Case study 1

The Tasmanian Dam Case, 1983

The South West Wilderness area of Tasmania is remote and rugged, has unique wildlife and flora, and is a UNESCO World Heritage Area. The Gordon-below-Franklin Dam project was planned to generate hydroelectricity and much-needed employment in the area by damming the Gordon River. After a failed attempt to stop other dams at Lake Pedder in the 1970s, the Tasmanian Green movement protested against the Gordon Dam project. The pro-dam lobby also protested that it should go ahead in a state with high unemployment that needed more power.

The state government supported the dam but the federal Australian government and environmental groups did not. The case went to the Australian High Court which decided (in a judge vote of 4 to 3) that the federal government was correct. The dam did not go ahead.

This was a landmark case. When UNESCO designates a World Heritage Site, this has no binding force on any government. The overriding of state law by federal law, in support of the UNESCO designation, can be seen as a commitment to upholding environmental aims.

Large parts of Australia's main national environmental law, the Environment Protection and Biodiversity Conservation Act 1999, depend for their constitutional validity on the decision in the Tasmanian Dam Case.

▲ Figure 6 Lake Pedder

Environmental impact assessments (EIAs)

Environmental law underpins the environmental impact assessments (EIAs) that many governments require. These assessments investigate and report on the environmental impacts of projects before they can be carried out. EIAs are vital in enabling sustainable development to occur.

Effectiveness of environmental laws

When a law is passed it is written into the **statute book** of a country, which is a collection of all the existing laws of that country. The effectiveness of the law depends on the strength of legislation and its enforcement. This can be diluted by various parties which lobby politically for weaker laws, or if enforcement is not firm enough; there has to be enough funding for regulators to identify and fine those who break the laws. Environmental laws are less effective if funding is low or if the laws are not generally accepted by civil society.

Economic systems incentivize financial profit for shareholders and directors over the long-term interests of the common good. If the polluter-pays principle is not enforced (see subtopic HL.b), the majority can suffer.

Case study 2

UK sewage in watercourses

Several decades ago, England and Wales became the only countries in the world to have a fully privatized water and sewage disposal system. There are 11 water companies, many owned by overseas companies. The UK's antiquated sewerage system is inadequate. Water companies have failed to invest to protect coastal and river environments. Instead, they rely on a network of around 18,000 licensed sewer overflows to routinely discharge raw sewage into rivers and the ocean. The sewer system carries rainwater and wastewater from homes along the same pipes to water treatment works. When there is too much rainfall, water companies are allowed to discharge overflow from these pipes into the rivers and seas. In total, water firms in England discharged raw sewage into rivers 372,533 times in 2021. Only 14% of these rivers are in good biological health.

No new reservoirs have been opened since the water industry was privatized in 1991 and since that time the population of the UK has expanded by over 10 million.

The UK government water services regulation authority (OFWAT) regulates the companies but focuses on security and quality of supply and on keeping bills to consumers as low as possible. Over five years, there have been fines for pollution of GBP 250 million;

however, some GBP 50 billion has been paid to shareholders of these companies since privatization. The water companies self-monitor and report sewage discharges to OFWAT but the government does not have the funding to do its own inspecting in enough breadth.

▲ Figure 7 Sewage outfall pipe discharging into a lake

Questions

1. Explain why the water companies discharge raw sewage into English waterways.

2. Explain the potential results of sewage discharges into waterways.

3. Discuss the ways in which English water companies could improve the quality of water for consumers and the environment.

▲ Figure 8 A country's written constitution

TOK

How true is it that factual evidence may be abused, dismissed or ignored in politics?

International environmental law

Causes of climate change, its consequences and solutions inevitably cross national borders, so environmental law frequently requires international discussion and regulation.

The UN held one of the first international summits on environmental problems and issues in 1972. Since then, governments at multiple levels—along with human rights groups, non-governmental organizations (NGOs), and other international groups—have developed various frameworks, policies and treaties to safeguard natural resources, protect the environment, and forge connections between nations. International environmental lawyers have played a significant role in this.

Organizations working on environmental law at an international level include the UNEP, CITES, Greenpeace, ClientEarth and WWF.

How do these laws work?

A nation's **constitution** is a set of principles and rules by which a country is organized and governed. In many nations (e.g. India, Singapore, the USA) the constitution is one document, ratified at the time the nation came into being.

Constitutional law is the body of law that defines the relationship between different entities within a nation, most commonly the judiciary, the executive and the legislature bodies.

- **Environmental constitutionalism**—where environmental rights and obligations are incorporated into national constitutions—has become a widespread phenomenon. Internationally, more and more environmental cases are effectively addressed by nations' constitutions. Climate change issues are increasingly being addressed in this manner.

- **Climate constitutionalism** is increasing to address the global threat created by the climate crisis. Around the world, 150 countries had taken action on environmental constitutionalism as of 2017, and in 2020, the London School of Economics identified 11 countries with "climate clauses" in their constitutions. They are Algeria, Bolivia, Côte d'Ivoire, Cuba, Dominican Republic, Ecuador, Thailand, Tunisia, Venezuela, Vietnam and Zambia. There may be more now. Most of these clauses are broad aspirations and pledges to action, not targets. Rights-based provisions are relatively scarce but Tunisia's constitution guarantees all citizens the right to participate in the protection of the climate.

More examples of international laws

International law in general sets out the rules that apply to relationships between states. It also sets out rules on many issues that states have agreed are of international importance. It provides an essential framework for addressing transboundary issues of pollution and resource management.

The ASEAN Agreement on Transboundary Haze Pollution (AATHP)

This is a legally binding environmental agreement signed in 2002 by the 10 states of the Association of Southeast Asian Nations (ASEAN) to reduce pollution by haze in Southeast Asia.

Haze occurs when suspended dust and smoke particles accumulate in relatively dry air; PM$_{2.5}$ particulates do most damage to lungs. Haze from forest fires is a real problem when deforestation for agriculture, mainly in Sumatra and Kalimantan, is wind-blown over many ASEAN countries. The south-west monsoon blows the haze over Thailand, Singapore, Brunei and the Malay peninsula. Enforcement of the agreement is difficult, however, as it partly depends on educating small farmers to encourage sustainable practices.

Indonesia signed the agreement late and it is the third largest GHG emitter country in the world. Some 75% of its emissions are from deforestation. Singapore has imposed strict rules on multinational palm oil companies that invested in Indonesia. Greenpeace called for tighter regulations to control the overseas operations of big food companies, highlighting that contract farming through subsidiaries of multinational corporations was a major factor behind the high number of haze hotspots.

The FAO International Plan of Action (IPOA) to prevent, deter and eliminate Illegal, Unreported and Unregulated fishing (IUU)

The International Plan of Action (IPOA) is voluntary within the framework of the UN Food and Agriculture Organization (FAO) Code of Conduct for Responsible Fisheries.

Illegal, unreported and unregulated fishing (IUU) is an issue around the world. IUU occurs in most fisheries, and accounts for up to 30% of total catches in some. The objective of the IPOA is to prevent, deter and eliminate IUU fishing by providing all states with comprehensive, effective and transparent measures by which to act, including through appropriate regional fisheries management organizations established in accordance with international law.

> Connections: This links to material covered in subtopics 6.3 and 8.3.

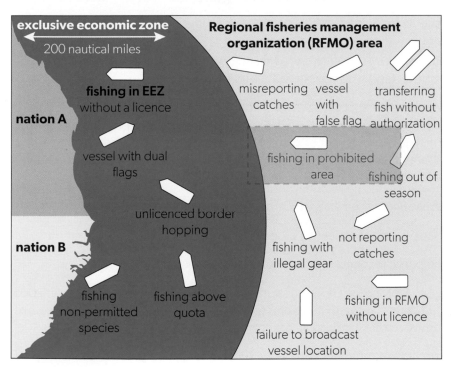

▲ Figure 9 US graphic on IUU fishing

Adapted from: U.S. Government Accountability Office

ATL Activity 4

ATL skills used: thinking, communication, research

Consider Figure 9 and your studies on sustainability and aquatic food production systems.

1. Review the ways in which IUU fishing may occur. How many are shown in Figure 9?

2. Which of these do you think are easier to regulate and why?

3. Describe ways in which nations A and B could ensure protection of their own EEZs (exclusive economic zones).

4. Explain why sustainable fishing is so difficult to achieve.

What does the UN do?

UN conferences produce international conventions (agreements) that are legally binding and protocols that may become legally binding to all signatories, though their development can be very slow and challenging.

There are difficulties involved in negotiating international agreements and the agreements have mixed results. Difficulties include:

- the complexity of the agreements

- rapidly evolving scientific knowledge

- pressures on individual governments from internal stakeholders with differing interests

- conflicts between countries over "differentiated responsibilities"

- financing commitments of high-income countries (HICs) towards low-income countries (LICs)

- general geopolitical conflicts

- potential economic impact of agreements.

But as people recognize that climate change is happening and affecting them, it becomes a more urgent issue.

The rule of law in environmental matters is essential for equity in terms of the advancement of Sustainable Development Goals (SDGs), the provision of fair access by assuring a rights-based approach, and the promotion and protection of environmental and other socio-economic rights.

The United Nations Environment Programme (UNEP) is responsible for coordinating responses to environmental issues within the UN system. Conflicts over natural resources and environmental crimes intensify the problems. At least 40% of internal conflicts over the last 60 years have a link to natural resources.

The Montreal Protocol of 1987 was an international treaty to phase out production of ozone depleting substances (ODS). Since then there have been nine revisions and the Kigali Amendment of 2016 phased down hydrofluorocarbons (HFCs) too as they are potent greenhouse gases (GHGs). The Montreal Protocol was widely adopted because:

- replacement chemicals (HFCs) could be produced

- there was a clear scientific issue of ozone depletion

- the public recognized and understood the risks.

The Kyoto Protocol on GHG emissions, adopted in 1997, came into force in 2005 with 192 parties. It committed states to reduce GHG emissions by anthropogenic activities which are causing climate change.

A second round of targets in the Doha Amendment in 2012 had lower take-up by countries, with the USA not ratifying the Amendment and others dropping out. By 2020, 147 states had agreed to the Doha Amendment.

TOK

At what level do political leaders and officials have different ethical obligations and responsibilities when making decisions that have such large impacts?

Connections: This links to material covered in subtopics 1.3, 3.3, 6.2, 6.3, 7.1 and 7.2.

Connections : This links to material covered in subtopic 3.3.

The Paris Agreement of 2015 was negotiated by 196 countries and signed by 195 (Iran, a major emitter, did not sign). The goal of this Agreement is to keep the global temperature increase to 2°C above pre-industrial levels and to reach net zero in carbon emissions by 2050. Each year there are conferences to assess progress and agree goals. These are the COP conferences.

Unfortunately, zero carbon targets are far more complex to agree and implement than those for phasing out ODS; they vary depending on many factors and some countries are able to fund emission reductions in other countries, thus meeting their own targets. As a result, progress towards the Paris Agreement targets has been slow.

International conventions and courts

Conventions

International agreements can generate institutions or organizations to aid their implementation. There is a range of conventions covered in the ESS course. Specific agreements are the most successful, for example:

- the Convention on International Trade in Endangered Species (CITES)

- the International Union for Conservation of Nature (IUCN).

Courts and tribunals

International courts and tribunals have had some role in the development of international environmental law.

- The International Court of Justice (ICJ), which has its seat in The Hague, is the principal judicial organ of the UN and rules on cases of discrimination and disputes between nations.

  For example, the Silala River in the Atacama Desert flows across the Chilean and Bolivian borders and its status as an international waterway was disputed. This centred on which country could take the water in a time of mega-drought. The court did not need to rule in the end as both parties agreed there that it was an international waterway and they would share the waters.

- The International Tribunal for the Law of the Sea (ITLOS) is an independent judicial body established by the 1982 UN Convention on the Law of the Sea.

- The European Court of Justice (ECJ) is the supreme court of the European Union (EU). It makes sure EU law is applied in the same way across the EU and that EU institutions and countries abide by EU law.

  One case was concerned with a Belgian law that prohibited waste that originated in member states from being dumped or stored in Wallonia (an area of Belgium). The ECJ ruled that waste was "goods" because it was the subject of a commercial transaction; therefore the Belgian law was counter to EU free movement of goods.

One of the difficulties faced by international judiciary bodies is how to evaluate appropriate compensation and damages for infringements of environmental law. Who decides the values is also complicated.

There are calls for a specialized International Court for the Environment.

▲ Figure 10 ICJ seal

▲ Figure 11 ECJ logo

Chagos Islands

A case brought before the ITLOS court is the dispute over the Chagos Islands in the Indian Ocean. The Chagos Archipelago is a group of atolls and 60 islands about 500 km south of the Maldives. They are a British Indian Ocean Territory (BIOT) because they were transferred to Britain after the Napoleonic Wars. Between 1967 and 1973, the UK expelled the indigenous peoples on Diego Garcia, the largest and only inhabited island of the Chagos, to allow the USA to build an airbase there. Many of these 1,500–2,000 people were sent to the Seychelles or Mauritius.

Mauritius gained independence from the UK in 1968, and has since claimed the Chagos Archipelago as Mauritian territory, arguing that it was coerced into giving up the Chagos Islands. The International Court of Justice ruled in 2019 that this action and the UK's continued administration of the Chagos was illegal. The UK dismissed this ruling as not legally binding.

In 2010, the UK announced the establishment of the Chagos Protected Marine Area—the world's largest marine reserve (larger than France or California). In 2015 the Permanent Court of Arbitration, a non-UN intergovernmental organization, ruled that this violated international law.

In 1992, the Maldives supported the UK's claim over the Chagos. In 2023, after the marine area was announced and fishing was prohibited in this area, they supported Mauritius' claim instead. The Maldives have a claim to fishing areas around the Chagos Islands within overlapping EEZs and fishing is 67% of Maldivian exports. At the time of writing, the UK still controls the Chagos and the airbase operates. Local peoples have not returned and have not received compensation. International law is complicated.

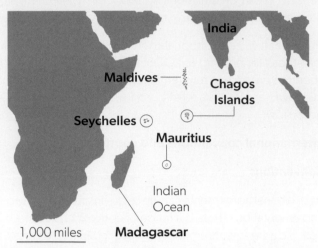

▲ Figure 12 Map showing the Chagos Islands, the Maldives and Mauritius in the Indian Ocean

Questions

1. Why did the UK make the indigeneous people of the Chagos Islands leave?

2. Why did the UK government believe it had the right to do this?

3. Which three countries are now disputing their claims to the Chagos and why?

4. Who benefits from and who suffers in the current dispute?

5. Discuss in class the ethical and justice issues in this case.

What is personhood?

Personhood is a term in law and in philosophy which describes the status of being a person. Granting legal personhood to natural entities can result in stronger environmental protection. Only a legal person can have rights, protections and legal liability in law. The law has a long history of granting legal personhood to corporations and other organizations, which means they have legal standing and can be sued or brought to court.

Examples of personhood are:

* the Ganges and Yamuna Rivers in India

* the Whanganui River (Te Awa Tupua) in New Zealand which is revered by the Maori people

- Mar Menor, a coastal saltwater lagoon in Spain—the first ecosystem to gain personhood

- Great ape personhood, first granted in 2007 by the Balearic Islands and then later by other countries.

▲ Figure 13 Mar Menor, Spain

▲ Figure 14 A silverback gorilla, part of the Great ape family

This is complicated too. Should personhood be granted to women, children, embryos, slaves, indigenous peoples, non-Great ape animals, aliens, A.I., mountains, scenic views, specific ecosystems, bacteria?

Indigenous knowledge systems do not recognize the distinction between humans and nature, or the differences between anthropocentric and ecocentric environmental value systems.

Linking environmental law and environmental economics

Environmental laws may be enacted but it is difficult to gain the agreement of all stakeholders and to ensure compliance with the laws. Similarly, environmental economics (see subtopic HL.b) attempts to value ecosystem services in financial terms but who makes the valuation and using what criteria? By integrating the two approaches, environmental sustainability may become more likely. For example, giving value to the cost of an oil spill can have an impact on the law which then imposes proportional fines.

TOK

At what level can the practices of indigenous cultures be judged with any validity by applying the moral values of another culture?

Check your understanding

How can environmental law help to ensure the sustainable management of Earth systems?

1. Define law and environmental law.

2. Describe the three levels at which environmental laws can be drafted.

3. Explain how national environmental laws may and may not be effective, with examples.

4. Discuss an example of a UN law or treaty that is effective and one that is less effective.

5. Evaluate the benefits of giving personhood to four entities of your choice. Two of these should be non-living.

- How can environmental economics ensure sustainability of the Earth's systems?
- How do different perspectives impact the type of economics governments and societies run?

Understandings

1. Economics studies how humans produce, distribute and consume goods and services, both individually and collectively.

2. Environmental economics is economics applied to the environment and environmental issues.

3. Market failure occurs when the allocation of goods and services by the free market imposes negative impacts on the environment.

4. When the market fails to prevent negative impacts, the polluter-pays principle may be applied.

5. "Greenwashing" or "green sheen" is where companies use marketing to give themselves a more environmentally friendly image.

6. The tragedy of the commons highlights the problem where property rights are not clearly delineated and no market price is attached to a common good, resulting in overexploitation.

7. Environmental accounting is the attempt to attach economic value to natural resources and their depletion.

8. In some cases economic value can be established by use, but this is not the case for non-use values.

9. Ecological economics is different from environmental economics in that it views the economy as a subsystem of Earth's larger biosphere and the social system as being a sub-component of ecology.

10. While the economic valuation of ecosystem services is addressed by environmental economics, there is an even greater emphasis in ecological economics.

11. Economic growth is the change in the total market value of goods and services in a country over a period and is usually measured as the annual percentage change in GDP.

12. Economic growth is influenced by supply and demand, and may be perceived as a measure of prosperity.

13. Economic growth has impacts on environmental welfare.

14. Eco-economic decoupling is the notion of separating economic growth from environmental degradation.

15. Ecological economics supports the need for degrowth, zero growth or slow growth and advocates planned reduction in consumption and production, particularly in high-income countries.

16. Ecological economists support a slow/no/zero growth model.

17. The circular economy and doughnut economics models can be seen as applications of ecological economics for sustainability.

What is economics?

Economics studies how humans produce, distribute and consume goods and services both individually and collectively. It focuses on supply and demand of resources and the outcomes of market interaction. It analyses the choices that individuals, businesses, governments and nations make to allocate resources. If demand is greater than supply, there will be shortages and prices rise in a free market. If supply is greater than demand, there is surplus and prices tend to fall.

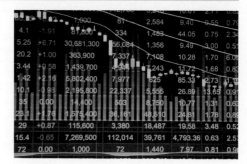

▲ Figure 1 A stock market chart

Case study 4

Water in Central Asia

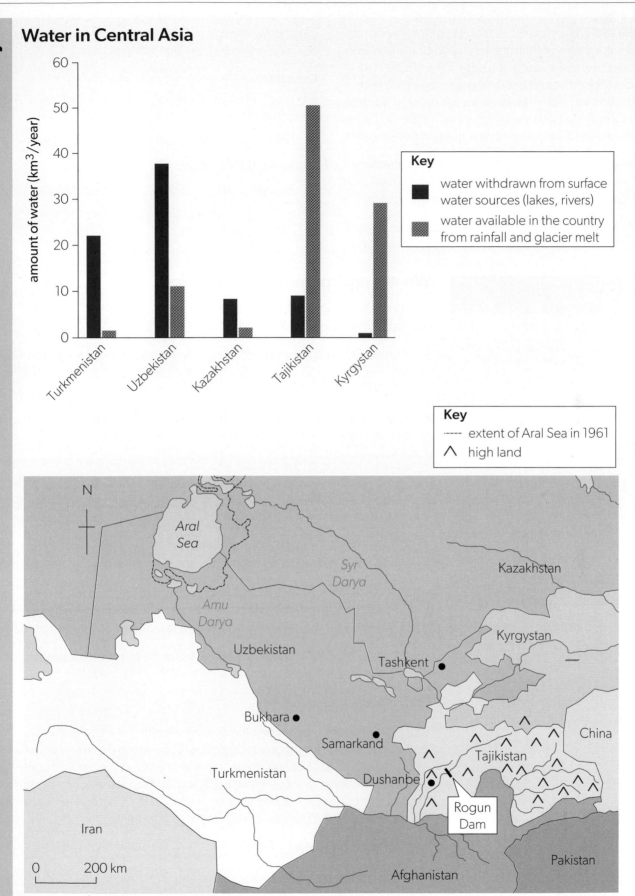

Key

■ water withdrawn from surface water sources (lakes, rivers)

▨ water available in the country from rainfall and glacier melt

Key

----- extent of Aral Sea in 1961

∧ high land

▲ Figure 2 Graph and map showing the relationship between demand for and supply of water in the countries of Central Asia

Central Asia is warming faster than the global average and glaciers are melting faster because wind-borne dust from the dried bed of the Aral Sea lands on the glaciers and reduces their reflectiveness (albedo) so they melt faster. River basins and water stores go across national borders. Although transboundary agreements are in place, they are not effective and some states view them as not equitable.

Where reliance is mostly on rainfall, droughts threaten agriculture and water supply. Water demand is increasing for irrigation and for the increased human population.

The World Bank has estimated that slow-onset climate change impacts in the region could result in 2.4 million climate migrants by 2050. These people are known as environmental refugees (see subtopic 8.1).

Questions

1. Discuss how water supply imbalances in Central Asia could be more equitably distributed.

2. For a region that you live in or know, investigate:

   a. water stores and use

   b. any transboundary issues relating to water

   c. ways in which water supply can be made more secure.

TOK

To what degree are political judgements a type of ethical judgement?

What is environmental economics?

Environmental economics is economics applied to the environment and environmental issues.

- It studies the economic effects of environmental policies around the world and it is concerned with the design and implementation of environmental policies.

- It developed as a result of environmental damage caused by economic activities and aims to enhance sustainable development.

- Its main focus is on the efficient allocation of environmental and natural resources and how alternative environmental policies deal with environmental damage, such as air pollution, water quality, toxic substances, solid waste and global warming.

- Environmental economics aims to deal with issues such as inefficient natural resource allocation, market failure, negative externalities and management of public goods.

- Technocentrics have the perspective that science and technology will enable environmental economics to work within the current economic framework.

- More ecocentric perspectives tend to value the approach of ecological economics (see later).

Economics and sustainability

Economic growth has negative impacts on environmental welfare: the use of resources or production of goods causes pollution which creates a net welfare loss on society at no cost to the producer. It may also lead to increased consumption of non-renewable resources, higher levels of pollution, global warming and loss of habitats. The challenge that countries face is how to develop economically—to lift the poorest out of the poverty trap—but to do this sustainably.

Polluter-pays principle

The polluter-pays principle is the commonly accepted idea that those who produce pollution should bear the costs of managing it to prevent damage to human health or the environment. For instance, a factory that produces a potentially poisonous substance as a by-product of its activities is usually held responsible for its safe disposal. In international environment law, this is mentioned in principle 16 of the 1992 Rio Declaration on Environment and Development.

Environmental economics has created solutions such as quotas, fines, taxes, tradeable permits and carbon neutral certification. These solutions ensure the polluter pays so as to limit the burden on society.

Scope of environmental economics

There are several important questions in environmental economics:

1. **What causes environmental challenges in terms of economics?** The concept of market failure is premised on the fact that markets for environmental goods—such as unpolluted air, a clean environment or scenic views of nature—are either non-existent or incomplete. There is likely to be no efficient allocation of environmental resources in the way that economists measure this.

2. **What is the monetary cost of environmental degradation through pollution, deforestation and loss of biodiversity?**

3. **What is the value of developments in the prevention and eradication of environmental harm?** How to measure and estimate the variables in questions 2 and 3 are vital in environmental economics.

4. **How can economic incentives and environmental policies be effectively designed to improve environmental quality and stop environmental damage?** Both economic incentives and environmental policies and regulations must be evaluated to find out if the intended objectives are achieved.

Environmental economics encompasses the following five concepts.

1. Sustainable development

Sustainable development is defined by the UNEP as "development that meets the needs of the present without compromising the ability of future generations to meet their own needs". The concept analyses the role of economic development in supporting sustainable development.

The four basic components of sustainable development are economic growth, environmental protection, social equity and institutional capacity.

2. Market failure

Market failure occurs when the allocation of goods and services by the free market imposes negative impacts on the environment. In an ideally functioning market, the forces of supply and demand are balanced, with a change in one side of the equation leading to another change that maintains the market's equilibrium. In a market failure, however, something interferes with this balance. For example, a factory that generates pollution while producing goods creates a net welfare loss on society at no cost to the factory.

Market failure occurs when a market is unable to efficiently allocate scarce resources at a given price because conditions for laws of supply and demand are not met. An example is an "environmental good" such as clean oceans. It is difficult to price the value of clean seas and oceans, and there exist no markets for clean water bodies where trading depends on the degree of cleanliness. This is a standard case of market failure.

> Connections: This links to material covered in subtopics 1.3, 3.3, 4.4, 5.2, 6.2, 6.3, 7.1, 7.2, and 8.3.

▲ Figure 3 What is the value of clean seas and oceans?

TOK

Is it true that technology has had an impact on how we perceive and filter data and information?

When the market fails to prevent negative impacts, the polluter-pays principle may be applied.

3. Externalities

Externalities are inadvertent consequences of economic activity that affect people over and above those directly involved in the activity. Externalities are another form of market failure. They can be negative or positive.

A **negative externality** creates unplanned outcomes that are harmful to the environment or to the general public. An example is industrial pollution, which results in unclean air and water and other health risks. The polluters may not incur any costs to address the pollution, even though their activities harm the environment and negatively affect the surrounding community.

A **positive externality** is a benefit to other people not directly involved in its generation. A community nature park can benefit people outside the community who visit family and friends in the area and would not have contributed to its development. People who benefit from an economic resource without contributing to its establishment are called "free riders".

▲ Figure 4 Negative (factory polluting) and positive (urban park for all) externalities

▲ Figure 5 An open cast copper mine

ATL Activity 5

ATL skills used: thinking, research

Figure 5 shows an open cast copper mine—a significant negative externality.

Investigate how the mining company could turn this into something positive for the community when they have finished mining the copper.

4. Valuation

Environmental accounting is the attempt to attach economic value to natural resources and their depletion. It is difficult to agree a consensus value with all stakeholders. There is a need for a common framework to measure the contribution of the environment to the economy and the impact of the economy on the environment.

The valuation of ecological resources is complex, as it is difficult to assign value to intangible benefits such as clean air and an unpolluted environment. But valuation is an important aspect of environmental economics as it helps us to evaluate a variety of options to manage challenges when using environmental and natural resources.

It is difficult to value resources that offer multiple benefits—for example, mountains may prevent flooding, provide scenic beauty, direct river flow patterns, and provide fertile soils for agriculture.

Environmental resources can be assigned values depending on use and non-use methods. It is far easier to assign value to a product in use, by observing what consumers are willing to pay to use it.

This is **use value**—when people benefit from direct use of the good. Opportunity cost (the price of missing out on the next best alternative) pricing, replacement cost, and pricing techniques can be employed in the "use" method.

▲ Figure 6 What is the value of cultural heritage?

▲ Figure 7 What is the value of natural scenery?

Non-use value is the value that people assign to economic goods they have never used and will never use. The concept is most commonly applied to the value of natural and built resources. Non-use value may include the intrinsic value of a species or the potential value of future use by forthcoming generations.

The contingent valuation technique is used for the "non-use" method. This uses a survey to measure what consumers would be willing to pay for a product or a common good they do not use or enjoy. It also considers how much people would accept as compensation payment in return for the destruction of a common good. For example, how can we value the cost of and compensation required for the *Exxon Valdez* oil spill of 1989 or the BP *Deepwater Horizon* oil spill of 2010?

5. Cost–benefit analysis

Humans have unlimited wants but live in a world with limited means. Economists study how people make decisions when faced with scarcity (when resources given to one end are not available to meet another). Under conditions of scarcity, there is an **opportunity cost** of any action.

Cost–benefit analysis (CBA) involves weighing the benefits arising from a policy against the perceived costs. The "best" policy is one which leads to the greatest surplus of benefits over costs.

CBA starts with a base policy where no changes are made to the status quo. Benefits are instances where human well-being is improved; costs decrease human well-being.

TOK

Look at Figures 6 and 7.

Decide on a price for each image:

- individually
- in small groups
- as a class.

To what extent is it economically valuable to put a price on natural resources?

ATL Activity 6

ATL skills used: thinking, research

Tankers or oil rigs spilling oil is not common but when it happens, it is an environmental disaster.

The *Exxon Valdez* ran aground in Prince William Sound on Bligh Reef, Alaska in 1989, spilling its cargo of crude oil. The oil eventually covered over 2,000 km of coastline. The spill had long-term environmental and economic impacts and litigation followed.

The BP *Deepwater Horizon* was an oil drilling rig in the Gulf of Mexico. In 2010, methane from drilling rose up into the rig, ignited and exploded. Oil leaked from the deep sea drill hole (1,600 m-deep) and flowed for 87 days. It was the world's largest accidental spill. The damage to species, the environment, the economy and human health was massive. Civil and federal criminal charges were brought and settlement payments were demanded.

Research the consequences of these oil spills in terms of:

1. legal actions and results

2. economic activity by communities affected

3. economic penalties paid by the companies involved

4. environmental effects, both short and longer term.

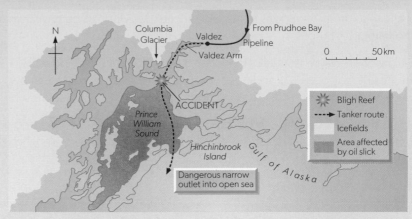

▲ Figure 8 *Exxon Valdez* oil spill

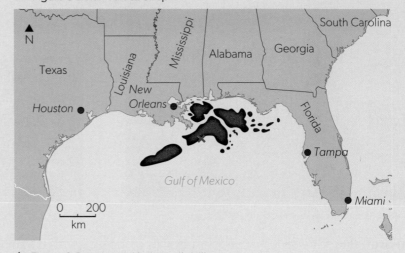

▲ Figure 9 *Deepwater Horizon* oil spill

Greenwashing

"Greenwashing" or "green sheen" is where companies use marketing to give themselves a more environmentally friendly image. Companies also produce, and charge more for, products that appear to be—but are not—less bad for the environment than alternatives.

Carbon offsetting: greenwashing or an effective market?

Connections: This links to material covered in subtopics 1.1 and 6.3.

Carbon offsetting is either a reduction in greenhouse gas (GHG) emissions or an increase in carbon storage that is used to compensate for emissions that occur elsewhere. Examples of this are:

- paying for methane gas capture in landfill sites
- creating new wind farms
- forest conservation or tree planting
- installing clean cookstoves
- creating solar farms
- creating dams for hydropower
- land restoration.

Carbon offsetting occurs when a polluting company or country buys carbon credits to make up for the GHGs it has emitted. The money should be used to fund action somewhere in the world that removes the same amount of carbon from the air, or prevents other GHG emissions.

But carbon offsets do not effectively reduce climate change because they only reduce carbon dioxide (CO_2) if the projects are additional and permanent.

They are measured in tonnes of CO_2 equivalents and are bought and sold through international brokers, online retailers, and trading platforms on the global carbon offset market. Table 1 summarizes some of the key problems with carbon offsetting.

Problem	Explanation
Carbon offsetting is not sustainable	**Environmentally:** Carbon offsets do not work towards the core issue of reducing CO_2 emissions. There are not enough offsets for all CO_2 emissions, and not all carbon offset projects are realized. **Economically:** Carbon offsets further an economic gap between the world's rich and poor. **Socially:** Carbon offsets are often used as greenwashing and they maintain an economic gap between the world's rich and poor.
Carbon offsetting is not ethical	Carbon offsets do not work towards the core issue of reducing CO_2 emissions; poorer countries are paid to offset carbon while "richer" countries continue to emit; carbon offset projects are often used as greenwashing.
Carbon offsetting is not good for the environment	Carbon offsets do not effectively reduce climate change, and there are not enough offsets for all of our CO_2 emissions.
Different projects have different effectiveness rates	Direct CO_2 removal is the most effective project category, followed by renewable energy, energy efficiency and, lastly, carbon sequestration.

▲ Table 1 Problems with carbon offsetting

The tragedy of the commons

The commons are those things we own together. Some are abstract like a language; others are tangible, such as fishing rights or common land. But if nobody takes responsibility for something, it will nearly always be abused. The tragedy of the commons (see subtopic 1.3) highlights the problem where property rights are not clearly delineated and no market price is attached to a common good, resulting in overexploitation. Often the solution involves private ownership or government control but there is another option if certain rules are followed.

▲ Figure 10 Swiss pasture where shared ownership can work

▲ Figure 11 Elinor Ostrom (1933–2012)

Elinor Ostrom was an American political economist who was awarded a Nobel Prize in Economics for her work. She showed that the use of resources by groups of people can be rational and prevent resource depletion without government intervention. Her theory stated that local communities are the best at managing their natural resources (as they are the ones that use them) and that all regulation on the use of resources should be done at the local level—not by a higher, central authority that does not have direct interaction with the resources. Ostrom looked at use of pastures in Africa, forests in Nepal, fishing waters in Indonesia and Maine, and pastures of a small alpine village in Switzerland. She developed eight design principles for responsibly managing shared resources.

The Ostrom principles are:

1. **Commons need to have clearly defined boundaries.** In particular, who is entitled to access to what? Unless there is a specified community of benefit, it becomes a free for all.

2. **Rules should fit local circumstances.** There is no one-size-fits-all. Rules should be dictated by local people and local ecological needs.

3. **Many users should be involved in decision-making.**

4. **Commons must be monitored.** Once rules have been set, communities need a way of checking that people are keeping them and are accountable if not.

5. **Sanctions for those who abuse the commons should be graduated.** Banning creates resentment so have systems of warnings and fines, as well as informal reputational consequences in the community.

6. **Conflict resolution should be easily accessible.** Resolving issues should be informal, cheap and straightforward.

7. **Commons need the right to organize.** A higher local authority must recognize them as legitimate.

8. **Commons work best when nested within larger networks.** Some things can be managed locally, but some might need wider regional cooperation —for example, an irrigation network might depend on a river that others also draw on upstream.

Connections: This links to material covered in subtopics 1.3, 3.2 and 4.3.

This sounds straightforward yet the rules are frequently not applied. Consider the Grand Banks fishery collapse, hunting of great whales or bison nearly to extinction, and hunting of the dodo, passenger pigeon and many other species to extinction.

Ecological economics

While environmental economics is based on traditional (classical) economic theories, ecological economics is transdisciplinary and addresses the interdependence of human economies and natural ecosystems both in space and over time. The economic valuation of ecosystem services is addressed by both environmental economics and ecological economics, but there is an even greater emphasis in ecological economics.

It focuses on:

- sustainable scales

- fair distribution, and equity between generations

- efficient resource allocation

- the value of natural capital as well as physical, human and financial capital

- environmental justice.

These factors all contribute to human well-being and sustainability.

Ecological economics views the economy as a subsystem of Earth's biosphere and the social system as a sub-component of ecology. Inputs to this economic system are solar energy and natural resources. Outputs are goods, waste and lower-grade energy as heat. This linear model is challenged in "new" economic thinking by the circular economy and the doughnut economy models. The concept that HICs should pay LICs not to deplete their natural assets is one element of ecological economics.

The precautionary principle

The precautionary principle was first set out in a European Commission communication adopted in February 2000, which defined the concept and envisaged how it would be applied. It is an approach to risk management, where, if it is possible that a given policy or action might cause harm to the public or the environment and if there is still no scientific agreement on the issue, the policy or action in question should not be carried out. The precautionary principle may only be invoked if there is a potential risk and may not be used to justify arbitrary decisions.

Economic growth

Gross domestic product (GDP) is the monetary measure of all goods and services produced by a country in a given period of time. Per capita GDP is a more accurate assessment of living standards but does not take into account inequalities in the actual distribution of income.

Economic growth is influenced by supply and demand and may be seen as a measure of prosperity (a condition in which people are successful and thriving in the economic sense). This linear economy approach does not usually take into account waste, pollution and issues that lead to environmental degradation. Prosperity has long been considered a good thing but global economic growth using more natural resources than are available is not possible. The question now is whether economic growth can ever be sustainable.

Economic growth has both negative and positive impacts on environmental welfare.

- Negative: Rising incomes associated with economic growth may lead to increased consumption of non-renewable resources, higher levels of pollution, global warming and the potential loss of habitats.

- Positive: Higher incomes could mean individuals and society use resources to protect the environment and mitigate environmental problems such as pollution.

Key term

Gross domestic product (GDP) measures economic growth by the annual percentage increase in the total market value of goods and services in a country over a period.

Connections: This links to material covered in subtopic 7.3.

▲ Figure 12 SDG 8

SDG 8 is about promoting inclusive and sustainable economic growth, employment and decent work for all. But the COVID-19 pandemic precipitated the worst economic crisis in decades and reversed progress towards decent work for all. The following figures come from the Sustainable Development Goals Report 2022:

• The real GDP for LICs is projected to rise by 4.0% in 2022, and 5.7% in 2023—still below the 7% target under the 2030 Agenda.

• In 2021, global output per worker rebounded sharply, rising by 3.2%; however, productivity in LICs declined by 1.6%.

• The average worker in a HIC produced 13.6 times more output than the average worker in a LIC in 2021.

• Worldwide, 160 million children (63 million girls and 97 million boys) were engaged in child labour at the beginning of 2020.

Eco-economic decoupling

This term refers to breaking the link between "environmental bad" effects and "economic good" effects. Decoupling may be absolute or relative and can be measured by decoupling indicators. A decoupled economy can grow without also increasing environmental pressures. Environmental pressure is often measured using emissions of pollutants and decoupling is often measured by levels of emissions against economic output.

Eco-economic decoupling is the notion of separating economic growth from environmental degradation. In many economies, economic growth increases pressure on the environment. If an economy can grow and at the same time reduce its consumption of clean water, other natural resources and energy from fossil fuels and not degrade the environment, it has "decoupled".

However, adopting the doughnut economy model changes perspectives and requires a total rethink of this viewpoint.

Connections: This links to material covered in subtopic 8.3.

Pressure on the environment is often measured by emission of pollutants as this is fairly easily measurable. If emissions reduce but economic growth increases, is this decoupling or are there other measurements to be made?

Decoupling indicators, like all other types of indicator, shed light on particular aspects of a complex reality but leave out other aspects. For example, the decoupling concept lacks an automatic link to resilience of the ecosystem.

Examples of absolute long-term decoupling are rare, but recently some industrialized countries have decoupled GDP growth from CO_2 emissions. However, this is often achieved by out-sourcing emissions—for example, importing fuels from other countries or buying carbon credits. A country may then claim to have decoupled when, in reality, degradation is still occurring.

Slow/no/zero growth

Ecological economists support a slow/no/zero growth model. Rather than focusing on GDP, they address the extent to which the ecological footprint of a country is sustainably balanced by its biocapacity.

Conventional economic theory has been that economic growth is not only desirable but is the best sign of a healthy economy. Other economists, including Kate Raworth of the doughnut model, say that economic stagnation (lack of or slow growth) can actually be a sign of prosperity. Stagnation is an indication that a country has neared its ceiling for economic success or shifted the material basis of its economy from manufacturing to services. In this view, it is not necessarily a bad thing if growth slows or stops during a stable economic period. Higher productivity can be achieved not only by increased GDP but also by higher efficiency using fewer inputs, working fewer hours, and using fewer resources. However, when this happens, unemployment rises which raises issues of equity.

(ATL) Activity 7

ATL skills used: thinking, communication, research

Review the circular economy and doughnut economics models, covered in subtopic 1.3. These models can be seen as applications of ecological economics for sustainability. Their effectiveness in addressing the sustainable activity of a society varies.

The doughnut model involves living sustainably. However, a country can claim to be sustainable if it imports most goods and exports waste products elsewhere—the problems are passed on to someone else. Clearly that is not globally sustainable.

There have been attempts to quantify the doughnut economic models for different countries.

1. Search online for "doughnut economy by country by Netlify".
2. Choose your country from the drop-down list. Hover over each segment of the doughnut to see more information.
3. Where has the ecological ceiling been exceeded?
4. Which social boundaries are within the green safe and just zone?
5. Where is social damage greatest?
6. Evaluate the model for your country.
7. Now study the model for two more countries that you know a little about—one HIC and one LIC. Describe and explain the differences you observe.

Check your understanding

How can environmental economics ensure sustainability of the Earth's systems?

How do different perspectives impact the type of economics governments and societies run?

1. Define economics, environmental economics and ecological economics.
2. Describe the impacts of market failure and how to mitigate the failure.
3. Describe and comment on one example of greenwashing of a product.
4. Explain why it is difficult to give a valuation in monetary terms to environmental services.
5. Define GDP and explain why it is difficult to make it compatible with sustainable development.
6. Explain how ecological economics differs from environmental economics.
7. Evaluate the possible effects of decoupling an economy from environmental degradation.
8. Evaluate the impacts of slow or zero growth (a) on equity and (b) on sustainability.

HL.c Environmental ethics

Guiding questions

- To what extent do humans have a moral responsibility towards the environment?
- How does environmental ethics influence approaches to achieving a sustainable future?

Understandings

1. Ethics is the branch of philosophy that focuses on moral principles and what behaviours are right and wrong.

2. Environmental ethics is a branch of ethical philosophy that addresses environmental issues.

3. A variety of ethical frameworks and conflicting ethical values emerge from differing fundamental beliefs concerning the relationship between humans and nature.

4. Instrumental value is the usefulness an entity has for humans.

5. Intrinsic value is the value one may attach to something simply for what it is.

6. The concepts of instrumental and intrinsic value are not exclusive.

7. An entity has "moral standing" if it is to be morally considered with regard to how we ought to act towards it.

8. There are three major approaches of traditional ethics: virtue ethics, consequentialist (for example, utilitarian) ethics and rights-based (deontological) ethics.

9. Virtue ethics focuses on the character of the person doing the action. It assumes that good people will do good actions and bad people will do bad actions.

10. Consequentialist ethics is the view that the consequences of an action determine the morality of the action.

11. Rights-based ethical systems focus on the actions and whether they conflict with the rights of others. There is debate about what these rights might be.

12. Some people hold the view that whatever is natural is correct or good. This position is contentious and is described as the "appeal to nature" fallacy.

13. Environmental movements and social justice movements have developed from separate histories but are increasingly seeking common goals of equitable and just societies.

Key terms

Ethics is the branch of philosophy that focuses on moral principles and what behaviours are morally right or wrong, just or unjust.

Morals are what you believe to be right and wrong.

Making ethical decisions on environmental issues

Humans make many **ethical** decisions with respect to the environment. For example:

- Should environmental research be subject to ethical constraints or is the pursuit of all knowledge intrinsically worthwhile?

- What environmental obligations do we have for future human generations?

- Is it right to cause the extinction of a species to benefit humanity?

- Should humans continue to cause deforestation and loss of biodiversity for our benefit, just one species of many?

- Do human rights and rights for other life exist in the same way that the laws of gravity exist?

We all have our own ideas about what is right and wrong and how we tell the difference. Our viewpoints are influenced by the communities in which we live (see subtopic 1.1). **Morality** is the belief that some behaviour is right and acceptable and that other behaviour is wrong.

Morality can be seen as something personal, whereas ethics relates to the standards of "good and bad" distinguished by a certain community or social grouping.

Moral standards are values that a society uses to determine reasonable, correct or acceptable behaviours. Some standards are universally accepted; for example, most societies believe killing is wrong, but some make exceptions to this in a just war or in self-defence.

ATL Activity 8

ATL skills used: thinking, communication, social, research

Look at the list of ethical questions at the start of this subtopic.

1. In groups, make a list of at least five more questions, thinking about your ethical relationship with your environment. For example: What do you use, what do you destroy, what do you do about pests, how do you view other lives?

2. For each question, consider your own perspective and values and your environmental value system (EVS). Do these change depending on the question?

Environmental ethics

Environmental ethics arose in the 1960s and 1970s when awareness of environmental issues increased. Most western ethical traditions focused on human–human actions and relationships only. They did not address the moral status of non-human or non-living environmental entities.

An ethical question for humanity now is: "What is our relationship with the natural environment?"

Different ethical frameworks and conflicting ethical values emerge from differing fundamental beliefs concerning the relationship between humans and nature.

Most current economic, political and cultural ideologies are based on anthropocentric values. The growth in size of the human population and of industrialization and exploitation of the natural world shows we live unsustainably and are damaging and destroying life on Earth. An ethical value system which sees nature as separate and there for humans to utilize would suggest a technocentric worldview. Viewing humans as stewards of nature but part of it suggests you would protect the environment and all life within it. If you see humans as part of nature in which all have intrinsic value and rights, you may hold an ecocentric environmental value system.

Instrumental and intrinsic values

In ethics there are two kinds of value—instrumental and intrinsic. You can value something (whether a living organism, an ecosystem or a non-living object) as an end in itself, or as a means to an end, or as both together.

TOK

From a TOK point of view, ethics focuses on knowledge questions about ethical issues, not on the ethical issues alone. It is about ethical considerations that impact inquiry in ESS. For example, "How can we know when we should act on what we know about the extinction of species?" The focus should be on the knowledge we are using to make decisions, not on extended discussions about the ethical theories.

If moral claims conflict, does it follow that all views about anthropogenic induced extinction are equally acceptable?

Key term

Environmental ethics is a branch of philosophy that addresses environmental issues.

Connections: This links to material covered in subtopic 1.3.

TOK

How much does the use of the terms "ecocentric", "anthropocentric" and "technocentric" convey ethical judgements?

Intrinsic value is the value that an entity has in itself, for what it is, or as an end in itself. A landscape, mountain, lake, ocean or single living insect may be valued for itself, regardless of whether a human sees it as beautiful or aesthetically pleasing—regardless even of whether a human ever sees it at all.

Instrumental value is the usefulness an entity has for humans. It is the value that something has as a means to a desired or valued end. Money has instrumental value, not in itself—because it is just a piece of paper, a pile of scrap metal or a computer print-out—but because of what it buys for us. Ecosystems have instrumental value for us because they provide goods such as food or water, or services such as waste processing, carbon storage or water purification.

An entity may have both intrinsic and instrumental value at the same time. Forests that are beautiful in their own right are also useful to humans. An individual animal in a zoo may attract the public but also has its own intrinsic value.

How much is one whale worth?

Whales have an important role in the overall health of the marine environment. They play a significant role in capturing carbon from the atmosphere. Each great whale sequesters an estimated 33 tonnes of CO_2 on average. Baleen whales eat krill (shrimp-like animals), small fish and phytoplankton. Phytoplankton absorb a massive amount of carbon from the atmosphere. Without whales, krill would probably eat much of the free-floating phytoplankton on the ocean's surface, resulting in a marked acceleration in climate change.

In 2019, the International Monetary Fund published a paper that proposed assigning a value of USD 2 million for each great whale to account for their role in carbon removal. That USD 2 million figure equates to the value of 1,000 large commercial trees. In addition to their carbon storage role, sperm whales have value due to ambergris (undigested waste that they excrete), which is used in the perfume market, especially to create fragrances like musk. There is also the ecotourism value of whale-watching around the world. Overall, the current population of whales is "worth" more than USD 1 trillion.

But what is the intrinsic value of whales? How can this be calculated? If you have seen a whale in the ocean, how can you value that? If we attach a dollar value to an animal, does that make its preservation more likely? If we reduce ecosystems to services they provide for us, and reduce species to an economic valuation, we miss both the aesthetic valuation and the intrinsic value of a whale for its own sake. Whales have been sources of awe, beauty, admiration and fear. And now they can wash up on beaches, dead and with stomachs full of plastic waste.

▲ Figure 1 Humpback whale breaching

 **ATL** **Activity 9**

ATL skills used: thinking

Consider your perspective on how to value:

1. whales
2. penguins
3. fleas
4. tigers
5. viruses
6. your favourite view
7. a green space near you.

TOK

To what extent does the beauty or cuteness of an organism play a role in how we perceive its value?

▶ Figure 2 A mosquito and a kitten

Moral standing—who counts?

An individual has moral standing for us if we believe that it makes a difference, morally, how that individual is treated, apart from the effects it has on others. But what about the moral standing of non-humans? And when does a fertilized egg have moral standing? The moral standing of many animals will be strongly influenced by culture and religion.

Human beings interact with non-human animals in a variety of ways. Some animals, like pet dogs, cats, birds and hamsters, live in our homes under our care. Others, like mice, rats, spiders and mosquitoes often live in our homes as unwelcome residents. Domestic animals, like cows and chickens, are bred for humans to eat their flesh, milk, eggs or other products. Other animals are experimented on by humans in laboratories for medical research, used for transport, hunted by humans for sport, or used for entertainment purposes in circuses, zoos and films.

Doctors look after the physical welfare of patients and believe it is morally wrong to mistreat them. They do this because they care. Farmers look after their animals because otherwise the herd may become ill or not grow fast enough. If farmers do this for the sake of the animals, the animals have moral standing in the farmers' eyes. If it is only because the animals would not provide income or food for the family, then animals have no moral standing with the farmer. If animals have a moral standing, this means it is objectionable to treat them in certain ways, such as using them in experiments for the benefit of human beings or causing them suffering. They should at least be treated with the respect a living being deserves.

▲ Figure 3 Doctor checking a patient

Anthropocentrists may believe that only humans have moral standing. But what about the Great apes and other animals that can reason? Should their moral standing be considered? That only human beings ultimately count in morality does not mean we do not have obligations to other species. Anthropocentric views hold that it is immoral to destroy plants or animals needlessly. This is because in doing so we are destroying resources that may provide significant benefits to ourselves or to future human generations. But non-humans count only to the extent that the welfare of human beings is affected.

The question "can animals suffer?" arose in the 18th century and the animal rights movement of the 1970s developed from this. In this movement, "species discrimination" is equivalent to other forms of discrimination. Animals have rights morally, not only for utility reasons. We should respect their individual lives.

In the 20th century, this was extended to all living things because all have interests and therefore moral rights. For example, plants grow towards light and show self-preservation (e.g. some plants have thorns or toxins to stop herbivores eating them).

It is a short step then to the moral rights of entire ecosystems. Aldo Leopold was a US naturalist who campaigned for wildlife and wilderness areas. He first put forward the ecocentric view of a "land ethic" that gives all nature moral standing. This includes soil, water and all living things. But it still raises the question of who deserves more consideration when interests are competing.

▲ Figure 4 A herd of goats near a water hole on the African savanna

▲ Figure 5 Bust of Aristotle

Similarly, what about the moral standing of future generations? Do humans alive today have obligations towards humans who will be living in the future, irrespective of the benefits to humans today? There are many questions here and the ethical dilemmas are large.

Three approaches of traditional ethics

There are three major approaches of traditional ethics or ethical decision-making frameworks:

* virtue ethics—based on whether you are a good person

* consequentialist ethics—based on outcomes

* rights-based (deontological) ethics—based on duty or rights.

1. **Virtue ethics** focuses on the character of the person doing the action, not on the action alone. It is person-based rather than action-based and assumes that good people do good actions and bad people do bad actions. There are duties which do not depend on consequences but the approach is to develop the character to act on those duties.

    Aristotle (384–322 BCE) wrote the first treatment of ethics in western civilization: the Nicomachean Ethic. This is virtue ethics but virtue here means the excellence of a thing. The modern philosopher Alasdair MacIntyre proposed three questions as at the heart of moral thinking:

    * Who am I?

    * Who ought I to become?

    * How ought I to get there?

    "Virtues" are attitudes, dispositions, or character traits that enable us to be and to act in ways that develop our potential. They enable us to pursue the ideals we have adopted. Honesty, courage, compassion, generosity, fidelity, integrity, fairness, self-control and prudence are all examples of virtues. Respect, compassion and responsibility are virtuous approaches to the natural world.

2. **Consequentialist ethics** focuses on the consequences of actions. It is the utilitarian view that the consequences of an action determine the morality of the action. Actions with good consequences are good actions. Actions with bad consequences are bad actions. Morally good actions are those that result in the greatest common good. The intention of an action does not affect its morality; it is simply a matter of the outcome. It asks, "How will my actions affect others?" and then assesses the impact of actions on a "least common denominator" such as happiness or wealth.

    The motto for this approach is, "The greatest good for the greatest number"—but what is good?

    In business, people do a cost–benefit analysis. This is a utilitarian approach to ethics, in which the common denominator is usually money. Everything from the cost of a product to the cost of a human life is given a monetary value. Utilitarians can be considered "consequentialists" because they look at the consequences of actions to determine whether any particular act is right or wrong.

Bulldozing a house to build a motorway, felling a forest to build an airport, and redistributing money to the poorest through taxes are consequentialist actions—but do they all serve the greater good?

Representative democracies work in this way and democratic governments are naturally majoritarian. But some will suffer.

3. **Rights-based** (deontological) **ethics** focuses on the duty or rights of the entities involved and whether or not there is conflict between them. "Deontological" means relating to the study (or science) of duty. The best ethical action is one which protects the ethical rights of those who are affected by the action—but there is debate about what these rights might be and whether they include the rights of animals or other lifeforms. Rights-based ethics also depends on your beliefs. If you follow a religion that allows you to eat meat, is it right or wrong to kill an animal for food?

Immanuel Kant is the founding deontological (duty-based) ethical theorist. Kant lived in 18th-century Prussia (1724–1804). To Kant, all humans must be seen as inherently worthy of respect and dignity. He argued that all morality must stem from such duties: a duty-based ethic. Consequences such as pain or pleasure are irrelevant.

Kant did not believe that humans could predict future consequences with any degree of certainty. Basing decisions on a guess about future consequences appalled him. He believed that if we used our unique (among the higher animals) facility of reason, we could determine our ethical duty, but regardless of whether or not doing our duty would make things better or worse (and for whom).

▲ Figure 6 Statue of Immanuel Kant in Kaliningrad

Individuals are never merely a means to an end. They are ends in themselves. Therefore, a duty could not be sacrificed to the greater good because all have to be respected.

The "appeal to nature" fallacy

The **appeal to nature** is a logical fallacy that occurs when something is seen to be good because it is perceived as natural, or bad because it is perceived as unnatural. Natural means existing or produced in nature—not artificial. But whether this includes things made or caused by humans depends on your interpretation and perspective. A fallacy is a mistaken belief based on unsound arguments or a failure in reasoning. If you believe that "natural" things are always good, you follow the appeal to nature fallacy. But is this a good ethical guide? There are natural things that may not be good. Diseases are natural. Mosquitoes spread many diseases. Viruses are natural.

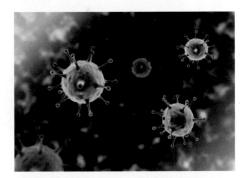

▲ Figure 7 COVID-19 coronaviruses

(ATL) Activity 10

ATL skills used: thinking, communication, research

What is nature and what is natural? The word "natural" can be a loaded term.

1. Find examples of natural products in advertising. They may be foods, cosmetics, medicines, cleaning products, clothing, furniture or fuels.

2. Make a collage, poster or graphic display of at least 15 "natural" things that you find.

3. Are they all good for you or do some do harm?

Smallpox virus eradication

Smallpox is the only human disease to have been eradicated so far. Smallpox was a viral disease caused by the variola virus. It caused mouth ulcers and skin blisters. Those who survived were left with scars over their body and face and were often blinded. The death rate from the virus was about 30% and between 300 and 500 million people died of it in the 20th century. It was found in Egyptian mummies dated 1500 BCE. There was never a cure. Kings and queens died of it and Mozart and Abraham Lincoln were infected. In 1507 it reached Hispaniola then Mexico after Columbus' voyages. Indigenous peoples in the Americas had no immunity to it and died with up to a 90% death rate.

Edward Jenner created a vaccine to smallpox in 1796 having seen that milkmaids who contracted cowpox were then immune to smallpox.

The World Health Organization (WHO) launched a smallpox eradication programme in 1959 and, with international efforts of mass inoculation, smallpox was declared eradicated in 1980.

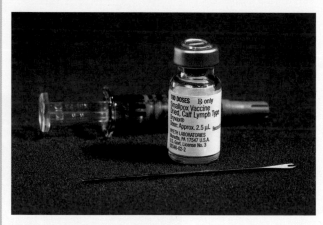

▲ Figure 8 Smallpox vaccine

◀ Figure 9 This young girl in Bangladesh was infected in 1973

However, research for effective vaccines for smallpox continues in case it is ever used as a bioterror weapon. Samples of the virus are kept for research in secure labs, but could they ever fall into the hands of terrorists?

Questions

1. Which ethical approach do you think would apply to the decision to eradicate smallpox?

2. Do you agree that it was the right thing to do?

3. Do you think that the smallpox virus had moral standing?

4. Do you think all human disease should be eradicated if medically possible in future?

5. What could be possible consequences if this happened?

6. Bears and tigers kill a few people per year. Would you advocate eradicating bears and tigers?

7. Does language matter: eradicate vs exterminate vs make extinct?

Environmental and social justice movements

Environmental movements are social movements in which individuals or groups have a common interest in protecting the environment. They lobby, educate and may show activism in attempting to bring change to government and global policies and practices.

Social justice movements promote fairness and equity across societies. They may be focused on racism, poverty, opportunity in education or employment or other inequalities.

Both types of movement came from different needs but are similar in that they are against exploitation. This could be environmental exploitation or exploitation of minority groups; in both, there is the assumption that humans or some humans are superior to nature or to others.

TOK

"Research for effective vaccines for smallpox continues because of a bioterror threat."

"Samples of the smallpox virus are kept to create a bioterror weapon."

How can we judge which version of the "truth" is correct or acceptable?

(ATL) Activity 11

ATL skills used: thinking, communication, social, research, self-management

1. Discuss and list in class examples of environmental movements and of social justice movements, both locally and globally.

2. Research an environmental movement and a social justice movement of your choice. Consider ones you may know already and that are of relevance to your life.

3. List the goals of each movement you have researched.

4. Evaluate the success of each movement.

Check your understanding

To what extent do humans have a moral responsibility towards the environment?

How does environmental ethics influence approaches to achieving a sustainable future?

1. State your ethical value system.

2. Discuss if you consider intrinsic value to be universal or selective.

3. Discuss if you consider some entities not to have moral standing.

4. Explain which ethical approach you tend to use and why.

5. Explain whether you think all that is natural is good or not.

6. Justify your ethical values with respect to future generations.

7. Comment on the use of environmental ethics in achieving sustainability.

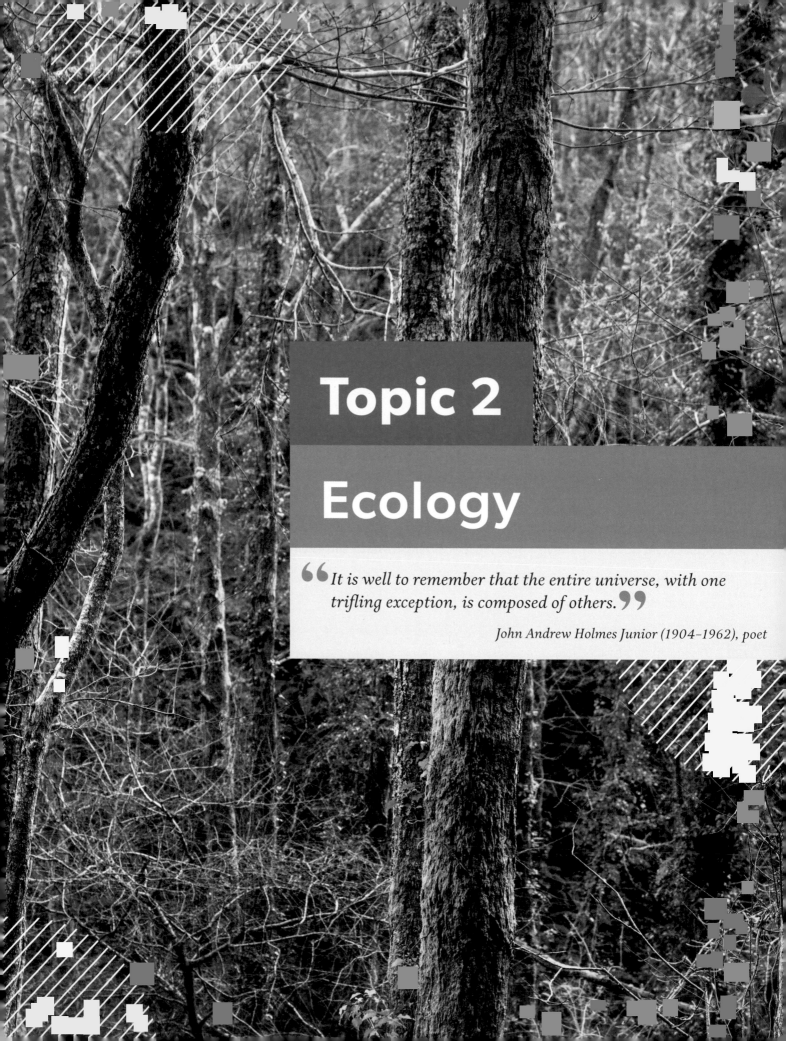

Topic 2

Ecology

“ *It is well to remember that the entire universe, with one trifling exception, is composed of others.* ”

John Andrew Holmes Junior (1904–1962), poet

2.1 Individuals, populations, communities, and ecosystems

Guiding question

- How can natural systems be modelled, and can these models be used to predict the effects of human disturbance?

Understandings

1. The biosphere is an ecological system composed of individuals, populations, communities, ecosystems.
2. An individual organism is a member of a species.
3. Classification of organisms allows for efficient identification and prediction of characteristics.
4. Taxonomists use a variety of tools to identify an organism.
5. A population is a group of organisms of the same species living in the same area at the same time, and which are capable of interbreeding.
6. Factors that determine the distribution of a population can be abiotic or biotic.
7. Temperature, sunlight, pH, salinity, dissolved oxygen and soil texture are examples of abiotic factors that affect species distributions in ecosystems.*
8. A niche describes the particular set of abiotic and biotic conditions and resources upon which an organism or a population depends.
9. Populations interact in ecosystems by herbivory, predation, parasitism, mutualism, disease and competition, with ecological, behavioural and evolutionary consequences.
10. Carrying capacity is the maximum size of a population determined by competition for limited resources.
11. Population size is regulated by density-dependent factors and negative feedback mechanisms.
12. Population growth can either be exponential or limited by carrying capacity.
13. Limiting factors on the growth of human populations have increasingly been eliminated, resulting in consequences for sustainability of ecosystems.
14. Carrying capacity cannot be easily assessed for human populations.
15. Population abundance can be estimated using random sampling, systematic sampling or transect sampling.*
16. Random quadrat sampling can be used to estimate population size for non-mobile organisms.*
17. Capture–mark–release–recapture and the Lincoln index can be used to estimate population size for mobile organisms.*
18. A community is a collection of interacting populations within the ecosystem.
19. Habitat is the location in which a community, species, population or organism lives.
20. Ecosystems are open systems in which both energy and matter can enter and exit.
21. Sustainability is a natural property of ecosystems.
22. Human activity can lead to tipping points in ecosystem stability.
23. Keystone species have a role in the sustainability of ecosystems.
24. The planetary boundaries model indicates that changes to biosphere integrity have passed a critical threshold.
25. To avoid critical tipping points, loss of biosphere integrity needs to be reversed.
*Understandings 7 and 15–17 will be covered in subtopic 2.6.

26. There are advantages of using a method of classification that illustrates evolutionary relationships in a clade.
27. There are difficulties in classifying organisms into the traditional hierarchy of taxa.
28. The niche of a species can be defined as fundamental or realized.
29. Life cycles vary between species in reproductive behaviour and lifespan.
30. Knowledge of species' classifications, niche requirements and life cycles help us to understand the extent of human impacts upon them.

What's what in ecology?

Taxonomy is the naming, describing and classifying of organisms both living and extinct. Each **species** is given a binomial (two part) name. Taxonomy organizes groups or types into a hierarchy. By naming species according to universal rules of taxonomy, we can easily communicate information. If you say you saw a hairy, brown, large animal with a tail and large ears, I may think of several possible examples. If you say you saw a "horse" or, better, *Equus caballus*, I know exactly what you are talking about.

To classify and name species, scientists may:

- make comparisons with reference collections (e.g. museums, herbaria, zoos, reference books)

- use a dichotomous key—a series of statements, each with two choices, which narrow down to identify a single species

- complete DNA surveys.

Examples of species include humans, giraffes, Scots pine trees and aardvarks (Table 1). Each species is given a binomial (two part) scientific name that comprises the genus name followed by the species name. Scientific names are always underlined when handwritten and put in italics when printed. The genus name is given first with a capital letter.

Common name	Scientific or binomial name
Human	*Homo sapiens*
Giraffe	*Giraffa camelopardalis*
Scots pine	*Pinus sylvestris*
Aardvark	*Orycteropus afer*

▲ Table 1 Common and binomial names for four species

Snails of one species in a pond form a **population**, but snails of the same species in another pond are a different population. A road or river may separate two populations from each other and stop them interbreeding. This may cause speciation.

Population density is the average number of individuals in a stated area (e.g. gazelles km^{-2}, bacteria cm^{-3}). The size of a population is affected by:

- natality (birth rate)

- mortality (death rate)

- migration, which includes immigration (moving into an area) and emigration (moving out of an area).

The natural environment includes the physical (abiotic) environment. Many populations of different species (a **community**) may share the same **habitat** and interact with each other (Figure 1).

Connections: This links to material covered in subtopic 2.6.

Key term

A **species** is a group of organisms (living things) that share common characteristics and are able to interbreed and produce fertile offspring.

A **population** is a group of organisms of the same species living in the same area at the same time, and which are capable of interbreeding.

A **community** is a group of two or more populations of species living in the same area at the same time.

A **habitat** is the environment in which a species normally lives.

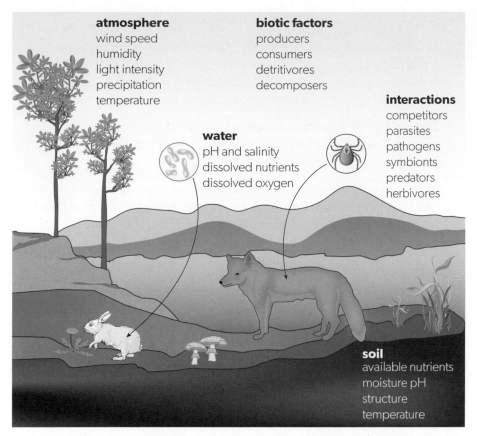

atmosphere
wind speed
humidity
light intensity
precipitation
temperature

biotic factors
producers
consumers
detritivores
decomposers

interactions
competitors
parasites
pathogens
symbionts
predators
herbivores

water
pH and salinity
dissolved nutrients
dissolved oxygen

soil
available nutrients
moisture pH
structure
temperature

▲ Figure 1 Abiotic and biotic factors and simple interactions within an **ecosystem**

A **niche** is a particular set of conditions needed by an organism or population. This includes:

- **biotic factors**—every relationship the organism has, where it lives, how it responds to resources available, to predators and to competitors, and how it alters these biotic factors

- **abiotic factors**—how much space there is, and availability of light, water and air.

No two species can inhabit the same **ecological niche** in the same place at the same time. If many species live together, they must have slightly different needs and responses so they are not in the same niche. For example, lions and cheetahs both live in the same area of the African savanna but they hunt different prey. Lions typically take down bigger herbivores such as zebra and Cape buffalo. Cheetahs focus on smaller antelopes such as Thompson's gazelles and impalas.

Key term

An **ecosystem** is made up of the organisms and their physical environment and the interactions between the living and non-living components within them.

A **niche** describes the particular set of abiotic and biotic conditions and resources upon which an organism or a population depends.

Biotic factors are the living components of an ecosystem—all organisms, their interactions and their waste that directly or indirectly affect another organism.

Abiotic factors are non-living, physical factors that influence organisms and ecosystems (e.g. temperature, sunlight, pH, salinity, pollutants).

Connections: This links to material covered in subtopic 3.1.

Key term

An **ecological niche** is the role of a species in an ecosystem.

▲ Figure 2 A lion preying on a zebra; a cheetah preying on an impala

▲ Figure 3 **Kirtland's warbler**

▲ Figure 4 **Prickly pear cactus,** *Opuntia*
stricta

Here are two examples of niches for named species:

1. Kirtland's warbler (*Setophaga kirtlandii*), also known as the jack pine warbler, is one of the rarest songbirds in North America. Because of habitat changes in the forests—due to felling rotations and fire management—they nearly became extinct but have made a comeback.

    These birds only breed in habitats that have young and large jack pine forests on sandy soils, with trees that are 6–20 years old and 2–4 m tall, where there is low ground cover and no tree canopy.

    As the trees grow, the dense nature of the forest opens up and lower branches die and break off. Kirtland's warblers then gradually abandon the habitat. They migrate in winter to the Bahamas.

2. Common prickly pear cactus (*Opuntia stricta*) is adapted to very dry desert conditions with hot days and cold nights. It grows in open and dry areas where soil is well-drained and sandy or gravelly as its roots need to be dry in winter so they do not rot. Native to the Americas, it has been spread by humans to southern Europe, Africa and Australia. In Australia, *Opuntia* grows very well and has become an invasive weed.

    Opuntia is adapted to habitats with low rainfall and cold nights by:

    *   storing water in fleshy pads

    *   having a thick waxy cuticle to reduce water loss

    *   modifying leaves into spines to reduce water loss by transpiration and to protect against being eaten

    *   having shallow roots that capture rainfall

    *   having antifreeze chemicals in its cells.

An activity investigating a local ecosystem is described in subtopic 2.6.

Population interactions

No organism can stay the same: it grows, eats, ages and dies. All habitats change too. Animals enter and leave, plants grow and shade the ground, water flows in and out. Animal migration may change a habitat greatly. Plagues of locusts can devastate all vegetation in their path, including crops. In 2013, a severe locust plague hit Madagascar with many swarms, each with over one billion locusts. The rice crop, livestock and rare wild animals were at risk. Only aerial spraying of insecticides stopped some of the damage. Fire, natural disasters and human activities all change ecosystems. Interactions between individuals, populations and communities change ecosystems too.

Interactions between organisms (e.g. predation, herbivory, parasitism, mutualism, disease, competition) are termed biotic factors. Interactions between species affect the population dynamics of those species and also affect the carrying capacity of the environment for those species.

Competition

Every organism in an ecosystem has some effect on every other organism in that ecosystem. Also, any resource in any ecosystem exists in a limited supply. When these two conditions apply jointly, competition takes place.

Intraspecific competition is between members of the same species. When the numbers of a population are small, there is little competition between individuals for resources. As long as the numbers are not too small for individuals to find mates, population growth will be high. Figure 5 summarises the effects of intraspecific competition on a seagull colony on an oceanic outcrop.

As the population grows, so does the competition between individuals for the resources, until eventually the carrying capacity of the ecosystem is reached. In this situation, the stronger individuals often claim a larger share of the resources.

Some species deal with intraspecific competition by being territorial (e.g. deer). An individual deer or a pair holds an area and fends off rivals. The biggest territory is held by individuals that are the most successful reproductively. They thus have access to more resources and will continue to be more successful at breeding.

Intraspecific competition tends to stabilize population numbers. It produces a sigmoid or logistic growth curve which is S-shaped (see Figure 20).

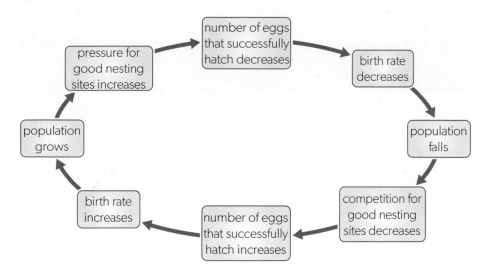

TOK

The **butterfly effect** is a term from chaos theory and refers to small changes that happen in a complex system and lead to seemingly unrelated results that are impossible to predict. It was first used in meteorology by Edward Lorenz in 1972 in a talk entitled "Does the flap of a butterfly's wings in Brazil set off a tornado in Texas?" Since then, it has been applied to systems other than the weather (e.g. asteroid travel paths, human behaviour).

One risk in applying the butterfly effect to complex environmental issues is that we might then think nothing can be done to improve things. But there is order in systems, however complex they are. As yet, there is no evidence to show that even many butterflies flapping their wings can affect weather patterns.

How do we decide whether or not all versions of an idea or ideas are equally acceptable?

◀ Figure 5 **Competition within a seagull colony**

Interspecific competition is between members of different species. Individuals of different species could be competing for the same limiting resource in the same area. When two species occupy the same niche or compete for the same resource, this is niche overlap. Interspecific competition may result in:

- a balance, in which both species share the resource, or

- one species totally out-competing the other.

This is the principle of competitive exclusion. Examples of both outcomes can be seen in a garden that has become overrun by weeds. A number of weed species coexist together, but the original domestic plants may be totally excluded. Competition reduces the carrying capacity for each of the competing species, as these species use the same resource(s).

In a temperate deciduous woodland, light is a limiting resource. Plant species that cannot get enough light will die. This is especially true of small flowering plants on the woodland floor. They are shaded by trees and by shrubs and bushes as well. Beech trees have very closely overlapping leaves, resulting in an almost bare woodland floor in beech woods.

But even in woods shaded by trees, flowers manage to grow. Carpets of snowdrops, primroses and bluebells are an integral part of all northern European deciduous woodlands in the spring. The key to the success of these species is that they grow, flower and reproduce before the shrub and tree species burst into leaf. They avoid competing directly with species that would outcompete them for light. To do this, these small plants complete the stages of their yearly cycle that require the most energy—and therefore the greatest photosynthesis—when there is less competition.

▶ Figure 6 Snowdrops flowering in a temperate woodland in spring

Types of interaction

Predation

Predation is when one animal—the **predator**, kills and eats another animal—the **prey**. Examples include lions eating zebras and wolves eating moose. All predators are **carnivores**. Be aware that not only do animals eat other animals (e.g. lion preying on zebra), some plants (insectivorous plants) consume insects and other small animals.

Look at the example of the Canadian lynx preying on snowshoe hare in subtopic 1.2 as an example of negative feedback control.

Key terms

A **predator** is a consumer that preys on other animals.

Prey are the animals a predator eats.

Carnivores are consumers that eat other animals.

Herbivory

Herbivory is defined as an animal **(herbivore)** eating a green plant. Some plants have defence mechanisms against this, for example thorns or spines (some cacti), a stinging mechanism (stinging nettles) or toxic chemicals (poison ivy). Herbivores may be large (e.g. elephants, cattle) or small (e.g. larvae of leaf miner insects that eat the inside of leaves) or in between (e.g. rabbits).

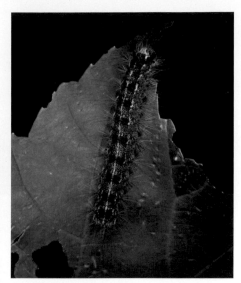

▲ Figure 7 Poplar sawfly larva (*Trichiocampus viminalis*) eating an aspen leaf (*Populus tremula*) in Glen Affric, Scotland, UK

▲ Figure 8 Cattle for meat production in pasture in Sao Paolo State, Brazil

Parasitism

Parasitism is a relationship between two species in which one species (the **parasite**) lives in or on another (the host). The parasite gains its food from the host. Normally parasites do not kill the host, unlike in predation. However, high parasite population densities can lead to the host's death. Parasites may be animals, plants, fungi or bacteria. Examples of animal parasites are vampire bats, fleas and intestinal worms.

Mutualism

Mutualism is a relation between two species in which both benefit.

Lichens are examples of mutualism. A **lichen** is a close association of a fungus underneath and a green alga on top. The fungus benefits by obtaining sugars from the photosynthetic alga. The alga benefits from minerals and water that the fungus absorbs and passes on to the alga.

The oxpecker bird feeds on parasites and insects on the skin of rhinos, Cape buffalos and other large mammals and may give an alarm call to them if there is danger. Both species benefit.

▲ Figure 9 Dodder, a parasitic plant living on another plant

Another example is the relationship between leguminous plants (beans, clover, vetch, peas) and nitrogen-fixing bacteria (Rhizobium). The bacteria live inside root nodules in the legumes. They absorb nitrogen from the soil and make it available to the plant in the form of ammonium compounds. The plants in turn supply the bacteria with sugar from photosynthesis. This mutualistic relationship enables legumes to live on very poor soils. As a consequence, leguminous plants are among the earliest pioneer species during succession on poor soil. Clover is also often used to increase the nutrient content of agricultural soil.

Key terms

Herbivores are consumers that eat plants.

Parasites live on or in living hosts and get their food from the host.

Figures 10–12 show mutualistic interactions between organisms.

▲ Figure 10 Yellow-billed oxpeckers on Cape buffalo—an example of mutualism

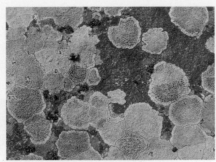

▲ Figure 11 A lichen on rock

▲ Figure 12 Nitrogen-fixing root nodules on a legume root

Saprotrophism

A **saprotroph** is an organism that feeds on dead organic material. Without such organisms, nutrients would not be recycled. (In your reading, you may see the term "saprophyte", which is now considered obsolete.)

Scavengers, detritivores and decomposers also gain their nutrients from dead organic matter but on different scales.

▲ Figure 13 Vulture feeding on a carcass

▲ Figure 14 A bracket fungus feeding on dead wood

▲ Figure 15 Earthworm in soil—a detritivore

Scavengers

A **scavenger** is an animal that feeds on carrion (decaying flesh). Examples are vultures, hyenas and raccoons.

Detritivores and decomposers

Detritivores and **decomposers** gain their nutrients from dead organic matter. The difference is that:

- detritivores eat (ingest) their food and then use enzymes to break it down (e.g. earthworms, crabs, flies, millipedes)

- decomposers secrete enzymes to break down their food but do not "eat" it (e.g. bacteria).

Key terms

Saprotrophs are organisms that live on dead or decaying organisms and get their food from these.

Scavengers mostly eat decaying biomass and are usually carnivores.

Detritivores and **decomposers** break down dead organic materials to get their food. This means they recycle organic matter.

Disease

A pathogen is an organism that causes harm to its animal or plant host. This harm is a disease as the host's function is negatively affected. Pathogens include bacteria, protists, viruses and fungi. Often the higher the density of a species, the easier it is for disease to spread between individuals. Examples of plant disease are potato blight and ergot. Examples of animal disease are rabies and tuberculosis (TB).

ATL Activity 1

ATL skills used: thinking, communication

1. Copy and complete Table 2.

Type of interaction	Species A	Species B	Example
Competition—neither species benefits, both species suffer	−	−	
Predation—one species kills the other for food	+	−	
Parasitism—the parasite benefits at the cost of the host	+	−	
Mutualism—both species benefit	+	+	
Saprotrophism			
Decomposition			
Disease			

▲ Table 2 Interactions between species
Adapted from: Six Red Marbles and OUP

2. What four things do all organisms need to survive? (Think back to your first biology lesson.)

3. What is the difference between interspecific and intraspecific competition?

4. What effect does intraspecific competition have on the individuals of a species?

Limits to population sizes

Limiting factors prevent a community, population or organism growing ever larger. They slow the rate of population growth as it approaches the **carrying capacity** of the system. There are many limiting factors which restrict the growth of populations in nature. These factors may be biotic or abiotic.

Biotic factors include lack of food; predation, herbivory or parasitism; difficulty in finding a mate; and competition between or within species.

Abiotic factors include lack of non-living resources such as oxygen, light, water or phosphate. Temperature can also be a limiting factor—for example, low temperature in the tundra freezes the soil and limits water availability to plants.

Key terms

Limiting factors are factors which slow down growth of a population as it reaches its carrying capacity.

Carrying capacity is the maximum number of a species (the maximum "load") that can be sustainably supported by a given area.

Key terms

Density-dependent limiting factors cause a population's growth rate to change (usually decrease) with increasing population density. They are biotic—for example, disease, predation, competition within a species for food or space.

Density-independent limiting factors change the size of a population regardless of its density. They are abiotic— for example, forest fires, earthquakes, floods, pollution.

Density-dependent limiting factors

Most **density-dependent** limiting factors decrease the growth rate of a population as its size increases. This is an example of negative feedback. Sometimes a population size briefly overshoots its carrying capacity but then it is reduced or even crashes. This is boom and bust of populations and is cyclical. There are four main density-dependent limiting factors.

- **Predation:**

  Many prey in an area attract more predators because prey are easier to catch (e.g. lambs in a field attract wolves, lions follow herds of gazelle). Prey numbers decrease due to predation, while predator numbers increase because there is a lot of food. Eventually, there will be too many predators and not enough prey to feed them all. Predators starve and their numbers fall. This can lead to cycles of high and low numbers of prey and predator (for example, snowshoe hare and lynx—see subtopic 1.2).

- **Competition within the population:**

  When a population reaches a high density, there are more individuals trying to use the same quantity of resources. This can lead to competition for food, water, shelter, mates, light and other resources needed for survival, and reproduction. A good example of this is provided by lemmings which emigrate if population density is too high.

- **Disease and parasites:**

  Disease is more likely to break out and result in deaths when more individuals are living together in the same place. Parasites are also more likely to spread under these conditions.

- **Waste accumulation:**

  High population densities can lead to the accumulation of harmful waste products that kill individuals or impair reproduction, reducing the population's growth.

Density-independent limiting factors

Density-dependent factors that tend to regulate the size of a population around the carrying capacity. In contrast, **density-independent** factors affect a population in the same way regardless of its size.

Density-independent limiting factors include:

- extremes of temperature

- events such as wildfires or floods.

Population fluctuations

The natural world is complex, and density-dependent and density-independent factors may operate at the same time. Density-dependent factors may keep a population near carrying capacity but a storm or fire will reduce its size rapidly.

Population cycles

Some populations undergo cyclical oscillations in size. **Cyclical oscillations** are repeating rises and drops in the size of the population over time. A graph of population size over time for a population with cyclical oscillations would look roughly like the wave in Figure 16—though probably not quite as tidy.

Where do these oscillations come from? In many cases, oscillations are produced by interactions between populations of at least two different species. Predation, parasitic infection, and fluctuation in food supply may create the oscillations.

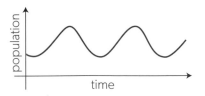

▲ Figure 16 Graph of oscillating population numbers over time

Lemmings

Lemmings are small rodents. Their habitat is the Arctic tundra. They are herbivores, eating grass and moss. They do not hibernate in winter but live in tunnel systems in large groups. Their population sizes oscillate greatly. They are very fast breeders and every four years or so numbers drop sharply before rising again. Lemmings migrate in spring to higher forested ground to breed and return to the tundra in winter.

Lemmings are predated on by owls, skuas, foxes and stoats. While owls, foxes and skuas have a varied diet, the stoats rely on eating lemmings. When the lemming population increases, there is more food for stoats and the stoat population increases with a time lag. This leads to increased predation so lemming populations fall, leaving the stoats with less food. As a result, stoat populations then fall after a time lag (Figure 18).

▲ Figure 17 A lemming (left) and its predator, a stoat (right)

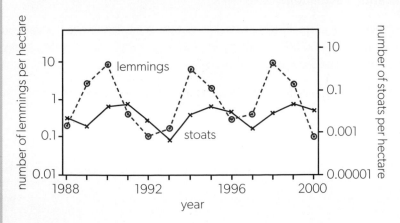

◀ Figure 18 Lemming and stoat population oscillations

115

Population growth

J-curves

Over time, the size of a population changes. Imagine you could take one bacterial cell, put it in a suitable supply of nutrients and then count the number of cells every hour. Bacterial cells reproduce asexually by cell division: one cell divides into two, and each of those cells divides into two, and so on.

If there are no **limiting factors** slowing growth, the number of cells will increase geometrically: 2, 4, 8, 16, 32, 64 and so on. This is called **exponential** or **geometric growth** and gives a **J-shaped curve** on a graph (Figure 19).

J-curves show a "boom and bust" pattern. The population grows exponentially at first and then, suddenly, collapses. These collapses are called diebacks. Often the population exceeds the carrying capacity before the collapse occurs (overshoot). A J-shaped population growth curve is typical of microbes, invertebrates, fish and small mammals.

It is important to remember that carrying capacity can be exceeded in the short term but this is not sustainable.

S-curves

When there are limiting factors, the rate at which the population size increases slows down until it reaches carrying capacity. This leads to an S-shaped curve (Figure 20).

S-curves start with exponential growth, unaffected by limiting factors. However, above a certain population size, the growth rate gradually slows down, finally resulting in a population of constant size.

The graph in Figure 20 shows population growth for a colony of yeast grown in a constant but limited supply of nutrient. During the first few days, the colony grows slowly as it starts to multiply (lag phase). Then it grows very rapidly as the multiplying colony has a plentiful nutrient supply (exponential phase). Eventually the population size stabilizes (stationary phase) because the number of yeast cells that can exploit the limited resources is fixed. Any more yeast cells and there is not enough food to go around. The population size stabilizes at the carrying capacity (K) of the environment. This is the maximum number or load of individuals that an environment can carry or support. The area between the exponential growth curve and the S-curve is called **environmental resistance**.

Population fluctuation

S- and J-curves are idealized curves. In practice, many limiting factors act on the same population and the resulting population growth curve normally looks like a combination of an S-curve and a J-curve.

Population size usually fluctuates around the carrying capacity of the environment.

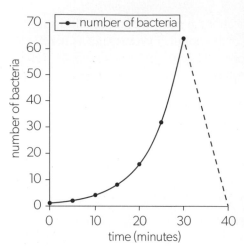

▲ Figure 19 A J-curve of exponential growth in a bacterial population over time

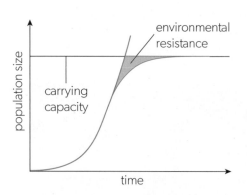

▲ Figure 20 S-shaped growth curve of a population

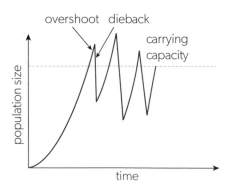

▲ Figure 21 Fluctuation of population size around the carrying capacity

Disappearing reindeer on St Matthew Island

St Matthew Island is a small, isolated island in the Bering Sea (Figure 22). It is now uninhabited by humans and is part of the USA. In August 1944, 24 female and 5 male reindeer were taken to the island by the US coastguard. The reindeer were intended to be a source of food for the coastguards should the island be cut off from supply ships.

Reindeer are not native to Alaska but their natural habitat is the Eurasian tundra. They eat lichen which was very abundant on St Matthew Island when the reindeer arrived. Lichen is a symbiotic combination of an alga and a fungus and is high in sugars. It is a good food source for herbivores. The body mass of the reindeer became greater than that of wild herds elsewhere. Between 1944 and summer 1963 the reindeer numbers were recorded as having increased to 6,000 (Figure 23).

But they ate all the lichen and most starved to death in the very severe winter of 1963–4. Numbers crashed to 42 in that winter. All but one of the survivors were female. Reindeer skeletons were everywhere.

Lichen is extremely slow-growing. There are no reindeer on St Matthew Island now. However, on a similar island (St Paul Island) about 400 reindeer survive even though they ate all the lichen there too. Their diet on St Paul Island is now grasses and grass roots.

It is thought that island ecosystems are not complex and therefore density-dependent factors did not operate as they would in a larger, more complex ecosystem. The drop in reindeer numbers was due to lack of food supply.

This pattern of boom and bust of populations introduced to isolated islands is seen elsewhere.

Look back at the snowshoe hare and lynx interaction in subtopic 1.2 and the lemming–stoat interaction in Case study 1.

Questions

1. Explain how the reindeer population changes are similar to and different from the other examples.
2. Explain which type of density-dependent factors are working in these three examples.
3. Explain what the reindeer herd on St Paul Island learned to do which the St Matthew Island herd did not.
4. Evaluate the advantages and disadvantages of introducing a population to a remote island.

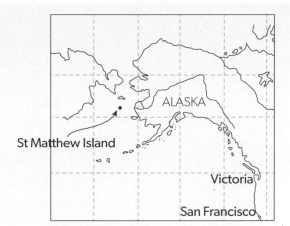

▲ Figure 22 Map showing the position of St Matthew Island

Klein, D. R. (1968). The introduction, increase, and crash of reindeer on St. Matthew Island. The Journal of Wildlife Management, 32(2), 350. doi:10.2307/3798981

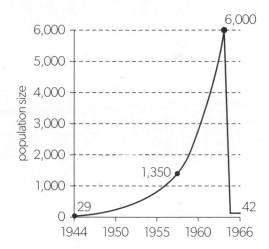

▲ Figure 23 Assumed population of the St Matthew Island reindeer herd, with four data points

Adapted from: Klein, D. R. (1968). The introduction, increase, and crash of reindeer on St. Matthew Island. The Journal of Wildlife Management, 32(2), 350. doi:10.2307/3798981

Practical work on estimating populations can be found in subtopic 2.6.

AHL lens

Evaluate the ethics of introducing reindeer to St Matthew Island.

Human carrying capacity

Remember that **carrying capacity** is the maximum number (or "load") of a species that can be sustainably supported by a given area.

Like all other populations, the human population has limiting factors but we have temporarily eliminated most of them. And we are living beyond the limits of sustainability.

Human ecological niche

By making tools, controlling energy sources, controlling our environment and expanding use of resources, humans have been able to live successfully in many different ecosystems on Earth. Such technologies have allowed us to live almost anywhere, including desert and Arctic biomes. Unlike other organisms, we have constructed our own niche.

Difficulties in measuring human carrying capacity

By examining the resources available and the requirements of a given species, it should be possible to estimate the carrying capacity of the environment for that species. This is problematic in the case of human populations for many reasons.

- Humans use a far **greater range of resources** than any other species. This means it is not just a case of working out what we eat and drink and what space we need for a house.

- We **substitute** resources with others if the original resource runs out. We may burn coal instead of wood, use solar energy instead of oil, eat mangoes instead of apples.

- Depending on lifestyles, culture and economic situation, human **resource use varies** from individual to individual, country to country. Money buys stuff so the more money is available, the more demand there tends to be for resources.

- Large predators that eat humans (e.g. sabre-toothed tigers) have largely been eliminated.

- We strip the environment of resources we need. This degrades the environment but temporarily provides us with resources.

- We **import resources** from outside our immediate environment so we cannot judge how many people a local environment can support.

- **Developments in technology** lead to changes in the resources we use. This can mean we use less because machines become more efficient. But it may mean we use more because we can exploit new resources (e.g. shale oil, EV transport).

While importing resources increases the carrying capacity for local populations, it does not increase the global carrying capacity. In fact, it may reduce carrying capacity. For example, the availability of cheaper imported food may force local farmers to reduce their prices. As a result, they will need to reduce costs and may choose to farm in ways that damage the local environment.

If the environment is degraded (e.g. by soil erosion), the land will become less productive and will produce food for fewer people than before. Also, at the moment, importing food involves burning fossil fuels in transport.

All these variables make it practically impossible to make reliable estimates of carrying capacities for human populations.

Ways to change human carrying capacity

Ecocentrists try to reduce their use of non-renewable resources and minimize their use of renewable ones. Some even try to "drop off the grid". This means they aim to become self-sufficient. For instance, they may use solar cells or wind energy for their electricity, recycle rainwater and grey water for their water supply, and grow their own food.

Technocentrists argue that the human carrying capacity can be expanded continuously through technological innovation and development. Algae and cyanobacteria may be a sustainable food source in the future as they have fast growth rates in the right conditions, are high in protein and fibre, and can be cultivated in small areas.

Using the remaining oil twice as efficiently would mean it lasts twice as long as it would have done otherwise. But that is only if the global population stays the same. The UN's estimate of human population size in 2050 is 9.7 billion, so efficiency in oil use would have to increase dramatically.

Conventional economists argue that trade and technology increase the carrying capacity. Ecological economists say that this is not the case. They argue that technological innovation can only increase the efficiency with which natural capital is used. Increased efficiency, at a particular economic level, may allow load on the ecosystem to increase but carrying capacity is fixed and once reached cannot be sustainably exceeded. The other difficulty with technology is that it may appear to increase productivity (e.g. energy-subsidized intensive agriculture gives higher yields) but this cannot be sustainable and long-term carrying capacity may be reduced (e.g. by soil erosion).

Connections: This links with material in subtopic 1.3.

Reuse, recycling, remanufacturing and absolute reductions

Humans can reduce their environmental demands (and thereby increase human carrying capacity) by reuse, recycling, remanufacturing and absolute reductions in energy and material use.

Reuse: the object is used more than once. Examples include reuse of soft drink bottles (after cleaning), furniture and pre-owned cars.

Recycle: the object's material is used again to manufacture a new product. Examples include:

- the use of plastic bags to make plastic fence posts for gardens or fleeces to wear

- recycling of aluminium: melting used aluminium to make new objects only takes a fraction of the energy required to obtain aluminium from aluminium ore so much energy can be saved by recycling.

▲ Figure 24 **Reuse, reduce, recycle poster**

Connections: This links to material covered in subtopic 1.3.

Remanufacture: the object's material is used to make a new object of the same type—for example, manufacturing new plastic (PET) bottles from used ones.

Absolute reduction: absolute reduction means using fewer resources (e.g. less energy or less paper). This increases carrying capacity, but the advantage is often eroded by population increase.

Remember that changes in birth rates and death rates do not change the carrying capacity. Carrying capacity is what the land can provide, and reducing the birth rate does not change that.

Limits to human carrying capacity

In 1798, the human population was about one billion. At that time, the economist Thomas Malthus wrote, "The power of the population is infinitely greater than the power of the Earth to produce subsistence for man."

In 1976, the population was 3.5 billion and environmentalist Paul Ehrlich warned that:

- "famines of unbelievable proportions" would arise and

- feeding a population of 6 billion (exceeded in 1999) would be "totally impossible in practice".

So far these predictions of disaster have been wrong and human carrying capacity may continue to increase. However, some people argue that pandemics, famine, natural disasters and climate change are already controlling human population size.

TOK

Considering use of resources, are there situations where ignorance or lack of knowledge is an excuse for unethical behaviour?

Ecosystem stability

Sustainability is a natural property of ecosystems which are open systems (exchanging energy and matter). Inputs balance outputs in a steady-state equilibrium.

Tropical rainforests cover less than 2% of the Earth's surface but contain an estimated 50% of biodiversity on land. Large trees in the forest may be 1,000 years old and average canopy age is 300 years. This complexity has many niches, habitats and communities and has made these rainforests stable systems. The more-or-less constant temperature and rainfall and high insolation (sunlight) mean plant growth is constant and food is available. These rainforests have high inputs and outputs and high storages. Undisturbed rainforest is a crucial carbon sink (store) of about 25% of all carbon stored on land.

These forests are constantly changing with fires, tree death and fall, small-scale clearing and landslides. This dynamic is healthy and can increase diversity by forming more ecological niches.

▲ Figure 25 Tropical rainforest

Deforestation of rainforests

In degraded rainforests, logging, burning and agriculture occur. The dense tree canopy is lost and the layered structure reduced. The forest floor is exposed to sunlight and rain, and is more likely to dry out or wash away. Loss of habitats means diversity decreases.

When the equilibrium reaches a tipping point, a new equilibrium is reached. This changes the ecosystem, often irreversibly. Loss of tree cover means less transpiration of water vapour and therefore less cooling and less rainfall. The remaining forest may then be vulnerable when conditions change.

Keystone species

A keystone species is one that plays a critical role in maintaining the structure of the ecosystem in which it lives (subtopic 3.3). Examples of keystone species include the following.

- Sea otters eat sea urchins in kelp forests. If there are no sea otters, the urchins eat only the holdfast (anchor) of the kelp so the fronds float away.

- Beavers make dams which turn a stream into a swampy area. Swamp-loving species then move in. Without the beaver and its dams, the habitat changes.

- Elephants in the African savanna remove trees which allows grasses to grow.

You might view beavers and elephants as engineers!

Biosphere integrity threshold

Review the planetary boundaries model. It indicates that changes to **biosphere** integrity have passed a critical threshold (Figure 26). This is due to disturbance of ecosystems and consequential loss of habitats.

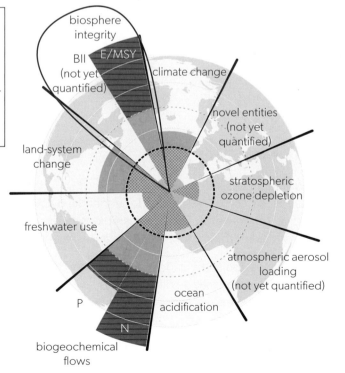

Key:
- ⊠ below boundary (safe)
- ▨ in zone of uncertainty (increasing risk)
- ▬ beyond zone of uncertainty (high risk)
- E/MSY = extinctions per million species years
- P = phosphorus
- N = nitrogen

biosphere integrity

BII (not yet quantified)
E/MSY
climate change
novel entities (not yet quantified)
land-system change
stratospheric ozone depletion
freshwater use
atmospheric aerosol loading (not yet quantified)
P
ocean acidification
N
biogeochemical flows

◀ Figure 26 Planetary boundaries model
Adapted from: Azote for Stockholm Resilience Centre, Stockholm University. Based on Rockström, J., W. Steffen, K. Noone, Å. Persson, et.al. 2009 (CC BY-NC-ND 3.0)

Rates of extinction (subtopic 3.1) and decline in population sizes of various species are evidence that the boundary has been crossed.

Review tipping points (subtopic 1.2). Once a species is extinct, it has gone. But before then, actions can be taken to restore habitats and protect niches for a particular species. There are success stories, and many people are active throughout the world in protecting vulnerable species.

Connections: This links to material covered in subtopic 3.1.

What is a clade?

A **clade** is a taxonomic group which contains one ancestor and all its descendants. Think of it as a branch of a tree with the common ancestor where the branch splits off.

A cladogram is an evolutionary tree that shows the ancestral relationships among organisms. Sometimes it is called a phylogenetic tree. In the past, cladograms were drawn based on similarities in phenotypes or physical traits among organisms. Today, similarities in DNA or amino acid sequences among organisms are used to draw cladograms.

The oldest common ancestors within a clade are close to the trunk of the evolutionary tree. Newly evolved species are farthest from the tree trunk. Cladograms represent hypotheses about the evolution of organisms. Because of this, they may change as more information is known.

Traditional taxonomy (classification of organisms) is usually based on how organisms look. It uses these taxa (groups):

- domain
- kingdom
- phylum
- class
- order
- family
- genus
- species.

The difficulty of classifying organisms according to their physical features is that some organisms have evolved independently to have the same feature. This is called convergent evolution. For example, bats, insects and birds can all fly, but their last common ancestor may not have had that ability.

Clades are nested and a clade may have a few or thousands of organisms in it. That means a clade may be different populations of a species, different species or different clades, each composed of many species.

(ATL) Activity 2

ATL skills used: thinking, communication

Remember: a clade is one ancestor and all its descendants.

1. State which box (1, 2 or 3) is **not** a clade and give the reason.

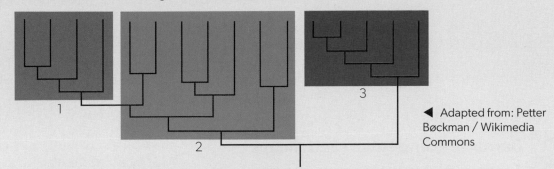

◀ Adapted from: Petter Bøckman / Wikimedia Commons

Use these diagrams to help you to answer question 1.

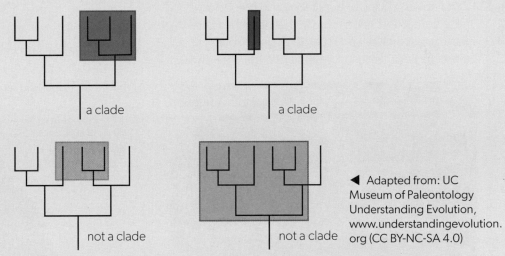

a clade

a clade

not a clade

not a clade

◀ Adapted from: UC Museum of Paleontology Understanding Evolution, www.understandingevolution.org (CC BY-NC-SA 4.0)

2. Explain why each of the following images shows a clade.

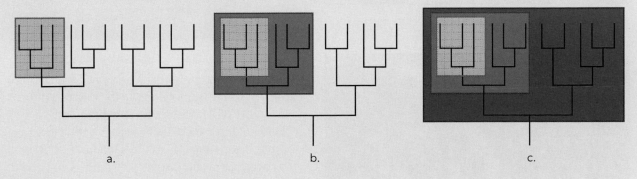

a.

b.

c.

▲ Adapted from: UC Museum of Paleontology Understanding Evolution, www.understandingevolution.org (CC BY-NC-SA 4.0)

Niches

Remember, a niche is where an organism best fits into an ecosystem. No two species can occupy the same ecological niche.

There are two types of niche.

- A **fundamental niche** is the full range of theoretical conditions and resources in which a species could survive and reproduce with no competition.

- A **realized niche** is the actual conditions and resources in which a species exists due to biotic interactions including competition from other species.

Fundamental niches are usually larger than realized ones because they do not include pressure from predators or other species. One individual may have different niches over its life cycle. For example, a tadpole lives in water, has gills and only eats plants. When the tadpole metamorphoses into a frog, it lives out of water, has lungs and eats other animals.

Barnacle competition and niches

A classic field experiment by Joseph Connell in 1961 considered two species of barnacle named *Chthamalus stellatus* and *Semibalanus balanoides*. These barnacles live on rocks in the intertidal zone off the coast of Scotland. They look similar but *Semibalanus* is larger and there are other differences.

▲ Figure 27 *Chthamalus stellatus*—Poli's stellate barnacle. Average diameter is about 5 mm

▲ Figure 28 *Semibalanus balanoides* (common rock barnacle). Average diameter is about 10 mm

Barnacles are small crustaceans. As larvae, they can swim. Once they are attached to rocks as adults, they stay there. *Semibalanus* is found on lower rocks which are covered by the sea more often and *Chthamalus* is found on higher rocks. But the larvae settle all over the rocks.

The question was this. Is the separation of the species due to competition between them or do they have different fundamental niches?

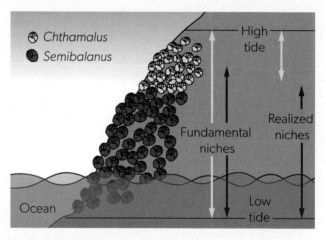

▲ Figure 29 Distribution of the two barnacle species

Connell moved the barnacles around, removed one species then the other and put the two together. He found that if he removed *Semibalanus*, *Chthamalus* survived on the lower rocks. However, if both species were put on the lower rocks *Chthamalus* was overgrown and displaced. *Semibalanus* could not survive on the higher rocks which were exposed to the air for longer when the tide went out.

Alone, *Chthamalus* can live all over the rocks in its fundamental niche but this overlaps the realized niche of *Semibalanus*.

(ATL) Activity 3

ATL skills used: thinking, communication, research

Brown and green anoles

Anoles are lizards that live in trees in south-eastern USA, and the Pacific and Caribbean islands. The green and brown anoles are different species. Green anoles are native to North America. Brown anoles are native to the Caribbean islands but arrived on the mainland in the early 1900s. Since then, the green anoles are found higher in the forest canopy and brown lower down. Habitat loss to urban areas is also affecting green anole populations.

▲ Figure 30 (a) Brown male anole displaying a dewlap—a flap of skin and (b) a green anole—a different species

1. Describe the realized and fundamental niche of the green anoles.

2. Describe the threats that both anole species face.

3. Predict the relative population sizes of the green and brown anoles in future.

K- and r-strategists' reproductive strategies

Species can be roughly divided into **K- and r-strategists** or **K- and r-selected species**. K and r are two variables that determine the shape of the population growth curve:

*   K is the carrying capacity

*   r describes the shape of the exponential part of the growth curve.

K- and r-strategies describe the approach of different species to getting their genes into the next generation and ensuring the survival of the species.

Different species vary in the amount of time and energy they use in raising their offspring. There are two extreme approaches: K-strategies and r-strategies. These strategies describe the approaches used by different species to get their genes into the next generation and ensure the survival of the species.

The main features of the two strategies are summarized in Table 3. Be aware, however, these two strategies are the extremes of a continuum of reproductive strategies, and many species show a mixture of these characteristics.

K-strategists	r-strategists
Have long lives	Have short lives
Grow more slowly	Grow rapidly
Mature later	Mature early
Reproduce more slowly	Reproduce quickly
Produce fewer large offspring	Produce many eggs or small offspring
Invest large amounts of energy in parental care or protection	Do not use energy in raising young after hatching
Most offspring survive	Most offspring die
Are adapted to stable environments	Are adapted to unstable environments
Have population sizes usually close to the carrying capacity (K), hence their name	Have fast reproductive and growth rates so may exceed the carrying capacity, with a population crash as a result
Appear at later stages of succession	Can colonize new habitats quickly (pioneers)
Are niche specialists	Are niche generalists
Are usually predators, at higher trophic levels	Are usually prey, at lower trophic levels
Are regulated mainly by internal factors	Are regulated mainly by external factors
Out-compete r-strategists in stable, climax ecosystems	Make opportunistic use of short-lived resources
Examples: humans, other large mammals, trees, albatrosses	Examples: invertebrates, fish, annual plants, flour beetles, bacteria

▲ Table 3 Typical characteristics of K- and r-strategists

ATL Activity 4

ATL skills used: thinking, communication

1. List characteristics of organisms which are r-strategists.

2. List characteristics of K-strategists.

3. Explain which species are most likely to be regulated by density-independent factors (e.g. weather).

4. Describe and explain the shape of the survivorship curve for K-selected species.

Connections: This links to material covered in subtopic 3.2.

A survivorship curve shows the fate of a group of individuals of a species (Figure 31). Curve type II is quite rare, showing species that have an equal chance of dying at any age, e.g. the hydrozoan *Hydra* and some species of birds.

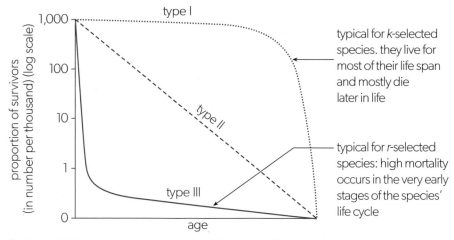

▲ Figure 31 Three hypothetical survivorship curves. Note that the y-axis is logarithmic

Only by knowing more about the niches of a species and its life cycle can we work out how to protect and conserve them. For example, the giant panda is the rarest large bear on Earth. There are estimated to be about 1,800 in the wild in south-west China. After years of declining numbers due to habitat loss and poaching, numbers are now increasing. These animals have been upgraded from endangered to vulnerable on the IUCN Red List.

Giant pandas are an umbrella species protecting many others in their habitats (multi-coloured pheasants, the golden monkey, the takin and crested ibis). Their diet is bamboo but because they have inefficient digestion they need to eat for 10–16 hours a day to get enough energy. They are solitary animals and need to find mates but habitat fragmentation makes that difficult. Corridors linking isolated forests can help. The Chengdu Panda Base studies and protects the pandas.

Check your understanding

In this subtopic, you have covered:

- terms in ecology

- why and how we classify organisms

- populations and their interactions

- carrying capacity

- keystone species and tipping points

- what clades are

- more on niches

- r- and K-selected species

- human impacts on ecosystems.

How can natural systems be modelled, and can these models be used to predict the effects of human disturbance?

1. Explain why it is important to classify species.

2. Explain why understanding about interactions between species helps species conservation.

3. Explain why sampling is used instead of gathering data from the whole population or ecosystem (see subtopic 2.6).

4. Discuss the limits to sizes of populations.

5. Discuss the limits to the size of global human population.

6. Explain why using evolutionary relationships in taxonomy has advantages.

7. Describe fundamental and realized niches.

8. Describe strategies in reproductive behaviour of r- and K-selected species.

❯❯ Taking it further

- Carry out an ecological investigation on natural and disturbed ecosystems, using the application of skills explored in this subtopic. Secondary data can be used as a comparison.

- Advocate about biodiversity loss.

- Take part in citizen science projects that collect data on species distributions and abundance.

2.2 Energy and biomass in ecosystems

Guiding questions

- How can flows of energy and matter through ecosystems be modelled?
- How do human actions affect the flow of energy and matter, and what is the impact on ecosystems?

Understandings

1. Ecosystems are sustained by supplies of energy and matter.
2. The first law of thermodynamics states that as energy flows through ecosystems, it can be transformed from one form to another but cannot be created or destroyed.
3. Photosynthesis and cellular respiration transform energy and matter in ecosystems.
4. Photosynthesis is the conversion of light energy to chemical energy in the form of glucose, some of which can be stored as biomass by autotrophs.
5. Producers form the first trophic level in a food chain.
6. Cellular respiration releases energy from glucose by converting it into a chemical form that can easily be used in carrying out active processes within living cells.
7. Some of the chemical energy released during cellular respiration is transformed into heat.
8. The second law of thermodynamics states that energy transformations in ecosystems are inefficient.
9. Consumers gain chemical energy from carbon (organic) compounds obtained from other organisms. Consumers have diverse strategies for obtaining energy-containing carbon compounds.
10. Because producers in ecosystems make their own carbon compounds by photosynthesis, they are at the start of food chains. Consumers obtain carbon compounds from producers or other consumers, so form the subsequent trophic levels.
11. Carbon compounds and the energy they contain are passed from one organism to the next in a food chain. The stages in a food chain are called trophic levels.
12. There are losses of energy and organic matter as food is transferred along a food chain.
13. Gross productivity (GP) is the total gain in biomass by an organism. Net productivity (NP) is the amount remaining after losses due to cellular respiration.
14. The number of trophic levels in ecosystems is limited due to energy losses.
15. Food webs show the complexity of trophic relationships in communities.
16. Biomass of a trophic level can be measured by collecting and drying samples.
17. Ecological pyramids are used to represent relative numbers, biomass or energy of trophic levels in an ecosystem.
18. Pollutants that are non-biodegradable, such as polychlorinated biphenyl (PCB), dichlorodiphenyltrichloroethane (DDT) and mercury, cause changes to ecosystems through the processes of bioaccumulation and biomagnification.
19. Non-biodegradable pollutants are absorbed within microplastics, which increases their transmission in the food chain.
20. Human activities, such as burning fossil fuels, deforestation, urbanization and agriculture, have impacts on flows of energy and transfers of matter in ecosystems.
21. Autotrophs synthesize carbon compounds from inorganic sources of carbon and other elements. Heterotrophs obtain carbon compounds from other organisms.
22. Photoautotrophs use light as an external energy source in photosynthesis. Chemoautotrophs use exothermic inorganic chemical reactions as an external energy source in chemosynthesis.
23. Primary productivity is the rate of production of biomass using an external energy source and inorganic sources of carbon and other elements.
24. Secondary productivity is the gain in biomass by consumers using carbon compounds absorbed and assimilated from ingested food.
25. Net primary productivity is the basis for food chains because it is the quantity of carbon compounds sustainably available to primary consumers.
26. Maximum sustainable yields (MSYs) are the net primary or net secondary productivity of a system.
27. Sustainable yields are higher for lower trophic levels.
28. Ecological efficiency is the percentage of energy received by one trophic level that is passed on to the next level.
29. The second law of thermodynamics shows how the entropy of a system increases as biomass passes through ecosystems.

Energy in systems

Energy in all systems is subject to the **laws of thermodynamics** and the **principle of conservation of energy**. This means that the total energy in any isolated system (e.g. the entire universe) is constant. Only the form that this energy takes can change. For example, photosynthesis transforms light energy to chemical energy; respiration transforms chemical energy to heat or kinetic energy (Figure 1).

In a power station, one form of energy (e.g. from coal, oil, nuclear power, moving water) is converted or transformed into electricity. In your body, food provides chemical energy which you convert into heat or kinetic energy.

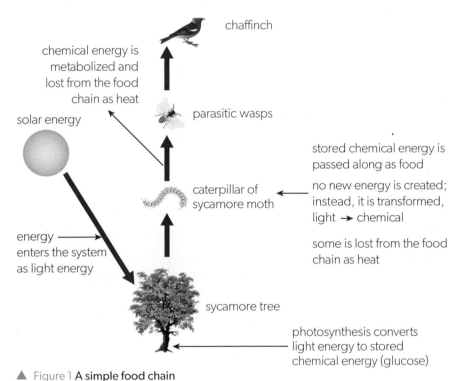

▲ Figure 1 **A simple food chain**

Figure 2 shows what happens to the Sun's energy when it reaches the Earth.

Respiration and photosynthesis

Three key ecological concepts are vital to your understanding of how everything else works: **photosynthesis**, **respiration** and **productivity**.

- Living things respire all the time, both in the dark and in the light.

- Photosynthesis occurs in plants, algae and some microorganisms. It requires water, carbon dioxide and sunlight.

- Water reaches the leaves of plants from the roots by transpiration.

- Photosynthesis and respiration transform energy and matter in ecosystems (Figure 3).

Key term

The **first law of thermodynamics** is the **principle of conservation of energy**. This states that as energy flows through ecosystems, it can be transformed from one form to another, but cannot be created or destroyed.

Connections: This links to material covered in subtopic 1.2.

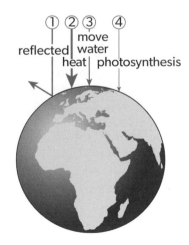

▲ Figure 2 When the Sun's energy reaches the Earth, about 30% is reflected back into space (1), around 50% is converted to heat (2), and most of the rest powers the hydrological cycle: rain, evaporation, wind (3). Less than 1% of incoming light is used for photosynthesis (4)

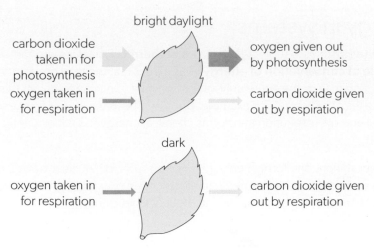

▲ Figure 3 Respiration and photosynthesis in a leaf

Respiration

All living things must respire to get energy to stay alive. If they do not do this, they die. **Cellular respiration** involves breaking down food, often in the form of glucose, to release energy which is used in living processes.

These processes are: movement, respiration, sensitivity, growth, reproduction, excretion, nutrition. Some people remember these processes by their first letters, which spell MRS GREN.

Respiration can use oxygen (aerobic respiration) or not (anaerobic respiration).

In aerobic respiration, energy is released and used, and the waste products are carbon dioxide and water. All living things respire all the time, in the light and in the dark, whether they are plants, animals, bacteria or fungi. Animals always respire, whether they are asleep or awake.

Aerobic respiration can be summarized as:

$$glucose + oxygen \longrightarrow energy + water + carbon\ dioxide$$
$$C_6H_{12}O_6 + 6O_2 \longrightarrow energy + 6H_2O + 6CO_2$$

Much of the energy produced in respiration is **heat energy** and is released (dissipated) into the environment. Heat is generated because respiration is not 100% efficient at transferring energy from substrates (e.g. carbohydrate) into the chemical form of energy used in cells. Heat generated within an **individual organism** cannot be transformed back into chemical energy and is ultimately lost from the body. This increases the **entropy** of the system, while the organism maintains a relatively high level of organization (low entropy).

Photosynthesis

In **photosynthesis**, light energy is converted to chemical energy, which is in the form of glucose. This is transformation of energy from one state to another.

The leaves of plants contain chloroplasts with the green pigment chlorophyll. In these chloroplasts, the energy of sunlight is used to split water and combine it with carbon dioxide to make food. The food is in the form of glucose, a sugar. Glucose is then used as the starting point for the plant to make every other molecule that it needs.

Key terms

Cellular respiration is the process by which glucose is broken down to release energy for living processes. Waste products of aerobic respiration are water and carbon dioxide.

Note the difference: Cellular respiration and photosynthesis are metabolic processes in cells. Be careful not to confuse cellular respiration with breathing, which is the physical process of taking air into and out of lungs. Breathing can also be called respiration. Here we are using the term "respiration" to be cellular respiration.

An **individual organism** is one individual living thing and a member of a species.

Photosynthesis is the process by which organisms make their own food from water and carbon dioxide using energy from sunlight.

In complex chemical pathways within their cells, plants:

- add nitrogen and sulfur to make amino acids and then proteins

- rearrange carbon, hydrogen and oxygen and add phosphorus to make fatty acids and lipoproteins which make up cell membranes.

Photosynthesis produces the raw material for making biomass. Animals are totally dependent on the chemicals produced by plants. Humans can make most of the molecules they need through the processes of eating, digesting and rebuilding. Food is eaten and then digested into its constituent molecules. These molecules are then used to make enzymes, fats and proteins.

Amino acids are the building blocks of proteins. There are twenty amino acid molecules but humans can only make eleven of these. The other nine are called essential amino acids and have to be eaten. They are found in meat, fish, dairy, soy, quinoa and buckwheat.

The waste product of photosynthesis is oxygen. This is useful because oxygen is used in aerobic respiration.

Photosynthesis can be summarized as:

$$\text{carbon dioxide} + \text{water} \xrightarrow[\text{chlorophyll}]{\text{light energy}} \text{glucose} + \text{oxygen}$$

$$6CO_2 + 6H_2O \xrightarrow[\text{chlorophyll}]{\text{light energy}} C_6H_{12}O_6 + 6O_2$$

When all the carbon dioxide that plants produce in respiration is used up in photosynthesis, the rates of the two processes are equal and there is no net release of either oxygen or carbon dioxide. This usually occurs at dawn and dusk when light intensity is not too high. This point is called the **compensation point** of a plant and is when the plant is neither adding biomass nor using it up to stay alive. It is just maintaining itself. This is important to remember when you learn about succession and biomes.

The **second law of thermodynamics** is about the quality of energy. As energy is transferred through food chains, some chemical energy is transformed into heat energy. Heat energy is less useful to living things than chemical energy or light energy. Heat is lost to the environment which means energy is lost from the food chain. Respiration causes the biggest loss. This is why energy transfers are never 100% efficient.

Efficiency is defined as the useful energy or work (output) produced by a process divided by the amount of energy consumed during the process (input).

$$\text{efficiency} = \frac{\text{work or energy produced}}{\text{energy consumed}}$$

$$\text{efficiency} = \frac{\text{useful output}}{\text{input}}$$

Multiply by 100% to express efficiency as a percentage.

(ATL) Activity 5

ATL skills used: thinking

Both respiration and photosynthesis are systems—biochemical processes.

You can draw systems diagrams for them with inputs, outputs, storages and flows which are transformations or transfers of energy and matter.

Draw a systems diagram for each of respiration and photosynthesis.

Where can you link the two?

Key term

The **second law of thermodynamics** states that as energy is transferred or transformed in a system, it is degraded to a less useful form of energy such as heat energy and entropy (disorder) increases.

Food chains and trophic levels

Almost all energy on Earth comes from the Sun so solar energy (solar radiation) is the start of almost every food chain. Some deep ocean vents give out heat from the core of the planet and some organisms get their energy from this through a process known as chemosynthesis. But most life forms get their energy from the Sun.

TOK

There is a philosophical implication to the inefficiency of energy transfers. According to physics, the fate of all the energy that exists today in the universe is to degrade into high entropy heat. When all energy has turned into heat, the whole universe will have a balanced temperature and no process will be possible because heat cannot turn into something of higher entropy. This is referred to as the thermal death of the universe.

To what extent is imagination crucial in the creation of this hypothesis?

Key terms

A **consumer** (heterotroph) gains its food from other organisms.

A **trophic level** is the position that an organism occupies in a food chain, or a group of organisms in a community that occupy the same position in food chains.

A **producer** (autotroph) makes its own food by photosynthesis.

A food chain shows the feeding relationships between species in an ecosystem. This means it shows the flow of energy from one organism to the next. Arrows connect the species, usually pointing towards the species that is the **consumer** of the other. The direction of the arrows shows the direction of transfer of biomass and energy (see Figures 1 and 4).

ATL Activity 6

ATL skills used: thinking, research

Copy and complete Table 1 to give two examples of each type of feeder.

Feeder	Example 1	Example 2
herbivore		
carnivore		
predator		
prey		
parasite		
saprotroph		
decomposer		
detritivore		

▲ Table 1 Types of feeder

Organisms are grouped into **trophic** (or feeding) **levels** (Greek for food is *trophe*). Trophic levels usually start with a primary **producer** (plant) and end with a carnivore at the top of the chain—a top carnivore.

▲ Figure 4 grass ⟶ grasshopper ⟶ dormouse ⟶ barn owl

Examine Table 2 to ensure you understand the hierarchy of trophic levels.

Name of group	Trophic level	Nutrition: Source of energy	Function
Primary producers (PP) Green plants	1st	Autotrophs: Make their own food from solar energy, CO_2 and H_2O	• Provide the energy requirements of all other trophic levels • Habitat for other organisms • Supply nutrients to the soil • Bind the soil and stop soil erosion
Primary consumers (PC) Herbivores	2nd	Heterotrophs: Consume PP	• Keep each other in check through negative feedback loops (see subtopic 1.2)
Secondary consumers (SC) Carnivores and omnivores	3rd	Heterotrophs: Consume herbivores and other carnivores and may consume PP	• Disperse seeds • Pollinate flowers • SC and TC remove old, weakened and diseased animals from the population
Tertiary consumers (TC) Carnivores and omnivores	4th	Heterotrophs: Consume herbivores and other carnivores and may consume PP	
Decomposers Bacteria and fungi		Saprotrophs: Obtain their energy from dead organisms by secreting enzymes that break down organic matter	Provide a crucial service for the ecosystem: • Break down dead organisms • Release nutrients back into the cycle • Control the spread of disease
Detritivores Snails, slugs, blow fly maggots, vultures		Saprotrophs: Derive their energy from detritus or decomposing organic material—dead organisms, faeces or parts of an organism, e.g. shed skin from a snake, a crab carapace	

▲ Table 2 **Hierarchy of trophic levels**

▲ Figure 5 Decomposer fungi in a woodland

Food webs

It would be very unusual to find an ecosystem with only a simple food chain. Most ecosystems involve many organisms, each of which may eat several other species.

It is possible to construct food chains for an entire ecosystem, but this creates problems. Food chains illustrate only a direct feeding relationship between one organism and another in a single hierarchy. The reality is very different. The diet of almost all consumers is **not** limited to a single food species. So, a single species can appear in more than one food chain.

A further limitation of food chains is that a species may feed at more than one trophic level. For example, voles are **omnivores**—they eat (insects) and plants. Humans also eat plants and animals. The animals that humans eat are usually herbivores. So, we would have to list all the food chains that contained voles or humans twice to show them at both the second and third trophic levels.

The reality is that ecosystems contain a complex network of interrelated food chains—a **food web**.

ATL Activity 7

ATL skills used: thinking, research, self-management

The food chains below are based on real food chains at Wytham Woods (a European Oak woodland) in Oxford, UK. Some pioneer ecologists worked here in the 1920s. In these four different food chains, only ten species are listed and some of them are in more than one food chain. If we listed all the species in the wood and their interactions in every food chain, the list would run for many pages.

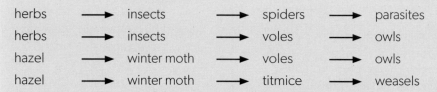

herbs	→	insects	→	spiders	→	parasites
herbs	→	insects	→	voles	→	owls
hazel	→	winter moth	→	voles	→	owls
hazel	→	winter moth	→	titmice	→	weasels

1. Construct a food web from this data.

2. In a table, summarize the differences between a food chain and a food web.

ATL Activity 8

ATL skills used: thinking

1. Draw the longest food chain in the food web in Figure 6.

2. Name two species that are found at two trophic levels.

3. If all lions die, explain what might happen to:

   a. hunting dogs   b. impala.

4. If there is a great increase in the zebra population, explain what might happen to:

   a. hunting dogs   b. producers.

5. If a pesticide is used to kill locusts, what may happen to baboons?

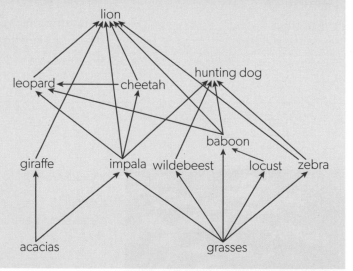

▶ Figure 6 Food web on the African savanna

The earliest food webs were published in the 1920s by Charles Elton (on Bear Island, Norway) and Alister Hardy (on plankton and herring in the North Sea). Elton's food web (Figure 7) looks complex but scientists have suggested that food web are not this simple. For example, polar bears are a keystone species and eat other animals as well as seals. Many species will change their diets depending on available food sources. But Elton was a pioneer in ecology and wrote about food chains and webs, niches, population cycles and pyramids of numbers before others understood their significance. He was one of the first to be concerned about the impact of invasive species on ecosystems.

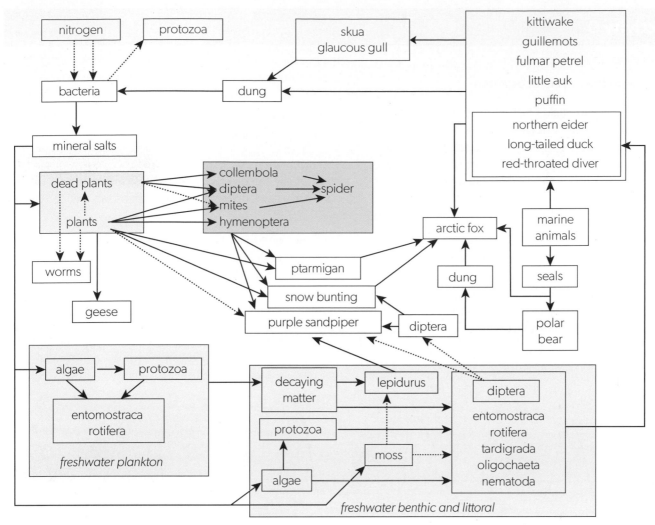

▲ Figure 7 One of the first food webs observed by Elton on Bear Island, Norway

Efficiency of energy transfer

The number of trophic levels in ecosystems is limited due to energy losses. Some of the energy released from respiration is lost to the organism as heat loss. This is especially the case for warm-blooded animals such as humans. This means the heat energy is not available to higher trophic levels. This limits the length of food chains.

The efficiency of transfer from one trophic level to the next (trophic efficiency) is considered, on average, to be about 10%. As always, things are not quite as straightforward as they at first appear. The 10% rule is a generalization and a helpful aid to our understanding of energy flow, but there are considerable variations. Trophic efficiencies generally range from 5% to 20%. This means only 5% to 20% of primary producer biomass consumed is converted into consumer biomass.

A community of small mammals in a grassland ecosystem may have a trophic efficiency of only 0.1% as they are warm-blooded, have a high metabolic rate and large surface area compared to their volume, and so lose a great deal of energy in respiration and heat.

In the oceans, zooplankton feeding on phytoplankton may have a trophic efficiency of 20% and consume most of the producer biomass. Cold-blooded animals have much slower assimilation rates than warm-blooded animals.

(ATL) Activity 9

ATL skills used: thinking, research

Carnivores in the tundra ecosystem

There are several species of bear in the tundra. Polar bears live further north, but are also found in the tundra searching for food. The Kodiak is the largest bear in the Alaskan tundra. It is usually a brown colour. Brown bears are not especially fierce and seldom eat meat. Wolves are the top predators of the tundra. They travel in small family groups (packs) and prey on caribou and other large herbivores that are too slow to stay with their groups.

Some wolves change to a bright white colour in the winter. Otters live near rivers and lakes so they can feed on fish. Shrews are the smallest carnivores of the tundra. Even bats are found in the tundra during the summer. They feed on the swarms of insects that fill the air.

The primary production is not sufficient to support animal life if only small areas of tundra are considered. The large herbivores and carnivores are dependent on

the productivity of vast areas of tundra and have adopted a migratory way of life. Small herbivores feed and live in the vegetation mat, eating the roots, rhizomes and bulbs.

1. Draw a food web for the tundra with **only** the animals mentioned here.

The populations of small herbivores such as lemmings show interesting fluctuations that also affect the carnivores dependent on them, such as the arctic fox and snowy owl.

The blue squares on the graph in Figure 8 represent the appearance and frequency of snowy owls after almost exponential population increases of lemmings. There is then a lag period of about two years before lemming numbers increase again.

2. Why do you think the snowy owls only appear when lemming numbers have fallen? (Hint: think about climate and decomposers.)

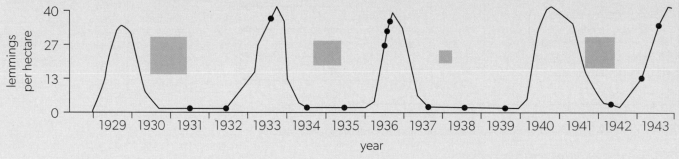

▲ Figure 8 Snowy owl and lemming numbers in the tundra from 1929 to 1943

Trophic inefficiencies occur because:

- not everything is eaten (if it were, the world would not be green as all plants would be consumed)

- digestion is inefficient (food is lost in faeces because the digestive system cannot extract all the energy from it)

- heat is lost in respiration

- some energy assimilated is used in reproduction and other life processes.

Calculating efficiency of transfer

To calculate the efficiency of transfer between two trophic levels, use:

$$\text{efficiency \%} = \frac{\text{biomass of higher trophic level}}{\text{biomass of lower trophic level}} \times 100$$

ATL Activity 10

ATL skills used: thinking

Look at this food chain:

phytoplankton ⟶ zooplankton ⟶ herring ⟶ seal

Table 3 shows the amount of biomass contained within each trophic level.

Trophic level	Organism	Total biomass (kg)
1	Phytoplankton	22
2	Zooplankton	1.8
3	Herring fish	0.26
4	Seal	0.019

▲ Table 3 Biomass at each trophic level

Calculate the efficiency of transfer between trophic levels 1 and 2, 2 and 3, and 3 and 4.

Maximum sustainable yield

Maximum sustainable yield (MSY) is a concept that assumes a population and a species produce surplus biomass (i.e. more biomass than is required to replace the population or species). This surplus can be harvested by humans.

MSY is the net productivity of a species or a trophic level that can be harvested without reducing future supply. It is usually applied to the fishing industry which aims to fish at MSY levels (see Figure 9 and subtopic 4.3).

Measuring biomass

The biomass of a trophic level can be measured by collecting and drying samples of that trophic level. For example, collect a sample of grasses, dry the sample to evaporate any water and weigh the sample. Units are usually $g\ m^{-2}$ or $Mg\ ha^{-1}$.

Productivity

Gross productivity (GP) is the gain in biomass of an organism. In a producer (mostly plants) gross productivity is called gross primary productivity (GPP). In consumers it is called gross secondary productivity (GSP).

All living things—whether producers or consumers—use some of the energy they take in to respire and to carry out metabolic processes (chemical reactions within cells). **Net productivity (NP)** is defined as gross productivity minus losses (Figure 10). A short equation for this is:

$$NP = GP - R \text{ or } NPP = GPP - R$$

where R = respiratory loss

In ecology, we usually talk about productivity and not production. This means we know the area or volume and the time period to which we refer.

Net productivity (NP) results from the fact that all organisms must respire to stay alive so some of the total energy gain is used in staying alive instead of being used to grow.

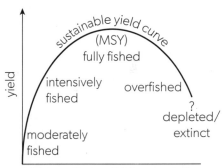

▲ Figure 9 MSY graph showing yield against fishing effort

Adapted from: Food and Agriculture Organization of the United Nations (FAO) (CC BY-NC-SA 3.0 IGO)

Key terms

Productivity is the conversion of energy into biomass over time. It is the rate of growth or biomass increase in plants and animals. It is measured for plants as mass per unit area per unit time (e.g. gram per square metre per year, $g\ m^{-2}\ yr^{-1}$).

Gross productivity (GP) is the total gain in biomass by an organism.

Net productivity (NP) is the amount remaining after losses due to respiration.

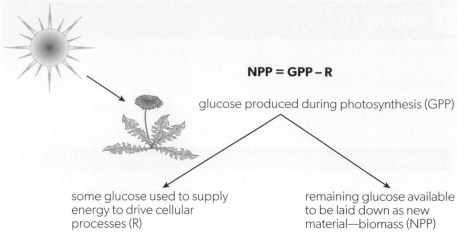

NPP = GPP − R

glucose produced during photosynthesis (GPP)

some glucose used to supply energy to drive cellular processes (R)

remaining glucose available to be laid down as new material—biomass (NPP)

▲ Figure 10 Net primary productivity (NPP)

Ecological pyramids

Ecological pyramids include pyramids of numbers, biomass and productivity. They are quantitative models, usually measured for a given area and time.

Pyramids are graphical models of the quantitative differences between amounts of living material stored at each trophic level of a food chain.

- They allow easy examination of energy transfers and losses.

- They give an idea of what feeds on what and what organisms exist at the different trophic levels.

- They help to demonstrate that ecosystems are systems that are in balance.

All pyramids may be represented as in Figure 11.

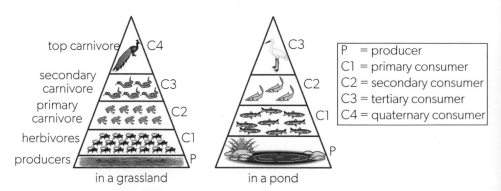

top carnivore — C4
secondary carnivore — C3
primary carnivore — C2
herbivores — C1
producers — P

in a grassland

C3
C2
C1
P

in a pond

P = producer
C1 = primary consumer
C2 = secondary consumer
C3 = tertiary consumer
C4 = quaternary consumer

▲ Figure 11 Ecological pyramids

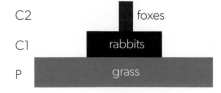

C2 — foxes
C1 — rabbits
P — grass

▲ Figure 12 Pyramid of numbers, showing producers (P), primary consumers (C1) and secondary consumers (C2)
Adapted from: Six Red Marbles and OUP

Pyramids of numbers

A **pyramid of numbers** (Figure 12) shows the number of organisms at each trophic level in a food chain at one time—the **standing crop**. The units are number per unit area.

The length of each bar gives a measure of the relative numbers. Most pyramids are broad at their base and have many individuals in the producer (P) level. But some may have a large single plant such as a tree as the producer so the base is one individual which supports many consumers (Figure 13).

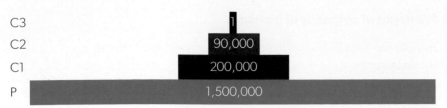

▲ Figure 13 Pyramids of numbers for a grazing ecosystem (left) and an oak tree (right)
Adapted from: Six Red Marbles and OUP

Advantage of pyramids of numbers

- This is a simple, easy method of giving an overview and is good for comparing changes in population numbers with time or season.

Disadvantages of pyramids of numbers

- All organisms are included, regardless of their size, so a pyramid based on an oak tree (Figure 13) is inverted (has a small base and gets larger as it goes up the trophic levels).

- Do not allow for juveniles or immature forms.

- Numbers can be too great to represent accurately.

Pyramids of biomass

A **pyramid of biomass** contains the biomass (mass of each individual × number of individuals) at each trophic level (Figure 14). Biomass is the quantity of (dry) organic material in an organism, a population, a particular trophic level or an ecosystem. Practical work on pyramids of biomass is covered in subtopic 2.6.

A pyramid of biomass uses units of mass per unit area, often grams per square metre ($g\ m^{-2}$) or kilograms per cubic kilometre of water ($kg\ km^{-3}$). A pyramid of biomass is more likely to be a pyramid shape (Figures 14 and 15) but there are some exceptions, particularly in oceanic ecosystems where the producers are phytoplankton (unicellular green algae). Phytoplankton reproduce fast but are present only in small amounts at any one time. As a pyramid represents biomass at one time only (e.g. in winter), the phytoplankton bar may be far smaller than that of the zooplankton which are the primary consumers (Figure 14).

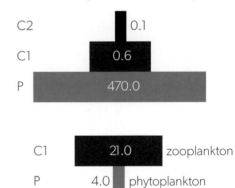

▲ Figure 14 Pyramids of biomass (units $g\ m^{-2}$)

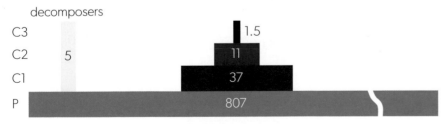

▲ Figure 15 Pyramid of biomass for a lake (units $g\ m^{-2}$)

Advantage of pyramids of biomass

- Overcome some of the problems of pyramids of numbers.

Disadvantages of pyramids of biomass

- Only use samples from populations, so it is impossible to measure biomass exactly.

- Organisms must be killed to measure dry mass.

The time of year at which biomass is measured affects the result. For example, the biomass of algae changes significantly during the year so the shape of the pyramid would depend on the season. The giant redwood trees of California have accumulated their biomass over many years but algae in a lake at the equivalent trophic level may only need a few days to accumulate the same biomass. A pyramid of biomass will not show these differences.

Pyramids of total biomass accumulated per year by organisms at a trophic level are usually pyramidal in shape. But two organisms with the same mass do not necessarily have the same energy content. A dormouse stores a large amount of fat, around 37 kJ g^{-1} of potential chemical energy, but a carnivore of equivalent mass contains larger amounts of carbohydrates and proteins, around 17 kJ g^{-1} potential energy. Some organisms contain a high proportion of non-digestible parts such as the exoskeletons of marine crustaceans.

Pyramids of numbers and biomass are snapshots at one time and place. Pyramids for the same food web in the same ecosystem may vary with season and year. In the spring, there will be more producers growing; in autumn, there may be more consumers living on the producers. As you have seen, pyramids of numbers may sometimes be inverted (Figure 13).

Pyramids of energy

A **pyramid of energy** (or **productivity**) shows the rate of flow of energy or biomass through each trophic level. It shows the energy or biomass being generated and available as food to the next trophic level during a fixed period of time. Unlike pyramids of numbers and biomass, which are snapshots at one time, these pyramids show the flow of energy over time. They are always pyramid-shaped in healthy ecosystems because they must follow the second law of thermodynamics. They are measured in units of energy or mass per unit area per period of time, often kilojoules per square metre per year (kJ m^{-2} yr^{-1})—see Table 4. Productivity values are rates of flow, whereas biomass values are stores existing at one particular time.

Pyramid	Units
Numbers (standing crop)	N m^{-2}
Biomass (standing crop)	g m^{-2}
Productivity (flow of biomass or energy)	g m^{-2} yr^{-1} J m^{-2} yr^{-1}

▲ Table 4 Pyramid units
Note the notation: N = numbers, g = grams, J = joules

Advantages of pyramids of energy

- Allow for rate of production over time.

- Allow comparison of different ecosystems.

- Can add solar radiation input.

- Never have inverted pyramids of energy.

Disadvantages of pyramids of energy

- Need to measure growth and reproduction over time.

- Still have issues of consumers at more than one trophic level and where to put decomposers and detritivores.

Data-based questions

Draw and label pyramids of numbers, biomass and energy, using the data in Table 5. Comment on these pyramids.

	Numbers	Biomass (kg m^{-2})	Energy (1000 kJ m^{-2} yr^{-1})
Primary producers	100,000	2,500	500
Primary consumers	10,000	200	50
Secondary consumers	2,000	15	5
Top consumers	500	1	–

▲ Table 5

ATL Activity 11

ATL skills used: thinking, communication, research

An ecosystem consists of one oak tree on which 10,000 herbivores are feeding. These herbivores are prey to 500 spiders and carnivorous insects. Three birds are feeding on these spiders and carnivorous insects. The oak tree has a mass of 4,000 kg, the herbivores have an average mass of 0.05 g, the spiders and carnivorous insects have an average mass of 0.2 g, and the three birds have an average mass of 10 g.

1. a. Construct a pyramid of numbers.

   b. Construct a pyramid of biomass.

   c. Explain the differences between these two pyramids.

2. Explain whether the energy "loss" between two subsequent trophic levels is in contradiction with the first law of thermodynamics.

3. Assuming ecological efficiency of 10%, 5% and 20% respectively (see Figure 16), what will be the energy available at the tertiary consumer level (4th trophic level), given a net primary productivity of 90,000 kJ m^{-2} yr^{-1}? What percentage is this figure of the original energy value at the primary producer level?

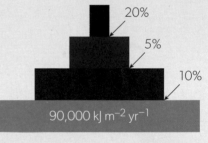

▲ Figure 16

Consequences of pyramids and ecosystem function

The way ecosystems work leads to: concentration of toxic substances in food chains; limited length of food chains; vulnerability of top carnivores; bioaccumulation and biomagnification.

Many chemicals are **biocides**. If a chemical in the environment (e.g. a **pesticide** such as DDT (dichlorodiphenyltrichloroethane) or PCBs (polychlorinated biphenyls) or a heavy metal such as mercury) breaks down slowly or does not break down at all, plants may take it up and animals may take it in as they eat or breathe. If they do not excrete or egest it, the chemical accumulates in their bodies over time. If the chemical stays in the ecosystem for a prolonged period of time, the concentration builds up. This is **bioaccumulation**. Eventually, the concentration may be high enough to cause disease or death.

Key terms

Biocides are substances or microorganisms that destroy, deter or render harmless living things. Examples include: disinfectants, antiseptics, preservatives, pesticides.

Bioaccumulation refers to the increasing concentration of non-biodegradable pollutants in organisms or trophic levels over time (as more are absorbed).

141

Key term

Biomagnification refers to the increasing concentration of non-biodegradable pollutants along a food chain (due to the loss of biodegradable biomass through, for example, respiration).

If a herbivore eats a plant that has the chemical in its tissues, the amount of the chemical that is taken in by the herbivore is greater than that in the plant that is eaten—because the herbivore grazes many plants over time. If a carnivore eats the herbivores, it too will take in more of the chemical than each herbivore contained because it eats many herbivores over time. In this way the chemical's concentration is magnified from trophic level to trophic level. This is **biomagnification**. While the concentration of the chemical may not affect organisms lower in the food chain, the top trophic levels may take in so much of the chemical that it causes disease or their death.

Data-based questions

In this food web, the smaller fish (minnows) eat plankton (microscopic plants and animals) in the water. Minnows are eaten by the larger fish called pickerel. These are eaten by herons, ospreys and cormorants; herons eat the minnows as well. The numbers give the percentage concentration of DDT.

1. How many trophic levels are in this food web?

2. How many times more concentrated is the DDT in the body of the cormorant than in the water? Explain how this happens.

3. In which species does bioaccumulation occur?

4. In which species does biomagnification occur?

5. Seals and penguins in Antarctica and polar bears in the Arctic have been found with pesticides in their tissues. The nearest land is over 1,000 km away. Discuss how the pesticides may have reached them.

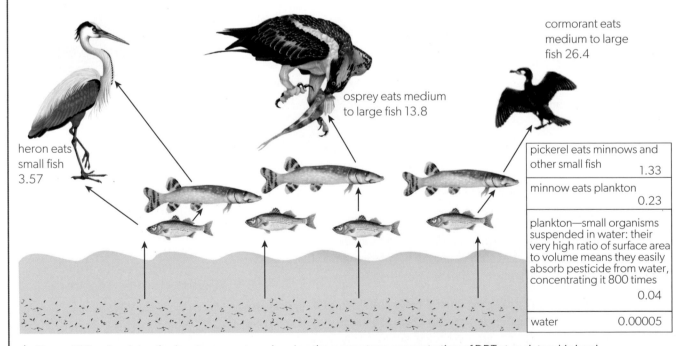

cormorant eats medium to large fish 26.4

osprey eats medium to large fish 13.8

heron eats small fish 3.57

pickerel eats minnows and other small fish	1.33
minnow eats plankton	0.23
plankton—small organisms suspended in water: their very high ratio of surface area to volume means they easily absorb pesticide from water, concentrating it 800 times	0.04
water	0.00005

▲ Figure 17 Food web in a freshwater ecosystem showing the percentage concentration of DDT at each trophic level

A serious problem with pesticides is how long they last in the environment once they are sprayed. Some decompose into harmless chemicals as soon as they touch the soil. Glyphosate (first sold by Monsanto as Roundup®)is one of these: once it touches the soil, it is inactivated. Others are persistent and do not break down in this way. Neither do they break down inside the bodies of organisms. They enter the food web and move through it from trophic level to trophic level. They are non-biodegradable—persistent organic pollutants or POPs. Many early insecticides such as DDT, dieldrin and aldrin fall into this group and they are stored in the fat of animals.

Human impact on flows of energy and transfer of matter

Just about all human activity has an effect on ecosystems. Biomass is lost to ecosystems when we build on land, grow crops and cut down trees. Food webs are disrupted when ecosystems are destroyed or degraded.

Loss of plants means less photosynthesis so less of the Sun's energy is captured and turned into chemical energy. Solar panels and solar farms do capture sunlight and convert it to electricity but efficiency is about 20% at best.

More on energy sources

Animals are known as **heterotrophs** or heterotrophic organisms, to distinguish them from plants **(autotrophs)**. Heterotrophs range from fungi and some bacteria and single-celled organisms to all animals and some parasitic plants.

Green plants are not the only organisms that can make their own food. Others include algae such as the green algae you see in ponds or brown algae which are seaweeds. There are also photosynthetic bacteria living in soil or water. These all use light as the external source of energy. They are called **photoautotrophs**.

There are other sources of energy such as chemical energy which some organisms use as their energy source. These organisms are called **chemoautotrophs**. These producers are the basis of food webs. Examples of chemoautotrophs include:

- Bacteria that live in deep sea vents where no sunlight penetrates and pressure and temperature are extremely high. Hydrogen sulfide and other chemicals provide the energy source.

- Two types of bacteria in the nitrogen cycle. One type gains its energy from oxidizing ammonia to nitrite (*Nitrosomonas*) and the other converts nitrite to nitrate (*Nitrobacter*).

AHL

Key terms

Primary productivity is the rate of production of biomass using an external energy source and inorganic sources of carbon and other elements.

Gross primary productivity (GPP) is the total gain in energy or biomass per unit area per unit time by green plants. It is the energy fixed (or converted from light to chemical energy) by green plants by photosynthesis. But some of this is used in respiration.

Net primary productivity (NPP) is the total gain in energy or biomass per unit area per unit time by green plants after allowing for losses to respiration. This is the increase in biomass of the plant: how much it grows and the biomass that is potentially available to consumers (animals) that eat the plant.

More on productivity

Gross primary productivity (GPP) is the total gain in energy or biomass per unit area per unit time by green plants. It is the energy fixed (or converted from light to chemical energy) by green plants by photosynthesis. Some of this energy is used in respiration.

Net primary productivity (NPP) is the total gain in energy or biomass per unit area per unit time by green plants after allowing for losses to respiration. This is the increase in biomass of the plant—how much it grows—and is the biomass that is potentially available to consumers (animals) that eat the plant.

GPP is very hard to measure as much of the light energy converted to chemical energy (glucose) is used up by plants in respiration almost as soon as it is produced. A more useful way of looking at production of plants is the measurement of net primary productivity (NPP).

Net primary productivity is simply plant growth and is the basis for food chains because it is the quantity of carbon compounds sustainably available to primary consumers.

The consumers may be herbivores in a natural food web or farmers, foresters and all humans in managed ecosystems whether agriculture (food) or silviculture (wood).

The units usually used for productivity are kg carbon m^{-2} $year^{-1}$ (kilograms of carbon per square metre of ecosystem per year).

An ecosystem's NPP is the rate at which plants accumulate dry mass (actual plant material). The glucose produced in photosynthesis has two main fates.

- Some provides for growth, maintenance and reproduction (life processes) with energy being lost as heat during processes of respiration.

- The remainder is deposited in and around cells as new material and represents the stored dry mass—this store of energy is potential food for consumers within the ecosystem.

This means that NPP represents the difference between the rate at which plants photosynthesize, GPP, and the rate at which they respire. This accumulation of dry mass is usually termed biomass and provides a useful measure of both production and utilization of resources (NPP = GPP – R, see Figure 10).

The amount of biomass produced varies:

- in space—some biomes have much higher NPP rates than others (e.g. tropical rainforest vs tundra)

- in time—many plants have seasonal patterns of productivity linked to changing availability of basic resources such as light, water and warmth.

Net secondary productivity (NSP)

As with plants, not all energy that goes into the herbivore is available to make new biomass.

- Only food that crosses the wall of the alimentary canal (gut wall) of animals is absorbed and used to power life processes (**assimilated food energy**). Some of the assimilated food energy is used in cellular respiration to provide energy for life processes. Some is removed as nitrogenous waste, in most animals as urine. The rest is stored in the dry mass of new tissue.

- Some ingested plant material passes straight through the herbivore and is released as faeces (egestion). This is not absorbed and provides animals with no energy.

Total food ingested, including food that is egested, is the measure of **gross secondary productivity (GSP)**. Therefore, **net secondary productivity (NSP)** can be thought of in the same way as net primary productivity (Figure 10).

net productivity of herbivores (net secondary productivity) =

energy in food ingested – energy lost in egestion – energy used in respiration

Only a very small percentage of the original NPP of plants is turned into secondary productivity by herbivores (Figure 18) and it is this secondary productivity which is available to consumers at the next trophic level. This change of primary productivity to secondary productivity follows the general conditions of energy transfer up trophic levels.

Carnivores—animals that eat other animals—are next in the trophic ladder. Secondary consumers are those that eat herbivores and tertiary consumers are those whose main source of energy is other carnivores. The ability of carnivores to assimilate energy follows the same basic path as that of herbivores, though secondary and tertiary consumers have higher protein diets—meat, which is more easily digested and assimilated.

Carnivores

- On average, carnivores assimilate 80% of the energy in their diets and egest less than 20%.

- Usually they have to chase moving animals so higher energy intake is offset by increased respiration during hunting.

- Biomass is locked up in the prey foods (e.g. non-digestible skeletal parts, such as bone, horn and antler) so they have to assimilate the maximum amount of energy that they can from any digestible food.

Herbivores

- Assimilate about 40% of the energy in their diets and egest 60%.

- Graze static plants.

Key terms

Secondary productivity is the gain in biomass by consumers, using carbon compounds absorbed and assimilated from ingested food. It can be estimated as ingested food minus faecal waste. (Faecal matter is not included as it is material that has remained undigested and unabsorbed.)

Gross secondary productivity (GSP) is the total energy or biomass assimilated (taken up) by consumers and is calculated by subtracting the mass of faecal loss from the mass of food eaten.

Net secondary productivity (NSP) is the total gain in energy or biomass per unit area per unit time by consumers after allowing for losses to respiration. There are other losses in animals as well as to respiration but respiration is the main one.

NSP is calculated by subtracting respiratory losses (R) from GSP.

GSP = food eaten – faecal loss

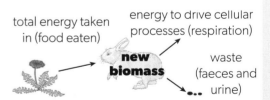

▲ Figure 18 **Net productivity of a herbivore**

Maximum sustainable yield

The MSY is the net primary or net secondary productivity of a system.

It is the largest crop or catch that can be taken from the stock of a species (e.g. a forest, or a shoal of fish) without depleting the stock. What is removed is the increase in production of the stock while leaving the stock to reproduce again.

Sustainable yields are higher for lower trophic levels. This is because less energy has been lost as there are fewer links in the food chain.

Eating meat or fish from higher trophic levels is expensive in terms of energy but these foods have high energy values. Muscle is mostly protein and fat is high in energy content. They are energy-dense foods.

For humans, eating a plant-based diet or from lower trophic levels is a more sustainable practice as we are then primary consumers not secondary or tertiary consumers. Less energy is wasted as there are fewer trophic levels.

Key term

Ecological efficiency is the percentage of energy received by one trophic level that is passed on to the next level.

Ecological efficiency

We need to know two quantities to establish assimilation and productivity efficiencies.

- What proportion of the NPP from one trophic level is assimilated by the next?

- How much of this assimilated material is turned into the tissues of the organism and how much is respired?

For an animal raised for meat these questions are:

- How much of the grass that an animal eats can it **assimilate** (absorb into its body)? This will determine how many animals the farmer can put in a field.

- How much of what is assimilated is used for **productivity** (turned into meat)? On a commercial farm this will determine the profits.

$$\text{efficiency of assimilation} = \frac{\text{gross productivity}}{\text{food eaten}} \times 100$$

$$\text{efficiency of biomass productivity} = \frac{\text{net productivity}}{\text{gross productivity}} \times 100$$

The length of food chains

As a rule of thumb, only 10% of the energy in one trophic level is transferred to the next—the **trophic efficiency** is 10%. A major part of the energy is used in respiration to keep the organism alive and is finally lost as heat to the environment. This is a result of the second law of thermodynamics, which states that energy is degraded to lower quality and finally to heat. More is lost because herbivores destroy more plant material than they eat (e.g. by trampling on it) or they reject it because it is too tough, old or spiky. Some material is not eaten at all and some dies and decomposes before it can be eaten. The 90% loss of energy from one trophic level to the next means there is very little energy available after about four trophic levels in terrestrial ecosystems and five in aquatic ecosystems.

ATL Activity 12

ATL skills used: thinking, communication, research

1. Consider the assimilation efficiencies in Table 6.

   a. Why do carnivores have a relatively high assimilation efficiency? (Think about the food they eat.)

   b. Do you think ruminant herbivores would be at the top or bottom of the range for herbivores? Why?

   c. Why does the giant panda have such a low assimilation efficiency? (Hint: its diet is mainly bamboo shoots.)

2. Copy Figure 19 and add the energy storages and transfers in Figure 20.

Organism	Assimilation efficiency
Carnivore	90%
Insectivore	70–80%
Herbivore	30–60%
Zooplankton feeding on phytoplankton	50–90%
Giant panda	20%

▲ Table 6

▲ Figure 19

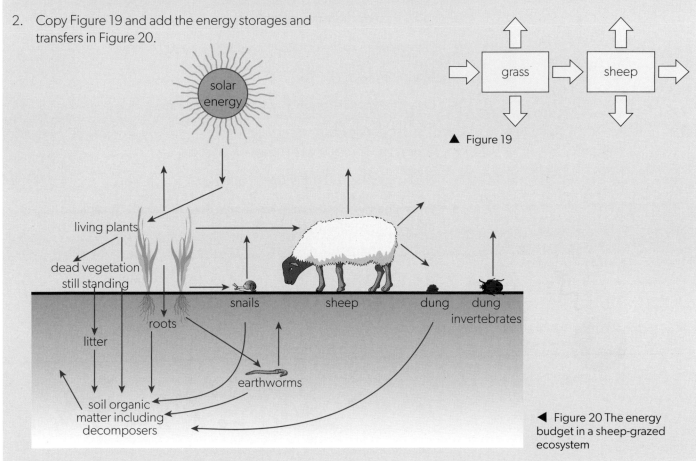

◀ Figure 20 The energy budget in a sheep-grazed ecosystem

Why top carnivores are in trouble

Top carnivores are vulnerable because of the loss of energy from each trophic level. There is only so much energy available and that is why big, fierce animals are rare. It is hard for them to accumulate enough energy to grow to a large size.

It is often the highest trophic level in a food chain that is the most susceptible to alterations in the environment. The UK population of the peregrine falcon fell dramatically in the late 1950s, probably due to agricultural chemicals such as DDT accumulating and then magnifying in the food chain. This appeared to cause egg-shell thinning and reduced breeding success. These chemicals were banned and from the mid-1960s, the peregrine population began to slowly recover despite persecution on grouse moors and the threat from egg collectors.

The top of the food chain is always vulnerable to the effects of changes further down the chain. Top carnivores often have a limited diet so a change in their food (prey) has a knock-on effect. Their population numbers are low because of the fall in efficiency along a food chain, therefore their ability to withstand negative influences is more limited than for species lower in the food chain with larger populations. They have lower resilience.

More on entropy

The **second law of thermodynamics** shows that the **entropy** of an isolated system not in equilibrium will tend to increase over time.

Entropy is a measure of the amount of disorder in a system. An increase in entropy arising from energy **transformations** reduces the energy available to do work.

- Entropy refers to the spreading out or dispersal of energy.

- More entropy = less order.

- Over time, all differences in energy in the universe will be evened out until nothing can change.

- Energy transformations are never 100% efficient.

- The biggest loss of heat is in respiration.

- When energy is used to do work, some energy is always dissipated (lost to the environment) as waste heat.

This process can be summarized by a simple diagram showing the energy input and outputs (Figure 21).

energy = work + heat (and other wasted energy)

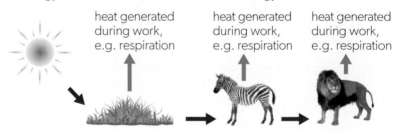

▲ Figure 22 Loss of energy to the environment in a food chain

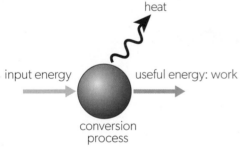

▲ Figure 21 Diagram to show the second law of thermodynamics

In the example in Figure 22, the energy spreads out so the useful energy consumed by one trophic level is less than the total energy at the level below.

- Depending on the type of plant, the efficiency at converting solar energy to stored sugars is around 1–2%.

- Herbivores (e.g. the zebra) on average only assimilate (turn into animal matter) about 10% of the total plant energy they consume. The rest is lost in metabolic processes and in escaping from the lion. This changes the stored chemical energy in its cells into useful work (running). But during the attempted escape some of the stored energy is converted to heat and lost from the food chain.

- A carnivore's efficiency is also around 10%. Carnivores also metabolize stored chemical energy, in this case trying to catch the herbivore.

- As energy is dispersed to the environment, there is a reduction in the amount of energy passed on to the next trophic level.

- The carnivore's total efficiency in the chain is $0.02 \times 0.1 \times 0.1 = 0.0002\%$.

Life is a battle against entropy and without the constant replenishment of energy, life cannot exist. Imagine rowing upstream, as shown in Figure 23. Stop for a moment and you will be swept back downstream by the current of entropy.

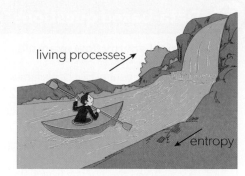

▲ Figure 23 A representation of life against entropy

Simple example of entropy

The situation in Figure 24 obeys the second law of thermodynamics, since the tidy room of low entropy becomes untidy, a situation of high entropy. In the process, entropy increases spontaneously.

▲ Figure 24 (a) Tidy room—low entropy vs (b) untidy room—high entropy

- Solar energy powers photosynthesis.

- Chemical energy, through respiration, powers all activities of life.

- Electrical energy runs all home appliances.

- The potential energy of a waterfall turns a turbine to produce electricity.

These are all high-quality forms of energy, because they power useful processes. They are all ordered forms of energy. Solar energy reaches us via photons in solar rays; chemical energy is stored in the bonds of macromolecules like sugars; the potential energy of falling water is due to the specific position of water, namely that it is high and falls. These ordered forms have low disorder, so low entropy.

On the contrary, heat may not power any process; it is a low-quality form of energy. Heat is simply dispersed in space, being capable only of warming it. Heat dissipates to the environment without any order; it is disordered. In other words, heat is a form of energy characterized by high entropy.

Data-based questions

The classical energy flow example

Silver Springs, in central Florida, is famous among ecologists as the place where Howard T. Odum researched energy flow in the ecosystem in the 1950s. Odum (1924–2002) was a pioneer ecologist working on ecological energetics. He was the first to attempt to measure an energy budget—he measured primary productivity and losses by respiration. (Later, near the end of a long and illustrious career, he and David Scienceman developed the concept of **emergy** (embodied energy) which is a measure of the quality and type of energy and matter that go into making an organism.)

Figure 25 shows the energy flows and biomass stores measured by Odum at Silver Springs. This simple community consists of algae and duckweed (producers); tadpoles, shrimps and insect larvae (herbivores); water beetles and frogs (first carnivores); small fish (top consumers); and bacteria, bivalves and snails (decomposers and detritivores). Dead leaves also fall into the water and spring water flows out, exporting some detritus.

1. Why does the width of the energy flow bands become progressively narrower as energy flows through the ecosystem?

2. Suggest an explanation for the limit on the number of trophic levels to four or five at most in a community.

3. How is the energy transferred between trophic levels?

4. Insolation (light) striking leaves is 1,700,000 units but only 410,000 are absorbed. What happens to the unabsorbed light energy?

5. A further 389,190 units escapes from producers as heat. Why is this?

6. Calculate the efficiency of net primary productivity.

7. Draw a productivity pyramid from the data given.

8. Would it be possible to draw a biomass pyramid from the data given?

9. Does the model support the first law of thermodynamics? Show your calculations.

10. How does the diagram demonstrate the second law of thermodynamics?

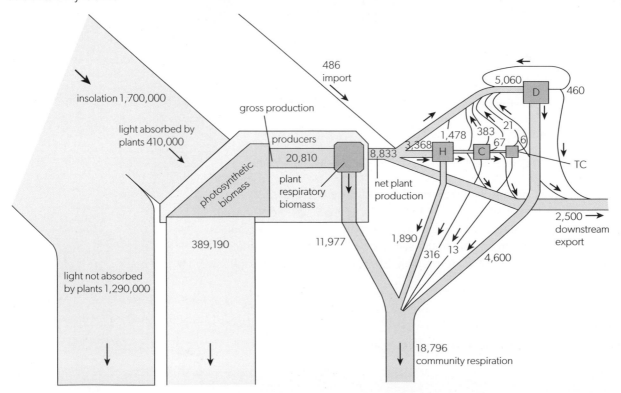

▲ Figure 25 The energy flow values in Silver Springs community. Units kcal m^{-2} yr^{-1} (1 kcal = 4.2 J)

Data-based questions

The data in Table 7 refer to carbon (in biomass) flows in a freshwater system at 40° N latitude.

1. From the data, write word equations and calculate:

    a. net productivity of phytoplankton

    b. gross productivity of zooplankton

    c. net productivity of zooplankton

    d. % assimilation of zooplankton

    e. % productivity of zooplankton.

2. Two more energy flow diagrams are shown in Figures 26 and 27.

    a. Copy Figure 26 and draw a rectangle on the diagram to show the ecosystem boundary.

    b. Explain why the storage boxes reduce in size as you go up the food chain.

    c. Name three decomposers and explain how they lose heat.

	$g\,C\,m^{-2}\,yr^{-1}$
Gross productivity of phytoplankton	132
Respiratory loss by phytoplankton	35
Phytoplankton eaten by zooplankton	31
Faecal loss by zooplankton	6
Respiratory loss by zooplankton	12

▲ Table 7

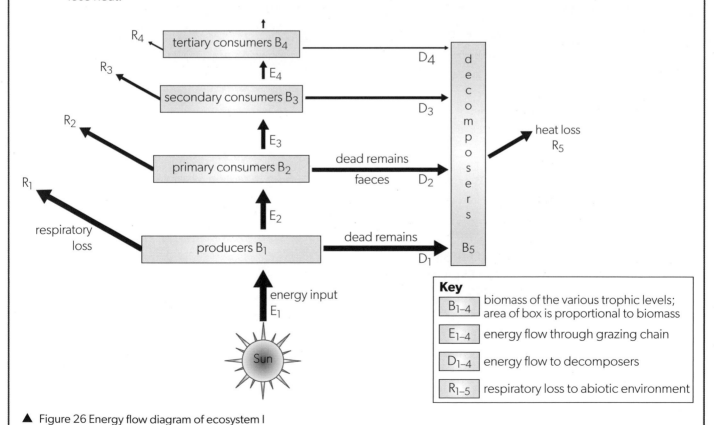

▲ Figure 26 Energy flow diagram of ecosystem I

Key

B_{1-4}	biomass of the various trophic levels; area of box is proportional to biomass
E_{1-4}	energy flow through grazing chain
D_{1-4}	energy flow to decomposers
R_{1-5}	respiratory loss to abiotic environment

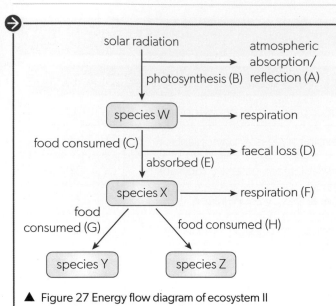

▲ Figure 27 Energy flow diagram of ecosystem II

d. For ecosystem II, identify from Figure 27 the letter(s) referring to the following energy flow processes and explain what happens to this energy at each stage as it passes through the ecosphere:

i. loss of radiation through reflection and absorption

ii. conversion of light to chemical energy in biomass

iii. loss of chemical energy from one trophic level to another

iv. efficiencies of transfer

v. overall conversion of light to heat energy by an ecosystem

vi. re-radiation of heat energy to the atmosphere.

Check your understanding

In this subtopic, you have covered:

- energy flow through ecosystems

- photosynthesis and respiration processes

- first and second laws of thermodynamics

- food chains and webs and trophic levels

- ecological pyramids

- pollutants in food webs

- human impacts on energy transfers

- more detail on autotrophs

- more on productivity

- maximum sustainable yields

- ecological efficiency

- entropy.

AHL

How can flows of energy and matter through ecosystems be modelled? How do human actions affect the flow of energy and matter, and what is the impact on ecosystems?

1. Explain the processes of respiration and photosynthesis.

2. Draw systems diagrams of these processes.

3. Review the two laws of thermodynamics and explain how energy degrades.

4. Define GP and NP.

5. Sketch pyramids of numbers, biomass and productivity for a lake and label the trophic levels.

6. Explain how some pollutants accumulate in food chains.

7. Explain why and how energy is lost in a food chain.

8. Explain biomagnification and bioaccumulation, with reference to named examples.

9. Discuss examples of loss of photosynthetic activity through human activities.

10. Define primary and secondary productivity, net and gross primary productivity.

11. Explain the significance of (a) maximum sustainable yield; (b) ecological efficiency; and (c) entropy.

AHL

≫ Taking it further

- Using primary or secondary data, study the impact of pollution on an ecosystem and the effect on food chains, for example, the effect of a sewage overflow on aquatic communities. (Consider health and safety, and ethical issues.)

- Advocate for the planetary health diet in your community based on the second law of thermodynamics.

- Contribute to an ecological citizen science programme.

2.3 Biogeochemical cycles

Guiding question

- How do human activities affect nutrient cycling, and what impact does this have on the sustainability of environmental systems?

Understandings

1. Biogeochemical cycles ensure chemical elements continue to be available to living organisms.
2. Biogeochemical cycles have stores, sinks and sources.
3. Organisms, crude oil and natural gas contain organic stores of carbon. Inorganic stores can be found in the atmosphere, soils and oceans.
4. Carbon flows between stores in ecosystems by photosynthesis, feeding, defecation, cellular respiration, death and decomposition.
5. Carbon sequestration is the process of capturing gaseous and atmospheric carbon dioxide and storing it in a solid or liquid form.
6. Ecosystems can act as stores, sinks or sources of carbon.
7. Fossil fuels are stores of carbon with unlimited residence times. They were formed when ecosystems acted as carbon sinks in past eras and become carbon sources when burned.
8. Agricultural systems can act as carbon stores, sources and sinks, depending on the techniques used.
9. Carbon dioxide is absorbed into the oceans by dissolving and is released as a gas when it comes out of a solution.
10. Increases in concentrations of dissolved carbon dioxide cause ocean acidification, harming marine animals.
11. Measures are required to alleviate the effects of human activities on the carbon cycle.
12. The lithosphere contains carbon stores in fossil fuels and in rocks, such as limestone, that contain calcium carbonate.
13. Reef-building corals and molluscs have hard parts that contain calcium carbonate that can become fossilized in limestone.
14. In past geological eras, organic matter from partially decomposed plants became fossilized in coal, and partially decomposed marine organisms became fossilized in oil and natural gas held in porous rocks.
15. Methane is produced from dead organic matter in anaerobic conditions by methanogenic bacteria.
16. Methane has a residence time of about 10 years in the atmosphere and is eventually oxidized to carbon dioxide.
17. The nitrogen cycle contains organic and inorganic stores.
18. Bacteria have essential roles in the nitrogen cycle.
19. Denitrification only happens in anaerobic conditions, such as soils that are waterlogged.
20. Plants cannot fix nitrogen so atmospheric dinitrogen is unavailable to them unless they form mutualistic associations with nitrogen-fixing bacteria.
21. Flows in the nitrogen cycle include mineral uptake by producers, photosynthesis, consumption, excretion, death, decomposition and ammonification.
22. Human activities such as deforestation, agriculture, aquaculture and urbanization change the nitrogen cycle.
23. The Haber process is an industrial process that produces ammonia from nitrogen and hydrogen for use as fertilizer.
24. Increases in nitrates in the biosphere from human activities have led to the planetary boundary for the nitrogen cycle being crossed, making irreversible changes to Earth systems likely.
25. Global collaboration is needed to address the uncontrolled use of nitrogen in industrial and agricultural processes and bring the nitrogen cycle back within planetary boundaries.

Biogeochemical cycles

All the **biogeochemical cycles** have both organic (when the element is in a living organism) and inorganic (when the element is in a simpler form outside living organisms) phases. Both phases are vital: the efficiency of movement through the organic phase determines how much is available to living organisms.

The major reservoir for all the main elements is as inorganic molecules in rock and soils. Flow in this inorganic phase tends to be much slower than the movement of these nutrients through the organic phase in organisms.

The major biogeochemical cycles are those of water, carbon, nitrogen, sulfur and phosphorus, all of which follow partially similar routes.

The impact of human activities is affecting the sustainability of the cycles. Biogeochemical cycles have **stores**, **sinks** and **sources**.

Key terms

Biogeochemical cycles are cycles of chemicals between biological and geological storages.

Stores (storages) remain in equilibrium with the environment, with equal amounts absorbed and released.

Sinks are where there is net accumulation of the element.

Sources are where there is net release of the element.

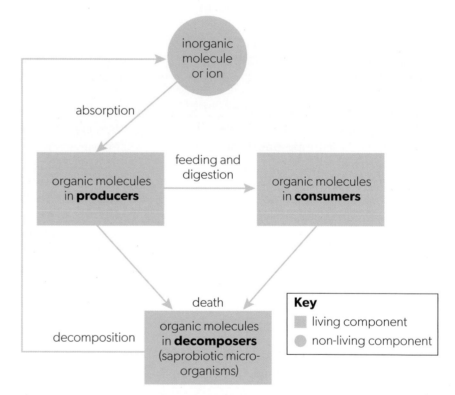

▲ Figure 1 Basic cycle of all nutrient cycles

Carbon cycle

The carbon cycle is the circulation of carbon through living and non-living systems on Earth. Carbon is found in four main stores: the soil, living things (biomass), the oceans and the atmosphere. Carbon not in the atmosphere is stored in carbon dioxide sinks (soil, biomass and oceans) as complex organic molecules or dissolved in seawater.

Carbon cycles between living (biotic) and non-living (abiotic) chemical cycles. It is fixed by photosynthesis and released back to the atmosphere through respiration. Carbon is also released back to the atmosphere through the combustion of fossil fuels and biomass.

Connections: This links to material covered in subtopics 1.2 and 6.2.

When dead organisms decompose, when living organisms respire, and when fossil fuels are burned, carbon is oxidized to carbon dioxide and this, water vapour and heat are released. By photosynthesis, plants recapture this carbon—**carbon fixation**—and lock it up in their bodies for a time as glucose or other large molecules.

When plants are harvested and cut down for food, firewood or processing, the carbon is released again to the atmosphere. As humans burn fossil fuels and cut down trees, they are increasing the amount of carbon in the atmosphere and changing the balance of the carbon cycle. Carbon can remain locked up for long periods of time, e.g. in the wood of trees or as coal and oil.

Human activity has disrupted the balance of the global carbon cycle (carbon budget) through increased combustion, land use changes and deforestation.

Residence time is the average period that an atom remains in a store. Without human interference (i.e. mining and burning) the residence time of carbon in fossil fuels would be measured in hundreds of millions of years. Fossil fuels are a carbon sink but become a carbon source when burned.

Carbon stores

Life on Earth is based on carbon, which may be:

- organic (complex carbon molecules)

  - organisms (biomass) in the biosphere—living plants and animals

  - stored as fossil fuels—crude oil, natural gas

- inorganic (simple carbon molecules)

  - locked up or fixed into solid forms and stored as sedimentary rocks

  - in the oceans where carbon is dissolved or locked up as carbonates in the shells of marine organisms

  - in soil

  - as carbon dioxide in the atmosphere.

Most of the carbon on Earth—about 65,500 billion metric tonnes—is stored in rocks and sediments (limestone, chalk, fossil fuels) and this is locked up for millions of years. The rest is in oceans, the atmosphere, soil, fossil fuels and living things. All fossil fuels contain carbon because they are fossilized life forms.

Carbon is an essential element in living systems, providing the chemical framework to form molecules that make up living organisms. The molecules of organic compounds are built from chains of carbon atoms to which atoms of the other elements (mainly hydrogen, oxygen, nitrogen and sulfur) are attached.

Carbon makes up around 0.037% of the atmosphere as carbon dioxide and is present in the oceans as carbonate and bicarbonates, and in rocks such as limestone and coal.

Connections: This links to material covered in subtopics 1.2, 5.1, 5.2 and 6.2.

Carbon flows

Carbon flows between stores in ecosystems by:	In inorganic stores, carbon flows by:
• photosynthesis • respiration • feeding • defecation • death and decomposition.	• fossilization • combustion • dissolving.

(ATL) Activity 13

ATL skills used: thinking, communication, research

1. Copy the carbon cycle diagram in Figure 2 and complete the blank boxes with the names of the processes.

2. Review subtopic 1.2 on transfers and transformations, then colour transfer and transformation processes in different colours.

3. Colour biotic and abiotic stores in different colours.

4. Add the inorganic stores to your diagram.

5. Add a key.

carbon dioxide
in the atmosphere

A

D

B

plants

respiration

C

fossil fuels

animals

death

▶ Figure 2 Basic carbon cycle

There is a fast carbon cycle (biotic, Figure 3) and a slow one (abiotic). The fast carbon cycle is a lifespan long. The slow carbon cycle is measured in millions of years.

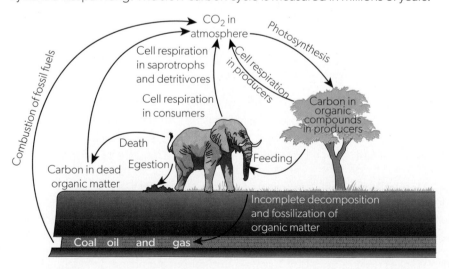

▶ Figure 3 Fast carbon cycle involving living things

Adapted from: U.S. DOE http://genomicscience.energy.gov/

If there were only inorganic stores of carbon, it would take millions of years for carbon to move between rocks, soil, oceans and atmosphere via transfer and transformations of precipitation and weathering processes. For example, rain falls on limestone and forms a weak acid—carbonic acid. This dissolves the rocks, releasing calcium ions that flow to the sea in rivers and there combine with bicarbonate ions to form calcium carbonate—chalk or limestone. Carbon returns to the atmosphere via volcanic eruptions. This is the inorganic **slow carbon cycle**.

The carbon budget

The amount of carbon on Earth is finite and we have a rough idea of where it goes. It is not easy to get accurate figures on global carbon exchange but there are (slightly different) estimates in Figures 4 and 5.

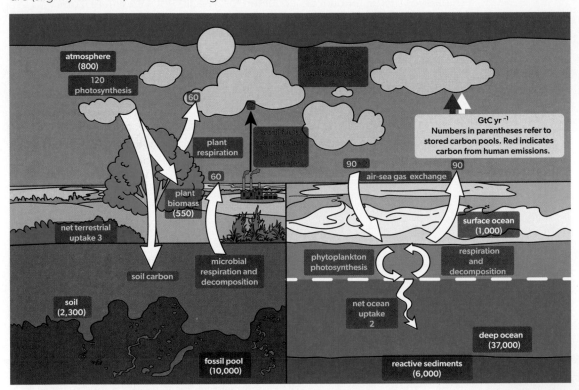

◀ Figure 4 **The fast carbon cycle with sizes of sinks, stores and flows in gigatonnes of carbon (GtC)**
Adapted from: U.S. DOE, Biological and Environmental Research Information System

The fast carbon cycle in Figure 4 involves living things in ecosystems and shows the movement of carbon between land, atmosphere and oceans. Yellow numbers are natural flows. White numbers indicate stored carbon. Red are human contributions. All are in gigatonnes of carbon per year GtC yr^{-1}. A gigatonne is 1 billion metric tonnes, or 10^{12} kg.

Humans and the carbon cycle

Without human interference, the carbon cycle is in balance over the long term. The carbon in fossil fuels would leak slowly into the atmosphere through volcanic activity over millions of years in the slow carbon cycle. But by burning coal, oil and natural gas, we accelerate the process, releasing carbon that took millions of years to accumulate into the atmosphere. In this way we move carbon from the slow cycle to the fast cycle.

Carbon dioxide levels in the atmosphere have risen from pre-industrial levels of about 280 ppm (parts per million) to 414 ppm in February 2023 and are still rising. Find out what they are now.

Each year:

- about 9 GtC enter the atmosphere
- about 5 GtC of this stays in the atmosphere
- about 3.4 GtC are taken up by plants

ATL Activity 14

ATL skills used: thinking, research, self-management

Study Figure 4 and state:

1. the largest terrestrial stores of carbon

2. the largest oceanic stores of carbon

3. the amount of carbon added to the atmosphere per year by human activity

4. the amount of this carbon that is fixed by phytoplankton in oceans

5. the amount of carbon that is fixed by terrestrial plants

6. the total amount of carbon added to the atmosphere each year by anthropogenic activity.

TOK

A lot of numbers have been stated in this section so far. They are estimates calculated from data collected. Should we trust this knowledge?

What is the role of inductive (specific to general conclusions) and deductive (testing an existing theory) reasoning in this scientific inquiry and prediction?

- diffusion of carbon dioxide into the oceans and uptake by oceanic phytoplankton accounts for about 2.5 GtC

- new growth forests fix about 0.5 GtC.

This leaves a large amount unaccounted for. We are not sure where it goes because of the complexity of the system. The amounts of carbon (in GtC) in other reservoirs are:

- atmosphere 860

- standing biomass 550

- soils 2,300

- surface oceans 1,000.

Since the pre-industrial period, we have added 200 GtC to the atmosphere. Nearly half the carbon humans emit per year through various processes is not fixed or sequestered.

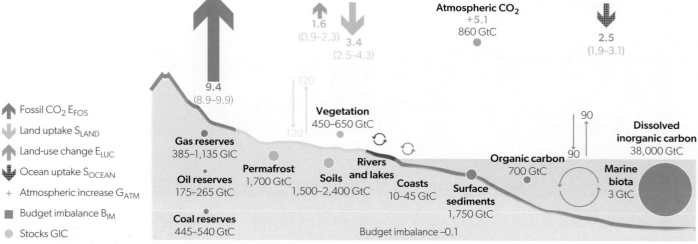

Fossil CO_2 E_{FOS}

Land uptake S_{LAND}

Land-use change E_{LUC}

Ocean uptake S_{OCEAN}

+ Atmospheric increase G_{ATM}

Budget imbalance B_{IM}

Stocks GIC

▲ Figure 5 Anthropogenic activities affecting the carbon cycle, 2010–2019. Numbers are average figures in GtC per year

Friedlingstein, P. et al. (2020) 'Global Carbon Budget 2020', Earth System Science Data, 12(4), pp. 3269–3340. doi:10.5194/essd-12-3269-2020.

Agricultural and forest systems and carbon

Forests and agricultural crops can act as carbon stores, sources or sinks. The carbon source–sink–store balance is dynamic.

- Regenerative agricultural methods like crop rotation, cover crops and no till promote the role of soil and vegetation as a carbon sink.

- Drainage of wetland, monoculture and heavy tillage promote the role of soil and vegetation as a carbon source.

- Cropping over a longer timescale (e.g. timber production) and the subsequent use of harvested products will also affect whether a crop is a store, sink or source.

- Forests have moderated climate change caused by humans by absorbing about 25% of the carbon emitted by human activities.

- A forest is a **carbon source** if it releases more carbon than it absorbs, for example when trees burn or when they decay after dying (due to old age or fire, insect attack or other disturbance).

- A forest is a **carbon sink** if it absorbs more carbon from the atmosphere than it releases.

Carbon is absorbed from the atmosphere through photosynthesis. It is then deposited in forest biomass (trunks, branches, roots and leaves), in dead organic matter (litter and dead wood) and in soil. This is **carbon sequestration**.

The carbon cycle of forests—a dynamic process

Trees, like all other plants, fix atmospheric CO_2 through photosynthesis and convert it to biomass and other materials necessary for metabolism. Above ground, most of a forest's long-term carbon storage occurs as woody biomass. Some of that carbon becomes soil organic carbon through addition and decomposition of fallen branches, leaf litter and dead roots. The remaining carbon is released back to the atmosphere by respiration and the decomposition of soil organic matter.

Storing carbon in woody biomass is good because it is a stable, long-term carbon pool. Even if a forest is no longer sequestering additional carbon or is sequestering it at low rates, the carbon previously sequestered in biomass is preserved for a long time because wood decomposes very slowly and tree roots prevent erosion and subsequent oxidation of soil organic carbon.

In young trees, respiration and losses of carbon to the atmosphere are low; therefore, most of the carbon fixed through photosynthesis is converted to biomass and sequestered. As trees age, respiration increases because energy is needed to replace dying tissues and a lower proportion of carbon fixed through photosynthesis is converted to biomass and sequestered.

Eventually a tree no longer sequesters additional carbon but instead maintains a constant quantity of carbon. This steady-state condition occurs when the carbon gained from photosynthesis and the carbon lost from respiration are equal. Different tree species reach a steady state at different times, somewhere between 90 and 120 years. Research indicates that late successional forests have more stable steady-state carbon pools because of a larger biomass of fungi associated with their roots below the zone of decomposer fungi found in the oxygen-rich upper soil layers.

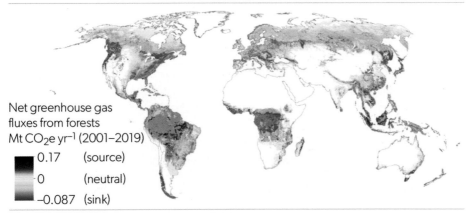

Net greenhouse gas fluxes from forests Mt CO$_2$e yr^{-1} (2001–2019)

- 0.17 (source)
- 0 (neutral)
- −0.087 (sink)

GLOBAL FOREST WATCH ⬢ **WORLD RESOURCES INSTITUTE**

▲ Figure 6 Forests: carbon sink or source?

Look at Figure 6. Where are forests a source of carbon? Why?

Oceans and carbon

Oceans sequester another 25% of carbon from the atmosphere and most is in the upper ocean. But burning fossil fuels releases inorganic carbon faster than oceans can absorb it. Adding more carbon dioxide from the atmosphere acidifies the oceans, with major consequences for ecosystems and marine animals. This is because small decreases in pH can interfere with calcium carbonate deposition in mollusc shells and coral skeletons.

Alleviating the effects of humans on the carbon cycle

There are four main ways to alleviate the impact of human activities.

1. **Low carbon technology** products are energy sources that produce low levels of GHGs (greenhouse gases). They include renewable energy technologies, such as wind turbines, solar panels, biomass systems, carbon capture and storage (CCS) and nuclear power. They are not zero carbon as GHGs are emitted in their construction, transport and maintenance but they produce less carbon than burning fossil fuels. The expectation is that the global economy will continue to grow but carbon emissions will fall.

2. **Reducing fossil fuel burning** is an obvious measure. This can be done through low carbon technologies and also by energy efficiency. Insulating homes and using LED lights, smart thermostats and electric vehicles all reduce fossil fuel use. "Green" electricity production using renewables is a great marketing tool for energy companies and makes environmental sense. Our need for electricity to power our technologies will only increase.

3. **Biological sequestration** can be increased by increasing forests. On average, growing forests capture twice the carbon they emit, whereas croplands or grasslands capture about 25% of carbon emissions. Soil also sequesters carbon, and peat bogs and swamps contain more than 600 GtC. This is more than all other vegetation types yet natural peatland is threatened with drainage, fire and removal.

The following forestry practices help to store carbon.

*   **Afforestation** is the planting of trees where trees have not grown (in the last 100 years). A common type of afforestation is the planting of short-rotation woody crops like hybrid poplar. These species grow very quickly and as a result sequester large amounts of carbon in a short time. They are planted and harvested within a short time frame—10 to 15 years—and the biomass is sold for paper or other processed wood products. Wood pellets are used in power stations as a source of fuel.

*   **Reforestation** is the re-establishment of trees on land that had been deforested in the last 100 years.

*   **Forest management** can maximize carbon storage—for example, by lengthening time between harvests, selective thinning for increased stocking, and planting fast-growing species.

*   **Management practices** not to clear fell the forest but leave some trees and allow natural regeneration.

Agricultural practices also affect carbon storage.

▲ Figure 7 A conifer plantation in Europe—fast-growing trees capture carbon

Connections: This links to material covered in subtopic 5.2.

4. **Artificial (or geological) carbon sequestration**—there are several ways to use technology to capture carbon emissions at point of production and then compress and store them in underground or under sea geological formations or rocks.

- **Carbon capture and storage (CCS) and bioenergy with carbon capture and storage (BECCS)**

  Sustainably sourced biomass-generated energy (bioenergy) can be carbon neutral because plants absorb CO_2 from the atmosphere as they grow. This offsets CO_2 emissions released when the biomass is burned as fuel to create electricity in power plants. When sustainable bioenergy is paired with CCS, it becomes a source of negative emissions, as CO_2 is permanently removed from the carbon cycle.

  CCS is the same but the energy source is fossil fuels, so it cannot be as low carbon as BECCS.

- **Graphene production** requires CO_2 as a raw material. Although limited to certain industries, it is used heavily in the production of devices such as smartphones or computer processors.

- **Engineered molecules** are a fairly new science. Scientists can change the shape of molecules to form new compounds by capturing carbon from the air. In practice, this could present an efficient way of creating raw materials while reducing atmospheric carbon.

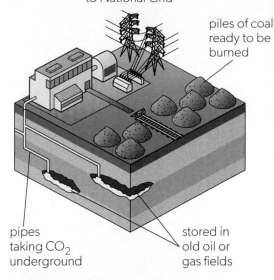

electricity distributed to National Grid

piles of coal ready to be burned

pipes taking CO_2 underground

stored in old oil or gas fields

▲ Figure 8 Diagram to show carbon capture and storage

Data-based questions

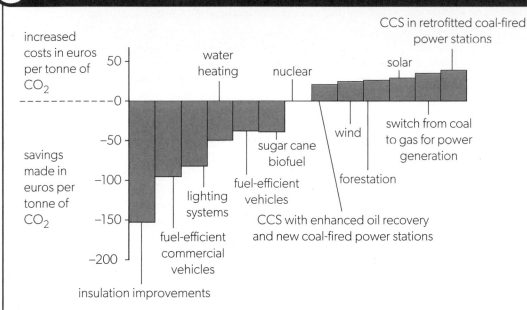

▲ Figure 9 Relative cost effectiveness, in euros, of different methods of reducing carbon emissions

Use the data in Figure 9 to answer these questions.

1. Which is the most cost-effective method of reducing carbon emissions?

2. Why do you think this is?

3. What are the disadvantages of using sugar cane as biofuel?

4. Why does nuclear power generation have a zero cost here?

5. As wind and solar energy production become more widespread, what do you think will happen to the cost of these?

6. Why is retrofitting CCS more expensive than CCS in new oil or coal power stations?

A little more on the carbon cycle

More on carbon stores

- The lithosphere carbon stores are limestone rocks and fossil fuels.

- Limestone—calcium carbonate—is the largest store of carbon.

- Limestone can be laid down by minerals precipitating out of seawater but most limestone is the fossilized remains of reef-building corals and mollusc shells which are made of calcium carbonate.

- Fossil fuels are the partially decomposed remains of plants (coal) and marine organisms (oil and gas) which were compressed by heat and pressure over millions of years between 10 and 180 million years ago.

- Fossil fuels take millions of years to form.

▲ Figure 10 Coal mining, crude oil (being cleaned up after a spill) and peat (cut and drying to be used as a fuel)

Residence time of carbon stores (how long something stays in one place) varies. For limestone it can be hundreds of millions of years if it is not mined; for the atmosphere 5 years; for the biosphere 13 years; and for the oceans 350 years.

Peat as a carbon sink

- Peatlands are very good carbon sinks because they absorb and store huge amounts of carbon – twice as much as all forests, even though peat occupies only 3% of global land area.

- Peat is partially decomposed decayed vegetation which forms in waterlogged conditions. It does not fully decompose because conditions are too acid and anaerobic (no air).

- It takes a year for 1 mm of peat to accumulate in temperate regions but less time in the tropics.

- When burned or drained, peat becomes a carbon source.

- Farmers and foresters have drained and then planted peatlands because they are fertile.

- Peatland degradation is most acute in the tropical peatlands of Southeast Asia, primarily in the islands of Borneo and Sumatra, where over two-thirds of all peat swamps have been cleared and drained since 1990.

Methane

- Methane molecules contain one carbon and four hydrogen atoms—CH_4.

- It is released when burning fossil fuels and from other activities (Figure 11).

- Methane is a potent GHG, with a much greater global warming potential than carbon dioxide.

- It stays in the atmosphere—residence time—for about 10 years and is then oxidized to carbon dioxide.

- Methane is also the main contributor to ground-level ozone pollution (see subtopic 5.1).

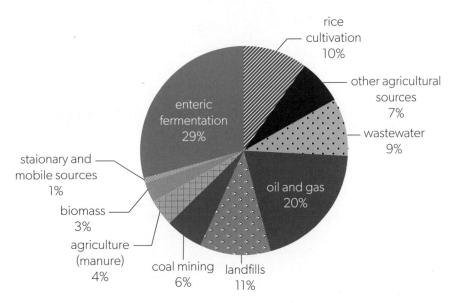

▲ Figure 11 Estimated anthropogenic methane emissions by source
Adapted from: Institute for Governance and Sustainable Development

Main sources of methane emissions

Methane emissions are of two sorts: natural, 40%, and anthropogenic (human-made), 60%.

Natural:	Anthropogenic:
- decay of plant material in anaerobic wetlands (swamps) by methanogenic bacteria	- fossil fuel burning
- seepage of gas hydrates from underground deposits in melting permafrost	- anaerobic digestion of biofuels
- volcanoes	- rice paddies
- digestion of food by ruminant animals	- cattle farming (called enteric fermentation in Figure 11)
- digestion of food by termites	- landfill emissions from decaying waste.
- wildfires	
- ocean sediments.	

Case study 4

Sources of methane

Methane is more than 80 times more powerful than carbon dioxide in its warming effect over 20 years. At least 25% of global warming is driven by methane from human activity.

There are three main sources of methane:

1. Permafrost blow holes

▲ Figure 12 Permafrost blow hole in Siberia

Inside the Arctic circle permafrost in Russia are some strange, large holes. They can be up to 50 m deep and probably formed in the last 10–15 years. An explosion causes them so soil, ice and rock are forced into the air. Some researchers think they are cryovolcanoes spewing ice not lava but most currently think they are due to the emission of gases which build up as ice layers melt in the ground. Inside the craters are higher levels of methane and carbon dioxide. There are various theories on how these gases are released and one is that the frozen methane hydrate melts, the gas builds up and then ruptures the surface.

Surface air temperatures in the Arctic are warming twice as fast as the global average, which is increasing the amount of permafrost thaw during the summer months.

The amount of these methane hydrates under the permafrost is not clear but there are possibilities of mining it. While methane is a source of energy, the dangers of methane escape are clear.

2. Rice paddies

Rice is the staple food of more than half the human population. Methane release from rice growing is about 20% of anthropogenic methane emissions.

▲ Figure 13 Planting rice in Vietnam

Methane is released because rice growing requires flooded paddy fields and the lack of oxygen causes anaerobic conditions so bacteria emit methane. Draining the fields in the middle of the growing season stops most methane release. This also saves water for irrigation and can increase yield. Many farmers in China have done this for 20 years now. The downside is that this practice increases nitrous oxide emissions—another GHG.

3. Ruminant methane

▲ Figure 14 The cow, like all ruminants, has four stomachs

Ruminants, including cows, sheep and goats, have four stomachs to digest tough grass. One of these, the rumen, contains bacteria which break down the grass cellulose in a digestive process called enteric fermentation. This releases methane which the animals release by belching. Estimates vary, but livestock are thought to be responsible for up to 14% of all GHG emissions from human activities. Ways to reduce this are:

- feed additives and supplements such as legumes or oils

- give vaccines to remove the gut microorganisms that release methane

- practise selective breeding for low methane producing animals

- feed silage (fermented grass), not other cereals.

Nitrogen cycle

Bacteria are essential in the nitrogen cycle. The nitrogen cycle is in three stages: nitrogen fixation, nitrification and denitrification. Nitrogen stores or sinks are:

- organisms

- soil

- fossil fuels

- the atmosphere

- water.

All living organisms need nitrogen as it is an essential element in proteins and DNA.

The nitrogen cycle has organic and inorganic stores.

Organic stores (containing carbon in long chains in plants and animals) include:	Inorganic stores (not in plants or animals) include:
- proteins - amino acids - DNA - other carbon compounds containing nitrogen in living organisms and in dead organic matter.	- nitrogen in the atmosphere - ammonia and other nitrogen compounds (nitrites and nitrates) in soil and water - nitrogen locked in sedimentary rocks on the ocean floors, mostly as ammonium salts, e.g. mica, feldspars, clay minerals.

Flows in the nitrogen cycle are transfers and transformations.

Transformation flows are:	Transfers are:
- photosynthesis - decomposition - ammonification - denitrification - nitrogen fixation - nitrification.	- consumption - excretion - death - mineral uptake by producers.

ATL Activity 15

ATL skills used: thinking, communication, social, research, self-management

Look at Figure 11 again. Research how landfills and wastewater also produce methane emissions.

In small groups of three or four, draw a poster or produce a podcast about the problems caused by methane and what we can do to reduce the problems.

Key term

Nitrogen fixation is conversion of nitrogen from the atmosphere into ammonia; some bacteria that fix nitrogen are free-living, others are found in root nodules of plants.

▲ Figure 15 **Lightning transforms nitrogen gas to nitrate**

Key terms

Nitrification is conversion of ammonia to nitrates; this occurs in two stages: ammonia to nitrites then nitrites to nitrates.

Denitrification is conversion of nitrates to nitrogen.

Decomposition is the conversion of amino acids into ammonium.

Nitrogen is the most abundant gas in the atmosphere, making up 78% of it. But it is unavailable to plants in this form. For plants to take up nitrogen, it must be in the form of ammonium ions (NH_4^+) or nitrates (NO_3^-). Animals eat plants and so take in their nitrogen in the form of amino acids and nucleotides.

Nitrogen fixation

Atmospheric nitrogen, N_2, cannot be used by plants or animals. But bacteria convert it to ammonium ions NH_4^+ which are in a form that plants can use. This process is **nitrogen fixation**.

This conversion from gaseous nitrogen to ammonium ions can happen in one of the following five ways.

1. By nitrogen-fixing bacteria free-living in the soil (*Azotobacter*).

2. By nitrogen-fixing bacteria living symbiotically in root nodules of leguminous plants (e.g. *Rhizobium*). The plant provides the bacteria with sugars from photosynthesis, the bacteria provide the plant with nitrates.

3. By cyanobacteria (sometimes called blue-green algae) that live in soil or water. Cyanobacteria are the cause of the high productivity of Asian rice fields, many of which have been productive for hundreds or even thousands of years without nitrogen-containing fertilizers.

4. By lightning also causing the oxidation of nitrogen gas to nitrate which is washed into the soil.

5. By the industrial **Haber process**—a nitrogen-fixing process used to make fertilizers. Nitrogen and hydrogen gases are combined under pressure in the presence of iron as a catalyst (speeds up the reaction) to form ammonia.

Nitrification

Some bacteria (e.g. *Nitrosomonas*) in the soil are called **nitrifying** bacteria and convert ammonium ions (NH_4^+) to nitrites (NO_2^-). Others (e.g. *Nitrobacter*) convert the nitrites to nitrates (NO_3^-) which are then available to be absorbed by plant roots.

Denitrification

Denitrifying bacteria (*Pseudomonas denitrificans*), in waterlogged and anaerobic (low oxygen level) conditions, reverse this process by converting ammonium, nitrate and nitrite ions to nitrogen gas which escapes to the atmosphere.

Decomposition

Decomposition of dead organisms supplies the soil with much more nitrogen than nitrogen fixation processes. Important organisms in decomposition are animals (insects and worms, among others), fungi and bacteria. They break down proteins into different ions: ammonium ions, nitrite ions and finally nitrate ions. These ions can be taken up by plants which recycle the nitrogen.

Assimilation

Once living organisms have taken in nitrogen, they assimilate it or build it into more complex molecules. Protein synthesis in cells turns inorganic nitrogen compounds into more complex amino acids which then join to form proteins. Nucleotides are the building blocks of DNA and these too contain nitrogen.

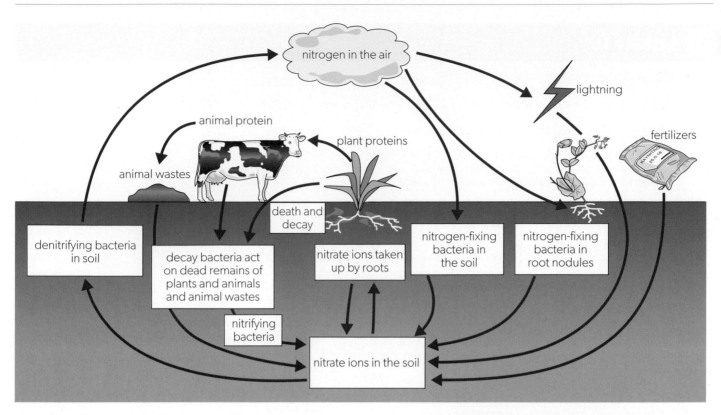

▲ Figure 16 **The nitrogen cycle**

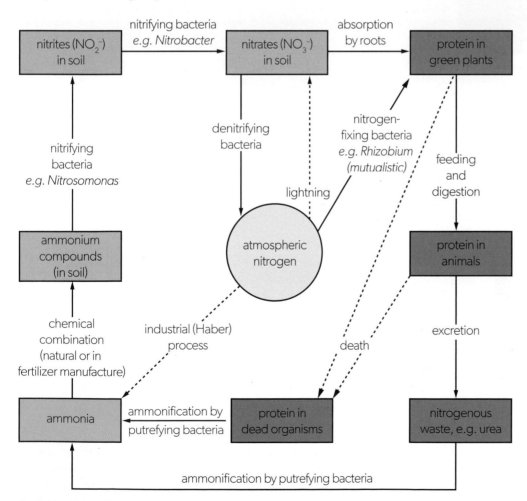

▲ Figure 17 **Another diagram of the nitrogen cycle**

(ATL) Activity 16

ATL skills used: thinking, communication, self-management

1. Draw your own systems diagram of the N cycle, including as many organisms or colours as you can.

2. Add stores of ocean, soils and atmosphere.

3. Make a key to show stores and flows, transfers and transformations.

4. Try to scale the stores and flows according to the data in Tables 1 and 2.

Rocks and sediments	190,400,120 (deep, unavailable)
Atmosphere	3,900,000
Ocean	23,348
Soils	460
Land plants	14
Land animals	0.2

▲ Table 1 Stores of N, in ×10^{15} grams

Biological fixation	on land: 140 in oceans: 15
Fixation by lightning	<3
Denitrification	on land: 200 in oceans: 110 by human activity: 1,000
Decomposition and assimilation in plants	1,200

▲ Table 2 Flows of N, in ×10^{15} grams

5. Compare and contrast Figures 16 and 17 as models showing the nitrogen cycle.

Connections: This links to material covered in subtopics 1.2, 2.1, 2.5, 5.2, 5.3 and 6.2.

▲ Figure 18 **Nitrogen-fixing root nodules**

Natural nitrogen fixation

Nitrogen is often a limiting factor in plant growth so any advantage a plant gets in access to nitrogen is a good thing. The only way that plants can use atmospheric nitrogen is by forming mutualistic associations with nitrogen-fixing bacteria (e.g. *Rhizobium*).

Legumes are plants that have pods, e.g. peas, beans, chick peas, soy beans. They have root nodules (Figure 18) where *Rhizobium* live in a symbiotic relationship with the plant. *Rhizobium* bacteria can only fix nitrogen if they are in nodules on a legume root but other free-living bacteria can fix it anywhere. Alfalfa and clover are the best nitrogen-fixing crops.

The legume gains nitrogen in a form it can use and the bacteria gain sugars made in photosynthesis from the plant. This is a mutualistic relationship.

This is a huge advantage to farmers because legumes do not need additional nitrogen fertilizers and can grow on nitrogen-poor soils.

Cereals (e.g. rice, wheat, maize) cannot fix nitrogen in this way but current work in genetic engineering is researching how to introduce nitrogen-fixing genes into cereals. All genes code for proteins and the *Nif* genes code for proteins that fix atmospheric nitrogen. They are found in free-living bacteria and in *Rhizobium*. Adding them to cereals by genetic engineering could mean a reduction in the need for artificial fertilizers.

Denitrification only happens in anaerobic conditions, such as soils that are waterlogged. In waterlogged, anaerobic soils, plant growth is reduced or stopped and nitrates are washed away by denitrification and leaching. In these soils, insectivorous plants (such as pitcher plants and sundews) can capture and digest insects and use them as a nitrogen source.

Artificial nitrogen fixation—the Haber process

This is an industrial process which uses high pressure and temperature and an iron catalyst to combine nitrogen from air with hydrogen (from natural gas— methane, CH_4) to make ammonia (Figure 20). Over 200 million tonnes (Mt) of ammonia is made in the Haber process per year.

The ammonia is sold to farmers as ammonium phosphate, nitrate or sulfate (synthetic fertilizers). About half the world population relies on food fertilized by synthetic fertilizers. The ammonium nitrate global market is over 50 million tonnes.

▲ Figure 19 A pitcher plant *Nepenthes* sp. on Mt Kinabalu, Borneo. The pitcher attracts insects which fall into the well, cannot escape and are dissolved by enzymes. The plant gains nitrogen from the insects

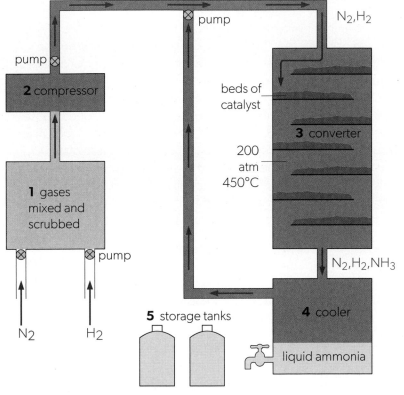

◀ Figure 20 The Haber process. Liquid ammonia is highly toxic and burns the skin

Advantages of the Haber process

- Produces inorganic fertilizer high in nitrogen—mostly ammonium nitrate and urea

- Nitrogen increases crop yield enormously

Disadvantages of the Haber process

- Expensive because the pressure is so high in the vessel and pipes where the reaction occurs so they have to be strong and well-maintained

- Inefficient—only about 15% of the gases are combined to form ammonia each time so they are recycled

- Uses 3–5% of the global natural gas supply and produces 1.4% of global carbon emissions

TOK

What criteria should be used to make ethical decisions about the use of advances in technology in the physical world?

Humans and the nitrogen cycle

Human activities such as deforestation, agriculture, aquaculture and urbanization change the nitrogen cycle. Nitrogen (and phosphorus) are both essential elements for plant growth, so fertilizer production and application are the main concerns.

Human activities convert around 120 million tonnes of nitrogen from the atmosphere into reactive nitrogen, mainly as fertilizer to help feed the world. Much of this new reactive nitrogen is emitted to the atmosphere in various forms rather than taken up by crops. When it is rained out, it pollutes waterways and coastal zones or accumulates in the terrestrial biosphere. Similarly, a relatively small proportion of phosphorus fertilizers applied to food production systems is taken up by plants; much of the phosphorus mobilized by humans also ends up in aquatic systems. These systems can become oxygen-starved as bacteria consume the blooms of algae that grow in response to the high nutrient supply. This causes eutrophication and ocean dead zones (see Figure 21).

It is easy for humans to alter the cycle and upset the natural balance. When people remove animals and plants for food for humans, they extract nitrogen from the cycle. Much of this nitrogen is later lost to the sea in human sewage. But people can also add nitrogen to the cycle in the form of artificial fertilizers, made in the Haber process, or by planting leguminous crops with root nodules containing nitrogen-fixing bacteria. These plants enrich the soil with nitrogen when they decompose. The soil condition also affects the nitrogen cycle. If it becomes waterlogged near the surface, most bacteria are unable to break down detritus because of lack of oxygen but certain bacteria can. Unfortunately, they release the nitrogen as gas back into the air. This is called denitrification. Excessive flow of rainwater through a porous soil, such as sandy soil, will wash away the nitrates into rivers, lakes and then the sea. This is called leaching and can lead to eutrophication.

A significant fraction of the applied nitrogen and phosphorus makes its way to the sea, where it can push marine and aquatic systems across ecological thresholds of their own. One regional-scale example of this effect is the decline in the shrimp catch in the Gulf of Mexico's "dead zone" caused by fertilizer transported in rivers from the US Midwest.

Reactive nitrogen pollutes waterways and coasts and, in nitrous oxide form, exacerbates global warming. Synthetic fertilizers, leguminous crops (soybeans, peanuts, alfalfa), many types of manufacturing, and fossil fuel burning industries and vehicles all produce reactive nitrogen.

Summary of human activity affecting the nitrogen cycle

1. Ocean dead zones

Rain washes nitrates in the soil into watercourses and oceans. This feeds algal blooms which choke aquatic life as they decompose. Dead zones are formed. There are over 400 such zones in the oceans, including a very large one in the Gulf of Mexico (Figure 21).

> Connections: This links to material covered in subtopic 4.4.

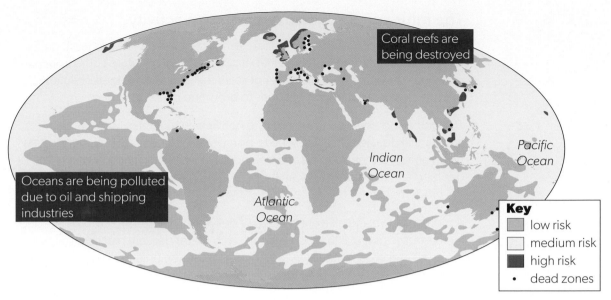

▲ Figure 21 Ocean dead zones and polluted zones

2. Air pollution

Nitrogen oxides are emitted by power stations and the internal combustion engine in cars. These form particulate matter—small particles that can be breathed in and so damage health.

Ammonia in the air makes it more alkaline. Peat bogs need acidic conditions and peat degradation is occurring with high ammonia levels.

3. Ozone depletion

Excess nitrate can be converted to nitrous oxide, N_2O. At high altitudes this reacts with UV light and breaks down ozone. Nitrous oxide molecules have a 120-year residence time in the atmosphere.

4. Climate change

Nitrous oxide is a GHG and is the third largest contributor to global warming after carbon dioxide and methane. A molecule of nitrous oxide is 300 times more potent at this than one molecule of carbon dioxide.

5. Soil acidification

Nitrates in soil, if not taken up by plants, can combine with magnesium and calcium. This makes the soil too acidic and plants are unable to take up nutrients as well.

Planetary boundary for biogeochemical flows

The **planetary boundary** for the nitrogen cycle is calculated in millions of tonnes per year removed from the atmosphere; the background level is 0, the boundary is set at 35, and we are already at 121. Phosphorus is a mineral that is mined for use in fertilizers, detergents, pesticides, steel production and even toothpaste. It is measured in millions of tonnes per year entering the ocean. The background level is −1, the boundary is 11, and currently 8.5 to 9.5 million tonnes end up each year in the ocean where it depletes oxygen levels, harming marine life. The planetary boundaries model is illustrated in subtopic 1.3, Figures 12 and 13.

The future of nitrogen use by human activity

As mentioned above, nitrogen fertilizer production uses 3–5% of the global natural gas supply. Energy for the Haber process could in future come from renewable resources but this does not solve the issues of nitrogen pollution. And the human population continues to grow and needs to be fed.

The global food system is vulnerable to conflicts, pandemics, weather patterns and natural disasters. In 2022, due to disruption in Ukraine and an energy crisis when energy prices doubled and trebled, prices of wheat and fertilizers also doubled and trebled. Food scarcity and suffering will continue unless the world develops more food resilience.

A more effective and efficient process of delivering fertilizers to crops is needed. Increasing costs are actually driving this, with better delivery practices, use of biosolids and organic fertilizers, and better cultivation practices in different parts of the world. As yet, however, there is no global agreement on reduction of fertilizer use.

(ATL) Activity 17

ATL skills used: thinking, communication, social, research

Intensive use of chemical fertilizers helped fuel the four-fold expansion of the human population over the last century (the Green Revolution onwards) and will be crucial for feeding over eight billion people by 2050.

The advantages and disadvantages or inorganic (artificial) and organic fertilizers are summarized in Tables 3 and 4. Organic fertilizers are derived from organic matter—either from animals (bone meal, blood meal, fish meal, shellfish, decomposed dead material, animal wastes) or plants (compost, seaweeds, legumes).

Advantages of inorganic fertilizers	Disadvantages of inorganic fertilizers
• Fast-acting on plants	• Can act too quickly and lead to fertilizer burn of crops
• Exact amount applied can be calculated	• Overfertilization produces weak stems, too large dark green leaves, slow root growth
• Increase crop yields greatly	• High levels of ammonium are toxic to plants
• Excess crop can be fed to livestock, increasing their production	• Soil degradation as no organic matter added
• Easy to transport, handle and apply for farmers— granules, powder or liquid rather than organic manures	• Require several applications
• Were inexpensive (until 2022 when prices tripled)	• Nitric acid is formed in the atmosphere which contributes to acid rain
	• About 50% of fertilizer spread is not taken up by plants; excess is leached away into rivers and seas causing eutrophication and algal blooms
	• Nitrous oxide—the third most important GHG— is released

▲ Table 3 Advantages and disadvantages of inorganic fertilizers

Advantages of organic fertilizers	Disadvantages of organic fertilizers
• Overapplication is rare	• Often more expensive to produce
• Stay in the soil longer so fewer applications needed	• Bulky, so transport and application costs are higher
• Prevent soil degradation and can restore organic matter content to soil	• Require animals to produce enough organic matter (or is that an advantage?)
• Retain water	
• Slowly release nutrients and lead to soil sustainability	
• Also have other nutrients	

▲ Table 4 Advantages and disadvantages of organic fertilizers

Work in groups of three or four. Considering various environmental value systems, suggest and justify the best course of action—in terms of use of organic and inorganic nitrogen fertilizers—for:

1. a subsistence farmer who needs more crop yield to feed their family, currently using subsidized inorganic fertilizer

2. a large commercial crop farmer with low margins and a small workforce

3. the state of Sikkim, India when deciding to become 100% organic in 2016

4. the UK government subsidizing the growing of nitrogen-fixing plants and green manures while also delaying a ban on use of urea fertilizer which emits ammonia (2002).

Check your understanding

In this subtopic, you have covered:

- what biogeochemical cycles are
- carbon cycles
- how carbon is stored
- effects of ocean acidification
- how humans are altering the carbon cycle
- what can be done about this

- more about how carbon is stored
- methane from peat bogs and permafrost
- nitrogen cycles
- how humans are affecting the nitrogen cycle
- what can be done about this.

How do human activities affect nutrient cycling and what impact does this have on the sustainability of environmental systems?

1. Describe the carbon cycle briefly with a diagram.

2. List the processes by which carbon flows through its stores.

3. Define carbon sequestration.

4. Explain how humans are affecting the carbon cycle.

5. Evaluate the means by which humans alleviate their impact on the carbon cycle.

6. Describe where carbon is stored in the lithosphere and biosphere.

7. Explain how human activity releases more methane into the atmosphere.

8. Describe the nitrogen cycle briefly with a diagram.

9. Explain how humans are affecting the nitrogen cycle.

10. Comment on the crossing of the planetary boundary for biogeochemical flows.

❯❯ Taking it further

- Update a display monthly with the latest carbon dioxide ppm in the atmosphere.

- If your school has green fields (playing fields, grounds) find out if they are fertilized and, if so, with what type of fertilizer. Are organic fertilizers used instead of inorganic? If not, can you provide an argument for change?

- Explore issues of justice for local communities when the local environment is overexploited for financial gain.

2.4 Climate and biomes

Guiding questions

- How does climate determine the distribution of natural systems?

- How are changes in Earth systems affecting the distribution of biomes?

Understandings

1. Climate describes atmospheric conditions over relatively long periods of time, whereas weather describes the conditions in the atmosphere over a short period of time.
2. A biome is a group of comparable ecosystems that have developed in similar climatic conditions, wherever they occur.
3. Abiotic factors are the determinants of terrestrial biome distribution.
4. Biomes can be categorized into groups that include freshwater, marine, forest, grassland, desert and tundra. Each of these groups has characteristic abiotic limiting factors, productivity and diversity. They may be further classed into many subcategories (for example, temperate forests, tropical rainforests and boreal forests).
5. The tricellular model of atmospheric circulation explains the behaviour of atmospheric systems and the distribution of precipitation and temperature at different latitudes. It also explains how these factors influence the structure and relative productivity of different terrestrial biomes.
6. The oceans absorb solar radiation and ocean currents distribute the resulting heat around the world.
7. Global warming is leading to changing climates and shifts in biomes.

8. There are three general patterns of climate types that are connected to biome types.
9. The biome predicted by any given temperature and rainfall pattern may not develop in an area because of secondary influences or human interventions.
10. The El Niño Southern Oscillation (ENSO) is the fluctuation in wind and sea surface temperatures that characterizes conditions in the tropical Pacific Ocean. The two opposite and extreme states are El Niño and La Niña, with transitional and neutral states between the extremes.
11. El Niño is due to a weakening or reversal of the normal east–west (Walker) circulation, which increases surface stratification and decreases upwelling of cold, nutrient-rich water near the coast of north-western South America. La Niña is due to a strengthening of the Walker circulation and reversal of other effects of El Niño.
12. Tropical cyclones are rapidly circulating storm systems with a low-pressure centre that originate in the tropics and are characterized by strong winds.
13. Rises in ocean temperatures resulting from global warming are increasing the intensity and frequency of hurricanes and typhoons because warmer water and air have more energy.

Biomes

Abiotic factors determine terrestrial **biome** distribution—precipitation, temperature and insolation are the main ones.

For any given temperature and rainfall pattern, one natural ecosystem type is likely to develop in different parts of the world.

Each biome has characteristic limiting factors, productivity and biodiversity.

Key term

A **biome** is a group of comparable ecosystems that have developed in similar climatic conditions wherever those conditions occur.

Why biomes are where they are

The **climate** is the major factor that determines what grows where and therefore what lives where. The other important factor is the terrain or morphology—slope, aspect and altitude.

Climate is made up of general **weather** patterns, seasons, extremes of weather and other factors. The two most important factors are temperature and precipitation (rain and snowfall). The climate is hotter nearer the equator and generally gets cooler as we go towards the poles (increase in latitude).

As well as orbiting around the Sun, the Earth rotates and is tilted at 23.5° on its axis. It takes 365 days (and a quarter) for the Earth to go once round the Sun and this gives us a year and our seasons (Figure 1).

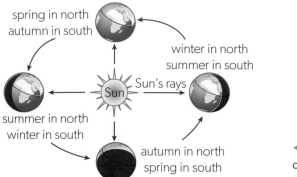

Figure 1 **How the Earth's tilt causes seasons**

Latitude (distance north or south from the equator) and **altitude** (height above sea level) both influence climate and biomes. It generally gets colder as you increase either latitude or altitude. So, there is snow on Mt Kilimanjaro, the Himalayas and the Andes at their higher altitudes. They have alpine or polar biomes even though they are at lower latitudes (nearer the equator).

Ocean currents and winds distribute surplus heat energy at the equator towards the poles. The oceans absorb solar radiation and ocean currents distribute the resulting heat around the world. Air moving horizontally at the surface of the Earth is called wind. Winds blow from high to low pressure areas. Winds cause the ocean currents but the water is responsible for transferring the heat.

Water can exist in three states—solid (ice and snow), liquid (water) and gas (water vapour). As it changes from state to state it either gives out or takes in heat. This is called latent heat. As water changes from solid to liquid (melts) to gas (evaporates), it takes in heat because more energy is needed to break the molecular bonds holding the molecules together. As water changes from gas to liquid (condenses) to solid (freezes), it gives out heat to its surroundings. These changes distribute heat around the Earth. Water is the only substance that occurs naturally in the atmosphere that can exist in all three states within the normal climatic conditions on Earth.

Increasing temperature causes increasing evaporation so the relationship between precipitation and evaporation is also important. Plants may be short of water even when it rains or snows a lot if the water evaporates straight away (deserts) or is frozen (tundra and poles). So we must also consider the **precipitation to evaporation ratio (P/E ratio)**. This is simple to calculate— see Table 1.

Key terms

Climate describes atmospheric conditions over relatively long periods of time.

Weather describes the conditions in the atmosphere over a short period of time.

P/E ratio is approximately 1 when precipitation is about the same as evaporation. In this case, the soils tend to be rich and fertile.

Tundra: Norway	Desert: Jordan
• 75 cm of snow falls per year • 50 cm is lost by evaporation	• 5 cm of rain falls per year • 50 cm is lost by evaporation
P/E ratio is 75/50 or 1.25	P/E ratio is 5/50 or 0.1
P/E ratio is much greater than 1	**P/E ratio is far less than 1**
• It rains or snows a lot and evaporation rates are low. • There is leaching in the soil when soluble minerals are washed downwards.	• Water moves upwards through the soil and then evaporates from the surface. • This leaves salts behind and the soil salinity increases to the point that plants cannot grow (salinization).

▲ Table 1 Comparison of P/E ratio in Norway and Jordan

Different biomes have differing amounts of productivity due to limiting factors. These are the raw materials or the energy source (light) for photosynthesis. For example, solar radiation and heat may be limited at the south pole in winter, while water is in limited supply in a desert. All food webs depend on photosynthesis by producers to provide the initial energy store. If producers cannot photosynthesize to their maximum capacity, other organisms will not have enough food.

Productivity is greater at low latitudes (nearer the equator), where temperatures are high all through the year, sunlight input is high and precipitation is also high. These conditions are ideal for photosynthesis. Towards the poles, the temperatures and the amount of sunlight decline. This means the rate at which plants photosynthesize is lower, and so both GPP (gross primary productivity) and NPP (net primary productivity) values are lower. In the terrestrial areas of the Arctic, Antarctic and adjacent regions (i.e. at high latitudes), certain conditions tend to cause a reduction in photosynthesis and lower productivity values. These conditions are:

- low temperatures

- permanently frozen ground (permafrost)

- long periods in winter when there is perpetual darkness

- low precipitation (cold air cannot hold as much moisture as warm air).

In desert areas (e.g. the Sahara and much of Saudi Arabia) and semi-arid areas (e.g. central Australia, south-west USA), the absence of moisture for long periods lowers productivity values severely, despite high temperatures and abundant sunlight. Temperate deciduous forests would become temperate rainforest if precipitation were higher, and temperate grassland if it were lower. However, these are generalizations and variations are considerable. In a few sheltered and favourable places in Greenland and south Georgia—in the Arctic and sub-Antarctic, respectively—productivity values close to those of mid-latitude forest have been recorded.

Connection: This links to material covered in subtopics 1.3, 2.2 and 6.2.

Biome types

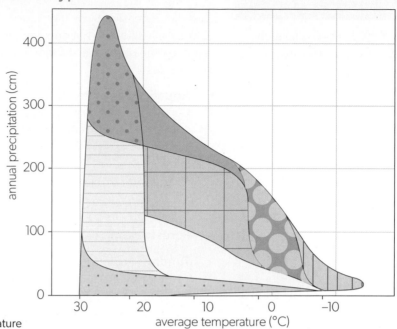

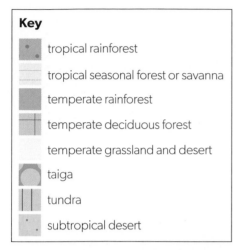

Key

- tropical rainforest
- tropical seasonal forest or savanna
- temperate rainforest
- temperate deciduous forest
- temperate grassland and desert
- taiga
- tundra
- subtropical desert

▲ Figure 2 **Graph of precipitation against temperature first plotted by Robert Whittaker, an American ecologist, to show biome distribution**

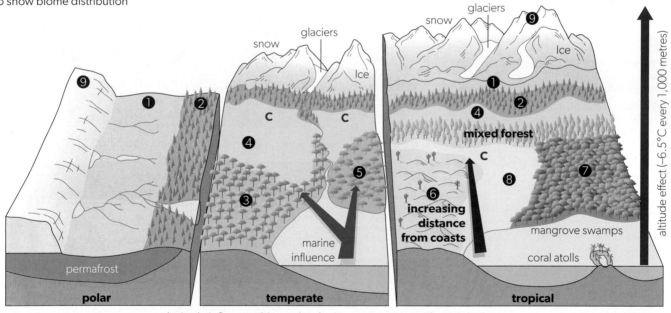

Latitude influences biomes by altering temperature and precipitation

arctic ← → **equator**

Key

Biomes seasonally lacking in heat and/or water
- ❷ coniferous forest
- ❸ temperate deciduous forest
- ❹ temperate grassland
- ❺ mediterranean
- ❽ tropical grassland (savanna)

Biomes permanently lacking in heat and/or water
- ❶ tundra
- ❻ hot desert
- ❾ mountain or alpine

Biomes promoting growth all year round
- ❼ tropical rainforest

Marine influence The sea cools nearby land in the hot season and warms it during the cold season. This reduces annual temperature range and increases precipitation.

Continentality (C) Away from the sea, the land heats up in the hot season and cools quickly in the cold season. This increases the annual temperature range and reduces precipitation.

▲ Figure 3 **Diagram showing how latitude, altitude and oceans influence biomes**

How many biomes are there?

Opinions differ slightly on the number of biomes. This is because they are not a natural classification but one devised by humans. However, it is possible to group biomes into several major types with sub-categories in each type (see Figures 2–4).

- **Aquatic**—freshwater (swamp forests, lakes and ponds, streams and rivers, bogs) and marine (rocky shore, mud flats, coral reef, mangrove swamp, continental shelf, deep ocean)

- **Deserts**—hot and cold

- **Grassland**—tropical or savanna and temperate

- **Chaparral**—hot and dry coastal (with woody shrubs)

- **Forests**—tropical, temperate and boreal (taiga)

- **Tundra**—Arctic and alpine

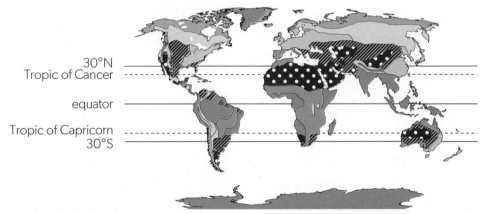

Key:
- tropical forest
- savanna
- desert
- polar and high-mountain ice
- chaparral
- temperate grassland
- temperate deciduous forest
- coniferous forest
- tundra (arctic and alpine)

30°N Tropic of Cancer

equator

Tropic of Capricorn 30°S

▲ Figure 4 **Terrestrial biome distribution**

ATL Activity 18

ATL skills used: thinking, research

1. Copy and complete the table, giving an example of each biome type.

Biome	Named example
tropical rainforest	
hot desert	
tundra	
temperate forest	
deep ocean	
temperate grassland	

2. Explain the terms: (a) latitude; and (b) altitude.

3. Explain why temperature decreases with increasing altitude.

4. Explain why there is more insolation at the equator.

5. Explain why there are seasons.

ATL skills used: thinking, communication, research

1. Go to earthobservatory.nasa.gov/biome/graphindex.php

2. Click on Enter Mission For Advanced Users and try the Great Graph Match. Use the information here and your knowledge of world geography.

3. Now research climate data for where you are living. Find data on average precipitation and temperature throughout the year. On a spreadsheet or graph paper, prepare a set of axes like the one below with scales suiting your data. Use it with your climate data to produce a combined graph, making precipitation a bar chart and temperature a line graph.

4. In which biome are you living?

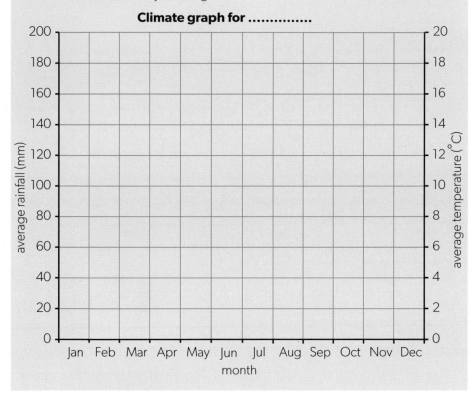

Climate graph for ....

Climate change and biome shift

With increasing mean global temperature and changes in precipitation, there is evidence that biomes are moving. There is general agreement that the climate is changing in these ways:

*   global average temperature increase of 1.5°C to 4.5°C by 2100 (according to the Intergovernmental Panel on Climate Change)

*   greater warming at higher latitudes

*   more warming in winter than summer

*   some areas becoming drier, others wetter

*   stronger storms.

These changes are happening very fast, within decades, and organisms change slowly, over many generations through evolutionary adaptation. All they can do to adapt to fast change is to move and that is what they are doing.

These moves are:

- towards the poles where it is cooler

- higher up mountains where it is cooler—500 m of altitude decreases temperature about 3°C

- towards the equator where it is wetter.

Examples of biomes shifting are:

- in Africa in the Sahel region, woodlands are becoming savannas

- in the Arctic, tundra is becoming shrubland.

Plants migrate very slowly because seeds are dispersed by wind or animals. Animals can migrate faster over longer distances (e.g. albatross, wildebeest, whales). But there are obstacles to migration, which may be:

- natural (e.g. mountain ranges, seas)

- caused by human activities (e.g. roads, agricultural fields and cities).

Animals may not be able to cross these obstacles and species are becoming extinct at fast rates (see topic 3).

Connections: This links to material covered in subtopics 3.1 and 3.2.

Areas known as **hotspots** are predicted to have a high turnover of species due to climate change. Examples include:

- the Himalayas—sometimes called the third pole—as species can move no higher than the land mass

- equatorial Eastern Africa—with a very drought-sensitive climate

- the Mediterranean region

- Madagascar

- the North American Great Plains and Great Lakes.

Up to one billion people live in regions which are vulnerable to biome changing. But these changes can also bring new opportunities for exploitation of resources.

- Drilling for oil under the Arctic Ocean is becoming possible with the decrease in sea ice.

- The North-West passage for ships between the north pole and North America could become a trade route without icing up.

- Under-sea mining of the Arctic regions is becoming possible.

Biomes in detail

Tropical rainforest

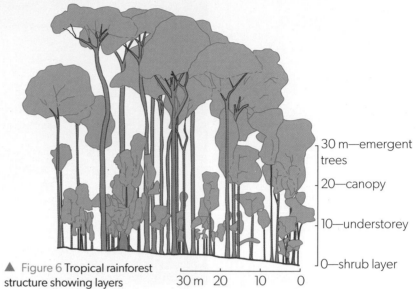

▲ Figure 5 A tropical rainforest in Borneo

▲ Figure 6 Tropical rainforest structure showing layers

30 m—emergent trees

20—canopy

10—understorey

0—shrub layer

30 m   20   10   0

What	Hot and wet areas with broadleaved evergreen forest.
Where (distribution)	Within 5° north and south of the equator.
Climate and limiting factors	High rainfall (2,000–5,000 mm yr^{-1}). High temperatures (26–28°C) and little seasonal variation. High insolation as near equator. Precipitation and evaporation are not limiting but rain washes nutrients out of the soil (leaching) so availability of nutrients may limit plant growth.
What's there (structure)	Amazingly high levels of biodiversity—many species and many individuals of each species. Plants compete for light and so grow tall to absorb it so there is a multi-storey profile to the forests with very tall emergent trees, a canopy of others, an understorey of smaller trees and a shrub layer under. This is called stratification. Vines, climbers and orchids live on the larger trees and use them for support (epiphytes). In primary forest (not logged by humans), so little light reaches the forest floor that few plants can live here. Nearly all the sunlight has been intercepted before it can reach the ground. Because there are so many plant species and a stratification of them, there are many niches and habitats for animals and large mammals can get enough food. Plants have shallow roots as most nutrients are near the surface so they have buttress roots to support them.
Net productivity	Estimated to produce 40% of NPP of terrestrial ecosystems. Growing season all year round, fast rate of decomposition and respiration and photosynthesis. Plants grow faster. But respiration is also high, and for a large mature tree in the rainforest, all the glucose made in photosynthesis is used in respiration, so there is no net gain. However, when rainforest plants are immature, their growth rates are huge and biomass gain very high. Rapid recycling of nutrients.
Human activity	The problem is that more than 50% of the world's human population lives in the tropics and subtropics and one in eight of us lives in or near a tropical rainforest. With fewer humans, the forest could provide enough resources for the population but there are now too many exploiting the forest and it does not have time to recover. This is not sustainable. In addition, commercial logging of valuable timber (e.g. mahogany) and clear-felling to convert the land to grazing cattle all destroy the forest.
Issues	Logging, clear-felling, conversion to grazing. Tropical rainforests are mostly in low-income countries and have been exploited for economic development.
Examples	Amazon Rainforest, Congo Basin in Africa, Borneo Rainforest.

▲ Table 2 Features of a tropical rainforest biome

Deserts

▲ Figure 7 **A desert in south-west North America**

What	Dry areas which are usually hot in the day and cold at night as skies are clear and there is little vegetation to insulate the ground. There are tropical, temperate and cold deserts.
Where (distribution)	Cover 20–30% of the Earth's surface about 30° north and south of the equator where dry air descends. Most are in the middle of continents. (Some deserts are cold deserts, e.g. the Gobi Desert.) The Atacama Desert in Chile can have no rain for 20 years or more. It is one of the driest places on Earth.
Climate and limiting factors	Water is limiting. Precipitation (P) less than 250 mm per year. Usually evaporation (E) exceeds precipitation: E > P.
What's there (structure)	Few species and low biodiversity but what can survive in deserts is well-adapted to the conditions. Soils are rich in nutrients as they are not washed away. Plants are drought-resistant and mostly cacti and succulents with adaptations to store water and reduce transpiration (e.g. leaves reduced to spines, thick cuticles to reduce transpiration). Animals too are adapted to drought conditions. Reptiles are dominant (e.g. snakes, lizards). Small mammals (e.g. kangaroo rats) can survive by adapting to be nocturnal (i.e. coming out at night and staying in a burrow in the heat of the day), or reducing water loss by having no sweat glands and absorbing water from their food. There are few large mammals in deserts.
Net productivity	Both primary (plants) and secondary (animals) are low because water is limiting and plant biomass cannot build up to large amounts. Food chains tend to be short because of this.
Human activity	Traditionally, nomadic tribes herd animals such as camels and goats in deserts as agriculture has not been possible except around oases or waterholes. Population density has been low as the environment cannot support large numbers. Oil has been found under deserts in the Gulf States and many deserts are rich in minerals including gold and silver. Irrigation is possible by tapping underground water stores or aquifers so, in some deserts, crops are grown. But there is a high rate of evaporation of this water and, as it evaporates, it leaves salts behind. Eventually these reach such high concentrations that crops will not grow (salinization).
Issues	Desertification—when an area becomes a desert through overgrazing, overcultivation or drought or all of these (e.g. the Sahel).
Examples	Sahara and Namib in Africa, Gobi in China.

▲ Table 3 **Features of a desert biome**

Temperate grasslands

▲ Figure 8 **Temperate grassland**

What	Fairly flat areas dominated by grasses and herbaceous (non-woody) plants.
Where (distribution)	In centres of continents 40–60° north of equator.
Climate and limiting factors	P = E or P slightly >E. Temperature range high as not near the sea to moderate temperatures. Clear skies. Low rainfall, threat of drought.
What's there (structure)	Grasses, wide diversity. Probably not a climax community as arrested by grazing animals. Grasses die back in winter but roots survive. Decomposed vegetation forms a mat, which contains high levels of nutrients. Burrowing animals (e.g. rabbits, gophers), kangaroos, bison, antelopes. Carnivores—wolves, coyotes. No trees.
Net productivity	600 g m^{-2} yr^{-1} so not very high.
Human activity	Used for cereal crops. Cereals are annual grasses. Black earth soils of the steppes rich in organic matter and deep so ideal for agriculture. Prairies in North America are less fertile soils so have to add fertilizers. Called world's bread baskets. Plus livestock—cattle and sheep that feed on the grasses.
Issues	The Dust Bowl in the 1930s in America—ecological disaster occurred when overcropping and drought led to soil being blown away on the Great. Overgrazing reduces them to desert or semi-desert.
Examples	North American prairies, Russian steppes in northern hemisphere; pampas in Argentina, veld in South Africa (30–40° south).

▲ Table 4 **Features of a temperate grassland biome**

▲ Figure 9 **Temperate forest**

Temperate forests

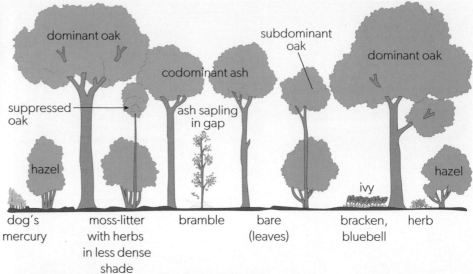

▲ Figure 10 **Temperate forest structure in Europe**

Key term

Climax vegetation is the dominant and stable community of plant species at the end of a succession.

What	Mild climate, deciduous forest.
Where (distribution)	Between 40° and 60° north and south of the equator.
Climate and limiting factors	P > E. Rainfall is 500–1500 mm per year, colder in winter. Winters freezing in some (Eastern China and NE USA), milder in western Europe due to the Gulf Stream. Temp range –30°C to +30°C. Summers cool.
What's there (structure)	Fewer species than tropical rainforests. For example, in Britain, oaks, which can reach heights of 30–40 m, become the dominant species of the **climax vegetation**. Other trees, such as elm, beech, sycamore, ash and chestnut, grow a little less high. Relatively few species and many woodlands are dominated by one species (e.g. beech). In the USA, there can be over 30 species per km². Trees have a growing season of 6–8 months, may only grow by about 50 cm a year.
	Woodlands show stratification. Beneath the canopy is a lower shrub layer varying between 5 m (e.g. holly, hazel and hawthorn) and 20 m (e.g. ash and birch). The forest floor, if the shrub layer is not too dense, is often covered in a thick undergrowth of brambles, grass, bracken and ferns. Many flowering plants (e.g. bluebells) bloom early in the year before the taller trees have developed their full foliage. Epiphytes (e.g. mistletoe, mosses, lichens and algae) grow on the branches. The forest floor has a reasonably thick leaf litter that is readily broken down.
	Rapid recycling of nutrients, although some are lost through leaching. The leaching of humus and nutrients and the mixing by biota produce a brown-coloured soil. Well-developed food chains in these forests with many autotrophs, herbivores (e.g. rabbits, deer and mice) and carnivores (e.g. foxes). Deciduous trees give way to coniferous towards polar latitudes and where there is an increase in either altitude or steepness of slope. P > E sufficiently to cause some leaching.
Net productivity	Second highest NPP after tropical rainforests but much lower because of leaf fall in winter, so reduced photosynthesis and transpiration and frozen soils when water is limiting. Temperatures and insolation lower in winter too as further from the Sun.
Human activity	Much temperate forest has been cleared for agriculture or urban developments. Large predators (e.g. wolves, bears) virtually wiped out.
Issues	Most of Europe's natural primary deciduous woodland has been cleared for farming, for use as fuel and in building, and for urban development. Some that is left is under threat (e.g. US Pacific Northwest old-growth temperate and coniferous forests). Often mineral wealth under forests is mined.
Examples	US Pacific Northwest.

▲ Table 5 **Features of a temperate forest biome**

Arctic tundra

▲ Figure 11 **Arctic tundra**

▲ Figure 12 **Distribution of Arctic tundra (shown in yellow)**

What	Cold, low precipitation and long, dark winters. 10% of Earth's land surface. Youngest of all the biomes as it was formed after the retreat of the continental glaciers only 10,000 years ago. Permafrost (frozen soil) present and no trees.
Where (distribution)	Just south of the Arctic ice cap and small amounts in the southern hemisphere. (Alpine tundra is found as isolated patches on high mountains from the poles to the tropics.)
Climate and limiting factors	Cold, high winds and little precipitation. Frozen ground (permafrost). Permafrost reaches to the surface in winter but in summer the top layers of soil defrost and plants can grow. Low temperatures so rates of respiration, photosynthesis and decomposition are low. Slow growth and slow recycling of nutrients. Water, temperature, insolation and nutrients can be limiting. In the winter, the northern hemisphere, where the Arctic tundra is located, tilts away from the Sun. After the spring equinox, the northern hemisphere is in constant sunlight. For nearly three months, from late May to August, the Sun never sets. This is because the Arctic regions of the Earth are tilted towards the Sun. With this continuous sunlight, the ice from the winter season begins to melt quickly. During spring and summer, animals are active, and plants begin to grow rapidly. Sometimes temperatures reach 30°C. Much of this energy is absorbed as the latent heat of melting of ice to water. In Antarctica, where a small amount of tundra is also located, the seasons are reversed.
What's there (structure)	No trees but thick mat of low-growing plants—grasses, mosses, small shrubs. Adapted to withstand drying out with leathery leaves or underground storage organs. Growing season may only be eight weeks in the summer. Animals also adapted with thick fur and small ears to reduce heat loss. Mostly small mammals such as lemmings, hares and voles. Predators such as Arctic fox, lynx, snowy owl. Most hibernate and make burrows. Simple ecosystems with few species. Often bare areas of ground. Low biodiversity—900 species of plants compared with 40,000 or more in the Amazon Rainforest. Soil poor, low inorganic matter and minerals.
Net productivity	Very low. Slow decomposition so many peat bogs where most of the carbon is stored.

▲ Table 6 **Features of an Arctic tundra biome**

Human activity	Few human activities but mining and oil. Nomadic groups herding reindeer.
Issues	Fragile ecosystems that take a very long time to recover from disruption. May take decades to recover if you even walk across it. Mining and oil extraction in Siberia and Canada destroy tundra. Many scientists feel that global warming caused by greenhouse gases may eliminate Arctic regions, including the tundra, forever. The global rise in temperature may damage the Arctic and Antarctic more than any other biome because the Arctic tundra's winter will be shortened, melting snow cover and parts of the permafrost, leading to flooding of some coastal areas. Plants will die, animal migrating patterns will change, and the tundra biome as we know it will be gone. The effect is uncertain but we do know the tundra, being the most fragile biome, will be the first to reflect any change in the Earth. Very large amounts of methane are locked up in tundra ice in clathrates. If these are released into the atmosphere then there will be a huge increase in greenhouse gases (clathrates contain 3,000 times as much methane as is in the atmosphere now and methane is more than 20 times as strong a greenhouse gas as carbon dioxide).
Examples	Siberia, Alaska.

▲ Table 6 **Features of an Arctic tundra biome (cont.)**

Deep ocean

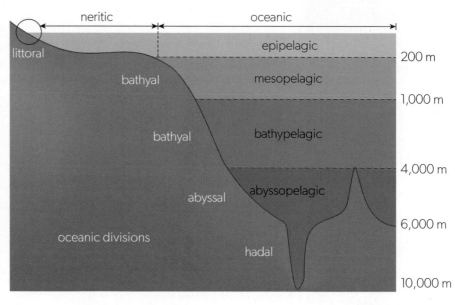

▲ Figure 13 **Deep ocean animals—fangtooth fish, tubeworms**

▲ Figure 14 **Deep ocean divisions**

What	The ocean and seafloor beyond continental shelves.
Where (distribution)	65% of the Earth's surface. Most is abyssal plain of the ocean floor—averaging 3.5 miles deep.
Climate and limiting factors	Pressure increases with depth, temperature variation decreases to a constant −2°C at depth. Light limiting below 1,000 m—there is none.
	Nutrients—low levels and low primary productivity but some dead organic matter falls to deep ocean floors.
What's there (structure)	Top 200 m—some light for photosynthesis so phytoplankton and cyanobacteria live here and they and algae are the main producers. They are eaten by zooplankton, fish and invertebrates (e.g. squid, jellyfish).
	200–1,000 m deep—as pressure increases with depth, fish here are muscular and strong to resist pressure. Very little light reaches here so large eyes, reflective sides and light-producing organs on their bodies. Many are red which absorbs shorter wavelengths of light that penetrate further.
	1,000–4,000 m deep—higher diversity here, always dark. Fish are black with small eyes, bristles and bioluminescence (i.e. creating their own light to hunt or avoid predators). Very little muscle, large mouths.
	4,000 m to bottom—huge pressures, constant cold. Mostly shrimps, some fish, jellyfish, tubeworms on bottom.
	Bottom surface—fine sediments made up of debris from above—plankton shells, dead organisms, whale and fish skeletons. Also mud and volcanic rocks in mid-ocean ridges. Where volcanoes erupt, there are hydrothermal vent communities high in sulfides where chemosynthetic bacteria gain their energy from the sulfur. These producers support communities of crabs, tubeworms, mussels, and even octopus and fish.
Net productivity	Low.
Human activity	Minimal but rocks rich in manganese and iron could be a resource.
Issues	Pollution from run-off from rivers, sewage, ocean warming due to climate change.
Examples	Arctic, Atlantic, Pacific Oceans.

▲ Table 7 Features of a deep ocean biome

Biome	Net primary productivity (g m^{-2} yr^{-1})	Annual precipitation (mm yr^{-1})	Area (10^6 km^2)	Plant biomass (10^9 T)	Mean biomass (kg m^{-2})	Animal biomass (10^6 T)	Solar radiation (W m^{-2} yr^{-1})
Tropical rainforest	2,200	2,000–5,000	17.0	765	45	330	175
Temperate forest	1,200	600–2,500	12.0	385	32.5	160	125
Boreal forest	800	300–500	12.0	240	20	57	100
Tropical grassland (savannas)	900	500–1,300	15.0	60	4	220	225
Temperate grassland	600	250–1,000	9.0	14	1.6	60	150
Tundra and alpine	140	<250	8.0	5	0.6	3.5	90
Desert (rock, sand, ice)	90	<250	24.0	0.5	0.02	0.02	75
Deep ocean	20–300	Variable	352	1,000+	Very low	800–2,000	Variable

▲ Table 8 Comparison of biomes

(ATL) Activity 20

ATL skills used: thinking, communication

1. State what deciduous means.

2. Define net primary productivity (NPP).

Use data in Table 8 and Figure 15, and information in the text, to answer the following questions.

3. Explain which biome has the highest NPP per m² per year and why.

4. Explain which biome has the largest NPP and why.

5. Explain why NPP is low in tundra, deserts and deep oceans.

6. Explain why there are no large deciduous forests or tundra in the southern hemisphere.

7. Explain why trees in temperate biomes are deciduous.

8. Describe how the number of tree species and their distribution differs in temperate and tropical rainforests.

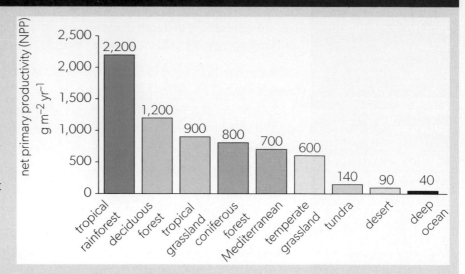

▲ Figure 15 NPP of major biomes

9. State the main factors causing the distribution of biomes.

10. Explain which biome(s) is/are most threatened and why.

11. Explain which biome(s) has/have been most changed by human activity and why.

12. Describe the effects climate change may have on biome distribution.

Connections: This links to material covered in subtopics 1.2, 4.1, 6.1 and 6.2.

Climate

The equator gets the most direct sunlight and energy because the Sun's rays reach the equator at a higher angle, almost 90 degrees (Figure 16). Much less solar energy reaches the poles because the angle is lower and snow and ice reflect more. The difference in the amount of solar energy between the poles and the equator drives atmospheric circulation in three cell types called the Hadley cell, the Ferrel cell and the polar cell.

Circulation of the atmosphere and circulation of the oceans are how heat is redistributed around the Earth (Figures 17 and 18). Energy from the Sun drives this circulation due to uneven distribution of solar energy—greatest at the equator, least at the poles. This is known as the tricellular model (Figures 17 and 18).

▶ Figure 16 Solar radiation hitting the Earth. You can see this effect if you shine a torch beam at a flat surface and an angled surface
Adapted from: Six Red Marbles and OUP

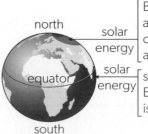

the rays hit the Earth at a more acute angle so are spread over a greater surface area

solar radiation hits the Earth at 90° angle so is most intense

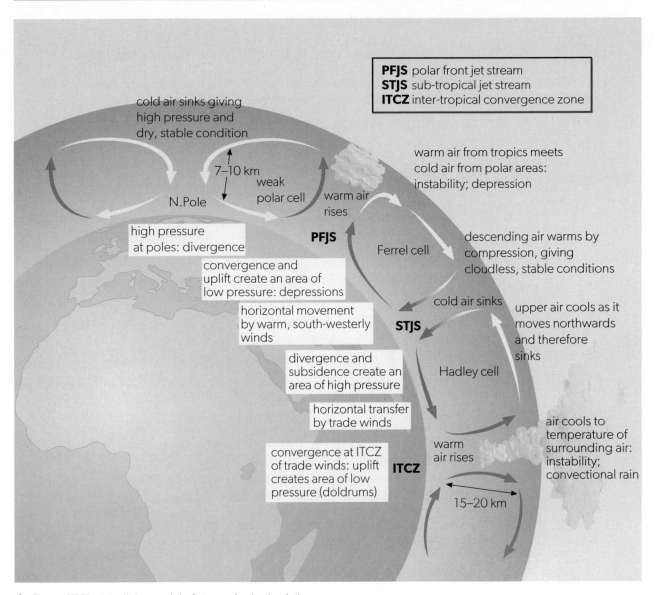

cold air sinks giving high pressure and dry, stable condition

7–10 km

N.Pole

weak polar cell

warm air rises

high pressure at poles: divergence

PFJS

convergence and uplift create an area of low pressure: depressions

horizontal movement by warm, south-westerly winds

divergence and subsidence create an area of high pressure

horizontal transfer by trade winds

convergence at ITCZ of trade winds: uplift creates area of low pressure (doldrums)

ITCZ

PFJS polar front jet stream
STJS sub-tropical jet stream
ITCZ inter-tropical convergence zone

warm air from tropics meets cold air from polar areas: instability; depression

Ferrel cell

descending air warms by compression, giving cloudless, stable conditions

cold air sinks

STJS

upper air cools as it moves northwards and therefore sinks

Hadley cell

warm air rises

air cools to temperature of surrounding air: instability; convectional rain

15–20 km

▲ Figure 17 The tricellular model of atmospheric circulation

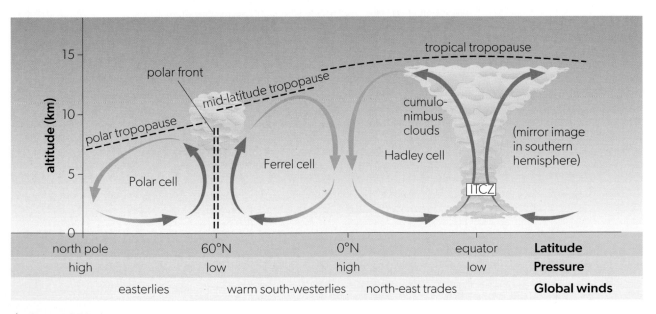

tropical tropopause

polar front

mid-latitude tropopause

15

cumulo-nimbus clouds

10

altitude (km)

polar tropopause

(mirror image in southern hemisphere)

Hadley cell

5

Ferrel cell

Polar cell

ITCZ

0

north pole		60°N		0°N		equator	**Latitude**
high		low		high		low	**Pressure**
	easterlies		warm south-westerlies		north-east trades		**Global winds**

▲ Figure 18 Tricellular model showing latitudes, winds and sizes of cells

You should know the following facts.

- Water exists in three states—solid, liquid and gas (i.e. water vapour).

- Hot air holds more water vapour than cold air. It has more gaseous water molecules in it (has a higher relative humidity).

- When liquid water gains energy the molecules can escape from the liquid and become a gas. This gas mixes with other gases in the air.

- **Latent heat of condensation** is energy released when water vapour condenses to form liquid droplets.

- The **Coriolis force** is caused because the Earth rotates on its axis and deflects air to the right in the northern hemisphere and to the left in the southern hemisphere.

- The **intertropical convergence zone (ITCZ)** is a band of low pressure, clouds and usually thunderstorms near the equator. It is called the doldrums by sailors because there is no wind.

Hadley cell

Hadley cells are convection cells, covering the area from 0° latitude to 30° north or south. (The same things happen either side of the equator, so there are two Hadley cells.)

At the equator, air is warmed by the Sun and gains energy. Liquid water is also warmed and some becomes water vapour in the air. The molecules of air then move more quickly and move apart. This means the air is less dense—there is more space between molecules—and so hot air rises.

As it rises, it cools and some water vapour in the air falls as rain, releasing its latent heat. Below these areas, there is a low-pressure zone with less air.

The hot air is forced towards the poles (northwards and southwards) away from the equator, so it cools as it receives less of the Sun's energy. As it cools, the molecules move more slowly and take up less space. The air becomes more dense and so falls. This increases air pressure because there is more air where the air is more dense and falling.

Air moves from higher to lower pressure areas (from areas where there is more to areas where there is less) so moves back towards the equator, replacing the air that rose as it got hotter.

Easterly trade winds are found below the Hadley cells.

What does this mean for biomes and winds?

At the equator is a belt of high rainfall and high insolation. Temperature and precipitation determine climate and the climate is hot and wet all year. Here are tropical rainforests. In the oceans are strong winds which sailors call the trade winds as they allowed sailing ships to cross oceans more easily. They blow from the north-east in the northern hemisphere and the south-east in the southern hemisphere (Figure 19).

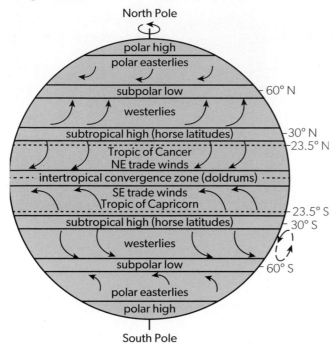

◀ Figure 19 Wind names and directions in northern and southern hemispheres

Adapted from: United States Geological Survey (USGS)

Ferrel cell

Ferrel cells are convection cells, weaker than the Hadley and polar cells.

Not all air at 30° north or south flows back to the equator. Some flows away from the equator, forming the Ferrel cells.

This air descends at 30° latitude and returns poleward at ground level. It deviates towards the east as it does so, which is why westerly trade winds are found beneath the Ferrel cells.

In the upper atmosphere of the Ferrel cell, air moves towards the equator, deviating towards the west. Both deviations are driven by conservation of angular momentum (as in Hadley cells and polar cells).

Polar cell

The polar cells are found between 60° and 90° north and south. They are similar to Hadley and Ferrel cells—weaker than Hadley cells but stronger than Ferrel cells.

Air at 60° north is still warm and moist enough to rise, flow northwards, cool and fall again. This leads to a cold, dry, high-pressure area at the North Pole. Air then flows back to complete the cell. The reverse is true at the South Pole.

Due to the Coriolis force, air moving towards the North Pole is deflected towards the east, while air moving towards the South Pole is deflected towards the west.

ATL Activity 21

ATL skills used: thinking, communication, social, research, self-management

Read the following instructions before you start. Work in groups of two or three.

1. Taking Figure 20 as a template, draw the Earth and the cells of the tricellular model (three in the northern hemisphere and three in the southern hemisphere).

2. Accurately mark the latitude lines at 0, 30° and 60° north and south.

3. Draw on your Earth the outline of the continent where you are living. Add any neighbouring continents that fit on your Earth diagram.

4. Draw in and label the biomes on your land area.

5. Using Figures 17–20, adjust your biomes to show the effects of the different wind directions on biome distribution.

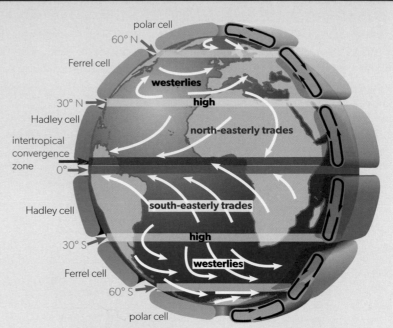

▲ Figure 20 Idealized diagram of the six cells of the tricellular model of atmospheric circulation
Kaidor / Wikimedia Commons, adapted from Jet Propulsion Labratory / NASA (CC BY-SA 3.0)

TOK

What is the role of imagination and intuition in the creation of the tricellular model?

The tricellular model shows there are three general climate types—tropical, temperate and polar—which correspond with the cells of the tricellular model (Table 9). As latitude increases, there is less energy, less heat, less rainfall and less productivity.

Cell	Climate type	Biome	Precipitation	Temperature	Vegetation
Hadley	tropical: equatorial	tropical rainforest	high	high	tropical rainforest
	tropical: seasonal	deciduous rainforest	high but lower than tropical rainforest	high	deciduous seasonal forest
Ferrel	temperate: maritime	temperate forest	moderate	cooler summer, warmer winter as moderated by oceans	various—conifer or deciduous trees, grasses, shrubs
	temperate: continental	temperate forest	moderate	warm summer, cold winter	various—conifer or deciduous trees, grasses, shrubs
Polar	polar: Arctic, Antarctic	tundra and ice cap	low	low	lichen, mosses, grasses

▲ Table 9 Three climate types correspond with the cells of the tricellular model

Human influence on biome distribution

The biome predicted by any given temperature and rainfall pattern may not develop in an area because of secondary influences or human interventions. Clearly humans have altered the vegetation types of biomes—clearing rainforests for grasslands; clearing temperate forest for crops, livestock and human habitation.

Ocean currents and energy distribution

Ocean currents are movements of water both vertically and horizontally. Currents move in specific directions and some have names. There are both surface currents and deep-water currents. These currents have an important role in the global distribution of energy. Without an understanding of ocean currents, we cannot understand global atmospheric energy exchanges.

Surface currents are found in the upper 400 m of ocean and are moved by the wind. The Earth's rotation deflects them and increases their circular movement.

Deep-water currents are also called thermohaline currents. They make up 90% of ocean currents and cause the oceanic conveyor belt (Figure 21).

*   They are due to differences in water density caused by (a) salt and (b) temperature.

*   Warm water can hold less salt than cold water so it is less dense and rises.

*   Cold water holds more salt so it is denser and sinks.

*   When warm water rises, cold water comes up from depth to replace it. These are called **upwellings**.

*   When cold water rises, it is replaced by warm water in **downwellings**.

*   In this way, water circulates.

ATL Activity 22

ATL skills used: thinking, social, research

1. Research which biome you are living in now.

2. Describe the climate type—rainfall, temperature, latitude, altitude, other factors.

3. Explain if you live in a natural biome or one with human influences.

4. Give reasons for the human influences in your biome.

5. To what extent has biodiversity changed in your biome?

6. Discuss your findings in class.

Key terms

An **upwelling** occurs when cooler, deeper water rises towards the surface.

A **downwelling** occurs when surface water is forced downwards, where it may take oxygen to deeper water.

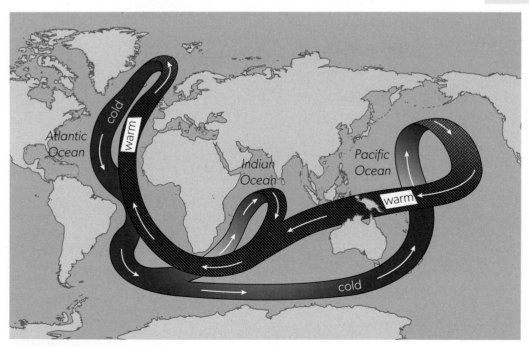

▲ Figure 21 The great oceanic conveyor belt moves water around the oceans in a constantly moving system of deep-ocean circulation which is driven by temperature and salinity

Cold ocean currents run from the poles to the equator, for example:

- the Humboldt Current (off the coast of Peru)

- the Benguela Current (off the coast of Namibia in western Africa).

Warm currents flow from the equator to the poles, for example:

- the Gulf Stream (in the North Atlantic Ocean)

- the Angola Current (off the coast of Angola).

Ocean currents and climate

The ocean plays a role in our climate by absorbing most of the incoming radiation—mostly in equatorial regions. Land does absorb sunlight but not as much as water. Water has a higher **specific heat capacity** than land. Specific heat capacity is the amount of heat needed to raise the temperature of a unit of matter by 1°C. Water masses heat up and cool down more slowly than land masses. As a result, land close to seas and oceans has a milder climate with moderate winters and cool summers.

Ocean currents (see Figure 22) also affect local climate.

- The warm Gulf Stream/North Atlantic Drift moderates the climate of north-western Europe, which otherwise would have a sub-Arctic climate.

- The cold Benguela Current comes under the influence of prevailing south-west winds and moderates the climate of the Namibian Desert.

- The Humboldt Current affects the climate in Peru (see ENSO below).

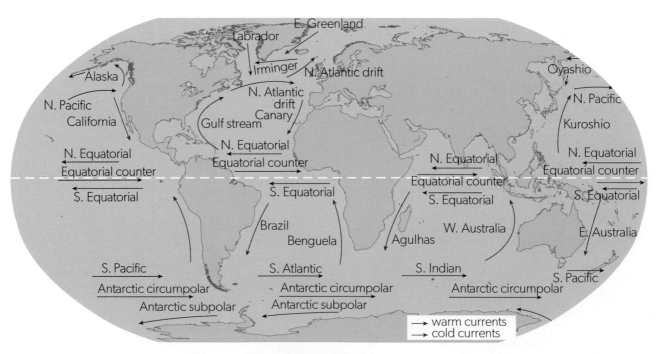

▲ Figure 22 **The ocean currents**

El Niño Southern Oscillation

The El Niño Southern Oscillation (ENSO) is the fluctuation (variation) in wind direction and sea surface temperature in the equatorial Pacific Ocean. It has three stages:

- neutral

- El Niño, the "warm phase"

- La Niña, the "cool phase".

El Niño and La Niña refer to the ocean temperature. During normal conditions in the Pacific Ocean, the trade winds blow westwards along the equator. This moves warm water from South America towards Asia and Australia. To replace that warm water, cold water rises from the depths. This is upwelling.

El Niño and La Niña are two opposing climate patterns that change the normal conditions. This is the El Niño Southern Oscillation (ENSO) cycle. El Niño and La Niña can both have impact globally with wildfires, floods and droughts. Changes between El Niño and La Niña occur irregularly in 2–7 year cycles and each lasts about 9–12 months. The changes are hard to predict. El Niño often arrives around Christmas time (hence the name—El Niño means "little boy" in Spanish).

Walker circulation

As well as the north–south circulations of the tricellular model, there are weaker east–west circulations. Across the equatorial Pacific, the east–west circulation is called the Walker circulation or Walker cell (Figure 23).

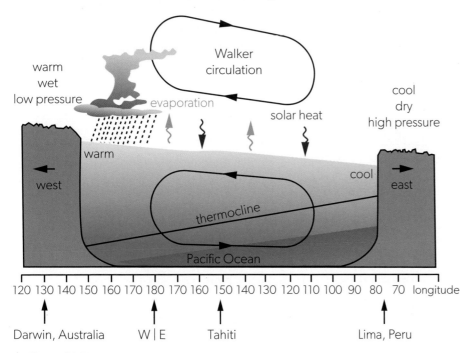

▲ Figure 23 **The Walker circulation**

Adapted from: PAR / Wikimedia Commons

A thermocline is a vertical temperature gradient in a body of water. As depth of water increases, temperature decreases but in a thermocline the temperature decreases rapidly with increasing depth.

In the neutral phase with no El Niño or La Niña:

- low-pressure areas form over the eastern Pacific (Australia, Asia)

- high-pressure areas form over the western Pacific (Peru, Central America)

- winds near the surface blow from east to west—the trade winds

- these winds blow warmer water towards the west and warm the air there

- that warm air rises and makes clouds, so rain falls

- the dryer air then travels back towards the west, creating a loop

- the cooler, dry air then sinks along the Pacific coast.

As the winds blow east to west and warmer surface water is blown westwards, colder water rises to fill the space (upwelling). This is blown westwards too and then descends in downwellings to the sea floor, where it flows back towards the east (see subtopic 4.1). The Pacific can be 60 cm higher in the west than the east in the neutral phase.

In the El Niño phase:

- trade winds weaken and reverse (Figures 24 and 25)

- warm water is pushed east to west, back to the coast of South and Central America

- the jet stream (strong winds around 8–11 km above the Earth's surface, blowing from west to east) moves south causing major changes in rainfall patterns globally.

In the La Niña phase:

- trade winds are even stronger than in the neutral phase, pushing more warm water from east to west, towards Australia and Asia (Figure 26)

- upwelling increases on the coast of South America, bringing cold, nutrient-rich water to the surface

- the cold waters in the Pacific push the jet stream northward

- results are the opposite to El Niño, for example drought in southern USA and heavy rains and flooding in the Pacific Northwest and Canada

- there can be more severe cyclones.

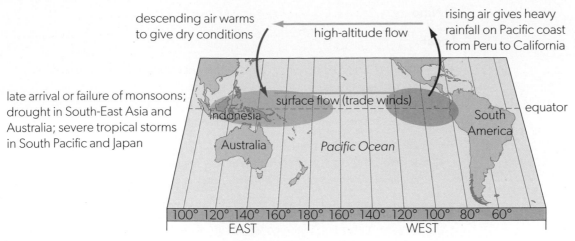

▲ Figure 24 **Atmospheric circulation during the El Niño phase**

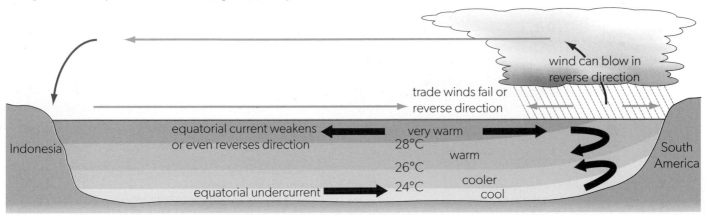

▲ Figure 25 **A cross-section through the Pacific Ocean during the El Niño phase**

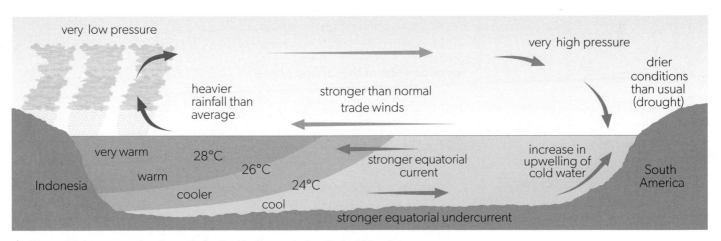

▲ Figure 26 **A cross-section through the Pacific Ocean during the La Niña phase**

The Walker circulation also drives other cells across the Indian Ocean and the equatorial Atlantic.

Local and global effects of ENSO

ENSO impacts:

- ocean temperatures

- the speed and strength of ocean currents

- global weather patterns.

El Niño

Local

- Without the upwelling which brings nutrients from depth to the surface waters, there are fewer phytoplankton. Phytoplankton are the basis of marine food chains. They are eaten by clams and small fish, which in turn are eaten by larger fish or marine mammals. All have less food.

- Fishing industries of Peru, Ecuador and Chile catch anchovy, sardine, mackerel, shrimp, tuna and hake. All fish have less food if there is no upwelling. Anchovy stocks collapse.

- Sea birds starve due to lack of food; massive death of sea birds.

- Storms and flooding in the coastal plain of Peru.

- Rising convection currents bring increased precipitation (Figure 24) in different places depending on the phase of ENSO. Higher rainfall in Ecuador and Peru causes coastal flooding and erosion. Crops are destroyed. Lower rainfall leads to droughts in Australia and Indonesia.

Global

- The warm ocean water that moves east during El Niño events contains tremendous amounts of energy. The amount of energy is large enough to alter major air currents like the jet streams, so El Niño affects global weather.

- Droughts occur in the Pacific north-west of the USA, and British Columbia (Canada). Forest fires are common in these areas.

- Heavy winters in higher latitudes of North and South America.

- Heavy storms often resulting in flooding in California and the Midwest of the USA, Central Europe and eastern Asia.

- Absence of the monsoon in India. The Indian population depends on the monsoon rains for its food production.

Some benefits of the El Niño phase

- Fewer hurricanes and other tropical cyclones in the north Atlantic.

- Milder winters in southern Canada and northern continental USA.

- Replenishment of water supplies in south-western USA.

- Less disease in some areas due to drier weather (e.g. malaria in south-eastern Africa).

La Niña

- Higher occurrences of fog due to increased upwelling along the coast of South America.

- Increase in food supply from upwellings off the coast of South America.

- Drought in southern USA, floods in north-west Pacific USA and Canada.

Connections: This links to material covered in subtopics 6.2 and 6.3.

Human impacts on ENSO

While ENSO is a natural phenomenon, its effects are exacerbated by human pressures on the environment. And with a bigger population living in regions affected by ENSO, humans are facing greater impacts. Examples include:

- fires leading to respiratory problems with smoke inhalation

- drought leading to crop failure causing famine or malnutrition

- flooding leading to loss of cropland, housing and life

- economic changes such as increases in the price of energy and food

- political and social unrest

- crash of fisheries

- insect population explosion leading to disease and plagues such as hanta virus, cholera, dengue fever and malaria.

The 1982–3 and 1997–8 El Niño events were particularly strong. They led to:

- severe drought in Australia, Indonesia and the Philippines

- typhoons in Tahiti

- flooding in central Chile

- drops in fish catches.

Case study 5

Anchovy fishery

The anchovy fishery off the coast of Peru is extremely rich because of the occurrence of an upwelling when cold nutrient-rich waters come up from the ocean depths. The majority of the Peruvian anchoveta catch is processed in large plants that reduce the fish to fishmeal or oil to be used as animal feed, fertilizer, and other industrial products. The fishery has yielded greater catches than any other single wild fish species in the world, with annual harvests between 3.14 and 8.32 million tonnes throughout the 2010s. But catch size fluctuates with ENSO phases and is reducing annually.

Normally, productivity in the oceans is quite low because of one of two limiting factors: light level and nutrient concentration.

- In the upper water levels, light intensity is high but nutrient levels are low and limit productivity. The available nutrients are taken in by phytoplankton. Organisms that are not eaten by some other organism, die and sink.

- The nutrients stored in these sinking organisms end up on the ocean bottom. The lower water layers are therefore nutrient rich. But the absence of light makes photosynthesis impossible.

▲ Figure 27 One anchovy—about 8 cm long

West of the Peruvian coast, the prevailing eastern trade winds push the surface water westward. This water is replaced by cold, nutrient-rich water from the deep Humboldt Current, which originates in the Antarctic region and follows the South American coast to the north. The appearance of nutrient-rich water at the surface allows for high productivity, hence the high numbers of fish and their predators, the sea birds.

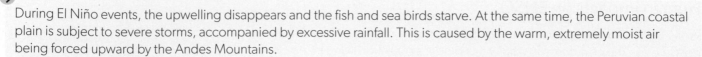

During El Niño events, the upwelling disappears and the fish and sea birds starve. At the same time, the Peruvian coastal plain is subject to severe storms, accompanied by excessive rainfall. This is caused by the warm, extremely moist air being forced upward by the Andes Mountains.

Questions

1. Research fishing rights offshore in Peruvian waters.

2. Determine if there is a legal limit to the amount of anchoveta caught each year.

3. Explain if it is more energy efficient to produce fishmeal or to eat whole anchovy.

4. Discuss if the fishery industry should be compensated in years that the catch falls.

▲ Figure 28 **Hurricane Florence in 2018 seen from the international space station**

Tropical cyclones

Cyclones are:

* one of the most dangerous natural hazards to humans

* rapidly circulating storm systems with a low-pressure centre that originate in the tropics

* characterized by strong winds with wind speeds over 119 km/h

* called different things depending on where they originate: hurricanes if they occur in the Atlantic and north-east Pacific; typhoons in the north-west Pacific; or cyclones if formed over the south Pacific and Indian Oceans (Figure 29)

* given a name so they can be identified—the first storm of the year starts with an A and so on.

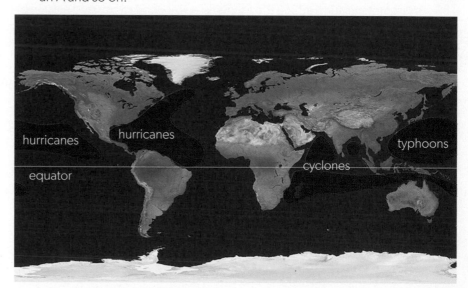

▲ Figure 29 Tropical cyclone distribution

How are cyclones formed?

* In the equatorial zone over the sea, air is heated by very warm seawater in the Hadley cell and rises quickly. This air cools as it rises and is pushed aside by more warm air rising. This causes convection currents and daily thunderstorms (Figure 30).

- In cyclones, thunderstorms group together leading to very low pressure at the surface, causing a depression. The depression strengthens and deep thunderclouds form around a central area known as the eye of the storm.

- Energy is released as warm water vapour condenses (latent heat) causing clouds and rain. This release of energy warms air around it, decreasing air pressure.

- More air rises to fill the low pressure area, so more moist warm air evaporates from the sea. This creates a self-sustaining heat engine.

- Winds circulate anti-clockwise in the northern hemisphere and clockwise in the southern hemisphere.

- Cyclones may be 10 km high and 600 km wide and move at 25–100 km/h.

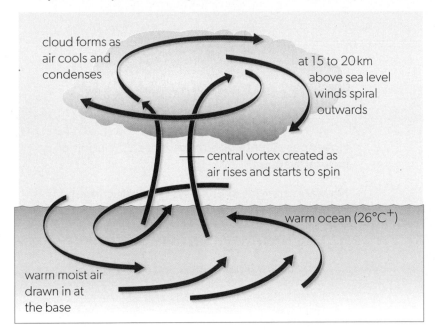

cloud forms as air cools and condenses

at 15 to 20 km above sea level winds spiral outwards

central vortex created as air rises and starts to spin

warm ocean (26°C$^+$)

warm moist air drawn in at the base

◀ Figure 30 **Cyclone formation**

Intensity and frequency of tropical cyclones

Increases in ocean temperatures resulting from global warming are increasing the intensity and frequency of cyclones because warmer water and air have more energy.

- Typhoons have long been recorded in the north-west Pacific. In 1957, one struck land near Hong Kong, killing 10,000 people.

- The most deadly tropical cyclone ever recorded (Great Bhola cyclone) hit Bangladesh and east Bengal in 1970, killing approximately 300,000 people as a result of a storm surge in the Ganges delta.

- Katrina (2005) was the most costly hurricane recorded. It caused 1,500 deaths and an estimated USD 108 billion of damage in Louisiana and Mississippi.

- In 2008, cyclone Nargis hit Myanmar causing a storm surge in the Irrawaddy delta and 138,000 people died.

- Sandy (2012) was the second most costly hurricane on record causing USD 71 billion of damage on the eastern coast of the USA.

- Typhoon Haiyan (2013) was the most powerful ever recorded and hit Southeast Asia, particularly the Philippines, causing loss of life.

AHL

Tropical cyclones are increasing in intensity and frequency but the evidence for this is complex and mixed. Here is what is known.

- Sea levels are rising which will cause more coastal flooding when cyclones hit land.

- Cyclone rainfall rates will increase as anthropogenic warming increases evaporation and moisture content of the atmosphere.

- In the north-west Pacific there seems to be a shift of cyclones to higher latitudes.

- In the Atlantic there appear to be more tropical storms but this may be a result of better reporting.

What is clear is that, with rising human population numbers, more of the population live in or near areas where cyclones hit land. This increases the risk of loss of life and damage to property.

TOK

How might the beliefs and interests of a person influence their conclusions about human impacts on natural ecosystems?

Check your understanding

In this subtopic, you have covered:

- climate and weather

- biome distribution

- atmospheric circulation

- ocean circulation

- biome shifting

- more detail on biomes

- ENSO

- cyclones

- anthropogenic activity causing changes in weather patterns.

AHL

How does climate determine the distribution of natural systems?

How are changes in Earth systems affecting the distribution of biomes?

1. Discuss what are the major factors influencing climate.

2. Explain why biomes are where they are.

3. Explain the effect of latitude and altitude on biome distribution.

4. Describe how humans are influencing this distribution.

5. Explain the ENSO and Walker circulation.

6. Describe the impacts on ecosystems of El Niño and La Niña phases of the ENSO.

7. Evaluate the evidence for human activity increasing cyclone intensity and frequency.

AHL

Taking it further

- Explore the effect of climate change on your local or regional biome and produce a presentation which explains the cause and effect of any shift.

- Raise awareness and fundraise for communities impacted by severe hurricanes, earthquakes or typhoons.

2.5 Zonation, succession and change in ecosystems

Guiding question

- How do ecological systems change over time and over space?

Understandings

1. Zonation refers to changes in community along an environmental gradient.
2. Transects can be used to measure biotic and abiotic factors along an environmental gradient in order to determine the variables that affect the distribution of species.
3. Succession is the replacement of one community by another in an area over time due to changes in biotic and abiotic variables.
4. Each seral community (sere) in a succession causes changes in environmental conditions that allow the next community to replace it through competition until a stable climax community is reached.
5. Primary successions happen on newly formed substratum where there is no soil or pre-existing community, such as rock newly formed by volcanism, moraines revealed by retreating glaciers, wind-blown sand or waterborne silt.
6. Secondary successions happen on bare soil where there has been a pre-existing community, such as a field where agriculture has ceased or a forest after an intense firestorm.
7. Energy flow, productivity, species diversity, soil depth and nutrient cycling change over time during succession.
8. An ecosystem's capacity to tolerate disturbances and maintain equilibrium depends on its diversity and resilience.
9. The type of community that develops in a succession is influenced by climatic factors, the properties of the local bedrock and soil, geomorphology, together with fire and weather-related events that can occur. There can also be top-down influences from primary consumers or higher trophic levels.
10. Patterns of net productivity (NP) and gross productivity (GP) change over time in a community undergoing succession.
11. r- and K-strategist species have reproductive strategies that are better adapted to pioneer and climax communities, respectively.
12. The concept of a climax community has been challenged, and there is uncertainty over what ecosystems would develop naturally were there no human influences.
13. Human activity can divert and change the progression of succession leading to a plagioclimax.

Key terms

Succession is the process of change over time in an ecosystem involving pioneer, intermediate and climax communities.

Zonation is the change in community along an environmental gradient due to factors such as changes in altitude, latitude, tidal level or distance from shore, coverage by water.

Succession and zonation

Do not confuse **succession** with **zonation**.

- Succession is how an ecosystem changes over time.

- Zonation is how an ecosystem changes along an environmental gradient (e.g. a gradient in altitude, latitude, tidal level, soil, or distance from water source).

Zonation	Succession
A sequence of vegetation and organisms in space showing a gradual change in distribution.	Dynamic and temporal (takes place over long periods of time).
Caused by an abiotic gradient such as changes in elevation, latitude, tidal level, soil horizons or distance from a water source.	Caused by progressive changes through time (e.g. as vegetation colonizes bare rock).
For example, temperature decreases with increasing altitude, altering the abiotic factors and so changing the species composition.	One community changes the environmental conditions so another community can colonize the area and replace the first through competition.

▲ Table 1 **Differences between zonation and succession**

Zonation

For each species, there is an ecological niche. Each niche has boundary limits and outside these, the species cannot live. Many abiotic and biotic factors influence these limits. Here are the most important ones on mountains (Figure 1).

- Temperature—which decreases with increasing altitude and latitude.

- Precipitation—on mountains, most rainfall is at middle altitudes so deciduous forest grows. Higher up, the air is too dry and cold for trees.

- Solar insolation—more intense at higher altitudes so plants have to adapt, often with red pigment in their leaves to protect themselves against too much insolation.

- Soil type—in warmer zones, decomposition is faster so soils are deeper and more fertile. Higher up, decomposition is slow and soils tend to be acidic.

- Interactions between species—competition may crowd out some species and grazing may alter plant composition. Mycorrhizal fungi may be very important in allowing trees to grow in some zones.

Human activities alter zonation. For example, road building on mountains may allow tourism into previously inaccessible areas. Deforestation and agriculture change previously undisturbed areas.

Connections: This links to material covered in subtopic 2.1.

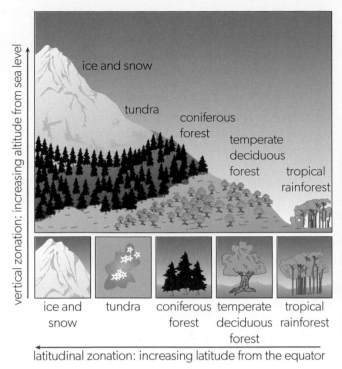

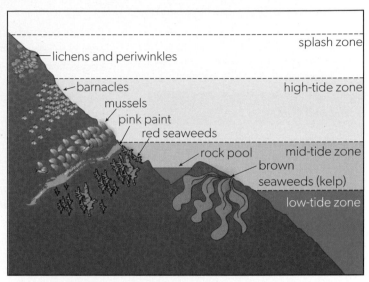

▲ Figure 1 **Zonation with increasing altitude on a mountain, and with increasing latitude from the equator**

▲ Figure 2 **Zonation of species on a rocky shore due to increasing exposure to air higher up the shore**

Graphical representation of zonation is often by a kite diagram where the width of the "kites" corresponds to the number of that species (Figure 3).

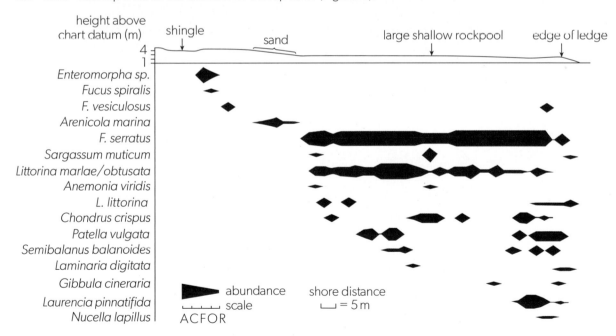

ACFOR = Abundant, Common, Frequent, Occasional, Rare

▲ Figure 3 **Kite diagram showing zonation of species on a rocky shore**

Succession

Succession is the replacement of one community by another in an area over time due to changes in biotic and abiotic variables. The process can take hundreds of years.

- Primary succession occurs on bare ground, where soil formation starts the process. Secondary succession occurs where soil is already formed but the vegetation has been removed (e.g. by a forest fire).

- Early in succession, in **pioneer communities**, gross primary productivity (GPP) is high and respiration is low, so net primary productivity (NPP) is high as biomass accumulates.

- In later stages, while GPP may remain high, respiration increases so NPP may approach zero and the productivity:respiration ratio (P:R) approaches 1.

- A **climax community** is reached at the end of a succession when species composition stops changing. There may be several states of a climax community depending on abiotic factors.

- The more complex the ecosystem (higher biodiversity, increasing age), the more stable it tends to be.

In agricultural systems, humans often deliberately stop succession when NPP is high and crops are harvested.

- Humans also interrupt succession by deforestation, grazing with animals or controlled burning.

- Sometimes the ecosystem recovers from this interruption and succession continues; sometimes the interruption is too great and the system is less resilient and so succession is stopped.

- Species biodiversity increases as succession continues, falling a little in a climax community. The higher the diversity, the higher the resilience.

- Mineral cycling also changes over the succession, increasing with time.

- The pollen record in peat bogs gives a record of succession of plant species.

- Each **seral community** changes the environmental conditions and this allows the next seral stage to replace it through competition.

Primary succession

Bare land occurs due to volcanic activity, moraines left by glacier retreat or when wind or water leave silt or sand on a surface. However, this land does not stay bare for long.

Plants very quickly start to colonize the bare land and, over time, an entire plant community develops. The change is directional as one community is replaced by another. This process is **primary succession**.

Key terms

A **pioneer community** is the first community that grows on bare ground.

A **climax community** is one that has reached a stable stage of a limited number of species.

Seral communities are stages in succession.

Primary succession is the colonization of bare ground or rock with no existing living things.

Primary succession:

- occurs on a bare inorganic surface

- involves the colonization of newly created land by organisms

- occurs as new land is either created or uncovered (e.g. river deltas, after volcanic eruptions, on sand dunes).

Table 2 and Figure 4 show the stages of primary succession.

Bare, inorganic surface ↓	A lifeless abiotic environment becomes available for colonization by pioneer plant and animal species. Soil is little more than mineral particles, nutrient-poor and with an erratic water supply.
Stage 1: Colonization ↓	First species to colonize an area—called **pioneers**—are adapted to extreme conditions. Pioneers are typically *r-selected species* showing small size, short life cycles, rapid growth and production of many offspring or seeds. Simple soil starts from windblown dust and mineral particles.
Stage 2: Establishment ↓	Species diversity increases. Invertebrate species begin to visit and live in the soil, increasing humus (organic material) content and water-holding capacity. Weathering enriches soil with nutrients.
Stage 3: Competition ↓	Microclimate continues to change as new species colonize. Larger plants increase cover and provide shelter, enabling **K-selected species** to become established. Temperatures, sunlight and wind are less extreme. Earlier pioneer *r*-species are unable to compete with *K*-species for space, nutrients or light, and are lost from the community.
Stage 4: Stabilization ↓	Fewer new species colonize as late colonizers become established, shading out early colonizers. Complex food webs develop. *K*-selected species are specialists with narrower niches. They are generally larger and less productive (slower growing) with longer life cycles and delayed reproduction.
Climax community	The final stage or **climax community** is stable and self-perpetuating. It exists in a steady-state dynamic equilibrium. The climax represents the maximum possible development that a community can reach under the prevailing environmental conditions of temperature, light and rainfall.

▲ Table 2 Stages of primary succession. Stages 2, 3 and 4 are intermediate

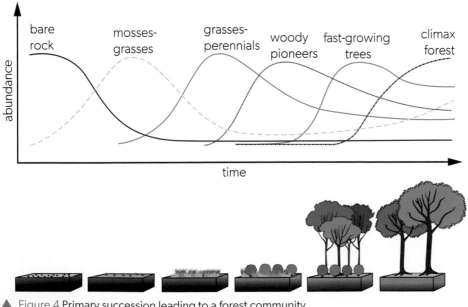

▲ Figure 4 Primary succession leading to a forest community

Adapted from: The Open University

Surtsey Island—an example of primary succession

Surtsey is an island about 1 km² in area off the southern coast of Iceland. It was formed by a volcanic eruption starting in 1963 and continuing until 1967. Its size is reducing due to wave erosion. It was declared a nature reserve and is studied as an example of primary succession on bare rock.

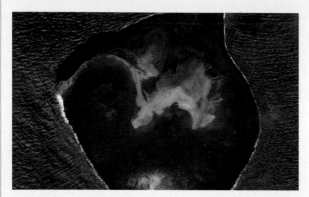

▲ Figure 5 Surtsey Island

There are now around 89 bird species and about 335 species of invertebrate. The flora of Surtsey now includes moss, lichens and 60 species of vascular plant.

Order of observation of living things:

- insects in 1964
- a vascular plant
- mosses and lichens
- seals breeding in 1987
- 20 species of vascular plant by 1987
- birds nesting—fulmar and guillemot first
- earthworms in 1993
- slugs, spiders, beetles in 1998
- a willow shrub in 1999
- 69 species by 2008 (compared with 460 species on mainland Iceland)
- a golden plover nesting in 2009
- about 2–5 new species each year.

Key term

Secondary succession is a succession started by an event such as forest fire or flood when seeds that are dormant may be in the soil.

Secondary succession

Where an already established community is suddenly destroyed—for example, by fire or flood or by human activity such as ploughing or deforestation—a shortened version of succession occurs.

This **secondary succession** occurs on soils that are already developed and ready to accept seeds carried in by the wind. There may also be dormant seeds left in the soil from the previous community. This shortens the number of stages the community goes through (Figure 6).

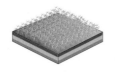 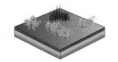 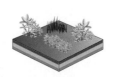 

time

agricultural land kept in an artificial seral stage; crops such as wheat act as a grassland

after being abandoned wild grasses from wind-blown and dormant seeds in the ground take over

with time, small shrubs start to colonize the grassland

eventually trees establish leading to the development of a climax community on mature soils

▲ Figure 6 Stages of secondary succession in abandoned agricultural land

Examples of secondary succession have been studied in abandoned farmland in North Carolina in the USA. The farmland had become infertile because farmers did not return enough nutrients to the soil after crops had been taken, and through wind erosion. As the land became unproductive and uneconomical to farm, farmers simply abandoned the land. This process created a patchwork of former farmland that was of various ages following abandonment.

(ATL) Activity 23

ATL skills used: thinking, communication, social, research, self-management

Work as a class.

1. Find a local example of an area that has been cleared of vegetation. Clearance could be by fire, flooding or human activity removing vegetation—leaving either bare soil (primary succession) or an area that has been harvested, cut or thinned (secondary succession). You could clear an area yourself, with permission.

2. Map the area using an online mapping tool of your choice. Consider scale, a key to mark types of vegetation, slope, aspect and other factors.

3. Try to find satellite data (using Google Earth or an alternative mapping site), or historical data to investigate how the area has changed over time. Record the changes.

4. Return every few weeks and record what is growing. Use a suitable sampling technique. What changes do you observe? Record these changes on your mapping tool to show how succession occurs.

Species diversity in successions

In the early stages of succession, there are only a few species within the community. As the community passes through subsequent stages, the number of species increases. Very few pioneer species are ever totally replaced as succession continues. The result is increasing diversity (i.e. more species). This increase tends to continue until a balance is reached between possibilities for new species to establish and existing species to expand their range, and local extinction.

Evidence following the eruption of the Mt St Helens volcano in 1980 has provided ecologists with a natural laboratory to study succession. In the first 10 years after the eruption, species diversity increased dramatically but after 20 years very little additional increase in diversity occurred (Figure 7).[1]

Disturbance

Early ideas about succession suggested that the climax community of any area was almost self-perpetuating. This is unrealistic as all communities are affected by periods of disturbance to a greater or lesser extent. Even in large forests, trees eventually age, die and fall over, leaving gaps. Other communities are affected by flood, fire, landslides, earthquakes, hurricanes and other natural hazards. All of these events create gaps that can be colonized by pioneer species within the surrounding community. This adds to both the productivity and diversity of the community (Figure 8).

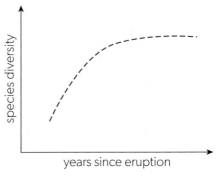

▲ Figure 7 Generalilzed graph of species diversity over time, following Mt St Helens eruption

[1] Carey, Susan, John Harte & Roger del Moral. 2006. Effect of community assembly and primary succession on the species-area relationship in disturbed systems. *Ecography* 29:866-872

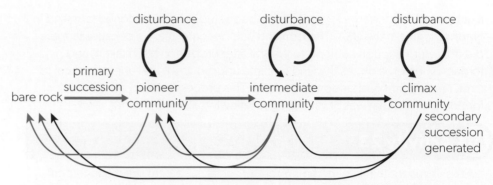

disturbance can send any seral stage back to an earlier seral stage or create gaps in a later community that then regenerate, increasing both productivity and diversity of the whole community

▲ Figure 8 **Effects of disturbance in a succession**

Changes in a succession

During a succession the following changes occur:

* The size of organisms increases with trees creating a more hospitable environment.

* Energy flow becomes more complex as simple food chains become complex food webs.

* Soil depth, humus, water-holding capacity, mineral content and cycling all increase.

* Biodiversity increases as more niches (lifestyle opportunities) appear and then falls as the climax community is reached (Figure 9).

* NPP and GPP rise and then fall.

* Productivity:respiration ratio falls.

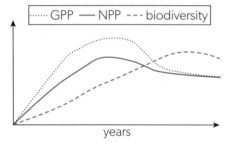

▲ Figure 9 **Changes in GPP, NPP and biodiversity in a succession**

Factor	Pioneer	Climax
NPP	high	low
GPP	low	high
Complexity	low	high
Biodiversity	low	high
Soil depth	shallow	deep
Organism strategy	*r*-strategist	*K*-strategist
Energy flow	simple	more complex

▲ Table 3 **Table of changes during succession**

Primary productivity varies with time (Figure 10). When plants first colonize bare ground, it is low as there are not many plants and they are starting from a seed. It rises quickly as more plants germinate and the biomass accumulates. When a climax community is reached (stable community of plant and animal species), productivity levels off. This is because the rate at which energy is fixed by the producers is approximately equal to the rate at which energy is used in respiration and emitted as heat.

In early stages, **gross primary productivity** is low due to the initial conditions and low density of producers. The proportion of energy lost through community respiration is relatively low too, so **net productivity** is high (i.e. the system is growing and biomass is accumulating).

In later stages, with an increased producer, consumer and decomposer community, gross productivity continues to rise to a maximum in the climax community. However, this is balanced by equally high rates of respiration, particularly by decomposers. So net productivity approaches 0 and the Productivity:respiration (P:R) ratio approaches 1.

During succession, GPP tends to increase through the pioneer and early stages and then decreases as the climax community reaches maturity. This increase in productivity is linked to growth and biomass.

Early stages are usually marked by rapid growth and biomass accumulation—grasses, herbs and small shrubs. GPP is low but NPP tends to be a large proportion of GPP as, with little biomass in the early stages, respiration is low. As the community develops towards woodland and biomass increases, so does productivity. But NPP as a percentage of GPP can fall as respiration rates increase with more biomass.

Studies have shown that standing crop (biomass) in succession to deciduous woodland reaches a peak within the first few centuries. Following the establishment of mature climax forest, biomass tends to fall as trees age, growth slows and an extended canopy crowds out ground cover. Also, older trees become less photosynthetically efficient and more NPP is allocated to non-photosynthetic structural biomass such as root systems (Table 4).

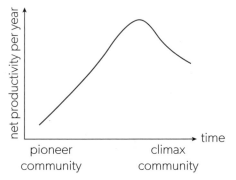

▲ Figure 10 **Productivity changes in a succession**

Early stage	Middle stage	Late stage
low GPP but high percentage NPP	GPP high	trees reach their maximum size
little increase in biomass	increased photosynthesis	ratio of NPP to R is roughly equal
	increases in biomass as plant forms become bigger	

▲ Table 4 Biomass accumulation and successional stage in deciduous woodland

r- and *K*-strategists

Reread the section on *r*- and *K*-strategies (in subtopic 2.1).

In pioneer communities, *r*-strategist species dominate. They produce large numbers of offspring so they can colonize new habitats quickly and make use of short-lived resources. Later in succession, *K*-strategist species produce a small number of offspring, which increases their survival rate and enables them to survive in long-term climax communities. For example, in a temperate woodland, birch trees, brambles and alder are pioneers, and oak or beech trees are long-lived climax species.

Influences on development of a community

Exactly what community develops in a succession depends on many variables:

- climatic factors

- properties of the local bedrock and soil—parent rock may lead to very acidic or alkaline soils

- geomorphology—steep slopes restrict soil development; lack of drainage leads to waterlogging

- fire and weather-related events

- presence of primary consumers or top carnivores.

TOK

What kinds of explanations do natural scientists offer?

Case study 7

Grey Wolves in Yellowstone

Yellowstone National Park in western USA was created in 1872 but wildlife was not protected. By 1926, the wolves were locally extinct. In 1995, they were reintroduced. Wolves are a **keystone species and apex predator** (see subtopic 2.1).

Once the wolves were gone, their prey species, elk, increased greatly. There were so many elk that they overgrazed the trees. The park service started trapping and moving the elk and eventually killing them. In the late 1960s, elk-killing stopped and populations rose again.

With no wolves, coyote populations also increased and they prey on pronghorn antelope.

▲ Figure 11 Grey wolf

Much debate raged around reintroducing the grey wolf. Finally in 1995, 21 wolves were captured in Alberta, Canada and taken to Yellowstone. In 2020 there were 123 wolves in the park. As expected, since 1995 elk numbers have declined.

Wolves kill 22 elk per wolf per year. Elk have changed their behaviour and moved to less hospitable areas of the park, with less food. They have produced fewer offspring.

Wolves prey on coyotes too and outcompete them for other foods. Coyote numbers have declined and they have moved to steeper slopes where wolves cannot run as fast. Both wolves and coyotes eat each other's pups.

Coyotes prey on foxes. With fewer coyotes, fox numbers have increased. Foxes prey on deer and hares, rodents and ground-nesting birds. With more foxes, there are fewer hares and young deer. Decreased populations of fox prey animals have led to changes in numbers of their food plants and insects.

Increased wolf populations have led to increased beaver populations, as elk have moved away from eating the willows which beavers rely on for winter food. More beavers build more dams, reducing run-off and erosion and creating ponds for fish habitat. Wolf carcasses provide food for scavengers such as ravens, wolverines, bald eagles and grizzly bears.

Wolves also claim kills made by cougars so the cougars have moved back up the mountains.

This is a **trophic cascade**—a series of powerful indirect interactions that can control entire ecosystems. Trophic cascades occur when predators limit the density and/or behaviour of their prey and thereby enhance survival of the next lower trophic level.

Yellowstone offered a unique opportunity to observe the effect of reintroducing a key species. It is often the loss of a key species that changes an ecosystem.

→

Questions

1. Draw two food webs for Yellowstone:

   a. with wolves

   b. without wolves.

2. Research the role of sea otters in kelp forest (see subtopic 2.1 and many online sources) and make short notes.

3. a. Research the role of elephants in the savanna of Africa, referring to the Global Conservation website and other sites.

   b. Research numbers of elephants on the savanna now and the change over time. There is some relevant data in Table 5.

   c. Draw a graph of your results. Think carefully about what type of scale will be appropriate.

   d. State if these elephants are on the IUCN Red List and explain why.

Year	1800	1976	1987	1989	2003	2021
Estimated population	26 million*	1.34 million*	760k*	608k*	200k–430k	~350k

▲ Table 5 Global population size of the Savanna elephant over time
*These historical numbers are forest and savanna elephants combined, as they were not recognized as separate species until 2021

Climax community debate

Features of climax communities:

- high tolerance of disturbances

- moderate conditions

- high species diversity, complex food webs

- large size of organisms

- specialist niches

- NPP low, biomass and organic matter high

- low cycling of minerals and nutrient exchange.

All climax communities have these features but, in practice, a steady-state climax community is very rare. Some ecologists think it cannot exist and that there are many different possibilities for any climax community depending on the path of succession and factors influencing it.

Hypotheses

- Climate is the control—climate determines the climax community: same climate, same climax.

- Several environmental factors may dominate (e.g. slope, soil type, human interference) leading to different stable communities. Random events determine the climax community.

- All environmental factors are equally important—environmental gradients determine the community and a series of climax communities develop in parallel to each other.

Vera wood–pasture hypothesis

Historically, Europe after the last ice age was thought to be all primary forest. But the Dutch scientist Frans Vera hypothesized that the predominant ecosystem was actually semi-open wood and pasture kept open by grazing of tree seedlings by large herbivores.

This was a highly controversial hypothesis as it had been assumed that humans gradually cleared prevalent forests to grow crops and livestock.

Rewilding

Connections: This links to material covered in subtopic 3.3.

Rewilding is restoring ecosystems by allowing the natural world to restore degraded landscapes. This is not merely leaving things alone. It is providing the conditions for natural forest regeneration and reintroducing species that have been lost.

In many areas, rewilding involves large herbivores ranging fairly freely in large stretches of land, grazing among forest trees as well as in open pastures. In Europe in particular, large keystone species and top predators have been lost and rewilding involves reintroducing some of these. This is controversial as beavers, bison and wolves are being reintroduced in various sites.

Case study 8

Rewilding example

Knepp, southern England is a 1,000-hectare estate with farms that were not very profitable. Its owners decided to regenerate the land by taking out all but perimeter fences and having free-roaming pigs, deer, cattle and ponies. Since this started over 20 years ago, rare species have arrived and are breeding. Examples include nightingales, white stork, turtle doves, peregrine falcons and purple emperor butterflies. The estate faced local opposition at first because it appeared the land was overrun by weed species. However, the ecosystem established a balance of broadleaved woodland, wood pasture, rivers and streams, grassland and meadow. It still produces meat for sale and is now an educational and tourist site.

▲ Figure 12 Longhorn cattle grazing in wood pasture

Question

Visit the website of Rewilding Europe and read about some of their projects.

Choose one of these projects, or a rewilding project local to you, and write a short paragraph about its aims, challenges and successes.

Plagioclimax

Plagioclimax is an altered or reduced climax of a plant ecosystem caused by human activity. The plagioclimax may be caused by a variety of disturbances, such as:

- burning—forest clearance

- felling existing vegetation

- planting crops or trees

- grazing and trampling by livestock

- harvesting crops.

Most farmland is a plagioclimax. So are grouse moors, paths through woods and fields, coppiced woodland and anywhere humans stop a climax community forming. A particular example is sand dunes where humans trample across the dunes.

Case study 9

Sand dunes at Studland Bay

Studland Bay is on the southern coast of England in Dorset. Sand dunes have continued to be formed there since the 16th century.

Succession begins with a bare surface of sand. Vegetation colonizes the sand. The pioneer plants tend to be low growing—why? They have fat fleshy leaves with a waxy coating and can survive being temporarily submerged.

Later, the predominant plant species is marram grass on the seaward side due to its ability to cope with the environmental conditions. Like the other grasses, it has

▲ Figure 13 Sand dunes at Studland Bay, Dorset, UK

leaves which are able to fold to reduce their surface area. The leaves are waxy to reduce transpiration and can be aligned to the wind direction. It incorporates silica into its cell structure to give the leaves extra strength and flexibility.

As a result of the humus from the previous stages, a sandy soil develops. This is able to support "pasture" grasses and bushes. Species such as hawthorn, elder, brambles and sea buckthorn (which has nitrogen-fixing root nodules so can thrive in nutrient-poor soil) are present. As the scrub develops, shorter species are shaded out.

The oldest dunes have forest. First pines and finally oak and ash woodland grow. This is the climax vegetation for the area. Henceforward, species diversity declines due to competition.

In every case, vegetation colonizes in a series of stages. The final stage is in dynamic equilibrium with its climatic environment and is known as climax vegetation for that climate. In the British Isles, this is temperate deciduous forest. As succession develops, there are increases in vegetation cover, soil depth and humus content, soil acidity, moisture content and sand stability (Figure 14).

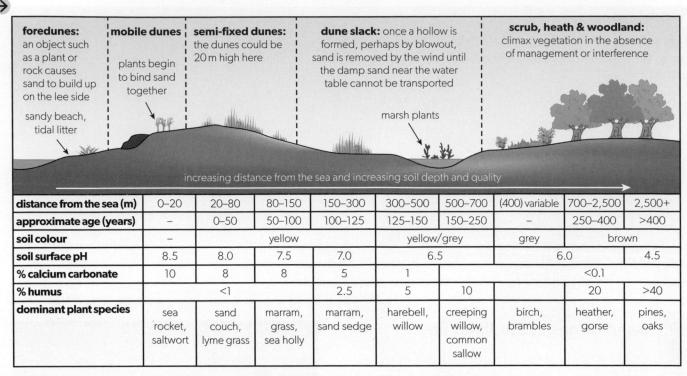

foredunes: an object such as a plant or rock causes sand to build up on the lee side

sandy beach, tidal litter

mobile dunes

plants begin to bind sand together

semi-fixed dunes: the dunes could be 20 m high here

dune slack: once a hollow is formed, perhaps by blowout, sand is removed by the wind until the damp sand near the water table cannot be transported

marsh plants

scrub, heath & woodland: climax vegetation in the absence of management or interference

increasing distance from the sea and increasing soil depth and quality

distance from the sea (m)	0–20	20–80	80–150	150–300	300–500	500–700	(400) variable	700–2,500	2,500+
approximate age (years)	–	0–50	50–100	100–125	125–150	150–250	–	250–400	>400
soil colour	–	yellow			yellow/grey		grey	brown	
soil surface pH	8.5	8.0	7.5	7.0	6.5		6.0		4.5
% calcium carbonate	10	8	8	5	1	<0.1			
% humus	<1			2.5	5	10		20	>40
dominant plant species	sea rocket, saltwort	sand couch, lyme grass	marram, grass, sea holly	marram, sand sedge	harebell, willow	creeping willow, common sallow	birch, brambles	heather, gorse	pines, oaks

▲ Figure 14 Diagram of natural sand dune succession

Questions

1. State where the calcium carbonate is from and why it decreases.

2. In the climax vegetation of oak and pine, there are fewer species due to competition. Explain what these species are competing for.

3. Explain why the percentage of humus increases.

4. Explain what happens to the succession if these sand dunes are in a popular tourist area and many people walk across the dunes to get to the sea.

(ATL) Activity 24

ATL skills used: thinking, communication, research

1. Find a local example of a plagioclimax, such as:

   • farmland with either crops or livestock

   • an ecosystem where the top carnivore or keystone species has disappeared

   • a managed woodland

   • a garden or park

   • a footpath through natural vegetation.

   a. Describe the community there.

   b. Describe the community that would develop if the human activity stopped.

Check your understanding

In this subtopic, you have covered:

- zonation and succession

- primary and secondary succession

- changes during succession

- net and gross productivity

- how succession is disrupted

- how succession is influenced

- primary productivity

- different reproductive strategies

- concept of a climax community

- plagioclimaxes.

AHL

How do ecological systems change over time and over space?

1. Describe the difference between succession and zonation.

2. Explain why zonation occurs.

3. Explain why succession occurs.

4. Describe what influences the type of climax community.

5. Describe what biotic factors change during succession and why they change.

6. Explain how pioneer and climax community species differ.

7. To what extent do humans affect succession?

AHL

>> Taking it further

- Research what was on your school site before the school was built (or what was there before a new block was built).

- Map how the school grounds have changed over time and how much green space remains.

- Produce an infographic or poster for the school to inform others about the fieldwork in which you have participated.

- Encourage your school to improve green spaces by tree planting, leaving grass to grow or creating wetland areas.

TOK

Feeding relationships can be represented by different models. How can we decide when one model is better than another?

What role does indigenous knowledge play in passing on scientific knowledge?

When is quantitative data superior to qualitative data in giving us knowledge about the world?

Controlled laboratory experiments are often seen as the hallmark of the scientific method. To what extent is the knowledge obtained by observational natural experiment less scientific than that from a manipulated laboratory experiment?

Why do we use internationally standardized methods of ecological study when making comparisons across international boundaries?

2.6 Practical work: Ecology lab and fieldwork

Why practical work?

The ESS guide recommends time to be spent in various areas of the course.

Activity	Time in hours
Practical activities (PA)	30
Collaborative sciences project (CSP)	10
Individual investigation (II)	10

▲ Table 1

This is the same for SL and for HL.

The practical work is an important aspect of the ESS course. It may be lab-based, classroom-based, or out in the field. Many parts of ESS can only be covered effectively by doing things, rather than reading about them. Such practical work is an opportunity for you to gain and develop skills and techniques.

The guide has sections called "Application of skills"; these are practical activities designed to cover the skills and techniques you need to know. They help you to understand the topics covered and should be fully integrated in the course.

Elsewhere in the book, these skills and activities are sometimes within topics. Here in subtopic 2.6, however, most of the ecology activities are together in one place. We suggest that you carry them out when the relevant content is covered in class.

There are also other activities here which will add to your understanding of ESS and be of use in your individual investigation (II).

Ecology skills list

To carry out your individual investigation (II), you will need ecology skills.

Table 2 shows the skills you need in laboratory work and field work. They are covered in several subtopics and the table shows where in the course they arise.

	Skill	Subtopics
Laboratory work	For carrying out laboratory experiments, students should be able to: • make appropriate quantitative measurements (e.g. counts, time, mass, volume, temperature, length, pH and concentration) • select and justify appropriate techniques, sampling strategies, apparatus and materials • carry out procedures for estimating biomass (dry weight) of plant matter only • carry out procedures for measuring gross and net, primary and secondary productivity and biological oxygen demand **(HL only)** • set up and utilize appropriate laboratory equipment and materials with safety and accuracy.	2.1 Use models that demonstrate feeding relationships, such as predator–prey. 2.2 Work out the efficiency of transfer between trophic levels. 2.2 Follow experimental procedures on how to find biomass and energy from biological samples (plant material only). 2.2 Use laboratory and field techniques for measuring primary and secondary productivity and work out GP and NP from data. 2.2 Consider protocols for determining primary productivity in ecosystems. Estimates can be based on photosynthesizing samples within a laboratory or, in the field, measuring change in biomass of samples (such as grassland) over time. 2.2 Create pyramids of numbers, biomass and energy from given data. 5.1 Sample two soils from the subsoil (B horizon): one from a local garden or field, and one from a natural ecosystem. Investigate texture, organic matter content, nitrogen, phosphorus, and potassium (NPK) concentrations, aeration, drainage and water retention. 5.1 Determine the amount of carbon in a dry soil sample by burning off the organic matter and calculating the change in mass.
Field work	For carrying out fieldwork, students should understand how to: • measure a range of abiotic factors (climatic, edaphic and aquatic) • identify flora and fauna using dichotomous keys, online databases and apps, and correctly use binomial nomenclature • use appropriate quadrat sampling for the estimation of abundance, population density, percentage cover, and percentage frequency of non-motile organisms • use capture–mark–release–recapture with the Lincoln index to estimate population size of motile organisms	2.1 Investigate a local ecosystem. 2.1 Use methods for measuring at least three abiotic factors in an aquatic or terrestrial ecosystem, including the use of data logging. 2.1 Know how to use dichotomous keys, applications and databases for the identification of species. 2.1 Use quadrat sampling to estimate abundance, population density, percentage cover and percentage frequency for non-mobile organisms and measure change along a transect. 2.1 Use the Lincoln index to estimate population size. Understand the assumptions made when using this method. 2.5 Investigate zonation along an environmental gradient using a transect sampling technique and a range of relevant abiotic measurements. Create kite diagrams to show distribution. 2.5 Use secondary data and a mapping database to recreate or map the changes through succession in a given area.

	3.1 Collect data in order to work out Simpson's reciprocal index for diversity.
• use transects to measure changes along an abiotic gradient • carry out sampling to collect data for calculating species diversity • carry out sampling to collect data for calculating biotic index (**HL only**).	3.2 Investigate the impact of human activity on biodiversity in an ecosystem by studying changes in species diversity along a transect laid perpendicular to a site of human interference or by randomly sampling within transects before and after the human activity. 4.4 Apply protocols for assessing biological oxygen demand and a named biotic index. 4.4 Use methods for measuring key abiotic factors in aquatic systems—dissolved systems, for example, dissolved oxygen, pH, temperature, turbidity, and concentrations of nitrates, phosphates and total suspended solids. Possible methods may include the use of oxygen and pH probes, a thermometer, a Secchi disc, and nitrate/phosphate tests.

▲ Table 2

This subtopic also includes practical work on the following skills, in addition to those in Table 2:

6.1 Investigate the impact of albedo or different greenhouse gases on the temperature of a closed system.

8.3 Plan an experiment to use an indicator species as a correlate for pollution in the local environment.

Guidelines

Take into account the following considerations when you do this work.

Use named examples to illustrate concepts or your arguments.

- Always give the full name of the animal or plant (e.g. not "fish" but "Atlantic salmon", not "tree" but "common oak tree").

- When you use a habitat or local ecosystem as an example, give as much detail as possible about it (e.g. not "beach" but "Tai Tam mangroves SSSI, in Tai Tam Harbour, Hong Kong Island, south-facing").

- All ecosystem investigations should follow the guidelines in the IB sciences experimentation policy and ESS guide. They may be more stringent than your local or national standards so check these guidelines carefully before designing an experiment.

The IB states the following:

- No experiments involving other people will be undertaken without their written consent and their understanding of the nature of the experiment.

- No experiment will be undertaken that inflicts pain on, or causes distress to, humans or live animals.

Topic 1 Concepts

1. Investigating systems: Set up an aquatic or terrestrial ecosystem in a bottle

There are links to this practical in subtopic 1.2 and throughout the course.

This could be used to practise the following IA skills:

- Research question (A)

- Method (C)

(Remember: no animals should be harmed)

A terrarium is a mini indoor garden that models a closed system.

To start this investigation, create a systems diagram to show the inputs and outputs of the terrarium.

The inputs will allow you to make a list of equipment you will need.

Materials

1. Plants:

    - Choose small plants that will not outgrow the terrarium and that grow well together—you may need to do some research. The plants selected will depend largely on where you live—mosses and ferns are good.

    - Terrariums tend to have low light levels because the plants are close together so make sure you choose plants that can tolerate such conditions.

    - Closed terrariums have high humidity because once the garden is set up it is sealed, so make sure the plants are tolerant of high humidity.

    - A terrarium community is hard to balance so choose plants that are easy to grow and inexpensive.

2. Container: there is a wide range of containers that you can use. The most usual is a glass jar with a tight-fitting lid. Make sure it is big enough for the plants and that there is plenty of room for root development. Or use a discarded plastic bottle or any suitable container. Be inventive.

3. Location: decide where you will put the terrarium—once it is up and running there should be no maintenance (if it works).

    - Make sure there is plenty of indirect light—direct light is too harsh and may cause extreme variations in temperature.

    - Keep the terrarium in a warm indoor environment where there are no significant variations in temperature.

4. Soil: use light potting compost that drains easily. Do some research to find out which soil is best for your plants and a terrarium.

▲ Figure 1 Recycling an old bottle for a terrarium

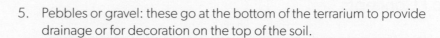

5. Pebbles or gravel: these go at the bottom of the terrarium to provide drainage or for decoration on the top of the soil.

6. Activated charcoal: this keeps the soil fresh.

7. Sheet moss tends to soak up excess water so it is useful to have in the bottom of the terrarium.

Method

1. Clean the container.

2. Line the bottom with layers of sheet moss, pebbles or gravel, and soil and activated charcoal.

3. Add the plants.

4. Leave the container open for a while to settle and establish the right amount of water.

5. Seal the container.

You have now made a terrarium (a terrestrial ecosystem). Observe what happens over time (once a week perhaps, at the start of a lesson) and record the visible changes.

Do your own research and consider setting up an aquatic ecosystem as a whole class (an aquarium). Remember: no animals must be harmed in any experiment so you will need to think carefully about how to do this.

2. Investigating systems: Investigation of feedback

Research models that demonstrate feeding relationships, for example, predator–prey.

1. TEDEd has a short feedback video which may help you to review the concept: search for "Feedback loops: How nature gets its rhythms" by Anje-Margriet Neutel.

2. Look for predator–prey feedback simulations on the web.

   Run your chosen simulation with different combinations of organisms and investigate questions such as:

   a. Which food web set-ups lead to a stable ecosystem?

   b. What happens if you remove one trophic level or one species?

3. If you set up a terrarium, can you work out the feedback loops there?

Topic 2 Ecology

In section 2.1.5 of the ESS guide, the application of skills statement is "to investigate a local ecosystem".

Carefully selecting an ecosystem that you can return to easily has many advantages:

- You get to know one ecosystem in some depth and can observe how it changes.

- You can cover many of the skills required by carrying out field work in this ecosystem.

- If it is not far from your classroom, you can visit often for short practical activities.

- It could be useful to visualize your ecosystem as an example when answering exam questions.

When studying ecosystems, there are simple questions to ask (Table 3).

Question	Example of technique to use
What is there?	sampling techniques, quadrats, basic taxonomy, classification and use of identification keys
Where are they?	quadrats, transects
How many are there?	quadrats, transects, Lincoln index, species diversity index
Why are they where they are?	measure abiotic factors

▲ Table 3

To create a food chain for a local ecosystem, you need to find out what is there by sampling, identify the sampled organisms, then draw the food chain.

If you then want to draw a food web, you need to investigate what the organisms eat and in which trophic level(s) they are found.

To work out pyramids of numbers, biomass or energy, you need to investigate and calculate productivity at each trophic level.

Biodiversity is hard to measure but a simple method is to find out the number of species and abundance of each species in an ecosystem. This can then be compared with other ecosystems.

Working out why organisms are where they are involves measuring abiotic and biotic factors in the ecosystem. These include ecological niches, competition, stages of succession, human activity and many other factors.

The techniques here show some of the ways in which ecosystems are studied.

Capturing small motile (mobile) animals

This could be used to practise the following IA skills:

- Method (C)

- Treatment of data (D)

You should select an area very close to where you live or go to school, because you will need to catch small motile animals, identify them and put them back.

It is difficult to count small animals because they move around. The first step is to catch the organisms. You will also need a key to help you to identify the organisms you are likely to catch.

Warning:

- Under no circumstances should any animal be stressed or killed during any investigation—there are humane ways to catch and count small animals.

- Make sure there are no venomous organisms in your local area.

- **Do not** handle any insects directly—move them with tweezers or a pooter.

Terrestrial ecosystems

There is a range of safe harmless techniques that can be used to catch insects.

1. Pitfall traps

A **pitfall trap** is ideal for catching insects and other small crawling animals that cannot fly away (see Figure 2). Insects can be attracted by decaying meat or sweet sugar solution (this must be covered so the insects do not fall in it and drown) and will fall into the trap.

Several of these traps can be placed around the study area. They should be checked at regular intervals (every 6 hours) and the species and number of each species recorded.

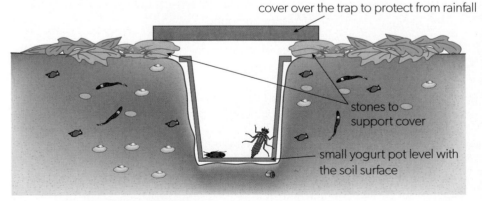

cover over the trap to protect from rainfall

stones to support cover

small yogurt pot level with the soil surface

▲ Figure 2 Diagram of a pitfall trap

2. Sweep net

Use a sweep net to capture many insects from vegetation by sweeping it slowly though the vegetation (Figure 3).

3. Tree beating

This method can find insects in tree branches or resting on leaves. Simply place a catching tray beneath a tree branch and gently tap the branch. The tray will catch anything that falls from the tree and you can log the species and their numbers.

4. Sheet trap

Hang a white sheet behind a light. Night-flying moths attracted to the light will settle on the sheet for you to observe.

5. Pooters

You can catch small insects and invertebrates using a pooter—a small jar with two tubes attached (Figure 4). You suck gently on one tube and the animal is pulled into the jar through the other. You cannot swallow the animal as there is gauze at the end of the mouthpiece tube!

Aquatic ecosystems

The organisms of most interest will be stream invertebrates and the most efficient way to catch them is through kick samples. Turning over stones is also effective.

Kick sampling is simple technique to loosen invertebrates, which then drift into a net (Figure 5).

1. Place a sweep net downstream from you.

2. Shuffle your feet into the streambed for 30 seconds.

3. Empty the contents of the net into a tray filled with stream water.

4. Use a pipette to sort the various insects into small plastic cups and record your results.

5. Repeat three times to ensure good results.

In aquatic systems, you can use nets of various sizes, and with various mesh sizes, to catch plankton, small invertebrates or larger fish. The nets can be towed behind boats or held in running water. Simple plastic sieves are also effective.

Warning:

- Some of these methods can be destructive so use them carefully.

- Always return all organisms to their habitats.

Identify organisms

Design your own dichotomous key for six or more organisms in your local ecosystem

Collect, draw or photograph six or more organisms from your local environment. Then create a dichotomous key to allow someone else to identify them.

▲ Figure 3 Using a sweep net

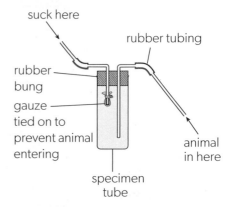

▲ Figure 4 A pooter

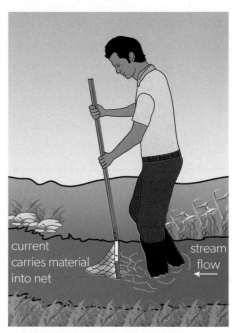

▲ Figure 5 Kick sampling for aquatic invertebrates

▲ Figure 6 Six organisms

An easy way to create keys is using a mind map.

Figure 7 shows a dichotomous key to identify the organisms in Figure 6.

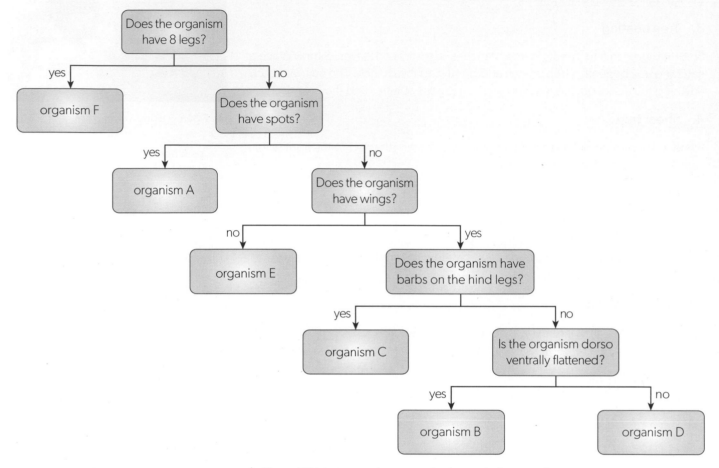

▲ Figure 7 Dichotomous key created using a mind map tool

Identify organisms A–F in Figure 6. Can you name them?

Identification databases

There are many online and paper field guides for all groups of organisms—insects, birds, mammals, beetles, flowering plants, fungi and many, many others.

Search for one relating to the ecosystem you are studying and attempt to identify the organisms you have found.

Collate all information:

- Identify producers: Sample the vegetation in the study area—this does not need to be detailed, you merely need to establish a list of all the primary producers in the area.

- While sampling the vegetation look for evidence of secondary consumers—such as faeces, burrows, footprints, scratch marks, nests, eggs.

- Identify consumers: Set pitfall traps, use a sweep net, do tree beating, observe and briefly capture the organisms.

- Identify the organisms (e.g. by using keys) and research the identified organisms' feeding habits.

- Draw simple food chains. If you have enough information on feeding habits, attempt to draw a food web.

Measuring biotic components of a system

1. Plant biomass

Measuring plant biomass is simple but destructive. It is usually best to measure above-ground biomass because it can be very difficult to collect the parts below the ground.

For low vegetation or grasses:

1. Place a suitably sized quadrat (see Figure 9).

2. Harvest all the above-ground vegetation in that area.

3. Wash it to remove any insects.

4. Dry it at about 60–70°C until it reaches a constant weight. Water content can vary enormously so all water should be removed. The mass given is dry weight.

5. For accurate results, repeat three to five times to obtain a mean per unit area.

6. Extrapolate the result to give the total biomass of that species in the ecosystem.

For trees and bushes:

1. Select the tree or bush you wish to test.

2. Harvest the leaves from three to five branches.

3. Repeat steps 3–6 in the above method.

2. Primary productivity

In **aquatic ecosystems** (both marine and freshwater) the **light and dark bottle technique** can be used to measure the gross and net productivity of aquatic plants (including phytoplankton). This is simple but has given us a good idea of the productivity of the oceans and of many lakes.

The productivity is usually calculated from the oxygen concentrations in the bottles. The procedure is as follows.

1. Take two bottles filled with water from the ecosystem—one made of clear glass; the other made of dark glass or covered to exclude light.

2. Measure the oxygen concentration of the water by chemical titration (Winkler method) or an oxygen probe, and record as mg oxygen per litre of water.

3. Place equal amounts of plants of the same species into each bottle.

4. Completely fill both bottles with water and seal with a cap. (No air should be present.)

5. Allow to stand and incubate for several hours. This incubation can take place in the laboratory or outdoors in the ecosystem under investigation.

6. Measure the oxygen level in each bottle and compare with the original oxygen level of the water.

In the light bottle, photosynthesis and respiration have been occurring. In the dark, only respiration occurs.

In terrestrial ecosystems, you can do a similar experiment with square "patches".

1. Select three equally sized patches with similar vegetation (e.g. grass).

2. Harvest the first patch (A) immediately and measure the biomass (see above).

3. Cover the second patch (B) with black plastic (so there is no photosynthesis, just respiration).

4. Leave the third patch (C) as it is.

5. After a suitable time period (depending on the season), harvest patches B and C and measure the biomass (as above).

6. Calculate GPP, NPP and R. Units of productivity are expressed as energy used, e.g. joules m^{-2} day^{-1}.

Measuring abundance

Having established what organisms are present, you need to find out how many there are.

Investigating the Lincoln index

You can use the capture–mark–release–recapture technique and the Lincoln index to estimate a population of mobile organisms.

This is a class activity.

1. Take 500 large beans, preferably light in colour—for example, large white lima beans or chickpeas.

2. Place the beans in a suitable container with a lid.

3. Each student takes a small handful of beans, counts them and marks them in some way (n_1). This number is recorded in a group data table (see Table 4).

4. Replace all the beans in the container. Shake the container vigorously to try to remove some of the marks.

5. Each student takes a second handful of beans and records in the table the total number of beans in this sample (n_2) and the number of beans that are marked (m_2).

6. Apply the Lincoln index formula to each row of data:

$$N = \frac{n_1 \times n_2}{m_2}$$

n_1 = number of animals or beans first marked and released

n_2 = number captured in second sample

m_2 = number of marked beans in second sample

N = total population

Student	n_1	n_2	m_2	N
1				
2				
3				
4				
5				
6				
7				

▲ Table 4

Note: If this process is used with a population of living organisms (e.g. woodlice or snails), the following assumptions are made:

- Mixing is complete—the marked individuals have spread throughout the population.

- Marks do not disappear.

- Marks are not harmful and do not increase predation by making the individual more easily seen.

- It is equally easy to catch every individual.

- There is no immigration, emigration, births or deaths in the population between the times of sampling.

- Trapping the organisms does not affect their chances of being trapped a second time.

Investigating the abundance of plants

There are various ways of assessing plant species abundance. Note that percentage frequency and percentage cover give an estimate of abundance but not actual population size.

- Density: mean number of plants per m^2.

- Percentage frequency: the number of occurrences divided by the number of possible occurrences. For example, if a plant occurs in 5 out of 100 squares in a grid quadrat, then the percentage frequency is 5%.

- Percentage cover: because plants spread out and grow, percentage cover is often measured instead of individual numbers. Percentage cover is an estimate of the area in a given frame size (quadrat) covered by the plant in question; it may help to divide the quadrat into smaller sections for this. Species may overlap or lie in different storeys in a forest, so the total percentage cover of all species within a quadrat may be well over 100% or much less if there is bare ground. You can estimate percentage cover either by comparing the sample area with Figure 8 and then grading the coverage on a scale from 0 to 5, or by using the ACFOR scale (Table 5).

▲ Figure 8

Percentage cover (%)	ACFOR scale	Score
50	Abundant	5
25–50	Common	4
12–25	Frequent	3
6–12	Occasional	2
<6, or single individual	Rare	1
absent		0

▲ Table 5 ACFOR percentage cover scale

Quadrats and transects

A **quadrat** is a frame of specific size (depending on what is being studied), which may be divided into subsections.

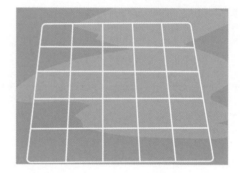

▲ Figure 9 Two examples of quadrats

Quadrat sampling can be used to estimate abundance, population density, percentage cover and percentage frequency for non-motile organisms and, in a line transect, measures change along the transect.

Question: How many quadrat samples, and of what size?

The size of the quadrat chosen will depend on the size of the organisms being sampled (Table 6).

Quadrat size	Quadrat area	Organism
10 × 10 cm	0.01 m²	Very small organisms such as lichens on tree trunks or walls, or algae.
0.5 × 0.5 m	0.25 m²	Small plants: grasses, herbs, small shrubs. Slow moving or sessile animals: mussels, limpets.
1.0 × 1.0 m	1 m²	Medium size plants: large bushes.
5.0 × 5.0 m	25 m²	Mature trees.

▲ Table 6

There is a balance to strike between increasing accuracy with increasing size and time available and the number of times a quadrat is placed.

Your choices will vary depending on the ecosystem, size of organisms and their distribution. The following simple method will help you to determine the appropriate number of samples.

- Increase the number of samples and plot the number of species found. When this number is stable, you have found all species in the area—so in Figure 10, eight samples are enough.

- Increase the size of the quadrat (e.g. from side length 10 cm, 15 cm, 20 cm and so on) and plot the number of species found. When this number reaches a constant, that is the quadrat size to use.

How to place quadrats

Quadrats can be placed randomly, continuously or systematically (according to a pattern).

1. **Random quadrat sampling** can be used to estimate population size for non-motile organisms.

   Random quadrats may be placed by throwing the quadrat over your shoulder, but we do not recommend this as it could be both dangerous and not random—you could decide where to throw.

   The conventional method is to use random number tables.

   1. Map out your study area.

   2. Draw a grid over the study area and number each square (Figure 11).

   3. Use a random number table to identify which squares you need to sample.

2. **Stratified random sampling** is used when there is an obvious difference within an area to be sampled and two or more sets of samples are taken.

   If the area for study has two distinctly different vegetation types and three areas for study—as in Figure 12—samples need to be taken in each area.

   1. Treat each area separately.

   2. Draw a grid for each area.

   3. Number the squares in each area (they can be the same or different numbers).

   4. Use a random number table to identify which squares you need to sample in each area.

3. **Continuous and systematic sampling** along a transect line.

   You might use this to look at changes in organisms as a result of changes along an environmental gradient (e.g. zonation along a slope, a rocky shore or grassland to woodland) or to measure the change in species composition with increasing distance from a source of pollution. Transect sampling is quick and relatively simple to conduct.

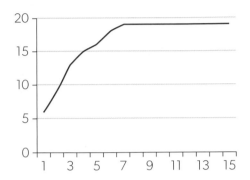

▲ Figure 10 Number of species and quadrat size

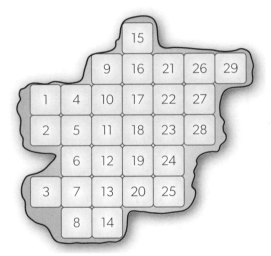

▲ Figure 11 Placement of quadrats

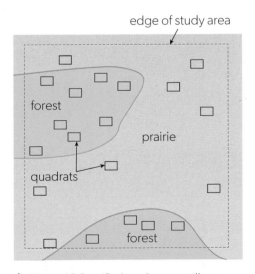

▲ Figure 12 Stratified random sampling

Key term

A **transect** is a sample path, line or strip along which you record the occurrence or distribution of plants and animals in a particular study area.

There are two main types of transect that could be useful to you.

1. A **line transect** consists of a string or measuring tape which is laid out in the direction of the environmental gradient. Species touching the string or tape are recorded.

2. A **belt transect** is a strip of chosen width through the ecosystem. It is made by laying two parallel line transects, usually 0.5 or 1 m apart, between which individuals are sampled.

Transect lines may be continuous or interrupted.

1. In a **continuous transect** (line or belt transect) the whole line or belt is sampled.

2. In an **interrupted transect** (line or belt) samples are taken at points along the line or belt. These points are usually taken at regular horizontal or vertical intervals. This is a form of systematic sampling. Quadrats are placed at intervals along the belt.

Note: You will need to sample along many line transects (at least three, preferably five) to obtain sufficient reliable data.

Measuring diversity

Key term

Species diversity is a function of the number of species and their relative abundance.

In other words, species diversity is the number of different species and the relative numbers of individuals of each species, i.e. species richness and species evenness.

Ecologists try to express diversity in numbers. The higher the number, the greater the **species diversity**. This makes it possible to compare similar ecosystems or to see whether ecosystems are changing in time. The most common way to turn diversity into a number is by using the **Simpson diversity index**.

Be careful. The name "Simpson diversity index" actually describes three related indices (Simpson's index, Simpson's index of diversity and Simpson's reciprocal index). Here we are using Simpson's reciprocal index, in which 1 is the lowest value (indicating there is only one species present) and a higher value means more diversity. The highest value is equal to the number of species in the sample. In the other indices, the value ranges from 0 to 1.

To calculate Simpson's reciprocal index, use the following formula:

$$D = \frac{N(N-1)}{\sum n(n-1)}$$

where: D = Simpson's reciprocal diversity index

N = total number of organisms of all species found

n = number of individuals of a particular species

Consider the example data in Table 7.

	Number of individuals of species		
	A	**B**	**C**
Ecosystem 1	25	24	21
Ecosystem 2	65	3	4

▲ Table 7 Diversity data for two ecosystems

The reciprocal diversity index of ecosystem 1 can be calculated like this:

$$N = 25 + 24 + 21 = 70$$

$$D = \frac{70 \times 69}{(25 \times 24) + (24 \times 23) + (21 \times 20)} = 3.07$$

In ecosystem 2, the diversity index is 1.22.

Both ecosystems have the same species richness (3) but the species in ecosystem 1 are more evenly distributed so species diversity, by this measure, is higher.

A high value for D indicates a highly diverse ecosystem—often a stable and ancient site. In contrast, low values for D are found in disturbed ecosystems such as logged forests. Pollution also results in low values for D. Agricultural land has extremely low values for D, as farmers try to prevent competition between their crops and other species (weeds). Be careful though: low values for D in Arctic tundra may represent ancient and stable sites as growth is so slow there and diversity is low.

Simpson's reciprocal diversity index is most often used for vegetation, but it can also be applied to animal populations.

Question: How diverse is the local ecosystem?

Answering this question will allow you to practise the IA skills of **Planning**, because you will need to plan a method.

▲ Figure 13 Sand dunes—a possible ecosystem for biodiversity measurements

Follow these steps to plan your investigation.

1. Select your study area and decide on a sampling strategy and quadrat size.

2. Design a data recording table to ensure you record all the relevant information. This might look like the example in Table 8—although, in a real investigation, you will need more than four quadrats.

Quadrat number	Species	Frequency of species	Total (n)
1			
2			
3			
4			

▲ Table 8

3. Go out and collect your data. If possible, collect data for a number of contrasting areas.

4. Calculate the Simpson's reciprocal diversity index for your study areas and compare the diversity.

5. To measure abundance, you can estimate percentage cover by comparing the sample area with Figure 8 and then grading it using the ACFOR scale (Table 5).

In some locations, the local ecosystem may not be suitable for investigation. If this is the case, you can calculate the diversity of cars in the school car parks or in local shopping areas.

1. Choose a suitable sampling area. You will not need a sampling strategy—just go to the car parks and count.

2. Decide how you are going to "speciate" the cars— for example, by colour or make.

3. Use a suitable table to record your data (e.g. Table 9).

Car park location or name	"Species" of car	Frequency of species	Total (n)
1			
2			
3			
4			

▲ Table 9

Investigating zonation and succession

It is good to start by walking around your local area or school grounds and observing what is there.

- Is there a playing field?

- Is there a footpath on soil rather than concrete?

- Does the ground slope?

- Is it more shady or more moist in one area than another and what difference does that make to the type and number of species living there?

There are various changes in abiotic factors and biotic factors that can occur.

- Over space: an environmental gradient—which is a trend in one or more abiotic or biotic components of an ecosystem (zonation).

- Over time: short-term diurnal cycles (day and night) or long-term changes (succession).

- Changes due to human activity: sewage effluent outfall, intensive agriculture.

Question: Investigate the changes that occur along an environmental gradient in your local area.

This could be used to practise any of the IA skills and could be included in your practical work. However, it is not suitable for a full IA unless there is an environmental issue and the societal element is strong.

1. Choose a suitable environmental gradient. The environmental gradient you select will depend on where you are doing the investigation. You could investigate:

    *   up or down a hill slope

    *   along a stream

    *   travelling away from a river or some other linear feature (road)

    *   in a line away from the sea or lake shore (from shallow water to land)

    *   through a woodland area from edge to centre.

2. Using an interrupted belt transect, lay out the transect line.

    *   Record the abundance of species in each quadrat using a suitable method.

    *   Repeat the transect three to five times.

3. Present your results in a kite diagram (see Figure 3 in subtopic 2.5) in which the height of the "kites" represents the abundance of the species. Kite diagrams are an excellent way to show the spatial distribution of a plant species, especially along an environmental gradient, succession or zonation.

Investigating the impact of human activity on species diversity

This could be used to practise the following IA skills:

*   Method (C)

*   Treatment of data (D)

*   Analysis and conclusion (E)

*   Evaluation (F).

You will need to consider the following points.

1. Choose a suitable area to sample—this should be somewhere that is affected by human activities.

2. Choose a suitable sampling method:

    a. Use a belt transect, laying your transect line perpendicular to the site of human interference. The obvious example of this is a footpath made through an ecosystem, e.g. across a school playing field, through a wooded area, in a park.

    b. In an area that you know will be affected by human activity, sample using random sampling with quadrats both before and after the activity. Remember, in a woodland area, you will have to vary the size of your quadrat to ensure you include the trees.

3. Decide whether you will collect data for plants, animals or both.

4. Choose an appropriate sampling strategy. If you wish to collect data on animals you can use methods described earlier.

5. Collect your data. As you do so, make notes about weather conditions or anything else that may affect your results. Take pictures as a record. Remember qualitative data.

6. Present your data in an appropriate way.

Measuring abiotic components of the system

Ecosystems can be roughly divided into marine, freshwater and terrestrial ecosystems. Each of these ecosystem types has a different set of physical (abiotic) factors that you can measure.

Marine ecosystems

Abiotic factors include salinity, pH, temperature, dissolved oxygen, wave action.

Whichever abiotic factor you choose, remember they vary over space and time so be careful to control one of these factors. Many abiotic factors can be measured using modern data loggers with interchangeable sensors or probes for pH, salinity, temperature or dissolved oxygen (Figure 14).

General method:

1. Select your abiotic variable and the appropriate probe.

2. Decide if you are going to measure the changes over time or space. Keep the one you are not changing constant.

    a. If you are measuring changes in temperature with depth, try to take all readings at the same time (this can be difficult).

    b. If you are measuring changes in salinity through time, take all measurements in the same place.

3. Repeat each reading at least five times so you can calculate a mean value and reduce errors.

Dissolved oxygen can be measured using a Winkler titration. A series of chemicals is added to the water sample and dissolved oxygen in the water reacts with iodide ions to form a golden-brown precipitate. Acid is then added to release iodine which can be measured, and is proportional to the amount of dissolved oxygen, which can then be calculated.

Freshwater ecosystems

Abiotic factors: turbidity, flow velocity, pH, temperature, dissolved oxygen.

The methods for measuring pH, temperature and dissolved oxygen are the same as for marine ecosystems.

Turbidity

Turbidity can be measured with optical instruments or by using a Secchi disc. High turbidity = cloudy water; low turbidity = clear water.

A Secchi disc is a white or black-and-white disc attached to a graduated rope (Figure 15). The disc is heavy to ensure the rope goes vertically down.

▲ Figure 14 Example of a probe with a pH sensor

▲ Figure 15 A Secchi disc

The procedure is as follows.

1. Slowly lower the disc into the water until it disappears from view.

2. Read the depth from the graduated rope.

3. Slowly raise the disc until it is just visible again.

4. Read the depth from the graduated rope.

5. Repeat in the same spot three to five times and calculate the mean to reduce errors. The mean reading is known as the Secchi depth.

For reliable results, follow a standard procedure.

- Always stand or always sit in the boat.

- Always wear your glasses or always work without them.

- Always work on the shady side of the boat.

Flow velocity

This is the speed at which the water is moving and it determines which species can live in a certain area. Flow velocity varies with the following factors.

1. Time: meltwater in the spring gives high flow rates; summer drought causes low flow rates.

2. Depth: surface water may flow more slowly than that in the middle of the water column.

3. Position in the river: inside bend has shallow slower-moving water; outside bend has deeper fast-moving water.

There are three basic methods for measuring flow velocity.

- Flow meter: these are generally expensive and can be unreliable as mixing water with electricity has its problems.

- Impellers: a simple mechanical device as shown in Figure 16. The impeller is mounted on a graduated stick. Its base should be placed on the river bed. The height of the impeller can be adjusted and the velocity measured at different depths, **but** it can only be used in clear shallow water, as you must be able to see the impeller. The impeller is held at the end of the side arm and lowered into the water facing upstream. The impeller is released and the time it takes to travel the length of the side arm is measured. Repeat three to five times for accurate results.

- Floats: The easiest way to measure flow velocity is to measure the time a floating object takes to travel a certain distance (Figure 17). The floating object should preferably be partly submerged to reduce the effect of wind. Oranges or tennis balls can be used as floats but the water needs to be deep enough for them. The method shown in Figure 17 should be repeated three to five times for accuracy.

  This method gives the surface flow velocity only. You can estimate the average flow velocity of a river by dividing the surface velocity by 1.25.

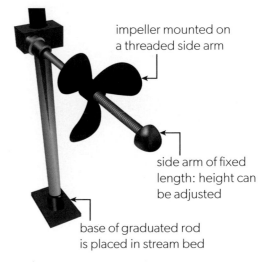

impeller mounted on a threaded side arm

side arm of fixed length: height can be adjusted

base of graduated rod is placed in stream bed

▲ Figure 16 Impeller

Warning: This method gives seconds per metre, **not** metres per second.

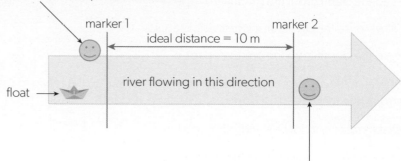

person 1 drops the float above the first marker and shouts "start" as it passes the marker

marker 1

ideal distance = 10 m

marker 2

river flowing in this direction

float →

person 2 starts the stopwatch on command from person 1 and stops it as the float passes marker 2, catching the float

▲ Figure 17 How to measure stream velocity

Terrestrial ecosystems

Abiotic factors: temperature, light intensity, wind speed, soil texture, slope, soil moisture, drainage and mineral content.

As with marine ecosystems, many of the abiotic variables of a terrestrial system can be measured using a data logger and an appropriate probe.

Air temperature

Temperature can be measured using simple liquid thermometers and min–max thermometers.

Wind speed

There is a variety of techniques used to measure wind speed.

- A revolving cup anemometer consists of three cups that rotate in the wind. The number of rotations per time period is counted and converted to a wind speed. Revolving cup anemometers can be mounted permanently or hand-held.

- A ventimeter is a calibrated tube over which the wind passes. This reduces the pressure in the tube, making a pointer move. It is easy to use and inexpensive.

- By observation of the effect of the wind on objects. In some countries, the observations can then be related to the Beaufort Scale (a scale of wind speed from 0 to 12).

Rainfall

Rainfall can be collected using a rain gauge (Figure 18). Some schools have an established weather station, in which case collecting rainfall data is easy. Many schools will not have a weather station but rain gauges are very easy to make and there are plenty of websites that can give you advice on how to make your own.

▲ Figure 18 A rain gauge

Once you have made your rain gauge:

1. place it in a suitable spot in the study area—somewhere away from the influence of buildings, trees and other obstacles that may affect rainfall

2. check it every 24 hours, at the same time every day. Pour rain into a graduated cylinder and record the daily amount of rainfall.

Measuring soil

Soil has a significant impact on plant growth and various aspects of the soil can be measured.

Soil texture (particle size)

Soil is made up of particles (gravel, sand, silt, clay) and the average size and distribution of these particles affect a soil's drainage and water-holding capacity. You will use different methods to measure the different types of particle—see Table 10.

Particle	How to measure
Gravel: very coarse, coarse and medium	Measure individually—simple, but time-consuming procedure.
Gravel: fine and very fine	Sieve through a series of sieves with different mesh sizes.
Sand: all sizes	
Silt and clay	Use sedimentation or optical techniques. Sedimentation techniques are based on the fact that large particles sink faster than small particles. Optical techniques use light scattering by the particles (light scattering is what makes suspensions of soil particles in water look cloudy). Both sedimentation and light scattering can be done using automated instruments but such instruments are expensive for secondary school use.

▲ Table 10 Measuring soil texture

You can take a sample of soil from each horizon, return to the lab and use a sediment settling technique.

1. Mix each soil sample with a known volume of water.

2. Pour each sample into a large measuring cylinder.

3. Record settling times and then compare depth and types of layers once the soils have settled.

Alternatively you can do field tests to assess the soil texture. You can find a number of "soil feel tests" online or you can print the key in Figure 19 and use it to assess soil texture.

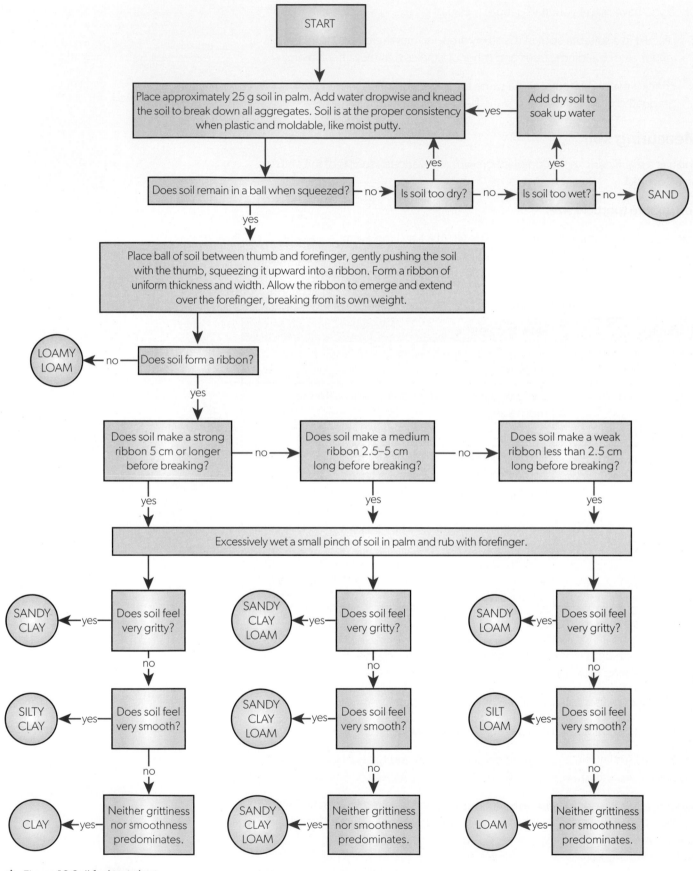

▲ Figure 19 Soil feel test chart.

Adapted from S J Thien, 1979, 'A flow diagram for teaching texture by feel analysis', *Journal of Agronomic Education*, 8: 54–55

Soil air

To find the volume of air in a soil sample:

1. Put 100 ml of water into a 250 ml measuring cylinder.

2. Add 100 ml soil into the 250 ml cylinder with water while stirring.

3. Measure the total volume of the mixture; it will be less than 200 ml.
   The decrease in volume is the volume of the air.

Soil moisture

This is the amount of water in the soil. It can be measured by drying soil samples.
Use a minimum of three to five samples to improve the accuracy of your results.

1. Place a sample of the soil in a crucible.

2. Weigh it and record the weight.

3. Dry the sample in a conventional drying oven or a microwave oven.

   - In a conventional oven, set the oven to 105°C—hot enough to dry the
     soil but not so hot as to burn off organic matter. Leave for 24 hours and
     weigh the sample; repeat this until the mass becomes constant. This
     could take several days.

   - In a microwave oven, place the sample in the microwave for 10 minutes.
     Weigh the sample, then return to the oven for 5 minutes. Repeat until the
     mass becomes constant.

Soil organic content

The organic content of a soil is made up of plant and animal residues in various
stages of decay and it has several functions:

- supplies nutrients to the soil

- holds water (like a sponge)

- helps reduce compaction and crusting

- increases infiltration.

Organic content can be determined by the loss on ignition (LOI) method.

1. Dry the sample as above and record the weight of the dry sample.

2. Heat the soil at high temperatures of 500–1,000°C for several hours.

3. Weigh the sample and repeat this until its mass becomes constant.

A minimum of three to five samples should be tested.

Consider the risks of using such high temperature and how to mitigate them.

Soil mineral content and pH

There is a wide range of soil nutrients essential for a fertile soil. NPK (nitrogen,
phosphorus and potassium) are the main ones and are found in many fertilizers.
These are easy to measure using soil testing kits available in many garden centres.

Soil pH can also be measured using a soil testing kit or a pH probe.

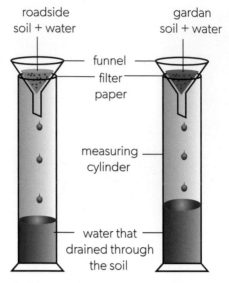

▲ Figure 20 Measuring soil water retention

▲ Figure 21 Measuring soil infiltration rate

Soil water retention and drainage

1. Take two funnels and line them with filter paper.

2. Add dry samples of equal weight of the two soils.

3. Pour 100 ml of water into each funnel (Figure 20).

4. When the dripping of water from the funnel stops, record the volume of water which drains into the measuring cylinder and the time taken.

5. Retention of water = 100 – volume in measuring cylinder.

Soil infiltration rate

This must be done in the field using the following method (**away** from where the soil samples were taken).

1. Select a suitable spot to measure drainage—it should be flat.

2. Take a short section of sturdy plastic tubing (drainpipe).

3. Knock the tubing about 15 cm into the soil (Figure 21).

4. Pour a set amount of water into the tube and time how long it takes for the water to drain away completely:

   a. If the soil has a high clay content or is compacted, drainage will be poor and it will take longer for the water to drain away (poorly drained).

   b. If the soil is sandy, the water will drain away quickly (well drained).

Investigating pollution

Water pollution

Biochemical oxygen demand

Biochemical (or biological) oxygen demand (BOD) is a measure of the amount of dissolved oxygen required to break down the organic material in a given volume of water through aerobic biological activity (by microorganisms). It is an indirect measure of organic material in water—dead plants and animals, manure, or even food.

This could be used to practise the following IA skills:

- Research Question and Inquiry (A)

- Strategy (B).

Select a range of water sources—such as various taps, streams, water fountains or lakes. The variety of sources will depend on where you are and what sources are available. Table 11 shows some examples of BOD values.

At each water source:

1. Prepare eight water collection bottles and make sure they are clean with **no detergent** in them.

2. Allocate two bottles to each water source and label them—for example: Tap in school, Bottle 1; Tap in school, Bottle 2; Water tank, Bottle 1; Water tank, Bottle 2; and so on.

3. Collect two identical samples of water from each water source.

    a. Bottle 1: Seal immediately (with no air space in the bottle) and then place in an incubator (at 20°C) for five days. After five days, use a suitable probe to measure the dissolved oxygen content of the water (mg dissolved oxygen l^{-1}).

    b. Bottle 2: Use a probe to measure the dissolved oxygen content and record this value (mg dissolved oxygen l^{-1}).

4. Calculate BOD = Day 5 reading – Day 1 reading (mg dissolved oxygen l^{-1})

Source of pollutant	BOD (mg dissolved oxygen l^{-1})
Unpolluted river	<5
Treated sewage	20–60
Raw domestic sewage	350
Cattle slurry	10,000
Paper pulp mill	25,000

▲ Table 11 Examples of BOD values

Trent biotic index for freshwater invertebrates

This could be used to practise the following IA skills:

- Method (C)

- Treatment of data (D)

- Analysis and conclusion (E)

- Evaluation (F).

The Trent biotic index is a 1–10 scale that gives a measure of the quality of an ecosystem by the presence and abundance of species living in it. It is an indirect measurement of pollution.

This index is based on the fact that some species tend to disappear and the species diversity decreases as organic pollution in a water course increases. Organisms are classified by tolerance levels. BOD is also measured.

1. Before carrying out any fieldwork, remember to follow the IB sciences experimentation guidelines. Fieldwork should have minimal impact on the environment and a risk assessment should be carried out to assess hazards and reduce potential risk.

2. Select one site upstream and one downstream of a point source of pollution.

3. Sample the invertebrates at each site, using kick sampling and hand nets.

4. Take several (five to seven) samples at each site.

5. Use a suitable guide to identify the species and calculate the index values. There are many such guides available online.

Air pollution

Investigating air pollution using a biotic indicator (lichens)

This could be used to practise the following IA skills:

- Method (C)

- Treatment of data (D)

- Analysis and conclusion (E)

- Evaluation (F).

A common source of air pollution is the combustion of fossil fuels so for this investigation we are using a busy road. Thermal power stations also burn fossil fuels but the distances over which the pollution can travel are very large.

Lichens grow in exposed places such as on rocks and tree bark. They absorb water and nutrients from rainwater so if the rain is polluted, the lichens can be damaged. You will have to look up the lichens common in your area, but as a general rule:

- crusty lichens are the most tolerant of pollution

- leafy lichens tolerate a little air pollution

- bushy lichens are intolerant of any air pollution and will not be present in polluted air.

▲ Figure 22 Types of lichens

1. Research your local area to find out which lichens are most common.

2. Find a suitable study site with an environmental gradient from polluted (e.g. roadside) to less polluted (e.g. woodland or forest).

3. Lay out a minimum of three transects, starting as close to the road as you can and moving into the forest.

4. Select a relevant sampling interval—this will depend on how far into the forest you go but every 10 m for 100 m is usually suitable.

5. At each point, select the three trees closest to the transect line.

   a. On the side of the tree closest to the road, measure a set distance (1 m) from the base of the tree.

b. Use a suitably sized quadrat (10 cm × 10 cm) and an appropriate method to assess the abundance of each species of lichen.

6. Repeat this along all three transects.

7. Throughout the investigation, make sure all the trees are the same species.

Investigating the impact of albedo or different greenhouse gases on the temperature of a closed system

Albedo is the proportion of sunlight reflected back from an object. Light colours have much higher albedo rates than dark colours. Artificial surfaces have much higher rates than natural ones. You can look up the actual differences for yourself.

To investigate the impact of albedo on the temperature of a closed system, follow these steps:

1. Select a range of surfaces, including a mix of natural and human-made—for example grass or other vegetation, glass, mirrors, bare soil, snow or ice, water, and concrete.

2. The easiest way to collect data is with a data logger and temperature sensor. This must be set up in a safe area where it will not be disturbed.

a. Set up the data logger and temperature sensor at a fixed height above the surface.

b. Record the temperature changes during sunlight hours.

c. Repeat for all surfaces you have selected.

3. If you do not have access to a data logger:

a. Set up an analogue thermometer at a fixed height above the selected surface.

b. Check the temperature every hour during sunlight hours.

c. Repeat for all surfaces you have selected.

4. Compare the differences in temperature and extrapolate to predict the possible effects of albedo on the global climate.

To investigate the impact of a greenhouse gas on the temperature of a closed system, follow these steps:

1. Set up identical soda bottles containing different gases:

- ambient air

- CO_2 in varying concentrations

- humid air

- dry air.

2. Place a thermometer in each of the bottles and seal them.

3. Check the temperature in the bottles regularly.

This investigation could also be done with coloured filters around the bottles to assess the impact of colour on heating.

Exam-style questions

1. The family Leporidae includes hares and rabbits.
    Figure 1 shows four species that can be found in
    western North America.

Species A: Length 55–66 cm

Species B: Length 35–40 cm

Species C: Length 46–63 cm

Species D: Length 25–29 cm

▲ Figure 1 Four species of the family Leporidae

1. a. Less than 30 cm in length .... **pygmy
        rabbit** (*Brachylagus idahoensis*)

    b. Greater than 30 cm in length .... go to 2

2. a. Has black tail .... **Black-tailed jackrabbit**
        (*Lepus califomicus*)

    b. Has a mostly white tail .... go to 3

3. a. Has short rounded ears .... **Nuttall's
        cottontail** (*Sylvilagus nuttallii*)

    b. Has ears at 2.5 times as long as wide ....
        white-tailed jackrabbit (*Lepus townsendii*)

▲ Figure 2 A dichotomous key for species A to D

   a. Use **Figures 1** and **2** to identify Species B
       and Species C. [1]

Sagebrush
(*Artemisia tridentata*)

Pygmy rabbit burrow

▲ Figure 3 Sagebrush
ecosystem without invasive
cheatgrass

Cheatgrass
(*bromus tectorum*)

▲ Figure 4 Sagebrush
ecosystem with invasive
cheatgrass

The sagebrush ecosystem provides a habitat for
pygmy rabbits.

   b. Suggest **one** reason why there might be a greater
       number of pygmy rabbits in the ecosystem shown
       in **Figure 3** than in the ecosystem shown in
       Figure 4. [1]

   c. Describe **one** method to determine the impact
       of invasive cheatgrass on sagebrush density. [3]

   d. Distinguish between the biodiversity of the
       sagebrush ecosystems in **Figures 3** and **4**. [2]

2. a. Describe biotic and abiotic factors with
        reference to a **named** ecosystem. [4]

    b. Using a system diagram, explain the transfer and
        transformation of energy as it flows through an
        ecosystem. [7]

3. a. Outline how species diversity and population
        size influence the resilience of an ecosystem. [4]

    b. Describe the similarities and differences in using
        a biotic index and a diversity index to assess
        ecosystems. [7]

4. a. Outline the role of the atmospheric system in the
        distribution of biomes. [4]

    b. Explain how human impacts on the atmosphere
        may influence the productivity of terrestrial
        biomes. [7]

5. Look at **Figure 5**.

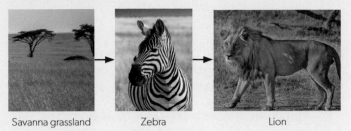

▲ Figure 5 Savanna food chain

Savanna grassland   Zebra   Lion

a. State the trophic level of the zebra. [1]

b. State how you could determine gross secondary productivity of the zebra. [1]

c. Explain how the second law of thermodynamics applies to this food chain. [2]

Biting flies **(Figure 6)** bite and drink the blood of zebras. They commonly carry diseases that can be fatal to zebras.

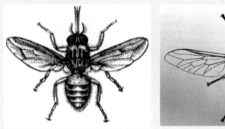

▲ Figure 6 Biting flies in the savanna

Second report of the Wellcome Research Laboratories at the Gordon Memorial College, Khartoum / Andrew Balfour / The Wellcome Trust (CC-BY 4.0)

d. State the type of relationship that exists between biting flies and the zebra. [1]

e. Zebra stripes may reduce the ability of the biting flies to land on the zebra. Describe how natural selection may have led to the evolution of zebra stripes in response to biting flies. [3]

6. a. Identify **four** ways to ensure reliability of the capture–mark–release–recapture method in estimating population size. [4]

b. Explain how the interactions between a species and its environment give rise to the S-shape of its population growth curve. [7]

7. Look at **Figure 7**.

a. Distinguish between **two** named biomes and the factors that cause their distribution. [4]

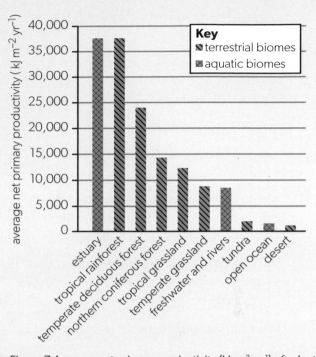

▲ Figure 7 Average net primary productivity (kJ m^{-2} yr^{-1}) of selected world biomes

b. Evaluate **one** method for measuring primary productivity in a named ecosystem. [7]

c. Discuss how human activities impact the flows and stores in the nitrogen cycle. [9]

8. a. Using **Figure 7**, identify an ecosystem that has an average net primary productivity above 30,000 kJ m^{-2} yr^{-1}. [1]

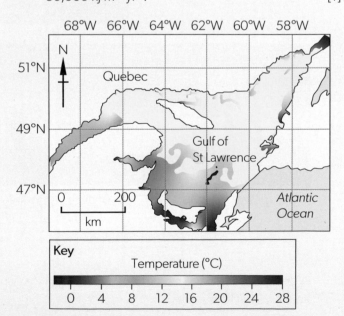

▲ Figure 8 Water surface temperature variation across the Large Ocean Management Area

Fisheries and Oceans Canada. Reproduced with the permission of © Her Majesty the Queen in Right of Canada, 2019

b. Suggest one reason for the zonation seen in Figure 8. [1]

c. Estuaries are some of the most productive ecosystems in the world, but only account for 3% of global productivity.

State **one** reason why this occurs. [1]

d. Outline why estuaries are highly productive ecosystems. [3]

9. a. Outline **two** reasons why the species within pioneer communities in **Figure 9** are more likely to be *r*-strategists than *K*-strategists. [2]

b. Outline **two** reasons why the climax community in **Figure 9** is more stable than the intermediate communities. [2]

c. Distinguish between zonation and succession. [1]

d. Outline **two** ways in which the food web is likely to change as a result of succession. [2]

e. Outline **two** ways in which the soil quality in the pioneer stages of the succession model shown in **Figure 9** will differ from that in the climax ecosystem. [2]

10. a. Identify **four** impacts on an ecosystem that may result from the introduction of an invasive species of herbivore. [4]

b. Explain how both positive and negative feedback mechanisms may play a role in producing a typical S population growth curve for a species. [7]

11. Outline **two** ecosystem services in a named biome. [4]

12. a. Outline how **four** different factors influence the resilience of an ecosystem. [4]

b. Explain how a community of trees in a woodland may be considered a system. [7]

13. a. Distinguish between the terms *niche* and *habitat* with reference to a named species. [4]

b. Suggest the procedures needed to collect data for the construction of a pyramid of numbers for the following food chain: [7]

plants ➡ snails ➡ birds

14. Describe the role of primary producers in ecosystems. [4]

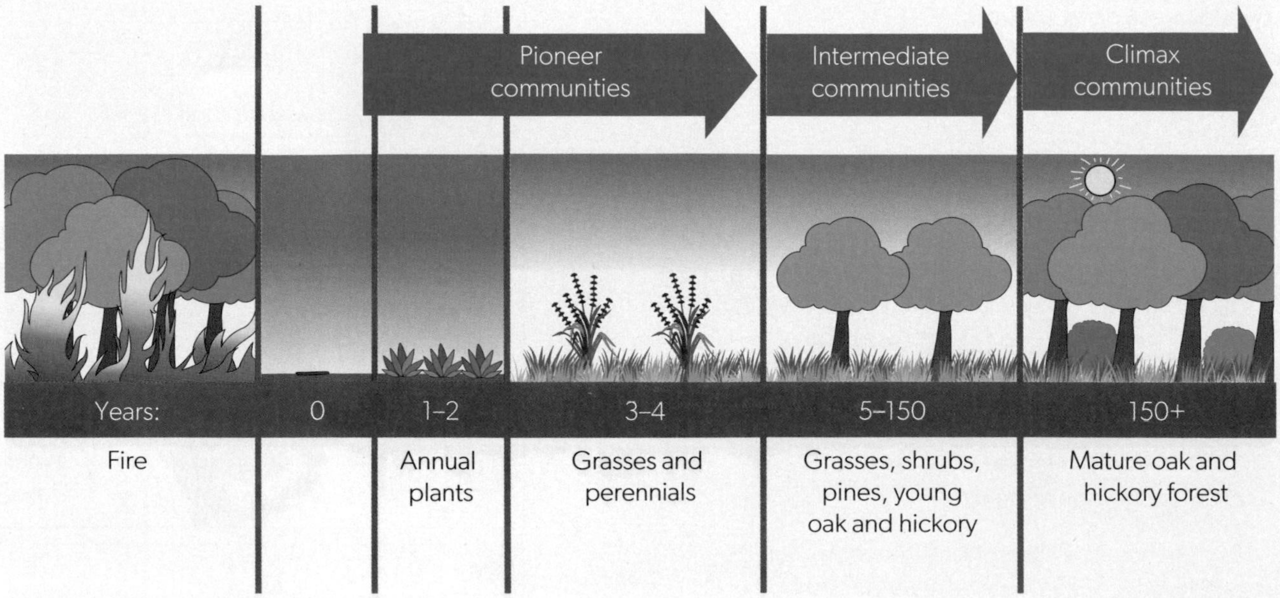

▲ Figure 9 Stages of succession following disturbance by fire

Adapted from: Katelyn Murphy

15. **Figure 10** shows the concentration of DDT at different trophic levels of a food chain.

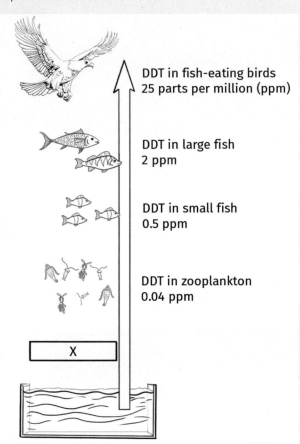

▲ Figure 10 Levels of concentration of DDT in food chain
Adapted from: toppr / © 2023 Haygot Technologies, Ltd.

a. State the main source of energy for the food chain in **Figure 10**. [1]

b. State the trophic level labelled **X** in **Figure 10**. [1]

c. Identify **one** use of DDT that has led to its presence in the environment. [1]

d. With reference to the concepts of bioaccumulation **and** biomagnification, outline how the concentration of DDT has changed along the food chain. [2]

e. State the relationship between large and small fish in **Figure 10**. [1]

f. Outline how this relationship may be of benefit to the populations of both species. [2]

16. a. Identify **four** ways in which solar energy reaching vegetation may be lost from an ecosystem before it contributes to the biomass of herbivores. [4]

b. Suggest a series of procedures that could be used to estimate the net productivity of an insect population in kg m⁻² yr⁻¹. [7]

17. Look at **Figure 11**.

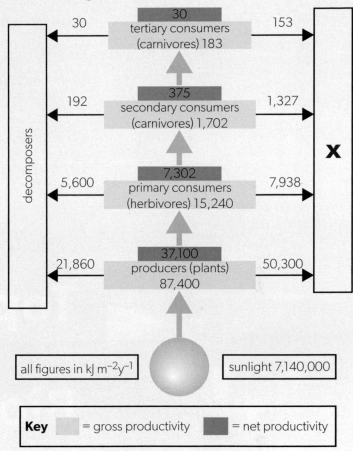

▲ Figure 11 The gross and net productivity at different trophic levels within the Silver Springs, Florida, ecosystem
Adapted from: John W. Kimball (CC BY 3.0)

a. State the process represented in the box labelled **X**. [1]

b. Define *net primary productivity*. [1]

c. Describe how the second law of thermodynamics operates in relation to the transfer of energy within the Silver Springs ecosystem. [2]

d. Distinguish between a pyramid of numbers and a pyramid of productivity. [2]

18. Outline why top carnivores are vulnerable to non-biodegradable toxins. [4]

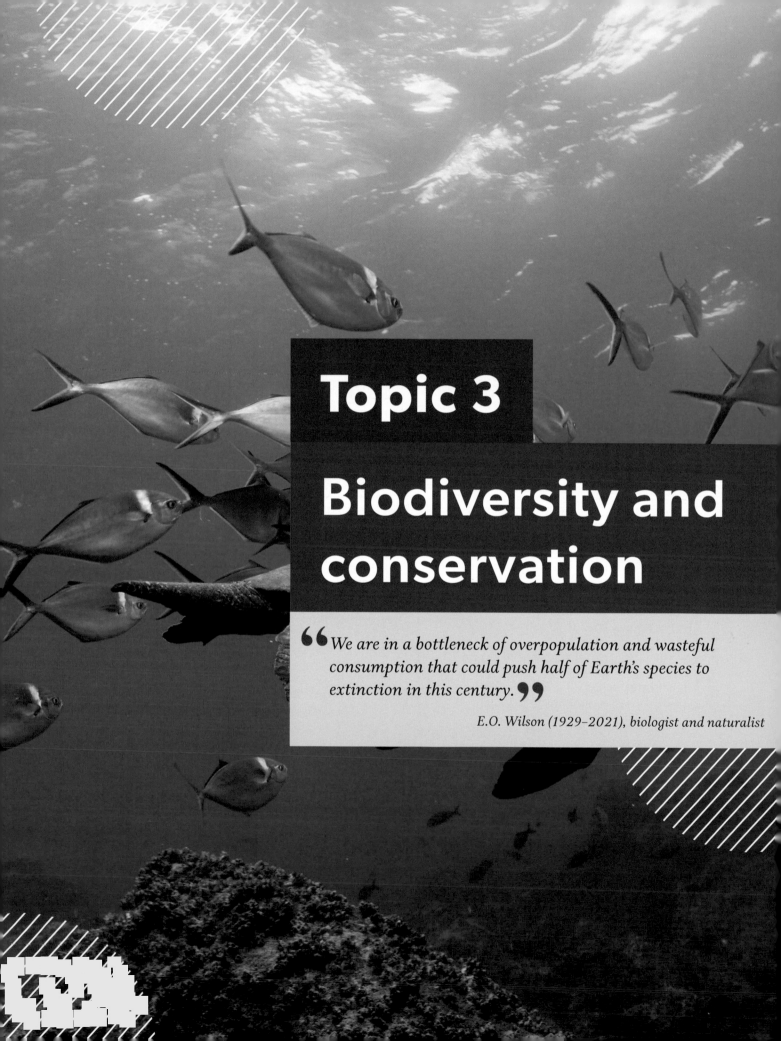

Topic 3

Biodiversity and conservation

> ❝ We are in a bottleneck of overpopulation and wasteful consumption that could push half of Earth's species to extinction in this century. ❞
>
> E.O. Wilson (1929–2021), biologist and naturalist

Guiding questions

- How can diversity be explained and quantified, and why is this important?
- How does the unsustainable use of natural resources impact biodiversity?

Understandings

1. Biodiversity is the total diversity of living systems and it exists at several levels.

2. The components of diversity contribute to the resilience of ecological systems.

3. Biodiversity arises from evolutionary processes.

4. Natural selection is the mechanism driving evolutionary change.

5. Evolution by natural selection involves variation, overproduction, competition for limited resources, and differences in adaptation that affect rates of survival and reproduction.

6. Speciation is the generation of new species through evolution.

7. Species diversity in communities is a product of richness and evenness.

8. Simpson's reciprocal index is used to provide a quantitative measure of species diversity, allowing different ecosystems to be compared and for change in a specific ecosystem over time to be monitored.

9. Knowledge of global and regional biodiversity is needed for the development of effective management strategies to conserve biodiversity.

10. Mutation and sexual reproduction increase genetic diversity.

11. Reproductive isolation can be achieved by geographical separation or, for populations living in the same area, by ecological or behavioural differences.

12. Biodiversity is spread unevenly across the planet, and certain areas contain a particularly large proportion of species, especially species that are rare and endangered.

13. Human activities have impacted the selective forces acting on species within ecosystems, resulting in evolutionary change in these species.

14. Artificial selection reduces genetic diversity and, consequently, species resilience.

15. Earth history extends over a period of 4.5 billion years. Processes that occur over an extended timescale have led to the evolution of life on Earth.

16. Earth history is divided up into geological epochs according to the fossil record.

17. Mass extinctions are followed by rapid rates of speciation due to increased niche availability.

18. The Anthropocene is a proposed geological epoch characterized by rapid environmental change and species extinction due to human activity.

19. Human impacts are having a planetary effect, which will be detectable in the geological record.

What is biodiversity?

Biodiversity is a complex concept covering the variety of life on Earth. It is:

- the numbers of species of different animals, plants, fungi and microorganisms in ecosystems—**species richness**

- the relative proportion of individuals of each species—**species evenness**

- three different but inter-related types of diversity: **habitat**, **species** and **genetic**.

TOK

The term "biodiversity" has replaced the term "nature" in much writing on conservation issues. Does this represent a paradigm shift?

Connections: This links to material covered in subtopics 1.2, 3.3, 6.2, 6.3 and 7.2.

Ecosystem **resilience** is the capacity of an ecosystem to withstand disturbance and still maintain the same functions of organisms working together to maintain a balanced sustainable ecosystem. Usually, high biodiversity indicates a healthy ecosystem, but low diversity is still possible in stable healthy ancient sites where conditions are harsh, for example, the Arctic.

Species diversity

Key term

Species diversity in communities is a product of two variables, the number of species (richness) and their relative proportions (evenness).

Species diversity is probably the aspect of biodiversity that you consider when you hear the term "biodiversity". It is not just the total number of organisms in a place—it is also the number of organisms within each of the different species. Species diversity is a product of species richness and species evenness.

- Species richness: the number or variety of species there is. This does not consider how common or rare each species is. This is a problem because rare species are less likely to maintain their presence in the area. To measure species richness, count the number of species present.
- Species evenness: the abundance or number of organisms of each species. This solves the weakness of only looking at richness because it takes into account the number of individuals that are present in each species. Species evenness is measured from 0 to 1 with 1 being the most even. High evenness is usually good. To measure species evenness, count the number of species and number of individuals in each species—their abundance.

ATL Activity 1

ATL skills used: thinking, research

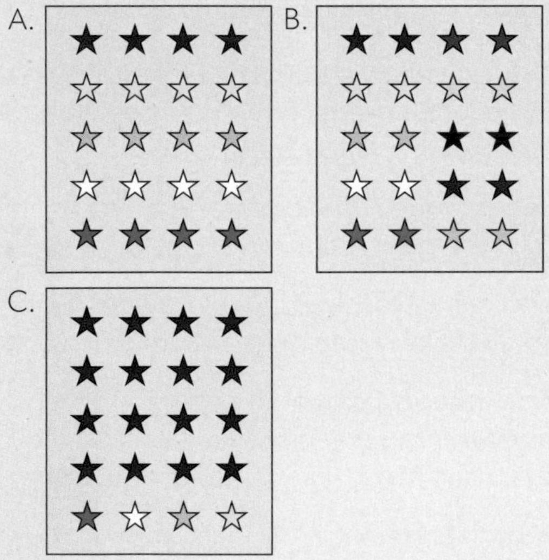

▲ Figure 1 Three boxes representing ecosystems

Figure 1 shows three boxes representing ecosystems. Each star represents one individual; the different colours are different species.

1. Which box has the highest species richness?
2. Which box has the highest species evenness?
3. Which box has the highest biodiversity?
4. Which ecosystem is most resilient?
5. Give examples of ecosystems that could be A, B and C.
6. Box C has abundant reds and only one of each of the other colours. What may happen to the equilibrium of the ecosystem in future?
7. Which box is most likely to have been disturbed by human activity?

Diversity indices

- A diversity index (*D*) gives a quantitative estimate of biological variability in space or time and describes and compares communities.
- We can compare two similar ecosystems, two communities within an ecosystem or the same ecosystem over time.

- When comparing communities that are similar, low diversity could be evidence of pollution, eutrophication or recent colonization of a site.

- The number of species present in an area is often used to indicate general patterns of biodiversity but only tells us part of the story.

- It is important to repeat investigations of diversity in the same community over time, to identify whether change is "natural" due to succession, or caused by human activity. Human activities may increase or decrease biodiversity, which would tell us if conservation efforts were succeeding.

- The value of D will be higher where there is greater richness (number of species) and evenness (similar abundance) with 1 being the lowest possible value in Simpson's reciprocal index.

- Several indices can be used to measure diversity; Simpson's is the best known.

Simpson's reciprocal diversity index

This is a numerical value that can be used to measure species richness and species evenness. The higher the number, the greater the species diversity. This makes it possible to compare similar ecosystems or to see whether ecosystems are changing over time.

Note: See subtopic 2.6 for how to use this. You do not need to memorize the formula.

(ATL) Activity 2

ATL skills used: thinking

Calculate Simpson's reciprocal diversity index to support or refute the statements made about diversity in Table 1.

Forest A	Forest B
15 different species	15 different species
100 individuals of one species 1 individual each of 14 other species	7 individuals of each of 15 species
Total individuals = 114	Total individuals = 105
Greater number of individuals	
Low species diversity because the individual species may not be able to breed and so may disappear	Greater species diversity—because the species are evenly spread and the populations more viable

▲ Table 1 Diversity in two forests

Habitat and genetic diversity

Variation in **genetic diversity** alters from habitat to habitat. Some habitats such as coral reefs and rainforests have high species diversity. Urban habitats and polar regions have much lower species diversity by comparison.

TOK

Biodiversity index is not a measure in the true sense of the word: it is merely a number (index) as it involves a subjective judgement on the combination of two measures, proportion and richness. Are there examples in other areas of the subjective use of numbers?

Key term

Genetic diversity is the range of genetic material present in a gene pool or population of a species.

Another way to measure diversity is to look at the amount of genetic diversity. Species are made up of both individuals and populations, both of which impact genetic diversity.

- Individuals: each individual in a species has a slightly different set of genes from any other individual in that species. So, more individuals means higher genetic diversity and thus a bigger gene pool.

- Populations: if a species is made up of two or more different populations in different places, then each population will have a different total genetic make-up. Therefore, it is important to conserve different populations of a species to maximize genetic diversity.

Not all species have the same amount of genetic diversity. A population with no genetic variation (in which every individual is genetically identical) cannot evolve in response to environmental or situational changes. A population with high levels of genetic variation is much more likely to include at least a few individuals carrying gene versions that provide protection from threats such as pathogens; such populations can evolve in response to new threats instead of going extinct.

Examples of species' genetic diversity

Cheetah (*Acinonyx jubatus*)

Cheetahs have very low genetic diversity. This, is probably due to a population bottleneck they experienced around 10,000 years ago. Their population size was greatly reduced and they barely avoided extinction at the end of the last ice age. In modern times, habitat encroachment and poaching have further reduced cheetah numbers, reducing genetic variation and leaving cheetahs even more vulnerable to extinction. Female cheetahs bear a single litter with multiple fathers which may be a behaviour to increase genetic diversity in the offspring.

Cavendish bananas (*Musa acuminata var.*)

Do you eat bananas? Have you ever seen them labelled as Fyffes or Chiquita? These bananas are identical clones (genetically identical) and unable to reproduce sexually. They grow in monocultures (one species) in plantations in tropical areas from South and Central America to Africa, India, and China. A fungal disease of bananas called Panama disease made a previous banana cultivar extinct, and this could also happen to the Cavendish variety due to a new strain of Panama disease called TR4. As yet, there is no similar replacement variety of banana.

What might be the results of an outbreak of Panama disease to (a) local societies dependent on growing bananas, (b) world trade in bananas, and (c) your diet?

European red fox (*Vulpes vulpes*)

The European red fox is found throughout Europe. It has a large genetic diversity. There are three distinct populations, each with a slightly different genetic make-up. This means there is genetic variety within the species and the species is sustainable. If one population were to disappear (e.g. wiped out by disease), others would survive.

▲ Figure 2 Cheetah in Masai Mara National Park, Kenya

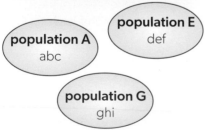

▲ Figure 3 European red fox and three populations with different genotypes

Domesticated dogs (*Canis familiaris*)

Dogs have been selectively bred by humans for over 15,000 years. Pleistocene wolves were the evolutionary ancestors of all dogs today. Dogs are used by humans for many purposes. They show a wide range of phenotypes from small dogs such as the Chihuahua (weight 1–3 kg) to the Great Dane (weight 54–90 kg). These are the same species and could therefore mate to produce fertile offspring.

ATL Activity 3

ATL skills used: thinking, research

Dogs have been selectively bred to develop particular physical traits.

Investigate:

- the roles dogs are bred for

- whether selective breeding is for economic, cultural or social reasons

- whether it matters if the dogs become feral

- whether it matters if the dogs have inherited inbred weaknesses.

Do you think such selective breeding of dogs is ethical?

For many conservationists, more genetic variation is a good thing. They would want to maximize genetic diversity. That means having many species and much variability within each species. Then species have a better chance of adapting to change in their habitats.

Humans can alter genetic diversity by artificially breeding (e.g. domesticated dogs) or genetic engineering of populations (e.g. altering the DNA of bacteria as in the manufacture of medicinal drugs). This reduces variation in their genotypes or even produces identical genotypes—clones. This can be an advantage if it produces a high-yielding crop or animal but a disadvantage if disease strikes and the whole population is susceptible. The loss of genetic variety resulting from the domestication of animals and plant breeding has increased the importance of "gene banks".

Tropical rainforests are high in **habitat diversity** because there are many ecological niches due to the layering of the forests. Tundra has a lower level of habitat diversity (see subtopic 2.4).

Biodiversity links to health, resilience and sustainability

Ecosystems vary considerably so they are difficult to compare. However, biodiversity is often used as a measure by which comparisons can be made. High biodiversity usually equates with a healthy, resilient, sustainable ecosystem (see Figure 4).

See also subtopic 1.2 on resilience and stability of ecosystems.

AHL lens

To what extent is human intervention by artificial selection or genetic engineering (including genetic engineering of humans) ethical?

Key term

Habitat diversity is the range of different habitats per unit area in a particular ecosystem or biome.

Connections: This links to material covered in subtopics 1.2, 3.3, 6.2 and 6.3.

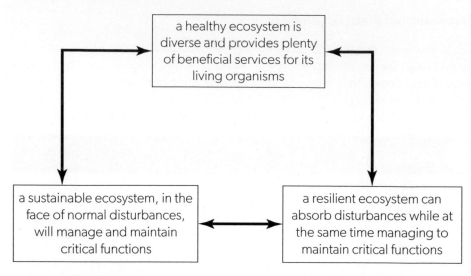

▲ Figure 4 The interrelationship between health, sustainability and resilience

Complexity of the ecosystem

The more complex a food web, the more resilient it is to the loss of one species or reduction in its population size. If one type of prey or food source or predator is lost, the others will fill the gaps left. This resilience of more complex communities and ecosystems is a good thing for biodiversity overall. But it may reduce species diversity as one species is lost completely but the community continues.

Stage of succession

When plants and animals colonize a bare piece of land, only a few pioneer species colonize it at first. As time passes and conditions become more favourable, species diversity increases until a climax community is reached. At this point, the species composition is stable.

In the early stages of succession, abiotic conditions are harsh and limiting factors are numerous (temperature, water supply, nutrients). Biodiversity is low. So, communities in young ecosystems that are undergoing succession tend to be more vulnerable to disturbances and less resilient.

Limiting factors and stability

If it is difficult for the organisms in an ecosystem to get enough raw materials for growth (e.g. water is a limiting factor in a desert), any change in, for example, water supply makes it even harder and species may disappear. If the abiotic factors required for life (water, light, heat, nutrients) are available in abundance, the system is more likely to manage if one is reduced.

Some areas of the world are more prone to environmental disasters and this impacts resilience of the ecosystems. Major disruptions such as volcanic eruptions and glaciations can wipe out large numbers of species. These are events to which there is no resilience.

Age of the ecosystem

As a rule, the older the ecosystem, the greater the biodiversity is likely to be. If an ecosystem is undisturbed for a long period of time, more habitats can develop allowing for greater diversification of species.

ATL Activity 4

ATL skills used: thinking, communication

Using Figure 5, explain what would happen to the food web if the following animals were removed:

1. the mouse

2. the rabbit

3. the owl.

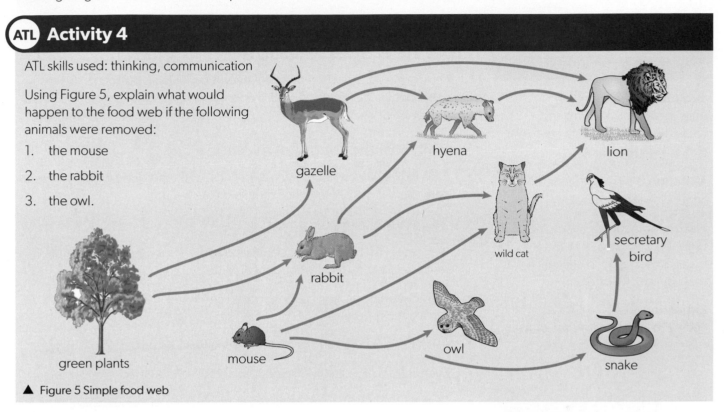

▲ Figure 5 Simple food web

Habitat diversity

Tropical rainforests are high in habitat diversity as there are many ecological niches due to the layering of the forests. A broad collection of habitats encourages a broad assortment of ecological niches and thus more species diversity. Tropical rainforests (vertical range), mountains (altitudinal range) and coral reefs all have a wide array of habitats.

High biodiversity has these advantages:

- resilience and stability due to the range of plants present, of which some will survive drought, floods, insect attack or disease

- genetic diversity, so resistance to diseases

- some plants there will have deep roots so can cycle nutrients and bring them to the surface making them available for other plants.

But high biodiversity does not always equate to a healthy ecosystem:

- diversity could be the result of fragmentation (break up) of a habitat or degradation when species richness is due to pioneer species invading bare areas quickly

▲ Figure 6 Vertical nature of the tropical rainforest

- managing grazing can be difficult as plant species have different requirements and tolerance to grazing

- some stable and healthy communities have few plant species so are an exception to the rule.

Key terms

Evolution is the change in traits (genetically determined characteristics) of a population over successive generations.

Mutation is a change in the DNA sequence of an organism.

Natural selection is the process by which populations of living organisms adapt and change. It is a mechanism of evolution.

Extinction is when a species ceases to exist after the last known individual of that species dies.

TOK

The theory of evolution by natural selection tells us that change in populations is achieved through the process of natural selection.

What is the difference between a convincing theory and a correct one?

How new species form

Charles Darwin proposed the theory of **evolution** which is outlined in *On the Origin of Species*, published in 1859. Evidence for this theory comes from the fossil record, discovery of the structure of DNA and mechanisms of **mutations**.

Summary of the theory of evolution

- Each individual is different (except identical twins and clones) due to their particular set of inherited genes and to mutations.

- This genetic diversity gives rise to variation within a population.

- Resources are limited for any population and there is competition for these resources, e.g. for food, light, space, water.

- Individuals with an advantage in a given environment are more likely to survive and reproduce.

- Variation is inherited so this advantage is likely to be passed to offspring.

- Each individual is slightly differently adapted (or fitted) to its environment.

- Over time these changes are inherited by offspring of the successful individuals and the whole population gradually changes. For example, a giraffe with a slightly longer neck than the other giraffes may be able to reach tree leaves that are out of reach to the others and so get more food. This gives that giraffe a competitive advantage. These small differences mean that some individuals will be more successful. They will survive to breed more than others and so pass their genes on to the next generation.

This is **natural selection**: individuals that are more adapted to their environment have an advantage and flourish and reproduce but those less adapted do not survive long enough to reproduce. The fittest survive—"**survival of the fittest**" is a term you may have heard.

Extinction is a natural process and eventually all species become extinct. The average lifespan for a species varies. Most mammals have a species lifespan of one million years, some arthropods ten million years or more. The rate at which extinctions occur is not constant and is made up of the background extinction rate plus mass extinctions when a sudden loss of species occurs in a relatively short time.

Speciation

In sexual reproduction, if a mutation occurs in the gametes (sperm and eggs), it can be passed on to the next generation. This leads to a new combination of genes and more variation. This is how a population in one habitat will gradually become better adapted to that habitat.

When populations of the same species become separated (e.g. by a barrier such as a river or mountain range), they cannot interbreed. If the environments they inhabit change, the populations may start to diverge and new species form. This is **speciation**. Separation of populations takes three main forms.

- **Geographic isolation:** populations of a species are physically separated. This may be by: a population being transported to an island; a population having such a large range that individuals only mate with those nearby; pollution in an area (e.g. toxic waste from mining), which stops most plants growing but some survive and adapt.

- **Temporal isolation:** some members of a population find a competitive advantage by being active at a different time of day (e.g. night-time) so they do not meet other members of the population to breed.

- **Behavioural isolation:** mating seasons are not synchronized or flowers mature at different times.

Over many generations, if a population is separated from others and isolated, the differences may increase to such an extent that, should the populations be reunited, they will be unable to interbreed. Then a new species has formed, and speciation has occurred.

The island rule

When a population becomes isolated on an island and evolves separately, there appears to be a tendency for small species to become much larger while large species tend to become much smaller when compared with populations of that species on the mainland. This suggests that limited resources and limited space on islands may lead to large animal species becoming smaller. But lower competition and absence of predators may allow small animal species to evolve into larger ones. Examples are Komodo dragons and the dodo (increase in size) and fossils of dwarf elephants (decrease in size).

Physical barriers

Species can develop into two or more new species if their population is split by a physical barrier, for example, a river, mountain range or ocean. The barrier splits the gene pool. The populations on either side of the barrier do not mix, so the genes on either side of the barrier do not mix. The two populations can therefore develop in different directions.

Continental drift and plate tectonics

The theory of plate tectonics explains how continents drifted apart as lithosphere tectonic plates float on a fluid-like lower layer. At plate boundaries where plates meet, there is often volcanic activity or earthquakes, mountain ridges or oceanic trenches. We think there were supercontinents lasting some 500–600 million years before breaking apart. Supercontinent Rodinia is one which broke up 760 MYA (million years ago). Then Pangaea formed (Figure 7) which broke up 200 MYA into Laurasia and Gondwana in the Triassic. Study Figure 7 and you can see the shapes of the continents today.

> ### Key term
>
> **Speciation** is the process by which new species form.

Connections: This links to material covered in subtopic 3.3.

▲ Figure 7 Supercontinent Pangaea
Adapted from: Hok / Wikimedia Commons (CC BY-SA 3.0)

▲ Figure 8 Large flightless birds (rhea, ostrich, emu, kiwi, tinamou, moa, elephant bird, cassowary)

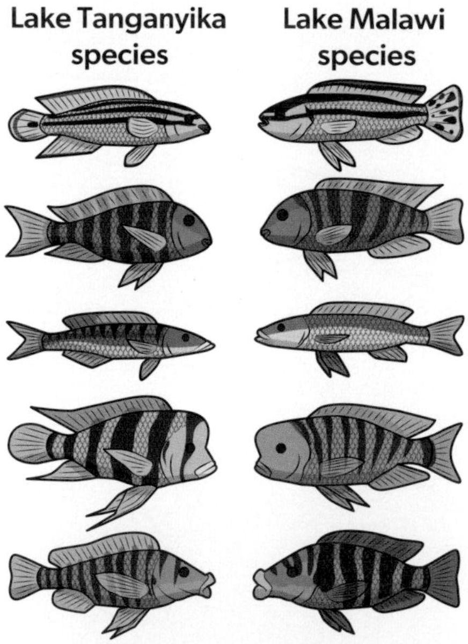

Lake Tanganyika species	Lake Malawi species

▲ Figure 9 Cichlid fish

Key term

Endemic species are only found in one geographical location and nowhere else on Earth.

Evidence for continental drift comes from:

- geological studies of similar rocks on different continents

- fossil evidence across continents

- similar groups on different continents, e.g. llamas and camels; kangaroos and cattle (marsupials in Australia demonstrate that this continent split off earlier than others); African and Indian elephants.

Here are three examples of speciation due to physical barriers.

1. Large flightless birds (Figure 8) only occur on the continents of Africa, Australia (including New Zealand) and South America. Continental drift describes how large supercontinents where the ancestor of these birds lived split up over time to form the continents we know today. These land masses were once part of Gondwana but they split off from Pangaea about 180 million years ago. Because of the long period of separate development, these large flightless birds are no longer closely related.

2. Australia (and Antarctica) split off from Gondwana approximately 140 million years ago. At that time, both marsupials and early placental mammals lived in the same area. In South America and Africa, the placental mammals prevailed and outcompeted the marsupials. In Australia, the marsupials outcompeted the mammals and now marsupials are mostly found in Australia and nearby Papua New Guinea (though a few still survive in South America). Native placental mammals are rare in Australia. These mammals came by sea (seals), air (bats) or were introduced by humans (dogs, cats, camels, rabbits and rats).

3. The cichlid fish in the lakes of East Africa (Figure 9) are one of the largest families of vertebrates:

- Lake Victoria: 3,170 species of cichlids (99% endemic)

- Lake Tanganyika: 126 species (100% endemic)

- Lake Malawi: 200 species (99% endemic).

These populations have probably been at least partially isolated from each other for around 100,000 years. They have had different selection pressures due to slightly different environments. So, the fish have adapted to the different environments they live in. As long as the population is large enough, they will continue to diverge from other populations, but some isolated populations may be too small and die out.

Separation of species by isolation is not the only factor that causes speciation or changes to a population of organisms. When continents come together, land bridges form.

Endemism on isolated islands

There are about 450,000 small islands on Earth, making up 5% of the total land area. Remote islands are often high in **endemic species** but low in species richness. This is because only a few individuals from a population may have reached the islands and then reproduced successfully. As they are then isolated, speciation occurs over time.

Reasons for high rates of endemism are:

- specialization to exploit the ecological niches on the island

- reduction of genetic diversity in the species as population numbers are often low

- no or little input from outside the island—few new individuals arrive and those that do may no longer be able to breed with the island species.

But there are downsides to endemism among island species:

- low population numbers lead to high extinction rates

- reduced genetic diversity in the gene pool so less chance of adaptation to new conditions

- no or little defence against predation as it was not needed in isolation

- vulnerable to arrival of invasive species which predate or outcompete them

- habitat destruction by humans or natural events and low resilience to changes.

As well as islands, isolated areas such as the mountains of the Ethiopian Highlands or Lake Baikal (which is far from any other large sea) also have high rates of endemism.

Examples of island endemism include:

- Darwin's finches or Galapagos swimming iguanas

- 8,000 endemic species in Madagascar

- 90% of Hawaiian island species

- 50% of all higher plants, mammals, birds, reptiles and amphibians in Mauritius.

Look at Case study 1. There is a theory that chimps and bonobos were one species which became separated into two populations north and south of the Congo River. About 2.5 million years ago, a drought in the southern region resulted in loss of food plants for gorillas which then died or migrated north. The southern "bonobo" population then had the remaining food source for themselves with no need to compete with gorillas or each other. They could form large social groups because there was enough food for them. In the north, gorillas and "chimps" had to share the niche for foods and competed. There was not enough food for larger groups to travel together so they developed weaker social bonds and more aggression because they had to fight for food to survive. The environmental pressure was different, leading to differentiation of the two populations into two species.

There is evidence that bonobos and chimps can mate and produce healthy and fertile offspring. This has occurred in captivity and there is DNA evidence that it has also occurred, if rarely, in the wild when an individual managed to cross the river and find a mate.

Chimps and bonobos demonstrate the process of speciation through geographical isolation but more research needs to be done. At what point are two species formed when populations become separated?

Bonobos and chimpanzees

There are two African apes which are the closest living relatives of humans. They share 98% of their genome with humans. These are the chimpanzee (*Pan troglodytes*) and the bonobo (*Pan paniscus*). Although they are similar physically, bonobos and chimpanzees differ in key social and sexual behaviours.

In the wild, chimpanzees are widespread across equatorial Africa including north of the Congo River. Bonobos live only south of the Congo River in the Democratic Republic of Congo in a relatively small and remote habitat. Bonobos are the rarest apes in captivity and little studied.

DNA sequences in bonobos diverged from those in chimpanzees around two million years ago which is relatively recent in terms of speciation. Bonobos are therefore closely related to chimpanzees.

Bonobos:

- males are less aggressive than chimpanzees and males and females are more playful

- female-dominant societies

- not observed to hunt cooperatively or use tools in the wild

- slender with pink lips

- limited geographical range.

▲ Figure 10 Female bonobo

Chimpanzees:

- male-dominant societies

- aggression between groups can be lethal

- use tools

- hunt cooperatively

- more robust bodies

- wider geographical range.

▲ Figure 11 Female chimpanzee

(ATL) Activity 5

ATL skills used: thinking, communication, social, research, self-management

Speciation can be due to geographical or behavioural barriers.

In small groups, research a contemporary example of speciation, for example:

- apple maggot/hawthorn fly (*Rhagoletis pomonella*)

- North Atlantic killer whale (*Orcinus orca*)

- a new, recent finch species in the Galapagos

- Northern and Mexican spotted owls

- polyploid wheat.

Present your findings as a poster or slide show.

Hotspots

Biodiversity is not equally distributed on Earth. Some regions have more biodiversity than others—more species and more of each species than in other areas. These are hotspots where there are also unusually high numbers of endemic species (e.g. the lemurs of Madagascar). Where these hotspots are is debated but there are currently about 36 of them globally.

Hotspots:

- include about 10 in tropical rainforest but also regions in most other biomes

- tend to be nearer the tropics because there are fewer limiting factors in lower latitudes

- are all threatened areas where 70% of the habitat has already been lost

- have more than 1,500 species of plants which are endemic in the habitats

- cover only 2.5% of the land surface

- tend to have large densities of human habitation nearby.

These hotspots are shown in Figure 12; between them they contain about 60% of the world's species so have very high species diversity.

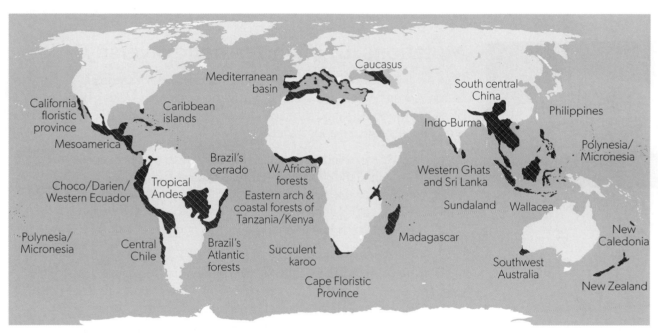

▲ Figure 12 Regions marked with red hatching are biodiversity hotspots

Critics say that naming hotspots can be misleading because they:

- focus on vascular plants and ignore animals

- do not represent total species diversity or richness

- focus on regions where habitats, usually forest, have been lost and ignore whether that loss is still happening

- do not consider genetic diversity

- do not consider the value of services, for example water resources.

But they are a useful model to focus our attention on habitat destruction and threats to unique ecosystems and the species within them.

ATL skills used: thinking, social, research

1. Research hotspots around the world and complete Table 2 for at least 10 hotspots.

2. Why is the number of plant species important for biodiversity?

3. What are the criteria for defining hotspots?

4. What is an endemic species?

5. For what reasons do you think hotspots are threatened areas?

6. Discuss how environmental sustainability may conflict with economic expansion and growth of a country.

Hotspot	Number of plant species	Number of endemic plant species

▲ Table 2

Mutations, DNA and genetic diversity

▲ Figure 13 Albinism in a hedgehog due to one mutation

DNA exists in every cell of every living organism. It contains the instructions needed for an organism to develop, survive and reproduce. This genetic code is called the genotype and although unseen, it determines the phenotype of an organism—its outward appearance.

For the body to be maintained, all cells must divide to produce replicas of themselves. During this division, errors may occur in the replication of DNA, leading to mutation. Such mutations can be in a single gene or multiple genes. Mutations are a natural occurrence but outside factors such as chemicals, infections, radiation, or human interference can also be a cause.

During sexual reproduction half the DNA from the egg cell combines with half the DNA from a sperm to give the full complement of the genetic code for the next individual. This process causes endless mixing of the genetic code of two individuals, which in a large population will increase genetic variation.

The resultant mutations:

*   may or may not have any effect on the individual organism

*   may be internal or may impact the physical appearance of the individual

*   may be beneficial or detrimental

*   are not intentional

*   may be small or large scale in their impacts, e.g. antibiotic-resistant bacteria.

Humans can speed up speciation by artificial selection of animals and plants and by genetic engineering but the natural process of speciation is a slow one.

Human impact on natural selection

Connections: This links to material covered in subtopics 3.2, 4.3, 5.2, 6.2 and 6.3.

ATL Activity 7

ATL skills used: communication, research

Table 3 shows human population growth since the year 1 CE.

Year	1	1000	1200	1400	1500	1600	1800	2000
Population (000,000)	188	295	393	390	461	554	990	6,143

▲ Table 3

1. Draw a line graph of the data in the table.

2. Find three estimates of the current global human population.

3. Evaluate the accuracy of these data sources.

The human population has grown immensely since about 1600. At time of writing (2023) the global population had just exceeded eight billion. With this growth comes a price—we are putting increasing pressure on the natural environment. We all know these stresses: overexploitation of natural resources, deforestation, global climate change, and hunting and poaching to name a few. What many people are not aware of is the ways in which humans have impacted natural selection and driven evolutionary changes in many species.

In agriculture the continuous use of pesticides, insecticides and herbicides has driven species to evolve resistance to these chemicals. In response, industry develops different pesticides in an attempt to keep pests under control and so the battle continues.

With advances in medicine, humans no longer have to suffer certain diseases and infections. However, over time pathogens such as bacteria, viruses, fungi and parasites develop resistance to the medication. A well-known example of this is methicillin-resistant *Staphylococcus aureus* (MRSA) which is very difficult to treat. Such resistance can result in the spread of diseases and cause illness or death.

Poaching in Gorongosa National Park, Mozambique

Elephants in Gorongosa are poached for their ivory tusks. Ivory has no intrinsic value, but it has significant cultural value as a status symbol. It has been carved and made into artworks for millennia. Its current value on the black market is estimated to be USD 1.44 billion a year. In Gorongosa, this activity has resulted in a 90% crash in the elephant population since 1970.

Poaching appears to have resulted in some unexpected evolutionary changes. In areas where there is intensive hunting, the African elephants have evolved to be tuskless. In this area before 1970, approximately 18.5% of the females were naturally tuskless—this is an evolutionary trait that made them unattractive to poachers. Now, however—according to research published in the respected journal *Nature* in 2021—33% of the females are tuskless. In this case, being tuskless is an evolutionary advantage.

▲ Figure 14 Elephant mother and calf

(ATL) Activity 8

ATL skills used: thinking, research

The genetic changes in the Gorongosa elephants were an indirect impact of poaching.

Methicillin-resistant *Staphylococcus aureus* (MRSA) is a pathogenic bacterium that is resistant to a range of antibiotics. This pathogen is very dangerous because it is extremely difficult to successfully treat MRSA infections. The development of drug resistance by *S. aureus* could be seen as a direct result of over-prescription of antibiotics.

1. To what extent do you think the evolution of (a) tuskless elephants and (b) antibiotic resistance was avoidable?

2. What could humans do to reverse these changes?

3. Should humans ever eradicate a disease or infection because we can?

4. Should humans keep specimens of disease pathogens such as smallpox? Is it ethical?

The geological timescale

The Earth formed about 4.6 billion (4.6×10^9) years ago. The first life forms are thought to have been simple cells like bacteria. Some 65 million years ago, the dinosaurs became extinct. The human species has been recognizably human for the last 200,000 years. To put this timescale into a perspective based on the 24-hour clock, consider geological time as the full 24 hours. Humans appeared at a few seconds to midnight.

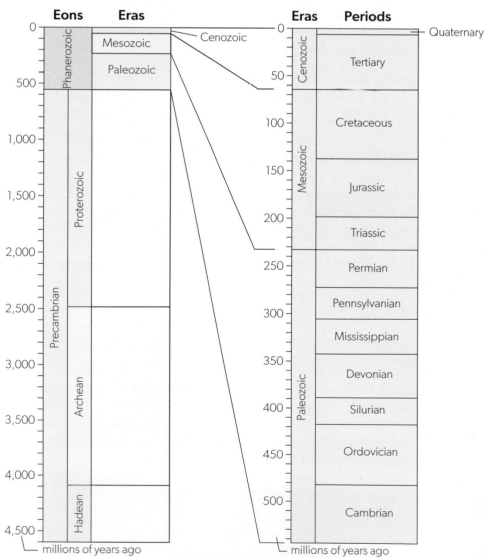

▶ Figure 15 The geological timescale

Adapted from: Six Red Marbles and OUP

The geological timescale is divided into sections by time. The names of these divisions do not need to be remembered but you may recognize a few.

Eons: longest time span of at least one billion years. These are marked by major events such as:

- the first signs of life on Earth in the appearance of single-celled organisms

- the appearance of multicellular organisms

- the development of complex, diversified life forms.

Eras: shorter than eons and determined by significant events such as:

- stabilization of the continents and increase in oxygen levels in the atmosphere

- breaking up of supercontinent Rodinia and evolution of sexual reproduction

- snowball Earth (when the Earth's surface was nearly all frozen)

- life on land and mass extinctions.

Periods: range from 1–100 million years in length. We are in the Quaternary period.

Epochs: smaller divisions spanning not more than 10 million years. These are marked by significant changes in fossils, which indicate environmental changes causing extinctions and the evolution of new species. The current epoch is the Holocene starting 11,000 years before the present, although many scientists argue that we have now entered a new epoch called "the Anthropocene" (see following section).

Fossils

A fossil is the preserved remains, impression, or trace evidence of organisms from the past. They were buried in sand or mud and are typically at least 10,000 years old. Bones and other hard body parts may be preserved. On rare occasions, the soft body parts may also be preserved.

▲ Figure 16 **Fossil of a dinosaur**

Fossils give us useful insights into the Earth's history. We can learn what the environment was like, how and when it changed, how a single landmass broke up and how the continents moved. Through fossils we can trace the evolution of species and construct a phylogenetic tree to show evolutionary relationships between species. (This model is sometimes called the "tree of life" model.) Fossils also help geologists to date rocks and create models of extinct animals.

Artificial selection

Natural selection is driven by nature and survivability in wild populations. Artificial selection is driven by humans and is most often seen in domesticated animals. Humans see certain traits as favourable, so they select seeds or animals with those desirable, inheritable traits and breed those organisms. This has been done for centuries, before we knew about evolution through genetics. Farmers in particular selected livestock and plants that gave higher yields or were more robust. Examples include cows that produce more milk and wheat that produces better ears. More recently, selective breeding has been used to produce a wide variety of cats and dogs as pets.

▲ Figure 17 **Domestic cat—bred for fluffiness**

Artificial selection has both advantages and disadvantages:

- It can increase productivity.

- Useful new varieties may be developed (e.g. the loganberry was created by cross-breeding the North American blackberry and the European raspberry).

- It is possible to produce plants with a higher resistance to pests or disease.

- Breeding organisms (particularly animals) for characteristics that humans want but that are not necessarily a survival advantage, can be seen as inhumane. Chickens without feathers cannot flap their wings and they are more prone to parasites and sunburn; hip dysplasia is common in many dog breeds (e.g. German Shepherd) due to consistent inbreeding to gain a particular posture.

- Breeding for selected traits can be a problem because those traits may not match changing environments on Earth.

- Selective breeding reduces resilience of a species because it restricts genetic diversity and therefore the ability of the species to adapt and survive.

> Connections: This links to material covered in subtopics 4.3 and 5.2.

Practical

Skills:
- Tool 2: Technology
- Inquiry 2: Collecting and processing data
- Inquiry 3: Concluding and evaluating

For hundreds of years humans have been selecting the "best" characteristics of organisms to improve agricultural yields. Choose a media format (podcast, video, newspaper article, creative writing) and create a short presentation.

1. Find a list of the different organisms humans have selectively bred to improve agricultural yield (economics).

2. Investigate the history of the artificial selection:
   a. How was selection done (e.g. any laws controlling experiments)?
   b. What changes took place in the organism (e.g. taller, shorter, fatter, faster growth)?
   c. What were the advantages and disadvantages of this selection process for:
      i. the organism   ii. humans?

Extinction

Extinction is the dying out of a species. There are two types of extinction which are determined by the rate at which extinction occurs. Both rates are estimated from the fossil record.

- Background extinction rate is the natural extinction rate of all species. Scientists think it is about one species per million species per year. This works out to between 10 and 100 species going extinct per year.

- Mass extinction is when species go extinct at a much faster rate than the background rate. This is generally seen as being over 75% of the world's species going extinct quickly. In the geological context, "quickly" means in fewer than 2.8 million years.

There have been five major mass extinctions in the geological record, spread over 500 million years. These mass extinctions show in the fossil record because the fossils are abruptly lost from the rock strata. Scientists think mass extinctions may be due to a rapid change of climate. This could be caused by a natural disaster (e.g. volcanic eruption, meteorite impact) which results in a change in climate conditions that many species cannot survive.

The Holocene extinction event

Most biologists now think we are in the sixth mass extinction. This is called the Holocene extinction event because the Holocene is the part of the Quaternary geological period that we are in now (see Figure 15). This extinction event started at the end of the last ice age about 9,000 to 13,000 years ago when large mammals such as the woolly mammoth and sabre-toothed tiger became extinct, probably through hunting. But the rate has accelerated in the last 100 years. While it may be partly due to climate change, the big difference is that it has been caused by one species—*Homo sapiens*.

> **Connections:** This links to material covered in subtopic 2.1.

We think there are about 5,000 mammal species alive today. Their background extinction rate would be one per 200 years but the past 400 years have seen 89 mammalian extinctions, much more than the background rate. Another 169 mammal species are listed as critically endangered or the "living dead". "Living dead" species are ones which have such small populations that there is little hope they will survive, or ones that have lost a species they depend on, such as a pollinator insect for a flowering plant or a prey species for a predator.

ATL Activity 9

ATL skills used: thinking

Lonesome George was the world's most famous reptile. He was a Pinta Island giant tortoise from the Galapagos Islands and, at over 100 years old, was the last of his subspecies. He was found in 1971 and no other tortoise had been recorded on the island since 1906. George's fellow tortoises were taken by sailors for meat or goats destroyed their habitat and food source. He was a symbol of what went wrong for many species whose fate is in the hands of humans.

1. Do you think it matters that Lonesome George was the last of his kind?

2. Is anyone to blame for this?

▲ Figure 18 Lonesome George: born before 1912, died 24 June 2012

3. What could have been done?

4. Is the title "Lonesome George" anthropomorphizing the tortoise?

5. Does this matter?

You may have heard of some of the earlier mass extinctions, particularly the one when the dinosaurs became extinct. This was the Cretaceous–Tertiary extinction, also called the K–T extinction or boundary. It happened about 65 million years ago. Most of the large animals on land and in the sea died out, along with small oceanic plankton, but most small animals and plants survived. The causes have been argued about for years but the general view now is that it was caused by the combination of the following events.

- A volcanic eruption: The volcanoes of the Deccan plateau in what is now India erupted for a million years at the time of the K–T boundary.

- A meteor impact put huge amounts of dust into the atmosphere: The evidence for this impact is the Chicxulub crater in the Yucatan peninsula, Mexico. The crater is 180 km in diameter and the igneous rock underneath contains high levels of iridium, a mineral common in extraterrestrial objects such as meteorites but not common on Earth.

- Climate change over a long period: the eruption and the meteor impact would have caused dust clouds to block much of the incoming solar radiation. Plants would have been unable to photosynthesize, so food webs would have collapsed.

All the mass extinctions are shown in Table 4.

Mass extinction	MYA (million years ago)	Geological period	Estimate of losses (a family contains up to 1,000 species)
Sixth	Now	Holocene	unknown
Fifth	65	Cretaceous–Tertiary	17% families and all large animals including dinosaurs
Fourth	199–214	End Triassic	23% families, some vertebrates
Third	251	Permian–Triassic	95% of all species, 54% of families
Second	364	Devonian	19% families
First	440	Ordovician–Silurian	25% families

▲ Table 4 The six mass extinctions

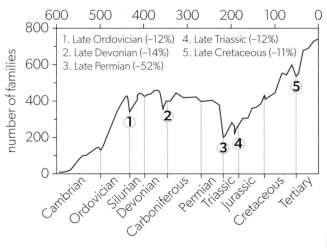

▲ Figure 19 The first five major extinction episodes of life on Earth shown by family diversity of marine vertebrates and invertebrates (after E.O. Wilson, 1988—see below)
Adapted from: Six Red Marbles and OUP

After each of the first five mass extinctions, there was a burst of adaptive radiation of the remaining species which adapted to fill the ecological niches left vacant (see Figure 19). But this is all within a geological and evolutionary timescale—so it happened over tens of millions of years.

Current extinction rates

Humans do not know how many species there are on Earth but a low estimate is around two million; about 1.4–1.8 million have been identified. At present, there are an estimated 200–2,000 extinctions each year. If the species estimate of two million is correct, this is mass extinction with a rate between 1,000 and 10,000 times the natural extinction rate. We do not even know what some of the species were.

Over the past 250–300 years, human activities have increased species extinction rates by as much as 1,000–10,000 times the background rate. Many scientists are calling this the sixth mass extinction. E.O. Wilson, a well-known biologist from Harvard, suggests that 30–50% of species could be extinct within 100 years. The rate is estimated to be about three species per hour. Habitat loss is the major cause of species extinction.

The rate of extinctions is not constant worldwide: it is far greater in certain areas called hotspots, which have a high concentration of different species. Up to 50% of animal and plant species are in one of the 36 hotspots which together make up only 2% of the land area on Earth. These areas are very vulnerable to habitat loss and many species within them are endemic. Tropical rainforests and coral reefs are particularly vulnerable and some species are more likely to become extinct than others—in particular, species that humans like to hunt or eat or wear and those that are dangerous to us or our crops.

Extinction rates are not linear. When half a habitat is lost, animals and plants remain in the other half. There are fewer of them but the species are still there. Only when nearly all the habitat has disappeared do extinction rates increase rapidly—so the current rate can only increase. However, scientists think that protecting just 5% of a habitat could preserve 50% of the species within it.

In the sixth mass extinction, humans have wiped out many large mammal and flightless bird species, such as woolly mammoths and ground sloths, and moas in New Zealand. In previous centuries, many extinctions were due to hunting for food—for example, dodos in Mauritius. Now most extinctions are due to habitat loss or degradation. One estimate suggests that 25% of all plant and animal species became extinct between 1985 and 2015—and the rate is accelerating.

The UN estimated that the human population reached six billion on 12 October 1999, seven billion on 12 March 2012 and eight billion on 15 November 2022. It is expected to reach 9.7 billion in 2050, reach a peak in 2085 of nearly 10.4 billion and then decrease. Humans alter the landscape on an unprecedented scale. Some organisms do well in the environments that we create (urban rats, domesticated animals, some introduced species) but most do not. We call the successful ones "weedy species", both animals and plants. Many weedy species will probably survive and thrive in the current mass extinction. But others, many never identified, are likely to die out. The question we should ask is whether humans are a weedy species or not. It has taken 5–10 million years to recover biodiversity after past mass extinctions—equivalent to more than 200,000 generations of humankind. To put this into context, there have been about 7,500 generations since the emergence of *Homo sapiens* and human civilization started only 500 generations ago.

The previous mass extinctions were due to physical (abiotic) causes over extended periods of time, usually millions of years. The current mass extinction is caused by humans (anthropogenic) so has a biotic cause and is accelerating rapidly over a few decades, not millennia. Humans are the direct cause of ecosystem stress because we:

- transform the environment—with cities, roads, industry, agriculture

- overexploit other species—in fishing, hunting and harvesting

- introduce alien species—which may not have natural predators

- pollute the environment—which may kill species directly or indirectly.

WWF, the World Wide Fund for Nature, produces a periodic report on the state of the world's ecosystems, called the *Living Planet Report*. The 2012 report was a grim account of loss and degradation, describing a 30% decline in the overall living planet index (a measure of ecosystem health) between 1970 and 2008. The figures in the 2022 report are worse, revealing an average 69% decrease in monitored wildlife populations since 1970. The worst hit areas are Latin America and the Caribbean with 94% losses; even the "best" areas—Europe and Central Asia—show an 18% loss. However, while major losses are seen in the tropics, temperate oceans and terrestrial ecosystems are seeing some improvements.

The living planet index measures trends in the Earth's biological diversity. The 2012 report followed populations of 2,500 vertebrate species (fish, amphibians, reptiles, birds, mammals), while the 2022 report looked at 32,000 species from all around the world to give a numerical index of changes in biodiversity (see Figure 20).

▶ Figure 20 Percentage changes in biodiversity (the living planet index) between 1970 and 2008

Adapted from: Our World in Data. Data from: WWF (2022) Living Planet Report 2022 - Building a nature-positive society. Almond, R.E.A., Grooten, M., Juffe Bignoli, D. & Petersen, T. (Eds). WWF, Gland, Switzerland.

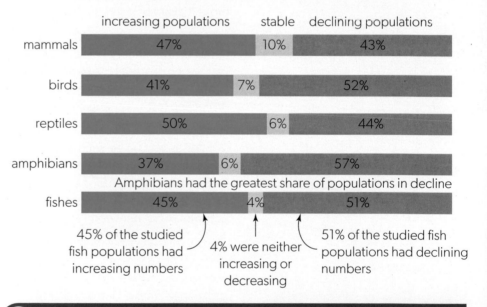

	increasing populations	stable	declining populations
mammals	47%	10%	43%
birds	41%	7%	52%
reptiles	50%	6%	44%
amphibians	37%	6%	57%
fishes	45%	4%	51%

Amphibians had the greatest share of populations in decline

45% of the studied fish populations had increasing numbers

4% were neither increasing or decreasing

51% of the studied fish populations had declining numbers

ATL Activity 10

ATL skills used: thinking, communication, research

Figure 20 shows data from the 2022 WWF *Living Planet Report*. Since 1970, there has been a reported 69% decline in wildlife populations.

1. In which three groups are the highest losses?

2. Suggest reasons for these losses.

3. Suggest with reasons which species may have increased population sizes.

4. To what extent are the data reliable?

There have been two stages to the sixth (current) mass extinction:

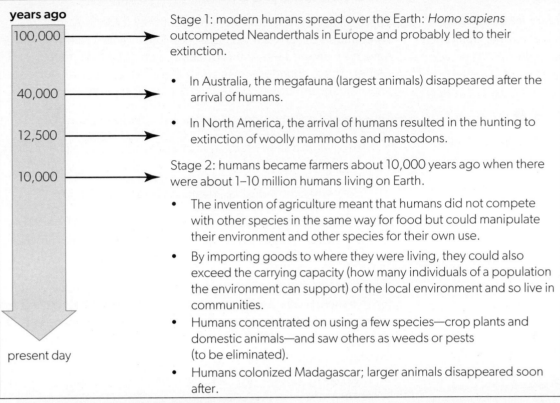

years ago

100,000 → Stage 1: modern humans spread over the Earth: *Homo sapiens* outcompeted Neanderthals in Europe and probably led to their extinction.

40,000 → • In Australia, the megafauna (largest animals) disappeared after the arrival of humans.

12,500 → • In North America, the arrival of humans resulted in the hunting to extinction of woolly mammoths and mastodons.

10,000 → Stage 2: humans became farmers about 10,000 years ago when there were about 1–10 million humans living on Earth.

• The invention of agriculture meant that humans did not compete with other species in the same way for food but could manipulate their environment and other species for their own use.

• By importing goods to where they were living, they could also exceed the carrying capacity (how many individuals of a population the environment can support) of the local environment and so live in communities.

• Humans concentrated on using a few species—crop plants and domestic animals—and saw others as weeds or pests (to be eliminated).

present day

• Humans colonized Madagascar; larger animals disappeared soon after.

▲ Figure 21 **The two stages of the sixth mass extinction**

ATL **Activity 11**

ATL skills used: thinking, research

Figure 22 shows human impacts on biodiversity.

1. Only in Africa did the large mammals survive. Why do you think this may be?

2. How have humans adapted to living in such a diverse range of climatic conditions?

3. What are the consequences of a higher rate of mass extinction for:

   a. humans
   b. the Earth?

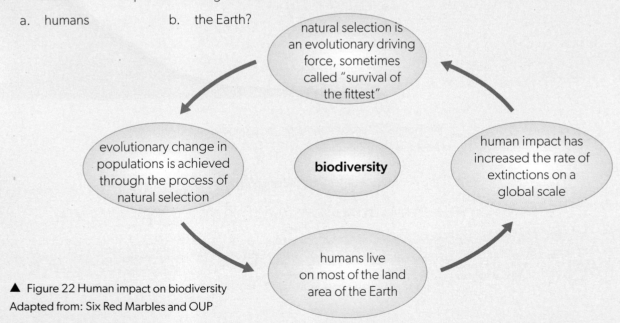

▲ Figure 22 **Human impact on biodiversity**
Adapted from: Six Red Marbles and OUP

The Anthropocene

There has been much discussion about the Holocene—was this the period when humans first made an impact on Earth? Many scientists argue that we have now entered the Anthropocene, with humans putting pressure on the planet. These changes are known as "golden spikes" because their effects are dramatic enough to be seen in the geological strata.

The start of the Anthropocene is debated. Was it:

- 1610, when a severe reduction in farming caused a dip in global carbon dioxide levels?

- 1760, when the Industrial Revolution began?

- 1950s, when the increase in the combustion of fossil fuels released large amounts of spherical fly ash particles which, being resistant to chemical degradation, appear in sediments and soil?

A study in *Nature* magazine showed that in 2020 the volume of artificial human-made materials was greater than that of all living beings for the first time ever. Is this what marks the start of the new geological Anthropocene era? On average, in 2022, each person on Earth produced anthropogenic mass greater than their own bodyweight every week.

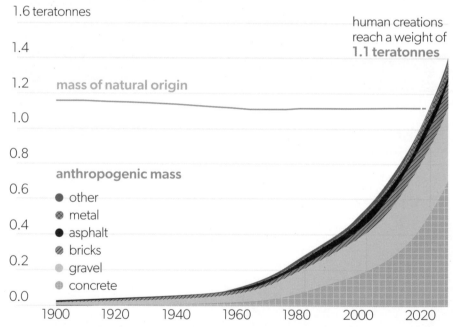

▲ Figure 23 The anthropogenic mass generated by humankind

Adapted from: Iberdrola

The Anthropocene is characterized by:

- technological progress during and since the Industrial Revolution

- population growth due to a better understanding of the causes of the high death rate plus improved methods of food production

- increases in production and consumption.

These three factors have allowed humans to evolve, leading to increased demand for all natural resources—minerals, water, energy sources, and land for farming, habitation and infrastructure.

Evidence for the Anthropocene

The impacts of human actions are numerous. Some are obvious and we experience them on a daily basis, like climate change. Other impacts can be less visible but still profound as evidence appears in the fossil record.

1. The geoscientist Jan Zalasiewicz thinks he has found the perfect marker for the start of the Anthropocene, and it is 16 July 1945 at 5.29. This is when the world's first atomic bomb was detonated by US scientists, releasing radioactive isotopes which will take millennia to decay. The identifying isotope is cesium 137. There is no natural source of this isotope, so if it is in the rocks, humans "put it there".

2. Climate change has been caused by the increased amount of carbon dioxide in the atmosphere from burning fossil fuels. In 2013, the concentration of carbon dioxide passed 400 parts per million (ppm)—a number once claimed to be the **tipping point**. In early 2023, it was 419.2 ppm.

    This is more than a 47% increase since the beginning of the Industrial Revolution, when it was 280 ppm. What is the concentration now?

3. The statistics for deforestation are staggering:

    *   Over 420 million hectares of forest have been lost since 1990.

    *   10 million hectares of forest were destroyed every year 2015–20.

    *   17% of the Amazon has been deforested—the tipping point is considered to be 20%.

4. In 2022, the International Union for the Conservation of Nature (IUCN) Red List listed over 42,100 species as threatened with extinction (more than a quarter of all assessed species).

> ### Key term
>
> A **tipping point** is a critical level beyond which the responses of a system are unpredictable and probably unstoppable.

▲ Figure 24 Deforestation

Practical

Skills:

*   Tool 2: Technology
*   Tool 3: Mathematics
*   Inquiry 3: Concluding and evaluating

Figure 25 and Table 5 show a small collection of data about the number of species that are under threat of extinction or that have become extinct since the end of 2002.

Produce an infographic for primary school children to show them what is happening and encourage them to increase biodiversity in their school.

▶ Figure 25 Groups of organisms threatened with extinction
Data from: IUCN Red List

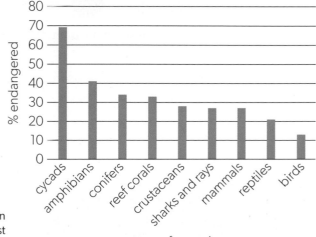

Organism	Date declared extinct	Cause
Pyrenean ibex	2000	Hunting for trophies of their huge horns caused population size reduction and a bottleneck with low genetic diversity. Eventually one female was left and she was killed by a falling tree (but another subspecies has been reintroduced)
Baiji river dolphin	2007	Rivers are overused for fishing, transportation and hydropower
Caribbean monk seal	2008	Humans eliminated their food source
Moorean viviparous tree snail	2009	An introduced predator, the rosy wolf snail, eradicated all native snails of Polynesia
West African black rhino	2011	Population lost by poaching for horn
Formosan clouded leopard	2013	Logging, loss of prey and poaching for pelts
Bramble Cay melomys	2015	Loss of habitat and rising sea levels
Spix's macaw	2016	Illegal pet trade and habitat destruction
Poo-uli	2019	Habitat destruction, spread of disease-carrying mosquitoes and invasive predators
Splendid poison frog	2020	Deforestation and habitat destruction

▲ Table 5 Some examples of organisms declared extinct since 2000

Check your understanding

In this subtopic, you have covered:

- types of biodiversity
- diversity and resilience
- natural selection and evolution
- speciation
- measuring biodiversity
- more on speciation
- human impact on speciation
- mass extinctions
- effects of the Anthropocene.

How can diversity be explained and quantified, and why is this important?

How does the unsustainable use of natural resources impact biodiversity?

1. Review and describe the three types of biodiversity.
2. Explain what Simpson's reciprocal diversity index measures.
3. Explain speciation by natural selection.
4. Explain what is happening in the sixth mass extinction.
5. Describe your worldview about other organisms.
6. Discuss your view about the ethics of extinction caused by human activity.
7. Review resilience and factors that make an ecosystem less resilient.
8. Explain how human activity reduces biodiversity.

⟩⟩ Taking it further

- Research one species that has gone extinct in your lifetime and prepare a three-minute talk for the class on its ecological role and reasons for its loss.
- Check if there is a citizen science and biodiversity project near you (these are local projects allowing citizens to identify and monitor plant or animal species in their local area).
- Create a podcast exploring the Age of the Anthropocene.
- Research reasons for tuskless elephants increasing in number in areas of civil conflict.
- Find local examples of the impact of inequality on knowledge of biodiversity.

3.2 Human impact on biodiversity

Guiding question

- What causes biodiversity loss and how are ecological and societal systems impacted?

Understandings

1. Biological diversity is being adversely affected by both direct and indirect influences.

2. Most ecosystems are subject to multiple human impacts.

3. Invasive alien species can reduce local biodiversity by competing for limited resources, predation and introduction of diseases or parasites.

4. The global conservation status of species is assessed by the International Union for Conservation of Nature (IUCN) and is published as the IUCN Red List. Status is based on number of individuals, rate of increase or decrease of the population, breeding potential, geographic range and known threats.

5. Assigning a global conservation status publicizes the vulnerability of species and allows governments, non-governmental agencies and individual citizens to select appropriate conservation priorities and management strategies.

6. Investigate three different named species: a species that has become extinct due to human activity; a species that is critically endangered; and a species whose conservation status has been improved by intervention.

7. The tragedy of the commons describes possible outcomes of the shared unrestricted use of a resource, with implications for sustainability and impacts on biodiversity.

8. Biodiversity hotspots are under threat from habitat destruction, which could lead to a significant loss of biological diversity, especially in tropical biomes.

9. Key areas that should be prioritized for biodiversity conservation have been identified on the basis of the international importance of their species and habitats.

10. In KBAs, there is conflict between exploitation, sustainable development and conservation.

11. Traditional indigenous approaches to land management can be seen as more sustainable but are facing challenges of population growth, economic development, climate change and a lack of governmental support and protection.

12. Environmental justice must be considered when undertaking conservation efforts to address biodiversity loss.

13. The planetary boundary "loss of biosphere integrity" indicates that species extinctions have already crossed a critical threshold.

Connections: This links to material covered in subtopics 4.4, 5.3, 6.2, 6.3, 7.2 and 8.3.

Threats to biodiversity

Most people will acknowledge that the primary threat to biodiversity is humans, the speed at which our populations are growing and the consequences this has for resource use. Humans need basic resources to survive such as food, water and shelter but we also want luxuries such as cars, computers and a plethora of other things. These resources are not being exploited sustainably and nature pays the price.

▲ Figure 1 Forest fires are both natural and human-induced

Natural hazards and environmental disasters

Naturally occurring events may have a negative impact on the environment (and humans). Natural events are seen as a hazard or even a natural disaster, when the negative impacts cause significant harm to human communities (see Table 1). Human activities can also have negative impacts at a massive scale, and such events are called environmental disasters. Examples include:

- Union Carbide cyanide gas leak, Bhopal, India (1984)

- nuclear power plant explosion at Chernobyl, Ukraine (1986)

- *Exxon Valdez* oil spill (1989)

- British Petroleum oil spill, Gulf of Mexico (2010).

ATL Activity 12

ATL skills used: thinking, communication, social, research, self-management

Work in pairs or groups. Each person chooses an example of an environmental disaster, a natural hazard or a natural disaster.

Research the causes and effects of your chosen event. Share your findings and discuss the similarities and differences.

Natural hazard	Example of natural hazard	Impact on human communities = natural disaster
Volcanic eruptions	Eruption of Mount St Helens in Washington State, USA in 1980	Reduced vast areas to wasteland, killed thousands of animals Cost USD 1 billion
Earthquakes	Haiti 2010 earthquake	Killed 160,000
Floods	Yangtze River floods in China in 1998	4,000 killed and 14 million people homeless
Wildfires	Bushfires, Australia, 2019–2020	Damaged approximately 18 million hectares of land 9,000 buildings and homes destroyed 400 deaths (directly or indirectly)
Hurricanes	Hurricane Ike, 2008	Caused extensive damage and many deaths across the Caribbean

▲ Table 1 Examples of recent natural hazards

Habitat loss

Habitat is the environment in which a species lives. The stability and variety of habitats are important to the maintenance of biodiversitvy: habitat loss is the major cause of loss of biodiversity due to human activities.

In many parts of the world, humans have destroyed or changed most of the original natural habitat to develop or build on the land.

- In the Philippines, Vietnam, Sri Lanka and Bangladesh, where human population levels are high, we have lost most of the wildlife habitat and most of the primary rainforest.

- In the Mediterranean region, only 10% of original forest cover remains.

- Madagascar is the only place where lemurs occur and there are many endemic species. Loss of their habitat is highly likely to make them extinct. The humid lowland forests where most endemic species live are degraded and fragmented due to shifting cultivation not leaving enough time for regeneration before clearing the land again. This is due to human population pressures. It is thought that about half the forests have been lost.

Habitats are lost when they are damaged or destroyed so much that they can no longer support the species that live there. The loss is generally to make way for human activities such as transportation routes, agriculture, mining, logging or or water supply (dams). Occasionally, nature destroys habitats through volcanic eruptions or extreme weather events such as tornadoes and hurricanes.

Habitat loss is a broad term that covers habitat destruction, degradation and fragmentation.

Habitat destruction

Destruction is usually caused by machinery—for example, bulldozers removing the vegetation. Such destruction can cause the loss of biodiversity and possibly the extinction of certain species. But sometimes it clears the way for succession and the evolution of new ecosystems. Life is resilient but humans destroy habitats at a scale and speed that exceeds nature's ability to heal.

Habitat degradation

This is not so much the removal of habitats but the downgrading of their quality to the extent that flora and fauna can no longer survive there. This is achieved through pollution, climate change and the introduction of invasive species.

This type of habitat loss can also cause problems for human populations because degrading the land makes it more susceptible to desertification and nutrient depletion of the soil. Erosion becomes more likely. None of this is good for natural habitats nor for agricultural production.

Habitat fragmentation

Habitats are changed from a single large coherent area into a patchwork of fragments, separated from each other by roads, towns, factories, fences, power lines, pipelines or fields. The fragments are isolated in a modified or degraded landscape and they act as islands within an inhospitable sea of modified ecosystems. This restricts the movement of animals and reduces the range over which the animals can roam. Some animals are at a higher risk of extinction and separation of breeding populations reduces genetic diversity.

Fragmented habitats suffer from edge effects where there are greater fluctuations of light, temperature and humidity than in the middle. As the areas get smaller, the edge effects impact a proportionally larger area. Greater edge effects also mean more invasion by pest species or the possibility that domestic and wild species come into contact and spread diseases between populations.

Pollution

Pollution is not just visible things such as rubbish and litter; it also includes hidden pollution. Most pollution is caused by human activities. It degrades or destroys habitats and makes them unsuitable to support the range of species that a pristine ecosystem can support.

Air pollution is the contamination of air with particles or poisonous gases. The most common causes are:

* partially combusted exhaust gases

* industrial by-products such as sulfur dioxide and carbon monoxide

▲ Figure 2 Habitat destruction for development

Connections: This links to material covered in subtopics 1.2 and 6.2.

- products of burning of plastics and rubber

- small particulate matter, PM$_{2.5}$, from forest fires, burning organic waste and wind-blown dust.

Air pollution causes breathing problems, and irritation of the eyes and respiratory system.

Light pollution occurs when artificial lights used by humans alter natural light levels. This disrupts ecosystems by altering plant growth patterns and disrupting animal navigation systems. If you look at the night sky in a city, you see very few stars because of light pollution. If you look up at night in an area that is a long way from human habitation, you will see a sky full of stars.

Noise pollution is when human activities cause an excessive amount of sound. This form of pollution affects humans mentally and physically, causing stress, hearing loss, sleep disturbance, high blood pressure and depression. It could have the same effect on animals but we are not sure about that. We do know that noise pollution disturbs the navigation, communication and feeding strategies of animals.

Soil pollution occurs when the soil is contaminated with abnormal levels of toxic substances. Some soil pollutants and their effects are shown in Table 2.

Type of pollutant	Source	Impact
Heavy metals	Medical waste, mining, agriculture and e-waste	Prolonged exposure affects liver, kidney, brain and blood
Polycyclic aromatic hydrocarbons (PHA)	Coal processing, vehicle emissions and cigarette smoke	Linked to cancers
Herbicides	Agriculture	Kill wild plants that are non-target plants
Insecticides		Kill non-target parasitic fungi
Fungicides		
Radioactive waste	Nuclear power stations	Reduces soil fertility by causing an imbalance in the nutrients
		Reduces organic matter and kills soil organisms
		Causes mutative diseases

▲ Table 2 Sources and impacts of some of the major soil pollutants

▲ Figure 3 Water pollution from point source pipes

Soil pollution can cause soil erosion. This happens because soil organisms die and the soil structure breaks down. The soil becomes crumbly and prone to wind and water erosion.

Water pollution contaminates all aquatic environments plus groundwater stores. Water is vital to all life on Earth, so contamination of water is a problem. Insecticides, pesticides and industrial waste can cause the death of aquatic organisms due to their toxicity. Fertilizers and detergents cause eutrophication which reduces oxygen levels in bodies of water and blocks the sunlight. The main sources of water pollution are industrial waste, detergents, agriculture, oil spills and road wash.

There are many measurements that can be taken to assess the levels of pollution:

1. Air: tropospheric ozone (Schoenbein test paper), particulate matter, air quality index

2. Light: use a light meter or lux meter (there are many apps)

3. Soil: pH, phosphorous, potassium, heavy metals

4. Water: acidity (pH), dissolved oxygen, dissolved solids, suspended sediment (turbidity), temperature

5. Noise: measured in decibels (dB) by a sound level meter.

See subtopic 2.6 for relevant practical work.

Overexploitation

When population densities were low, traditional customs and practices prevented overexploitation of species. However, the human population has grown massively and methods of hunting and harvesting have become many times more efficient than traditional practices. More and more humans are now overexploiting the environment. Rural poverty has increased and for many the choice is between starvation or eating bush meat (wild animal meat), being cold or felling the last tree for fuel. These are not really choices.

With population expansion came technological advances and increased wealth. Humans no longer catch animals just to eat. We are getting better and better at catching, hunting and harvesting.

*   Chain saws have replaced hand saws in the forests.

*   Factory ships with efficient sonar and radar find and process fish stocks.

*   Bottom trawling scoops up all species of fish whether humans eat them or not.

Humans "harvest" the natural environment for animals to fill the ever-increasing demands of the pet trade. Victims of this trade include many primates, birds and reptiles.

The Grand Banks off Newfoundland were once one of the richest fishing grounds in the world. Now they are fished out. If we exceed the **maximum sustainable yield (MSY)** of any species, the population is not sustainable (see subtopic 1.3). The difficulty is knowing what the MSY figure is and how much is left.

Introducing non-native (exotic or alien) species

Introduction of non-native species can drastically upset a natural ecosystem. Humans have done this through colonization of different countries, bringing their own crops or livestock. Sometimes exotic species are introduced by accident or by unexpected escapes from gardens or zoos.

TOK

There may be long-term consequences when biodiversity is lost. Should people be held morally responsible for the long-term consequences of their actions?

Key term

Maximum sustainable yield (MSY) is the maximum harvest we can sustainably take each year and still allow replacement by natural population growth.

▲ Figure 4 Rubber tapping, the collection of latex from rubber trees. Natural rubber is mostly collected by smallholders working on small plots in rainforest. They cut the bark (called tapping) and collect the fluid latex that drips out. One tree can be tapped for 30 years before being used as timber and others planted. This is a sustainable practice and MSY cannot be exceeded

Sometimes such introductions work well. Examples include:

- potatoes from the Americas to Europe

- rubber trees from the Amazon to Southeast Asia.

Sometimes introducing an exotic species is a disaster.

- Rhododendrons were introduced to Europe from Nepal by plant collectors because of the large and beautiful flowers. But they have escaped into the wild where they outcompete the native plants and are toxic.

- Dutch Elm disease entered Europe via imported American logs. The disease decimated elm tree populations.

- Sudden oak death was imported in the same way as Dutch Elm disease.

- Australia has been particularly unfortunate with rabbits, cane toads, red foxes, camels, blackberry, prickly pear and the crown of thorns starfish, to name just a few. The very different flora and fauna of Australia are well adapted to their environment but unable to compete with aggressive invasive species.

ATL Activity 13

ATL skills used: thinking, communication, research

Rabbits are not indigenous to Australia. In 1859, a few rabbits were shipped from Europe to Australia for sport and for meat. With no predator in Australia, they multiplied exponentially and after 10 years, there were estimated to be two million. The rabbits ate all the grass and forage so there was none for sheep and the farming economy collapsed. They also caused erosion as the topsoil blew away and probably caused the extinction of many Australian marsupials which could not compete with them.

It was so bad that in the period from 1901 to 1907, Western Australia built a rabbit-proof fence. It ran from north to south and was over 3,200 km long. Other control methods are shooting, trapping and poisoning.

In 1950, the **myxomatosis virus** was brought from Brazil and released to control the rabbits. This biological control measure caused an epidemic that killed up to 500 million of

the 600 million rabbits. But the rabbits were not eliminated because a very few were resistant to the virus. They bred and now rabbits are again a problem with the population recovering to 200–300 million within 40 years. In 1996, the government released **rabbit haemorrhagic disease** (RHD) in a second attempt to control the rabbits. But again the attempt failed. Neither of these biological control measures eliminated the rabbits.

1. Describe the factors that contributed to the success of rabbits in Australia.

2. Explain what is a biological control measure.

3. Research and describe the impact of the introduction of cane toads (*Rhinella marina*) and prickly pear (*Opuntia*) into Australia.

4. List what these three species have in common in terms of being a pest.

Spread of disease

Some diseases can decrease biodiversity. The last population of black-footed ferrets in the wild was wiped out by canine distemper in 1987 and the Serengeti lion population is reduced because of distemper. Diseases of domesticated animals can spread to wild species and vice versa, particularly if population densities are high.

In zoos, disease is a constant threat where species are kept close together. Diseases tend to be species specific (only affect one species). But if they mutate, they can infect across the species barrier and this is an ever-present threat. Recent examples include the following.

- Swine flu outbreak in 2010—swine flu is endemic in pigs and can sometimes pass from pigs to humans.

- Bird flu (avian flu) is a virus adapted to both birds and humans and a very pathogenic strain (H5N1) has spread through Asia to Europe and Africa since 2003.

- Foot-and-mouth disease (FMD) is a virus that affects all animals with cloven hooves (cattle, sheep, pigs, deer, goats) and spreads easily among these populations. It rarely infects humans. A major outbreak in the UK in 2001 led to the government slaughtering up to 10 million animals to try to halt the spread of FMD.

- The COVID-19 pandemic in humans is suspected to have crossed species from bats to pangolins or other animals and then to humans.

Poaching and the illegal pet trade

Poaching is the illegal trafficking or killing of wildlife. It is worth USD 20–216 billion a year. In 2013, the street price of rhino horn in Asia was USD 60,000–100,000 per kilogram. That is more than the price of gold. Poaching is a major threat to biodiversity. Due to the popularity of tiger products in Asia, there are now more tigers in captivity than in the wild. Illegal wildlife trading includes live animals, animal parts such as tusks, skins, organs, horns, shells (turtles), and plants and timber.

Tiger—endangered: Tiger numbers have fallen from an estimated 100,000 individuals a century ago to about 4,500 today. It is estimated that over 100 tigers are poached each year.

Rhino—critically endangered: Poaching of rhinos in South Africa grew 9,000% from 2007 to 2014 with over 7,000 rhinos poached in this time. In 2017 alone, 1,028 rhinos were lost in South Africa—a slight decline of 26 from the year before. Three of the world's five rhino species are critically endangered and the estimated global rhino population is 29,000.

Pangolin—critically endangered: Pangolins (the only scaled mammal) are traded for their meat and scales. It is estimated that over one million pangolins have been killed and illegally traded in the past 15 years, making the pangolin the most heavily trafficked wild mammal on the planet. Two of the eight species of pangolin are now critically endangered.

Sharks and rays—endangered or vulnerable: In our oceans many shark species are being driven to extinction due to massive levels of overfishing, estimated at over 100 million individuals killed annually. Over one-third of all sharks and rays are at risk of extinction due to overfishing.

African elephant—critically endangered: Between 2007 and 2014 there was a 30% decline in the African savannah elephant population, the majority of which were slaughtered for ivory. Central Africa populations suffered a 60% decline in the 10 years from 2002 to 2012.

▲ Figure 5 Baby black rhino

Sea turtle—threatened, endangered or critically endangered: Turtle populations are suffering due to slaughter for meat, eggs, skin and shell; nesting beaches being disrupted by human activity; being caught in fishing gear and drowned; and marine pollution.

Poaching supplies the illegal pet trade, estimated to be worth USD 23 billion a year. Trade records published by the Convention on International Trade in Endangered Species (CITES) show the large numbers of traded animals (probably for the pet trade):

- over 500 species of birds, especially parrots

- just under 500 species of reptiles such as lizards and snakes

- approximately 100 species of mammal—many of them carnivores and primates.

Climate change

Climate change alone has impacts on biodiversity but it can also exacerbate the changes caused by other factors. Greenhouse gases are a natural part of the atmospheric system but the pollutants (carbon dioxide, methane) added by human activities are the problem.

The Earth's climate has changed for millennia. Variations in solar activity, volcanic eruptions and changes in the orbit of the Earth have been the main causes, and organisms have adapted to the changes. But the human-induced changes that are now happening are occurring at a speed which is too fast for animals to adapt. The impacts of the changes are outlined below.

- The rising temperatures will alter ecosystems because the organisms that can survive there will change. Many animals are moving to cooler areas (either towards the poles or to higher elevations); this adds to the problems caused by pollution.

- Higher temperatures are reducing the amount of water vapour in the atmosphere. Since the end of the 20th century nearly 60% of vegetated areas are showing reduced growth rates.

- Higher temperatures cause rising sea levels as polar ice caps melt. This changes the salinity of seawater and of estuaries and salt marshes.

- Increasing ocean temperatures affect the mutualistic relationship between coral and zooxanthellae. Coral polyps provide protection and zooxanthellae photosynthesize and "feed" the coral. As ocean temperatures rise, the coral expels the zooxanthellae. This means corals lose their colour and the effect is called bleaching. The corals are technically still alive at this point, but their survival chances are poor. Coral reefs provide habitats for thousands of species—they have very high biodiversity. Studies have shown that coral bleaching and a decline in fish species diversity are linked.

- Increases in carbon dioxide levels in the atmosphere mean more carbon dioxide dissolves into the oceans causing ocean acidification. This makes it harder for shellfish and coral to extract calcium ions and carbonate ions from seawater. They need these ions to make hard calcium carbonate shells. Without hard shells, their survival chances are reduced. When nitrate pollution is added to the problem, the occurrence of marine algae blooms increases.

▲ Figure 6 Bleached coral

- Climate change also increases the intensity and frequency of fires, storms and drought, all of which put a significant strain on ecosystems.

> Connections: This links to material covered in subtopics HL.a, HL.b and HL.c.

(ATL) Activity 14

ATL skills used: thinking, communication, social, research

In small groups, copy and complete Table 3, conducting your own research to see if there are ethical or economic reasons for and against the actions that cause the events listed here.

Event	Ethical		Economic	
	Reasons for action	Reasons against action	Reasons for action	Reasons against action
Habitat loss				
Pollution				
Overexploitation				
Introducing non-native (exotic/alien) species				
Spread of disease				
Poaching and the illegal pet trade				
Climate change				

▲ Table 3

Total world biodiversity

How many species are there on Earth today? That seems like a straightforward question; we have explored just about every region of the Earth and logged and catalogued what is there so we should have an accurate answer. But, in fact, we have very little idea about how many groups of organisms there are and certainly no clear idea of how many are becoming extinct. A conservative estimate of species alive today (2023) is about 8.7 million species excluding bacteria.

- 6.5 million species are terrestrial and 2.2 million in the oceans.

- According to a study published by *PLOS Biology*, 86% of all species on land and 91% of species in the seas have yet to be discovered, described and catalogued.

- Two-thirds are in the tropics, mostly tropical rainforests.

- 50% of tropical rainforests have been cleared by humans. We are clearing, burning or logging about 1 million km^2 every 5–10 years of the original 18 million km^2. Many countries no longer have any primary forest (forest that is not degraded or destroyed by humans).

Only about 1.4–1.8 million species of organisms have been described and named, so are "known to science". How we reach an estimate depends on the species. It is easier to see large animals but harder for smaller ones. Big, furry animals grab our attention. It is relatively easy to identify the big animals or ones that do not run away. Most mammals and birds are known. It is also easier to find plants because they cannot move. But many groups of smaller organisms such as insects, nematodes, fungi and bacteria have not been found, identified and named.

Estimates are wildly variable, for instance, the overall range is 3–100 million. Some think that there are 70–90 million species on Earth, while others think there are about 8–10 million species. That is quite a wide margin of error.

Groups	Species found	Total estimated species	Percentage identified and named
Vertebrates	46,500	50,000	93
Molluscs	70,000	200,000	35
Arthropods (insects)	840,000	8,000,000	11
Arachnids	75,000	750,000	10
Crustaceans	30,000	250,000	12
Protozoa	40,000	200,000	20
Algae	15,000	500,000	3
Fungi	70,000	1,000,000	70
Plants	256,000	300,000	85

▲ Table 4 Numbers of identified species and total estimated species for various groups

(ATL) Activity 15

ATL skills used: thinking, communication, social, research

Beetles (Coleoptera) are the group with the most identified and named species—they make up about 25% of all named species (around two million). One way of finding and identifying insects is "fogging" the canopies of rainforest trees with short-lived insecticides. The organisms die, fall out of the trees and can be collected and counted, then the numbers are extrapolated.

1. Discuss if this is ethical in your view.

2. Investigate how biodiversity data is gathered in the country where you live.

3. Discuss how indigenous peoples can help to collect data about species in the area in which they live.

What makes a species prone to extinction?

Some species are more vulnerable than others to extinction. Even in the same ecosystem, some species survive habitat loss or degradation while others do not. The following factors all make a species more likely to be in danger of extinction. Often though, more than one pressure operates on an organism. The nearer an organism comes to extinction, the greater the number of operating pressures.

Narrow geographical range

If a species only lives in one place and that place is damaged or destroyed, the habitat has gone along with the organisms that live there. These species are then known to be "extinct in the wild". The Spix's macaw (also called the little blue macaw) has only 177 captive individuals left in the world.

Small population size or declining numbers—low genetic diversity

A small population has less genetic diversity and is therefore less resilient to change—it cannot adapt as well. A fall in the number of individuals causes more inbreeding. If the numbers drop too far, the species becomes known as the "living dead" or it goes extinct. Large predators and extreme specialists are commonly in this category—for example, snow leopard, tiger, Lonesome George (subtopic 3.1).

Low population densities and large territories

The individuals of some species, such as the giant panda, require a large territory or range over which to hunt or find food. This means they only rarely meet others of the species for breeding. If a city, road, factory or farm splits up the territory, finding a mate becomes very difficult, so they are less likely to survive as a species.

Few populations of the species

If there are only one or two populations of a species left, they are very vulnerable. If the populations are wiped out, the species will vanish, for example lemurs.

A large body

Top predators are rare. This is because only about 10% of the energy in the food chain is passed on at each trophic level. The rest is lost to the environment. The top predators in both aquatic and terrestrial ecosystems—such as tigers and killer whales—tend to be large animals with large ranges and low population densities and need a lot of food.

Low reproductive potential

Reproducing slowly and infrequently means the population takes a long time to recover. Whales fall into this category. Many of the larger species of sea bird only produce one egg per pair per year, and do not breed until they are several years old. Examples include gannets, albatrosses and some species of penguin.

Seasonal migrants

Species that migrate face many challenges. They have long and hazardous migration routes and they need intact habitats at both ends; if one habitat is destroyed, they arrive to find no food or habitat. Examples include swallows (migrate between southern Africa and Europe) and songbirds (migrate between Canada and Central and South America). Barriers on their journey can also prevent the species completing it (e.g. salmon trying to swim upriver to spawn).

Poor dispersers

Species that cannot move easily to new habitats are also in trouble. Plants rely on seed dispersal or vegetative growth to move. That takes a long time and climate change resulting in biome shift may mean the plant dies out before it can move. Non-flying birds (e.g. various flightless birds in New Zealand) are mostly extinct because they cannot escape introduced hunters or fly to another island.

Specialized feeders or niche requirements

The giant panda (1,864 left) mostly eats bamboo shoots in the forests of central China. The koala only eats eucalyptus leaves and lives in the coastal regions of southern and western Australia. Bog plants such as the sundew or pitcher plant can only survive in damp places. All suffer from habitat loss or degradation.

Edible to humans and herding together

Overhunting or overharvesting can eradicate a species quickly, especially if that species lives in large groups—for example, herds of bison in North America, shoals of fish, flocks of passenger pigeons. Modern technology means it is easier to find large herds and shoals so many can be caught.

▲ Figure 7 Giant panda

▲ Figure 8 Minke whale harvest

Also under threat from humans are tigers. Although they are not edible, all their body parts are used in traditional Chinese medicine and demand for them is very high.

Island organisms

Island organisms can be particularly vulnerable to extinction. Depending on the size of the island:

- populations tend to be small

- there is a high degree of endemic species

- genetic diversity tends to be low in small unique island populations

- islands tend to be vulnerable to the introduction of non-native predators against which endemic species have no defence mechanism.

The dodo is a classic example of an island species that became extinct.

Combined pressures

Many of the factors that make species prone to extinction are combined in certain species. Mammals such as tigers, leopards, bears and wolves are very vulnerable because they:

- are large

- need a large territory

- represent a danger to humans both directly and indirectly so are killed to "remove the danger"

- are hunted for "sport"

- are top carnivores in the food chain.

The minimum viable population size that is needed for a species to survive in the wild is a figure that scientists and conservationists consider. But there is no magic number. It depends on many factors such as genetic diversity in the individuals left, rate of reproduction, mortality rate, growth rate and threats to habitat.

For large carnivores, 500 individuals is sometimes thought of as the absolute minimum below which there is little hope for the species. Humans have hunted many big animal species to very endangered levels.

- There are thought to be 400 tigers in Sumatra and about 4,500 in the wild today (compared with 100,000 at the start of the 20th century).

- There are 10,000–25,000 blue whales left in the world today compared with 140,000 in the 1920s.

▲ Figure 9 Island species are prone to extinction, e.g. on Sipadan Island, Sabah

▲ Figure 10 Tiger hunting

The IUCN Red List

The International Union for Conservation of Nature (IUCN) is usually called the World Conservation Union. It is an international agency, founded in 1948 and brings together 83 states, 110 government agencies, more than 800 NGOs, and some 10,000 scientists and experts from 181 countries.

The IUCN's mission is to:

- influence, encourage and assist societies throughout the world to conserve the integrity and diversity of nature

- ensure that any use of natural resources is equitable and ecologically sustainable.

The IUCN monitors the state of the world's species through the Red List of Threatened Species. It supports the Millennium Ecosystem Assessment and educates the public, advises governments and assesses new World Heritage sites.

The Red List of Threatened Species is a list of species under varying levels of threat to their survival. The data is gathered scientifically on a global scale and species are put into one of the Red List categories according to their relative danger of extinction. Two of the categories are extinct and extinct in the wild. The lists are regularly updated. They are used by governments to inform policies on trade in endangered species and conservation measures. The IUCN Red List has 42,108 threatened species on it out of 150,000 species that have been assessed so far.

EXTINCT (EX)	when there is no reasonable doubt that the last individual has died; a taxon is presumed extinct when exhaustive surveys in known and/or expected habitats have failed to record an individual
EXTINCT IN THE WILD (EW)	when it is known only to survive in cultivation, in captivity or as a naturalized population (or populations) well outside the past range
CRITICALLY ENDANGERED (CR)	when it is considered to be facing an extremely high risk of extinction in the wild
ENDANGERED (EN)	when it is facing a very high risk of extinction in the wild
VULNERABLE (VU)	when it is facing a high risk of extinction in the wild
NEAR THREATENED (NT)	when it has been evaluated against the criteria and does not qualify for Critically Endangered, Endangered or Vulnerable now, but is close to qualifying for or is likely to qualify for a threatened category in the near future
LEAST CONCERN (LC)	when it is widespread and abundant
DATA DEFICIENT (DD)	when there is inadequate information to make a direct, or indirect, assessment of its risk of extinction based on its distribution and/or population status
NOT EVALUATED (NE)	when it has not yet been evaluated against the criteria

Increasing risk of extinction

▲ Figure 11 IUCN Red List categories

The Red List determines the conservation status of a species based on several criteria:

- population size
- degree of specialization
- distribution
- reproductive potential and behaviour
- geographic range and degree of fragmentation
- quality of habitat
- trophic level
- probability of extinction.

Critically endangered species

Carnaby's black-cockatoo (*Zanda latirostris*)

Geographic range

South-west of Western Australia

Description

- A large, black bird, threatened by destruction of its habitat.
- Estimated population (in late 2022) was 10,000–60,000 breeding individuals.
- Begins breeding at four years old and produces only two eggs each year.
- Survival rate is very low with one or no chicks surviving the first year.
- Live for 40–50 years.
- Protected by law and are officially listed as endangered.
- Numbers continue to decline and captive breeding is largely unsuccessful.

Ecological role

- Very specialized habitat requirements.
- Breed in holes in mature (over 100 years old) salmon gum trees using the same breeding holes each year.
- Feed in open heathland on seeds and insect larvae.
- Migrate to populated coastal areas in late summer and autumn.

▲ Figure 12 Carnaby's black-cockatoo

Known threats

- The cockatoo's habitat has been lost to wheat farming during the last 100 years and more recently to gravel mines, firebreaks and agriculture.

- Regeneration of mature trees for nesting is slow. As the young trees begin to grow they are grazed by rabbits and sheep.

- Nest hollows are cleared for firewood or in tidying up back yards.

- There is competition from invasive species for nesting sites.

- Poaching of prized specimens for bird collectors and illegal robbing of eggs and baby birds from nests is a lucrative activity. Enforcement of the law is difficult in remote areas.

Conservation measures

There is a national plan outlining recovery actions needed to prevent further decline of the Carnaby's black-cockatoo. The intentions include the following.

- Identify and manage crucial habitat areas across its range.

- Monitor habitat, threats and ongoing status of the populations.

- Research biology and ecology for proper management.

- Encourage community conservation efforts.

- Increase education and awareness of decision makers and establish joint management agreements for information sharing.

Rafflesia (*Rafflesia spp.*)

Geographic range

Jungles of Sumatra and Borneo, Southeast Asia

Description

- There are 15–19 species of this tropical parasitic plant.

Ecological role

- Parasitic on one genus of vine.

- Plants are single sexed (either male or female).

- Pollination must be carried out when the plants are in bloom (flowering) so a male and female in the same area must both be ready for pollination at the same time.

- Seeds are dispersed by small squirrels and rodents and they must reach a "host" vine.

▲ Figure 13 Rafflesia flower

Known threats

- Rafflesia plants are very vulnerable because they need very specific conditions to survive and carry out their life cycle—a narrow ecological niche.

- They are vulnerable due to deforestation and logging which destroy their habitat.

- Humans damage them and fewer plants means less chance of breeding.

Conservation measures

- In Sabah, Sumatra and Sarawak there are Rafflesia sanctuaries.

- There are many locating, protecting and monitoring programmes being set up.

- Education: there are "Save Rafflesia" campaigns.

Tiger (*Panthera tigris*)

Geographic range

East and Southeast Asia

Description

- One of the big cats.

- Carnivore.

- Territorial, solitary and endemic to regions with the highest human populations on Earth.

- At the beginning of the 20th century, it was estimated that there were around 100,000 wild tigers.

- In 2015 there were fewer than 8,000; in 7 years that number has dropped to 4,500—these are in: India: 3,000 (70% of the global population); Malaysia: 150; Indonesia: 400; Bangladesh: 300; and Vietnam and Russia about 200 each.

- There were eight sub-species, now there are five.

- Javan, Balinese and Caspian tigers became extinct during the 20th century.

- The Sumatran tiger is likely to go extinct in the next 50 years.

Ecological role

- Top carnivore—eats various deer and large herbivores.

Known threats

- Forests across Asia have been destroyed for timber or for conversion to agriculture, causing population fragmentation.

- Tigers in one area cannot mate with tigers in nearby areas, causing inbreeding and a weakening of the gene pool, so tigers are born with birth defects and mutations.

- Uncontrolled growth of the human population in LICs is the main cause of habitat destruction.

- Tiger poaching is illegal, but all their body parts (e.g. bones, whiskers, penis) can be sold on the black market (see Table 5). Tiger parts are commonly used in traditional Chinese medicine or sold as luxuries.

▲ Figure 14 A Siberian tiger in Philadelphia Zoo

Body part	Average price
Boned-out tiger	USD 50,000–60,000 (price is falling)
Skin	Up to USD 25,000
Bone	USD 1,250/kg
Tiger penis soup	USD 320/bowl
Pair of eyes	USD 170

▲ Table 5 What does it cost to buy a tiger?

Conservation measures

- The WWF and many other organizations have begun innovative and aggressive approaches, some beginning as early as the 1970s. The WWF works in nearly all 14 of the countries that tigers inhabit.

- Conservation sites such as Tiger Island have been set up.

- The tiger conservation plan includes: strengthening international treaties; supporting surveys; monitoring tiger populations; pushing for the enforcement of laws controlling the illegal trade in tiger parts; working with the traditional Chinese medicine community to find alternative products and reduce the use of tiger body parts; supporting anti-poaching efforts where tigers are most in danger; and using state-of-the-art ecology methods to target critical populations of tigers in certain areas. Tigers are also now covered under the strictest protection of the Convention on International Trade in Endangered Species (CITES).

However, the value and demand for tiger parts make such enforcement efforts very difficult. Without conservation efforts, the future of the tiger looks bleak and the current efforts do not seem to be enough. Many feel the extinction of the tiger is inevitable.

Recovered species

Australian saltwater crocodile (*Crocodylus porosus*)

Of the 23 species of crocodiles worldwide, 18 were once endangered but have since recovered sufficiently to be removed from the list. Many species are thriving.

Geographic range

Coastal Northern Territory of Australia

Description

- Bulky reptile with a broad snout; can grow up to 5 m long.

- Lays up to 80 eggs each year which take up to 3 months to hatch.

▲ Figure 15 Australian saltwater crocodile

- Matures at 15 years old.

- Was listed as a protected species in Australia in 1971.

- Protected under CITES which banned trade in endangered animals.

Ecological role

- Habitat is estuaries, swamps and rivers.

- Nests are built on riverbanks in a heap of leaves.

- The eggs are food for goannas, pythons, dingoes and other small animals.

- Older crocodiles eat young crocodiles, mud crabs, sea snakes, turtle eggs and catfish. Baby crocodiles eat tadpoles, crabs and fish.

- Top predator.

Known threats

- The saltwater crocodile was overexploited for skin (leather), meat and body parts through illegal hunting, poaching and smuggling.

- It was hunted for sport.

- It was deliberately killed because of attacks on humans.

Conservation measures

- To restore the populations a sustainable use policy was introduced and only allowed limited culling of wild populations.

- Ranching (collecting eggs and hatchlings from the wild and raising them in captivity) and closed-cycle farming (maintaining breeding adults in captivity and harvesting offspring at four years of age).

- The exploitation of farmed animals reduces the hunting of wild crocodiles.

- Visitors tour areas to see wild crocodiles, so they are now a valued species. The money tourists pay supports the local economy and conservation efforts. This policy was supported by the Species Survival Commission (SSC) of IUCN but was accused by others of treating crocodiles inhumanely.

Golden lion tamarin (*Leontopithecus rosalia*)

Geographic range

Atlantic coastal rainforests of Brazil

Description

- Small endemic monkey.

- Among the rarest animals in the world. In 2015 there were about 1,000 in the wild and 500 in captivity. In late 2022 numbers were estimated at 2,500–3,200 in the wild and 500 in captivity.

- Life expectancy of about eight years.

▲ Figure 16 Golden lion tamarin

Ecological role

- Omnivores.

- Live territorially in family groups in the wild in tropical rainforest canopy.

- Prey to large cats and birds of prey.

Known threats

- Only 2% of their native habitat is left.

- Poachers can earn USD 20,000 per skin.

- Predation is great in the wild and their food source is not dependable.

Conservation measures

- A captive breeding programme (breeding in zoos) for the last 40 years or more.

- Over 150 institutions are involved in this and exchange individuals to increase genetic diversity.

- Some are reintroduced to the wild but with only a 30% success rate as the habitat is threatened and predators (including humans) take many.

- It is unlikely that this species would have survived without captive breeding as it is estimated that nearly 40% of the current wild population is reintroduced.

- The long-term future is uncertain in the wild.

Extinct species

Tasmanian tiger or thylacine (*Thylacinus cynocephalus*)

Geographic range

Australia (for the last few hundred years, it was found only on the island of Tasmania, but before that it had existed on the mainland)

Description

- Was a marsupial.

- Similar in appearance to a wolf, but with a rigid tail.

- Strong fighter with jaws that could open more widely than those of any other known mammal.

- Life expectancy of 12–14 years.

- Gave birth to two to four young a year.

- Habitat was open forest and grassland.

▲ Figure 17 The last thylacine

Ecological role

- Became restricted to dense rainforest as the population declined.

- Lived in rocky outcrops and large, hollow logs.

- Were nocturnal.

- Typical prey were small mammals and birds, but they also ate kangaroos, wallabies, wombats, echidnas, and sheep.

Known threats

- Were outcompeted by dingoes on the mainland of Australia and became extinct there hundreds of years ago.

- In Tasmania they were hunted by farmers whose stock was the thylacine's prey.

- Private and government bounties were paid for scalps, leading to a peak kill in 1900. Bounties continued until 1910 (government) and 1914 (private), leading to an intense killing spree by hunting, poisoning and trapping.

- Shooting parties were organized for tourists' entertainment.

- The severely depleted population was affected by disease and competition from settlers' dogs.

- The last wild thylacine was killed in 1930 and the last captive animal died at the Hobart Zoo on 7 September 1936.

- The thylacine was legally protected in 1936.

- It was classified by IUCN as endangered in 1972 and is now listed as extinct.

Consequences of disappearance

- The thylacine was a carnivorous marsupial from a unique marsupial family.

- In Tasmania, where there were no dingoes, the thylacine was a significant predator.

- Introduced dogs have taken over the ecological role of the thylacine.

Dodo (*Raphus cucullatus*)

Geographic range

Island of Mauritius

Description

- Large endemic flightless bird.

Ecological role

- No major predators on Mauritius so the dodo had no need of flight.

- A ground-nesting bird.

▲ Figure 18 **Drawing of a dodo**

Known threats

- In 1505, Portuguese sailors discovered Mauritius and used it as a restocking point on their voyages to get spices from Indonesia.

- They ate the dodo as a source of fresh meat.

- Island became a penal colony (jail) and rats, pigs and monkeys were introduced.

- Dodo eggs were eaten by the invasive species and humans killed dodos for sport and food.

- Crab-eating macaque monkeys introduced by sailors stole dodo eggs.

- Conversion of forest to plantations also destroyed their habitat.

- Known to be extinct by 1681.

Consequences of disappearance

- Island fauna impoverished by loss of the dodo.

- Very few skeletons remain.

- It became an icon due to its apparent stupidity (it had no fear of humans, never having had need to fear predators) and its untimely extinction. "Dead as a dodo" is now a common saying.

(ATL) Activity 16

ATL skills used: thinking, communication, research

1. Define biodiversity.

2. Explain what is an endangered species. List several examples.

3. Identify and discuss the two main reasons for extinction of species.

4. Discuss whether it matters if a species becomes extinct.

5. To what extent should effort be made to preserve a species and why?

6. List four examples of human activity that threaten species.

7. Examine the case studies of organisms on the previous pages and the list of characteristics that make species more prone to extinction.

    a. Make a table to compare the characteristics shared by these species.

    b. Describe what they have in common.

The tragedy of the commons

This metaphor illustrates the tension between the common good and the needs of the individual, and how they can be in conflict. If a resource is seen as belonging to all, we all tend to exploit it and overexploit it if we can. This is because the advantage to the individual of taking the resource (be it fish, timber, minerals, apples) is greater than the cost to the individual because the cost is spread among the whole population.

In the short term, it is worth taking all the fish you can because, if you do not, someone else will. This assumes that humans are selfish and not altruistic, and it has caused much debate among economists and philosophers. The solution is often regulation and legislation by authorities which limit the amount of common goods available to any individual. This may be by permit, agreed limits or cooperation to conserve the resource.

Exploitation of the oceans is a good example of the tragedy of the commons. Read Case study 4 in subtopic 4.3, on exploitation of the Grand Banks cod.

Another example is marine plastic pollution. We all use plastic products on a regular basis in our daily lives. Plastic is a very convenient material used for items such as packaging, cutlery, straws, bags, textiles and electronics. The problem is that we do not manage plastic waste disposal well and marine ecosystems are paying the price.

Individuals use plastic without considering the overall consequences of the massive amount of garbage that is generated. Poor management of domestic waste and careless disposal of the plastic items mean that much of it ends up in the gyres that have developed in international waters. It is easy for us to throw away a straw or a plastic bag because it does not impact us as individuals. But the "commons" in this case is the international waters where much of the waste plastic ends up. Who is responsible for cleaning up international waters? Everybody? Nobody?

▲ Figure 19 Plastic garbage affects local waters and beaches as well as the oceanic gyres

Practical

Skills:

* Tool 2: Technology
* Inquiry 1: Exploring and designing
* Inquiry 2: Collecting and processing data
* Inquiry 3: Concluding and evaluating

Research your local area or the country you live in and choose a conservation area that you are passionate about.

1. Give a description of the area with pictures.

2. What role does the area play—social, ecological, recreational, economic?

3. What are the pressures on the area?

4. Is the area being used sustainably?

5. What conflicts are there over the conservation of the area?

AHL

Connections: This links to material covered in subtopics HL.a, HL.b and HL.c.

Hotspots under threat

By definition, hotspots (subtopic 3.1) are under threat because they are areas that have lost 70% of their original natural vegetation, mostly because of human activities. "Why does biodiversity matter?" The answer is simple: without biodiversity there would be nothing to support life on Earth. An ecosystem is not sustainable without plants and animals of every size from the biggest living animal (blue whale) to the smallest microorganisms. The absence of a healthy ecosystem means there will be no food to eat, no clean water to drink and no fresh air to breathe. Not to mention the inherent spiritual, cultural and aesthetic value that nature brings.

Many of these benefits are obvious to most people but others are less widely known.

- Pollinators are crucial to 33% of the world's crop production; no bees, birds and insects = no pollination.

- Microbes maintain soil health as they break down dead organic matter and release the minerals in it for plant uptake.

- The vast wetlands (swamps and marshes) regulate the flow of water and decrease the risk of flooding.

- Vegetation cleans the air we breathe and acts as a carbon sink because it absorbs carbon dioxide.

- Mangroves and coral reefs act as coastal defences and decrease coastal erosion.

- Many medicines come from nature (e.g. aspirin is from willow bark and morphine is from opium).

- Being in nature improves mental health through stress reduction. It also lowers blood pressure.

The 36 biodiversity hotspots cover about 2.5% of the Earth's surface and yet provide 35% of the ecosystem services. Part of the problem is that 27 of these hotspots are in LICs. Five are in North and Central America and four are in Europe and Central Asia.

Vulnerability of tropical rainforests

▲ Figure 21 Large tree with buttress roots in the Amazon Rainforest

Review the following key facts about tropical rainforests:

- contain over 50% of all species of plants and animals on Earth

- cover 6% of the land area on Earth

- one hectare of rainforest may have 300 species living in it

- produce about 40% of the oxygen that animals use (the lungs of the Earth)

- have high species diversity and habitat diversity

- contain 50% of the world's timber (the world's biggest resource after oil).

▲ Figure 20 Hummingbird drinking nectar

(ATL) Activity 17

ATL skills used: thinking, communication, social, research, self-management

How can you raise awareness in your local community about the existence and benefits of global biodiversity hotspots? Be creative!

Malaysia's tropical rainforests are a biodiversity hotspot. They contain 8,000 flowering plant species of 1,400 genera including 155 genera of the huge dipterocarp trees (e.g. mahogany). These trees have enormous commercial importance. Denmark is a similar size to the Malay Peninsula and has 45 species of mammals compared with Malaysia's 203.

Rainforests are so diverse because of the many ecological niches they provide. High levels of heat, light and water are present all year. This means that photosynthesis is rapid and is not limited by lack of raw materials. The four vertical layers of the rainforest allow for many different habitats and niches which are filled by diverse species. Growth occurs all through the year, so food can be found at any time. Some rainforests are old in both living and geological terms. The lack of disturbance may mean that the system has had time to become more complex.

The fast rate of respiration and decomposition means that the forests appear to be very fertile with high levels of biomass in standing crop (trees and other plants). This lush appearance made them attractive to humans as potentially rich agricultural land. But appearances are deceptive. The majority of nutrients are held in the plants, not in the soil or leaf litter. Once the trees are cleared for timber and the rest of the vegetation burnt, the main store of nutrients has gone. The absence of vegetation exposes the soil to heavy rainfall which washes away any remaining nutrients and soil, and fertility reduces rapidly. So, more forest is cleared to get short-term fertility for crop growth and, of course, more timber.

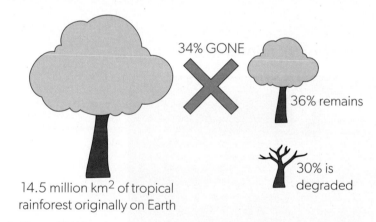

34% GONE

36% remains

30% is degraded

14.5 million km² of tropical rainforest originally on Earth

▲ Figure 22 What has happened to the tropical rainforests of the world?

Rainforest is being lost at a massive rate. Approximately 1.5 hectares of rainforest is cleared every second.

Commercial activities are not the biggest threat to the rainforests—individual people are. At least two billion humans live in the wet tropics and many of these rely on the rainforests for subsistence agriculture (shifting cultivation). Shifting cultivation is sustainable at low population densities. People clear a small area of forest, and grow crops for two or three years then move on when the soil is exhausted. This works if there is enough time; about 100 years is needed to allow the forest to regenerate fully before it is cleared again. Forest areas cleared and then not managed do regrow in time. But it is estimated that it takes 1,000 years for the biodiversity of the primary forest (before logging or clearing) to be recovered and the secondary forest that grows up in its place is impoverished in many ways. The increasing human population means that forest is cleared too quickly and there is a gradual degradation of nutrients and biodiversity. This sets up a positive feedback cycle that accelerates the degradation.

Much is written and spoken about rainforests because of their biodiversity and their vulnerability. Of course, clearance of the forest does not mean that nothing then grows on that land. There may be animals grazing grasslands, oil palm or soy plantations, subsistence agriculture and urban development. But much biodiversity is lost in these areas.

Key biodiversity areas

Key biodiversity areas (KBAs) are "sites contributing significantly to the global persistence of biodiversity", in terrestrial, freshwater and marine ecosystems.

Key biodiversity areas are the most important places as regards species and their habitats. They are where organizations such as the IUCN are focusing their conservation efforts. KBAs are mapped, monitored and conserved to try and protect the most critical areas in a wide variety of biomes (Figure 23).

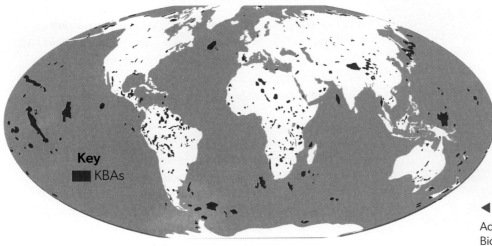

Key
■ KBAs

◀ Figure 23 KBAs (IUCN 2016)

Adapted from: The World Database of Key Biodiversity Areas. BirdLife International (2022). Developed by the KBA Partnership. Basemap: Esri, DigitalGlobe, GeoEye, i-cubed, USDA FSA, USGS, AEX, Getmapping, Aerogrid, IGN, IGP, swisstopo, and the GIS User Community.

An area is defined as a KBA if it has one or more of the following criteria.

The major categories are:

• threatened biodiversity or geographically restricted biodiversity

• ecological integrity

• biological processes

• irreplaceability through quantitative analysis.

Most of these criteria are assessed using the IUCN parameters for conservation status:

• number of mature individuals

• range and area of occupancy

• extent of suitable habitat

• number of localities

• genetic diversity.

Case study 2

Mundaring, Kalamunda Key Biodiversity Area

Location: Perth area, western Australia

Size: 13,710 ha

Human population (Mandurah): 91,300 (2011)

Vegetation: mainly scrub, woodland and Mediterranean forests

Native birds

Baudin's black-cockatoo: critically endangered

Carnaby's black-cockatoo: endangered

The rest of the birdlife is colourful, varied and not at risk. It includes red-capped parrot, western rosella, red-winged fairywren, western spinebill, western thornbill, western yellow robin, white-breasted robin, red-eared firetail, peregrine falcon, forest red-tailed black-cockatoo, long-billed black-cockatoo.

Native animals include numerous frogs, kangaroos, possums, various reptiles (snakes and lizards), echidna and bandicoots.

▲ Figure 24 Western rosella

Conservation efforts

The community has an extensive local biodiversity strategy: *Natural Area Planning in the Shire of Kalamunda*. This can be accessed online. Community involvement is key and people are encouraged to:

- maintain birdbaths especially during the dry season when drought can be an issue; small sticks or pebbles are included to help small animals climb out

- not feed wildlife because it is bad for the health of the wildlife and disturbs normal behaviours; it also encourages invasive species like galahs that outcompete the indigenous birds for nesting sites

- enjoy the natural environment in parks, access to natural wetlands and woodlands

- conserve natural resources.

Conflicts between conservation and land use

The Mundaring, Kalamunda KBA has several land uses that could be seen to conflict with conservation and the maintenance of biodiversity. The dryland agriculture across much of the region requires irrigation—thus taking water away from the natural habitat areas.

Water supply in the area is mainly from groundwater; there are approximately 170,000 residential bores in the Perth area (individuals sinking wells or boreholes to access the groundwater). The increase in extraction from these groundwater sources is causing replenishment issues, exacerbated by climate change (low precipitation levels). This causes deficit in the natural environment and puts pressure on the wildlife. To try to combat this issue a groundwater replenishment scheme is being constructed to extend the extraction of recycled water, treat it to drinking water standards and pump it into the groundwater stores.

Other than that, the pressure of increased population means that residential areas are expanding and dryland agriculture and grazing take over more and more of the natural environment.

Conflicts in KBAs

Biodiversity in all areas of the world is under threat and there is a conflict between exploitation and conservation, much of which is caused by population growth.

Threats include the following.

- Habitat loss and fragmentation: caused by the human need for additional land to grow food, store water and build infrastructure; 37% of the Earth's land (not including Antarctica) is used for food production.

- Invasive species: brought into an area accidently (in cargo) or deliberately to control pests or as pets.

- Overexploitation of natural resources: for economic development and the extraction of a huge range of resources, from uranium to water.

- Pollution: many of our activities produce pollutants.

Traditional indigenous land management

Indigenous peoples and nations:

- are distinct from other societies both socially and culturally

- have a historical connection with pre-invasion and pre-colonial societies

- share collective ancestral ties to specific lands and natural resources

- may have been removed from their ancestral lands.

There are many indigenous cultures across the planet, all of which have different approaches to land management. Such cultures are unique and making generalizations shows a lack of sensitivity to this. For the purposes of this discussion, we will consider the Australian Aboriginal (indigenous) People.

> Connections: This links to material covered in subtopics 5.2 and 7.2.

settlement improvements, e.g. dust control, firewood collection, water management

cultural resources, e.g. hunting, gathering, burning, sharing of knowledge and ceremonies

Indigenous individuals and groups

reduction of danger, e.g. weed and animal control, fire management, revegetation

commerce and economy, e.g. pastoral practices, art, bush harvests

environment and natural resources, e.g. plants, animals, shelter and water, protection of endangered species

▲ Figure 25 The different aspects of indigenous land management

Indigenous land management is complex and holistic, involving many aspects of life—see Figure 25. It takes into account the 50,000-year relationship indigenous societies have had with their land and the sea. To many of the Aboriginal People, fire was a friend and the land was a patchwork of areas that were burnt and re-grown. The intuitive use of fire allowed the accurate prediction of plant growth which in turn attracted the animals they hunted. The land was managed to provide food and drink in a sustainable way. They managed the land in a holistic fashion that suited certain species like kangaroos, and benefitted other animals, birds, insects, reptiles, snakes and marsupials. The landscape maintained its complexities and supported the communities for many generations. Such management takes deep knowledge of the land, in particular to:

- control fires so that the food source for a whole community is maintained (there is very little evidence of deep burns)

- develop specific patterns for specific areas of land where they knew the preferences of different animals (e.g. kangaroos like short grass so they managed the land accordingly).

Many of these management practices have been under threat for decades. Between the late 1890s and the 1970s, Aboriginal People in Australia lost their traditional territories and ways of life. This had a profound impact on children who lost their sense of identity and were alienated from their culture and language. Indigenous communities were pushed off the land by European settlers and there was little or no government support or protection.

Now, Aboriginal People have an increased level of formal involvement. Funding is being embraced and indigenous communities are taking on significant projects across Australia, providing crucial land management in many remote areas.

ATL Activity 18

ATL skills used: thinking, communication, social, research

According to the World Bank, there are 476 million indigenous peoples worldwide and 6% of the global population is indigenous.

Study the facts below and discuss if the situation is ethical.

- 19% of indigenous people live in extreme poverty.

- Life expectancy is 20 years lower than for non-indigenous people.

Investigate the indigenous peoples in your area (or somewhere else of your choice).

1. Do they have any rights?

2. Do they have access to their ancestral lands?

3. Are there any environmental laws that protect their land rights?

Connections: This links to material covered in subtopics 1.1 and HL.a.

Planetary boundaries

These nine boundaries (discussed in subtopic 1.3) show how far we can safely "push" the planet before the impact of anthropogenic changes will be unpredictable. We are operating at high risk as regards biogeochemical flows and biosphere integrity. Additionally, we are beyond what is considered safe in climate change and land-use system change. By 2022, five boundaries had been crossed (see Figure 13 in subtopic 1.3).

Rockström et al. highlighted loss of biodiversity as the boundary that is furthest beyond the safe operating space. This is measured using extinction rates, but they are not a good indicator at global levels. To determine biodiversity loss boundaries, we need to consider planetary large-scale responses which may impact the ability of the Earth system to sustain human societies. Three facets of biodiversity could be used to establish the boundary:

- genetic library of life: the evolutionary development and diversification of species

- functional diversity

- biome condition and extent.

Scientists around the world continue to research ways to enhance and improve delineation of the boundaries. DIVERSITAS (an international programme of research on biodiversity) concluded that "biodiversity's role in supporting a safe operating space for humanity may lie primarily in its interactions with other boundaries, suggesting an immediate area of focus for scientists and policymakers". Therefore, managing biodiversity is crucial to prevent other boundaries going from "safe" to "high risk". The biosphere interacts through energy, nutrient and material cycles, thus impacting (and being impacted by) climate change, biochemical flows and freshwater use.

> *Species diversity in communities is a product of two variables, the number of species (richness) and their relative proportions (evenness).*
>
> Stockholm Resilience Centre

Connections: This links to material covered in subtopic 1.3.

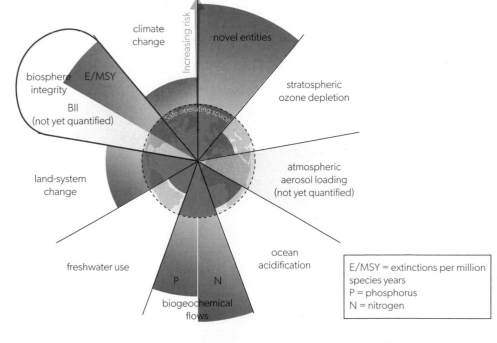

◄ Figure 26 Planetary boundaries model with biosphere integrity overshoot circled

Adapted from: Azote for Stockholm Resilience Centre, Stockholm University. Based on Richardson et al. 2023, Steffen et al. 2015, and Rockström et al. 2009 (CC BY-NC-ND 3.0)

Check your understanding

In this subtopic, you have covered:

- how humans impact biodiversity

- natural and anthropogenic threats to biodiversity

- problems caused by habitat loss, pollution, overexploitation, introduced species, poaching and climate change

- how much biodiversity there is in the world

- why some species are more prone to extinction than others

- what the IUCN is and the work it is doing

- biological hotspots, what they are, and which ones are most under threat

- key biodiversity areas and the conflicts surrounding them

- planetary boundaries and biodiversity.

AHL

What causes biodiversity loss and how are ecological and societal systems impacted?

1. Explain how biodiversity is being directly and indirectly adversely affected.

2. Discuss how alien species arrive in an area and are managed.

3. Outline how the IUCN defines the conservation status of a species.

4. Discuss differences between the perspectives of governments, agencies and individuals in conservation.

5. Describe examples of species that have been overharvested.

6. Discuss case studies about a species that is now extinct, one that is critically endangered and one that has recovered due to conservation efforts.

7. Examine the impact of human activity on biodiversity along a transect (see subtopic 2.6).

8. Evaluate the consequences of biodiversity distribution for conservation.

9. List two key biodiversity areas that have been prioritized and describe their importance for global biodiversity.

10. Discuss the conflicts that arise in key biodiversity areas.

11. Discuss the threats to an example of sustainable traditional indigenous land management practice.

12. Discuss an example of an indigenous or marginalized community that has been forcefully relocated due to conservation efforts.

13. Evaluate the claim that species extinctions caused by anthropogenic actions may lead to a tipping point in the whole Earth system.

AHL

⟫ Taking it further

- Assess the tensions between exploitation, sustainable development and conservation in a local ecosystem or protected area.
- Join a citizen science or voluntary agency to participate in gathering knowledge of local and regional biodiversity.
- Volunteer in a local NGO for wildlife rehabilitation.
- Raise awareness of indigenous land rights.

3.3 Conservation and regeneration

AHL

Guiding questions

- How can different strategies for conserving and regenerating natural systems be compared?
- How do worldviews affect the choices made in protecting natural systems?

Understandings

1. Arguments for species and habitat preservation can be based on aesthetic, ecological, economic, ethical and social justifications.
2. Species-based conservation tends to involve *ex situ* strategies and habitat-based conservation tends to involve *in situ* strategies.
3. Sometimes a mixed conservation approach is adopted, where both habitat and particular species are considered.
4. The Convention on Biological Diversity (CBD) is a UN treaty addressing both species-based and habitat-based conservation.
5. Habitat conservation strategies protect species by conservation of their natural environment. This may require protection of wild areas or active management.
6. Effective conservation of biodiversity in nature reserves and national parks depends on an understanding of the biology of target species and on the effect of the size and shape of conservation areas.
7. Natural processes in ecosystems can be regenerated by rewilding.
8. Conservation and regeneration measures can be used to reverse the decline in biodiversity to ensure a safe operating space for humanity within the biodiversity planetary boundary.
9. Environmental perspectives and value systems can impact the choice of conservation strategies selected by a society.

10. Success in conserving and restoring biodiversity by international, governmental and non-governmental organizations depends on their use of media, speed of response, diplomatic constraints, financial resources and political influence.
11. Positive feedback loops that enhance biodiversity and promote ecosystem equilibrium can be triggered by rewilding and habitat restoration efforts.
12. Rewilding projects have both benefits and limitations.
13. The success of conservation or regeneration measures needs to be assessed.
14. Ecotourism can increase interdependence of local communities and increase biodiversity by generating income and providing funds for protecting areas, but there can also be negative societal and ecological impacts.

Why conserve biodiversity?

The relatively new discipline of conservation biology has brought together experts from various disciplines in the last 40 years. All are concerned about the loss of biodiversity and want to act together to save species and communities from extinction. Their rationale is that:

- diversity of organisms and ecological complexity are good things
- untimely extinction of species is a bad thing
- evolutionary adaptation is good
- biological diversity has intrinsic value and we should try to conserve it.

Here are some commonly used acronyms that you will need to understand this area.

- UN   United Nations
- IUCN   International Union for Conservation of Nature

TOK

Before you read about why it is important to conserve biodiversity, take a few moments to think about it. Make a list of reasons why you think we should conserve biodiversity and reasons why you think we should not. These questions may help you to make your lists.

- Do humans need other species? In what ways?

- Do other species exist for human use? Do they only exist for human use?

- Do other species have a right to exist? Does a great ape have more rights than a mosquito?

Are your reasons based on rational thought or emotion or both? Does this affect how valid they are?

- UNEP   United Nations Environment Programme

- CITES   Convention on International Trade in Endangered Species

- UNDP   United Nations Development Programme

- WWF   World Wide Fund for Nature

- WRI   World Resources Institute

- NGO   non-governmental organization

- GO   governmental organization

- SDG   sustainable development goals

There are many reasons to conserve biodiversity—economic, ecological, social, ethical and aesthetic.

The value of biodiversity

Biodiversity can be valued in a number of ways, but it is often considered as:

- goods—direct benefits from ecosystems such as food, water and wood

- services—indirect environmental services from ecosystem processes such as nutrient cycling, biogeochemical cycling and climate regulation.

Direct benefits

This is about the value of biodiversity through consumption. Such benefits relate to things that can be harvested and consumed like food or medicine.

Food sources

Rice, wheat and maize (corn) provide over half the world's food. In the 1960s, wheat stripe rust disease wiped out a third of the yield in the USA. The crops were saved by the introduction of resistant genes from a wild strain of wheat in Turkey.

Maize was particularly vulnerable to disease prior to the 1970s because it was almost all genetically identical worldwide. Now there are more than 700 different commercial hybrids of maize and it contains 20 times more genetic diversity than humans. This is partly because, in the 1970s, a perennial maize was found in a few hectares of threatened farmland in Mexico. The plants contained genes that confer resistance to four of the seven major maize diseases and could give the potential of making maize a perennial crop that would not need sowing.

Biodiversity of food crops is important because:

- we need to preserve old varieties in case we need them in the future

- we need to ensure there is enough genetic diversity for some of the crop to survive

- we know pests and diseases can wipe out non-resistant strains

- breeders are only one step ahead of the diseases and require wild strains from which they may find resistant genes.

▲ Figure 1 Maize (corn) variation

Natural products

- Many of the medicines, fertilizers and pesticides we use are derived from plants and animals.

- Guano (sea bird droppings) is high in phosphate and is used locally by farmers in some LICs.

- Oil palms give us oil for everything from margarine to toiletries, but oil palm plantations are a major driver of deforestation.

- Rubber (latex) is from rubber trees, linen from flax, rope from hemp, cotton from cotton plants and silk from silkworms.

- Honey, beeswax, rattan, natural perfumes, leather and timber are all from plants or animals.

Indirect benefits

These are the non-consumptive resources that nature provides—they are valued for social, ethical, cultural, ecological and aesthetic reasons.

Environmental services

We are just beginning to give monetary value to environmental services (see subtopic 8.2). If a value can be placed on these processes, then we can quantify the natural capital of biodiversity.

- Soil aeration depends on worms.

- Fertilization and pollination of some food crops depend on insects.

- Ecosystem productivity gives our environment stability and recycles materials. Plants capture carbon and store it in their tissues. They release oxygen which nearly all organisms need for respiration.

- Agriculture captures 40% of the productivity of the terrestrial ecosystem.

- Soil and water resources are protected by vegetation.

- Climate is regulated by the rainforests and vegetation cover.

- Waste is broken down and recycled by decomposers.

The preservation of as many species as possible in natural or semi-natural habitats may render the environment more stable. This makes the environment less likely to be affected by the spread of disease (plant, animal or human) or some other environmental catastrophe. Natural, multi-layered, forest with many tree species was removed from tropical islands and replaced with a single species—coconut plantations. This was followed by serious damage when the islands were affected by tropical storms.

Scientific and educational value

We investigate and research the diversity of plants and animals. The Encyclopaedia of Life (eol.org) intends to document all the 1.8 million species that are named on Earth. Humans do this because we believe we should know.

Biological control agents

Some species help us to control invasive species without the use of chemicals. For example, the prickly pear cactus (*Opuntia stricta*), introduced to Australia in 1840, was a pest plant covering millions of hectares of farm lands. In 1924, a moth (*Cactoblastis cactorum*) whose caterpillars eat prickly pear was introduced and by 1932 most of the land was clear of the cactus.

Gene pools

Wild animals and plants are sources of genes for hybridization and genetic engineering.

Future potential for even more uses

With new discoveries to come, there will be many more practical reasons to appreciate biodiversity. Some species can act as environmental monitors—an early warning system. Miners used to take canaries into the mines with them to warn of dangerous gases. If the birds died, the miners knew there was a toxic gas around and they should get out. Indicator species, such as lichens, can show air quality.

Human health

- The first antibiotics (such as penicillin) were obtained from fungi.

- A rare species of yew (*Taxus brevifolia*) from the Pacific Northwest of the USA produces a chemical called paclitaxel which is used to treat several types of cancer. It became the best-selling cancer drug ever manufactured.

- The rosy periwinkle (*Catharanthus roseus*), from the Madagascan forest, is curing some children with leukaemia or lymphomas.

Human rights

If biodiversity is protected, indigenous people can continue to live in their native lands. If we preserve the rainforests, indigenous tribes can continue to live in them and continue to make a livelihood.

Recreation

Biodiversity is often the subject of aesthetic interest. People rely on wild places and living things in them for spiritual fulfilment and outdoor activities.

Ecotourism

Many people take vacations in areas of outstanding natural beauty and national parks. This brings extra finance to an area and provides employment.

Ethical or intrinsic value

Each species has a right to exist—a bioright unrelated to human needs. Biodiversity should be preserved for its own sake and humans have a responsibility to act as stewards of the Earth.

Biorights self-perpetuation

Biologically diverse ecosystems help to preserve their component species, reducing the need for future efforts to conserve single species.

▲ Figure 2 Area of outstanding natural beauty

ATL Activity 19

ATL skills used: thinking, research

It is relatively easy to put a price on the direct benefits of the goods provided by the environment:

- the global honey market in 2022 was USD 9.01 billion

- the global timber market in 2021 was USD 285 billion.

Although these numbers are hard to imagine, they are accurate.

1. Try to find the global value of soil, the water cycle or any other indirect benefits of biodiversity.

2. What are the advantages and disadvantages of putting a monetary value on environmental services?

ATL Activity 20

ATL skills used: thinking, communication, social

1. Read the reasons to preserve biodiversity in the text. Sort them under the headings: Economic reasons, Ecological reasons, Social reasons, Ethical reasons and Aesthetic reasons. Do some fit in more than one column?

2. Under the same headings, list reasons not to preserve biodiversity.

3. Discuss your lists with your fellow students.

Conservation and preservation of biodiversity

Many people use these two terms interchangeably, but they have different meanings.

Conservation biologists do not necessarily want to exclude humans from reserves or from interacting with other organisms. They will even consider the harvesting or hunting of species as long as it is sustainable. They recognize that it is very difficult to exclude humans from habitats and development is needed to help people out of poverty. But they want to ensure this development is not at the expense of the environment. A conservation biologist would look for ways to create income for local people, e.g. through ecotourism or by allowing managed access to a reserve to enable education (anthropocentric viewpoint).

Preservation biologists have an ecocentric viewpoint which puts value on nature for its intrinsic worth, not as a resource that humans can exploit. Deep green ecologists argue that whatever the cost, species should be preserved regardless of their value or usefulness to humans. So, according to preservation biologists, the smallpox virus should not be destroyed even though it causes disease in humans. Preservation is often a more difficult option than conservation.

Efforts to conserve and preserve species and habitats require citizens, conservation organizations and governments to act in various ways. Because conservation is mostly seen as for the "public good", politicians act to change public policy to align with conservation aims. But there is tension between what

Key terms

Conservation biology is the sustainable use and management of natural resources.

Preservation biology attempts to exclude human activity in areas where humans have not yet encroached.

is seen as good for the economy and human needs and what is seen as good for the environment. However, these goals can coincide and this is being recognized more and more. Swiss Re Institute's stress-test analysis suggests that global GDP will lose up to 18% from climate change if no mitigative strategies are employed. According to the World Economic Forum, if the Paris Agreement targets are hit and the temperature rises by less than 2°C, the global GDP losses will be about 4.2%.

There is a danger in thinking that you, one individual, can have little or no impact on a global issue. Do not think that. You can have an effect locally in a specific place as an individual and perhaps a larger effect as a member of a group. The group may act locally or globally or both, and major changes have happened in thinking about and acting for the environment in the last few decades. You may have heard the slogan, "Think globally, act locally". It was first used in the early 1970s and is as true today as it was then. Everyone's environmental conscience starts at home, whether in using less plastic packaging and fossil fuel or acting to help a local nature reserve. You may not be personally able to save an endangered species but you can act to save some form of biodiversity.

Sustainable development means meeting the needs of the present without negatively impacting the needs of future generations and biodiversity (WRI/IUCN/UNEP 1992). It has been applied to conservation projects such as providing local people with the money to set up a national park. However, beware of "greenwashing" (hiding unsustainable practices by claiming they are "green"). Examples of unsubstantiated claims about products include:

- convincing people that a product is environmentally friendly when it isn't

- convincing people that a product has positive impacts on the environment when it doesn't.

For instance, when a mining company has extracted all the minerals it economically can, it may say it is being environmentally friendly by bulldozing the spoil heaps and adding grass seed. But so much more than that must be done to be sustainable.

Approaches to conservation

There are three basic approaches to conservation:

- species-based

- habitat-based

- a mixture of both.

Conservation is more successful when it involves research, adequate funding and the support of the local community.

Species-based conservation

This type of conservation focuses on conserving the species but does not look at conserving the habitat in which it lives. Here are five examples of species-based conservation approaches.

▲ Figure 3 This popular slogan has been widely used in many contexts. It is particularly appropriate in the context of climate change

TOK

There are various approaches to the conservation of biodiversity. To what extent is our decision to act based on emotions?

1. CITES

Many species are becoming endangered because of international trade. CITES is an international agreement between governments to address this problem. Governments sign up to CITES voluntarily and must write their own national laws to support its aim—"to ensure that international trade in specimens of wild animals and plants does not threaten their survival". Threatened species from elephants to turtles, orchids to mahogany are covered by CITES.

CITES has dramatically reduced the trade in endangered species of both live animal imports (e.g. tortoises) and animal parts (e.g. elephant tusk for ivory, rhino horn).

▲ Figure 4 The CITES logo

Species are grouped in the CITES appendices according to how threatened they are by international trade.

- **Appendix I**—species cannot be traded internationally as they are threatened with extinction.

- **Appendix II**—species can be traded internationally but within strict regulations ensuring sustainability.

- **Appendix III**—a species included at the request of a country which then needs the cooperation of other countries to help prevent illegal exploitation.

The appendices include some whole groups, such as primates, cetaceans (whales, dolphins and porpoises), sea turtles, parrots, corals, cacti and orchids and many separate species or populations of species.

CITES is a great example of what can be done voluntarily to conserve species. About 5,000 animal species and 28,000 plant species are on its lists. Since 1975, it has been one of the most effective international wildlife conservation agreements in the world.

(ATL) Activity 21

ATL skills used: thinking, research

Visit their official websites to answer the following questions about the WWF and Greenpeace.

1. What does WWF stand for?

2. Why do you think WWF has a panda as its logo?

3. For both WWF and Greenpeace, find out:

    a. who they are

    b. what they do

    c. where they work

    d. how they work.

4. Find the nearest WWF project to where you live.

5. What are Greenpeace's main biodiversity campaigns?

2. Captive breeding and zoos

Zoos, aquaria and other captive breeding facilities keep many examples of species, but they cannot keep every species. Some species are not easy to breed in captivity (e.g. cheetahs, Yangtze giant softshell turtle and whooping cranes). Zoos and other captive breeding facilities need to build and maintain a healthy population with good genetic diversity. Therefore, zoos keep detailed records of which animals breed together so that genetic lines can be tracked. They have breeding programmes and exchange animals with each other to widen the gene pools for reproductive success.

Zoos have had a bad press and sometimes this is justified when animals are confined in small cages or treated with cruelty. Most zoos are open to the public. This means they need to keep examples of megafauna (giraffe, elephant, hippo, rhino, polar bear) because these are what people will pay to see. The best zoos look after their animals very well, and many have educational centres. Aquaria may keep large cetaceans (e.g. killer whales, porpoises), fish and invertebrate species.

Pros of captive breeding	Cons of captive breeding
Saves animals from extinction and other threats	Some animals suffer in captivity, e.g. they suffer from obsessive-compulsive disorder, depression and anxiety
Educates the population about conservation	Conditions may not be suited to the animals
Helps generate funds for research and conservation	They are expensive to run
Animals may be able to be reintroduced into the wild	Reintroduction can be hard or impossible
Good for scientific research	Can lead to inbreeding which reduces genetic diversity
May provide employment for local people	Needs global cooperation, which is difficult

▲ Table 1 The pros and cons of captive breeding

▲ Figure 5 The downside of zoos

▲ Figure 6 The upside of zoos

3. Reintroduction

Even if a species is bred successfully in captivity, programmes to reintroduce wild populations or establish new ones are expensive and difficult. A few are successful, e.g. the Californian condor, Przewalski's horse in Mongolia and the black-footed ferret in Wyoming, USA. The best programmes are highly worthwhile and the best hope of saving a species that is extinct in the wild or in severe decline. Success is more likely if there are incentives for local people to support a reintroduction programme. Released animals may need extra feeding or care or even be recaptured if their food supply runs out or they are in danger of dying. The reintroduction of the golden lion tamarin to the remaining Atlantic Coast Forest of Brazil has been a rallying point for local people and reintroduction of the Arabian oryx in Oman has given employment to the local Bedouin tribes.

However, many programmes are not successful and are seen as an unnecessary waste of money, poorly run or unethical. Often the less successful programmes are those where the animal has become habituated to humans. Orphaned orang-utans have to be taught how to climb and socialize with each other. It takes patience and perseverance to get them to live in the wild after years in captivity and many do not make the move successfully.

Rare plants can be reintroduced after raising seedlings in controlled conditions. But they may be dug up by collectors once planted out, outcompeted by other plants, or eaten by herbivores.

Sometimes it is impossible to reintroduce a species to its native habitat because that habitat has gone. Then the only way to keep that species is in captivity —animals in zoos, game farms and aquaria, plants in arboretums, botanic gardens and seed banks.

Frozen zoos are stores of animal tissue (e.g. sperm, eggs, embryos, skin) that is cryopreserved (frozen at very low temperature for long-term storage, often in liquid nitrogen at $-196°C$). In theory, animals which have become extinct in the wild could be raised from this material and then bred in zoos or reintroduced to nature reserves. In practice, there are many issues involved in this process. For one example, look up the Frozen Ark Project online.

ATL Activity 22

ATL skills used: thinking, communication, social

We can watch wildlife on any number of devices and through most media. It is not the same as being there and watching a real elephant feed on the African savanna. For many people watching elephants in the wild is not an option but we have zoos. Discuss these questions in class.

1. Are zoos ethical?

   a. Should zoos be just for research and breeding?

   b. Consider the conflict between the ethics and the economics of zoos.

   c. Do we have the right to capture and cage other species even if we treat them well?

2. Do animals have intrinsic rights?

3. Why should we keep the last few individuals of a species in a zoo if there is no habitat left for them?

4. Botanical gardens and seed banks

The largest botanical garden in the world is the Royal Botanical Gardens in Kew, London. These gardens grow 25,000 plant species (10% of the world's total) and about 10% of these are threatened in the wild.

Around the world, there are some 1,500 botanical gardens which grow plants, identify and classify plants, and carry out research, education and conservation.

▲ Figure 7 Seeds in storage in a seed bank, Kew Gardens

Seed banks are where frozen or dried seeds are stored for many years. They are gene banks for the world's plant species and an insurance policy for the future. If the plant species is lost in the wild, the seeds may be preserved for future use and repopulation. Up to 100,000 plant species are in danger of extinction. Seed banks are a way of preserving the genetic variation of a species. Some crop plants which are widely grown have only a few varieties but seed banks contain many more varieties of the species.

There are seed banks around the world, holding national and international collections. Who owns the seeds in these banks is a matter of concern. Seed banks generally require high levels of energy and technology, so they tend to be in high-income countries. But they may contain seeds from many countries. A seed bank in Svalbard, Norway is funded by the Global Crop Diversity Trust as well as the Norwegian government. It is built within the permafrost so needs no power to keep the seeds frozen. This is the ultimate safety net against loss of other seed banks through civil strife or war.

5. Flagship species

These species are popular and can capture our imagination. Most of the flagship species are large and furry but they may not have a significant role in the ecosystem. If they disappeared, the rest of the community would be little affected. But these species have instant appeal, so they are used in fundraising from the public. The funds are then used to protect the habitat which will include other species that may be under more threat than the flagship species.

There are disadvantages to naming flagship species:

* They take priority over others.

* If they become extinct, the message is that we have failed.

* They may be in conflict with local peoples, e.g. human-eating tigers.

▲ Figure 8 Ring-tailed lemur—a flagship species of Madagascar

Some conservationists use the term "umbrella species". If they can gain support to conserve that species, this will also help other species in the same habitat—those under the umbrella. A species can be both keystone (see below) and umbrella, e.g. the lemurs of Madagascar.

6. Keystone species

A keystone species is one that plays a critical role in maintaining the structure of the ecosystem in which it lives.

All species are not equal. Some species have a greater effect on their environment than others, regardless of their abundance or biomass. They act like the keystone in an arch, which holds the arch together. The disappearance of a keystone species from an ecosystem can massively imbalance the ecosystem (far more than the loss of other species)—such a loss can even destroy the ecosystem. This impact is far greater than and not proportional to the number or biomass of the keystone species.

The difficulty for researchers is to identify the keystone species. These species tend to be predators or engineers in the ecosystem. A small predator can keep a herbivore population in check. Without the predator, the herbivores increase and eat all the producers, causing the loss of all food in the ecosystem.

Examples of keystone species include the following.

- Sea otters eat sea urchins in kelp forests. If there are no sea otters, the urchins eat the holdfast (anchor) of the kelp and it floats away.

- Beavers are engineers, making dams which turn a stream into a swampy area. Swamp-loving species then move in. Without beavers, the habitat changes.

- Elephants in the African savanna are also engineers, removing trees so the grasses can grow.

 Practical

Skills:

- Tool 2: Technology

Choose an animal species that is threatened. Browse the IUCN Red List or WWF website if you cannot make up your mind.

For your chosen species, find the following:

- an image of your chosen species

- the Red List category of your species (critical, endangered or vulnerable)

- the global distribution of your species (find a map if possible)

- its estimated current population size

- the CITES appendix of your species

- the threats facing your species (ecological, social and economic pressures)

- suggested conservation strategies to remove the threat of extinction

- how local people and government actions can help your species to recover.

Now make a video or animation about your chosen species, including the information you have found.

Share with the class.

Habitat-based conservation

This is a management practice that is essential to biodiversity preservation. It is very important for migratory species because they need more than one habitat and their habitats need to be joined by a safe migratory route. The aims of habitat conservation are:

- conserving, protecting and restoring habitats

- preventing species extinction

- reducing or preventing fragmentation of habitats, to preserve the range of a species.

Mixed approach—nature reserves

The location of a protected area (nature reserve) within a country is a significant factor in the success of the conservation effort. It is essential to consider the land use around a reserve because the interactions do not stop at the boundary. Land use outside the area may disrupt flows and impact the biodiversity within a reserve. Therefore, the proximity to urban centres or other intensive human activities must be considered.

In the past, many nature reserves were set up on land that no-one else wanted. It may have been poor agricultural land, land far away from high human population density or land that was degraded in some way. This all means that early reserves may not have been large enough or appropriate to the needs of the species they were aiming to protect.

Designing protected areas

When planning a protected area, conservationists now ask:

- How large should it be to protect the species?

- Do the species that need protection need a large reserve?

- Is it better to have one larger or many smaller reserves?

- What about the edge effects?

- How many individuals of an endangered species must be protected?

- What is the best shape?

- If there are several reserves, how close should they be to each other?

- Should reserves be joined by corridors or kept separate?

The large or small debate is known as the "SLOSS" debate ("single large or several small").

Single large (SL)	Several small (SS)
Contains sufficient numbers of a large wide-ranging species—top carnivores	Provide a greater range of habitats
Minimizes edge effects	More populations of a rare species
Provides more habitats for more species	Danger of a natural or human-made disaster (fire, flood, disease) wiping out the reserve and its inhabitants is reduced as some reserves may escape the damage
Both used for education and so further long-term goals of conservation through education	

▲ Table 2 Comparison of single large or several small conservation areas

The consensus is that the ideal size for a reserve depends on the size and requirements of the species it is intended to protect. Sometimes only small is possible—and that is better than nothing. In reserves of all sizes, **edge effects** should be minimized so that the circumference-to-area ratio is low (Table 3).

Edge effects occur at **ecotones** (where two habitats meet and there is a change near the boundary). There are more species present in ecotones because there are species from each habitat plus some opportunists. So, there is increased predation and competition. There is a change in abiotic factors as well, e.g. more wind or precipitation. Long thin reserves have a large edge effect, while circular ones have the least. But, in practice, the shape is determined by what is available and most reserves are irregular in shape. Division of the area by fences, pylons, roads, railways or farming should be avoided if possible because this fragments the habitat. But sometimes it is easier for governments to put road and rail links across national parks because there is less opposition than from privately owned land.

	Better	Worse	Why?
A			bigger better than smaller: includes more species, allows bigger populations, more interior per edge
B			intact better than fragmented: larger single population, no dispersal problem, more interior per edge
C			close better than isolated: easier to disperse among patches, allows easier recolonization if a local patch loses all individuals
D			clumped better than in a row: shorter distance to other reserves
E			connected with corridors better than not connected: facilitates dispersal
F			round better than any other shape: decreases amount of edge

▲ Table 3 Possible shapes of protected areas

Corridors are strips of protected land that link reserves. They allow individuals to move from reserve to reserve and so increase the size of the gene pool or allow seasonal migration. This has worked well in Costa Rica where two parks were linked by a 7,700 hectare corridor several kilometres wide. At least 35 species of bird use the corridor. But corridors have disadvantages. For example, disease in one reserve may be spread to the other and it may be easier for poachers and hunters to kill animals, because it is harder to protect the corridor than the reserves. Exotic or invasive species may also get into a reserve via corridors.

The Man and the Biosphere (MAB) programme reserves have buffer zones (transitional zones around the core reserve) where possible (Figure 9). Some farming, extraction of natural resources (e.g. selective logging) and experimental research is allowed in the buffer zone. The core reserve is undisturbed and species that cannot tolerate disturbance should be safer there.

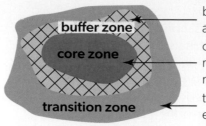

buffer surrounds the core area and is used for activities compatible with ecological ideals

core protected sites with minimal activity restricted to conserving biological diversity, monitoring, research

transition for sustainable development, e.g. farming, small settlements and communities

▲ Figure 9 Zones of a conservation area
Adapted from: BodoInfoexpress

ATL Activity 23

ATL skills used: thinking, communication, research, self-management

1. Using named examples, explain the difference between a nature reserve and habitat conservation.

2. Explain the land use conflicts that may arise in a local or global named botanical garden or park.

What is happening globally?

Connections: This links to material covered in subtopics 1.1, 1.2, 1.3, 2.1, 2.2, HL.a, HL.b and HL.c.

A range of activists, societies and organizations are working towards conservation of biodiversity. Some organizations are national and some international, and they employ a variety of strategies.

Convention on Biological Diversity

The Convention on Biological Diversity (CBD) is the international legal instrument for "the conservation of biological diversity, the sustainable use of its components and the fair and equitable sharing of the benefits arising out of the utilization of genetic resources". It has been ratified by 196 nations.

The CBD is governed by the Conference of the Parties (COP). Every government or organization that has ratified the treaty meets every two years to review progress, set priorities and commit to plans of action. This assists governments to implement the CBD programmes and coordinate with each other to collect and spread information.

The convention covers everything directly or indirectly related to biodiversity, including ecosystems, species and genetic resources. It deals with all aspects of humanity from culture to politics and everything in between. It also includes some key protocols, including the following:

- The Cartagena Protocol on Biosafety aims to protect the health of both humans and the environment from any adverse effects of biotechnology.

- The Nagoya Protocol aims to see equitable sharing of any benefits that arise from the study and use of genetic resources.

Bioreserves

In 1970 UNESCO set up the Man and the Biosphere programme, an intergovernmental scientific programme combining natural and social sciences. These biosphere reserves aim to:

- unite biodiversity conservation with sustainable use

- improve human livelihoods

- safeguard ecosystems—natural and managed

- enhance the relationship between people and the environment

- promote innovative approaches to economic development that encourage social, cultural and environmental sustainability.

There are 738 bioreserve sites in 134 countries around the world; 22 of these sites are transboundary.

Case study 3

Glacier Bay and Admiralty Island Biosphere Reserve, USA

▲ Figure 10 Satellite image of Glacier Bay basin

Description

- Established: 1986

- Surface area: 1,515,015 ha

- Located in south-east Alaska and made up of two areas: Glacier Bay and Admiralty Island.

- Retreating and advancing glaciers have shaped the landscape for about 20 million years.

- Since the last advance (200 years ago) the ice has retreated nearly 100 km.

Ecology

- Succession can be seen in this reserve.

- Pioneer plants are closest to the glacier: alder, cottonwood, lichens, moss, soapberry and willow.

- More mature forest (western hemlock and Sitka spruce) occurs further away.

- Admiralty Island has large stands of productive, mature temperate rainforest.

- Subtidal meadows, freshwater marshes and shrubland are found in both areas.

Socio-economics

- Humans have been around for approximately 10,000 years.

- Tlingits occupied the area in the 18th and 19th centuries—they were artistic people making Chilkat robes, basket weaving and totem poles.

- European settlers had sporadic settlements based on mining, logging, fishing and fur-trading.

- Human impact is limited so succession and recolonization can be studied.

- Heavy local involvement in development and review of management plans.

- Tourists arrive 80% on cruise ships; others hike, camp and kayak.

- Managed hunting and fishing are allowed.

- It is a UNESCO World Heritage site.

Ecosanctuary

An ecosanctuary is a place where wildlife can be preserved and where multi-species mammalian pest control is implemented. This is to allow ecosystem recovery through community involvement. In the late 1970s, authorities in New Zealand tried to evaluate the impact of pest mammals on native wildlife. Pest mammals include introduced invasive mammals such as rats (brown and black), cats, dogs and hares.

Many ecosystems have suffered reduced biodiversity caused by introduced predators. The chances of recovery for such ecosystems are limited if the predator remains in the area. It is very difficult to remove predators from an area, so fenced ecosanctuaries have been introduced. The predators are removed from an area and fences are erected. The aim is to reduce predation by decreasing the number of invasive predators, and also to reduce reinvasion by the predators. This is a costly exercise, but it allows the naturally occurring population numbers to recover.

Rewilding

Connections: This links to material covered in subtopics 5.2 and 6.3.

Rewilding (see also subtopic 2.5) is a strategy whereby the land is allowed to go back to its natural state. The aim is to restore ecosystems to the point where nature takes over and human intervention is no longer needed. The benefits are tackling climate change (carbon sequestering), improving biodiversity and improving human health and well-being (Figure 11).

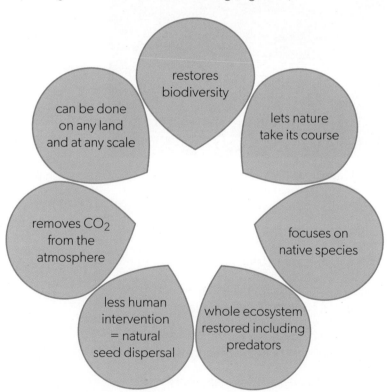

▲ Figure 11 Aspects of rewilding

Rewilding allows natural processes to be re-established and missing species to be replaced whenever possible. This includes replacing the top predators. The ecosystem then functions as a whole, balance is restored and nature determines the end result.

Rewilding is possible in any degraded system such as wetlands, forests and grasslands. Campaign group Rewilding Britain calculates that rewilding six million hectares of land could sequester 47 million tonnes of CO_2 per year. That is a lot, and every action that reduces climate change is welcome. See also the reintroduction of the grey wolf to Yellowstone, in subtopic 2.5.

Environmental value systems and conservation

As we saw in subtopic 1.1, there are different ways to view the value of the environment (Figure 12). It is likely that the three EVSs work in conjunction with each other.

ecocentric—the natural world has intrinsic value and is the most important

technocentric— technology will resolve all environmental issues

anthropocentric— humans are the most important element

▲ Figure 12 EVS perspectives

ATL Activity 24

ATL skills used: thinking, communication, social

Consider everything you have read so far in this topic and complete the following tasks.

1. Review your personal EVS (subtopic 1.1).

2. Make a table with the headings Ecocentric, Anthropocentric and Technocentric. List the various conservation strategies in the appropriate columns. Select the column that is the closest fit to your EVS.

3. Discuss which EVSs are most important in the various aspects of conservation.

Community support

Conservation International works with local communities and indigenous peoples on conservation projects. This people-centred approach inspires local individuals to take ownership of the project to protect biodiversity and support traditional lifestyles. In many LICs, conservation is closely intertwined with people. Local communities that live close to protected areas often exploit the natural resources there (sustainably or otherwise). For example, local people may hunt tigers to capitalize on the high prices on the black market or hunt and eat bush meat to survive. But if a community-based conservation project can provide an alternative income, there is a chance of protecting the natural resource.

TOK

Is it ethical to take away a community's traditional way of life and their livelihood without providing other opportunities for them?

Baboon Sanctuary, Belize

In 1985, Dr Rob Horwich went to Belize to study the black howler monkey (called baboons by the local population). It was apparent that habitat loss was such a serious issue that both the monkeys' habitat and the monkeys would soon disappear. Dr Horwich worked with a local landowner, Fallet Young. They reached out to other local landowners and farmers to educate them about managing their lands in a way that would protect howler monkey habitat. The programme was a huge success.

- 120 landowners signed voluntary pledges to change land management practices to protect howler monkey habitat.

- There is a sanctuary run by local women leaders which attracts international visitors.

- Guided nature hikes and a local museum bring in income.

- Small restaurants and B&Bs have been established.

- Howler monkey habitat is protected.

- Black howler monkey numbers have increased by thousands.

▲ Figure 13 Howler monkeys

Adequate funding

Funding for any project is central—you must have the money to start the project. The fact is that most threats to natural and cultural heritage are based on economics. Issues such as pollution, climate change, poaching, overexploitation and deforestation have a human element and can possibly be controlled if there is financial support. But they cannot be tackled without including the local economy.

Funding comes from a range of sources.

- Earned income: making money from sustainable use of the environment—such as ecotourism, food and fibre collection.

- Public funding including government grants: for example, in 2022 the US Environmental Protection Agency (EPA) had USD 76.49 billion to distribute to various conservation projects.

- Private funding: commercial banks, private individuals and companies can donate funds to support biodiversity conservation through projects, policies and technology.

- Non-profit organizations: this list is almost endless. You probably know of the WWF and the Wildlife Conservation Society. Maybe less well-known are the Durrell Wildlife Conservation Trust and the Charles Darwin Foundation.

Education and awareness

There are many aims of wildlife conservation programmes under the umbrella of education and raising awareness. Here are some.

- To ensure future generations can experience the natural world in its fullest glory and diversity.

 **Activity 25**

ATL skills used: thinking, social, research

In small groups, research a local conservation group and make short notes on its:

- aims

- purposes

- achievements

- funding and organizational structure.

- To develop understanding of the holistic and complex nature of ecosystems, including interactions within ecosystems and how ecosystems react to external human influences.

- To develop critical thinking skills in order to make informed decisions about what actions to take—not just decisions about whether to conserve species or ecosystems.

Appropriate legislation

Environmental conservation laws provide structures and support for wildlife management strategies. They cover every form of conservation from species to habitats, both terrestrial and aquatic. Some of these laws are broad and on the face of it not linked to conservation, for example the Clean Air Act (UK 1956), Ecological Solid Waste Management Act (Philippines 2000) and Clean Water Act (USA 1948).

Maybe the most famous legislation is CITES, an international agreement to stop the trade in endangered species.

Environmental justice

Environmental justice is discussed in subtopic 1.3. It is the right of all people to live in a pollution-free, efficiently protected environment, with equitable access to natural resources, regardless of race, gender or socio-economic status.

(ATL) Activity 26

ATL skills used: thinking, social, self-management

As a class, discuss these questions.

1. What constitutes "appropriate" in this context?

2. To what extent does "appropriate" differ for ecocentrists and technocentrists?

3. How differently will LICs and MICs view "appropriate"?

4. We probably cannot save all the endangered species. How do we decide which ones it is appropriate to save?

5. How does the Clean Air Act aid conservation of plants and animals?

Case study 5

Maasai tribes of Kenya and Tanzania

Ecotourism is seen as a sustainable kind of tourism because it is based on underdeveloped natural resources. People enjoy seeing the untouched wilds of many areas of the world, actively participating in conservation and supporting local communities.

Kenya and Tanzania were colonized by the British who quickly recognized the abundant natural resources in these countries. The British colonists thought the Maasai pastoralist way of life conflicted with the wildlife in the areas in which the Maasai lived. The Maasai were removed to facilitate the creation of national parks. This disastrous practice was continued by the country's own governments. Parks and reserves were developed without any consultation with the local tribespeople. The Maasai land has been repeatedly taken away from them and they have had no compensation and not been involved in any decision-making. They are now fighting for their land and their rights.

It is we Maasai who have preserved this priceless heritage in our land. We were sharing it with the wild animals long before the arrival of those who use game only as a means of making money. So please do not tell us that we must be pushed off our land for the financial convenience of commercial hunters and hotel-keepers. Nor tell us that we must live only by the rules and regulations of zoologists... If Uhuru (independence) means anything at all, it means that we are to be treated like humans, not animals. **99**

Edward ole Mbarnoti, a Maasai leader

Question

Review the information about Maasai land rights in subtopic 1.3. Then research the current status of Maasai land rights.

Scientific research

Scientific research about biodiversity must be the supporting structure for any decisions made about conservation.

- It investigates interspecies dynamics of complex ecosystems and that knowledge can be applied to other ecosystems.

- A better understanding of biodiverse ecosystems can tell scientists how biodiversity could be regenerated and increased in unbalanced ecosystems.

- It can assess emerging threats to ecosystems thus raising awareness and improving decision-making processes.

- It will develop science-based management of ecosystems to support wildlife.

How conservation organizations work— IGOs, GOs and NGOs

Organizations that work to conserve or preserve biodiversity and the environment may be local or global or both. Most large international organizations have national branches and regional and local offices. They fall into three categories depending on their constitution and funding:

- Intergovernmental organizations (IGOs): composed of and answering to a group of member states (countries); also called international organizations; e.g. the UN, the IPCC.

- Governmental organizations (GOs): part of and funded by a national government; highly bureaucratic; research, regulation, monitoring and control activities, e.g. the Environmental Protection Agency of the USA (EPA), the Environmental Protection Department of China.

- Non-governmental organizations (NGOs): not part of a government; not for profit; may be international or local and funded by altruists and subscriptions; some run by volunteers; very diverse; e.g. Friends of the Earth, Greenpeace.

Table 4 summarizes the key features of IGOs, GOs and NGOs.

▲ Figure 14 The UN logo

	IGO and GO	NGO
Use of media	• Media liaison officers prepare and read written statements • Control or work with media (at least one TV channel propagates official policy in even the most democratic regimes) so communicate decisions, attitudes and policies more effectively to the public	• Use footage of activities to gain media attention • Mobilize public protest to put pressure on governments • Gain media coverage through variety of protests (e.g. protest on frontlines or sabotage); sometimes access to mass media is hindered (especially in non-democratic regimes)
	• All provide environmental information to the public on global trends, publishing official scientific documents and technical reports gathering data from a plethora of sources	
Speed of response	• Considered slow—they are bureaucratic and can take time to act as they depend on consensus, often between differing views • Directed by governments, so sometimes act against public opinion	• Can be rapid—usually its members have already reached consensus (or they would not have joined in the first place)
Political diplomatic constraints	• Considerable—often hindered by political disagreement especially if international • Decisions can be politically driven rather than by best conservation strategy	• Unaffected by political constraints—can even include illegal activity • Idealistic or driven by best conservation strategy; often hold the moral high ground over other organizations and may be extreme in actions or views
Enforceability	• International agreements and national or regional laws can lead to prosecution	• No legal power—use of persuasion and public opinion to pressure governments
Public image	• Organized as businesses with concrete allocation of duties • Cultivate a sober or upright, measured image based on scientific or business-like approach	• Can be confrontational or radical in their approach to environmental issues such as biodiversity
	• All lead and encourage partnership between nations and organizations to conserve and restore ecosystems and biodiversity	
Legislation	• Enforce their decisions via legislation (may even be authoritarian sometimes)	• Serve as watchdogs (suing government agencies or businesses who violate environmental law)
	• All seek to ensure that decisions are applied	
Agenda	• Provide guidelines and implement international treaties	• Use public pressure to influence national governments or lobby governments over policy or legislation • Buy and manage land to protect habitats and wildlife
	• All may collaborate in global, transnational scientific research projects; may provide forum for discussion	

▲ Table 4 Comparison of IGOs, GOs and NGOs

(ATL) Activity 27

ATL skills used: thinking, communication, research

1. Copy and complete Table 5.

International agency or organization	Full name of IGO, GO or NGO	Logo	Aims	Actions
IUCN				
Greenpeace				
Wildlife Conservation Society				
US EPA				
UNEP				
WWF				
UNESCO				

▲ Table 5

The organizations in Table 5 operate in different ways. In their dealings with people, some work at government level while others work with local people "in the field". To bring about change, some work conservatively by careful negotiation, while others are more radical and draw attention to issues using the media.

2. Place each organization from the table on the axes in Figure 15 and add any others that you know of locally and nationally.

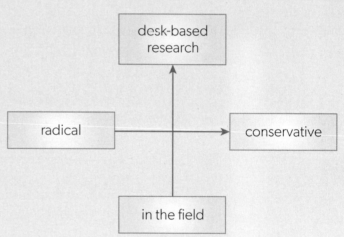

▲ Figure 15 Categorizing different organizations

International conventions on biodiversity— the good news

The consolidation of freedom of information via the worldwide web means that individuals, communities and nations have more power than ever before. Here is how this is reflected in the environmental arena.

- International cooperation was formalized in the UN and its many specialist agencies which have a major impact.

- United Nations Environmental Programme (UNEP) set up the Intergovernmental Panel on Climate Change (IPCC).

- UNEP drove the Montreal Protocol for phasing out the production of CFCs. Countries are forming international economic groupings (e.g. ASEAN, the EU, the African Union, OPEC). Some of these organizations are working for sustainable development and environmental protection.

- International organizations are complemented by NGOs. These often start locally as "grassroots" movements but are now having global impacts (e.g. Greenpeace, WWF).

In the past, institutional inertia (an inability to get going) has been a block to change. Now there is power in the huge numbers of people who want change and vote for politicians who say they will deliver it. This power may be enough to slow down or even reverse the degradation of our planet by human activities.

1980: The World Conservation Strategy

The World Conservation Strategy (WCS) was published by the IUCN, UNEP and WWF. It was a ground-breaking achievement which presented a united, integrated approach to conservation for the first time. It called for international, national and regional efforts to balance development with conservation of the world's living resources. Its aims were to:

- maintain essential ecological processes and life-support systems

- preserve genetic diversity

- ensure the sustainable utilization of species and ecosystems.

Many countries adopted the WCS and developed their own strategies for addressing national issues.

In 1982, the UN World Charter for Nature was adopted. It had the following principles.

- Nature shall be respected and its essential processes shall not be impaired.

- The genetic viability on Earth shall not be compromised; the population levels of all life forms, wild and domesticated, must be at least sufficient for their survival, and to this end necessary habitats shall be safeguarded.

- All areas of the Earth, both land and sea, shall be subject to these principles of conservation; special protection shall be given to unique areas, to representative samples of all the different types of ecosystems and to the habitats of rare or endangered species.

- Ecosystems and organisms, as well as the land, marine and atmospheric resources that are utilized by humans, shall be managed to achieve and maintain optimum sustainable productivity, but not in such a way as to endanger the integrity of those other ecosystems or species with which they coexist.

- Nature shall be secured against degradation caused by warfare or other hostile activities.

In 1991, *Caring for the Earth: A Strategy for Sustainable Living* was updated and launched in 65 countries. It stated the benefits of sustainable use of natural resources, and the benefits of sharing resources more equally among the world population.

1992: Rio Earth Summit and later

At the Rio Earth Summit in 1992, world leaders agreed on a sustainable development agenda called Agenda 21, and the Earth Council Global Biodiversity Strategy. The aim of the strategy was to help countries integrate biodiversity into their national planning. The three main objectives were:

- conservation of biological variation

- sustainable use of its components

- equitable sharing of benefits from utilizing genetic resources.

Agenda 21 was intended to involve action at international, national, regional and local levels. Some national and state governments have legislated or advised that local authorities implement the plan locally, as recommended in Chapter 28 of the document. Such programmes are known as Local Agenda 21 (LA21).

2002: The Johannesburg Summit on sustainable development

This was supposed to consolidate the Rio Earth Summit but little action came out of its deliberations.

2005: World Summit, New York

This outlined a series of global priorities for action and recommended that each country prepare its own national strategy for the conservation of natural resources for long-term human welfare. National conservation strategies have been written as a result of the international meetings. Integrating conservation with development is now a high priority—this was not the case a few decades ago.

2010: Biodiversity target

Adopted by EU in 2001 to halt biodiversity decline by 2010. This has largely failed.

2013: Rio+20

Considerations on how to build a green economy and how to improve cooperation for sustainable development resulted in a non-binding paper, *The Future We Want*.

2015: Sustainable Development Goals (SDGs)

The SDGs built on the success of the Millennium Development Goals (MDGs) but are broader in nature. They are time-bound and measurable goals for combating poverty, hunger, disease, illiteracy, environmental degradation and discrimination against women. The aim was to achieve the goals by 2030.

2023: SDGs midpoint

Although gains have been made since 2015 on achieving SDG targets, there have also been reverses with many causes including the COVID-19 pandemic, conflicts and climate change. In 2023 many UN conferences were held including COP28, the 78th UN General Assembly and a Water Conference.

2030: end of 15-year plan for SDGs

Annual meetings monitor progress towards the SDGs and reports are written. Progress towards meeting targets has happened but is slow. Look at the most recent SDG report to review current progress.

ATL Activity 28

ATL skills used: communication, social, research, self-management

The SDGs are broad and cover many aspects of human existence. We can all play a part in progressing towards the goals. Work in small groups.

1. Design a new logo for any or all of the goals. Be creative!

2. Make a mind map to show what you do to help to achieve these goals.

3. Remind yourself of the SDGs. Clicking on a goal on the UN SDG website will take you to an overview from the latest SDG report, including targets and indicators.

4. Figure 16 shows how one action to achieve SDG 11 also contributed to SDGs 12, 13, 14 and 15. Draw a similar diagram for SDG 12 by selecting one target in SDG 12 and showing how it will support four other SDGs. For example, if you select "reduce food waste" as an action, which other goals would this support?

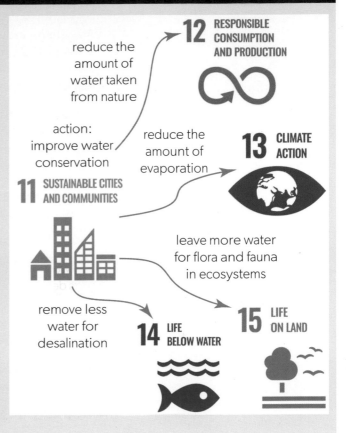

▶ Figure 16 Links between SDGs 11, 12, 13, 14 and 15

2002–25: The UN Environmental Programme World Conservation Monitoring Centre

This is a collaboration between UNEP and the World Conservation Monitoring Centre (WCMC), a UK charity. It has four areas of impact.

*   **Nature conserved:** working to ensure sustainability in the international trade in wildlife, promote nature connectivity, and support governments and others to strengthen networks of protected and conserved areas.

*   **Nature restored:** sharing knowledge and insights on the importance of restoring land and ocean ecosystems, as well as the opportunities to improve sustainability and resilience of agricultural systems.

*   **Nature-based solutions:** strengthening the use of nature-based solutions to climate change and championing the dependence of human health on the health of the natural world.

*   **Nature economy:** identifying leverage points to transform global economic systems, including through providing data platforms and metrics to help governments, businesses and investors understand impacts and dependencies on nature.

More about rewilding

Remind yourself about rewilding (subtopic 2.5).

All systems, including ecosystems, are subject to negative and positive feedback (subtopic 1.2). Positive feedback is when a change in the environment sets up a cycle that perpetuates that change. This can cause a complex, nonlinear response that results in an alternative stable state.

Sometimes, these changes are gradual and smooth. An increase in plant biomass during succession improves the abiotic conditions (temperature, water holding capacity). This means growth conditions improve and more complex plant life is established. This leads to more biomass.

In other cases, a critical threshold is reached, and the shifts are sudden and often catastrophic (e.g. eutrophication). What emerges is an alternative stable state. In the case of eutrophication, that alternative is not favourable.

Yellowstone National Park provides an example showing a slow cascade of positive feedback (see subtopic 2.5 and Figure 17). When wolves were reintroduced:

- elks were more cautious, so aspens were re-established because they were no longer overgrazed

- predation by the wolves controlled coyote populations so different carnivores were able to move in

- smaller herbivore numbers increased due to a reduction in the number of coyotes

- the wolves hunt throughout the year and leave carrion which can be scavenged by other carnivores.

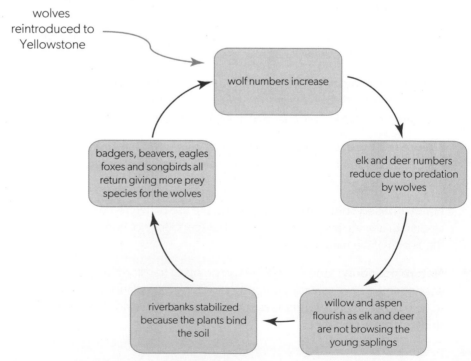

▲ Figure 17 Positive feedback in Yellowstone National Park

Benefits and limitations of rewilding

Benefits

- Beneficial against climate change.

- Increases and stabilizes bee populations—one of the most important pollinators.

- Introducing predators has a cascade effect on the whole food web.

- Changes in the food web can lead to more plant growth, reducing soil erosion and thus stabilizing some areas for more plant growth.

- In some areas, rewilding can raise awareness of the importance of these wild areas.

- Increases in ecotourism (e.g. the reintroduction of wolves to Yellowstone is projected to give a gain of USD 5 million for businesses in the area).

Limitations

- Loss of agricultural land and thus food production may reduce.

- Loss of land for development—roads, housing.

- Reintroduction of predators (e.g. wolves, lynx) presents a danger to farm stock and this has a significant economic impact.

- Reintroduction of herbivores can also be a problem (e.g. beavers in Scotland forage local croplands for food).

- Rewilding is seen as a high-risk strategy due to the unpredictability of the projects. Studies have shown that the survival of reintroduced plants is sometimes only 16–52%.

Success or failure

In the past, the criteria to assess the success or failure of conservation projects were:

- changes in biodiversity

- number of conserved species

- total area covered.

These criteria are relatively straightforward to measure because they are quantifiable. However, they do not take into account the rest of the environment, political factors or social factors. More importantly, there is no consideration of conflicts that may arise. Any or all of these factors may enhance or threaten the success of a conservation project. Looking at the projected results in a multifaceted way helps when assessing the outcomes of the project and improves learning for future projects. Effective evaluation of a project considers the initial state and changes during and after the project.

Practical

Skills:

- Tool 2: Technology
- Inquiry 2: Collecting and processing data
- Inquiry 3: Concluding and evaluating

Success of a conservation strategy can be evaluated at three levels:

- Did the project succeed as planned?
- Was the project well received by the communities impacted?

- Was this the best way to conserve nature?

1. Look up the Hinewai Reserve project in New Zealand.

   a. List its goals.

   b. Describe the ecosystem there.

   c. Explain the successes of the reserve.

2. Research a local or regional conservation project and give a report on whether or not it has been a success using these criteria: success as planned; reception; conservation achievement.

Case study 6

The Green Belt Movement (GBM)

In 1977, this conservation project was founded in Kenya to empower communities, especially rural women, to protect the environment. The founder of the Green Belt Movement was Professor Wangari Mathai. She received the Goldman Environmental Prize (1991) and the Nobel Peace Prize in 2004 "for her contribution to sustainable development, democracy and peace".

The emphasis on rural women is because it is the women who spend hours in the natural environment searching for food, water and firewood. With increased deforestation, the women had to travel further. This is a problem because they had less time to tend the crops and to look after the children.

On World Environment Day in 1977 seven seedlings were planted. Then the movement planted thousands of seedlings in long rows to form green belts of trees. Progress was not easy and there were efforts to suppress the movement between 1989 and 1999. The movement prevented construction in green spaces in Nairobi and protested against privatization of the Karura Forest. Private guards were hired to stop the protestors entering the forests which led to bloodshed as confrontations became violent.

▲ Figure 18 Professor Wangari Mathai

These problems did not stop the movement and over 5,000 tree nurseries have been established. Hundreds of thousands of women are involved and more than 51 million trees have been planted in Kenya. The movement is linked to sustainable development, peace and democracy. The project has broadened its efforts to include:

- tree planting and watershed management
- gender livelihood and advocacy
- climate change
- mainstream advocacy.

The GBM is a "grassroots" environmental movement. Grassroots movements are started by citizens, often local, who advocate for change and understand local issues, context and culture.

Question

1. Research one other grassroots environmental movement. The Global Citizen website has some examples.

2. Prepare a two-minute presentation to your class, describing a grassroots movement you would like to start.

Ecotourism—is it what it claims to be?

According to the World Tourism Organization, ecotourism is "tourism directed toward exotic, often threatened, natural environments, intended to support conservation efforts and observe wildlife". Is it what it claims to be, and is it sustainable?

What does ecotourism claim or aim to do?

- Create partnerships that empower the local communities and unite conservation with those communities.

- Be well designed and constructed with low-impact facilities that minimize the following impacts: behavioural—collecting souvenirs from the environment; physical—trampling, touching wildlife, picking flowers; social—disturbing local communities or imposing ideas from "home".

- Establish and maintain environmental and cultural awareness.

- Be positive for everyone and everywhere involved.

- Benefit conservation efforts through: increasing awareness of the environment; contributing financially to the local economy and conservation efforts; raising sensitivity about the political and social environment of the host country; raising awareness of indigenous peoples, their rights and their beliefs.

Tourism is increasing annually so these aims are ever more important. The old saying that "what people don't know about, they won't care about" suggests that what people do know about, they may care about. As people become more aware of and involved in ecotourism, they get to know more about the environment and its problems. This gives rise to the hope that they will then care about the environment and act positively to save it.

Does ecotourism deliver?

As ecotourism develops and learns it may inspire more research to improve management of the activities.

There are of course negative impacts of ecotourism, particularly if areas or tours are badly managed.

- A habitat may be visited too frequently for it to survive. For example, tourists may harvest souvenirs, trample plants causing compacting and soil erosion, or disturb animals.

- Ecotourist areas may expand and spill over into surrounding habitats, causing fragmentation and habitat loss.

- Ecotourists may cause stress just by their presence in the animals' habitat. For example, they may disrupt natural behaviours such as mating. Stress to plants and animals may also result from pollution by noise, light and heat.

- There must be community agreement for such intrusions—this is not always considered. This must be discussed in advance to avoid conflict between the local community and the ecotourist companies and their clients.

▲ Figure 19 An ecotourism sign

- Wildlife and humans may come into conflict. For example, tourists may discard food that animals take or there could be road collisions with animals involved.

- Animals may become acclimatized to human presence. This increases the risk of poaching and dependence on tourists for food.

- Incoming visitors may bring invasive species with them—either ones that escape into the environment or pathogens that impact the indigenous population.

- Traditional lifestyles may be impacted as indigenous people become tour guides.

(ATL) Activity 29

ATL skills used: thinking, research

Think about somewhere you have always wanted to go. Research online and identify an ecotourism option for that place. Read the reviews and, using all that you have learned so far, assess your chosen destination and the ecotourism option.

Consider the following questions:

1. Does it meet ecotourism criteria?

2. Are there negative aspects? What are they?

3. How much is the local population involved?

Check your understanding

In this subtopic, you have covered:

- why biodiversity is important and why we should conserve it

- the numerous direct and indirect benefits to humans of biodiversity

- how biodiversity is being conserved or preserved

- the different approaches to conservation

    - species-based conservation (e.g. CITES)

    - habitat conservation

- the different global projects that are going on

- the link between EVSs and conservation

- how the different conservation methods work

- how many different organizations and agreements there are

- new approaches to conservation such as rewilding

- effectiveness of ecotourism.

TOK

To what extent is international scientific collaboration important in biodiversity conservation?

AHL

How can different strategies for conserving and regenerating natural systems be compared?

How do worldviews affect the choices made in protecting natural systems?

1. Discuss the arguments for species and habitat preservation.

2. Understand that species and habitat preservation decisions are based on a wide range of criteria.

3. Describe the *in situ* and *ex situ* strategies of species-based, habitat-based and mixed approach conservation measures.

4. Support your descriptions with named examples.

5. Explain what is necessary for effective conservation of biodiversity in nature reserves and national parks, including the shape and size of the areas.

6. Describe an example of a rewilding project.

7. Use examples of individual, collective, national or international efforts to demonstrate that the measures taken to conserve and regenerate biodiversity can be at many different levels.

8. Explain how environmental perspectives and value systems can impact the choice of conservation strategies selected.

9. Compare and contrast international, governmental and non-governmental organizations in terms of their success with biodiversity conservation and restoration projects.

10. Using named examples, explain how positive feedback loops that enhance biodiversity and promote ecosystem equilibrium can trigger rewilding and habitat restoration.

11. Discuss the benefits and limitations of rewilding projects.

12. Evaluate the success of conservation or regeneration measures.

13. Evaluate the use of a named ecotourism project.

AHL

▶▶ Taking it further

- Use secondary data to assess the success of a rewilding project.
- Create a questionnaire to assess the impact of ecotourism in your area.
- Create a questionnaire to assess the level of awareness of conservation projects in your area.
- Investigate the role of an NGO in a conservation project.
- Visit a rewilding project or protected area and raise awareness about the project.
- Volunteer in a local conservation project, for example, the removal of an invasive species or putting up bird boxes.

Exam-style questions

1. a. Define biodiversity. [1]

   b. With reference to **Figure 1** identify **three** factors that could explain the high biodiversity in Ecuador. [3]

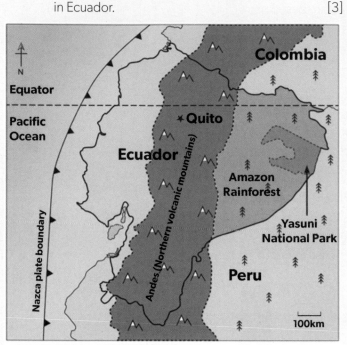

▲ Figure 1 Map to show the location of Yasuni National Park in Ecuador, a globally significant high biodiversity area

2. Look at **Figure 2** and **Table 1**.

▲ Figure 2 Examples of entanglement of marine species

Species	Status
Leatherback	Vulnerable
Flatback	Data deficient
Kemp's ridley	Critically endangered
Olive ridley	Vulnerable
Green	Endangered
Hawksbill	Critically endangered
Loggerhead	Vulnerable

▲ Table 1 Sea turtle species and status on IUCN Red List

Data from: IUCN 2023. The IUCN Red List of Threatened Species. Version 2022-2. <https://www.iucnredlist.org>

   a. Calculate the percentage of sea turtle species from **Table 1** that are critically endangered. [1]

   b. State **two** factors that are used to determine the conservation status of a species. [2]

   c. Identify **two** strategies for fisheries management that could improve the conservation status of sea turtles. [2]

3. a. Distinguish between the concept of a "charismatic" (flagship) species and a keystone species using named examples. [4]

   b. Explain the role of **two** historical influences in shaping the development of the environmental movement. [7]

   c. Discuss the implications of environmental value systems in the protection of tropical biomes. [9]

4. a. Outline **two** ecosystem services in a named biome. [4]

   b. Explain the causes, and the possible consequences, of the loss of a named critically endangered species. [7]

   c. Using examples, discuss whether habitat conservation is more successful than a species-based approach to protecting threatened species. [9]

5. a. Outline the factors that contribute to total biodiversity of an ecosystem. [4]

    b. Explain how ecological techniques can be used to study the effects of human activities on the biodiversity of a named ecosystem. [7]

    c. To what extent are strategies to promote the conservation of biodiversity successful? [9]

6. **Table 2** gives the approximate number of bird species found at different altitudes in tropical South America.

Altitude (m)	Number of species
0–500	2,000
500–1,000	1,950
1,000–1,500	1,550
1,500–2,000	1,100
2,000–3,000	950
3,000–4,000	500
4,000–5,000	200

▲ Table 2 Data from a diagram in Gaston K and Spicer J, *Biodiversity: An Introduction*, Blackwell Science, 1998

    a. Describe and explain the relationship between altitude and number of species shown in **Table 2**. [3]

    b. Define habitat diversity and species diversity. [2]

    c. Outline **three** characteristics that an area should have if it is to be designated a nature reserve or similar protected area. [3]

    d. An area of forest has been made a nature reserve. It is surrounded by farmland with several towns. Describe some of the changes that might occur in the area following its protection in this way. [2]

    e. Briefly describe a named protected area or nature reserve that you have studied and explain how it has been managed to protect its biodiversity. [3]

7. a. Explain the term biodiversity. [2]

    b. List **four** arguments for the preservation of biodiversity. [2]

    c. Describe the processes that may lead to the formation of new species. [3]

    d. Discuss the relative advantages and disadvantages of a species-based conservation strategy compared to the use of reserves. [4]

    e. **Figure 3** shows the layout of various conservation reserves. The reserves represent "islands" containing protected ecosystems surrounded by unprotected areas affected by human activities. For each pair (1) to (4), explain why the areas represented in column **A** might be considered better for conservation than the areas in column **B**. [4]

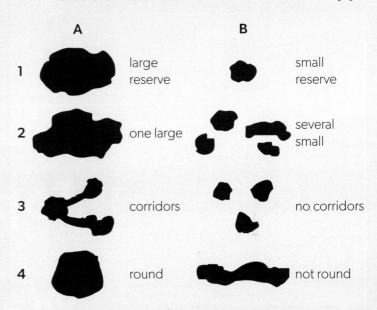

▲ Figure 3 Layout of conservation reserves

8. **Table 3** gives the number of flowering plant species for several tropical regions in the Americas, together with the area of each region in km².

Region	Surface area in km²	Estimated total number of species
Amazon Basin	7,050,000	30,000
Northern Andes	383,000	40,000
Atlantic coastal forests of Brazil	1,000,000	10,000
Central America including Mexico	2,500,000	19,000

▲ Table 3 Data from: Andrew Henderson and Steven Churchill, 'Neotropical plant diversity', *Nature* (1991), Vol. 351, pp. 21–22. © Nature

    a. Which region has the greatest number of species per unit area? [1]

    b. Which region has the lowest number of species per unit area? [1]

    c. Explain the range of biodiversity shown in the data. [2]

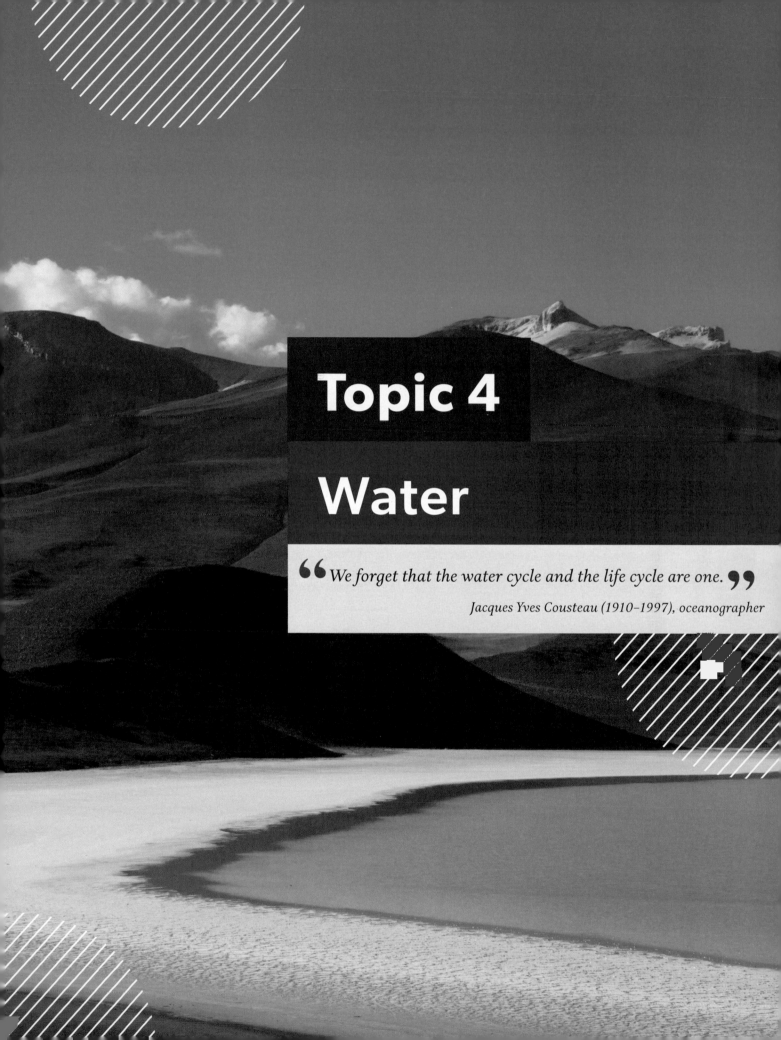

Topic 4

Water

 66 *We forget that the water cycle and the life cycle are one.* 99

Jacques Yves Cousteau (1910–1997), oceanographer

Guiding question

- How do water systems support life on Earth, and how do they interact with other systems, such as the carbon cycle?

Understandings

1. Movements of water in the hydrosphere are driven by solar radiation and gravity.

2. The global hydrological cycle operates as a system with stores and flows.

3. The main stores in the hydrological cycle are the oceans (96.5%), glaciers and ice caps (1.7%), groundwater (1.7%), surface freshwater (0.02%), atmosphere (0.001%), organisms (0.0001%).

4. Flows in the hydrological cycle include transpiration, sublimation, evaporation, condensation, advection, precipitation, melting, freezing, surface run-off, infiltration, percolation, streamflow and groundwater flow.

5. Human activities, such as agriculture, deforestation and urbanization, can alter these flows and stores.

6. The steady state of any water body can be demonstrated through flow diagrams of inputs and outputs.

7. Water has unique physical and chemical properties that support and sustain life.

8. The oceans act as a carbon sink by absorbing carbon dioxide from the atmosphere and sequestering it.

9. Carbon sequestered in oceans over the short term as dissolved carbon dioxide causes ocean acidification; over the longer term, carbon is taken up into living organisms as biomass that accumulates on the seabed.

10. The temperature of water varies with depth, with cold water below and warmer water above. Differences in density restrict mixing between the layers, leading to persistent stratification.

11. Stratification occurs in deeper lakes, coastal areas, enclosed seas and open ocean, with a thermocline forming a transition layer between the warmer mixed layer at the surface and the cooler water below.

12. Global warming and salinity changes have increased the intensity of ocean stratification.

13. Upwellings in oceans and freshwater bodies can bring cold, nutrient-rich waters to the surface.

14. Thermohaline circulation systems are driven by differences in temperature and salinity. The resulting differences in water density drive the ocean conveyor belt, which distributes heat around the world and thus affects climate.

The Earth's water budget

Viewed from space, the presence of water on Earth is obvious—hence the term "Blue Planet". Although 70% of the Earth's surface is covered with water, you can see in Figure 1 that the percentage of freshwater is very low—around 3%.

Key term

The **water budget** is a quantitative estimate of the amounts of water in stores and flows of the water cycle.

Activity 1

ATL skills used: communication

Using the data given, construct a table to list freshwater stores on Earth in decreasing order of size.

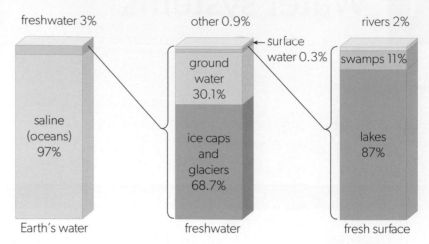

▲ **Figure 1** Distribution of the Earth's water

The atmosphere holds even less water.

- Only 0.001% of the Earth's total water volume is water vapour in the atmosphere.

- According to the US Geological Survey, if all the water in the atmosphere rained down at once, it would only cover the ground to a depth of 2.5 cm.

The time it takes for a molecule of water to enter and leave a particular part of the system is known as turnover time. Turnover times are very variable:

- oceans: 37,000 years

- ice caps: 16,000 years

- groundwater: 300 years

- rivers: 12–20 days

- atmosphere: only 9 days.

Water can be considered to be either a renewable resource or a non-renewable resource, depending on where it is stored. This is shown in Figure 2.

Key term

An **aquifer** is a body of porous rock or sediment saturated with groundwater.

Renewable	⟶	Non-renewable
Atmosphere	Groundwater **aquifers**	Oceans
Rivers	In aquifers, it takes longer than a human lifetime to replenish the water extracted	Ice caps

Water can easily move from being renewable to being non-renewable if poorly managed

▲ Figure 2 Status of water storages

The water (hydrological) cycle

The most fundamental question is "What drives the water cycle?"

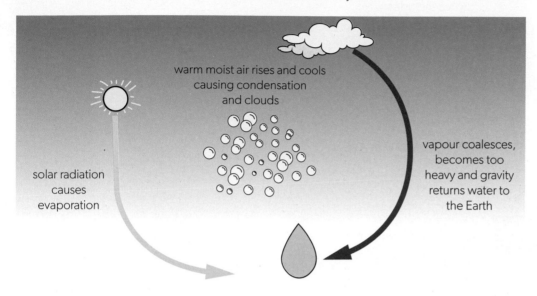

▲ Figure 3 Simple diagram to show the forces that drive the water cycle

The simple diagram in Figure 3 shows that energy from solar radiation and the force of gravity drive the water cycle. The water cycle (Figure 4) drives the Earth's weather systems.

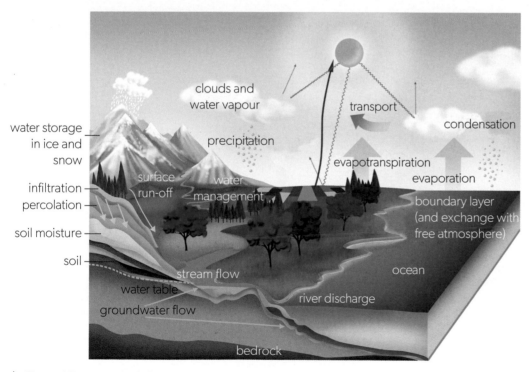

▲ Figure 4 The water cycle

The water cycle consists of stores (or storages) of water and the flows of water between the various stores. These flows may be transfers or transformations.

Stores include:

- oceans

- soil

- groundwater in aquifers

- lakes

- rivers and streams

- the atmosphere

- glaciers and ice caps.

Transfers (movements in the same state) are:

- advection: wind-blown movement

- flooding

- surface run-off: water running over the surface, not in a channel

- infiltration: when water runs into and through soil

- percolation: water running into and through rocks

- stream flow and current: water running over the surface in defined channels.

Transformations (changes of state between solid, liquid and vapour) are:

- evaporation: liquid to gas (water vapour)

- transpiration: plants lose water vapour through stomata

- evapotranspiration (EVT): a combination of evaporation and transpiration which can be measured and includes plant transpiration and evaporation from the soil below it

- condensation: water vapour to liquid

- freezing: liquid to solid snow and ice

- melting: solid ice to liquid water

- sublimation: solid snow and ice to water vapour.

Water is nearly unique as it is one of very few materials that exist naturally in all three phases—solid, liquid, gas (also called states of matter)—at surface pressures and temperatures. It can move directly from one phase to another—see Figure 5.

TOK

The hydrological cycle is represented as a systems model. To what extent can systems diagrams effectively model reality, given that they are only based on limited observable features?

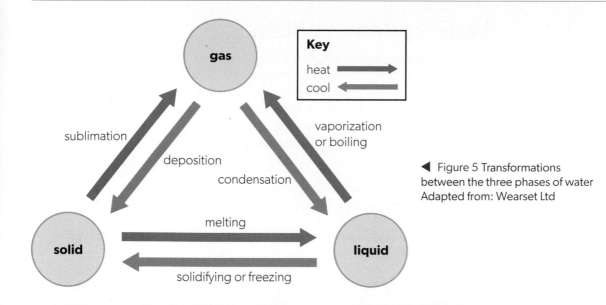

◄ Figure 5 Transformations between the three phases of water
Adapted from: Wearset Ltd

ATL Activity 2

ATL skills used: thinking, communication

1. Draw a systems diagram of the water budget and cycle, showing the stores and flows given in Table 1. Make the sizes of the store boxes, and the widths of the flow arrows, proportional to the volumes. Label all stores and flows.

Stores	Water volume (km³ × 10³)
Snow and ice	27,000
Ground water and aquifers	9,000
Lakes and rivers	250
Oceans	1,350,000
Atmosphere (variable)	13
Soil	35,000
Flows	**Water volume (km³ × 10³)**
Precipitation over oceans	385
Precipitation over land	110
Ice melt	2
Surface run-off	40
Evapotranspiration from land	70
Evaporation from sea	425

▲ Table 1 Stores and flows in the water cycle

2. Table 1 includes six stores of water.

   List them in order of decreasing size (largest first) and calculate the percentage of the total hydrosphere in each store.

3. Explain which of these stores humans can use and calculate what percentage of all water this is.

Water cycle as a system

The water cycle on a global scale is a closed system. There are no inputs or outputs of matter, only energy. However, on a local scale—e.g. a lake (Figure 6)—the system is open, with inputs and outputs of both energy and matter. Imagine a sink full of water: if you take out the plug but run the tap at just the right speed of water delivery, the volume will remain the same and the system is in a steady state. (Please do not try this—it is wasteful.)

Why does this matter? As with maximum sustainable yield (MSY) in ecosystems, (see subtopics 2.2 and 4.3), our water sources have a maximum sustainable rate at which the water can be harvested. This applies to lakes, aquifers and rivers.

> Connections: This links to material covered in subtopics 1.2, 2.2 and 4.3.

Inputs of water
precipitation
throughflow in the soil
groundwater flow through rocks
overland flow and river flow
glacier melt

Outputs of water
evaporation
infiltration into the soil
percolation into the rocks
overflow (during flooding)
river outflow

▲ Figure 6 Inputs to and outputs from a lake system

Many of the inputs and outputs have significant seasonal variation. If we are to use natural water sources sustainably, we must not harvest above the sustainable maximum. Each input and ouput can be measured and an MSY calculated, although—as always— these are estimates that do not allow for one-off events such as an overly hot summer that increases evaporation rates.

 Practical

Skills:

- Tool 1: Experimental techniques

- Tool 4: Systems and models

- Inquiry 2: Collecting and processing data

Set up your own mini lake and control inputs and outputs.

Set-up 1

1. Put 500 ml of water in five bowls of different sizes.

2. Leave all five bowls in the same place.

3. After one week, measure the amount of water in each container.

What will this tell you?

Set-up 2

1. Add 300 ml of water to a 500 ml beaker.

2. Each day, read off the amount of water in the beaker and then add 10 ml of water.

3. Repeat this every day for two weeks

What will this tell you?

Human impact on the water cycle

Connections: This links to material covered in subtopics 1.2 and 5.2.

Consider the myriad ways humans use water, and the impact these uses have on the water cycle. In addition, human changes to the landscape interrupt the movement of water in the following ways.

- Withdrawals—for domestic use, agriculture and industry.

- Discharges—by adding pollutants to water, e.g. chemicals from agriculture, fertilizers, sewage.

- Changing the speed at which water can flow and where it flows, including: urbanization—by building roads and channelling rivers underground or into concreted areas; canalizing—straightening large sections of rivers in concrete channels to facilitate more rapid flow through sensitive areas; and with dams, barrages and dykes—to make reservoirs.

- Diverting rivers or sections of rivers: many are diverted away from important inhabited areas to prevent flood damage; some are diverted towards dams to improve storage.

- Deforestation—clearance of forests for human use such as agriculture, industry, housing, infrastructure.

The following are examples of major changes caused by humans.

- Aral Sea—intense irrigation has almost stopped river flow into this inland sea and lowered the sea's level. Over the last 50 years the sea's area has reduced by 75% and the volume of water it holds has reduced by 90%.

- Ganges basin—deforestation increases flooding as precipitation is not absorbed by vegetation.

Deforestation

Trees and forests play a vital role in the water cycle:

- Trees store water in their canopy, branches and roots, so forests act as reservoirs for water. They release this water slowly through transpiration—one tree in the tropical rainforest can transpire about 378 litres per day. Transpiration from trees supplies clouds with moisture for formation of rain.

- The layered structure of the trees dissipates some of the power of storms, keeping rainfall patterns more regular.

Tree roots binds the soil and slow down water infiltration into the soil, reducing soil erosion and maintaining the soil water store.

When trees are growing in a forest, precipitation returns water to the land, plants take up the water via their roots and the cycle continues. For example, in the Amazon, the clouds created by the rainforest move westward and provide essential water to Central Brazil, Paraguay, Uruguay and northern Argentina. Deforestation disrupts this movement of water.

Deforestation also disrupts river flow and water volume. Research suggests that the River Amazon is at a tipping point: with only 81% of the forest left intact, the hydrological cycle could collapse and the forest turn to grassland or desert.

▲ Figure 7 Deforestation and soil erosion

Deforestation also impacts groundwater stores. If precipitation is depleted, there is less water to recharge aquifers and that store decreases—this impacts river flows and water supplies for humans.

Urbanization

More and more humans live in cities and there are more of us than ever before. This means more buildings and fewer natural permeable surfaces for water to infiltrate into. Natural water flows are disrupted and flash floods are more common. In January 2020, Jakarta in Indonesia had very heavy overnight rainfall (400 mm) which caused the major rivers to overflow. The resultant flooding killed 66 people and displaced 60,000.

- Flood peaks tend to be higher and more frequent in urban areas than in rural areas.

- According to Morgan Stanley analysts, the cost of flood damage in China could reach USD 77 billion by 2030.

- Stormwater is collected by impermeable surfaces like rooftops, roads and footpaths and flows directly into waterways, so floodwater can arrive in minutes not hours.

- Water flushing (high-speed movement of water over surfaces) means that water flows over impermeable surfaces, "cleaning" them and carrying pollutants to the rivers.

- There is very limited water storage in urban areas because the soil is sealed under impermeable surfaces. The removal of natural vegetation decreases the biomass store and limits interception and evapotranspiration. Surfaces are smooth so there are few natural surface depressions to store water, and human-made drainage systems move water away quickly.

▲ Figure 8 Urban flooding

Practical

Skills:

- Tool 4: Systems and models

- Inquiry 2: Collecting and processing data

1. On your journey to school observe and record what the surfaces of the land you pass are made of.

2. Figure 9 shows how to measure the rate of infiltration of a surface. The method is described in subtopic 2.6.

   Set up this experiment on different surfaces to see which ones allow water to pass through and how fast it infiltrates.

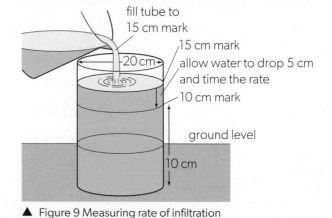

fill tube to 15 cm mark
15 cm mark
20 cm
allow water to drop 5 cm and time the rate
10 cm mark
ground level
10 cm

▲ Figure 9 Measuring rate of infiltration

Agriculture

The use of water in agriculture—for growing fresh produce and sustaining livestock—impacts stores and water quality. In many areas of the world, rainfall is a major supply of water for agriculture but this is not always the case, especially in summers when demand outstrips supply.

Water used in agriculture must be extracted sustainably to ensure a minimum environmental flow that provides sufficient water for rivers, lakes and wetlands. From the socio-environmental side, this means the extraction rate must take account of stakeholders both upstream and downstream. Key sources of water for agriculture include the following.

- Groundwater: to use this sustainably, the extraction rate must be lower than the annual recharge rate.

- Surface water such as rivers, streams, canals and reservoirs.

- Precipitation: obviously this is not extraction, but it is a source for agriculture and the impact of run-off into streams is important. Sometimes rainwater is collected in cisterns.

Water is used for irrigation, livestock and other farmyard activities such as washing animals, machinery and equipment. According to the UK Water Partnership, annual extraction of water by the UK agricultural sector is 120 to 150 million $m^3 yr^{-1}$, half of which is for irrigation. Over-extraction can cause springs, rivers and marshlands to dry up.

Run-off from agriculture is a problem as it washes excess nutrients and pesticides into local aquatic systems. Excess nitrogen and phosphorus cause degradation of the water quality and can cause eutrophication (subtopic 4.4).

The soil in agricultural fields is compacted by farm machinery and at certain times of the year there is a lack of vegetation cover. The compaction reduces infiltration rates and amounts, thus reducing soil water storage, and the lack of cover increases wind erosion. Eroded soil ends up in rivers, decreasing storage capacity and increasing the chances of flooding.

ATL Activity 3

ATL skills used: thinking, communication, research

Visit the AQUASTAT website, which is maintained by the Food and Agriculture Organization.

Review the country statistics for Jordan and use the information to answer these questions.

1. Describe where most of the population live and explain why.

2. Summarize the agricultural industry of Jordan.

3. List the water supply sources available to Jordan and state whether they are sustainable or not.

4. Describe the political and environmental issues around Jordan's water supplies.

5. Explain the water management strategies that Jordan is adopting.

(ATL) Activity 4

ATL skills used: thinking, communication, social, research, self-management

1. To what extent is water unique?

2. Make a list of everything you use water for.

3. Discuss your list with your classmates and rank the uses in order of importance.

4. Explain in your own words the terms in Table 2.

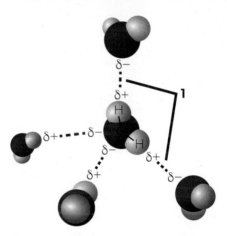

▲ Figure 10 Model of water molecules showing hydrogen bonds (dotted lines) between them

▲ Figure 11 Surface tension can support some invertebrates

Uniqueness of water

All life on Earth depends on water for survival. Luckily, it is the most abundant resource on Earth, and it has certain important chemical and physical properties.

Chemical properties	Physical properties
Polarity	Density or phases
Cohesion or surface tension	Transparency
Adhesion	Gas solubility
Solvent	
Specific heat capacity	

▲ Table 2 The chemical and physical properties of water

Chemical properties

Polarity

A water molecule is made up of two hydrogen atoms and one oxygen atom. Water is a polar molecule (Figure 10) because of its shape. The net charge of a water molecule is zero but the hydrogen ends are slightly positive (δ^+) and the oxygen is slightly negative (δ^-). Weak hydrogen bonds form between the hydrogen atoms in adjacent water molecules, so the molecules attract each other. Water molecules are bent because the negative charge of the oxygen atom repels the positive charges of the hydrogen atoms. This bending causes the polarity of water.

Cohesion

Water molecules "stick together" because they attract each other. Hence, we get large and small bodies of water from oceans to puddles. This leads to surface tension that allows light objects (Figure 11) to stay on the surface of the water.

Adhesion

Water molecules are attracted to molecules that are not water. In plants, this allows capillary action to take place—water moves up the xylem from the roots to the leaves because water molecules are attracted to the walls of the xylem vessels.

Solvent

Water is able to dissolve such a wide range of solutes that it is often described as the universal solvent. This is a misnomer as water cannot dissolve all substances, e.g. lipids do not dissolve well in water.

It is a good solvent because it has both positively and negatively charged poles. For this reason, water is important in cellular processes where it breaks the bonds between molecules and dissolves them.

Specific heat capacity

The hydrogen bonds in water take a lot of energy to break, so a lot of energy is needed for water to change phase; it is also slow to heat up and cool down. Humans are approximately 60% water, so we rely on these slow changes to keep our body temperature relatively constant. These slow changes are important for aquatic life and for the effective functioning of enzymes in our bodies. The specific heat capacity of water is much greater than that of land which is why land cools faster and heats faster than water.

Physical properties

Water phases

Water is one of three naturally occurring materials that are present in all three phases on Earth under normal pressures and temperatures. The other two are germanium and silicon, which are elements (water is a molecule made up of two elements—oxygen and hydrogen). Solid water (ice) is less dense and therefore lighter than the liquid form, because the hydrogen bonds push the molecules further apart in ice than in water. This means that, if a body of water freezes over then the ice layer on top insulates the aquatic life below, which survives. It also explains why icebergs float.

Water transparency

Transparency indicates the clarity of water and is measured by a Secchi disc; the clearer the water, the deeper sunlight can penetrate. If water is more transparent it:

- can indicate a healthy aquatic ecosystem; pollution and high sediment load tend to decrease transparency

- determines the depth to which plants can photosynthesize and therefore grow

- is linked to greater surface warming and stronger thermal stratification

- ensures that predator–prey interactions can occur; predators cannot catch prey they cannot see.

Water transparency varies seasonally and is affected by algal growth, surface run-off, suspended sediments and turbulence.

Gas solubility

The two most important gases found dissolved in water are oxygen and carbon dioxide; carbon dioxide is 200 times more soluble in water than is oxygen. Both are essential to aquatic life; the fauna require oxygen while the flora require both carbon dioxide and oxygen. Gas solubility increases under higher pressures and lower temperatures.

Oceans

Oceans are central to our survival. They also support global economies; consider how much freight is moved around the world by the international shipping industry. And how many people spend holidays on the beach enjoying the ocean?

▲ Figure 12 Iceberg floating because ice is less dense than water

Connections: This links to material covered in subtopic 2.5.

Connections: This links to material covered in subtopic 2.4.

We live on a blue planet, with oceans and seas covering more than 70 per cent of the Earth's surface. Oceans feed us, regulate our climate, and generate most of the oxygen we breathe. 99

UNEP

ATL Activity 5

ATL skills used: thinking, communication, social

To what extent can we justify the threats we pose to the oceans? Discuss the threats listed, considering each threat as economic, ethical or social.

Threats to oceans

The oceans face many threats from human activity:

- eight million tonnes of plastic

- coral reef damage due to climate change

- overfishing

- nutrient overload, creating dead zones, and other pollutants

- acidification and warming.

Carbon sequestration and sinks

A carbon sink is a store—a reservoir of carbon—that may be natural or human made. Carbon sinks may:

- sequester (absorb and store) atmospheric carbon

- use physical or geological processes such as storing carbon in rock formations

- use biological mechanisms such as storing carbon in vegetation, animals, soil

- use technology such as carbon-capture storage.

It is generally accepted that there are four carbon sinks: atmosphere, biosphere, hydrosphere and lithosphere (Figure 13). The lithosphere includes the pedosphere (soil), which may be considered a fifth sink. The focus here is the hydrosphere, specifically the oceans and how they sequester carbon.

The oceans act as a significant carbon sink, absorbing 25 to 31% of the carbon dioxide produced by human activities each year. This is a two-way exchange or flux. Carbon dioxide is released from the oceans to the atmosphere (positive atmospheric flux) and also taken in by the oceans from the atmosphere (negative atmospheric flux). Think of these fluxes as an inhale and an exhale, where the net effect of these opposing fluxes determines the overall effect.

The polar regions are very important in the carbon cycle, because the lower temperatures mean that carbon dioxide dissolves more easily in water. Denser cold water sinks towards the depths, taking the dissolved carbon dioxide with it.

▲ Figure 13 The carbon sinks: atmosphere, biosphere, lithosphere and hydrosphere

Acidification of oceans

Seawater had an average pH of 8.2 before the Industrial Revolution. It is now 8.1—an increase of nearly 30% (remember, the pH scale is logarithmic). This ocean acidification is due to absorption of more carbon dioxide from the atmosphere and it upsets the chemical balance in the oceans and coastal waters. The change in pH slows the rate of calcification—the process that some marine life uses to make calcium carbonate, the raw material of skeletons and shells. Therefore oysters, clams, coral reefs, shrimps, and other crustaceans and shellfish are suffering from weaker skeletons.

The ocean carbon sink is moderating the increases in carbon dioxide produced by the combustion of fossil fuels. However, if the oceans warm up too much, ocean circulation will slow down. This will trap surface water at the surface, where it will become saturated with carbon—so carbon uptake will slow down. It all depends on the ocean carbon pump, which is made up of two pumps.

1. The **physical pump** is a slow ocean circulation taking several hundred years to complete one cycle. Carbon dioxide from the atmosphere diffuses into the ocean surface waters. It is then transferred to the deep ocean in zones where cold, dense surface waters sink.

2. The **biological pump** transfers surface carbon in the food web to the seabed. It depends on a healthy ecosystem with plenty of phytoplankton and microscopic algae. These producers extract carbon dioxide for photosynthesis from water and are the basis of the food chain. When organisms die, they may be eaten by scavengers but everything eventually ends up in the sediment on the sea bed. Under such conditions, geological processes take place, forming oil deposits and rocks such as chalk and limestone. These are extremely long-term carbon sinks.

Stratification

The ocean is a single body of saltwater but it is not uniform, it is stratified. The **stratification** between layers of water is due to the different physical properties (Figure 14) which act as a barrier to mixing. This is a natural phenomenon—it is usually vertical but may at times be horizontal. Mixing occurs slowly due to winds, tides and currents—the bigger the difference, the slower the mixing.

> ## Key term
>
> **Stratification** is when there are two or more distinct layers in a vertical column of water (e.g. salt water or freshwater).

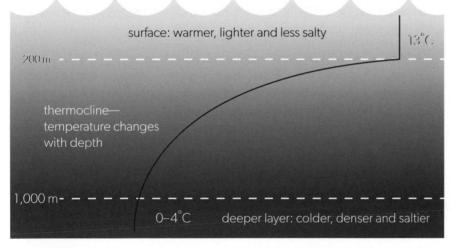

surface: warmer, lighter and less salty

13°C

200 m

thermocline—
temperature changes
with depth

1,000 m

0–4°C   deeper layer: colder, denser and saltier

◀ Figure 14 Ocean stratification of three layers: surface, thermocline, deeper (up to 10,000 m deep)

Surface layer

This upper "mixed layer" of the ocean is approximately 200 metres in depth; this varies according to season and latitude. It is subject to a great amount of turbulence created by:

• wind generating ripples that travel downwards and outwards

• currents moving in different directions and swirling the surface water around

• diurnal (daily) patterns of heating and cooling which cause vertical movement.

This is the layer that experiences constant gaseous exchange with the atmosphere. This exchange maintains the balance between the two layers.

However, balance is not instantaneous; in fact it can take 10–100 years to achieve. So the effect of changes in atmospheric carbon dioxide now may take a long time to become apparent.

In general, water temperature decreases with depth—with the important exception that water is most dense at 4°C. This means that water colder than 4°C will float above warmer water and a body of water will freeze from the surface down. This is vital to aquatic life, allowing for survival under insulating ice.

The bottom of the surface layer is determined by the sharp drop in water temperature and increase in density that marks the **thermocline**.

The thermocline

The two layers of ocean water are separated by the thermocline. The strength and depth of the thermocline change seasonally and with latitude—it is:

- fairly stable in the tropics

- deeper in the mid-latitudes in summer

- nearly absent at the poles.

Salinity, temperature and density

Salinity is the measure of dissolved salts in water. Salinity and temperature are inversely proportional. The colder the ocean, the more salty it is. Cold salty water is denser than warm freshwater and sinks below it.

The density of seawater increases with increasing depth because of increased pressure. As density increases, salinity increases because the mass of the water increases. This is a positive relationship.

Denser water sinks. It is mainly water temperature that determines where water sinks. Cold ocean water is at the poles, warm at the equator. High salinity and low temperature at the surface (at the poles) mean seawater sinks and flows along the ocean floor, driving the deep ocean currents.

More heat at the surface leads to more evaporation of seawater. This means salinity increases and water becomes more dense and sinks. But more heat also means density decreases so water rises.

Oxygen and nutrients

Changes in density, temperature and salinity result in stratification in seawater. There is also an impact on dissolved oxygen content and nutrient levels of the water. In summer, the warm, light surface waters accelerate the rate of plant photosynthesis, increasing the dissolved oxygen content of the shallow water. Lack of mixing means that this oxygen does not reach the deeper layers. Lack of mixing also keeps the nutrient-rich waters in the depths.

Colder water can hold more oxygen than warmer water because when temperature increases, the increased energy breaks weak bonds holding oxygen in the water so the oxygen escapes. Solubility of gases decreases with increasing temperature.

Key term

A **thermocline** is a thin transitional layer in which there is a very sudden change in temperature.

 ## Practical

Skills:
- Tool 2: Technology
- Tool 3: Mathematics
- Inquiry 2: Collecting and processing data
- Inquiry 3: Concluding and evaluating

Use data from a suitable database to analyse the relationship between water temperatures and salinity and oxygen concentrations. Use an appropriate statistical test.

The NASA Scientific Visualization Studio and the European Space Agency websites provide helpful overviews.

You can also visit the website of the USA National Oceanic and Atmospheric Administration (NOAA) and look for online research papers.

Global warming and ocean stratification

Recent studies have shown that the increased warming of the ocean's surface layers (due to global warming) is increasing stratification. This is not good news for marine life, or the Earth generally. The climate is warming which increases ocean stability. We often consider "stability" to be a good thing, but it has certain consequences in the context of the oceans.

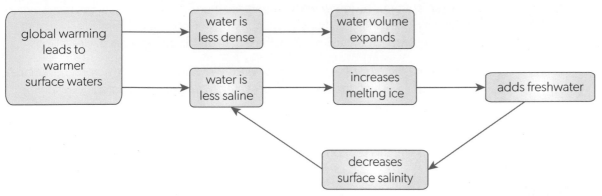

▲ Figure 15 **Positive feedback from global warming**

Figure 15 shows a positive feedback cycle by which increases in stratification drive global warming. Warmer water absorbs less gases, including carbon dioxide, so the oceans cease to be a carbon sink removing carbon dioxide from the atmosphere. This causes the Earth's surface to warm further, so the ocean surface also warms.

This warming of the ocean surface has a number of impacts, as warm water:

- absorbs less oxygen

- does not mix as easily with cooler deep ocean water

- increases coral bleaching

- favours the development of long-lasting and more intense hurricanes.

ATL Activity 6

ATL skills used: communication

Draw a positive feedback cycle to show the impact of global warming on ocean temperatures.

ATL Activity 7

ATL skills used: thinking, communication, social, research, self-management

Stratification of water occurs in deeper lakes, coastal areas and enclosed seas, as well as in the open ocean.

In small groups, carry out some research into this. Find a lake—locally or globally—that has stratification and look at the potential impacts on the area in terms of:

- the economy

- the cultural traditions of the people

- the ecosystem and biodiversity.

Connections: This links to material covered in subtopics 2.1, 2.4 and 4.3.

Upwellings and downwellings

Normally, productivity in oceans is quite low due to the limiting factors: light level and nutrient concentration. The **upper water levels** have good light intensity, but low nutrient levels which limit productivity. The nutrients are taken in by phytoplankton and travel along the food chain. Dead organisms that are not eaten by some other organism sink and the nutrients stored within them end up on the ocean bottom. The **lower water layers** are therefore nutrient-rich, but the absence of light makes photosynthesis impossible. Upwellings (Figure 16) solve this issue; cold nutrient-rich water flows from the ocean depths to the surface. Upwellings:

- are a natural process caused by winds blowing at an angle across the surface water, pushing the warm surface water away and dragging colder water up from below

- fertilize the surface waters and encourage phytoplankton and seaweed growth; this supports the rest of the marine food web, including birds

- generate the most fertile ecosystems—off the coast of Peru, a continuous upwelling nearly 26,000 square kilometres in size supports one of the richest fishing grounds in the world

- move fish and invertebrate larvae up from the depths; they are caught up in the ocean currents and swept offshore where survival chances are lower

- cool coastal regions which may encourage sea fogs—San Francisco is famous for these

- influence the El Niño Southern Oscillation (ENSO).

The reverse process, called downwelling, occurs when wind blowing on-shore causes surface water to build up along a coastline and forces it downwards, taking down dissolved oxygen. The water on the sea surface then becomes more dense than the water beneath. Most downwellings happen at the poles and usually lead to reduced productivity.

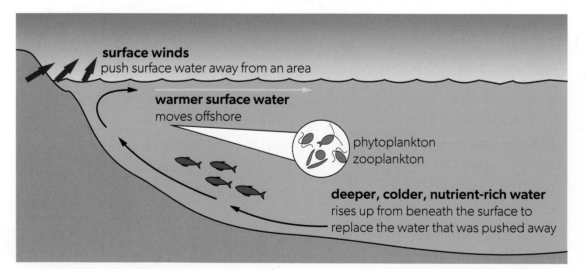

surface winds
push surface water away from an area

warmer surface water
moves offshore

phytoplankton
zooplankton

deeper, colder, nutrient-rich water
rises up from beneath the surface to
replace the water that was pushed away

▲ Figure 16 Upwelling

Adapted from: National Ocean Service / NOAA

Thermohaline circulation

Look back at subtopic 2.4 to remind yourself about ocean currents and energy distribution. Then study Figure 17.

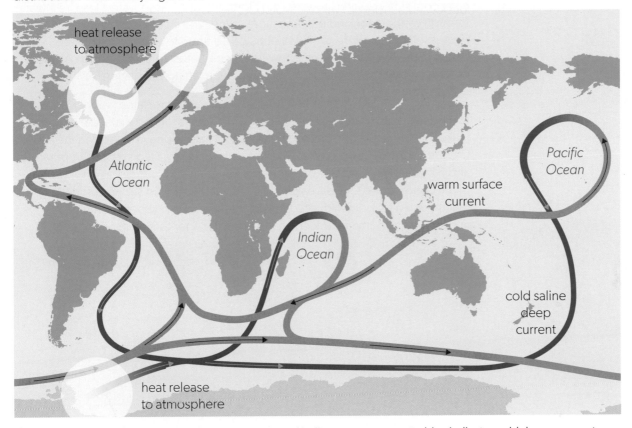

heat release to atmosphere

Atlantic Ocean

Pacific Ocean

warm surface current

Indian Ocean

cold saline deep current

heat release to atmosphere

▲ Figure 17 Conveyor belt system of ocean currents: red indicates warm currents, blue indicates cold deeper currents

The ocean is constantly on the move; some of these movements are visible, others are less obvious. The movements we rarely see are the ones that have an important role in the global distribution of energy. Without an understanding of these ocean currents, we cannot understand global atmospheric energy exchanges. There are two types of current.

- **Surface currents** (red in Figure 17) move through the oceans like giant rivers. They occur in the upper 400 m; they are moved by the wind and deflected by the Earth's rotation.

> Connections: This links to material covered in subtopic 6.2.

- **Deep underwater currents** (blue in Figure 17) are driven by the **thermohaline circulation**, and they cause the oceanic conveyor belt. They are massive, slow-moving currents that mix the ocean's water on a global scale.

Thermohaline circulation circulates water from the north to the south and back again; the term comes from "thermos" (heat) and "haline" (salt)—the two factors that determine seawater density. There are a few basic principles that drive this.

- Water always flows to the lowest point.

- Water density depends on the water's temperature and salinity.

- Cold water holds more salt, so it is denser and sinks.

- Warm water has lower salinity than cold water so is less dense and rises.

The North Atlantic conveyor belt

Rivers and melting ice caps bring low salinity, low density water into the North Atlantic from the Labrador current. Wind currents from the equator, however, carry cool surface waters towards the North Atlantic; these waters lose freshwater through evaporation and become more saline. The increased salinity and decreased temperature of these waters cause them to sink and form deep ocean currents which flow back towards the equator. This is the **North Atlantic conveyor belt** or the Atlantic Meridional Overturning Circulation (AMOC).

The North Atlantic conveyor belt has for 8,000 years circulated warmer water towards northern Europe. It takes about 1,000 years for one molecule of water to make the full circulation. Since 2004, oceanographers have measured the AMOC to observe if it is slowing down. The theory is that increased surface temperature and increased rainfall make ocean water lighter and so reduce the sinking of the conveyor belt, thus weakening it. As yet, there is no strong evidence that there is a slowdown due to climate change but models are complex and separating human-induced changes from longer-term shifts is not easy.

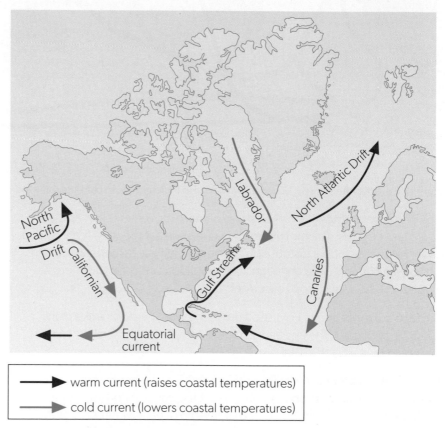

▲ Figure 18 Ocean currents with names

Check your understanding

In this subtopic, you have covered:

- the hydrological cycle and the water budget, the proportions of saline and freshwater

- the stores, transfers and transformations within the hydrological cycle

- how the hydrological cycle is a system

- human impacts on the water cycle, both direct and indirect

- how unique water is and how this makes it so versatile

- the oceans and how important they are to humans

- carbon sequestration by the oceans

- the way oceans are stratified and the links to global warming

- the action of ocean currents and energy distribution

- the strong links between the ocean currents and climate.

How do water systems support life on Earth, and how do they interact with other systems, such as the carbon cycle?

1. Describe the movement of water through the hydrological cycle and the forces that drive this cycle.

2. Construct and interpret systems diagrams of the hydrological cycle. Outline the stores, flows, transfers and transformations.

3. State the relative sizes of the main stores in the hydrological cycle.

4. Describe how human activities alter the flows and stores of the cycle.

5. Explain how a water body maintains a steady state using a flow diagram.

6. Outline the unique physical and chemical properties of water.

7. Explain how oceans act as a carbon sink.

8. Outline the process of ocean acidification.

9. Explain the causes and consequences of ocean stratification.

10. Explain the link between global warming, ocean acidification and stratification.

11. Explain upwellings and how they are linked to ENSO events.

12. Analyse the thermohaline circulation system and how it affects climate.

 AHL

>> Taking it further

- Find out if there is thought to be water on other planets or moons in our solar system and what this might mean for life there.

- Research situations where recent human activities have changed water stores and flows on Earth, sometimes with disastrous consequences.

Guiding questions

- What issues of water equity exist, and how can they be addressed?
- How do human populations affect the water cycle and how does this impact water security?

Understandings

1. Water security is having access to sufficient amounts of safe drinking water.

2. Social, cultural, economic and political factors all have an impact on the availability of, and equitable access to, the freshwater required for human well-being.

3. Human societies undergoing population growth or economic development must increase the supply of water or the efficiency of its utilization.

4. Water supplies can be increased by constructing dams, reservoirs, rainwater catchment systems and desalination plants, and by enhancement of natural wetlands.

5. Water scarcity refers to the limited availability of water to human societies.

6. Water conservation techniques can be applied at a domestic level.

7. Water conservation strategies can be applied at an industrial level in food production systems.

8. Mitigation strategies exist to address water scarcity.

9. Freshwater use is a planetary boundary, with increasing demand for limited freshwater resources causing increased water stress and the risk of abrupt and irreversible changes to the hydrological system.

10. Local and global governance is needed to maintain freshwater use at sustainable levels.

11. Water footprints can serve as a measure of sustainable use by societies and can inform decision-making about water security.

12. Citizen science is playing an increasing role in monitoring and managing water resources.

13. "Water stress" (like "water scarcity") is measure of the limitation of water supply; it not only takes into account the scarcity of availability but also the water quality, environmental flows and accessibility.

14. Water stress is defined as a clean, accessible water supply of less than 1,700 cubic metres per year per capita.

15. The causes of increasing water stress may depend on the socio-economic context.

16. Water stress can arise from transboundary disputes when water sources cross regional boundaries.

17. Water stress can be addressed at an industrial level.

18. Industrial freshwater production has negative environmental impacts that can be minimized but not usually eliminated.

19. Inequitable access to drinkable water and sanitation negatively impacts human health and sustainable development.

Water as a critical resource

Just how crucial is water and who has enough for daily life?

- According to the WHO (World Health Organization) each person needs 20 litres of water for daily functioning.

- 1 in 10 people on the planet (780 million people) have no access to clean drinking water.

- 2.3 billion people on the planet have no basic sanitation.

- 800 children (under the age of 5) die every day from water-related diseases such as cholera and diarrhoea.

- In rural areas of Africa, many women and children walk 6 km a day to collect 18 kg of water, and spend 200 million hours moving water every day.

- Estimated annual sales of water in the USA total USD 4 billion.

- 90% of natural disasters worldwide are water-related.

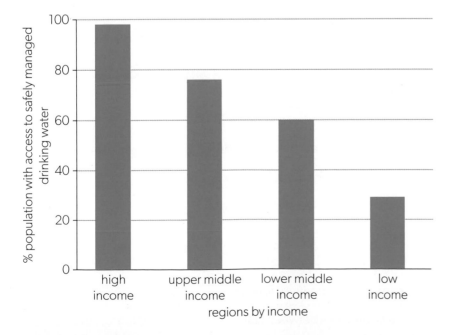

◄ Figure 1 Graph to show the percentage of people with access to safely managed drinking water

Data from: WHO/UNICEF Joint Monitoring Programme (JMP) for Water Supply, Sanitation and Hygiene (washdata.org) (CC BY-4.0)

- Daily per capita use of water in residential areas in litres: Japan = 286; Europe = 200; China = 180; Tanzania = 101.

- Over 263 river basins are shared by two or more countries, mostly without adequate legal or institutional cooperative arrangements.

- Quantity of water in litres needed to produce 1 kg of: cheese = 5,605; rice = 2,248; potatoes = 59; beef = 15,000.

ATL Activity 9

ATL skills used: thinking, communication, research

Use the facts above to answer the following questions.

1. Assuming there are 8 billion people alive on Earth, how many people globally have no clean drinking water?

2. Calculate:

   a. the daily per capita use of water in Tanzania as a percentage of the daily per capita use in Japan

   b. how many days of water use by one person in Europe it would take to produce one kilogram of beef

   c. how many kilograms of rice could be produced using the average annual water consumption of one person in China

   d. your water footprint—make a list of all the food you eat in a day and find out how much water it takes to produce it.

Uses of freshwater by humans

Although there is a lot of water on Earth, most of it is saline. It is possible to use desalination plants to remove the salt, but the energy costs are very high—10–13 kWh for every 3,800 litres of water. This sort of expense is only possible in wealthy countries which are water-stressed and near the sea, for example Australia, Saudi Arabia and the UAE. A major issue with desalination is that salt is a by-product. This increases the density of the water causing it to sink to the ocean-bottom ecosystems and damaging them. Unless we can develop technology to desalinate water cheaply, and dispose of the salt by-product safely, it is not a viable proposition globally.

Water is used both directly and indirectly in:

- domestic tasks: water used at home for drinking, washing, cleaning

- agriculture: irrigation, for animals to drink

- industry: including manufacturing, mining

- power generation: hydro power, cooling thermal power station water

- transportation: ships on lakes and rivers

- recreation: swimming, boating, fishing

- marking the boundaries between nation states (seas, rivers and lakes).

Connections: This links to material covered in subtopics 1.3 and 7.2, and in topic 8.

ATL Activity 10

ATL skills used: social, research, self-management

According to the World Bank in 2019, water usage by percentage is:

- agriculture 72%

- industry 16%

- domestic usage 12%.

The World Bank percentages are global averages but these values vary considerably between countries.

Work in groups of three. Each person chooses a country with a different level of development and investigates the water usage in that country. Then, in your groups, discuss and explain the differences.

Agriculture uses water for irrigation and to provide water for livestock; these usage rates are tens of times higher than domestic use. As human population expands, we need water to grow more food but, like food, it is not that there is not enough water worldwide, it is that the distribution is uneven. For example:

- Egypt imports more than half its food as it does not have enough water to grow it.

- In the Murray–Darling basin in Australia there is water scarcity for humans because so much water is used for agriculture.

Adding droughts, climate change, soil erosion and salinization to the story, you can see that water is a major issue for nations and international organizations and will become even more important.

Water availability and security

The WHO states that each human should have access to a minimum of 20 litres of fresh water per day. Agenda 21 says this should be 40 litres. Much of the world has access to far less than this, while other areas have far more.

Our sources of freshwater are:

- **surface freshwater**: rivers, streams, reservoirs and lakes

- underground **aquifers**.

An aquifer is a layer of porous rock (holds water) sandwiched between two layers of impermeable rock (does not let water through)—see Figure 2. They are filled continuously by infiltration of precipitation where the porous rock reaches the surface, but they are present only in some areas.

Water flow in aquifers is extremely slow (horizontal flows can be as slow as 1–10 metres per century). As a result, aquifers are often used unsustainably. Many aquifers are also "fossil aquifers"—meaning the recharge source is no longer exposed at the surface and so they are never refilled. These can never be used sustainably.

Key term

Water security is access to a sufficient quantity of clean water to live sustainably.

Connections: This links to material covered in subtopic 1.3.

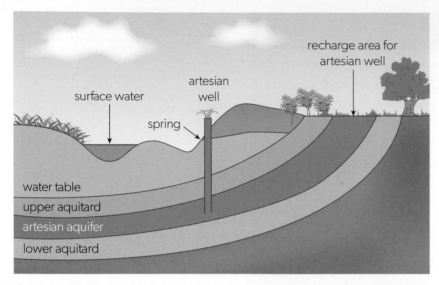

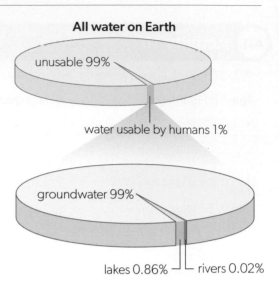

All water on Earth

▲ Figure 2 Aquifer structure. An aquitard is non-permeable rock that restricts water movement

▲ Figure 3 Proportion of water on Earth that is usable by humans

These freshwater sources are limited (Figure 3) and the UN has applied the term "water crisis" to our management of water resources today. There is not enough usable water and it can be very polluted. Up to 40% of humans alive today live with some level of water scarcity, a figure that will only increase.

Water scarcity

Water scarcity refers to the limited availability of water to human societies. This may be:

- physical scarcity—limited by the actual abundance of water present

- economic scarcity—limited by the available storage and transport systems.

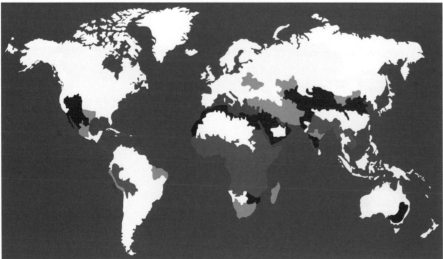

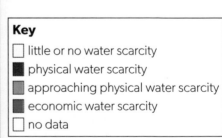

Key
- □ little or no water scarcity
- ■ physical water scarcity
- ■ approaching physical water scarcity
- ■ economic water scarcity
- □ no data

▲ Figure 4 Distribution of water scarcity

ATL Activity 11

ATL skills used: thinking

Refer to Figure 4 to answer the following questions.

1. List three regions with physical water scarcity.

2. What factors do they have in common?

3. List three regions with economic water scarcity.

4. What biomes are these regions in?

5. What factors do regions with no water scarcity have in common?

Factors that impact water availability and scarcity

Water supply across the globe is not even and there is a range of physical, social, economic and political factors that impact that supply.

If the **climate** has lower temperatures and high rainfall it creates water surpluses. Higher temperatures increase evaporation rates. In areas where precipitation is higher in the warmer summer months, there can be seasonal water deficits.

Figure 2 shows the importance of **geology** in water supply. The permeable rock in aquifers is a very good water supply and approximately 70% of south-east England gets its water from the chalk aquifer that underlies the area. However, permeable rock on the surface (limestone) limits surface water supplies.

Infrastructure is important in moving and treating water. It may be human made (e.g. reservoirs, pipelines, pumping stations or sewage treatment facilities) or natural. The human-made parts of this infrastructure are expensive. Lack of political will or poor financial support often means that water scarcity is an issue even in areas with a plentiful physical water supply.

Poverty is the root cause of scarcity in many places. Nearly 420 million people living in Africa lack basic drinking water supplies and 780 million lack sanitation services. This traps them in the never-ending cycle of poverty.

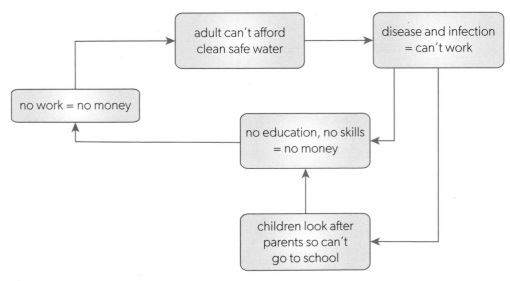

▲ Figure 5 Water scarcity and the cycle of poverty

The politics of water supply are complex because many major rivers run through several countries.

- The Danube River basin is shared by 19 countries and 81 million people.

- The Tigris and Euphrates Rivers carry water that is extracted by Iran, Iraq and Syria.

- The River Rhine rises in Switzerland, flows through Germany and then flows out to the North Sea in the Netherlands.

- The Nile (longest river in the world) flows through 11 countries before it flows into the Mediterranean Sea (Figure 6).

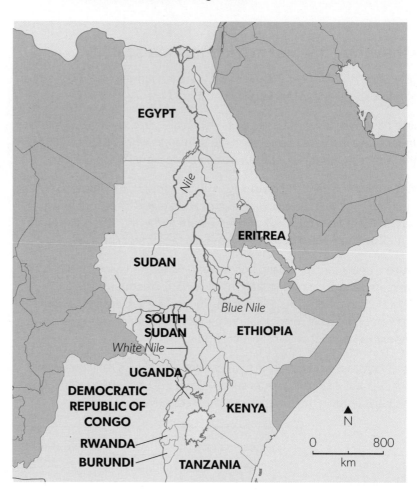

▲ Figure 6 River Nile and the countries it flows through

TOK

Aid agencies often use emotive advertisements to highlight issues of water security. To what extent can emotion be used to manipulate knowledge and actions?

Under these circumstances, communication must take place within and between countries. Water pollution in one country has detrimental impacts on other countries. Hydropolitics is the name given to politics surrounding the availability of water resources.

When more water is taken from an aquifer than is replaced by rainfall, it is called **over-abstraction**. This is often an issue in agricultural areas where a lack of water in the aquifer can cause subsidence of the land above as well as affecting the sustainability of such extraction.

Pollution is another human-induced issue that impacts water security. The sources of water pollution are covered in subtopic 4.4.

Sustainability of freshwater resource usage

Global freshwater consumption is increasing due to population growth and improvements in quality-of-life for many people. This increased freshwater use leads to two types of problems: water scarcity (discussed above) and water **degradation** (when water quality deteriorates, making it less suitable for use).

Sustainable use of resources allows full natural replacement of the resources exploited and full recovery of the ecosystems affected by their extraction and use. At the moment, humans are not using water sustainably—it is predicted that 700 million people could be displaced by intense water scarcity by 2030. There are many reasons for this.

- **Climate change** is causing water access issues. Certain areas of the world are impacted more than others—unfortunately, they are the areas that are already water stressed. Climate change disrupts rainfall patterns and changes monsoon rain cycles, exaggerating the inequality of supplies. Rising sea levels cause salination of groundwater, making it unfit to drink.

- **Rising temperatures** are blamed for the reduction in the flow of the Colorado River in the USA, making navigation into the Gulf of Mexico more difficult.

- **Slower flow rates** in the lower courses of rivers result in sedimentation, which makes the already shallow river even shallower and may extend deltas further into the sea.

- **Deforestation** reduces transpiration which reduces precipitation resulting in droughts.

- **Underground aquifers** are being exhausted by high pumping rates, causing a cone of exhaustion. This means they cannot be used anymore. This affects agriculture and causes building damage when the soil shrinks due to water loss.

- Freshwater is **contaminated** by human activities and natural events: fertilizers and pesticides are used in agriculture; industries release pollutants into surface water bodies; warm water outflow from power stations holds less oxygen than cold water, which negatively affects aquatic organisms; wastewater is often not dealt with effectively—on a global scale, 80% of the wastewater from households flows back into ecosystems without being treated; natural disasters such as drought and floods can contaminate or even destroy clean water sources, and the frequency and severity of these events is increasing.

- Water **waste** (dripping taps, over-watering lawns) can account for up to 40% of a city's lost water—143 billion litres a year.

- Irrigation often results in **salinization** of the soil (degradation), especially in dry areas. Irrigation water evaporates before it is used by the crops. This brings dissolved minerals to the surface, making the soil too saline (salty) for further agriculture.

- Many countries **mismanage** water supplies because they lack adequate infrastructure to deliver water to the people. According to the UN, the issue is that water infrastructure "is typically capital intensive, long-lived with high sunk costs. It calls for a high initial investment followed by a very long payback period."

Connections: This links to material covered in subtopics 1.2, 1.3 and 4.1.

367

Connections: This links to material covered in subtopic 1.3.

Methods of increasing water supply

Improving water access

Improved water access and availability can be achieved through increasing the actual supply or conserving water where it is used.

Increasing freshwater supply

This is a brief overview of the main techniques for increasing water supplies.

Dams and reservoirs

Dams are human-made structures that create reservoirs (artificial lakes). The water stored in the reservoir can be used in residential areas, farming and industry. The dams can produce HEP (hydroelectric power), and reservoirs are also used for recreation and transportation.

▲ Figure 7 Dam and reservoir water supply

Dams provide a range of social, economic and environmental advantages, including:

- water storage: helps regulate the water supply, particularly in the dry season

- flood control: flood mitigation is an essential role of dams—they can control the flow of floodwater by diverting it or storing it and releasing it slowly

- irrigation: helps farmers through water shortage periods, boosting crop production

- hydroelectric power (HEP) production: this is a clean energy source that, once built, produces no pollution

- recreation: people can enjoy swimming, skiing and boating on the reservoirs.

However, dams have several significant disadvantages. The construction of a dam has several drawbacks:

- the weight of the water can stress the Earth's surface and causes geological weaknesses which may result in dam collapse

- very costly and time-consuming to build

- use of concrete and building materials leads to emission of GHGs

- displaces or disturbs local people

- disrupts local ecosystems

- disrupts groundwater.

After construction, dams block the progression of water to other states or countries and stop migration routes of aquatic animals. Failure of the dam can have catastrophic consequences.

Rainwater harvesting systems

Rainwater is collected and stored instead of being allowed to run off. This can be done on a small or large scale, through two basic methods. Water may be collected and moved into deep pits, reservoirs or wells so it can recharge aquifers. This is artificial recharge that can supplement natural recharge. Alternatively, water can be collected from roofs or other hard surfaces and then directed into gutters and drains via a tank ready for use.

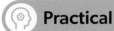

Practical

Skills:
- Inquiry 1: Exploring and designing
- Inquiry 2: Collecting and processing data
- Inquiry 3: Concluding and evaluating

Dams solve a lot of issues but may create others.

Select and research an example of a large-scale dam, either locally or globally. Working individually or in small groups, investigate the reasons for the dam being built, the impacts (both negative and positive), and the risks. Put together a case **for** and **against** the dam. Alternatively, set up a class debate about the pros and cons of a specific dam.

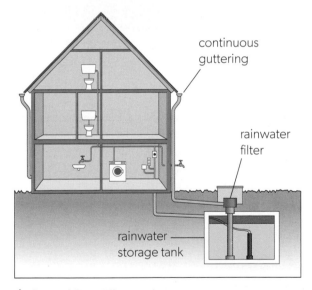

▲ Figure 8 Two different rainwater capture systems

There are many advantages of rainwater harvesting, most of which are logical. It:

- supplies relatively clean water and decreases demand

- reduces the need to import water

- doubles as a conservation method

- improves groundwater stores

- reduces surface run-off, thus reducing soil erosion and decreasing flood risk

- can be retrofitted to existing structures

- reduces the cost of infrastructure in urban areas

- is better for garden plants as rainwater does not have added chemicals

- uses technology that is simple, cheap and can be extended if necessary.

In Australia, 25% of households have a rainwater tank and it is estimated that rainwater harvesting provides 274 billion litres of water a year. This can reduce the cost of residential water by AUD 540 million. Outside urban areas, rainwater provides 109 billion litres (63%) of residential water. In Sydney rainwater harvesting provides more water than the desalination plant.

As ever, there are some disadvantages to be considered.

- Rainfall is not predictable.

- Many people do not have access to a proper storage system or the technical skills to install the equipment required.

- The systems need regular maintenance.

- There is a limit to the amount of water that can be stored.

- Poor installation can lead to mosquitoes using the water to lay eggs.

In addition to rainwater harvesting, in some areas of the world it is possible to use nets to collect dew and fog. The nets are one metre squared and positioned perpendicular to the wind direction. The nets catch droplets of fog or dew and then gravity takes the water down to collection containers. They are simple to set up and do not need to be checked regularly—just collect the water at the end of the day.

Desalination

This is the process of removing salt from seawater and, with declining rainfall and a growing population, it is becoming a much-needed secure water source in some areas of the world. There are two main techniques used to desalinate seawater. Reverse osmosis removes salts by moving water through a series of semi-permeable membranes. Thermal desalination uses heat to evaporate and condense water.

The major advantage of desalination is that it provides an accessible source of drinking water in areas where there is no other supply—for example, 70% of Saudi Arabia's water is supplied by desalination. Desalination can also be used in the drier areas of countries with a plentiful supply of water, and it reduces pressure on freshwater supplies.

However, desalination has very high setup and operational costs—building the plant can cost between USD 300 million and USD 2.9 billion, depending on the location. Running costs are high due to the energy requirements and as amounts of energy required vary, the cost of desalination varies accordingly. Disposal of the salt presents problems, as the brine water discharge lowers the amount of oxygen in water. This stresses or kills aquatic life in the receiving body of water. Also, the chemicals used in the plants, such as chlorine and hydrochloric acid, can be toxic in high concentrations.

▲ Figure 9 A desalination plant

Enhancing natural wetlands

This intervention by humans has multiple benefits. The major one is that healthy wetlands with rich natural biodiversity of flora and fauna improve water quality. They do this by filtering the water to remove sediments, nutrients and pollutants. At the same time wetlands provide wildlife habitats and are aesthetically pleasing. This capacity is being compromised by extreme weather conditions (caused by global climate change).

▲ Figure 10 Aerial photo of Okavango Delta, Botswana

Water conservation

Water conservation is increasingly important as more and more areas of the world face water scarcity and insecurity. Measures can be implemented at small-scale domestic levels or at larger scales in industry and agriculture.

Domestic level conservation

First, be aware of your water use and aim for water-efficient practices, as suggested here.

- Take shorter showers; some advertising campaigns advise that a shower should take no longer than a popular song (3 minutes).

- Turn off the tap while you are brushing your teeth, shaving, cleaning vegetables or hand washing the dishes.

- Install water-saving devices such as flow-controlled shower heads, aerators on taps and half-flush toilets.

- Only use a dishwater and washing machine for full loads.

- Keep a bottle of water in the fridge to avoid running the water until it is cold.

- Check all water devices for leaks—taps, pipes, toilets.

ATL Activity 12

ATL skills used: thinking, communication, research

Visit the website of the US Environmental Protection Agency (EPA), or any other website about the value of wetlands.

1. List the benefits of wetlands.

2. Draw a mind map to show the possible effects of these benefits.

> *Water conservation refers to the preservation, control and development of water resources, both surface and groundwater, and prevention of pollution.* 99
>
> *OECD*

ATL Activity 13

ATL skills used: thinking

Before you read on, list all the ways in which you could reduce your daily water usage.

- In the garden—only water grass and flowers when they need it; water during the cool part of the day to reduce evaporation; make sure sprinklers water plants and not pavements; plant drought-resistant plants; and use mulch around trees to slow evaporation.

- Do not use a hose to clean driveways and steps—use a broom.

- Educate children about the problems of water supply and discourage them from playing with hoses and sprinklers.

- Wash cars in car washes with a closed water system. Not washing a car in the street also reduces pollution from oil.

- Use grey water recycling—grey water is water from showers, baths, household laundry and kitchen sinks, which can be reused on site for flushing WCs and garden irrigation.

Grey water

If you live in a house with running water, usually only purified and treated freshwater enters your home in one pipeline. This water is used for all the processes in the home that need water, but not all of these processes need to use drinking water.

Black water or sewage contains human waste and may carry disease-causing bacteria and other organisms (e.g. worms).

Grey water can be very lightly used water. Do you keep the tap running when you clean your teeth or run a glass of water? If so, most of this water is perfectly clean yet goes down the pipes and mixes with black water in a shared sewerage system. All this water is then cleaned to the highest standards before being recycled back into our piped water system.

This is an enormous waste of clean water and the two types, black and grey, could be separated before they leave our homes. Grey water could be used to irrigate gardens, clean cars, flush WCs and so on. It must be used at once though, to avoid build-up of bacteria.

There are more official ways of reducing water usage, including **water metering**. A water meter is a simple device fitted to water pipes as they enter a residential or commercial building. This allows authorities to log how much water a building uses and charge accordingly. This raises awareness, because when people are charged for how much water they use, they are more careful about wastage. Figures suggest that meters cut consumption by 9–20%.

Another way to cut water consumption is through **rationing** or restrictions. This is where the use of water is restricted and certain uses are banned, e.g. watering lawns, car washing, or filling swimming pools. Restrictions may limit the volume of water available or the times it is available.

On 10 August 2022, *The Brussels Times* reported that across Europe 60% of the land was under drought warnings or alerts. Italy had the driest period in 70 years and in Verona "drinking water could only be used for essential daily activities such as cooking". Spain had the driest climate experienced there for 1,200 years.

▲ Figure 11 Water meter to measure consumption

Water conservation in agriculture

Due to rising population and the effects of global climate change (droughts, extreme temperatures) agriculture is under pressure. It is the world's greatest user of water and is expected to need up to 19% more by 2050. Thus, water conservation in agriculture is essential.

Irrigation uses a lot of water but about 20% of the water delivered to agricultural land is lost in distribution channels, while 10–15% is lost through overwatering. Various strategies can be used to address such losses. Careful timing of irrigation and accurate calculations of the amount of water needed by the crop can minimize water usage. This can be worked out by measuring soil moisture and plant growth rates.

Drip irrigation uses pipes to drip water slowly onto the roots of the plants; this method uses up to 50% less water than overhead irrigation. It also reduces surface run-off (and soil erosion), surface evaporation and overwatering. The initial cost of installation is high, but this is balanced by lower operating costs. Drip irrigation pipes can be installed underground and the amount and timing of water delivery controlled by a computer. This is the most efficient method but also the most expensive.

Gated pipe irrigation is an old irrigation technique whereby water is allowed to flow into irrigation ditches that border the field, saturating the soil. The water that enters the ditches is controlled by gates and sensors can be used to monitor soil moisture and control the gates automatically.

Estimates suggest that 40–60% of the water loss in agriculture is due to **evaporation**. It is better to have a few large reservoirs instead of numerous small ones. Water is also lost if it is moved around in open ditches; pipes are better.

▲ Figure 12 **Surface drip irrigation system**

As climate change brings more frequent and more severe droughts, farmers may need to switch to **drought-resistant crops**. These can be designed by traditional cross-breeding or genetic modification. Such crops require less water but still provide the required yield. Some good drought-resistant crops include beans, broccoli, corn and cucumber. Crops that use a lot of water are rice, wheat, cotton and alfalfa.

Composting, mulch and black plastic all improve water conservation.

- Compost: vegetable peelings, fruit waste, grass cuttings, dead organic matter in general.

- Mulch: can be made from compost—decaying leaves and plant remains.

- Black plastic: sheets or shredded black plastic used as mulch.

These three elements are spread on and around crop plants to suppress weeds, warm the soil and prevent evaporation.

Rainwater harvested from the roofs and gutters of greenhouses can be used as irrigation water for the plants growing in the greenhouses. The size of the storage tanks must be calculated to ensure an adequate supply of water.

Aquaponics is a relatively new farming technique for raising fish and growing vegetables. It is a sustainable method to produce more food using less water and less labour. Fish are raised in recirculating aquaculture tanks, with soilless plants above with their roots in the water. The nutrient-rich water from the fish tanks is a natural fertilizer for the plants and plants purify the water for the fish.

▲ Figure 13 Fresh organic vegetables grown using hydroponics

Water shortage

Water is scarce in many urban societies. Case study 1 describes the problems associated with water shortage in Jordan.

Case study 1

Jordan

Jordan is one of the 10 most water scarce countries in the world. The major surface water resources are the Jordan River and Yarmouk River, both shared with neighbouring countries. Water is also taken from the Disi aquifer which is 500 m below ground level under both Jordan and Saudi Arabia. Both countries extract water from it.

- Household water supply is intermittent and may be on for only 36 hours per week. In summer 2022, household taps ran dry for three weeks.

- Wealthier households buy water tanks which they put on their roofs and fill with bought water from private tankers..

- Poorer households cannot afford to buy water tanks or water.

- The population was 8 million in 2012, including more than 760,000 people registered with the UN as refugees. It is now estimated to exceed 11 million.

- All Jordan's water sources are near borders, with infrastructure issues moving it to urban areas.

- The River Jordan is nearly dry.

- The Disi aquifer is being drained at roughly twice the rate at which it can be replenished naturally and now accounts for about 60% of the country's water supply. Estimates are that at current extraction rates, it will be dry in 50 years.

- Jordan depends on winter rainfall which is not enough to meet demand (40% below normal).

- Jordan lacks natural lakes.

- Jordan is well below the absolute water scarcity threshold of 500 m³ per person set by the UN.

- A desalination plant is being built but will not operate for years.

- Water for irrigation is reduced by 50% and crops with low water demand allowed.

Solutions

- Public awareness campaigns on rainwater harvesting techniques, water storage and use are effective.

- The King Talal Dam created a reservoir across the Zarqa River which stores winter rainwater. Water from the dam irrigates about 17,000 hectares.

- Reuse of treated wastewater for irrigation has increased.

- Grey water use has increased.

- A desalination plant takes water from the Dead Sea and supplies Amman.

- Funding from the EU, USA, Japan and Germany, as well as the World Bank and UN, supports projects to increase water supply.

Freshwater and the planetary boundary

The freshwater planetary boundary (Figure 14) now has two aspects.

- Blue water measures humanity's use of groundwater, lakes and rivers.

- Green water measures the impact on the hydrological cycle—rainfall, soil moisture and evaporation. This is measured by assessing the soil moisture in the root zone of plants.

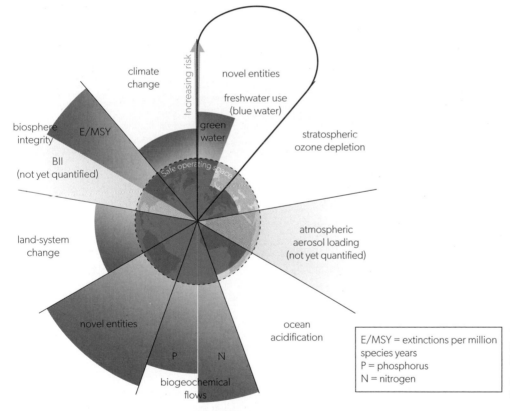

E/MSY = extinctions per million species years
P = phosphorus
N = nitrogen

◀ Figure 14 The planetary boundaries model, with the freshwater boundary highlighted

Adapted from: J. Lokrantz/Azote based on Steffen et al. 2015 (CC BY-NC-ND 3.0)

The concept of green water has been added because soil moisture conditions are seen to have far-reaching consequences for ecological, biochemical and hydrological dynamics. Abnormally dry and wet soils are now common in biomes that are not adapted to seasonal changes. In the Amazon, the drier conditions are pushing the biome towards the rainforest-to-savanna tipping point. This will release large amounts of carbon dioxide.

Johan Rockström, Professor at Stockholm Resilience Centre, observed that "This latest scientific analysis shows how we humans might be pushing green water well outside of the variability that the Earth has experienced over several thousand years during the Holocene period".

Acceptable freshwater use was set at a consumption rate of 4,000–6,000 km³ per year utilized but not returned as run-off. It is now thought that 4,000 km³ per year is close to the danger zone because it is very close to the point beyond which there would be catastrophic impacts.

In April 2022, scientists announced that the freshwater planetary boundary was the sixth one to be crossed. They also predicted that the situation will worsen. Lan Wang-Erlandsson from the Stockholm Resilience Centre stated that "We are profoundly changing the water cycle". This will destabilize the Earth system which will impact the health of the planet and reduce resilience.

According to the research, human overexploitation now requires immediate action if we hope to remain in a safe "operating space". We urgently need to address issues of overexploitation of water, deforestation, soil erosion, water pollution, climate change and land degradation.

ATL Activity 15

ATL skills used: communication, social, research, self-management

Work in a group of three.

1. Six results of human activity (listed below) have contributed to the crossing of the freshwater boundary. Divide them into three pairs.

   - water overexploitation
   - deforestation
   - soil erosion
   - water pollution
   - climate change
   - land degradation

2. Each person takes one pair of actions and researches:

   a. how the actions have contributed to the crossing of the freshwater planetary boundary

   b. what mitigating strategies can be taken by local communities and global leaders to avoid disaster.

Water footprints

The idea of an ecological footprint was discussed in subtopic 1.3. One aspect of the ecological footprint is the water footprint (Figure 15).

Water footprint is an environmental concept that measures the amount of freshwater consumed or polluted in litres or cubic metres per unit of time.

- It includes direct and indirect uses of a producer, consumer, item or service.

- It can be applied to individuals, communities, businesses, products or countries.

- It is a geographically explicit indicator.

▲ Figure 15 Water footprint?

The concept aims to raise awareness about how water is used. This is very important in the face of limited supply caused by increasing populations and climate change. It can be used to assess sustainability levels and inform decisions by governments. Like the freshwater planetary boundary, the water footprint has different elements (Figure 16).

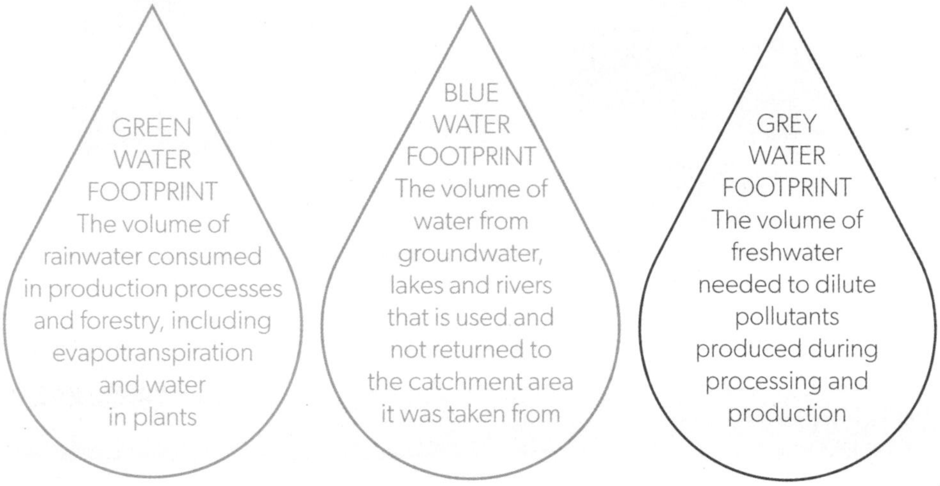

GREEN WATER FOOTPRINT
The volume of rainwater consumed in production processes and forestry, including evapotranspiration and water in plants

BLUE WATER FOOTPRINT
The volume of water from groundwater, lakes and rivers that is used and not returned to the catchment area it was taken from

GREY WATER FOOTPRINT
The volume of freshwater needed to dilute pollutants produced during processing and production

▲ Figure 16 Elements of the water footprint

Water footprint is important because it acknowledges that human impacts on the freshwater system are linked to the entire supply chain not just consumption of the product. Water problems are linked to the global economy because many countries import water-intensive goods from elsewhere. This "saves" their own water supplies but puts pressure on the water supply of the exporting countries. Worse, the exporting countries are usually countries that lack water, smart governance and knowledge of conservation techniques.

Individual water footprint

It is easy to estimate how much water you drink, and how much you use to shower, wash the dishes, and so on. However, your individual water footprint must also include the water needed to grow your food and produce the energy to make all the products you use.

 Practical

Skills:

- Tool 2: Technology
- Inquiry 3: Concluding and evaluating

Whole-class activity

1. As a class, find a range of freshwater footprint calculators online. Decide which one you will use by evaluating the strengths and weaknesses of each.

2. Working individually, complete the questionnaire. Write your score on a piece of paper and put it in a container or record the results in a shared document.

3. As a class, compare water footprints.

4. What questions in the questionnaire surprised you?

5. Discuss some ways you can reduce your water footprint.

Freshwater footprints for food

Figures 17 and 18 show the highest and lowest water-consuming food items.

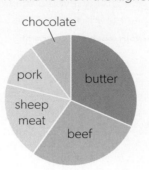

▲ Figure 17 Pie chart to show five foods that consume high amounts of water in litres/kg to grow

Data from: Institution of Mechanical Engineers

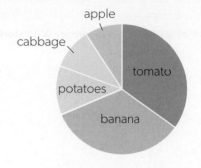

▲ Figure 18 Pie chart to show five foods that consume low amounts of water in litres/kg to grow

Data from: Institution of Mechanical Engineers

(ATL) Activity 16

ATL skills used: thinking, research

1. Can you compare Figures 17 and 18 in terms of the amount of water the foods consume in production?

2. Do these pie charts, without numerical labels, tell you anything about the amounts of water consumed?

3. Does the use of the same colours in each chart help or hinder interpretation?

4. Evaluate the use of pie charts to present this data.

1 kg of food item	Water consumed (litres)	1 kg of food item	Water consumed (litres)
Chocolate	17,196	Apple	822
Beef	15,415	Banana	790
Sheep meat	10,412	Potatoes	287
Pork	5,988	Cabbage	237
Butter	5,553	Tomato	214

◀ Table 1

5. Present the data in Table 1 using a different type of graph or chart.

6. The pie charts were produced using the data in the tables above. Discuss the values in terms of sustainability. Do your own research to find data for the volume of water consumed to produce some items you use on a daily basis such as fabric, a laptop, sports equipment, or anything that interests you.

Fierce national competition over water resources has prompted fears that water issues contain the seeds of violent conflict. If all the world's peoples work together, a secure and sustainable water future can be ours. 🙼

Kofi Annan (1938–2018), former UN Secretary-General

Local and global water governance

Water governance is a system whereby institutions meet the water needs of society by using resources in a controlled, accountable and responsible way. Good governance ensures the actions taken reduce risks and unlock new opportunities for water availability. It also ensures sustainable water use and environmental protection. This is usually achieved by political and institutional rules.

The Organisation for Economic Co-operation and Development (OECD) is an independent body that promotes economic and social well-being across the globe. It has a water governance initiative giving advice about policy and good governance. There is a policy forum to share good practices and a set of principles that constitute water governance.

ATL skills used: thinking, communication

Local and global governance is needed to maintain freshwater use at sustainable levels.

1. Research a local or global example of regulations restricting water use for a particular community. For example, in California in 2022 a regulation banned all use of drinkable water to irrigate decorative lawns or areas of grass.

2. Transboundary waters are aquifers or lake- or river-drainage basins shared by two or more countries. Often there is little or no cooperation or written agreements over water extraction.

    153 countries have territory within one or more transboundary waters—most basins are between only two countries, but 13 are shared between 5–8 nations; 5 are shared between 9–11 nations; and the Danube flows through 18 nations.

    List the issues that international sharing of water resources may cause.

Connections: This links to material covered in subtopic HL.a.

As water demand increases, the supply is decreasing in both quality and quantity. This means competition for water resources intensifies. When this competition involves political borders, diplomacy is required.

Competition for water can cause conflict so nations use discussion and treaties to stabilize relations. The first known treaty was in 2500 BCE, when Lagash and Umma (Sumerian cities) signed an agreement to end a dispute over water from the Tigris. Treaties used to focus on navigation and border demarcation. Nowadays there is a greater focus on use, development, protection and conservation of water resources.

Legal agreements over water have been maintained in the face of conflicts over other issues. Here are two examples.

*   Since 1957, Cambodia, Thailand, Laos and Vietnam have cooperated to share the waters and sustainable development of the Mekong River under the Mekong River Committee and then Commission. The aim is "to promote and coordinate sustainable management and development of water and related resources for the countries' mutual benefit and the people's well-being". Yet during this time, the Vietnam War was ongoing.

*   In 1999, 10 countries agreed on a framework for the Nile River Basin to promote equitable benefits from and use of the Nile water resources. The hope was to combat poverty and increase economic development.

In 1997, the UN Convention on Non-Navigational Uses of International Watercourses set out two essential principles that must underpin agreements for shared watercourses. First, resource use must be equitable and reasonable. Second, all parties must avoid causing harm to other users.

Most transboundary water resource agreements focus on hydropower and water utilization, with some focus on flood control, industrial provision, navigation and pollution.

ATL skills used: thinking, communication, social, research

You have just read about two examples of transboundary water agreements. Research one or two other agreements and answer these questions for each one.

1. List the countries involved.

2. State the starting date.

3. Describe the aims.

4. Discuss the outcomes.

Citizen science

Citizen science is when members of the public collect, report and analyse data to help scientific research. Monitoring water resources to ensure sustainable use is a good example of citizen science. The citizen (you) can work alone, in a group or collaborate with professional scientists.

Connections: This links to material covered in subtopics 1.3, 3.1, 3.2 and 7.1.

There are water-monitoring projects going on around the world. Table 2 lists a few of them.

Project	Location	Data started	Monitors
Brooklyn Atlantis	USA	2012	Water quality
Anecdata	Worldwide	2013	Water quality
SPLASSH	Worldwide	2013	Water quality, pollution
CyanoTracker	Worldwide	2014	Algal-bloom, water quality
AppEAR	Argentina	2015	Water quality
EyeOnWater Australia	Australia	2017	Water colour and quality
CrowdWater	Worldwide	2017	Hydrology

▲ Table 2 Examples of water-monitoring citizen science projects

Technology and innovation have made things easier for the citizen scientist. They make it easier for people to find projects, communicate with others, share ideas and learn new skills. Many researchers use crowdsourcing to get the public to help them collect data. This involves gathering information via the internet and through social media and smartphone apps.

Crowdsourcing allows researchers to collect large amounts of data that is of high enough quality to be useful. But there are some problems.

- Data is usually unstructured because the times and locations are random.

- Data could be biased.

- Data may require complex analysis.

Water stress

Water stress—like water scarcity—is a measure of the limitation of water supply. It takes into account the scarceness of availability but also the water quality, environmental flows and accessibility. So, a region with an ample supply (not suffering from water scarcity) may experience water stress because of low water quality.

Water stress is defined by the UN Food and Agriculture Organization (FAO) as the symptoms of water scarcity or shortage. It is usually caused by a combination of physical and economic factors:

- Physical: lack of water due to local environmental conditions, droughts or insufficient water resources.

- Economic: lack of water due to lack of adequate infrastructure or lack of investment.

Key term

Water stress is defined here as a clean, accessible water supply of less than 1,700 m³ per year per capita.

Causes of water stress

Causes may depend on socio-economic contexts. These include:

- increasing industrialization in an emerging economy

- over-abstraction due to increased population pressure in a low-income country

- transboundary disputes when water sources cross regional boundaries.

Water stress intensifies as demand intensifies or supply decreases in quantity or quality. Demand appears to be infinite as economies grow and resource usage accelerates. But supply may be inadequate due to:

- poor infrastructure—pipes and storage structures

- underestimation of demand by authorities

- physical stress—water availability may be limited due to lack of rainfall or lack of adequate natural stores

- the fact that water is a finite resource

- pollution and unsustainable extraction of reserves such as fossil aquifers (aquifers that do not refill).

Several countries across the Middle East, North Africa and South Asia have extremely high levels of water stress. Many, such as Saudi Arabia, Egypt, United Arab Emirates, Syria, Pakistan and Libya, have withdrawal rates well in excess of 100%—this means they are either extracting unsustainably from existing aquifer sources, or produce a large share of water from desalination.

Impacts of water stress

Water stress undermines sustainable living. The impacts are far-reaching and usually linked to poverty.

- A common solution to lack of **clean piped water** is to send girls to collect water. This can involve long distances and many hours of walking. The result is that girls do not attend school and therefore get no education to allow them to find paid employment.

- Using dirty water increases the risk of debilitating **waterborne disease** such as diarrhoea, malaria and schistosomiasis. The schistosomiasis parasite enters through the skin and causes chills, cough, fever and muscle aches. These symptoms persist for years and make the climb out of poverty impossible.

- Irrigation can improve crop yields by up to 400% so water insecurity can have a significant impact on food production.

- Water is central to **economic development** because it is used in all stages of industrial production, e.g. cleaning, cooling, transport and as a source of energy. Water insecurity within a country can cause industrial production problems and necessitate water imports, which are not cheap.

(ATL) Activity 19

ATL skills used: thinking, research

1. Research other impacts of water insecurity.

2. Categorize each one as Socio-economic, Political, Environmental or Cultural.

3. Find out about a country which has water insecurity.

   a. Why does it have insecurity?

   b. What are the impacts of this insecurity?

   c. Can people live sustainably under these conditions?

Socio-economic causes of water stress

The causes and impacts of water stress are outlined above but there is an underlying issue: what takes precedence over supplying a population with the water it needs? Unfortunately, many low-income countries (LICs) must make a choice between improving their economy by investing in industrialization or using the money to improve the infrastructure of the water supply.

The rising demand for freshwater is often due to industrialization and estimates suggest that industry's share of global water consumption will hit 21% by 2040. Power stations are a big consumer of water so there is a vicious circle here; see Figure 19.

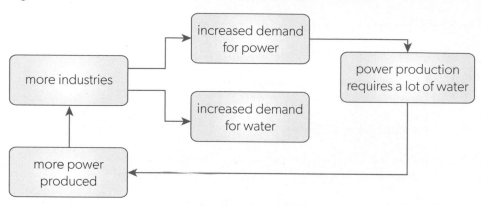

▲ Figure 19 Increased industrialization increases demand for water

Many industries use water as their waste disposal unit and dump pollutants into the closest water body. This may be deliberate or accidental—either way it means the water source is degraded.

LICs that are developing through industrialization often have water stress issues, high population growth rates and urbanization. In these countries, birth rates are usually twice as high as the global average due to limited family planning and lack of education for women.

The urbanization associated with a growing population causes additional water stresses. Urban areas change natural permeable surfaces into impervious ones, limiting freshwater stores in areas with high population density. The infrastructure to supply water is also lacking so the number of people without access to clean water increases.

Examples of countries with water stress or water scarcity

Three billion people experience severe water stress for at least one month each year. Over two billion people live in countries where water supply is inadequate. According to UNICEF, half of the world's population could be living in areas facing water scarcity by as early as 2025. UNICEF also reports that women and girls spend an estimated 200 million hours hauling water every day.

The WHO reports that 884 million people lack access to safe drinking water. Around 3.2 billion people live in agricultural areas with high water shortages, and approximately 73% of people affected by water shortages live in Asia.

(ATL) Activity 20

ATL skills used: thinking, communication, research

LICs want to develop through industrialization and often have a growing population. How could their governments improve access to freshwater for their people? Illustrate your response with named examples.

Consider legislation, ethical issues and environmental issues.

Case study 2

Four countries with water stress or scarcity

Afghanistan

The main water source is the Kabul River Basin, located in eastern Afghanistan. It joins the Indus River in neighbouring Pakistan. Most inflows are generated from snow melt in the Hindu Kush mountains, with their heavy snowfalls and many glaciers.

Water has become even scarcer in Afghanistan after decades of crisis caused by conflict, natural disasters, economic insecurity and climate change—including the worst drought in the last 27 years. About 8 out of every 10 Afghans drink unsafe water, and 93% of the country's children live in areas with high water scarcity and vulnerability. 80% of drinking water in Afghanistan is polluted. Low rainfall, irregular use of groundwater and insufficient infrastructure in Afghan cities are among the main causes of drinking water pollution.

Niger

Niger is one of the least developed countries in the world with 9.5 million people affected by conditions of extreme poverty. It has intense droughts, poor soil conditions and the gradual spread of the desert. It is completely within the Sahel, so the entire country is threatened by drought and desertification. In some regions, there is a chronic shortage of clean water, particularly during the warm months which regularly see extremely high temperatures. Only 56% of Nigeriens have access to a source of drinking water, and just 13% have access to basic sanitation services. In the areas that need help the most, violence and insecurity make it difficult for experts to assess the situation and work with communities to find solutions. The country is highly dependent on agriculture, so even minor climate shocks directly affect the livelihood of thousands of households. This situation is aggravated by the presence of armed conflicts at Niger's shared borders with Burkina Faso, Nigeria and Mali, causing massive displacement and increasing vulnerability.

Nepal

Nepal is home to 2.7% of the Earth's available freshwater due to the number of glaciers, rivers, springs, lakes, high levels of groundwater, and high amounts of rainfall. However, water stress has reached a crisis point, especially in the last 20 years.

While Nepal has many water sources, its network and infrastructure are not able to meet supply or demand. In 2020, the region of the capital city Kathmandu met less than 20% of its local water needs. Of the eight rivers that flow through Kathmandu, not one is clean.

The rapid urbanization of Kathmandu is one key reason that water has become scarce in the country. The city also faced infrastructural setbacks due to the earthquake of 2015. About 2.6 million Nepalese do not have access to a WC and sanitation-related diseases remain a major problem.

Qatar

Qatar is one of the most water-stressed countries in the world. Yet it has one of the highest domestic water consumption rates in the world: an estimated 450+ litres per person per day. This country is a desert without a single river to help sustain the population. The growing population and economy saw water use almost double between 2006 and 2013. Households require the most water, followed closely by agriculture. In just two decades, Qatar has gone from being a country with the lowest amount of emergency freshwater in the region to being one of the most water-resilient.

Until the mid-20th century, the country's only known source of water comprised several aquifers which were being depleted faster than they were refilled by rainwater. A low number of storms in the Arabian Peninsula led to the aquifers being depleted. This made Qatar's water security problems in the second half of the 20th century even worse.

To address its water supply issues, Qatar constructed and expanded desalination plants, which supply 99% of the country's water needs. It also collects 50% of all wastewater and treats nearly all of it. The treated water is used in agriculture and parks, injected into aquifers or stored in lagoons. The plants are expensive and consume a lot of energy during operation. They are not environmentally friendly. They are also vulnerable to natural crises and security conflicts.

Therefore, the Water Security Mega Reservoirs was started in 2015, to build large concrete reservoirs. The first phase will provide storage for seven days' drinkable water at 2026 usage levels; the second phase will provide seven days' drinkable water at 2036 usage levels. Only a very rich country such as Qatar can fund such a major project.

ATL Activity 21

ATL skills used: thinking, communication, social, research

1. List the long-term environmental problems of overuse of water.

2. Draw an annotated diagram of the water cycle to show overuse impacts.

3. Discuss the factors causing water stress in the four countries discussed in Case study 2.

4. Discuss a sustainable solution to water management for one of these countries.

5. Water stress can lead to conflict, particularly where sources are shared. Research the water politics of the Nile basin in north-east Africa. Alternatively, research a transboundary water politics issue of your choice.

   a. Evaluate the claims for water of the countries involved.

   b. Suggest solutions to any tensions.

ATL Activity 22

ATL skills used: thinking, research

1. Wars have been and will continue to be fought over water as both demand and stress increase. Why?

2. Shared water resources (lakes and rivers) are examples of the tragedy of the commons. Choose one of the examples in the case study or find another one that interests you. Research how the tragedy of the commons applies to this resource.

More ways to deal with water stress

Connections: This links to material covered in subtopic 3.2.

There are many strategies to increase water supply. The use of dams and reservoirs, rainwater harvesting systems, desalination, and enhancing natural wetlands were discussed earlier in this subtopic. The following section will discuss some additional techniques.

Water transfer schemes

Water stress is a complex issue, one aspect of which is unequal distribution of water around the globe. Some areas of the world have a plentiful supply of freshwater while others suffer from a severe shortage. Climate change is likely to intensify this inequity. One of the large-scale human responses is the physical transfer of water from a source or donor basin to a recipient basin.

These transfers involve extensive civil engineering works in the construction of dams and reservoirs for storage and pipes or canals for moving the water. Water may be moved within the same country or across international borders, which requires political negotiations.

Water transfers come with some specific advantages: they address water deficit issues in the recipient area and the dams and reservoirs can be used to generate HEP and irrigate agricultural land. However, they may reduce water availability in the donor region. Also, they are expensive and take a long time to complete.

Water transfer schemes need **pipelines or canals** to move water from the donor region to the recipient region. Pipes are simply tubes (usually of plastic) that carry water from one place to another. Large diameter pipes move water from treatment plants to community areas, then smaller diameter pipes move water around settlements. Canals are artificially constructed ditches to move water from one place to another.

Pipes have the following advantages over canals:

- stop losses from seepage and evapotranspiration

- avoid embankment breaches and overflows

- water is protected from pollution and particles

- avoid erosion by flowing water

- not impeded by vegetation or other obstacles

- lower operation and maintenance costs **but** higher installation costs

- between 95 and 100% efficient compared with 70% efficiency of canals.

▲ Figure 20 The problem with canals

Cloud seeding

Cloud seeding is a process in which silver iodide crystals are implanted into clouds to generate rain in a particular area. The crystals are sprinkled into the clouds from a plane. In the right conditions, cloud seeding provides the condensation nuclei that encourage the microphysical processes that produce rainfall. This can either cause rain to fall when it would not have done so naturally or it can increase the amount of rain that falls.

Certain conditions are needed for successful cloud seeding:

- accurate weather forecasting

- suitable clouds—deep and with temperatures between −10 and −12°C

- wind speeds of less than 30 m/sec

- wind in the right direction

- at least 50% cloud cover in the target area.

Cloud seeding works better in wet years!

In addition to these conditions, there are disadvantages:

- it is expensive

- the long-term impacts on the weather are unknown

- there have been no studies to assess the long-term impact of the chemicals used; higher levels of exposure to silver iodide may cause permanent skin problems.

Preliminary stage—removes large bulky solids by passing everything through smaller and smaller meshes and allowing debris to settle out

Primary—water is left in sedimentation tanks to allow any remaining solids to settle out as sludge. Grease and oil float to the top and are skimmed off

Secondary—biological methods are used to remove organic compounds

Tertiary—water is disinfected to make it drinking quality; this may involve UV lights or chemicals such as chlorine to kill microorganisms

▲ Figure 21 Water treatment process

Water treatment plants

Water treatment tends to be the responsibility of governments and is in place to some extent in most countries. The purpose is to remove harmful contaminants and improve taste, smell and appearance of drinking water (Figure 21).

Advantages to water treatment:

- reduces water waste because water discharged by human activity can be purified and reused

- the best way to return water to the natural cycle, maintaining the balance of the hydrological cycle

- removes harmful contaminants (chemicals, heavy metals, pathogens) that are detrimental to human health and the environment

- water can be used for a multitude of activities—domestic, agricultural and industrial—so less water is withdrawn from the natural resource base.

Disadvantages to water treatment:

- energy consumption is high: 2–3% of a high-income country's electrical power

- disposal of the sludge resulting from treatments is difficult

- large areas needed for treatment plants

- servicing is required on a regular basis

- smell pollution

- installation costs are high.

Aquifer storage

Look back at Figure 2 in this subtopic. With increasing water stresses, there is an increasing need for artificial groundwater recharge to increase aquifer storage. Engineering measures can replenish aquifers by diverting surface water via storage ponds and infiltration basins or water can be injected directly into the aquifer via wells. Rainwater harvesting often feeds aquifer recharging.

Advantages of artificial recharge:

- alleviates the problems of overextraction by increasing groundwater stores and keeping aquifers functional

- does not involve major constructional work so has none of the disadvantages of dams and reservoirs

- immune from most human-made and natural disasters

- increases vegetation by replenishing soil moisture reserves from below; this can decrease soil erosion and improve biodiversity

- improves conservation of wetlands

- can increase the amount of water available for irrigation, thus increasing the cropped area and yields

- in LICs, rural populations rely on wells and handpumps for drinking water. Water from aquifers is filtered so it is safer to drink, therefore recharging the aquifers is important.

Disadvantages of artificial recharge:

- potential for groundwater contamination as surface water may come from agricultural run-off or road run-off

- a minimum amount of water needs to be added to the aquifer to make it economically viable

- there has to be a sound understanding of the geological and hydrological background in the area to ensure success.

A focus on desalination

Desalination has been discussed earlier. Here we will consider some of the negative environmental impacts of the process.

One of the biggest issues with desalination is that it is energy intensive. It increases dependence on fossil fuels, increases greenhouse gas emissions, and escalates climate change impacts.

In addition, groundwater supplies in areas where desalination plants are located could be at risk from contamination by biological, chemical and mineral contaminants. This has implications for local water access and crop growth in agricultural areas. Whether or not this happens depends on how well the wastewater is treated and monitored.

By definition, desalination is taking the mineral salts out of seawater. These salts have to be disposed of somewhere (Figure 22). The brine that is created is corrosive and potentially deadly to vegetation and wildlife.

Water to be desalinated has good and bad minerals in it. We want to remove minerals such as arsenic, barium and lead but we want to keep calcium, magnesium and potassium. However, desalination does not differentiate between these minerals—they are all removed.

The intake pipes for desalination are a huge threat to marine life, which can be injured, trapped or killed by open intake pipe suction. The State Water Resources Control Board has estimated that open intake pipes in Californian waters kill 70 billion fish, larvae and other marine organisms each year.

▲ Figure 22 Salt stacks at a desalination plant in Trapani, Sicily

ATL Activity 23

ATL skills used: thinking, social, research

1. To what extent do the pros outweigh the cons in each method used to decrease water stress?

2. Discuss the ethical aspects of using these methods to decrease water stress.

3. Examine the sustainability of using these methods to decrease water stress.

Inequitable access to drinking water and sanitation

Inequitable access to drinking water and proper sanitation is a humanitarian crisis. Poor access to these basic needs undermines people's health through anaemia, schistosomiasis (parasitic worm), cholera, diarrhoea, malnutrition, malaria and death.

Such situations hit lower income families hardest and perpetuate poverty because treatments are unaffordable and work becomes impossible.

Case study 3

Peru

The following information comes from a study, published in the *International Journal for Equity in Health*, that investigated the socio-economic inequalities in access to drinking water between 2008 and 2018. In this context, safe drinking water is chlorinated water supplied by the public network.

Safe drinking water situation in South America:

- In 2015, 95% of the population had access to a drinking water source. However, 34 million people still lack access.

- Coverage in rural areas is much poorer.

- Access has improved in both urban and rural areas but it is worse in rural areas than large urban areas.

- Access for the poor is worse than for the rich, and poor families cannot afford household water disinfection methods.

Situation in Peru:

- Peru has the third best access to freshwater resources in South America.

- Only 87% of the population in Peru have access to chlorinated water.

- Access to safe drinking water for the wealthy is improving faster than for poorer people.

- 14% of the population drinks non-drinking water from rivers, springs and tankers.

- 85% of people in urban areas have access, compared with only 9% in rural areas.

- 77% of the water distributed in small cities is chlorinated.

- There is strong residential segregation by class and ethnicity, with impacts on water access and health.

	2008 (%)	2018 (%)
Households in extreme poverty	8.8	2.2
Households in poverty	22.4	14.4
Household access to safe drinking water	47	52
Access to safe drinking water in		
Small cities	29.2	29.1
Medium cities	49.1	52.4
Large cities	65.6	12.8

▲ Table 3 Facts and figures about access to water in Peru

⏺ ATL Activity 24

ATL skills used: thinking, communication, social, research, self-management

1. Read Case study 3, which reviews inequities in water supply due to location and income.

2. Discuss equity issues relating to water and sanitation for a named marginalized group within society. Differences may be due to ethnicity, gender, or other factors.

3. Research your own case study, including facts and figures.

4. Compare your example with the situation in Peru.

Check your understanding

In this subtopic, you have covered:

- how crucial water is to human life, especially freshwater

- water availability is not uniform and water insecurity can be a serious problem

- how freshwater can be used in a sustainable way

- water security can be improved by increasing the supply of freshwater

- water access and sustainable use can be improved through conservation efforts at all levels of usage

- the position of freshwater in the planetary boundary model

- water usage can be measured using water footprints for individuals, regions or countries

- water governance takes place at many different scales

- the causes of water stress and how to deal with it

- water can be the root cause of conflict

- there is lack of equity in access to water and sanitation.

What issues of water equity exist, and how can they be addressed?

How do human populations affect the water cycle, and how does this impact water security?

1. Define water security.

2. Describe the factors that impact access to freshwater.

3. Outline how water supply can be increased.

4. Outline water scarcity.

5. Compare and contrast the various water conservation strategies.

6. Explain how water scarcity can be addressed.

7. Discuss the concept of the freshwater planetary boundary and how crossing it may cause serious issues with the whole hydrological cycle.

8. Compare and contrast local and global governance measures for sustainable use of water.

9. Evaluate the use of water footprints as a measure of sustainability of water usage.

10. Analyse the use of citizen science as a method to gain accurate water quality data.

11. Compare and contrast water scarcity and water stress as measures of water supply.

12. Examine water stress.

    a.   What is it?

    b.   What causes it?

    c.   What are the consequences?

    d.   How can it be mitigated?

13. Discuss the negative environmental impacts of industrial freshwater production.

14. Examine the negative impacts of inequitable access to water and sanitation on human health and sustainable development.

AHL

>> Taking it further

- Use secondary data to evaluate water stress on a local, regional and global scale.

- Compare the water footprint for various activities such as: beef production vs soy bean production; synthetic cloth vs natural cloth; or a football vs a laptop.

- Investigate your own household water use for a weekend and compare it with your peers. Then produce a checklist to assess how well you conserve water.

- Produce an infographic, podcast or short video to raise awareness of ways to save water.

- Find a local citizen science group that is looking at water supply.

- Create a questionnaire to investigate water use in your local area and try to compare the water use of different socio-economic groups in your region.

4.3 Aquatic food production systems

Guiding questions

- How are our diets impacted by our values and perspectives?
- To what extent are aquatic food systems sustainable?

Understandings

1. Phytoplankton and macrophytes provide energy for freshwater and marine food webs.

2. Humans consume organisms from freshwater and marine environments.

3. Demand for foods from freshwater and marine environments is increasing due to the growth in human population and changes in dietary preferences.

4. The increasing global demand for seafood has encouraged use of unsustainable harvesting practices and overexploitation.

5. Overexploitation has led to the collapse of fisheries.

6. The maximum sustainable yield (MSY) is the highest possible annual catch that can be sustained over time, so it should be used to set caps on fishing quotas.

7. Climate change and ocean acidification are having impacts on ecosystems and may cause collapse of some populations in freshwater or marine ecosystems.

8. Unsustainable exploitation of freshwater and marine ecosystems can be mitigated through policy legislation addressing the fishing industry and changes in consumer behaviour.

9. Marine protected areas (MPAs) can be used to support aquatic food chains and maintain sustainable yields.

10. Aquaculture is the farming of aquatic organisms, including fish, molluscs, crustaceans and aquatic plants. The industry is expanding to increase food supplies and support economic development, but there are associated environmental impacts.

11. Productivity, thermal stratification, nutrient mixing and nutrient loading are interconnected in water systems.

12. Accurate assessment of fish stocks and monitoring of harvest rates are required for their conservation and sustainable use.

13. There are risks in harvesting fish at maximum sustainable yield (MSY) rate and these need to be managed carefully.

14. Species that have been overexploited may recover with cooperation between governments, the fishing industry, consumers and other interest groups, including NGOs, wholesale fishery markets and local supermarkets.

15. According to the UN Convention on the Law of the Sea (UNCLOS), coastal states have an exclusive economic zone stretching 370 km out to sea, within which the state's government can regulate fishing. Almost 60% of the ocean is the high seas outside these coastal zones, with limited intergovernmental regulation.

16. Harvesting of seals, whales and dolphins raises ethical issues relating to the rights of animals and of indigenous groups of humans.

Aquatic ecosystems and food webs

The primary producers in aquatic food webs include:

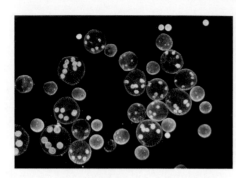

▲ Figure 1 Phytoplankton range in size from 1 μm to 10 mm

- **phytoplankton** (microscopic marine algae, diatoms and dinoflagellates)

- **macrophytes** (aquatic plants visible without a microscope)

- **cyanobacteria** (also called blue-green algae, aquatic and photosynthetic, found in all types of water).

Most of these producers photosynthesize, using sunlight to build the carbohydrates that feed through the rest of the food web. In the absence of sunlight around hydrothermal vents in the deep oceans, chemosynthesis is used.

Primary producers either float in open water or are anchored to substrate at the bottom. They provide 99% of primary productivity (see subtopic 2.2) in the oceans but less in freshwater systems.

Zooplankton are the primary consumers; they also float in the sea. These animals eat phytoplankton and their waste (dead organic matter—DOM), supporting the complex food webs of aquatic ecosystems.

Plankton are any organisms in water or air that cannot propel themselves against a current or wind. The term is defined by the niche of the organism—floating freely in currents—not the size. They are not necessarily microscopic, single-celled organisms: jellyfish are also plankton.

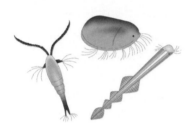

▲ Figure 2 Zooplankton

Marine ecosystems (oceans, mangroves, estuaries, lagoons, coral reefs, deep ocean floor) are usually very biodiverse with high stability and resilience.

Freshwater ecosystems (ponds, lakes, rivers, streams, bogs and wetlands) are heavily impacted by human activity and have undergone substantial alteration over time.

Aquatic organisms can be classified as:

- benthic—living on or in the bottom of the column of water

- pelagic—swimming or floating in the upper layers of water.

The aquatic organisms humans use

Humans harvest many aquatic species from the wild and from aquaculture (farming in water). We catch 2,370 different species of organism from the wild and 624 species produced by aquaculture (Figure 3).

Besides aquatic animals, humans also consume many aquatic plants. Wild rice, Chinese water chestnut, Indian lotus, water spinach, watercress and seaweeds (algae; green, brown and red) all provide humans with a wide range of products.

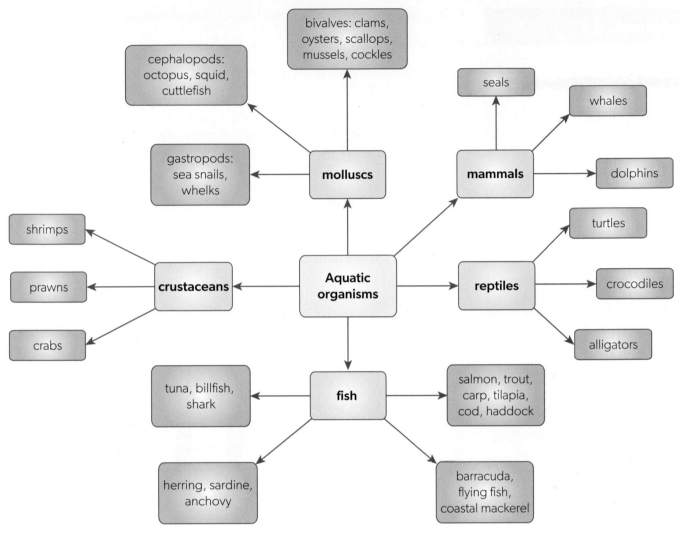

▲ Figure 3 Some of the aquatic animal species consumed by humans

The following list gives some examples of the uses of aquatic resources.

- Aquatic fauna are an important component of human diet in many cultures—15% of animal protein eaten by humans comes from fish. In Japan nearly half of the animal protein in the diet is fish.

- Fish products are a feedstuff for farmed fish and terrestrial livestock. About 25% of the global catch goes into fish meal and fish oil products to feed animals that humans farm for food.

- Fish oil, containing omega 3 fatty acids, is believed to help brain and eye development, fight inflammation, and help to prevent heart disease.

- Gelatine is a protein used to make ointments, capsules and cosmetics.

- Carotenoids are antioxidants that are believed to decrease the risk of certain cancers and eye disease.

- Chitin and chitosan have biochemical applications for drug and gene delivery, wound healing, tissue engineering and stem cell technology.

- Glucosamine treats inflammation and cartilage deterioration.

- Antifreeze from fish proteins can be used to prevent freezing injury in humans.

- Crushed fish scales can be used in cosmetics such as nail varnish, lipstick, highlighters, bronzers and eyeshadow.

- Algae are extremely versatile. They have potential as a biofuel. In China, Japan and Korea, algae aquaculture is big business—nori is the alga used in sushi wraps. In Canada and Ireland, dulce is eaten raw or fried; spirulina goes into smoothies; and blue-green algae is thought to have anti-cancer, antiviral, antibiotic, antioxidant and anti-inflammatory properties. Algae are a great source of natural pigments for food dyes and inks. Finally, they form a medium for bacterial cultures and gel electrophoresis in labs.

Demand for marine and freshwater food

Figure 4 shows a steady global increase in consumption of fish and seafood (shellfish, marine crustaceans) since 1961.

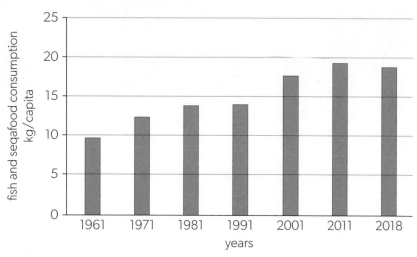

▲ Figure 4 Fish and seafood consumption

Data from: Based on free material from GAPMINDER.ORG, (CC BY 4.0)

The increase has not been uniform. Between 1961 and 2018, China, Indonesia, Malaysia, Cambodia and South Korea showed per capita increases of between 34% and 44%, while the Maldives increase was close to 74%.

Remember, an increase in per capita values cannot be attributed to an increase in a country's population. However, an increase in the population contributes to an overall increase in the total amount of fish and seafood consumed.

Factors increasing human consumption of fish and seafood

1. One major change is a shift in dietary choices. Fish is a good source of protein and fatty acids such as omega 3 along with calcium, iron and vitamin B12. It is low in saturated fats, carbohydrates and cholesterol, all of which increase the risk of heart disease.

2. Overall, society is getting wealthier. According to the World Bank, gross national income (GNI) was USD 482 in 1962 and by 2021 it was USD 12,023. This means that more people can afford imported fish and seafood from a diverse range of sources all year.

▲ Figure 5 Fish for sale in a Hong Kong market

3. Aquaculture production systems have improved with time, and this increases the supply. Aquatic animals produce millions of young and require less feed per unit of production than land animals. They are seen as a resource-efficient method of food production.

4. The distribution of aquatic products is becoming more efficient which means that great quantities of the product can reach the market in fresh condition.

5. Fisheries are becoming more and more effective and efficient at finding, hunting and processing fish and other seafoods.

Fisheries—industrial farming and hunting

Here are some key facts about fisheries.

- 90% of fishery activity is in the oceans and 10% in freshwater.

- Fisheries harvest shellfish (oysters, mussels and other molluscs, including squid and octopus) and vertebrates, both finfish (e.g. tilapia, tuna, salmon) and flatfish (e.g. plaice, halibut, turbot).

- Up to half a billion people make a livelihood in fisheries.

- Fish are a very important food for humans: three billion people gain 20% of their protein intake from fish and the rest of us gain about 15% of our protein from fish.

Key term

A **fishery** exists when fish are harvested in some way. It includes capture of wild fish (also called capture fishing) and aquaculture or fish farming.

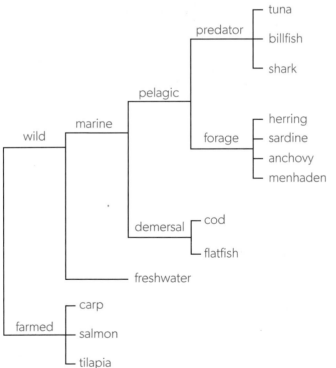

▲ Figure 6 Some of the species in the fishery industry

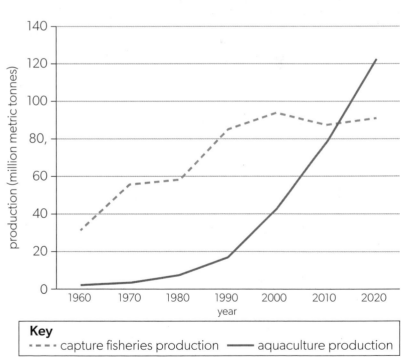

▲ Figure 7 World wild fish catch and aquaculture, 1960–2020
Data from: Food and Agriculture Organization (CC BY 4.0)

ATL **Activity 27**

ATL skills used: thinking, communication, research

Select two species of fish or seafood you have heard of and two you do not know. Research:

- their distribution worldwide

- their habitat

- estimated population numbers

- estimated harvest per year globally

- how they are used by humans—eaten directly by humans or converted to fishmeal for other consumers

- whether their harvesting is sustainable.

Create a poster or other visual display of your collected data.

Practical

Skills:

- Tool 1: Experimental techniques
- Tool 2: Technology
- Tool 3: Mathematics
- Tool 4: Systems and models
- Inquiry 1: Exploring and designing
- Inquiry 2: Collecting and processing data
- Inquiry 3: Concluding and evaluating

Investigate this hypothesis: As the level of development of a country increases, the amount of fish consumed increases.

1. Choose a minimum of 20 countries.

2. Find three secondary data sources for the:

    a. independent variable (IV): level of development of the country, measured by GDP per capita USD

    b. dependent variable (DV): kg of fish and seafood consumed per capita.

    The following websites may be useful:

    - World Bank Indicators, from World Bank Open Data
      Statista—Fisheries & Aquaculture
      Our World in Data—fish and overfishing

    - World Bank Indicators—GDP per capita
      Our World in Data—GDP

3. Determine how to graphically represent the data you have gathered.

4. Discuss the results with respect to the hypothesis.

5. Evaluate the study.

Unsustainable wild fishing industry

Fishing started as a hunter-gatherer ethos of fishers going to sea in primitive craft with simple fishing equipment to land a fish catch for their own family or community. In those days, the world fisheries were thought to be inexhaustible. Now, according to the World Bank, nearly 90% of the global marine fish stocks are fully exploited or overfished.

The global fish catch is no longer increasing (Figure 8). Demand is high and rising but fishers cannot find or catch enough fish despite improvements in technology.

Fish stocks are a resource under pressure, being exploited at an unsustainable level through overfishing. We are so good at finding and catching fish that, once found, fish populations can be depleted very quickly. Once we catch the larger specimens, we then catch smaller and smaller ones and do not leave individuals to mature and reproduce.

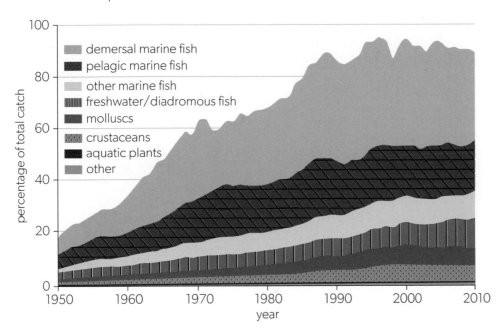

▲ Figure 8 Global wild fish capture in million tonnes, 1950–2010, as reported by the FAO (2019)

Humans are experts at finding and catching fish on an industrial scale.

- Commercial fishing is informed by the latest satellite technology, GPS navigation and fish finding (scanning) technology of military quality.

- Fishing fleets have become larger and, with modern refrigeration techniques, including blast freezing, they can stay at sea for weeks or an entire season.

- Within a fleet there will be a suite of vessels including fishing vessels, supply vessels and factory ships that process the catch at sea.

- Indiscriminate fishing gear takes all organisms in an area, whether they are the target species or not.

▲ Figure 9 Sharks and crushed mackerel on deck of a factory vessel

Many of the methods used to catch fish are highly destructive.

Bottom trawling: Trawlers drag huge nets over the seabed rather like clearcutting a forest. The nets are cone-like, with a closed end to hold the catch. They are dragged along the seafloor by fishing boats to catch fish that live deeper in the ocean. These nets destroy the seafloor by tearing up plants by the roots and demolishing animal burrows. They are also indiscriminate, catching both target and non-target species.

Gillnets: These do less harm to the environment because they are a stationary screen of netting hanging in the water. They do not damage the sea floor and the mesh size is set to avoid catching smaller, juvenile fish.

Purse seine nets: This open ocean method of fishing targets dense schools of pelagic fish like tuna and mackerel. A net curtain surrounds the school of fish and then the bottom is drawn in like closing a sack.

Blast or dynamite fishing: Fishers throw dynamite into the water to kill or stun the fish. This is illegal and highly destructive because the explosion destroys any nearby habitat—usually coral reefs. It is common is Malaysia, Indonesia and Lebanon. Apart from the damage to wildlife habitats, the improvised explosives often fail to detonate and present a danger to people in the water in that area. Accidents and injuries are common.

Ghost fishing: This is not really a method of fishing; it is more to do with the leftovers of fishing. Ghost gear is abandoned fishing gear that ends up in the aquatic environment. The ghost gear continues to do its job of catching fish and trapping animals.

Fish stocks are shrinking because industrialized nations subsidize their modern fleets by an estimated USD 35 billion a year. In addition, demand outstrips supply and MSY is exceeded annually in most fisheries.

The future of the fisheries industry

The world fish catch is between 90 and 95 million tonnes per year (which it has been since the 1990s), supplemented by aquaculture. In addition there are approximately 38 million tonnes of **bycatch**—marine species caught unintentionally while fishing for desired species. Bycatch is an issue because:

- the organisms thrown back into the ocean are dead, dying or seriously injured

- it includes turtles, dolphins, whales, marine mammals and even sea birds, some of which may be endangered

- it includes corals and sponges that are damaged thus damaging reefs and the habitats they provide

- it slows the recovery of stocks of non-target fish

- it changes the availability of prey animals, thus impacting the ecosystem

- it includes fish that are below the legal size limit in the allotted quota.

Water wars were discussed in subtopic 4.2. A significant number of serious international crises and near-war events have taken place over fishing and fishing rights in the last 60 years.

- 1970s: Iceland banned all foreign vessels from fishing in Icelandic waters. This led to three "Cod Wars" between Britain and Iceland.

- 1994: British and French fishers competed with Spanish fishers for tuna in the Bay of Biscay.

- 1995: turbot (also called halibut) war between Canada and Spain. Canada fired on and captured the crew of a Spanish fishing boat, the *Estai*, after chasing it from national into international waters. The *Estai* cut its trawl net when it was being pursued. The Canadians recovered the net from the seabed and found it had a mesh size smaller than was permitted, which caught the smaller turbot. Eventually, Canada and Spain agreed a solution which resulted in increased regulation of fisheries.

Case study 4

Newfoundland cod fisheries—a political, social and environmental disaster

The Newfoundland Grand Banks are a series of underwater plateaus off north-east Canada.

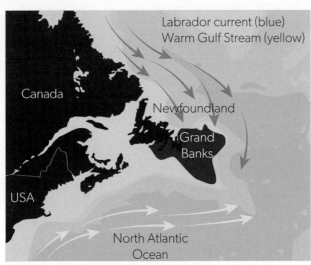

▲ Figure 10 The Grand Banks, where the cold Labrador Current and the warm Gulf Stream meet

Adapted from: British Sea Fishing

Over 500 years ago, Giovanni Caboto (John Cabot) sailed to the Grand Banks and commented that the sea was so full of cod that lowering a basket and pulling it up was all you needed to do to catch them. Because the Grand Banks are relatively shallow, sunlight reaches the seabed, allowing marine vegetation to grow and provide a home and food for small fish and crustaceans. These are eaten by larger fish such as swordfish, haddock and cod.

From the 1700s, sailing fishing vessels used lines to catch cod, which was preserved in salt. This was dangerous work as there was often fog and the small fishing boats collided or overturned. By the 1920s, steam-powered boats took over from sail. These were larger and used trawl nets dragged over the seabed to catch everything. This caused much damage. In the 1950s, diesel-powered ships and factory trawlers arrived. These were much larger still and could freeze the catch onboard, staying at sea for weeks. They came from the UK, Soviet Union, Spain, France, West Germany and other countries. The US and Canadian governments controlled only waters up to three nautical miles from their coastlines. Anyone with a boat could fish in the rest. Fish catch soared to unsustainable levels, meaning stocks could not be replaced as fast as they were removed.

In the 200 years from 1600 to 1800 an estimated 8 million tonnes of cod were taken. In the 15 years from 1960 to 1975, the same amount was taken by factory trawlers.

In 1977, the US and Canadian governments expelled foreign trawlers using the UN Convention on the Law of the Sea; the UNCLOS states that coastal states have an exclusive economic zone (EEZ) that extends 200 nautical miles out to sea.

This was a chance to reduce the volume of cod caught but there was little understanding of the need for this. Instead, the Canadian and US fleets expanded, catching just as much cod as the foreign fleets.

In the early 1980s, inshore fishing reported lower cod numbers. It was suspected that the cod were caught by trawlers offshore so did not reach the shallows to spawn (produce eggs).

The data collected by scientists at the time led them to think that about 16% of the total stock of cod was caught per year (sustainable levels). In fact it was 60% (unsustainable). Once it was known that cod stocks were in trouble, fishing quotas should have been established at a level to allow for replacement. Unfortunately, that would have caused major layoffs and economic hardship to the fishing industry, so quotas were set too high.

The inevitable happened and eventually a moratorium (ban) on cod fishing was put in place by the Canadian government in 1982. Cod was commercially extinct if not biologically extinct. By 1992, cod stocks were thought to be 0.3% of the original population level that Caboto noticed.

Some 30 years later, stocks of cod have not rebounded as expected. It is unclear why.

- Capelin (a small fish which is a staple food for cod) have been fished and turned into fishmeal (for animal feed). But capelin are predators of cod larvae so fewer capelin should mean more cod.

- Harp seals eat young cod and there are too many seals.

- Juvenile cod are found dead in high numbers.

- Fishing continues on the Grand Banks for shrimp, prawn and crab. Trawlers dredging for these cannot avoid catching cod as bycatch.

What is clear is that the loss of cod has altered the structure of the ecosystem.

By 2010 cod stocks had recovered a little, to about 10% of 1960s levels. This caused the fishery industry to demand cod fishing start again. A quota was allowed which was more than twice as high as recommended by research scientists. The quota was 13,000 tonnes, in contrast to the 1960s when 800,000 tonnes of cod were caught per year.

In 2019, Canada's Fisheries Act added further legislation—if stocks fell below a reference level, fishing would be kept at the lowest possible level (but not zero). But that amendment was not legally binding, so cod stocks keep fluctuating around low levels with minor recovery, more fishing, lower stocks, and so on.

Questions

1. Explain why the Grand Banks were so biologically productive.

2.a. Draw a simple food web for the Grand Banks ecosystem (pre-fishing industry).

   b. Now add the top predator.

3. Explain how we can measure cod stocks.

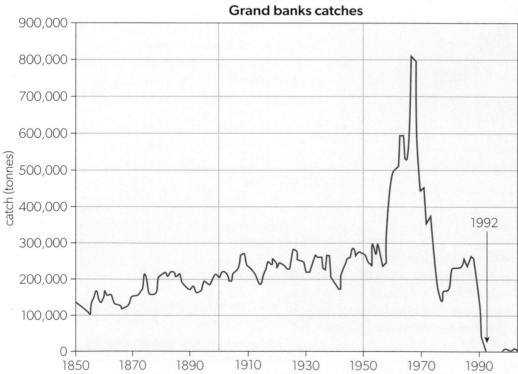

Grand banks catches

▲ Figure 11 Graph of Grand Banks fishery catch

Data from: Millenium Ecosystem Assessment

Sustainable fish?

The Grand Banks cod fishery collapse is not a one-off event. The story of stock collapse, ban, limited recovery, fishing starting again, collapse, ban, and so on is far too common.

It is an old tension between many factions: economics, livelihoods and ways of life of the societies involved, governments being lobbied by the industry, and science. Many factors are hard to measure and open to debate.

For centuries, the Newfoundland cod fishery was one of the world's most productive fisheries, yielding 800,000 tonnes of fish and employing 40,000 people at its peak in 1968. Then its stocks plummeted as a result of overharvesting and habitat damage. In 1992, the fishery was closed in an effort to save it. But it may have been too late: stocks are not expected to be out of the critical zone until 2024. (In the critical zone, there is still harm happening in the cod stock.) The federal department of fisheries reported in 2021 that mortality rates are rising and new immigration is slow. The sequence of minor recoveries of stocks, then fishing resuming, then stocks crashing again is one example of how mismanagement can reduce a productive fishery to virtually nothing.

This collapse was local in scale, but the issue is much larger. North Atlantic Ocean fisheries now catch half what they did 50 years ago, despite tripling their efforts. Many popular species—such as cod, tuna, flounder and hake—are now in serious decline. Cod stocks in the North Sea and to the west of Scotland are on the verge of collapse.

The deterioration of oceanic fisheries can be reversed. Granting fishers an ownership stake in fish stocks is one way to help them understand that the more productive the fishery, the more valuable their share. For example, fishers in Iceland and New Zealand have used marketable quotas, allowing them to sell catch rights, since the late 1980s. The upshot is smaller but more profitable catches and rebounding fish populations. The classic "tragedy of the commons" problem is averted.

Because of the complexity of marine ecosystems, some scientists are pushing for management of whole ecosystems rather than single species. Studies have shown that well-positioned and fully protected marine reserves, known as fish parks, can help replenish an overfished area. By giving fish a refuge to breed and mature in, reserves can increase the size and total number of fish both in the reserve and in surrounding waters. For example, a network of reserves established off Saint Lucia in 1995 has raised the catch of adjacent small-scale fishers by up to 90%. Preservation of nursery habitats like coral reefs, kelp forests and coastal wetlands is integral to keeping fish in the sea for generations to come.

Consumers can promote healthy fishery production by eating less fish and seafood and buying from well-managed, abundantly stocked fisheries. The *Seafood Lover's Almanac*, published by National Audubon Society's Living Oceans Program, is one valuable reference. For example, Chilean seabass stocks are on the verge of collapse and illegal fishing abounds, so it is on the list of fish to avoid. The list also distinguishes between wild Alaska salmon, which comes from a healthy fishery, and farmed salmon, which is fed meal made from wild fish and thus does not relieve pressure on marine stocks.

TOK

How can we decide what to do if experts disagree?

What shapes perspectives in the cod fishing industry?

How do you know when it is morally right to act?

Informative and truthful labels are needed to allow consumers to make wise purchasing decisions. The Marine Stewardship Council, a new independent international accreditation organization, has thus far certified seven fisheries as being sustainably managed with minimal environmental impact.

The capacity of the world's fishing fleet is now double the sustainable yield of fisheries. Ransom Myers and Boris Worm from Dalhousie University believe that the global fish catch may need to be cut in half to prevent additional collapses.

Policy tools that can help preserve the world's fish stocks include:

- reducing bycatch

- creating no-take fish reserves

- managing marine ecosystems for long-term sustainability instead of short-term economic gain.

If these policies are coupled with a redirection of annual fishing industry subsidies of at least USD 35 billion to alternatives such as the retraining of fishers, there could be a big payoff. It is difficult to overestimate the urgency of saving the world's fish stocks. Once fisheries collapse, there is no guarantee they will recover.

The list of unsustainable fisheries does change over time. Some species in some seas are greatly overfished. In others areas, regulations, monitoring and certifying have allowed fisheries to continue sustainably. The Marine Conservation Society, a charity based in the UK, produces a "Good Fish Guide" which provides information on fish to avoid and best choices if you do want to eat fish (Table 1).

One good news story is that of the Patagonian toothfish, also called Chilean seabass (*Dissostichus eleginoides*). This species lives in deep cold waters in the southern oceans. It was heavily overfished by illegal means (IUU—illegal, unreported and unregulated fishing) in the decades up to the late 1990s and stocks had collapsed. Campaigns to eliminate IUU had an effect, particularly the story of the Sea Shepherd ship which chased a pirate fishing ship for 110 days in 2014 before the pirate ship scuttled itself. Sea Shepherd is a direct action non-profit organization which aims to conserve the world's oceans and marine life. Read about the stopping of IUU fishing of the Patagonian toothfish on the Marine Stewardship Council website and elsewhere.

ATL Activity 28

ATL skills used: thinking, communication, social, research, self-management

Working in a small group, create a scientific article about an aquatic food consumption crisis of your choice. Make your article interesting and visual, as well as scientific.

Connections: This links to material covered in subtopics 1.3 and 2.2.

ATL Activity 29

ATL skills used: thinking, communication

Review the guidance from the Marine Conservation Society in Table 1.

1. Which of the species that are harvested unsustainably have you eaten?

2. How can you ensure you eat responsibly?

3. To what extent is it your responsibility to find out how these species are caught for consumption?

4. Do you think people would like to know how their food is caught and killed?

What not to eat

Species	Reason	Alternatives
Atlantic cod (from overfished stocks)	Species listed by World Conservation Union, IUCN. Some stocks close to collapse, e.g. North Sea	Line caught fish from Icelandic waters
Atlantic salmon	Wild stocks reduced by 50% in last 20 years	Wild Pacific salmon; responsibly or organically farmed salmon
Dogfish/spurdog	Trawling of spurdog destroys habitats and species is listed by IUCN as endangered	Line caught spurdog in USA
European seabass	Trawl fisheries target pre-spawing and spawning fish; high levels of cetacean bycatch	Line caught or farmed seabass
Grouper	Many species are listed by IUCN	None
Haddock (from overfished stocks)	Species listed by IUCN as vulnerable	Line caught fish from Icelandic and Faroese waters
Ling (*Molva spp*)	Deep-water species and habitat vulnerable to impacts of exploitation and trawling	None
Marlin	Many species listed by IUCN as endangered	None
Monkfish	Long-lived species vulnerable to exploitation. Mature females extremely rare	None
North Atlantic halibut	Species listed by IUCN as endangered; very slow-growing	Line caught Pacific species; farmed North Atlantic halibut
Orange roughy	Very long-lived, deep sea species. IUCN endangered. Can live over 100 years and accumulate toxins	None
Shark	Long-lived species vulnerable to exploitation. IUCN lists many as critically endangered	None
Skates and rays	Long-lived species vulnerable to exploitation	None
Snapper	Some species listed by IUCN, others overexploited locally	None
South African hake (*M. capensis*)	Species heavily overfished and now scarce	European hake (*Merluccius merluccius*); net, line caught or farmed is now sustainable
Sturgeon	Long-lived species vulnerable to exploitation. Five out of six Caspian Sea species listed by IUCN	None, although this species is now farmed
Swordfish	Species listed by IUCN. Gillnets also catch dolphins, sharks and turtles	Line caught in North Atlantic
Tuna	All commercially fished species listed by IUCN except skipjack and yellowfin are overfished. Bluefin particularly overfished and IUCN critically endangered. 5% or less of original biomass of Bluefin left in oceans	"Dolphin friendly" skipjack or yellowfin. Only choose pole and line caught
Warm-water or tropical prawns	High bycatch levels and habitat destruction	Responsibly farmed prawns only

▲ Table 1 What not to eat from the Marine Conservation Society

Maximum sustainable yield

The sustainable yield (SY) is the increase in natural capital (i.e. natural income) that can be exploited each year without depleting the original stock or its potential for replenishment. For commercial ventures, the **maximum sustainable yield (MSY)** is of interest. For a given population of fish, this means the maximum sustainable annual catch. MSY must be at a level that can be sustained through time and still allow the fish population to grow. It is a hypothetical equilibrium between fishing and fish numbers: an estimate of how many fish and of what size can be taken in any year so that the harvest is not impaired in subsequent years.

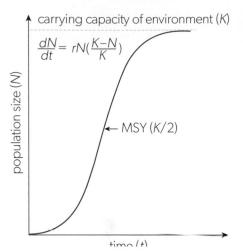

▲ Figure 12 S-shaped curve showing point of MSY

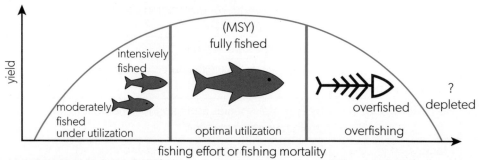

▲ Figure 13 Sustainable yield curve

ATL Activity 30

ATL skills used: thinking, research

In a world mostly committed to a free market—in a global context—have we the right to criticize or control those who use technology and labour efficiencies to produce more food at less cost? Without intensive agriculture, we could not feed the human population.

1. What is Fairtrade?

2. To what extent does Fairtrade help farmers out of poverty?

3. Do you think the label Fairtrade is there to make consumers in higher-income economies feel better or is it more meaningful?

4. What do we mean by ethical food?

The carrying capacity for a species depends on its reproductive strategy, its longevity, and the natural resources of the habitat or ecosystem.

Each breeding season or year, new individuals enter the population (either new offspring or immigrants). If the number recruited to (entering) the population is larger than the number leaving (dying or emigrating), there is a net increase in population. If the difference in population from initial size to new population size is harvested, the population will remain the same. This number is the MSY for the population.

$$SY = \frac{\text{total biomass at time } t + 1}{\text{energy}} - \frac{\text{total biomass at time } t}{\text{energy}}$$

An alternative formula is:

$$SY = \text{annual growth and recruitment} - \text{annual death and emigration}$$

In practice, harvesting the maximum sustainable yield normally leads to population decline and thus loss of resource base and an unsustainable industry or fishery.

ATL Activity 31

ATL skills used: thinking, communication, social, research, self-management

Research how two contrasting fisheries have been managed and relate your findings to the concept of sustainability (e.g. cod fisheries in Newfoundland and Iceland).

Technical issues that should be covered include improvements to boats, fishing gear (trawler bags), and detection of fisheries via satellites.

Management aspects should include use of quotas, designation of Marine Protected Areas (exclusion zones), and restriction on types and sizes of fishing gear (including mesh size of nets).

Present your findings in an appropriate way.

Data-based questions

The Inuit are indigenous aboriginal people of Northern Canada. The data in Table 2 come from a study of an Inuit fish farming community. The Inuit fish in the open sea but have also sectioned off a large fjord (a long narrow inlet of the sea) which they use for farming salmon and shrimps. The shrimps eat phytoplankton. Salmon and kawai (a wild fish) both eat shrimps.

1. Use the data in Table 2 to complete the diagram below. [6]

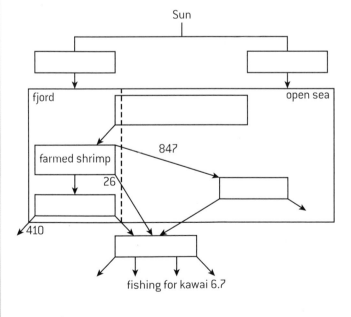

fishing for kawai 6.7

	All in kJ m^{-2} yr^{-1}
Insolation on fjord	185,000.0
Insolation on open sea	1,972,000.0
Farmed shrimp consumed by Inuit	26.0
Gross primary production by phytoplankton	3,470.0
Shrimp consumed by kawai	847.0
Respiratory loss by kawai (open sea)	572.0
Shrimp consumed by salmon (farmed)	461.0
Respiratory loss by salmon	410.0
Kawai consumed by Inuit	6.2
Salmon consumed by Inuit	4.3
Energy used in managing salmon farm	4.1
Energy used in fishing for kawai	6.7
Energy used in managing shrimp farm	14.0
Energy used in other human activities including trading furs	12.5

▲ Table 2

2. a. Define what is meant by the term "gross primary productivity" (GPP). [1]

   b. State how GPP differs from net primary productivity (NPP). [1]

   c. Identify the factors other than insolation which affect rates of GPP. [2]

3. Using the data in Table 2, determine whether salmon or kawai is more efficient at converting food into biomass. [3]

4. Compare the efficiency of aquatic food production systems with terrestrial food production systems. [3]

5. Calculations based on the data in Table 2 would suggest that farming and eating shrimp is the most energy efficient food source for the Inuit. Suggest why the Inuit continue to farm salmon. [1]

6. Suggest ways in which this indigenous food production system might differ from a large-scale commercial food production system. [3]

Climate change and ocean acidification

Climate change and water acidification are injurious to freshwater and marine ecosystems. The combined impacts of these two anthropogenic consequences augment each other. They also vary over time and space. It is possible that warming of the oceans may speed up calcification and thus counter the impact of the lower pH on reef production.

Climate change

Climate change causes a cascade of effects in the oceans.

- As oceans warm, the resulting thermal expansion causes sea level rises and alters the ocean currents.

- Melting ice on land and in the oceans causes sea level rises and reduces salinity.

- Warming of the oceans causes heat waves and acidification.

- Biodiversity in the oceans is reduced—many organisms are not adapted to higher temperatures.

- Changes in the oceans change the lives of human coastal communities: 600 million people (10% of the world's population) living close to the coast are just 10 metres above sea level, and half the world's megacities—accommodating two billion people—are on the coast.

- Increasing global temperatures increase the survival and transmission of tropical diseases.

- LICs are home to 95% of the world's fishers and their major source of food and income is the sea.

- The ocean is estimated to be worth USD 3–6 trillion/year in employment, ecosystem services and cultural services.

ATL Activity 32

ATL skills used: thinking, communication, research

Do some more research into climate change and the oceans.

Produce an infographic to show your findings.

Check out the UN Ocean Conference.

Ocean acidification

pH is measured on a scale from 1 to 14, with 1 being the most acidic, 7 neutral, and 14 the most alkaline. The ocean's surface pH is 8 (the same as baking soda) which is slightly alkaline. However, carbon dioxide forms carbonic acid in water, so as the oceans absorb carbon dioxide, they become more acidic with more hydrogen ions. The pH of seawater decreases and this is known as acidification. As the pH scale is a log scale, a drop of 0.1pH represents a 30% increase in acidity.

Increased acidity means carbonate ions are less abundant for marine animals that make their shells or exoskeletons out of calcium carbonate. These are crustaceans, corals, echinoderms, sea urchins and all shelled animals. Due to very slow ocean mixing, absorbed carbon dioxide is concentrated in the top 1,000 m where most animals live.

Ocean acidification also makes it more difficult for plankton to build their skeletons. Phytoplankton are the base of marine food chains so the ecosystem's food web is altered. Human food supplies are affected.

A lower pH also impacts fish reproduction as the eggs are sensitive to pH. This leads to reduced population growth or replacement.

A pH of 7.8 is predicted to be the point at which coral growth would stop. Reef habitats would start to disappear, decreasing storm protection. More storms would disrupt many habitats as well as the tourism industry.

▲ Figure 14 Sea urchin on a coral reef in the Maldivian archipelago

The Great Barrier Reef (GBR)

Climate change is the single biggest threat facing the Reef. ""

Great Barrier Reef Foundation

The GBR is a rich and complex natural ecosystem. The biggest threat to it is climate change.

Corals have a mutualistic relationship with zooxanthellae (microscopic algae that live within the coral and give it colour). A temperature increase of 1°C for four weeks induces heat stress in the coral and they expel the zooxanthellae. This is coral bleaching. The corals become transparent so the white skeletons can be seen. Coral bleaching may also be due to increased sun exposure, changes in water quality and severe low tides.

There have been four mass bleaching events on the GBR.

- 2016: Central third of the reef had severe bleaching; southern sector was less affected.

- 2017: Far north saw widespread severe bleaching with a 22% mortality rate; again the south was less severely affected.

- 2020: 1,036 reefs were surveyed and 60% had moderate to severe bleaching.

- 2022: The whole reef was hit—northern and central regions had extreme bleaching, while the south saw minor bleaching.

Climate change brings higher temperatures which escalate the frequency and intensity of severe weather events. Tropical cyclones generate high winds, heavy rain, huge waves and storm surges. The waves break over the corals, tear them up and reduce them to rubble. The waves also uproot mangrove trees. It can take the reef up to 10 years to recover from this sort of damage. But with climate change, the reef experiences up to 11 storms a year.

Higher water temperatures force marine species to migrate south towards cooler waters. This changes the dynamics of the entire ecosystem as competition shifts and food sources disappear. This impacts local coastal communities and tourism.

Practical

Skills:
- Tool 1: Experimental techniques
- Inquiry 1: Exploring and designing

You can study the impact of pH on calcium carbonate shells, so long as you use shells that have been discarded by the animals.

Plan an experiment to investigate the impact of pH on the disintegration of calcium carbonate shells.

Mitigation of unsustainable exploitation

Unsustainable fishing has drastic impacts on fish populations and the aquatic ecosystems in which they live. It also has wider implications for human societies. So what are the solutions?

Rights-based fishery management

The traditional approach to fishing is "first come first served". This means fishers catch as many fish as they can in the shortest time possible. A rights-based fishery management approach allows an entity (person, community or fishing boat) to fish in a specific place at a specific time. Each entity is guaranteed a set portion of the catch, but they have to stick to the given limits. These limits include how many fish they catch and when they can fish. This approach aims to balance socio-economic and ecological needs. This would stop the tragedy of the commons scenario that is currently happening. Each entity has a vested interest in maintaining the fishery for long-term sustainable gain rather than short-term profit.

In New England (part of the east coast of the USA), a catch-share was adopted to give the cod and haddock populations a chance to recover from overfishing. This system can result in a steadier income stream and encourages environmentally friendly, sustainable practices making for healthier fish populations. It allocates a harvest allowance or quota to each individual or company. The individual can decide how and when they fill their quota. They can even lease their quota to other fishers. But in this case, cod stocks continued to fall and the scheme closed. There are several hypotheses as to why this happened. The system depends on accurate information on stock levels before quotas are allocated. If the information is inaccurate and quotas are set too high, the fishery will continue to be unsustainable.

Fishing subsidies

To many, the idea of fishing subsidies is anathema (hateful). Why give money to an industry that is operating at a level that is destroying the resource it relies on? The subsidies given are for fuel, fishing gear and building new vessels. They are seen as an incentive to overfish at a time when estimates suggest there are already two and a half times more fishing fleets than are needed to meet current demand.

In June 2022, the World Trade Organization (WTO) negotiated the Fisheries Agreement, which aimed to:

- restrict subsidies

- decrease global overfishing and improve failing fish stocks

- preserve the world's oceans

- protect small local communities that rely on marine resources for their livelihoods

- support transparency and accountability of the bodies that issue subsidies (usually governments)

- allow subsidies for fishing activities that are helping to rebuild stocks

- encourage countries to cooperate to ban subsidies for illegal, unreported, unregulated fishing, and restrict subsidies for fishing that targets stocks already overfished.

Regulations

Regulations can be applied at any level though monitoring them is not easy. The colossal nets that are used in fishing have increased the problems of bycatch so placing regulations on net size, placement in the water, and fishing equipment used can reduce this issue.

Ban fishing in international waters

International waters are not owned by any country. This means they can be fished by any and all fishing operations. International waters are estimated to make up 58–77% of the oceans. In national water (under legislation of a country) protection laws have been established and they have been very effective at reducing overfishing. Therefore, imposing a ban on fishing in international waters is likely to be highly effective. However, convincing governments to adopt and enforce this policy would be a challenge.

In 2017, nine nations agreed to a 16-year ban on commercial fishing in the 3 million km² of the central Arctic Ocean. The FAO launched the Common Oceans Programme to regulate fishing in some areas of international waters. This programme banned fishing in critical areas and reduced tuna overfishing.

Protect predator species

Predators are essential for the maintenance of ecosystems. In marine ecosystems, predators, such as tuna, are prone to overfishing. Without the predator, overpopulation of prey species and algal blooms can become a problem.

And overfishing, bycatch, pollution and habitat loss are all thought to be major threats to one-third of shark species.

▲ Figure 15 Yellowfin tuna

Traceability and food labelling

This requires that food is labelled with information about where it has come from. Therefore the journey of fish from catch to market can be known by the consumer. The customer can make an educated decision about food choices. The labels are backed up by certificates and validation documents from trustworthy authorities.

ATL Activity 33

ATL skills used: thinking

1. State two reasons why we are overfishing the oceans.

2. Why has the world fish catch stalled?

3. What actions can be taken to reverse overfishing?

4. What actions can you take?

Marine protected areas

The UN World Database of Protected Areas lists 15,000 MPAs, totalling 27 million km^2 or 7.5% of the ocean. This is progress but a long way off the IUCN recommendation that 30% of the world's oceans should be protected by 2030.

To provide benefits to fisheries, successful MPAs across the globe share all or most of the following five key features.

- They are highly to fully protected.

- They are well enforced.

- They have been established for 10 years or more.

- They are large in size.

- They are isolated by deep water and sand.

The far-reaching advantages of MPAs are astounding and include the following.

- Protecting and restoring endangered species and ecosystems and so conserving biodiversity, which boosts resilience. MPAs have a 23% increase in biodiversity.

- Restoring key processes, e.g. carbon capture and water purification.

- Acting as an insurance policy if other measures fail.

- Enabling research and education.

- Attracting tourists, which supplements community livelihoods.

- Attracting larger fish to spawn in the protected waters. More spawning gives more offspring and these travel around the oceans and promote fisheries further afield, making wider areas more resilient.

Marine Protected Areas (MPAs) are established by governments to protect marine ecosystems from invasive human activity such as overfishing. These living laboratories provide research opportunities for marine biologists and oceanographers.

From a report in June 2019, from the American Progress website

- Enabling fish in MPAs to grow larger thus adding to the spawning potential. These fish are up to 28% larger than the fish in unprotected areas. This is important to sustainability of fisheries because larger females release more and larger eggs of higher quality than smaller females.

- Boosting fisheries' size as the density of organisms in the protected areas is one and a half times higher than in unprotected areas.

There is great economic value from MPAs. For every USD 1 spent, there are returns of USD 20. MPAs:

- benefit neighbouring fisheries

- combat global climate change

- establish and/or maintain storm buffers

- encourage ecotourism

- provide employment in the management of the MPA.

To be able to provide any of these benefits, there has to be back-up from other fisheries management strategies. MPAs need to cover large areas in which the ethos is well established and enforced. If these areas can be expanded, overfishing can be controlled in a greater area. As ever, governments need to be convinced to restrict fishing and enforce the rules.

Aquaculture

The range of organisms farmed in **aquaculture** is astonishing. According to the FAO, 494 recognized species are farmed worldwide (see Table 3).

Number of species	Examples
313 finfish	Tilapia bass perch
88 molluscs	Oysters, mussels, scallops, clams
49 crustaceans	Prawns, shrimps
6 marine invertebrates	Sea cucumbers, sea urchins
3 frogs	Marsh frog, giant swamp frog
2 aquatic turtles	Soft-shelled turtles
2 cyanobacteria	*Trichodesmium*

▲ Table 3 Number of species farmed globally in aquaculture

The FAO *State of the World Fisheries and Aquaculture 2022 Report* states that the growth of aquaculture pushed total fisheries production to 214 million tonnes in 2020. Just over 40% of that increase came from aquaculture. Humans may now eat more farmed fish than wild-caught fish for the first time since fishing began. If aquaculture is not sustainable, this cannot continue. The challenge is to make fish-farming sustainable.

(ATL) Activity 34

ATL skills used: thinking, research
Research a local or global MPA, to find out how it:

- supports aquatic food chains

- maintains sustainable yields.

Evaluate its results.

Key term

Aquaculture is the farming of aquatic organisms in either coastal or inland areas; it involves interventions in the rearing process to enhance production.

Connections: This links to material covered in subtopics 1.3 and 4.4.

▲ Figure 16 Fish farm in Norway

China produces nearly 40% of all farmed fish worldwide and most of that is carp or catfish. These are grown in rice paddies and their waste provides fertilizer for the rice. This system is used in many Southeast Asian countries and is of mutual benefit: the system produces rice and a healthy source of protein for the farmer.

Fish farming is becoming more sustainable because fishmeal now uses more trimmings and scraps which would have been wasted in the past; waste from livestock and poultry processing is also substituted for fishmeal. In addition, the US Department of Agriculture has proven that eight species of carnivorous fish can get enough nutrients from alternative sources without eating other fish. These fish are white sea bass, walleye, rainbow trout, cobia, Arctic char, yellowtail, Atlantic salmon and coho salmon.

Unfortunately, other aquaculture systems are less efficient. Shrimp and salmon are carnivores and are fed on fishmeal or fish oil produced from wild fish. Mangrove swamps have been replaced by fish farms—in the Philippines, two-thirds of the mangroves have been lost in 40 years.

Impacts of fish farms include:

- loss of habitats

- pollution (with feed, antifouling agents, antibiotics and other medicines added to fish pens)

- spread of diseases

- escaped species, including genetically modified organisms, may survive to interbreed with wild fish, or outcompete native species and cause the population to crash.

Land-based marine algae farms

This is a proposal by researchers from Cornell University, USA. A model used to predict yields suggests that a desert setting could become a future "breadbasket" that can meet global protein demand by 2050. The model looks at how aquaculture can help food supply through seawater-fed algae farms situated on land. With wild fish stocks over-stretched and constraints on aquaculture in coastal areas, this is a possible solution.

- It may meet future nutritional demands and increase food production by 56%, thus helping to feed the growing human population.

- It will enhance environmental sustainability.

- Protein-rich, nutrient-rich microalgae could provide food security for many people.

Connections: This links to material covered in subtopics 1.3 and 5.2.

- It is a good protein and nutrient supplement for vegetarians because it contains omega-3 fatty acids and certain critical amino acids and minerals.

- Marine algae are big carbon consumers, so the operations could be carbon negative.

- Algae grow 10 times faster than terrestrial crops, so an algae farm would not need added chemical fertilizers. This would reduce waterway pollution.

- To produce algae, carbon dioxide must be added to the water in the aquaculture ponds. There is a problem here because it is difficult to source the carbon dioxide for this purpose.

- Algae introduced into building materials such as cement absorbs carbon from the atmosphere.

Productivity of aquatic systems

Productivity is how much organic matter is produced by plants—mainly phytoplankton. So, what is needed for high productivity in an aquatic system? The basic components are:

- light, which means shallow water. Maximum productivity in the oceans is at about 0–80 m depth. At this depth chlorophyll levels are at the maximum due to high concentrations of phytoplankton biomass

- good nutrient levels, which are dependent on upwellings

- temperatures of around 20–30°C.

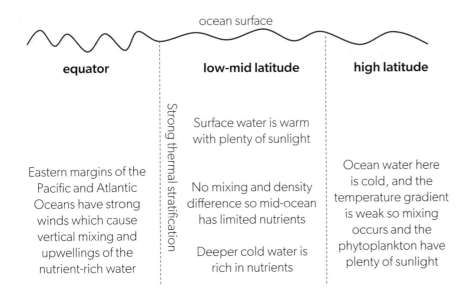

▲ Figure 17 Why different oceanic areas have different nutrient availability

AHL

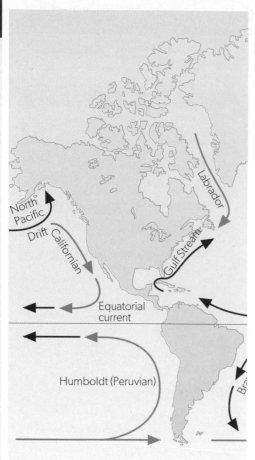

▲ Figure 18 Cold currents that drive upwellings in the eastern Pacific

Adapted from: David Russell Illustration, Peters & Zabransky Ltd, eMC Design Ltd

State(s) must set an allowable catch, based on scientific information, which is designed to maintain or restore species to levels supporting a maximum sustainable yield (MSY).

The UN Convention on the Law of the Sea (UNCLOS)

But productivity is dependent on more than these basic components:

- Surface areas of the ocean are nutrient-poor but warm and sunny.

- Deep areas are cold and dark but nutrient-rich.

- Without nutrient mixing, nutrient loading and stratification, ocean productivity would be much lower.

Coastal upwellings maintain high biodiversity and provide 20% of the oceans' fish harvest. These upwellings occur where major ocean currents flow parallel to the coast, e.g. the Humboldt current runs along the coast of Peru and the Californian current flows parallel to California (Figure 18).

The upwellings supply phosphorus and nitrogen (in the oceans as phosphate ions, ammonium ions, nitrates and nitrites). If these upwellings fail, the effects can be catastrophic—as seen when the Peruvian anchovy industry collapsed in the 1970s.

Fish stocks

To calculate how many fish can be caught for a sustainable fishery, we need:

- accurate assessment of MSY

- accurate assessment of current fish stocks and monitoring of harvest rates (for conservation and sustainability)

- cooperation between all stakeholders to ensure the recovery of any overexploited species.

Maximum sustainable yield assessment and management

The principal behind MSY is that if fish are removed from the population, this reduces pressure on the resources of the fish left behind. The remaining fish can benefit from less competition so there will be more food and habitat and reproduction rates can increase. Up to a point, the more fish harvested the greater the benefits for those remaining. However, there is a tipping point and if MSY is exceeded, reproductive potential is reduced and fish stocks decline. There are several reasons for this:

- The population dynamics of the target species are normally predicted (modelled) rather than the species numbers being quantitatively measured (counted).

- It is often impossible to be precise about the size of a population.

- Estimates are based on previous experience.

- Fishing fleets tend to harvest at higher rates than are necessary to meet demand—they fish to the max.

- Fish populations are dynamic, so age and sex ratios vary considerably. If the harvest takes reproductive females this will have a greater impact on future recruitment than if mature or old males are taken. Also, if immature fish are taken future recruitment rates will be impacted and populations will decline.

- Disease may strike the population.

If the stock is overfished, it is essential to set a timeframe that will allow the stock to recover and reach MSY. Within that timeframe, catch levels must be adjusted or fishing may have to be banned.

A much safer approach is to adopt the harvesting of an optimal sustainable yield (OSY). This usually requires less effort than MSY and maximizes the difference between total revenue and total cost. It has a much greater safety margin than MSY but may still have an impact on population size if there are other environmental pressures within a system. Fishing quotas are often set as a percentage or proportion of the OSY per fleet per year. The quota is set as a weight of catch not number of fish.

Assessing fish stocks and monitoring catches

Fish stocks must be accurately assessed and the catches carefully monitored if species are to be conserved through sustainable use. Fishery managers need to understand population dynamics, movements, breeding sites and feeding habits. Understanding the health of the fish stock involves biology, ecology, environment, fishing behaviour and knowledge of market values. Such evidence—combined with mathematical modelling, statistical analysis and computer science—allows people to develop harvest strategies and fisheries management.

Assessment

Assessment of fish stocks can be done through a range of methods.

- Tracking—marine biologists tag the fish electronically and map their movements.

- Marking and recapturing fish.

- Assessment models combine various fish-related factors to inform decisions about regulating and managing fish stocks. Surplus production models use population growth models to estimate annual stock and the surplus produced, thus giving the amount available to harvest. Virtual population analysis uses data on fish populations from previous decades (stock size, mortality and growth rates, and movements). This data is used to predict catch, fish size, age and composition. Statistical analysis then compares reality with predictions to get the best fit.

All monitoring techniques have a similar problem: fish swim in large schools or shoals, they are constantly moving vertically and horizontally, and they are hard to distinguish from each other. Apart from that, monitoring technology is not always available and when it is, it is expensive and time-consuming.

Connections: This links to material covered in subtopic 2.6.

Practical

Skills:

- Tool 1: Experimental techniques
- Inquiry 2: Collecting and processing data
- Inquiry 3: Concluding and evaluating

Capture–mark–release–recapture practical

A school of fish in a large lake is represented by pieces of scrap paper of different sizes in a large bag.

- Each student catches a handful of "fish" and marks them.

- Record the number of fish caught in a shared document.

- Put the fish back in the lake.

- Shake the lake to mix up the fish.

- Each student takes a second handful of fish.

- Record how many are caught and how many are marked.

- Calculate the Lincoln index and compare results.

Evaluate the accuracy of this method.

Monitoring

An important method of monitoring fishery catch is electronic or e-monitoring. A system of video cameras and sensors is installed on the fishing boats to record fishing activities. Such systems are now mandatory on commercial fishing fleets in many countries.

There are major advantages to this system.

- Video recordings can be viewed by the authorities to verify that the fishing logs are accurate as regards the amount and type of fish caught.

- It ensures that reports about interactions with threatened, endangered or protected species are reported.

- Fishers can prove they follow management arrangements and are being responsible operators.

- Over time, e-monitoring can improve data so that management decisions can be made to help protect all aquatic species.

Recovery of overexploited species

The problem with any solution to overexploitation of an aquatic species is that there are numerous stakeholders—people who have an interest in any projects. Many people can make a difference or even slow progress in the recovery—see Figure 19.

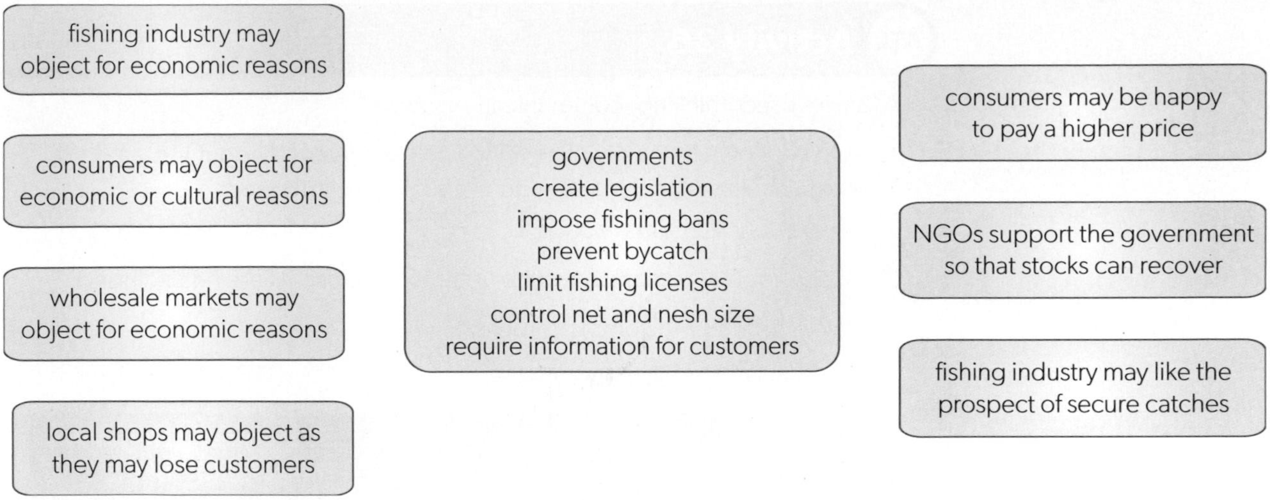

fishing industry may object for economic reasons

consumers may object for economic or cultural reasons

wholesale markets may object for economic reasons

governments
create legislation
impose fishing bans
prevent bycatch
limit fishing licenses
control net and nesh size
require information for customers

consumers may be happy to pay a higher price

NGOs support the government so that stocks can recover

fishing industry may like the prospect of secure catches

local shops may object as they may lose customers

▲ Figure 19 Some fisheries stakeholders and their thinking

Any recovery of a species involves the strategies discussed in "Mitigation of unsustainable exploitation": rights-based fishery management, fishing subsidies, regulations, bans on fishing in international waters, protection of predator species, traceability and food labelling, and MPAs.

In addition to commercial efforts, consumers can make a difference. Many websites give information and advice about what seafood is good for us *and* for the Earth's ecosystems. Remember the following key points.

- Avoid the big fish: tuna, marlin and shark are overexploited and have been for many years. Try to avoid eating these.

- Buy locally caught and sold fish: this means you know where it has come from, and less energy is used to store and transport it.

- Choose wild fish not farmed ones: although farmed fish are seen as a solution to overfishing, they are not always a better choice for the ecosystem. The spread of disease into wild populations of fish is a problem. Also, seals, sharks and whales are attracted to the aquaculture nets, get stuck in them and die. To farm 1 kg of salmon takes 5.5 kg of wild caught animals.

- Use seafood guides and smartphone apps designed to help people make good seafood choices, whether in the grocery store or in a restaurant. Some are specific to location, e.g. GoodFish is specific to Australia. Others, such as SeafoodCheck, are global. The SeafoodCheck app gives information about mercury toxicity levels of various seafood items.

- Ask the questions that matter: if you eat at a restaurant, ask them what seafood on the menu is harvested sustainably. This will make them stop and think—customer desires matter.

- Always opt for certified sustainable seafood: seafood marked with the blue fish tick is traceable to a sustainable fishery. Over 25,000 seafood products from all over the world carry this label.

- Encourage other people to think and act sustainably: we all have the power to set an example from how we act and what choices we make. You can influence people older and younger than you are.

ATL Activity 36

ATL skills used: thinking, communication, social

1. On your own, consider everything you have learned about accuracy and sustainability of the exploitation of aquatic environments.

2. Share your thoughts in pairs.

3. Copy Table 4 and add any other interest groups that are not included.

4. Work together to complete Table 4.

5. Present your ideas to the class.

	Strategies they would favour	Positives of the strategy	Strategies they would block	Negatives of the strategy
Governments				
Fishing industry				
Consumers				
NGOs				
Wholesale markets				
Local shops and supermarkets				

▲ Table 4

The Law of the Sea

Connections: This links to material covered in subtopic HL.a.

The UN follows the belief that everything to do with the ocean is interrelated and so approaches to law and order must be holistic. In December 1982, the UN Convention on the Law of the Sea (UNCLOS) was opened for signatures in Montego Bay, Jamaica. It had taken 14 years to compile through the collaboration of 150 countries. Legal, political, social and economic aspects were taken into consideration. UNCLOS provides a framework for development and exploitation of the resources of the oceans. It came into force in November 1994 and is now globally recognized as dealing with all maritime matters.

It is an extensive document that governs everything about ocean spaces:

- delimitation

- environmental control

- marine scientific research

- economic and commercial activities

- technology and settlement of disputes.

For a very brief summary, search online for "Oceans and Law of the Sea, UN".

Some of the key aspects of the Convention are that coastal states:

- have control over internal waters (right next to the coastline) where foreign ships may not travel and the country is free to set its own laws and regulate use

- have control over territorial waters up to 22 km from the coast. In this area, foreign ships can transit on "innocent passage" but not spy, fish or pollute

- have the rights to natural resources and economic activities in an exclusive economic zone (EEZ), which extends up to 370 km from the coast; this includes the water and the continental shelf

- must prevent and control marine pollution

- have control over any marine research in the EEZ.

The continental shelf is the extension of continents under the seas and oceans. Where continental shelf exists, it creates shallow water; there is plenty of light for producers to photosynthize and upwellings bring nutrient-rich water to the surface. Continental shelf has the richest fisheries with 50% of oceanic productivity in 15% of its area.

The width of the shelf averages 80 km but varies from almost zero (e.g. the coast of Chile and west coast of Sumatra where one tectonic plate is sliding under another) to nearly 1,600 km for the Siberian Shelf in the Arctic. The North Sea between the UK and mainland Europe is all continental shelf. The depth at which the shelf stops and the seabed slopes more steeply is remarkably constant at about 150 metres.

> Connections: This links to material covered in subtopic 3.1.

Outside the continental shelf are international waters which no one country controls. This leaves significant questions: who is allowed to fish there, who controls this and who cleans up when there is a pollution problem?

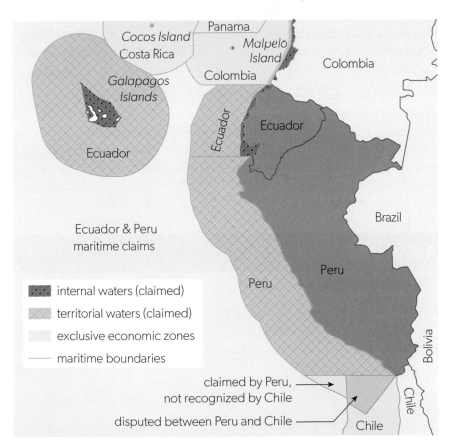

▲ Figure 20 UNCLOS delimitations of coastal waters around Peru

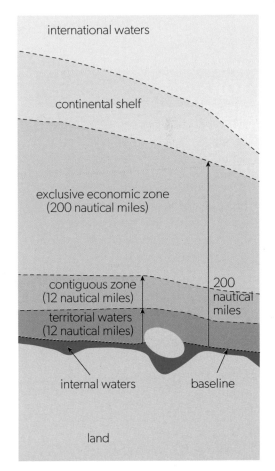

▲ Figure 21 Ocean zones according to UNCLOS

HL

Controversial harvesting

Some species of marine animals have been hunted to such an extent that a moratorium has been placed on their hunting.

-   The overexploitation of whale stocks resulted in a temporary ban on commercial whaling in 1986. The moratorium is still in place though some countries (e.g. Norway) still engage in commercial whaling and Iceland has a reservation that allows it to continue.

-   The USA has a complete ban on commercial hunting of all marine mammals, but indigenous peoples are permitted to hunt small numbers of ringed seals in a traditional way.

-   The Faroe Islands are a semi-autonomous region of Denmark. Hunting long-finned pilot whales and Atlantic white-sided dolphins is a tradition dating back to 800 CE. A school of the animals is driven ashore and beached, then killed. The meat and blubber were eaten through the winters and skins used for shoes. But today, the tradition is opposed by many Faroe Islanders and in campaigns by activists. In 2022, the Faroese government set a limit of 500 dolphins killed per year.

Seal hunting in Canada

In the 1970s, seal culls raised major protests in Europe; the images in the media were of baby seals being clubbed to death. At the time, Europe was the main market for seal products and they banned the import of all seal products by 2009, claiming "moral concerns" over the practice. There were also concerns over declining harp seal numbers, though by 2017 the Canadian Department of Fisheries and Oceans estimated the population to be close to 7.5 million.

▲ Figure 22 Harp seal

The media had sensationalized the commercial sealing without recognizing the fact that subsistence sealing had been in place for centuries. The subsistence farmers eat the meat and occasionally sell the fur to supplement income in winter when fishing is bad. The outright ban in Europe means that this survival line has gone. Anti-sealing campaigns were highly detrimental to the traditional sealers, who even received death threats.

The WWF and Greenpeace pushed for a broader perspective and acknowledgment of the connection between sealing and the Inuit cultures of Newfoundland and Labrador. They petitioned for the introduction of sustainable quotas. Sealers and their families still risk violence for staying connected to their culture.

Whaling in Indonesia

Whaling has been practised since at least 875 CE, mainly for meat, blubber and whale oil. The depletion of some species to near extinction led to a ban on international whaling by the 1980s. Modern-day need for whale meat is the subject of great debate. Countries such as Iceland, Japan and Norway want the ban on certain species lifted but anti-whaling countries and environmental activists oppose this.

Aboriginal peoples recognized by the International Whaling Commission (IWC) are permitted to hunt whales so long as it is part of their indigenous culture. There are catch limits to ensure the conservation of whales and hunters must report the number of whales killed. This allows tracking of whale populations. Aboriginal whaling is under threat by commercial whalers and poachers who are taking whales in large quantities; this makes it hard to maintain population numbers. In some places, whale watching has replaced whaling.

▲ Figure 23 A baleen whale

There are only two remaining whaling communities in Indonesia, in the islands of Lembata and Solor. Religious taboos require that the whole whale is used. Half the whale stays in the village and the rest goes to local market for barter. They primarily hunt sperm whales. Dolphins, manta rays and turtles are also hunted along with some sharks.

Whale hunts are very traditional in style with bamboo spears. The boats are small wooden outriggers 10–12 m in length and about 2 m wide with a sail of woven palm fronds. The whales are killed by a harpooner jumping on top of the whale and driving the harpoon in by hand. This method suits their natural resources and their culture so in 1973 when the FAO tried to modernize the hunt it was completely unsuccessful.

Surveys by the WWF have confirmed that the current level of hunting is sustainable and the whale stock is not in danger.

Check your understanding

In this subtopic, you have covered:

- aquatic ecosystems, their food webs and how humans impact them all

- the increase in demand for marine and freshwater food

- how the fishing industry has changed and unsustainable fishing methods

- maximum sustainable yield (MSY) and how it is useful

- the impacts of overexploitation of fish stocks

- the interconnection between climate change and the oceans, especially acidification

- the benefits of Marine Protected Areas (MPAs)

- the pros and cons of aquaculture

- the factors that impact the productivity of aquatic systems

- the risks of using MSY to establish catch rates

- the importance of accurate assessment and monitoring of fish stocks

- how or if overexploited species will recover

- who governs the oceans

- whether or not some harvesting of organisms should be allowed.

AHL

How are our diets impacted by our values and perspectives?

To what extent are aquatic food systems sustainable?

1. State what provides the energy for aquatic food webs.

2. Outline the different organisms humans consume from aquatic environments.

3. Discuss the reasons for the increase in demand for foods from aquatic environments.

4. Outline the causes and consequences of unsustainable harvesting practices and overexploitation of stocks.

5. Analyse the use of MSY for setting fishing quotas.

6. Explain how climate change and ocean acidification may cause the population collapse of aquatic organisms.

7. Outline how policy can mitigate unsustainable exploitation of aquatic ecosystems.

8. Describe how MPAs can support aquatic food chains.

9. Outline the environmental impacts of aquaculture.

$\rightarrow$

AHL

10. Outline the interconnection between productivity, thermal stratification, nutrient mixing and nutrient loading.

11. Discuss how fish stocks and harvest rates can be monitored accurately.

12. Explain the risks of harvesting at MSY.

13. Analyse how the various stakeholders cooperate to help overexploited species numbers to recover.

14. To what extent is the UN Convention on the Law of the Sea an effective method of controlling access to fishing zones?

❯❯ Taking it further

- Design an experiment to investigate the impact of ocean acidification on a shelled organism.

- Investigate fishing rates in various countries. Is there a change through time? Is there a difference between countries with different levels of development?

- Explore the role of protected aquatic reserves in conserving biodiversity.

- Compare the approaches of regional or national governments in the efficacy of protecting aquatic environments.

- Raise awareness of loss of access to local fishing due to international sale of fishing rights.

- Host a film show highlighting the tensions around consumption of fish.

- Raise awareness of marine certification programmes for fish consumption.

- Write emails to NGOs, highlighting changes that could alleviate environmental challenges around aquaculture.

- For a chilling look at the causes and problems of overfishing, watch the documentary *The End of the Line*.

Guiding questions

- How does pollution affect the sustainability of environmental systems?
- How do different perspectives affect how pollution is managed?

Understandings

1. Water pollution has multiple sources and major impacts on marine and freshwater systems.

2. Plastic debris is accumulating in marine environments. Management is needed to remove plastics from the supply chain and to clear up existing pollution.

3. Water quality is the measurement of chemical, physical and biological characteristics of water. Water quality is variable and is often measured using a water quality index. Monitoring water quality can inform management strategies for reducing water pollution.

4. Biochemical oxygen demand (BOD) is a measure of the amount of dissolved oxygen required by microorganisms to decompose organic material in water.

5. Eutrophication occurs when lakes, estuaries and coastal waters receive inputs of mineral nutrients, especially nitrates and phosphates, often causing excessive growth of phytoplankton.

6. Eutrophication leads to a sequence of impacts and changes to the aquatic system.

7. Eutrophication can substantially impact ecosystem services.

8. Eutrophication can be addressed at three different levels of management.

9. There is a wide range of pollutants that can be found in water.

10. Algal blooms may produce toxins that threaten the health of humans and other animals. Consider one example from freshwater and one from marine water.

11. The frequency of anoxic/hypoxic waters is likely to increase due to the combined effects of global warming, freshwater stratification, sewage disposal and eutrophication.

12. Sewage is treated to allow safe release of effluent by primary, secondary and tertiary water treatment stages.

13. Some species are sensitive to pollutants or are adapted to polluted waters, so these can be used as indicator species.

14. A biotic index can provide an indirect measure of water quality based on the tolerance to pollution, relative abundance and diversity of species in the community.

15. Overall water quality can be assessed by calculating a water quality index (WQI).

16. Drinking water quality guidelines have been set by the World Health Organization (WHO), and local governments can set statutory standards.

17. Action by individuals or groups of citizens can help to reduce water pollution.

Water pollution

Substances causing **water pollution** can be chemical or microbial. Both degrade the water quality enough to make it toxic to humans, other species and the environment. Water is vulnerable to pollution because it is a solvent to many substances. So, many of the substances that water comes into contact with will readily dissolve in it (Table 1).

Water pollution is an issue in both low- and high-income countries. Globally water pollution causes between 1.4 and 1.8 million deaths a year and a further 3.5 million people die from water-related diseases such as cholera, diarrhoea and dysentery. Nearly 790 million people do not have access to clean, safe drinking water.

Key term

Water pollution is the contamination of bodies of water by pollutants, either directly or indirectly.

Type	Pollutant	Example	Effects
Organic	Sewage	Human waste	Eutrophication
	Animal waste	Manure	Smell
	Biological detergents	Washing powders	
	Food processing waste	Fats and grease	
	Pesticides from agriculture	Insecticides, herbicides	Loss of biodiversity
	Chemicals from industry	PCBs, drugs, hormones, tributylin antifouling paint for boats	May be carcinogenic; growth-promoting hormones
	Pathogens	Waterborne and faecal pathogens	Disease
	Invasive species	Cane toads	Decimate native species
Inorganic	Nitrates and phosphates	Fertilizers	Eutrophication; change biodiversity
	Phosphates	Washing detergents	
	Heavy metals, e.g. Hg, Cd, Pb, As	Industry and motor vehicles	Bioaccumulation and biomagnification in food chains; toxic
	Hot water (thermal pollution)	Power stations	Changes physical properties of water; kills fish; changes biodiversity
	Crude oil	Industry	Floats on surface; contaminates sea birds; reduces oxygen levels
	Radioactive materials	Nuclear power stations	Radiation sickness
	Light	Cities, hotels on beaches	Disrupt turtle nesting sites
	Noise	Aircraft, speedboats, tankers	Upset whale navigation; change plant growth; upset bird cycles
Both	Suspended solids	Silt from construction sites	Damage corals and filter feeders
	Solid domestic waste	Household garbage including plastics	Plastics are especially bad, causing suffocation and starvation (Great Pacific Garbage Patch); microplastics; microbeads; ghost fishing nets

▲ Table 1 Examples of pollutants and their effects

Types of water pollution

Both seawater and freshwater can become polluted.

Pollutants can be:

- anthropogenic (created by human activities, including sound) or natural (e.g. volcanic eruptions, algal blooms)

- point source or non-point source

- organic or inorganic.

Sources of freshwater pollution include:

- agricultural run-off

- sewage

- industrial discharge

- solid domestic waste.

Sources of marine pollution include:

- rivers

- pipelines

- the atmosphere

- human activities at sea, both operational and accidental discharges.

Noise pollution

Sound travels faster in water than in air. Oceans are full of natural noises but human-made noise (from ships, oil exploration, seismic surveys, sonar) can be very disruptive. This is sonic pollution or noise pollution. Whales, dolphins and porpoises are particularly disturbed as they use echolocation to navigate, communicate and find mates. Whales can beach themselves when they are disorientated by noise. Other species including many fish and crustaceans are affected as well.

Case study 7

Urban water pollution and River Thames

Urban areas increase the amount and variety of pollutants that can enter water systems. The impermeable surfaces mean that more water moves quickly into rivers and other water bodies. Urban pollutants can harm or kill vegetation and fish, foul drinking water and make recreational areas unusable. These pollutants include:

- sediments from gardens and building sites

- motor vehicle waste such as oil, grease and other chemicals

- chemicals from lawn treatments such as pesticides and fertilizers

- biological waste from failure of septic tanks and pet waste

- road salts from ice clearance.

The River Thames has long been a source of water, fish, trade and transport for London. The Thames has been used for domestic purposes (drinking and cooking) and industrial purposes such as brewing, power generation and waste disposal.

Causes of River Thames pollution

By the 13th century, water supplies were dirty due to the disposal of household refuse and sewage.

By the mid-1800s, the population of London reached 2.5 million. With this increase came more factories and flushing toilets so the volume of domestic and industrial waste flushed into the Thames increased.

Impacts of River Thames pollution

- Decreased water quality was linked to increased spread of disease such as cholera.

- Freshwater and saltwater species, including the fisheries, decreased. This meant the loss of sole, cod, herring, whitebait, sprat, salmon and eel catches.

- The summer of 1858 was unusually hot and the smell from the Thames was so bad that politicians and Queen Victoria complained about the problem.

- In 1878, a passenger steamer sank where the sewers emptied into the Thames and people died from ingesting the polluted water.

- In the 1960s, the smell from the Thames was again significant in warm weather.

Solutions to River Thames pollution

- Newly established private water companies took water directly from the Thames and distributed it throughout London. So, in 1852 the Metropolis Water Act legislated the filtering of water through sand and shell filters.

- In 1859 a scheme by Joseph Bazalgette to pipe sewage away from London was started. Money was made available and 132 km of brick sewers and pumping stations were built. That sewer system still operates in the 21st century.

- "Bovril boats" were introduced to carry sewage sludge out to the Thames Estuary. These operated until 1998.

- Modern technologies are used to change sewage into fertilizer.

- In 1902, the Metropolitan Water Board was established to ensure the quality and cost of water was consistent.

- In 2015, the Thames Tideway Tunnel was started. The aim is to capture sewage from outflow pipes and redirect it to existing treatment facilities.

- In 2023, this tunnel was still due to open in 2025. Due to the COVID-19 pandemic, the cost rose to GBP 4.5 billion. The tunnel is 25 km long and should protect the River Thames from sewage for 100 years.

▲ Figure 1 The River Thames and surrounding urban land use today

Results of River Thames clean-up

Water quality has improved and there are now:

- 125 fish species and 400 invertebrate species

- salmon, grey seals and harbour porpoises

- keystone bird species such as herons, cormorants and Canadian and Egyptian geese

- bats, which had almost disappeared from the area.

Plastic pollution

Plastic is a synthetic organic polymer, made from petroleum. We all use it in various forms, probably every day. It is very useful—for example, for packing, household items, sports equipment, medical equipment and electronics. However, there are problems.

- 300–380 million tonnes of plastic are produced each year (estimates vary).

- 50% of the plastic produced is for single-use items—straws, shopping bags.

- Less than 10% of the plastic is recycled.

- 40 billion plastic bottles are used worldwide every month.

- In 2019, production and incineration of plastic added over 850 million metric tonnes of greenhouse gases to the atmosphere, bringing global climate catastrophe closer.

- Plastic made from oil takes a long time to break down, if it breaks down at all (Figure 2).

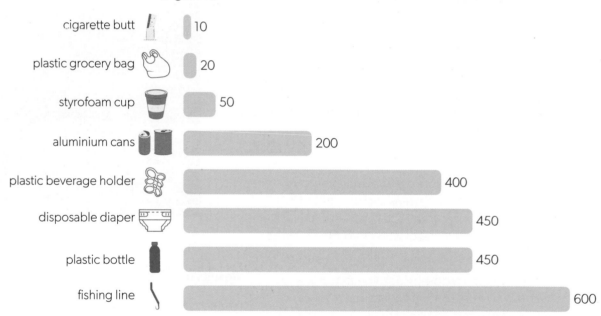

cigarette butt 10
plastic grocery bag 20
styrofoam cup 50
aluminium cans 200
plastic beverage holder 400
disposable diaper 450
plastic bottle 450
fishing line 600

▲ Figure 2 Estimated breakdown rates of types of plastic in the ocean. The exact number of years varies by type of product and marine conditions. Cigarette butts and grocery bags are an upper estimate

Biodegradable waste can be broken down by other living things and is mostly from biological sources, e.g. food waste, paper, human waste, manure, sewage, dead organisms.

Non-biodegradable pollutants are materials that do not decompose. Most plastics are made from crude oil and are composed of carbon and hydrogen. However, the polymers (long-chain molecules) which make plastics so useful and versatile cannot be broken down by decomposers. This is why they persist in the environment. Worse, although these polymers do not break down into molecules that can be reused by living things, they do break down into smaller and smaller pieces called **microplastics**.

Microplastics are also non-biodegradable. So are some biocides (such as DDT) and metals such as lead, arsenic and mercury. Synthetic fibres, particles of glass, silver (aluminium) foil and electronic waste are also non-biodegradable. These are everywhere and, once ingested, they stay in the body.

A small study in 2022 found that most of the humans sampled had microplastics in their blood. Half the samples contained PET plastic, which is used in drinks bottles. A third contained polystyrene, used for packaging food and other products. A quarter contained polyethylene, from which plastic carrier bags are made. More research is needed to see if microplastics cause harm to humans. But the younger you are, the more likely you are to accumulate more plastics because there are more and more in the environment.

Plastic in the oceans

Plastic litter in marine ecosystems comes from many sources, such as discarded fishing gear, bottles, bags, straws, food packaging, toys, cosmetic containers and microbeads. There is a huge amount of it and it is persistent—it breaks down into smaller and smaller pieces and microplastics but it never disappears. These pieces are taken in by marine organisms and enter the food chain.

There are estimated to be 5.25 trillion pieces of macro- and microplastic in the oceans, and around 8 million more are added every day. Plastic makes up approximately 80% of all marine debris; maybe the most iconic image of this environmental disaster is the Great Pacific Garbage Patch (GPGP). However, the plastic does not all remain on the surface. In fact, plastic trash impacts many different marine environments—the shorelines, the surface, the water column and the seafloor.

▲ Figure 3 Plastic trash littering the ocean

Issues of ocean plastics and microplastics

- Larger pieces entangle sea mammals (whales, porpoises etc.) and fish.

- Chemical contaminants of microplastics can be endocrine disruptors which mimic hormones and interfere with body systems, including reproduction, or cause cancers.

- Filter feeders ingest microplastics, which stay in their bodies.

- Larger organisms eat the filter feeders and, through biomagnification, microplastics are accumulated in the food chain.

- Humans eat marine organisms and therefore accumulate microplastics in their bodies.

Connections: This links to material covered in subtopic 2.2.

Ocean gyres are circulating ocean currents. There are five: the North Atlantic Gyre, the South Atlantic Gyre, the North Pacific Gyre, the South Pacific Gyre and the Indian Ocean Gyre.

Ocean garbage gathers in these gyres and the patches can be huge. For example, the GPGP covers 1.6 million km² between California and Hawaii. That is three times the size of France or twice the size of Texas (Figure 4).

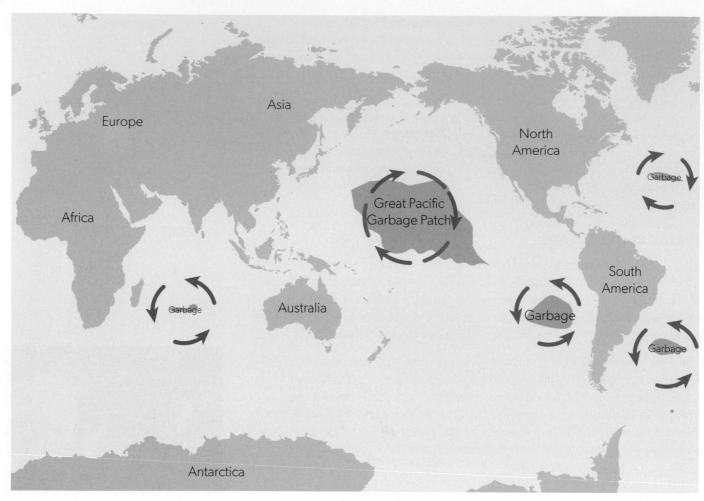

▲ Figure 4 Ocean gyres and garbage patches

Where does the plastic come from?

Some of the plastics in the oceans come from ocean-based activities such as the fishing industry and aquaculture, but the major source is on land. Some sources are obvious such as urban areas, littering, poor waste disposal and management, the construction industry and illegal dumping. A big contributor has been **microbeads**—very small solid plastic particles less than 1 mm in diameter, not degradable and insoluble. They were very common in a wide range of products especially cosmetics and other personal care products. Most countries have now banned or phased out use of microbeads but, as they do not degrade, they are still present in the environment. Next time you buy a body scrub, check out the ingredients list.

The combined effects of UV radiation, wind and currents cause plastics to break down into microplastics (under 5 mm) and nanoplastics (under 100 nm). These smaller plastics are easily ingested but **not** digested by marine animals. Larger pieces of plastic are mistaken for prey by sea birds, whales, fish and turtles, which can take in so much plastic that the body no longer has the capacity to feel hungry and the animals die of starvation (Figures 5 and 6). Plastic will also suffocate, entangle, and lacerate sea-life, reducing the ability to swim and increasing the chances of infection. The plastic also makes a good transport mechanism for invasive species.

▲ Figure 5 A turtle may mistake plastic bags for prey

▲ Figure 6 Plastics in the stomach of a dead albatross chick

Microplastics are found in drinking water. Some of the chemicals used in plastic production are known to be carcinogenic and detrimental to the endocrine system in humans. This causes neurological and immune disorders in humans and wildlife. These chemicals also biomagnify through the aquatic food chain and end up in humans because they are top predators.

Plastic also has an impact on the aesthetic value of the oceans. This is important for the physical and psychological well-being of humans and wildlife and is also important for the economy. In many coastal resorts, time and money must be spent clearing the beaches of plastic and other litter.

Management solutions

Plastic pollution is a problem because many countries lack appropriate facilities or management structures to deal with waste disposal and the quantity of plastics. Proper waste management needs landfill, incineration, recycling or circular economy infrastructure. If these are not in place or are poorly managed, plastic leaks into rivers and oceans. The following actions are designed to address this.

- International legislative frameworks should be strengthened or at least adhered to. In 1996, the Protocol to the London Convention (the London Protocol) was initiated.

- Producer responsibility is needed—putting the responsibility on the producers of plastic products to find alternative, low-cost solutions to reduce plastic usage in their products.

- Collaboration between research institutes, industries and governments is needed. The whole lifecycle of plastics needs to be updated, from design to infrastructure (new ways to dispose of and reduce microplastic waste and synthetic textiles) to household use.

- Additional funding is required for research and innovation to find solutions. Consumers and society need to take action to develop sustainable habits—the following are suggestions.

   - Reduce single-use plastic purchases, especially water in bottles, and use reusable containers instead.

   - Recycle and reuse plastic products.

(ATL) Activity 40

ATL skills used: communication, research

Research the GPGP.

1. How much garbage is estimated to be in the GPGP?

2. What types of plastic are there?

3. From which source is most of this plastic?

4. How deep is the garbage patch?

5. Is the size increasing or decreasing?

6. How persistent are the plastics in the GPGP?

7. Describe briefly how we find information on the GPGP.

8. What methods are being used to clean up the GPGP?

- Stop using products with microbeads.

- Buy second-hand items—without plastic if possible.

- Support plastic bag bans and use your own reusable bags.

- Stop buying mini-packs of snacks; instead, buy larger packs and use small reusable containers.

- Participate in local plastic clean-up projects or organize your own.

Many of these actions put pressure on the producers of plastic products and bring about change.

Stopping more plastic entering the oceans is not the only challenge. It is essential to clean up the existing garbage patch. There are several strategies already in place.

Ocean Cleanup is a Dutch non-profit organization dedicated to cleaning up the oceans. Two large ships tow a large floating net, three metres deep and in the shape of a large U, through the ocean. Plastic flows into a central retention zone and the vessels come together, pick up the retention zone and empty the plastic onto the decks of the ships.

The organization also has vessels that catch trash at the mouth of rivers to stop it entering the sea. The river water flows along a barrier that guides the trash onto a conveyor belt. The conveyor belt dumps it onto solar-powered shuttles that carry it to dumpsters on a barge. The barges then take the trash to the riverside where it is emptied into the waste management system. Ocean Cleanup has removed one million kilograms of trash from rivers in Southeast Asia, the Dominican Republic and Jamaica.

▲ Figure 7 Collected plastic bottles ready for treatment in Hanoi, Vietnam, 2016

In many areas of the world, a simple system of nets is being used to halt the flow of plastic down rivers so that it can be collected and disposed of properly.

Brief example of action on plastic

A letter coordinated by the environmental not-for-profit organization, City to Sea, called for the five biggest plastic polluters (Coca-Cola, PepsiCo, Nestlé, Unilever, and Procter & Gamble) to tackle the issue of single-use plastic pollution. The suggested solution is to stop single-use plastic and switch to affordable, accessible, refillable, reusable packaging.

The five top polluters could change their reputation from "big pollution" to "big solution". To do so, they should:

- **reveal** the real size of their plastic footprint and be accountable for reducing it

- set ambitious targets along with a clear action plan to prioritize these targets and **reduce** the amount of plastic they use

- **reinvent** packaging so it can be refilled and reused easily and cheaply.

This initiative could spread through collaboration with other companies to standardize reusable packaging.

Connections: This links to material covered in subtopic 2.6.

ATL Activity 41

ATL skills used: thinking, research

1. Investigate which products use microbeads and why. This may turn into an individual investigation.

2. How can your community or school help to reduce plastic use?

3. Investigate how else we can remove plastic from the oceans.

4. Investigate the big five further.

   a. Are they being ethical in their use of plastics?

   b. Should they receive a financial penalty for their actions?

   c. Should they be made to clean up their act?

TOK

To what extent should we care about our use of microplastics?

There is some good news, so do not be too pessimistic. Consumer pressures are very effective. Starch-based materials made from potato, cassava, corn or wood fibre are becoming more and more common in food packaging, envelopes and other uses. They are completely biodegradable.

Water quality

Water quality in aquatic systems can be assessed by sampling.

Direct measures look at abiotic factors in the rivers. Many of these abiotic factors are easily measured using scientific probes but some traditional methods are still used (Table 2).

TOK

A wide range of parameters are used to test the quality of water and judgements are made about cause and effect. To what extent can we know cause-and-effect relationships, given that we can only ever observe correlation?

Abiotic factor	Use of technology	Traditional instruments
Dissolved oxygen	Dissolved oxygen meter	Titration test
pH	pH probe	Colorimetry with indicator solutions or papers; litmus papers
Temperature	Temperature probe	Thermometer
Turbidity	Turbidity meter	Secchi disk
Concentration of nitrates	Nitrate sensor	Complete water testing kit—also tests for certain metals
Concentration of phosphates	High range phosphorus colorimeter	Field spectrophotometer or colorimeter
Concentration of specific metals	Magnetic resonance analyser	Requires laboratory tests
Suspended solids	Total suspended solids sensor or monitor	

▲ Table 2 Measurement of abiotic factors in rivers

As technology has developed, methods for assessing water quality have become more accurate, faster, more readily available and cheaper. New methods also avoid the risks of coming into contact with chemicals and substances that are toxic or harmful. Any data collected on water quality can be used to inform management strategies for reducing pollution.

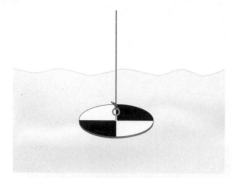

▲ Figure 8 Secchi disc

Adapted from: IFA Design, Plymouth, UK; Clive Goodyer; Q2A Media

The Secchi disc is a very traditional method to assess turbidity. Figure 8 shows how the disc is marked in black and white quarters so it is clearly visible in water. The disc is lowered into the water until it is no longer visible and the depth measured and recorded. This is repeated five times and the average calculated. High depth values indicate low turbidity.

Biochemical oxygen demand (BOD)

BOD is an indirect measure of the amount of organic matter in a sample. It is usually expressed in milligrams of oxygen consumed per litre of sample in five days at 20°C. It indicates the levels of organic pollutants in the water and it can be used to assess the proficiency of water treatment facilities.

Most water in the natural environment includes small amounts of organic compounds which are a food supply for aquatic microorganisms. In oxygenated water, the microorganisms use dissolved oxygen to break down the organic compounds to release energy for growth and reproduction. The available food acts as a limiting factor and the amount of microorganisms present in water is proportional to the food available. The metabolic activity creates oxygen demand that is proportional to the amount of organic compounds available. Sometimes, the speed of microbial metabolism is higher than the available oxygen can sustain and this can result in the death of fish and aquatic insects.

Key term

Biochemical oxygen demand (BOD) is a measure of the amount of dissolved oxygen required by aerobic biological microorganisms to break down the organic material in a given volume of water, at a certain temperature over a certain period of time.

Connections: This links to material covered in subtopic 2.6.

Practical

Skills:

- Tool 1: Experimental techniques
- Inquiry 1: Exploring and designing
- Inquiry 2: Collecting and processing data
- Inquiry 3: Concluding and evaluating

Investigation of water quality and abiotic factors in water samples

1. Compare a polluted and an unpolluted site (e.g. upstream and downstream of a point source of pollution).

2. Investigate water quality by comparing samples from the two sites in terms of:

   - dissolved oxygen
   - pH
   - temperature
   - turbidity
   - concentrations of nitrates
   - concentrations of phosphates
   - concentrations of specific metals
   - concentrations of total suspended solids.

(ATL) Activity 42

ATL skills used: thinking, communication

Four factories discharge effluent containing organic matter into rivers. Table 3 shows the volume of discharge into the river and the resulting biological oxygen demand.

Factory	Volume of effluent/ 1,000 l day^{-1}	BOD/g l^{-1}
A	14.0	27
B	1.0	53
C	3.0	124
D	0.8	33

▲ Table 3

1. Explain whether these pollution data are for point sources or non-point sources.

2. Which pollution source, point source or non-point source, is easier to regulate?

3. Explain which factory is adding most to the BOD of the river into which it discharges.

4. If factory C discharges water at a temperature of 50°C, describe three possible effects on the organisms in the river.

Eutrophication

Eutrophication is when excess nutrients are added to an aquatic ecosystem. If phosphate or nitrate concentrations have previously been limited, then algal blooms may develop with the extra nutrients. Eutrophication is a natural process but anthropogenic eutrophication is more common. It is caused by the release of detergents, sewage or agricultural fertilizers into water bodies.

When it is severe, eutrophication results in **dead zones** in oceans or freshwater where there is not enough oxygen to support life. In less severe cases, **biodegradation of organic material** uses up oxygen which can lead to anoxic (low oxygen) conditions and then anaerobic decomposition. This can release methane, hydrogen sulfide and ammonia, which are all toxic gases.

Sources of these nutrients may be point sources:	Non-point sources of pollution are harder to identify:
<ul><li>wastewater from cities and industry or animal feedlots</li><li>overflows from storm drains or sewers</li><li>run-off from construction sites.</li></ul>	<ul><li>run-off from crops and grassland</li><li>urban run-off</li><li>leaching from septic tanks</li><li>leaching from landfill sites</li><li>leaching from old mines.</li></ul>

Key term

Eutrophication can occur when lakes, estuaries and coastal waters receive inputs of nutrients (nitrates and phosphates) which result in an excess growth of plants and phytoplankton.

▲ Figure 9 Dead fish resulting from toxins or oxygen depletion in Lake Binder, Iowa

435

TOK

To what extent is it ethical for water companies to make water management decisions based on economic cost rather than ecosystem costs?

The process of eutrophication

1. Fertilizers wash into a river or lake.

2. High levels of phosphate in the water allow algae to grow faster (phosphate is often limiting).

3. Algal blooms (mats of algae) form, blocking out light to plants beneath them, which die.

4. More algae mean more food for the zooplankton and small animals that feed on them. These are food for fish which multiply as there is more food. Therefore, there are then fewer zooplankton to eat the algae.

5. Algae die and are decomposed by aerobic bacteria.

6. These bacteria use up oxygen in the water, so soon everything dies as food chains collapse.

7. Oxygen levels fall lower.

8. Dead organic material forms sediments on the lake or riverbed and turbidity increases.

9. Eventually, all life is gone and the sediment settles to leave a clear blue lake.

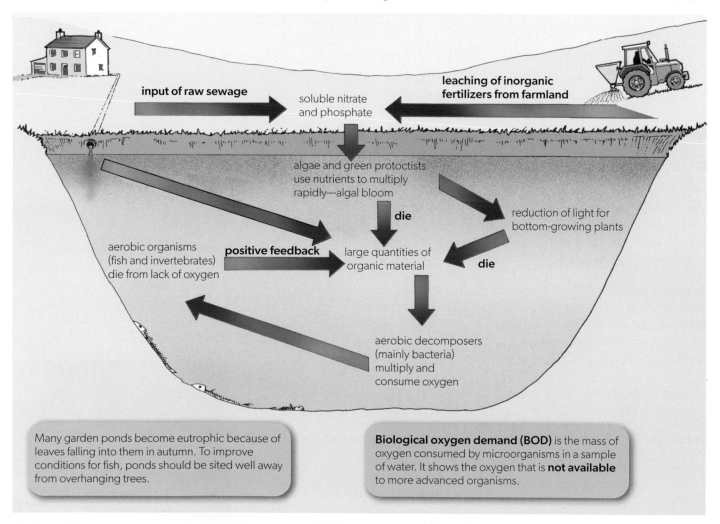

input of raw sewage

soluble nitrate and phosphate

leaching of inorganic fertilizers from farmland

algae and green protoctists use nutrients to multiply rapidly—algal bloom

die

reduction of light for bottom-growing plants

aerobic organisms (fish and invertebrates) die from lack of oxygen

positive feedback

large quantities of organic material

die

aerobic decomposers (mainly bacteria) multiply and consume oxygen

Many garden ponds become eutrophic because of leaves falling into them in autumn. To improve conditions for fish, ponds should be sited well away from overhanging trees.

Biological oxygen demand (BOD) is the mass of oxygen consumed by microorganisms in a sample of water. It shows the oxygen that is **not available** to more advanced organisms.

▲ Figure 10 Process of eutrophication in a nutrient-enriched pond or river

The excess nutrients are nitrates and phosphates. They come from detergents, fertilizers, drainage from intensive livestock rearing units, sewage and increased erosion of topsoil into the water.

Impacts of eutrophication

Anthropogenic eutrophication leads to unsightly rivers, ponds and lakes covered by green algal scum and duckweed. The water also gives off foul-smelling gases such as hydrogen sulfide (rotten egg smell). Other changes include:

- hypoxic or oxygen-deficient (anaerobic) water

- loss of biodiversity and shortened food chains

- death of higher plants (flowering plants, reeds)

- death of aerobic organisms—invertebrates, fish and amphibians

- increased turbidity (cloudiness) of water.

Impact on ecosystem services

When the over-abundant algae and other plants eventually decompose, they produce a large amount of carbon dioxide which lowers the pH of water causing acidification—impacts of ocean acidication were discussed in subtopic 4.3. This can reduce fisheries, both commercial and recreational.

The general degradation of water quality due to foul-smelling, noxious algal blooms causes:

- tainted drinking water supplies

- reduced recreational opportunities

- unattractive visual appearance

- decrease in success rates of predatory fish—those that rely on sight for hunting are impacted by the higher turbidity levels and changes in pH weaken chemosensory perception.

Noxious algal blooms which contain particular cyanobacteria (e.g. *Anabaena*) can cause gastrointestinal illness. This can present a health risk to both humans and cattle. These cyanobacteria also reduce the efficiency of the aquatic food chain as they are poor quality food for zooplankton.

Management strategies for eutrophic waters

The estimated cost of treatment of eutrophic water in the UK is USD 105–160 million per year and in the USA it is USD 2 billion per year.

Table 4 shows the three levels of strategy that can be used to manage pollution. Strategies that reduce production of pollutants have more widespread benefits and are less expensive to put into action than those that require removal of pollutants from the environment and restoration of the ecosystem.

ATL Activity 43

ATL skills used: communication, research

- Using Figure 10, draw an annotated systems model to illustrate the sequence of impacts and changes that take place during eutrophication.

- Identify the processes and state whether they are transfers or transformations.

- Indicate on your diagram an instance of positive feedback.

▲ Figure 11 Bright green algal bloom in water

Strategy for reducing pollution	Example of action
Altering or reducing the human activity that produces the pollutant	• Ban or limit detergents with phosphates (used to improve the performance of the detergent in hard water areas). • Use ecodetergents with no phosphates or new technology in washing machines. • Plant buffer zones between fields and water courses to absorb excess nutrients. • Stop leaching of slurry (animal waste) or sewage from their sources. • Educate farmers about more effective timing for fertilizer application.
Reducing the release of pollutants into the environment	• Treat wastewater before release to remove phosphates and nitrates. • Divert or treat sewage waste effectively. • Minimize fertilizer dosage on agricultural lands or use organic matter instead.
Removing pollutants from the environment and restoring ecosystems	• Pump air through lakes. • Dredge sediments with high nutrient levels from river and lake beds. • Remove excess weeds physically or by using herbicides and algicides. • Restock ponds or water bodies with appropriate organisms.

▲ Table 4 Three levels of strategy to manage pollutants

ATL Activity 44

ATL skills used: thinking, communication

1. Copy Table 4 and add a third column, headed Evaluation. Complete this column.

2. Consider a river where an outfall pipe discharges sewage—point source pollution.

   Copy the axes in Figure 12 and draw curves to show changes in:

   a. detritus (sewage)   e. oxygen concentration
   b. turbidity (suspended solids)   f. invertebrate and fish biodiversity
   c. bacterial growth   g. clean water species.
   d. BOD

   Make a key to your curves.

3. Now consider a river with high nutrient levels due to run-off from over-fertilized fields.

   Make a new copy of the axes in Figure 12, not including the "outfall" line. Draw curves to show changes in:

   a. nutrients (nitrate and phosphate)
   b. algal growth (bloom)
   c. detritus increase (due to higher plant death)
   d. bacterial growth
   e. BOD
   f. oxygen concentration
   g. invertebrate biodiversity
   h. clean water species.

   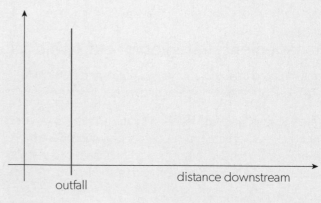

   ▲ Figure 12

   Make a key to your curves.

Harmful algal blooms

There are many different substances that pollute water. Some pollutants are more hazardous than others. Pollutants include organic matter (e.g. sewage), dissolved substances (e.g. tributyltin, an endocrine disrupting chemical), persistent chemicals that become biomagnified (e.g. PCBs), plastics and heat energy.

Excess of nutrients such as nitrates and phosphates can result in harmful algal blooms (HABs) and red tides; both of these have significant health impacts on humans and other animals.

Cyanobacteria in algal blooms

Cyanobacteria or blue-green algae occur in freshwater. They are varied in nature, occurring as single cells, colonies or thread-like strings. They can move through the water column and produce toxins called cyanotoxins. Coming into contact with water contaminated with these toxins is not difficult. This may be by:

- skin contact, when swimming or enjoying other water sports in contaminated water

- drinking the contaminated water

- breathing in droplets of the water

- eating fish or shellfish that live in the contaminated water

- eating blue-green algae supplements that come from the contaminated water.

The health implications for humans include gastrointestinal issues, liver damage, neurological damage, skin or eye irritation.

In Salto Grande dam, Argentina (2007), a young man was accidently immersed in an algal bloom of freshwater cyanobacteria. He experienced the symptoms shown in Figure 13.

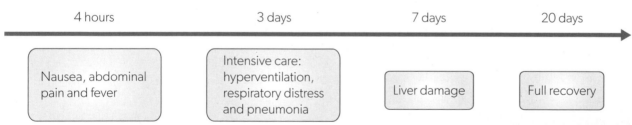

| 4 hours | 3 days | 7 days | 20 days |

Nausea, abdominal pain and fever

Intensive care: hyperventilation, respiratory distress and pneumonia

Liver damage

Full recovery

▲ Figure 13 Progression of symptoms after exposure to a bloom of freshwater cyanobacteria

Red tides

In coastal waters, algal blooms sometimes contain large numbers of dinoflagellates (e.g. *Gonyaulax*) which are red in colour and multiply rapidly. The large number of dinoflagellates makes the sea appear red. These "red tides" can be dangerous as the algae produce toxins that have similar effects on humans as cyanobacteria blooms.

The neurotoxins in red tides can cause the death of thousands of fish, which impacts marine food webs and causes closure of fisheries for shellfish and other seafoods. The dead fish wash up on the shores and the toxins work their way through the food chain by biomagnification. The foam from the red tides causes sea birds to lose the waterproof covering of their feathers which results in fatalities. Red tides have also been linked to the deaths of large numbers of whales and turtles.

▲ Figure 14 A red tide in the Gulf of Mexico

(ATL) Activity 45

ATL skills used: thinking, research

Find a website that shows satellite images of red tides and their development over time, such as NASA, NOAA, ESA (European Space Agency) or Science Direct.

How do sites like these inform scientists?

(ATL) Activity 46

ATL skills used: communication, social, self-management

In small groups, produce a piece of visual art (video, cartoon, podcast) about red tides or another water pollution problem in your area or region.

Key term

Indicator species are plants and animals that show something about the environment by their presence, absence, abundance or scarcity.

Anoxic and hypoxic waters

Algal blooms and red tides can cause anoxic (no oxygen) and hypoxic (less than 2–3 mg of oxygen per litre) water. Both compromise the survival of organisms and ecosystems. Such conditions are likely to become more common due to the following factors.

*   Eutrophication (see earlier section).

*   Freshwater stratification—caused by changes in saline and temperature gradients due to the excess of nutrients that cause eutrophication. This stratification disrupts vertical mixing, so oxygen-deficient waters are deprived of oxygen absorbed from the atmosphere.

*   Sewage disposal—this is one of the major causes of eutrophication but, with growing populations, the infrastructure to deal with human waste is prone to failure. This means more nutrients enter the freshwater systems.

*   Global warming—this causes loss of oxygen in water because warmer water holds less oxygen than cold.

Solutions to red tides

In the USA, red tides are estimated to cause loss of USD 82 million in income from tourism, fisheries and seafood revenue. The Gulf of Mexico has the largest dead zone in the USA caused by excess nitrates and phosphates from agriculture in the Mississippi River basin. There are also other dead zones, so solutions are vital.

Nutrient trading is a voluntary market-based reduction of nutrient use. It is a type of water quality trading scheme which allows those that can reduce nutrients at low cost to sell credits to those facing higher-cost nutrient reduction options. It could allow sources of pollution to meet their pollution targets in a cost-effective manner and create new revenue opportunities for farmers, entrepreneurs, and others who implement low-cost pollution reduction practices.

Red tide forecasts let people know how safe it is to bathe in certain areas around Florida. They use satellite images and data collected by citizen scientists using the HABscope system. (A HABscope is a miniature microscope that attaches to a smartphone and takes a video of water samples. These videos are then analysed to estimate the size, possibility of and potential severity of a bloom.)

In the Gulf of Maine, sediment samples are collected and the number of cysts that will form dinoflagellates is counted. This information goes into a model that predicts the number of cysts that will survive depending on anticipated salinity and ocean currents. Scientists can then forecast the likelihood of red tides.

Biotic indices and indicator species

Indicator species are the most sensitive to change so they act as early warning signs that something may have changed in an ecosystem. For example, canaries were once taken down coal mines in Britain because they are more sensitive than humans to poisonous gases (e.g. carbon monoxide, methane). If gases were present in the mine, the bird would die and so warn the miners in time for them to escape.

Invertebrates are used to estimate levels of pollution, as they are sensitive to decreases in oxygen concentration in water, caused by the action of aerobic bacteria as they decompose organic matter. The presence of certain indicator species that can tolerate various levels of oxygen is used to calculate a **biotic index**, a semi-quantitative estimate of pollution levels. Figure 15 shows the different indicator species in freshwater streams; these will vary by location.

A biotic index does not measure pollutants directly; instead, it measures the effect of pollutants on biodiversity. A scale (1–10) gives a measure of the quality of an ecosystem based on the presence and abundance of the species living in it. Biotic indices based on indicator species are usually used at the same time as BOD.

Key term

A **biotic index** indirectly measures pollution by assaying the impact on species within the community according to their tolerance, diversity and relative abundance.

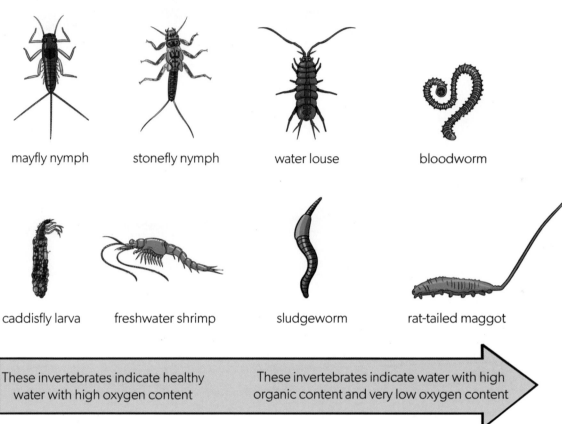

mayfly nymph   stonefly nymph   water louse   bloodworm

caddisfly larva   freshwater shrimp   sludgeworm   rat-tailed maggot

These invertebrates indicate healthy water with high oxygen content

These invertebrates indicate water with high organic content and very low oxygen content

▲ Figure 15 Indicator species of invertebrate found in temperate freshwater
Adapted from: Six Marbles and OUP

BOD gives a measure of pollution at the instant a water sample is collected whereas indicator species give a summary of recent history. A diversity index can also compare two bodies of water (see subtopic 3.1 for information about Simpson's diversity index).

The **Trent biotic index** is based on the presence or absence of indicator species as levels of organic pollution in a river increase. In highly polluted rivers with high levels of organic matter, levels of oxygen and light fall. Species diversity in these conditions is low. However, the number of individuals of species such as bloodworm and rat-tailed maggot may be high as they are tolerant of such conditions. Diversity decreases as pollution increases. Figure 16 shows some typical invertebrates that might be found above and at various points below a point-source of pollution such as a sewage outfall pipe.

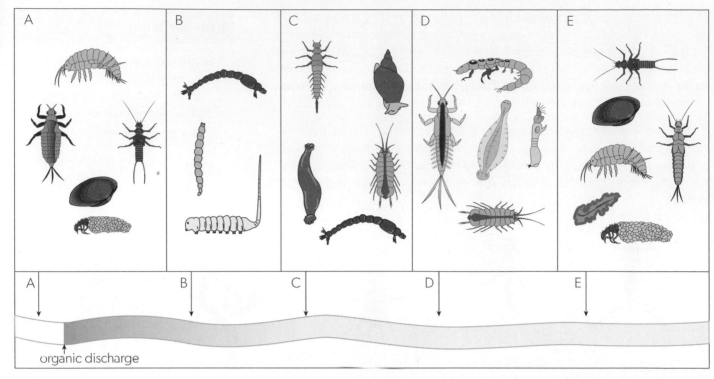

▲ Figure 16 Examples of invertebrates found in freshwater. Species that are more sensitive to pollution are not found in boxes B, C and D

Practical

Skills:

- Tool 1: Experimental techniques
- Inquiry 1: Exploring and designing
- Inquiry 2: Collecting and processing data
- Inquiry 3: Concluding and evaluating

Investigation of BOD

- Design an experiment to investigate the BOD of three water samples.
- Record and evaluate the data.

Investigation of biodiversity using a biotic index

- Design an experiment to investigate the biodiversity of a stream using a biotic index.
- Evaluate the uses of biotic indices in measuring aquatic pollution.

Sewage treatment

Sewage is wastewater plus human excrement. As population size increases, the need to treat sewage increases.

The major aim of wastewater treatment is to remove as much of the suspended solids as possible before the remaining water, called effluent, is discharged back into the environment.

Treatment also aims to eliminate pollutants and toxins, prevent the spread of diseases, allow water to be reused and allow clean water to be recycled into natural water resources.

There are three stages of treatment: primary, secondary and tertiary. But not all three are used for all sewage.

Primary sewage treatment

- Sewage is held in a settling tank. Heavy solids sink, lighter solids float.

- Scrapers push the solids (sludge) to the base of the tank.

- A hopper moves the sludge to different treatment areas.

- Solids are held back as the fluid portion moves on to secondary treatment.

Secondary sewage treatment

This stage removes 85% of the organic matter by using the bacteria in the sewage. Aerobic biological processes biodegrade contaminants in the waste. This makes it safer to release into local water systems.

- Biofiltration removes any additional sediment using sand filters and trickling filters. Sand filters the liquid. A trickling filter is a bed of stones up to six feet deep through which the fluid passes.

- Activated sludge treatment pumps sewage from the settling tank into an aeration tank and mixes it with air and sludge loaded with bacteria. The bacteria then break down the organic matter. This takes 30 hours, and it increases the oxygen saturation of the water.

Tertiary wastewater treatment

This stage raises the water quality so it is safe for domestic or industrial uses. Pathogens are removed at this stage to make the water safe to drink.

- Alum may be used to remove phosphorus particles—it is also good for removing any solids left after primary and secondary treatment.

- Chlorine is added to kill bacteria, viruses and some parasites (e.g. *Giardia*).

- Dechlorination is used to remove the chlorine as it is harmful to aquatic life.

ATL Activity 47

ATL skills used: thinking, communication, social

1. Draw an annotated flow diagram to show the processing of sewage.

2. Evaluate the process of sewage treatment.

3. Discuss the issues of providing sewage treatment to:

   a. a village close to you

   b. the capital city of the country you live in

   c. all the inhabitants of the country in which you live.

4. Discuss whether sewage is treated to the same standard throughout your country regardless of status of the inhabitants. Is it equitable?

Water quality index

A water quality index (WQI) is a single weighted average, consisting of the combined results of several individual water quality test parameters. A WQI represents the degree of contamination in a given water sample at a given time. It turns a complex set of water quality data into an easily understood number. It is like the UV index or the air quality index, giving the public a general impression of any potential problems. WQI values are controversial among scientists because a single number cannot possibly tell the whole story of the quality of the water. If the same index is used in different rivers or over time, the quality can be compared. But not if different indices are used.

There is no single WQI and they all use different parameters to contribute to the calculation of an index number. Possible parameters include:

- turbidity measured in Nephelometric Turbidity Units (NTUs) or metres depth with a Secchi disc
- dissolved oxygen (g l^{-1})
- BOD (g l^{-1})
- nutrients present, usually nitrogen (g l^{-1}) and phosphate (g l^{-1})
- bacteria present, e.g. total coliforms (# per ml) and faecal coliforms (# per ml^{-1})
- temperature (°C)
- pH
- electrical conductivity (S/m)
- oxidation–reduction potential (mV)—this shows how well the water can break down waste products.

Vernier water quality index

In the 1970s, the National Sanitation Foundation and over 100 water quality experts devised a standard WQI. This assesses nine parameters: temperature, pH, turbidity, total solids, dissolved oxygen, biochemical oxygen demand, phosphates, nitrates, and faecal coliforms. Each of these has a weighting factor based on how important it is to water quality.

To assess the Vernier WQI, numerical data is collected for each of the nine parameters. This gives a Q-value which is then multiplied by the weighting factor. These values are totalled, and the health of the water assessed on a scale of excellent, good, medium or average, fair and poor.

Drinking water quality

There are no accepted international standards for drinking water. Consider that fact for a moment—water is essential to the survival of every human being and water often contains harmful contaminants. And yet the quality standards for drinking water are variable and the permitted concentrations of certain constituents can vary by up to 10 times between standards.

The standards that exist in high-income countries are guidelines or targets but not requirements; they rarely have any legal basis and are not subject to enforcement, with the following exceptions.

- European Drinking Water Directive: member states must establish local legislation in each country; routine inspections are mandatory; penalties will be applied by the European Commission if this is ignored.

- Safe Drinking Water Act (United States) passed by Congress in 1974 and amended in 1986 and 1996: the Environmental Protection Agency (EPA) sets standards and monitors water supplies; 90 contaminants have maximum allowable levels.

Countries with guidelines include Canada, Australia, China and New Zealand. The WHO also publishes guidelines for countries but without any legislative framework to set standards.

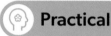

Practical

Skills:

- Tool 1: Experimental techniques
- Inquiry 1: Exploring and designing

Investigation of WQI for selected water sources

- Design an experiment to assess the WQI of three different water samples.

- You can devise your own WQI but make sure you investigate what parameters you can measure, and which ones are most important for water quality.

Key facts according to the WHO

- Globally, at least two billion people use a drinking water source contaminated with faeces.

- Some 829,000 people are estimated to die each year from diarrhoea as a result of unsafe drinking water, poor sanitation and poor hand hygiene.

- In 2017, over 220 million people required preventative treatment for schistosomiasis, caused by parasitic worms in water.

Case study 8

Water bottling plant: Angola

We buy bottled water with the assumption that it is safe to drink. After all, it has been processed in order to be sold. Is that true though?

A brief internet search will shows that there are water bottling plants in many different countries. But without international standards, can we be sure that there are no contaminants in the bottled water?

Table 5 gives some basic information about a water bottling plant in Angola and the water access there.

Water bottling plant	Water in Angola
Pre-engineered water bottling plant	59% of the population lack basic water services
A multi-purpose steel building with a production hall, storage unit, laboratory and office space	28 million people use contaminated water sources
Processes and packages 50,000 bottles of mineral water each year	Water must be carried to homes, mostly by women and girls
Uses 25 million litres of water a year	In 2019, World Vision piped water, drilled boreholes and rehabilitated water points for 16 communities

▲ Table 5

Questions

Consider the information in Table 5.

1. Would you drink the water from this bottling plant?

2. What regulations and standards are in place to ensure drinking water quality in Angola?

3. Investigate bottled water supply in your area or country. What regulations control the quality of the water?

Take action

We take water for granted because when we turn on the tap (faucet) we get water. However, as the global population continues to increase so does water consumption. Therefore, every person needs to take action to prevent water pollution and conserve the water we have so it remains abundant and safe for us and future generations. There are many ways to help.

Reduce consumption and waste disposal

Some of these suggestions are obvious, others less so.

- Make sure all litter is in rubbish bins so less goes into waterways.

- Mulch or compost grass and leaves—do not let it blow into drains.

Connections: This links to material covered in subtopic 1.3.

AHL

- Avoid garden chemicals if you can. If you do use them, make sure they are only used on the grass or flower beds.

- Wash cars and other outdoor objects on grass or gravel so the water soaks into the ground.

- Dispose of toxic chemicals properly, not down the drain or flushed into the toilet.

- Use phosphate-free detergents and biodegradable cleaners.

- Eat more organic food because that reduces pollution at the source (the farms).

- Reduce meat consumption—pastoral farming is extremely "water heavy".

- Get involved in local clean-up efforts on beaches and riversides.

- Reduce your use of motor vehicles: walk, ride a bicycle or use public transport.

- Save electricity by turning off lights as you leave the room.

- Use water-efficient appliances (taps, showers, washing machines).

- Do not flush medicines down the toilet.

- Do not pour oil into the drains—that includes motor oil, cooking oil, grease.

- Do not use single-use plastics.

Peaceful protest

There are different ways to protest environmental issues. Research the "Great Stink" when Queen Victoria and many others protested about the smell from the Thames.

You can protest about certain products by refusing to buy them. If everyone stopped buying items that contained single-use plastics, companies would find an alternative. In Australia, ocean activists united to protest the use of plastics and in response, seven out of the eight Australian states and territories have banned most single-use plastics.

Similar protests could be made about phosphate in detergents. Another way to make a difference is to support environmental charities and citizen science projects. Many charities work to protect river habitats, clean up pollution, plant trees or collect data for citizen science investigations. You can donate money or time—or both—to help the cause.

Check your understanding

In this subtopic, you have covered:

- water pollution: the multiple sources, impacts and types

- major issues with plastic pollution

- managements strategies for dealing with water pollution

- water quality, measurement, assessment and management

- the importance of dissolved oxygen in aquatic systems and what affects it

- eutrophication—causes, impacts and management

- the wide range of pollutants in water

- harmful algal blooms (HABs) and the threats they present to human health and other organisms

- global warming, freshwater stratification, sewage disposal and eutrophication combine to increase the problems of HABs

- how sewage is treated to make it safe to enter the hydrological cycle

- the use of indicator species and biotic indices to assess water quality

- the use of water quality indices (WQIs) to assess water safety

- the WHO and local governments' guidelines about drinking water quality

- what action can be taken to reduce water pollution.

How does pollution affect the sustainability of environmental systems?

How do different perspectives affect how pollution is managed?

1. Outline the multiple sources, impacts and management strategies associated with water pollution.

2. Compare and contrast the various management strategies needed to deal with the existing plastic pollution.

3. State the various methods used to assess water quality.

4. Outline BOD as a measure of water quality.

5. Describe the causes and impacts of eutrophication when it occurs in lakes, estuaries and coastal waters.

6. Construct a model to show the sequence of impacts and changes that takes place during eutrophication.

7. Analyse the impact of eutrophication on ecosystem services.

8. Compare and contrast the control of eutrophication using the three different levels of management.

9. State the wide range of pollutants that can be found in water.

10. Describe the impact of HABs/cyanobacteria in algal blooms on the health of humans and other animals.

11. Discuss the impact of HABs on aquatic food webs.

12. Explain how global warming, freshwater stratification, sewage disposal and eutrophication increase the frequency of HABs.

13. Discuss the treatment of sewage in the three water treatment stages.

14. Outline the value of indicator species and biotic indices in establishing water quality.

15. Describe the use of a water quality index.

16. Analyse the role of the WHO and local governments' statutory standards in assessing drinking water quality.

17. Evaluate the extent to which the action of individuals or groups of citizens is successful in reducing water pollution.

 Taking it further

- Assess the quality of water around your school.

- Use secondary data to investigate the effects of pollution on an aquatic system in your local area.

- Produce an information film or booklet about red tides or another water pollution issue in your local area.

- Visit a water treatment plant.

- Engage in plastic pollution clean-ups—in parks, roadsides, your school grounds.

- Organize a beach clean-up if you live near the coast.

Exam-style questions

1. Outline **one** method for measuring the impact of a build-up of dead organic matter in an aquatic ecosystem. [4]

2. Identify **four** strategies that can be used in the sustainable management of wild fisheries. [4]

3. Evaluate the sustainability of **two** water management strategies to improve access to freshwater resources in a society. [7]

4. Look at **Figures 1 and 2**. The Nadezhda plant may be a possible source of the water discolouration in the Daldykan River.

Daldykan River
Nadezhda plant
Leaking waste water
Slurry pipe
Tailings pond

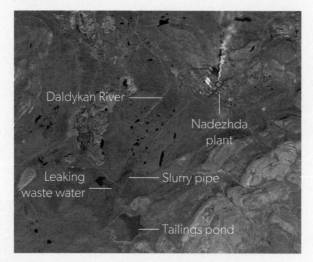

▲ Figure 1 The Nadezhda smelting plant in Norilsk opened in 1979
Adapted from: NASA Earth Observatory image by Jesse Allen, using Landsat data from the U.S. Geological Survey

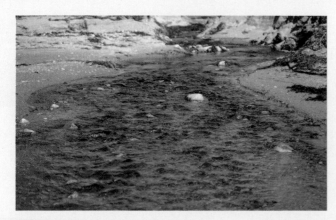

▲ Figure 2 Daldykan River in Norilsk

a. Describe a practical strategy using a biotic index to provide evidence that the Daldykan River is being damaged by effluent from the Nadezhda metal processing plant. [3]

b. i. When measuring levels of pollution, state **one** advantage of using a biotic index compared with measuring the pollutants directly. [1]

ii. When measuring levels of pollution, state **one** disadvantage of using a biotic index compared with measuring the pollutants directly. [1]

1995
Equator
2025
Equator
Key
water scarcity measured as water withdrawal/total available × 100
less than 10%
20% – 40%
10% – 20%
Over 40%

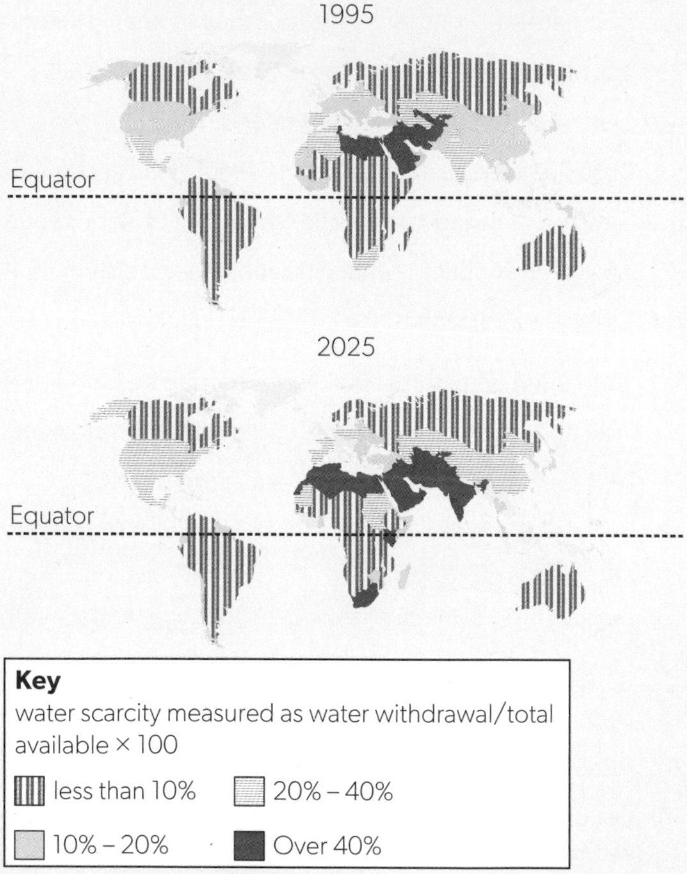

▲ Figure 3 Projected global water scarcity 1995–2025
Data from: GRID-Arendal (CC BY-NC-SA 3.0)

5. a. State the general pattern of change in global water scarcity predicted from 1995 to 2025, as shown in **Figure 3**. [1]

b. i. Identify **two** ways in which climate change may influence the predicted change shown in **Figure 3**. [2]

ii. Identify **two** possible human influences, not related to climate change, that may cause the changes in water scarcity predicted for 2025. [2]

c. Outline **two** reasons why some countries are unlikely to experience water scarcity. [2]

6. **Figure 4** shows global fish capture and production, from 1991 and projected to 2025.

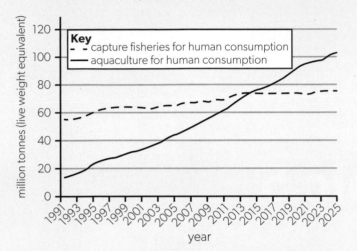

▲ Figure 4 Global capture fisheries and aquaculture production from 1991 and projected to 2025

Data from: Food and Agriculture Organization of the United Nations.

a. i. Using **Figure 4**, identify **one** reason for the trend shown in the curve for aquaculture. [1]

ii. Using **Figure 4**, identify **one** reason for the trend shown in the curve for capture fisheries. [1]

b. Outline **two** negative environmental impacts of aquaculture. [2]

c. Describe **two** strategies for the management of sustainable capture fisheries. [2]

7. Look at **Graphs A** to **C** in **Figure 5**.

a. Define biochemical oxygen demand (BOD). [1]

b. Outline how turbidity changes after the raw sewage discharge point in **Graph B**. [2]

c. Suggest how the population growth curve for algae in **Graph C** would appear if the pollutant had been nitrates and phosphates from fertilizer run-off. [3]

d. Outline why point source pollution is often easier to manage than non-point source pollution. [2]

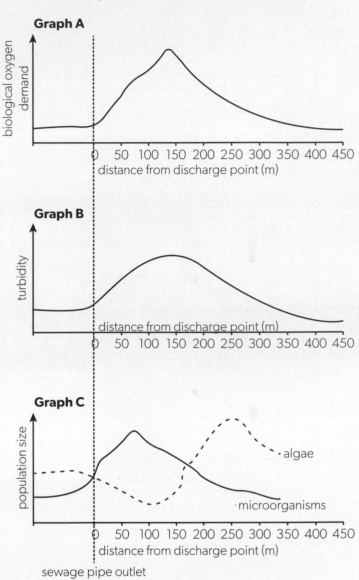

▲ Figure 5 The effects of organic pollution (raw sewage discharged from a pipe) on a stream ecosystem

Data from: Dr. Mel Zimmerman, Professor of Biology and Director of Clean Water Institute at Lycoming College. Adapted from Bartsch and Ingram (1975).

Topic 5

Land

" *Civilization as it is known today could not have evolved, nor can it survive, without an adequate food supply.* "

Norman Borlaug (1914–2009), Green Revolution agronomist

5.1 Soil

AHL

Guiding questions

- How do soils play a role in sustaining natural systems?
- How are human activities affecting the stability of soil systems?

Understandings

1. Soil is a dynamic system within the larger ecosystem that has its own inputs, outputs, storages and flows.

2. Soil is made up of inorganic and organic components, water and air.

3. Soils develop a stable, layered structure known as a profile, made up of several horizons, produced by interactions within the system over long periods of time.

4. Soil system inputs include those from dead organic matter and inorganic minerals.

5. Soil system outputs include losses of dead organic matter due to decomposition, losses of mineral components and loss of energy due to heat loss.

6. Transfers occur across soil horizons, into and out of soils.

7. Transformations within soils can change the components or the whole soil system.

8. Systems flow diagrams show flows into, out of and within the soil ecosystem.

9. Soils provide the foundation of terrestrial ecosystems as a medium for plant growth (a seed bank, a store of water and almost all essential plant nutrients). Carbon is an exception; it is obtained by plants from the atmosphere.

10. Soils contribute to biodiversity by providing a habitat and a niche for many species.

11. Soils have an important role in the recycling of elements as a part of biogeochemical cycles.

12. Soil texture defines the physical make-up of the mineral soil. It depends on the relative proportions of sand, silt, clay and humus.

13. Soil texture affects primary productivity through the differing influences of sand, silt, clay and dead organic matter, including humus.

14. Soils can act as carbon sinks, stores or sources, depending on the relative rates of input of dead organic matter and decomposition.

15. Soils are classified and mapped by appearance of the whole soil profile.

16. Horizons are horizontal strata that are distinctive to the soil type. The key horizons are organic layer, mixed layer, mineral soil and parent rock (O, A, B and C horizons).

17. The A horizon is the layer of soil just beneath the uppermost organic humus layer, where present. It is rich in organic matter and is also known as the mixed layer or topsoil. This is the most valuable for plant growth but, along with the O horizon, is also the most vulnerable to erosion and degradation, with implications for sustainable management of soil.

18. Factors that influence soil formation include climate, organisms, geomorphology (landscape), geology (parent material) and time.

19. Differences between soils rich in sand, silt or clay include particle size and chemical properties.

20. Soil properties can be determined from analysing the sand, silt and clay percentages, percentage organic matter, percentage water, infiltration, bulk density, colour and pH.

21. Carbon is released from soils as methane or carbon dioxide.

What is soil?

Soil is:

- a dynamic system with inputs, outputs, storages and flows

- a complex mixture of interacting components forming its own ecosystem with distinct soil organisms

- made up of inorganic components or mineral matter (rock fragments, sand, silt and clay) from weathering of parental rock, and organic components including living organisms and material from the decay of organisms.

Soil facts

- Soils store more carbon than the atmosphere and all of the world's plants and forests combined. 10% of the world's carbon dioxide emissions are stored in soil.

- Soil is a habitat for many organisms—there are more soil microorganisms in a tablespoon of healthy soil than there are people on Earth today. In some ecosystems, the below-ground biomass is greater than the above-ground biomass.

- Only about 1% of the microorganisms found in soil have been identified so far—our soils are one of the largest reservoirs of microbial diversity on Earth; this includes many single-celled organisms, such as bacteria and archaea, as well as certain fungi.

- 95% of food production relies on the soil: most plants grow in soil and we either eat these plants or eat animals that have eaten plants.

- Healthy topsoil is vital to our existence on this planet, but we are losing topsoil at an alarming rate—between 10 and 40 times faster than it is formed. Moreover, it can take 500 years to form one inch of topsoil.

- Earthworms are essential in soil—one earthworm can eat 5 tonnes of dry matter per hectare per year.

- We are only just beginning to find out about the complex relationships between soil organisms, mycorrhizal fungi, plant roots and the ways they can move nutrients around and even communicate with each other.

- As well as holding water and mineral nutrients that plants depend upon, soil acts as an enormous filter for any water that passes through it.

- Soils store and transfer heat, thus affecting atmospheric temperature, which in turn can affect the interactions between soil and atmospheric moisture.

- Soils are the part of the lithosphere where life processes and soil-forming processes both take place. They are the pedosphere (soil sphere). This is a thin bridge between biosphere and lithosphere and is acted upon and influenced by the atmosphere, the hydrosphere and the lithosphere.

- Soil is a highly porous medium typically with a 50:50 mix of solids and pore spaces. The pore spaces contain variable amounts of water and air.

- There are many soil types and they are classified using keys.

Activity 1

ATL skills used: thinking, communication

Make a systems diagram showing soil storages and flows. Add transfers and transformations. (It will be quite a complex model.)

Soil components and processes

Soils can act as carbon sinks, storages or sources depending on relative rates of input of dead organic matter and decomposition. A forest is an efficient carbon sink but you can see in Figure 1 that overall carbon storage is higher in wetlands, temperate grassland and boreal forests. How much carbon is stored in soil depends on the type of soil, the climate, and limiting factors of precipitation, aeration and temperature.

Connections: This links to material covered in subtopic 1.2.

- The colder it is, the slower the rate of decomposition as respiration slows down.

- The wetter it is, the higher the likelihood of anaerobic conditions—when aerobic respiration cannot occur.

- Although tropical forests have high biomass above ground, decomposition rates are fast and leaching of nutrients out of soils is frequent.

Storages	organic matter, organisms, nutrients, minerals from underlying rock, air and water
Inputs	<ul><li>organic material including leaf litter, manure, biomass</li><li>inorganic matter from parent material</li><li>precipitation</li><li>gases, air humidity</li></ul><ul><li>solar energy</li><li>guano</li><li>waterborne and windblown particles</li><li>anthropogenic: compost, fertilizer, agrochemicals, irrigation, salinization</li></ul>
Outputs	<ul><li>uptake by plants</li><li>soil erosion</li><li>loss of dead organic matter</li><li>loss through wind and water erosion</li></ul><ul><li>diffusion of gases</li><li>evaporation of water</li><li>loss of heat</li></ul>
Transfers	<ul><li>into and out of soils and across soil horizons</li><li>biological mixing</li><li>translocation (movement of soil particles in suspension)</li><li>leaching (minerals dissolved in water moved through soil)</li><li>infiltration</li></ul><ul><li>percolation</li><li>groundwater flow</li><li>aeration</li><li>erosion</li><li>leaching</li></ul>
Transformations	<ul><li>can change the whole system or the soil components</li><li>decomposition</li><li>weathering</li></ul><ul><li>nutrient cycling</li><li>salinization</li></ul>

▲ Table 1 The components and processes of the soil system
Adapted from: Six Red Marbles and OUP

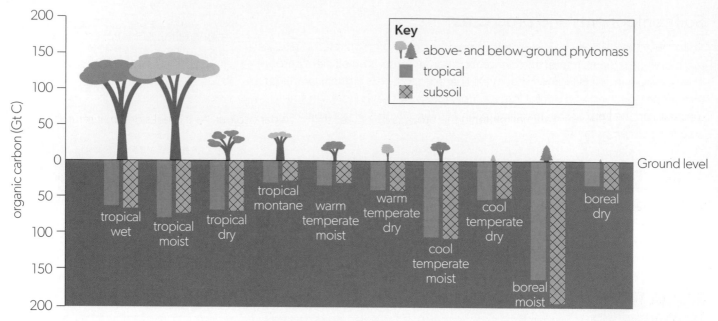

Figure 1 Carbon storage in different biomes, both above and below ground
Adapted from: Zac Kayler, Maria Janowiak, Chris Swanston / USDA FS Climate Change Resource Center

Connections: This links to material covered in subtopics 2.3 and 6.2.

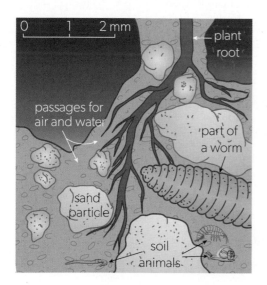

▲ Figure 2 Cross-section of soil

ATL Activity 2

ATL skills used: thinking, communication

Consider Figure 1. Phytomass is plant biomass.

1. Which biome has the largest below-ground carbon store and which three biomes have the smallest?

2. What are the major factors that determine rate of photosynthesis and therefore productivity in a biome?

3. Compare the total biomass in tropical moist forest and boreal dry forest. (Boreal forest has long, dry, cold winters and short summers.)

4. Explain why boreal moist and dry forest biomes have such different stores of phytomass.

5. Estimate the total organic carbon in GtC of tropical moist forest and warm temperate moist forest and explain the difference.

6. Explain why there is relatively little carbon stored in forest soils and much more in wetlands and tundra. (Hint: what do rates of decomposition and photosynthesis depend on?)

Figure 2 and Table 2 summarize the key components of a soil. The character of a soil depends on the exact mix of these components. In addition, the soils within any environment are the result of a mix of complex soil-forming processes. Climate, parent rock material, the shape of the land, the organisms living on and within it and time all contribute to and affect the type of soil.

Fraction	Constituents	Function
Rock particles	Insoluble—e.g. gravel, sand, silt, clay and chalk. Soluble—e.g. mineral salts, compounds of nitrogen, phosphorus, potassium, sulfur, magnesium	Provides the skeleton of the soil and can be derived from the underlying rock or from rock particles transported to the environment—e.g. glacial till
Humus	Plant and animal matter in the process of decomposition	Gives the soil a dark colour. As it breaks down, it returns mineral nutrients to the soil. Absorbs and holds a large amount of water
Water	Water either seeping down from precipitation or moving from underground sources by capillary action	Dissolved mineral salts move through the soil and are also available to plants. Rapid movement of water causes leaching of minerals. Large volumes of water in the soil can cause waterlogging, leading to anoxic conditions and acidification
Air	Mainly oxygen and nitrogen	Well-aerated soils provide oxygen for the respiration of soil organisms and plant roots
Soil organisms	Soil invertebrates, microorganisms and large animals	Soil invertebrates like worms help to break down dead organic matter into smaller particles. The small particles are then decomposed by soil microorganisms, recycling mineral nutrients. Larger burrowing soil animals help to mix and aerate the soil, e. g. moles

▲ Table 2 Constituents of soil
Adapted from: Six Red Marbles and OUP

Soil structure

Soils develop a stable, layered structure known as a profile, produced by interactions within the system over long periods of time.

Soil profiles have distinct horizons—O, A, E, B, C, R (see Figure 3)—with more organic matter at the top and inorganic bedrock at the bottom.

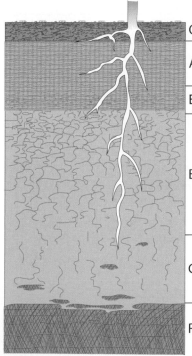

O leaf litter

A mineral horizon at the surface showing organic matter enrichment

E subsurface horizon showing depletion of organic matter, clay, iron, and aluminium compounds

B subsoil horizon showing enrichment of clay material, iron, aluminium, or organic compounds

C horizons of loosened or unconsolidated material

R hard bedrock

▲ Figure 3 A soil profile

Particle	Particle diameter (mm)
Clay	< 0.002
Silt	0.002–0.05
Sand	0.05–2.00

▲ Table 3 Clay, silt and sand particle types

The mineral portion of soil can be divided up into three particles based on size: sand, silt and clay (Table 3). Most soils consist of a mixture of these soil particles. Soil texture therefore depends on the relative proportions of sand, silt and clay particles. The range of particle sizes in soil types is quite large and particles over 2 mm diameter are gravel or stones. Accurate measurement of sizes depends on the instrument used and skill of the user. There will be scientific uncertainty in the measurements which are always inexact. However, the properties of sand, silt and clay are different enough for the particle size range to indicate the type of soil.

Soil functions

A medium for plant growth

Soils provide an anchor for plant roots, a seed bank, a store of water and almost all essential plant nutrients—the main ones are nitrogen, phosphorus and potassium or NPK (except carbon, which plants take in from the air as carbon dioxide). This is the soil fertility.

Provision of habitats

Soil communities have a large biodiversity in many niches. This includes microorganisms, animals and fungi, and many are unknown species. Earthworms burrow through soil by eating and excreting it, aerating it and allowing water to percolate.

Recycling of elements

There is a major input of dead organic matter from plant roots, leaf litter and dead organisms. These are broken down by detritivores (e.g. earthworms) into smaller fragments and decomposed by saprotrophs such as fungi and bacteria. They are then used by other organisms.

Water supply regulation

Water in soils is the solvent for nutrients and is taken in by plant roots. Weathering depends on water freezing and thawing. Above all, soils hold water in their pores as it infiltrates the soil. A compacted soil allows water to run off. An aerated soil absorbs the water, keeps a reservoir for plants and animals and allows the water to drain away more slowly. Long-term refilling of aquifers is by percolation through the soil.

A medium for housing, farming and landscaping

About 50% of people on Earth use soil to construct their houses. Soil mixed with water, straw or stones is an excellent building material, as long as it is kept dry. Road bases, housing and factories are built on firm foundations of soil. Soil can be pushed around to form gardens and parks or contoured on hillsides to form terraces, bunds to hold in rainwater for crops, or ditches to drain excess water or carry irrigation water.

▲ Figure 4 A mud house in rural Brazil

ATL **Activity 3**

ATL skills used: thinking, social, research

Wattle and daub, in building construction, is a method of constructing walls in which vertical wooden stakes, or wattles, are woven with horizontal twigs and branches, and then daubed with clay or mud. This method is one of the oldest known for making a weatherproof structure; it has been used for 6,000 years and still is in some parts of the world.

Cob as a building material is similar, being made of clay soils, straw and lime. Both methods provide high insulation levels, use local materials and are relatively simple to do.

In groups of two or three, research the use of mud or soil to make houses and compare it with the use of concrete, wood or bricks. What are the advantages and disadvantages of each in terms of sustainability? (See also subtopic 8.2, on urban planning.)

Soil texture

This depends on the relative proportions of sand, silt, clay and humus. Soil texture is an important property of a soil, as it determines the soil's fertility and primary productivity through the different influences of sand, silt and clay.

The proportions of these particles give soil its texture (Table 4). Most soils also contain particles that are larger than 2 mm in diameter (pebbles and stones), but these are not considered in a description of soil texture.

It is possible to feel the texture of moist soil if you rub it between your fingers.

- Sandy soils are gritty and fall apart easily.

- Silty soils feel slippery, like wet talcum powder, and hold together better than sandy soils.

- Clay soils feel sticky and can be rolled into a ball easily.

Most soils contain a mixture of different soil particles. If a soil contains fairly equal proportions of the three sizes of particle, it is said to be a **loam** (Figure 5). Loam soils are ideal for agriculture because:

- the sand particles ensure good drainage and a good air supply to the roots

- the clay retains water and supplies nutrients—so they are fertile

- the silt particles help to hold the sand and clay particles together and can be worked easily.

We get a clearer picture of the proportions of soil particles by drying out a soil sample and passing it through a series of standard sieves of decreasing mesh size, to separate the soil into clay, silt and sand "portions".

Alternatively you can place a sample of soil in a jam jar, fill it with water, shake it vigorously, then leave it to settle out. The heaviest particles will settle first (sand) and the finer particles will settle last (clay).

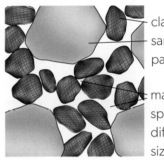

clay and sand particles

many air spaces of different sizes

▲ Figure 5 A loam soil with sand and clay particles

	Sandy soil	Clay soil	Loam soil
Composition (%)			
• Sand	100	15	40
• Silt		15	20
• Clay		70	40
Mineral content	High	High	Intermediate
Potential to hold organic matter	Low	Low	Intermediate
Drainage	Very good	Poor	Good
Water holding capacity	Low	Very high	Intermediate
Air spaces	Large	Small	Intermediate
Biota	Low	Low	High
Primary productivity	Low	Quite low	High

▲ Table 4 Comparison of three soil types

▲ Figure 6 Humus with earthworms

The importance of humus

Humus contributes significantly to the texture of soils in which it is abundant. It is a dark brown or black substance with loose, crumbly texture, formed by the partial decay of dead plant material and lying beneath the leaf litter. It affects primary productivity by influencing:

- mineral nutrient retention versus leaching

- water retention versus drainage

- aeration versus compaction or waterlogging.

Managed soil systems

"Managed" here means modified by humans to produce food or resources that we use. In agriculture, soils are managed to increase crop yield by:

- adding compost to increase nutrients and organic matter in soils

- adding inorganic fertilizers to increase nutrients available for crops

- treating with chemicals to kill pests (pesticides) or unwanted plants (herbicides); such chemicals may leave residues in the soil

- irrigating if water is scarce; this leads to increased salinity as the water supplied contains salts which are left behind when the water evaporates

- draining if water is in excess, usings ditches or subsoil pipe networks

- cultivating to increase aeration.

Soil profiles

What does a soil look like?

If you dig a trench in the ground, the side of the trench creates a soil profile—a cross-section. This profile changes as it goes down from the surface towards the underlying base rock. It is a record of the processes that have created the soil, its mineral composition, organic content, and chemical and physical characteristics such as pH and moisture.

Horizons

In cross-section, soils have a profile which is modified over time as organic material leaches (washes) downwards and mineral materials move upwards. These processes sort the soil into distinctive horizons (zones or levels), which are often visible in the soil (Figure 3).

Figure 7 shows some of the processes involved in the formation of soil horizons. The top layer of the soil is often rich in organic material while the lower layers consist of inorganic material. The inorganic material is derived from the weathering of rocks. Within this, materials are sorted and layers are formed by water carrying particles either up or down—known as **translocation**. In hot, dry climates, where precipitation < evaporation (P < E), water evaporates at the soil surface and water from lower soil layers moves upwards. When doing so, it dissolves minerals and takes them to the surface, where the minerals are left behind when the water evaporates. This also happens after irrigation and is called **salinization**. In colder and wetter climates, when P > E, water flows down in the soil, dissolving minerals and transporting them downwards. This is **leaching**.

O horizon: Many soils contain an uppermost layer of newly added organic material that comes from organisms that die and end up on top of the soil. There, fungi, bacteria and many different decomposers and detritivores will start to decompose the dead material.

A horizon: Upper layer. In many soils, this is where humus builds up. Humus forms from partially decomposed organic matter and is often mixed with fine mineral particles. Often decomposition is incomplete and a layer of dark brown or black organic material is formed—the **humus layer**. In normal conditions, organic matter decomposes rapidly through the decomposer food web, releasing soluble minerals that are then taken up by plant roots. Waterlogging reduces the number of soil organisms as they die without oxygen, which results in a build-up of organic matter and can eventually lead to the formation of peat soils.

B horizon: This is the layer where soluble minerals and organic matter tend to be deposited from the layer above. In particular, clay and iron salts can be deposited in this horizon.

C horizon: This layer is mainly weathered rock from which the soil forms.

R horizon: Parent material (bedrock or other medium).

Not all soils contain all three A, B and C horizons; sometimes there may be no distinct layering. In some cases, we cannot dig deep enough to find the C layer. In intensive agricultural systems only B and C horizons may remain as the organic matter in litter and crops has been removed.

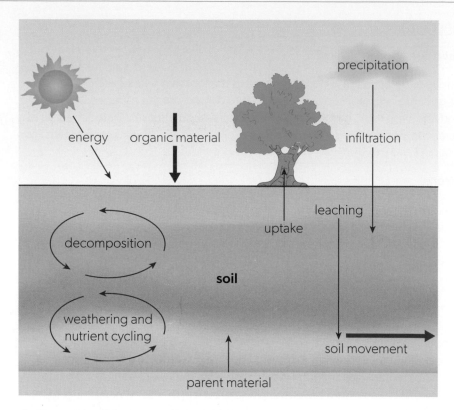

▲ Figure 7 Simplified model of the soil system

Classification of soils

Soils are classified and mapped by the appearance of the whole soil profile.

Soils and biomes are closely linked because climate and vegetation are two major factors that determine how soils form and therefore specific soil types are associated with each biome.

The US classification defines 12 soil types, which are given different names ending in –isol. There are also Canadian and FAO (Food and Agriculture Organization of the UN) classification systems so soil type naming is very complex indeed.

Here are four examples.

1. **Mollisols** (black or brown earths or chernozems)—Figure 8:

    *   temperate grassland soils that make up about 7% of the world's ice-free land surface

    *   formed under grass in climates with a moderate to pronounced seasonal moisture deficit

    *   have a dark-coloured surface horizon, relatively high in organic matter

    *   are base-rich throughout and therefore quite fertile

    *   extensive soils on the steppes of Europe, Asia, North American Great Plains and South America.

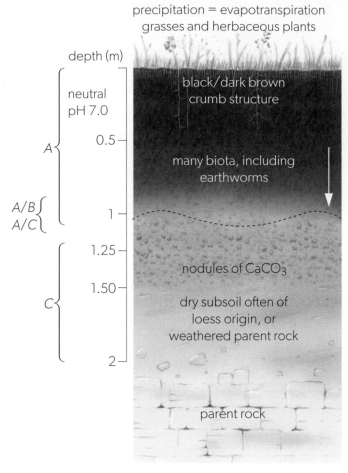

precipitation = evapotranspiration
grasses and herbaceous plants

depth (m)

neutral
pH 7.0

0.5

1

1.25

1.50

2

A

A/B
A/C

C

black/dark brown
crumb structure

many biota, including
earthworms

nodules of CaCO₃

dry subsoil often of
loess origin, or
weathered parent rock

parent rock

thick sod cover/organic
matter

accumulation of mull humus
and bases (Ca, Mg, Na, K) and
some Fe, Al and Si

slight leaching after
spring snowmelt and
summer storms

indistinct boundary; possibly
an absence of a B horizon

calcification

▲ Figure 8 Mollisol soil type, typical of continental grassland biome

2. **Alfisols**—Figure 9:

- deciduous forest soils with leaching (downward movement) of minerals and clay where they retain moisture and nutrients

- very productive soils for crops, along with mollisols

- in temperate and humid regions under hardwoods or where forest is cleared for agriculture.

3. **Oxisols** (or laterite soils)—Figure 10:

- most oxisols occur in humid tropical and subtropical zones

- vegetation is tropical rainforest, scrub and thorn forest, or savanna on flat to gently sloping uplands

- highly weathered soils on old landscapes subjected to shifting cultivation for many years

- acidic soils, dominated by low activity minerals, such as quartz, kaolinite and iron oxides

- tend to have indistinct horizons

- occur on land surfaces that have been stable for a long time

- very low natural fertility and low capacity to retain added lime or fertilizer.

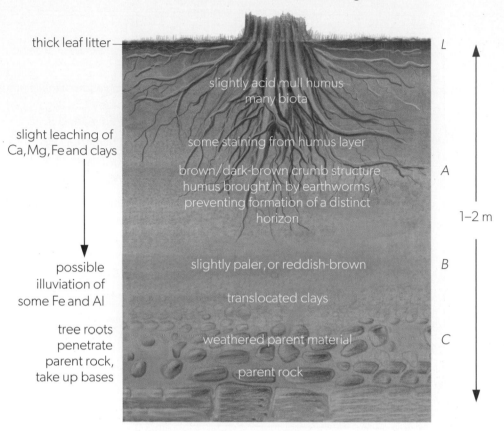

deciduous woodland with undergrowth

thick leaf litter

L

slightly acid mull humus
many biota

slight leaching of
Ca, Mg, Fe and clays

some staining from humus layer

A

brown/dark-brown crumb structure
humus brought in by earthworms,
preventing formation of a distinct
horizon

1–2 m

possible
illuviation of
some Fe and Al

slightly paler, or reddish-brown

B

translocated clays

tree roots
penetrate
parent rock,
take up bases

weathered parent material

C

parent rock

▲ Figure 9 **Alfisol soil type, typical of deciduous forest biome**

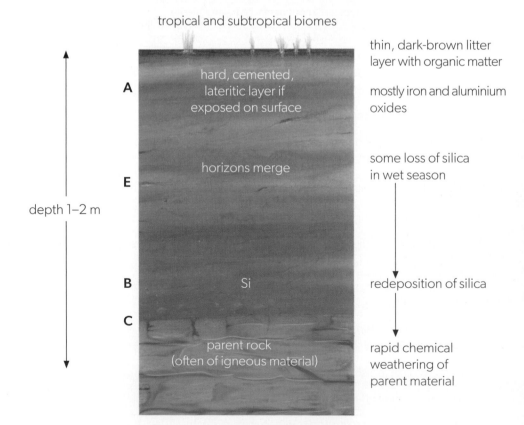

tropical and subtropical biomes

thin, dark-brown litter
layer with organic matter

A

hard, cemented,
lateritic layer if
exposed on surface

mostly iron and aluminium
oxides

E

horizons merge

some loss of silica
in wet season

depth 1–2 m

B

Si

redeposition of silica

C

parent rock
(often of igneous material)

rapid chemical
weathering of
parent material

▲ Figure 10 **Oxisol soil type, typical of tropical rainforest biome**

4. **Gelisol** (or cryosol)—Figure 11:

- in tundra biomes with permafrost
- waterlogged or frozen soils
- slow decomposition
- not fertile.

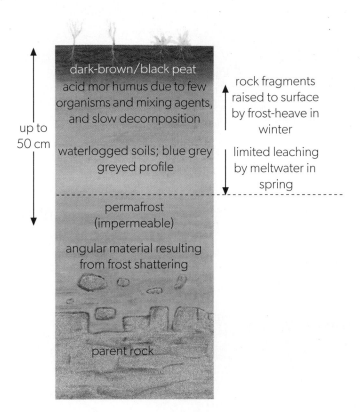

▲ Figure 11 Gelisol, typical of tundra biome

ATL **Activity 4**

ATL skills used: thinking, communication

1. Copy and complete Table 5.

Biome	Soil type	Human activity on these soils
temperate deciduous forest	alfisol	clear forest and grow wheat/other cereals
tundra	gelisol	minimal as frozen or waterlogged with low fertility
temperate grassland		
	oxisol	

▲ Table 5

2. Explain why mollisols and alfisols are good soils for agricultural crops.

3. Explain why oxisols and gelisols are poor soils for agricultural crops.

4. Compare and contrast the two pairs of soils in 2 and 3, considering properties of their biomes.

Factors influencing soil formation

Soils form through interactions of five major factors (Figure 12):

- time
- climate
- organisms
- geomorphology (landscape)
- geology (parent material).

Parent rock (geology)
Affects:
- permeability (drainage)
- mineral content (nutrients)
- acidity or alkalinity
- depth, colour, texture

Climate
- Temperature affects rate of weathering
- Precipitation affects type of vegetation and movement of water within soil
- Affects rate of vegetation decay and formation of humus

Soil

Topography (relief)
- Altitude (height)
- Steepness of slope
- Aspect

Organisms
- Flora (type of vegetation)
- Fauna (type of animals)
- Microorganisms which affect rate at which humus forms, recycling of nutrients and mixing of air and humus in soil

Time
It takes 3,000 to 12,000 years to create a layer of soil deep enough to support agriculture

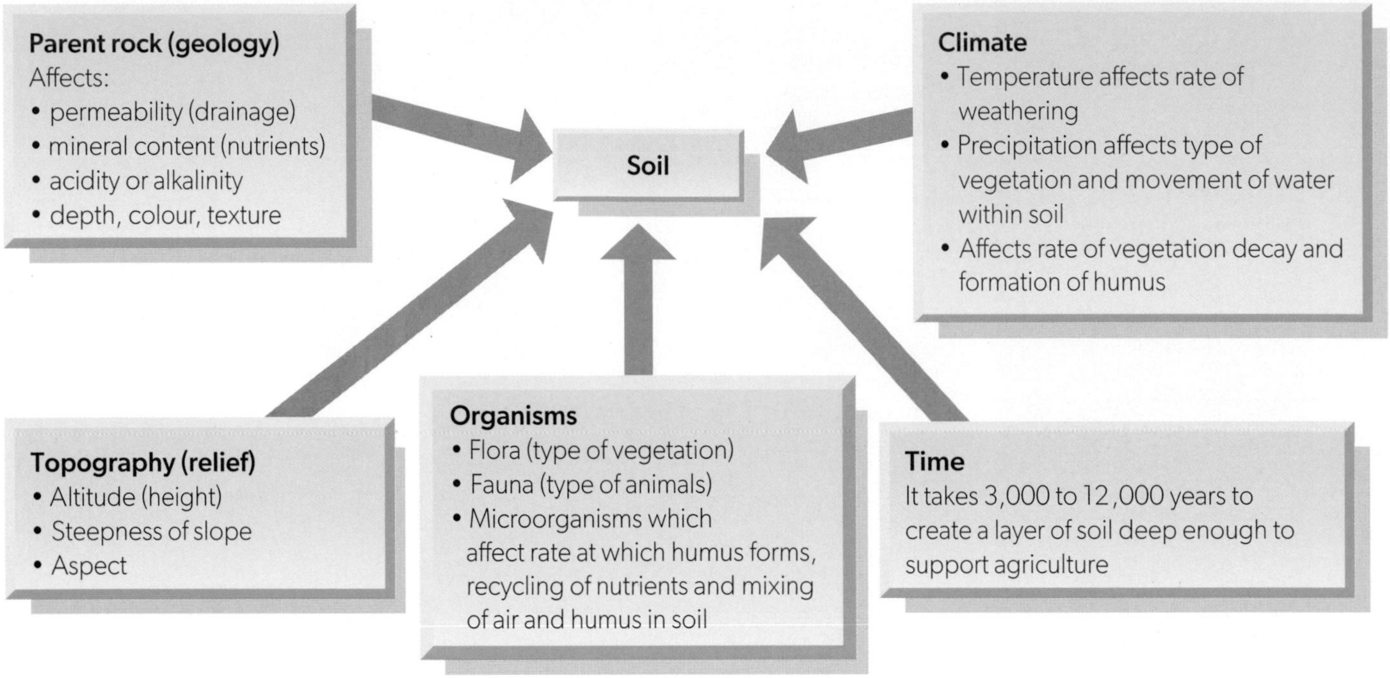

▲ Figure 12 Five factors influencing soil type

1. **Time**—soils can take years to form, from decades to thousands of years. This depends on the other factors but most soil cycles are long.

2. **Climate**— this is the dominant factor as both precipitation and temperature affect rates of reaction. Rates of leaching and evaporation depend on climate. Water deficiency can lead to oxisols (lateritic soils) or excess can lead to loss of fertility as nutrients are carried away.

   Waterlogged soils, where the soil is saturated and water cannot drain away, result in plant roots dying as all pores are filled with water not air; denitrification and soil acidity increase.

3. **Organisms**—each soil is its own ecosystem. Some have nitrogen-fixing bacteria, some have fungi that accumulate phosphorus; plant roots anchor soil and decomposition builds up organic matter. Soil animals burrow and allow gases and water to circulate in the soil.

4. **Topography** (relief of the landscape)—the slope, elevation and orientation (N, S, E or W) affect the soil. On steep slopes, rain runs off instead of percolating (soaking into) the soil, so there is more erosion and soils are shallow. Higher altitudes are cooler so soil activity is slower. Slopes facing the Sun's path are drier and warmer.

5. **Geology** (parent material)—rock, the parent material, is the source of soil minerals and most plant nutrients. Weathering and transportation (by erosion or deposition) of this material forms different soils. Weathering can be physical (rock breaking into smaller pieces by freezing then thawing) or by chemical reactions, e.g. soluble minerals dissolve in water or oxidize in air.

a. Calcareous parent materials are sedimentary rocks high in calcium carbonate derived from plankton, corals and molluscs (chalk, limestone). Soils formed from these rocks are alkaline with a higher pH. They are not very fertile as they generally have low organic matter and low nitrogen and phosphorus.

b. Volcanic parent materials (e.g. lava, basalt, granite) result in young soils called andisols. These are very fertile as the rock weathers to release minerals such as magnesium and potassium which can be limiting factors for plant growth. Soils in volcanic areas are therefore desirable for farming and extra layers of ash produced from volcanic eruptions add more natural fertilizer.

▲ Figure 13 Chalk cliffs weather to form alkaline, thin, and low-fertility soils

The Ring of Fire volcanic soils

- Volcanic soils cover 1% of the land surface of the Earth but can support 10% of the population due to their high fertility.

- Indonesia is situated on the Pacific Ring of Fire (Figure 14), a chain of volcanoes that stretches from Sumatra through Java and Bali to Timor, and constitutes the most dangerous of the world's tectonic areas.

- Indonesia's volcanic eruptions have affected the world's climate:

  - The volcano of Lake Toba erupted 74,000 years ago causing a 6-year volcanic winter as ash in the atmosphere cooled the Earth.

  - The Mount Tambora eruption in 1815 caused a year with no summer in Europe and the Krakatau eruption of 1883 cooled the Earth.

  - In 2014, Mount Sinaburg erupted with pyroclastic flow. The ash—250 million tonnes of it—contains high levels of calcium, magnesium, potassium and phosphate.

- Volcanic eruptions can also sequester carbon from the atmosphere, resulting in highly fertile soils.

 **Activity 5**

ATL skills used: research

1. Research the geology of the area where you live.

2. Identify the soil types that are formed from these parent rocks.

3. What type of soils are there in your local area?

4. What crops are grown in these soils?

5. What type of fertilizers are added to agricultural land in your area?

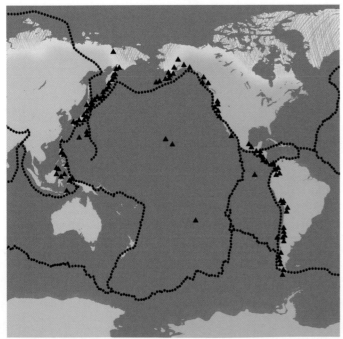

◀ Figure 14 Ring of Fire volcanoes. Dotted lines show boundaries between tectonic plates; small triangles indicate volcanoes

More on sand, silt and clay soils

Soil properties can be determined from analysis of the sand, silt and clay percentages, organic matter, percentage water, infiltration, bulk density, colour and pH. A soil triangle (Figure 15) shows how soils are named according to sand, silt and clay percentages.

Cation-exchange capacity

This is a measure of how many cations (positively charged ions) can be retained on surfaces of soil particles. Sand and silt are derived from quartz and have a low cation-exchange capacity (CEC). Clays are complex silicates that have a much greater CEC that increases availability of positively charged minerals.

CEC is important as it influences the soil's ability to capture and keep soil nutrients. Potassium, ammonium, magnesium and calcium are all available to plants as cations K^+, NH_4^+, Mg^{2+}, Ca^{2+}.

High CEC soils retain more nutrients and loss of applied fertilizer is lower than in low CEC soils.

Porosity and permeability

Porosity (the amount of space between particles) and permeability (the ease with which gases and liquids can pass through the soil) are features of soil texture. Soils with a very fine clay particle texture have lots of micropores. The total (combined) pore space is large but the soil has a low permeability as water molecules easily fill these micropores and become adhered (stuck) to the clay surface—the water is trapped as a film around the clay particles. In contrast, sandy soils have fewer but larger macropores, with a smaller total space. These spaces are too large for water's adhesive properties to work so sandy soils drain well.

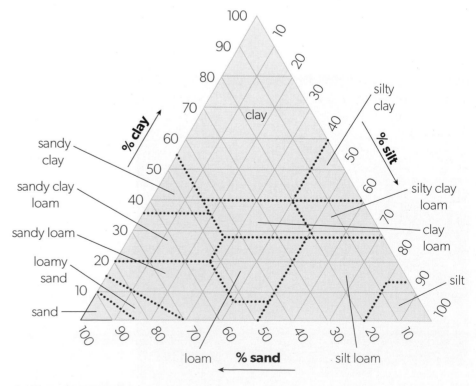

▲ Figure 15 Soil triangle

The low permeability of clay means that it can lock dissolved minerals between the pores, making it hard for plant roots to access them. Strangely, the result can be a soil that is rich in minerals but has low fertility.

Acidification of soils

The features of clay soils also encourage high acidity, which has a great effect on the chemical characteristics. As the soil absorbs more water, clay particles begin to fill up with positive hydrogen ions (H^+). This binds soil water tightly to clay and makes the soil more acidic. It also reduces the amount of other positive ions that can bind, allowing the important nutrients potassium, magnesium and ammonium to be lost through leaching. Another effect of decreasing soil pH is that, as soil pH decreases, ions of aluminium and iron—both plant toxins—start to become more available to plants.

Acidification has had a major impact on forestry in northern Europe where acid rain caused by industrial pollution has made the soil more acid. This in turn has meant more available aluminium and iron ions in the soil, causing damage to evergreen forestry through needle death.

Soil sustainability

Soil fertility

Fertile soil is a non-renewable natural resource. Once it is lost, it cannot be replaced quickly.

About one-third of all soils have been degraded and all the topsoil where intensive agriculture is practised could be gone within 60 years.

Soil formation takes a very long time. The natural soil renewal rate is about 1 tonne ha^{-1} yr^{-1} in natural ecosystems under the best conditions (wet, temperate climate), equivalent to 0.05–0.1 mm per year. And this is only after the initial chemical and physical weathering has occurred and fine material and soil organisms are present. So this figure actually represents the natural rate at which soils regain their fertility. As a consequence, soil use often exceeds soil formation.

Fertile soil has enough nutrients for healthy plant growth. The main nutrients are nitrates, phosphates and potassium (NPK, see Figure 16) and there are many micronutrients that plants also need. These nutrients can be leached out of soil or removed when a crop is harvested. They have to be replaced in agricultural soils via chemical fertilizer, growing legumes, crop rotation or through the application of organic matter (e.g. manure, compost).

Topsoil is the A horizon of soil, just beneath the uppermost organic humus layer where it is present. Topsoil is rich in organic matter and is the most valuable for plant growth but, along with the O horizon, is also the most vulnerable to erosion and degradation, with implications for sustainable management of soil.

Topsoil has more oxygen, organic matter, microorganisms and nutrient recycling than lower soil horizons, so it is where there is most root growth and other biological activity.

Intensive farming removes this layer and requires the addition of fertilizers to add nutrients that are taken away in harvesting the crops.

> *To be a successful farmer one must first know the nature of the soil.*
>
> Xenophon (circa 430–354 BCE), Ancient Greek philosopher and historian, student of Socrates

potash: for flower head or fruit

nitrate: for leaf and stem

phosphate: for root system

▲ Figure 16 NPK nutrients and organic fertilizer for plant growth

TOK

What value systems can you identify in the causes and approaches to resolving the issues addressed in soil conservation?

To what extent might the solutions to soil loss alter your predictions for the state of human societies and the biosphere some decades from now?

The soil system may be represented by a soil profile. Since a model is strictly speaking "false", how can it lead to knowledge?

Many of the issues threatening soil health are related to agriculture and the pressure to meet rising food demand, such as:

- degradation and decline caused by intensive practices

- compaction under heavy machinery and inappropriate land cultivation practices

- damage to biodiversity caused by monocultures and other land use and management practices

- pollution from chemicals (e.g. pesticides, heavy metals, pharmaceuticals, plastics)

- land abandonment and neglect.

Soil as a carbon source

Soils emit greenhouse gases—carbon dioxide, methane and nitrous oxide. The carbon is fixed again in the carbon cycle but agriculture and human activity disrupt this. The top metre of soil contains three times the carbon of the atmosphere in natural systems but now carbon is released faster than it is replaced.

This carbon may be released through increased decomposition due to global warming, agricultural practices, drainage of wetlands or other human activity. Increased carbon emissions may lead to a tipping point where increasing temperatures lead to breakdown of methane clathrates in underlying geological structures (see subtopic 6.2).

Check your understanding

In this subtopic, you have covered:

- what soil is and what it is made from

- soil is a system with transfers and transformations

- functions of soil

- soil profiles and horizons

- factors influencing soil types

- how carbon is released from soils.

How do soils play a role in sustaining natural systems?

How are human activities affecting the stability of soil systems?

1. Describe the soil system.

2. List the components of soil.

3. List the processes that occur in soils.

4. Describe the functions of soils.

5. Explain how a soil forms and how a soil degrades.

6. Explain how soil texture affects primary productivity.

7. Describe a soil profile with horizons.

8. Explain the five factors that influence soil formation.

9. For two soil types, explain their composition and relative importance for agriculture.

10. Comment on the sustainability of soil.

AHL

 Taking it further

- Watch well-researched documentaries on soil sustainability and soil fungi. Examples are *Kiss the Ground* (2020) and *Fantastic Fungi* (2019) but there may be more recent productions available.

- Discuss in class how humans can reduce soil loss and degradation.

- Research how local farms manage and value soil in relation to agriculture, climate change, biodiversity and overall sustainability.

- Host a documentary film festival about soil and food.

- Start an organic vegetable garden with compost bins.

5.2 Agriculture and food

- To what extent can the production of food be considered sustainable?

Understandings

1. Land is a finite resource, and the human population continues to increase and require feeding.

2. Marginalized groups are more vulnerable if their needs are not taken into account in land-use decisions.

3. World agriculture produces enough food to feed eight billion people, but the food is not equitably distributed and much is wasted or lost in distribution.

4. Agriculture systems across the world vary considerably due to the different nature of the soils and climates.

5. Agricultural systems are varied, with different factors influencing the farmers' choices. These differences and factors have implications for economic, social and environmental sustainability.

6. Nomadic pastoralism and slash-and-burn agriculture are traditional techniques that have sustained low-density populations in some regions of the world.

7. The Green Revolution (also known as the Third Agricultural Revolution in the 1950s and 1960s) used breeding of high-yielding crop plants—combined with increased and improved irrigation systems, synthetic fertilizer and application of pesticides—to increase food security. It has been criticized for its sociocultural, economic and environmental consequences.

8. Synthetic fertilizers are needed in many intensive systems to maintain high commercial productivity at the expense of sustainability. In sustainable agriculture, there are other methods for improving soil fertility.

9. A variety of techniques can be used to conserve soil, with widespread environmental, economic and sociocultural benefits.

10. Humans are omnivorous, and diets include fungi, plants, meat and fish. Diets lower in trophic levels are more sustainable.

11. Current global strategies to achieve sustainable food supply include reducing demand and food waste, reducing greenhouse gas emissions from food production and increasing productivity without increasing the area of land used for agriculture.

12. Food security is the physical and economic availability of food, allowing all individuals to get the balanced diet they need for an active and healthy life.

13. Contrasting agricultural choices will often be the result of differences in the local soils and climate.

14. Numerous alternative farming approaches have been developed in relation to the current ecological crisis. These include approaches that promote soil regeneration, rewilding, permaculture, non-commercial cropping and zero tillage.

15. Regenerative farming systems and permaculture use mixed farming techniques to improve and diversify productivity. Techniques include the use of animals like pigs or chickens to clear vegetation and plough the land, or mob grazing to improve soil.

16. Technological improvements can lead to very high levels of productivity, as seen in the modern high-tech greenhouse and vertical farming techniques that are increasingly important for supplying food to urban areas.

17. The sustainability of different diets varies. Supply chain efficiency, the distance food travels, the type of farming and farming techniques, and societal diet changes can all impact sustainability.

18. Harvesting wild species from ecosystems by traditional methods may be more sustainable than land conversion and cultivation.

19. Claims that low-productivity indigenous, traditional or alternative food systems are sustainable should be evaluated against the need to produce enough food to feed the wider global population.

20. Food distribution patterns and food quality variations reflect functioning of the global food supply industry and can lead to all forms of malnutrition (diseases of undernourishment and overnourishment).

Food facts

- There is enough food in the world but an imbalance in its distribution.

- A crop is a plant that can be grown and harvested for food or profit.

- Crop production globally has increased due to both increasing area of land in use and increased crop yields.

- A food staple is a food that makes up the dominant part of a population's diet, e.g. cassava, maize, plantains, potatoes, rice, sorghum, soybeans, sweet potatoes, wheat and yams.

- Cereals are grain-producing grasses, such as wheat, rice, maize and millet.

- Cereals account for more than half of the world's harvested area, and more of the Earth's surface is covered by wheat than by any other food crop.

- China, the USA and India are the three largest cereal-producing countries.

- Of the 2.7 billion tonnes of cereal produced each year, about 41% is for food for humans, 35% for animal feed, and the remaining 24% is processed for industrial use or biofuels, used as seed or wasted.

- Food production is closely linked with climate, culture, tradition and politics.

- Rice is the primary crop and food staple of more than half the world's population. Asia is the world's largest rice-producing and rice-consuming region. Rice is also becoming an increasing food staple throughout Africa.

- Yams are a major staple in West Africa, where they are consumed mainly as "fufu", a gelatinous dough. Fufu can also be made from cassava and plantains.

- Of all the mammals on Earth, 96% are livestock and humans. Only 4% are wild mammals.

- Livestock numbers far exceed human numbers—chicken (23 billion), cattle (1.5 billion), sheep (1.2 billion), ducks (1.2 billion), goats (1 billion) and pigs (1 billion).

- Meat consumption is increasing—three times as much is produced now than 50 years ago.

Agriculture facts

- Land is a finite resource and the human population continues to increase and require feeding.

- Half of the world's habitable land is used for agriculture.

- Historically, food production was localized—farmers produced food for their families or communities. Over time, however, agricultural production has also become much more international. There is now a global food system and large amounts of food are traded internationally.

Abundance does not spread, famine does.

Zulu proverb

This is a sad hoax, for industrial man no longer eats potatoes made from solar energy; now he eats potatoes partly made of oil.

Howard T. Odum (1924–2002), an American ecologist, referring to the common perception that modern agriculture has freed society from limits imposed by nature, when in fact it is highly dependent on non-renewable fossil fuels

- Different food production systems (e.g. terrestrial and aquatic) have different efficiencies and impacts and make different demands on the environment.

- Not all agricultural land can be used to grow crops; some of it is too steep or has poor soils.

- Approximately 77% of the world's agricultural land is devoted to raising animals, including cropland devoted to animal feed and pasture for grazing land.

- Our current use of the soil in which we produce food is not sustainable.

- Food production is responsible for 25% of global GHG emissions.

- Meat and dairy tend to have larger carbon footprints than crop production.

- 70% of global freshwater withdrawals are used for agriculture.

Key terms

Agribusiness is the business of agricultural production including farming, seed supply, breeding, chemicals for agriculture, machinery, food harvesting, distribution, processing and storage.

Commercial agriculture (or farming) is large-scale production of crops and livestock for sale.

Subsistence agriculture (or farming) is farming for self-sufficiency to grow enough for a family.

The global food system

Until a few decades ago, people in different parts of the world ate very different diets. Food did not travel far from where it was grown and most people ate a fairly monotonous and seasonal diet. Now, many of us can buy any foods at any time of year as they are flown or shipped around the world. But the diversity of our foods has decreased. Wheat, maize (corn), rice and soybeans account for 60% of all foods grown in agriculture. A handful of countries (USA, Argentina, Brazil, France, Ukraine) grow 75% of the world's maize. Five countries (Thailand, Vietnam, India, Pakistan, USA) sell 77% of the world's rice and five (USA, France, Canada, Russia, Australia) sell 65% of the world's wheat. According to the FAO, about 40% of the global population now rely on imported food as their own countries do not grow enough.

(ATL) Activity 6

ATL skills used: thinking, communication, research

Read the food and agriculture facts and study Figure 1.

1. What is a crop?

2. What is a food staple?

3. Calculate the quantity (in billion tonnes) of cereal produced each year that is used directly for human food.

4. Explain why 29% of the land surface of the Earth is not inhabited.

5. Explain why livestock farming takes up more land area than crops.

6. Explain why meat production has a larger carbon footprint than crop production.

7. Evaluate a plant-based food diet for humans.

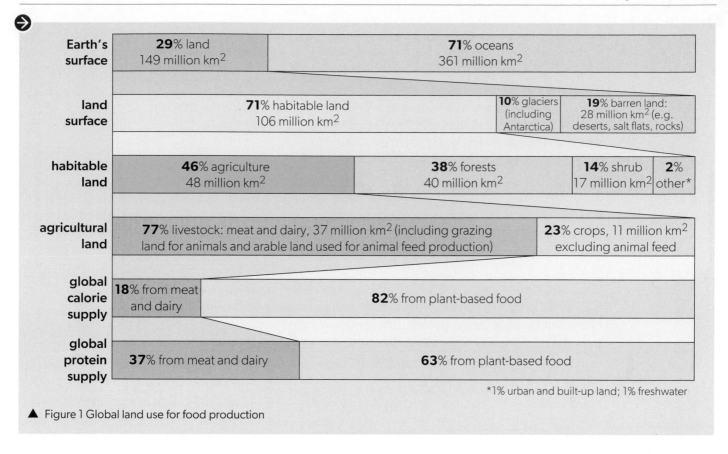

▲ Figure 1 Global land use for food production

Factors influencing sustainability of agriculture

Table 1 shows the two extremes of agriculture but notice that many of these factors are inter-related and difficult to separate.

Connections: This links to material covered in subtopics 1.1 and 1.3.

Factor	Which is more sustainable?	
	Commercial farming—mostly in HICs	**Subsistence agriculture—mostly in LICs**
Agribusiness (commercial vs subsistence)	**Commercial farming** is epitomized by many of the practices in this column but could still function with more environmentally friendly practices.	**Subsistence agriculture** tends to be more associated with this side of the equation.
Scale of farming	Large-scale farming tends to rely heavily on machinery, chemicals and extensive use of fossil fuels.	Small-scale farming tends to be more labour intensive but may still rely on chemicals to boost production.
Industrialization	HICs have many people working in industry and so they must be provided with food from large-scale commercial farming.	LICs have limited industry so job opportunities are limited and people may have to grow their own food.
Mechanization	Use of lots of heavy machinery can damage the soil and uses a lot of fossil fuels.	Use of draft animals (e.g. donkeys, oxen) or human power is less stressful on the soil and can add manure. They are powered by plants so no burning of fossil fuels.
Fossil fuel use	A heavy dependence on fossil fuels is using a finite resource which produces large amounts of pollution.	Use of manual labour or draft animals does not cause these problems.

Seed, crop or livestock choices	Farming systems sometimes grow crops or keep animals that are not indigenous to the area and this can create the need for irrigation, greenhouses and imports of feedstuffs.	Selecting organisms that are indigenous is less likely to create these problems.
Water use	Some agricultural systems have very heavy water demands and require large-scale irrigation solutions which divert water from people and may cause localized water supply problems and a drop in the water table.	Also requires water, which can be used unsustainably.
Fertilizers, pest control	Growing the same crop continually on the same land requires chemicals to support the soil and control pests. These often make their way into the local ecosystems—terrestrial and aquatic.	Crop rotation, biological pest control and other environmentally sound practices cause fewer problems but many subsistence farmers use large amounts of pesticides if they can afford them.
Antibiotics	Keeping animals in close quarters (often inside) causes the spread of diseases and this requires large amounts of antibiotics, often used routinely. If these make it into the local ecosystem they can cause super-bugs.	Free-range animals tend to be healthier and in less need of antibiotics.
Legislation	Large-scale commercial interests are controlled by legislation so may pollute less.	Small-scale operations often go un-noticed by legal bodies.
Pollinators	Many commercial crops, e.g. almonds in California, require pollination by bees and other insects. Honey bee hives are brought in to provide this but honey bee colony collapse disorder is killing many bees. Some crops have no pollinators left and humans have to do this by hand.	With a more biodiverse farm, pollinators have different habitats and there are usually enough insects to pollinate crops.

▲ Table 1 Factors influencing sustainability of agriculture
Adapted from: Six Red Marbles and OUP

Marginalized populations

Marginalized populations experience discrimination and exclusion due to unequal economic, cultural, social and political relationships. They may include ethnic minorities, indigenous groups, women farmers, low-caste farmers and LICs.

ATL Activity 7

ATL skills used: thinking, communication, social, research, self-management

Either individually or in groups of two or three, research two marginalized groups either locally or globally who are affected by land use policies and practices.

- One should be an indigenous culture, e.g. the Sámi of northern Europe, Maori of New Zealand, Lakota of the USA, Aymaras of Bolivia or one of your choice.

- One should be a group of low status or low income within a society. This could be related to gender,

race, religion, educational status, age, physical ability or other factors.

Answer the following questions and present your findings in a suitable form.

1. Who are they and where do they live?

2. How do their lifestyles and culture differ from the mainstream society there?

3. What land use or occupation issues are there?

4. What can be done to reduce or remove their vulnerable status?

Food waste

The world produces enough food for eight billion people or more. But distribution is unequal and there is much waste. It is estimated that about one-third of food produced is lost or wasted.

Loss occurs in the harvest, slaughter or catch. Waste in the food chain is in production, storage, processing, sales or consumption. Also food is redirected to animal feed, energy production as biofuels, or fertilizer.

Sustainable Development Goal (SDG) 12 is about ensuring sustainable consumption and production patterns, which is key to sustaining the livelihoods of current and future generations. It aims to "halve per capita global food waste at the retail and consumer level, and reduce food losses along production and supply chains by 2030".

According to the UN:

*   In 2020, an estimated 13.3% of the world's food was lost after harvesting and before reaching retail markets.

*   An estimated 17% of total food available to consumers (931 million metric tonnes) is wasted at household, food service and retail levels.

*   Food that ends up in landfill generates 8–10% of global greenhouse gas emissions.

If you are interested in finding out more about food waste, search online for the FAO document *The State of Food and Agriculture: Moving forward on food loss and waste reduction.*

▲ Figure 2 Sustainable Development Goal 12 (SDG 12)

Different agricultural systems

Like all systems, agricultural systems have inputs, outputs, storages and flows (Figure 3). They vary considerably across the world due to the different nature of the soils and climates. You know from subtopic 5.1 that different biomes have different potential soil fertility and therefore different productivity.

Connections: This links to material covered in subtopics 2.4 and 5.1.

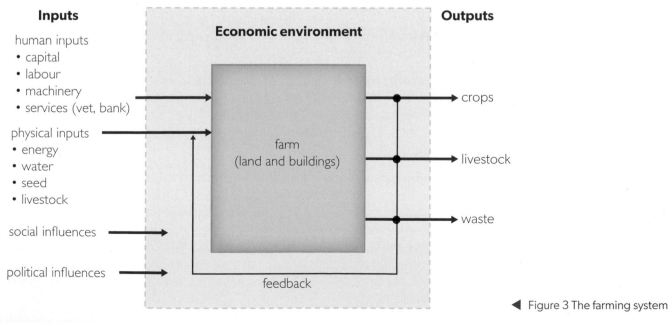

◀ Figure 3 The farming system

Types of farming system

Land aspect, altitude, latitude and slope all determine what can be grown. If there are large rocks near the surface, land cannot easily be cultivated. If soils are low in nutrients, contaminated, need irrigation or flood easily, these factors also limit what type of agriculture can be carried out. Also culture, tradition, politics and society's choice of food to eat influence what farmers grow. Subsidies, incentives or penalties may determine choice and farmers who sell their produce to a market have to run an efficient business as making a large loss could cause the business to fail.

Agricultural systems can be classified in a number of ways.

1. Outputs from the farm system—arable, pastoral or livestock, monoculture or mixed.

2. Reasons for farming—commercial or subsistence, sedentary or nomadic.

3. Types of inputs required for the farm system—intensive or extensive, irrigated or rain-fed, soil-based or hydroponic, organic or inorganic.

You may have heard of the following types of farming.

*   **Subsistence** farming is the provision of food by farmers for their own families or the local community—there is no surplus. Usually mixed crops are planted and human labour is used a great deal. There are relatively low inputs of energy in the form of fossil fuels or chemicals. With low capital input and low levels of technology, subsistence farmers are unlikely to produce much more than they need. They are vulnerable to food shortages as little is stored.

*   **Cash cropping** is growing crops for the market, not to eat yourself.

*   **Commercial** farming takes place on a large, profit-making scale, maximizing yields per hectare. This is often by a **monoculture** of one crop or one type of animal. High levels of technology, energy and chemical input are usually used with corresponding high outputs.

*   **Arable** farming is sowing crops on good soils to eat directly or to feed to animals or use as biofuels.

*   **Mixed** farming has both crops and animals. Animal waste is used to fertilize the crops and improve soil structure, and some crops are fed to the animals.

*   **Pastoral** farming is raising animals, usually on grass and on land that is not suitable for crops.

*   **Nomadic pastoralism** is a form of pastoralism in which people move cyclically with their herds to find fresh pasture or water. Herded animals may be cattle, sheep, goats, reindeer, camels or horses. About 40 million people practise this way of life, including Mongolian nomads, Aboriginal Australians, the San of Africa.

*   **Shifting cultivation** is a mode of farming long followed in the humid tropics of sub-Saharan Africa, Southeast Asia, and South America. In this practice of "slash and burn", farmers would cut the native vegetation and burn it, then plant crops in the exposed, ash-fertilized soil for two or three seasons in succession. This allows the farmed area to recover. Shifting cultivation is an example of arable, subsistence and extensive farming. It is the traditional form of agriculture in rainforest areas. 200–300 million people across 64 countries practise this form of agriculture.

▲ Figure 4 Deforestation of the Amazon rainforest by slash and burn in shifting cultivation

Connections: This links to material covered in subtopic 1.2.

Farming system	Shifting cultivation	Cereal growing	Rice growing	Horticulture and dairying
Example of where	Amazon rainforest	Canadian Prairies	Ganges Valley	Western Netherlands
Type	Extensive, subsistence	Extensive, commercial	Intensive, subsistence	Intensive, commercial
Inputs	Low—labour and hand tools	High use of technology and fertilizers	High labour, low technology	High labour and technology
Outputs	Low—enough to feed the family	Low per hectare, high per farmer	High per hectare, low per farmer	High per hectare and per farmer
Efficiency	High	Medium	High	High
Environmental impact	Low—if enough land to move to and time for forest to regrow	High—loss of natural ecosystems, soil erosion, loss of biodiversity	Low—padi rice has a polyculture, stocked with fish. Also grow other crops	High—greenhouses for salads and flowers are heated and lit In dairying, grass is fertilized, cows produce waste

▲ Table 2 Comparing some farming systems

- **Extensive farming** has low levels of inputs of fertilizers, pesticides, machinery and labour per unit of land and consequently lower productivity. It uses more land with a lower density of stocking or planting and lower inputs and corresponding outputs.

- **Intensive farming** maximizes productivity from a given unit of land with relatively high amounts of inputs and outputs per unit area. Animal feedlots are intensive. Inorganic fertilizer made using fossil fuels is key to maintaining soil fertility and crop yields.

- **Irrigated** agriculture requires huge amounts of water. It is the largest user of water globally, a trend encouraged by the fact that farmers in most countries do not pay for the full cost of the water they use. Agriculture irrigation accounts for 70% of water use worldwide and over 40% in many Organisation for Economic Co-operation and Development (OECD) countries. The alternative is to rely on rainfall for crops and animals and to capture excess in tanks or reservoirs.

- **Hydroponics** is the science of growing plants without soil by giving them mineral salts dissolved in water. The plant roots may be supported by an inert material such as gravel. It allows for more plants in a smaller space and no competition but is only used for salad crops or soft fruits. Less water is used but it is not used yet for large-scale staple crops.

- **Organic** farming avoids synthetic fertilizers or other chemicals and uses techniques such as crop rotation, green manures and compost to maintain a fertile soil. It uses biological pest control to reduce disease and pests.

- **Inorganic** farming uses synthetic fertilizers made by humans, which are often by-products of the petroleum industry.

Practical

Skills:

- Tool 2: Technology
- Tool 4: Systems and models
- Inquiry 2: Collecting and processing data
- Inquiry 3: Concluding and evaluating

1. Select one pair of contrasting farming systems from the list:

   - intensive or extensive
   - mixed or livestock based
   - subsistence or commercial
   - organic or inorganic

   - monocultures or diverse farms
   - family or corporate ownership
   - irrigated or rain-fed
   - soil based or hydroponic (no soil).

2. Research two examples of your chosen systems. They may be local or global. For example, you might investigate a cereal monoculture farm on the prairies and a mixed livestock or arable farm in a similar area.

3. Produce a detailed study of your selected farms using a medium of your choice. Your study needs to include:

   a. background information on where, what and how the farms operate

   b. crops or livestock grown

   c. inputs and outputs

   d. environmental impacts

   e. evaluation of sustainability of practices and soils.

▲ Figure 5 Nomadic pastoralism—reindeer herd near a Nenets chum in Yamal, northwest Siberia, Russia

Nomadic pastoralism and shifting cultivation

These two forms of extensive agriculture have existed since humans started farming and are successful when human population densities are low. But today there are great pressures from settled peoples on the nomadic way of life. When humans settle, they define their territories—homes, farm fields, towns or countries—and that leaves little room for wandering societies. There are also internal and external pressures on nomadic peoples to settle. Increasing modernization may mean they adopt modern conveniences which are not compatible with a nomadic way of life, or they are encouraged to settle by those settled societies which do not like their territories being traversed. This encouragement may take the form of coercion or persuasion.

(ATL) Activity 8

ATL skills used: thinking, communication

Figure 6 shows how the various forms of agriculture are carried out in certain regions of the world.

Use the figure and what you have learned to answer the questions.

1.   Give possible reasons for the distribution of these types of agriculture.

2.   Why do central Australia and southwest China have little or limited agricultural use?

3.   What is the dominant biome near the equator and why?

4.   What climatic conditions does rice grow in?

5.   Why is fishing concentrated near coastlines?

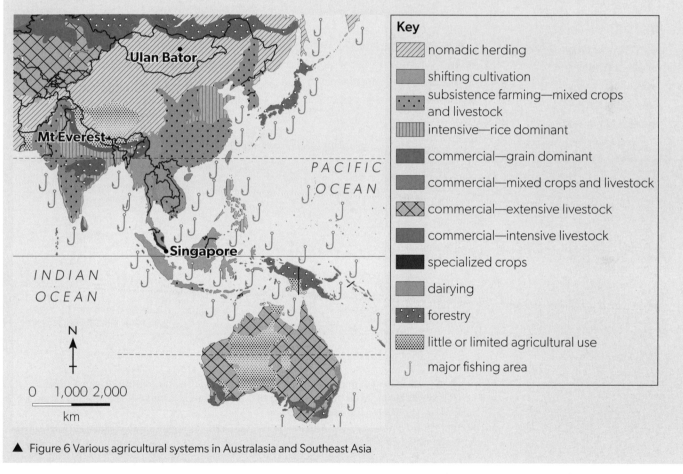

▲ Figure 6 Various agricultural systems in Australasia and Southeast Asia

The Green Revolution

From the 1940s to the 1960s, plant breeding of wheat and rice and then other cereals was undertaken to produce varieties that were less prone to disease, had shorter stalks (so they did not fall over in rain) and gave higher yields. This was done by artificial selection of the varieties and individual plants that had the traits wanted and the results were spectacular in terms of crop yields. (It was before we knew about the possibility of or could carry out genetic engineering—switching genes from one species to another.)

Connections: This links to material covered in subtopics 1.3, 2.3 and 6.2.

In Mexico, wheat yields increased such that the country became self-sufficient in wheat and then exported the surplus. In India, the IR8 variety of rice, a high-yielding variety (HYV), gave five times the yield of older varieties with no added fertilizer and ten times with fertilizer. This led to a fall in the cost of rice and to India exporting the surplus. However, in Africa, there was little difference in crop yields.

Some have been critical of the Green Revolution varieties, as the result has been increased use of inorganic fertilizer, irrigation, mechanization and pesticides (if farmers can afford them). While some say that the poor have become poorer because of this, there is little doubt that the HYVs have allowed humans to produce enough food for the global population (see Figure 7).

However, the increased use of fertilizers, irrigation and chemicals has caused eutrophication, salinization and the accumulation of chemicals in food chains. The Green Revolution has also reduced genetic diversity in crops as most farmers use the new varieties. Farmers were encouraged to take out loans to buy the new seeds and products. Improved productivity depended on fixing nitrogen into synthetic fertilizers, making crop production fossil-fuel dependent. These measures often replaced traditional practices and the Green Revolution has been criticized for its economic and environmental consequences (Figure 8).

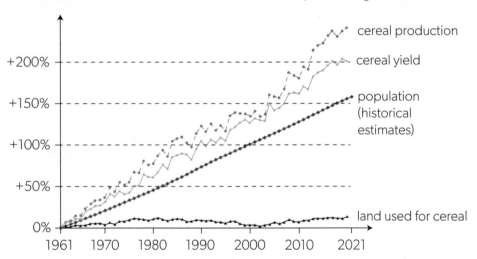

▲ Figure 7 Change in cereal production, yield, land use and population since 1961 (All figures are indexed to the start year of the timeline. This means the first year of the time-series is given the value 0%)
Our World in Data based on World Bank; Food and Agriculture Organization of the United Nations (CC BY 4.0)

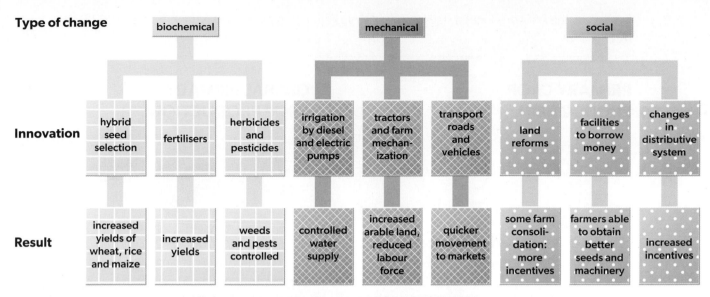

Figure 8 Key changes due to the Green Revolution and their impacts on farming

ATL Activity 9

ATL skills used: thinking, communication, research

Read about the Green Revolution and use the text and Figure 8 to answer these questions.

1. What was the Green Revolution?

2. Look at Figure 7. If land use is more or less constant from 1961 to 2021, explain how cereal yield can increase.

3. Describe the relationship between harvested area and cereal production, as shown in Figure 7.

4. State the percentage increase in cereal yields from 1961 to 2021.

5. Calculate the average percentage increase per year of (a) cereals and (b) population.

6. Explain what the difference in percentage increases in your answer to question 5 means in terms of limiting factors in human population growth.

7. For what reasons would cereal production be higher than cereal yield?

8. Evaluate the impacts of the Green Revolution.

Every year, the FAO produces a statistical yearbook. Figure 9 shows some key facts from the 2022 book.

PRIMARY CROP PRODUCTION

 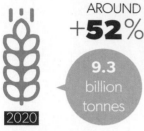

AROUND +**52**%

9.3 billion tonnes

2000 2020

The production of primary crops was 9.3 billion tonnes in 2020, 52% more than in 2000.

GLOBAL PRIMARY CROP PRODUCTION

OTHER PRIMARY CROPS

SUGAR CANE
MAIZE
WHEAT
RICE

Four crops account for about half of global primary crop production: sugar cane, maize, wheat and rice.

VEGETABLE OILS PRODUCTION

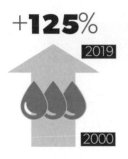

+**125**%

2019

2000

The production of vegetable oils went up 125% between 2000 and 2019, driven by a sharp increase in palm oil.

MEAT PRODUCTION

OTHERS

2020

+**45**%

CHICKEN MEAT

2000

337 MILLION TONNES

337 million tonnes of meat were produced in 2020, 45% more than in 2000, with chicken meat representing more than half the increase.

MEAT PRODUCTION BREAKDOWN

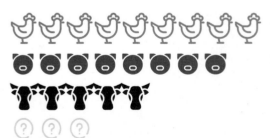

Chicken meat was the most produced type of meat in 2020.

CEREALS TRADE

LARGER EXPORTERS

EUROPE

AMERICAS

ASIA

LARGER IMPORTER

Cereals are the most traded commodity by quantity in 2020: the Americas and Europe are the largest exporters and Asia the largest importer.

▲ Figure 9 FAO yearbook facts, 2022

FAO. 2022. World Food and Agriculture – Statistical Yearbook 2022. Rome. https://doi.org/10.4060/cc2211en (CC BY-NC-SA 3.0)

RISE OF HUNGER

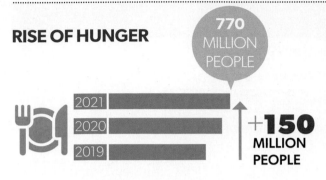

770 MILLION PEOPLE

2021
2020
2019

+150 MILLION PEOPLE

Hunger is still on the rise, with almost 770 million people undernourished in 2021, 46 million more than in 2020 and 150 million more than in 2019.

PREVALENCE OF UNDERNOURISHMENT

ASIA

AFRICA

MOST OF THE UNDERNOURISHED PEOPLE

HIGHEST PREVALENCE OF UNDERNOURISHMENT

While most of the undernourished people live in Asia, Africa has the highest prevalence of undernourishment.

PREVALENCE OF FOOD INSECURITY

In every continent, the prevalence of moderate or severe food insecurity is slightly higher for women than for men, with the largest differences found in Latin America and the Caribbean.

DIETARY ENERGY SUPPLY

ASIA

OTHER REGIONS

2000

Dietary energy supply went up in all regions since 2000, with the fastest increases in Asia.

INCREASE OF OBESITY

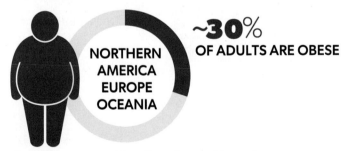

NORTHERN AMERICA EUROPE OCEANIA

~30% OF ADULTS ARE OBESE

Obesity has increased in all regions. Almost 30% of adults in Northern America, Europe and Oceania are obese.

▲ Figure 9 FAO yearbook facts, 2022 (cont.)
FAO. 2022. World Food and Agriculture – Statistical Yearbook 2022. Rome.
https://doi.org/10.4060/cc2211en (CC BY-NC-SA 3.0)

ATL Activity 10

ATL skills used: communication, research

Review Figure 9, which shows some key facts from the 2022 FAO yearbook.

Research the facts in the yearbook of the current year or other sources.

1. Has production of the four primary crops (maize, wheat, rice, sugar cane) changed and what is it now?
2. Have oil and meat production increased? How have they changed?
3. How many people in the world are hungry now?
4. Where in the world are most people undernourished?
5. Have obesity levels changed worldwide?
6. Summarize the factors causing the wide disparity between levels of nutrition worldwide.

Making agriculture sustainable

Agricultural systems are simplified ecosystems. Farmers grow a crop and discourage other plant species or pests as they need maximum yield of the crop and little competition. Soil profiles are simplified as organic matter (the crop) is removed. In inorganic intensive arable cropping, farmers are aware that soil is becoming impoverished but there is little incentive to spend money improving their soils if the market pays no more for this.

Connections: This links to material covered in subtopics 2.3 and 2.4.

Soil conservation techniques

Growing crops removes most of the organic matter from the soil, and inorganic fertilizers introduce no new organic matter. Therefore, soil degradation is inevitable unless other practices are used. In sustainable agriculture, there are methods for improving soil fertility that do not involve inorganic fertilizers.

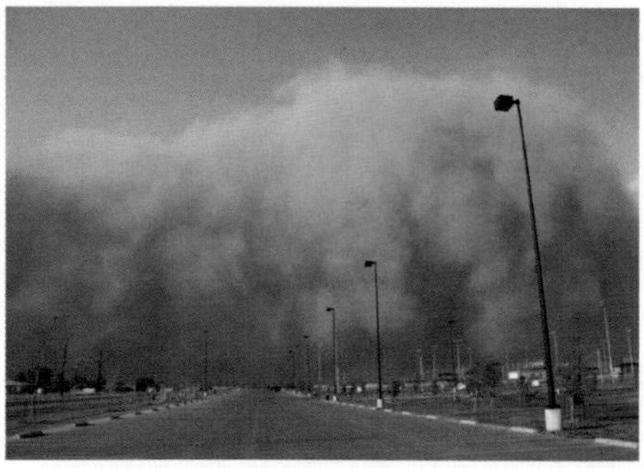

▲ Figure 10 Dust storm of dry soil blown by the wind —an example of soil erosion by wind

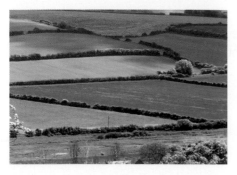

▲ Figure 11 Shelter belts in Saskatchewan, Canada

1. **Conservation from erosion by wind and water:**
   a. against erosion by wind:
      - planting tree or hedge windbreaks, or trees or bushes between fields (shelter belts—see Figure 11)
      - use of cover crops—crops grown not for sale but to protect the soil
      - stone walls and solid fencing
      - solid fencing

   b. against erosion by water:
      - terracing—often for rice growing
      - contour ploughing—plough along not down a contour line
      - bunding—soil embankments to slow or stop water flow
      - drainage systems—ditches, pipes, reservoirs
      - use of cover crops

- continuous ground cover—by an inert material or by mulch

- stone lines—placed along contours to slow water run-off.

◀ Figure 12 Deep channels develop on hillsides following rainfall—an example of soil erosion by water

2. **Conservation of fertility:**

- adding lime to increase pH of soils

- using organic materials such as compost, green manures, leaf litter

- selective harvesting of forest trees to remove the best and leave others to grow

- fallowing—when a field is not harvested for a year

- herbal mixed leys—a ley is temporary grassland and a herbal ley is a complex seed mixture of grasses, legumes and herbs which benefits livestock and increases soil fertility

- using mycorrhizal fungi—these form an association between the crop plant and the fungus. The plant provides sugars for the fungus; the fungus provides water and nutrients, particularly phosphorus, to the plant

- agroforestry—planting trees with agriculture on the same land, e.g. nut trees with sheep provides two products.

3. **Cultivation techniques:**

- avoid using marginal land which may be too steep or have poor soils

- avoid overgrazing or overcropping

- strip cropping—cultivate a field in long, narrow strips with crop rotation to reduce run-off and increase fertility

- mixed cropping

- crop rotations—change what crop is grown every year in a three or four year rotation

- reduced tillage (mechanical manipulation of soil to make it suitable for planting seeds)

- reduced use of heavy machinery.

(ATL) Activity 11

ATL skills used: research

Select four soil conservation techniques. Research their roles and benefits.

Human dietary shift

Humans have an almost unlimited choice of foods if they can afford it. Meat consumption is increasing as the world gets richer (Figure 9). Meat is an important source of nutrition for many people around the world. Global demand for meat is growing: over the past 50 years, meat production has more than tripled. The world now produces more than 340 million tonnes each year.

Animals that we may choose to eat mostly eat grass (ruminants such as cows, sheep, goats, buffalo) and much of that grass grows on soils that cannot support other crops. But meat production is energy intensive when, instead of grazing on grass, these animals are kept indoors or in pens and fed fishmeal or cereals. These are the feedlots of North America where animals are confined, fed high-protein diets and grow fast.

You know that more energy is lost in higher trophic levels of a food chain, so a dietary shift by humans towards eating more vegetables and less meat or fish may be more sustainable. The yield of food per unit of land area is greater in quantity and lower in cost with crops rather than livestock.

Whatever metric is used—e.g. protein or fat supply or calories per person—the world map looks much the same: the HICs are well-fed and LMICs and LICs less so.

If more people in HICs started eating less meat and fish and moved towards a more plant-based diet, there could be benefits both to their health and towards making agriculture more sustainable.

Sustainable food supply

Land area is finite and we are farming just about all of the land that can be farmed in some way. Therefore, the global strategies to achieve sustainable food supply have to focus on efficiencies. This means:

- reducing demand, food waste and GHG emissions from food production

- increasing productivity without increasing the area of land used for agriculture.

How can this be done?

1. Making plant-based meat substitutes affordable and desirable. These are meat analogues made from plants and usually have soy, pea, wheat gluten and various other pulses in them. Burgers are an example. However, they can have high sodium levels and controversial additives. So the question becomes: why not just eat the plants?

2. Reducing nitrogen loss to the atmosphere. Nitrogen is added to soils in fertilizers but is lost by leaching and by denitrification back to atmospheric nitrogen, N_2. Soil conservation measures, precision farming and effective application of nitrogen at the time of growth reduce this loss.

3. Reducing methane losses from ruminants and rice growing.

4. Extending shelf life for food to reduce wastage. Freezing, drying, irradiating, heat treatment and modifying the atmosphere of the product (e.g. bags of salad with an atmosphere of nitrogen) are all carried out.

5. Using genetic modification (GM) to boost yields. In this process, DNA of a crop plant is altered to change how it grows or make it resistant to a disease.

Connections: This links to material covered in subtopics 1.3 and 2.2.

Connections: This links to material covered in subtopics 1.3, 2.3, 3.1 and 6.3.

6. Using solar power to replace the Haber process which makes ammonia from nitrogen and hydrogen, and using sunlight and new catalysts called photocatalysts on farm sites. This technocentric approach may be as revolutionary as the Green Revolution.

Food security

Food security is physical and economic availability of food allowing all individuals to get the balanced diet they need for an active and healthy life. Food insecurity is the opposite.

Connections: This links to material covered in subtopic 7.1.

Ukraine in eastern Europe has very fertile soil (called mollisol, chernozem or black earth, and rich in humus). This allows Ukraine to produce 10% of the world's wheat, 15% of corn (maize), 13% of barley and over 50% of sunflower oil.

In 2021 after the Ukraine–Russia conflict started, and growing and exporting became difficult, food commodity prices in Ukraine were at a 10-year high and fuel prices increased hugely. This led to fertilizer prices tripling and food prices worldwide rising steeply.

In 2022, the World Food Programme (WFP), part of the UN, determined that conflict, COVID-19, the climate crisis and rising costs "combined to create jeopardy for up to 828 million hungry people across the world". These people face critical food insecurity.

According to the WFP, the following are four major causes of food insecurity.

- **Conflict** is still the biggest driver of hunger, with 60% of the world's hungry living in areas afflicted by war and violence. Events in Ukraine are further proof of how conflict feeds hunger, forcing people out of their homes and wiping out their sources of income.

- **Climate change** and disasters destroy lives, crops and livelihoods, and undermine people's ability to feed themselves.

- The economic **consequences of the COVID-19 pandemic** are driving hunger to unprecedented levels.

- **Costs** are also at an all-time high: WFP's monthly operating costs are USD 73.6 million above their 2019 average—a staggering 44% rise. The extra now spent on operating costs would have previously fed four million people for one month.

The hotspots of world hunger are in the Central American Dry Corridor and Haiti; in Africa in the Sahel, Democratic Republic of the Congo, the Tigray region and Sudan and eastwards to the Horn of Africa; Syria; and Yemen to Afghanistan.

People leave their home country to escape hunger, conflict, climate change or economic hardship. The number of refugees worldwide leaving their home country increased from 20.7 million at the end of 2021 to 35.3 million at the end of 2022.

When added to the number of internally displaced people (IDPs)—who are forced to move through hunger or conflict but do not leave their home country—the total figure is greater than 100 million people. This is about 1 in 78 people on Earth.

ⓐ Activity 12

ATL skills used: thinking, social, communication, research, self-management

The information about food insecurity above was correct in late 2022. Research the current food insecurity situation worldwide.

1. What is the current food price inflation level?

2. Which countries are facing severe food insecurity, and for what reasons?

3. Many LICs produce cash crops that are sold to HICs. This reduces food availability for the LIC's population.

   a. Why does this happen?

   b. What sociopolitical factors influence this?

   c. What ecological factors influence this?

   d. What are the international issues associated with this?

Contrasting agricultural choices are often due to differences in the local soils and climate. Where soils are the same type, local conditions, economics and cultural heritage also determine what is grown. Like wheat, rice grows in grassland regions, on mollisols. Cereal growing and ranching are also on these soils, in the steppe and prairies. Soy beans and cattle ranching are on the oxisols of tropical forests, such as the Amazon rainforest. Ranching and irrigated crops are on desert aridisols. Mixed arable and livestock are on temperate forest soils.

Case study 1

Ranching and soybean in the Amazon

Where? The Amazon basin is a huge area (6.5 million km²) in nine countries, comprising 5% of the Earth's land area. About 60% of the Amazon is in Brazil. This, the largest tropical rainforest on Earth, provides water for the region, regulates climate worldwide and provides habitats for many species.

What? Clearance of the Amazon rainforest for cattle ranching causes deforestation on a huge scale. Brazil exports more beef than any other country —about 20% of global beef exports; there are 230 million head of cattle in Brazil and their main diet is grass. In the 1960s there were five million head of cattle. Ranching accounts for 80% of deforestation in the Amazon basin and releases hundreds of millions of tonnes of GHGs.

Agriculture is the second main cause of deforestation. Traditionally, "slash and burn" shifting agriculture cleared a small area by fire and planted a crop for a few years, then moved on as the soil nutrients were depleted. With a small population, the land recovered, forest regrew, and the process was sustainable. Modern agribusiness—large-scale agro-industrial farming—is very different. Soy is the main crop—Brazil is the world's biggest soy exporter—but also grown are sugar cane and palm oil for biofuels, cotton and rice. Clearance for crops also occurs on the Cerrado—tropical savannah grassland.

▲ Figure 13 Map of South America showing the location of the Amazon rainforest

How? Forest is cleared by fire. Grazing density is low with less than one cow per hectare. Some workers were slave labour until 2008 when they were freed. More roads have allowed more clearance near them as transport becomes easier.

Why? Cattle ranching and soy production in this area are very profitable because forest and Cerrado land is very cheap and there is a lot of it.

Global demand for soy is increasing as grain becomes more expensive or subsidies elsewhere distort the market. For example, US government subsidies for corn-based ethanol production as a biofuel led US farmers to plant more corn and less soy. Then Brazilian farmers planted more soy as demand and price increased.

▲ Figure 14 Zebu, the predominant cattle in Brazil, at a recently logged ranch on the edge of the Amazon rainforest. Expanding cattle ranches are the biggest cause of deforestation in Brazil

What happens? Deforestation, ranching and agriculture alter the ecosystem, climate and weather patterns and cause biodiversity and habitat loss.

Cattle ranches are at risk of fire, cause soil erosion and silt up rivers.

Crop production involves fertilizer use, pesticides and soil degradation.

What has been done?

In 2006, Greenpeace reported the links between deforestation in the Amazon, soy and meat (between 70% and 90% of the world's soybean crop is used as animal feed). The industry responded with a moratorium which said that companies would not buy soy from soy traders supplied by farmers who cleared the rainforest after 2008, used slave labour or threatened Indigenous Lands. From 2004 to 2012, the clearing of trees in the Amazon fell by 84%. Each year the moratorium was renewed and in 2016 it was renewed indefinitely.

However, farmers continue to clear the land for other crops and deforestation reached a 15-year high in 2021, possibly encouraged by political leadership at that time.

What can be done?

Intensifying farming of cattle and crops would use less land for more productivity. But monocultures have hazards, including more mechanization in large fields with less labour required. Intensification requires more input and inorganic fertilizers have become far more expensive. Herbicides and insecticides used can damage aquatic life.

▲ Figure 15 Farmer spraying pesticide on a soy field

In Brazil, legislation could be used to limit the amount of any land holding that can be cleared of forest. Local governments should commit to reducing deforestation and GHG emissions, and act upon these commitments. They should also redirect investment to more sustainable use of forest products.

HICs could fund protection of the forests for LICs. Companies should not buy products resulting in deforestation. Individuals can boycott beef from deforested areas, eat less meat and aim to reduce carbon footprints.

Question

As a class, research the current state of the Amazon rainforest and Cerrado in terms of deforestation rates, crops grown, political actions taken and indigenous peoples' rights. Present your information in a form that can be shared with the class or the whole school.

Environmental law and environmental ethics

Most of the Amazon rainforest of Brazil is public property but "land grabbing" is when land is illegally registered as private property. Then, because there is more land for sale, the price of land falls.

Regulation is low in such a large area and it is too late to protect a forest once it has burned.

Indigenous peoples who may be evicted from land that is grabbed then move into pristine forest to survive, causing more to be cleared. What are their rights?

ATL Activity 13

ATL skills used: thinking, communication, research

Compare one pair of contrasting examples from one biome and soil type, for example:

- cereal and ranching in the mollisols of steppe and prairie
- ranching and irrigated crops in desert aridisols
- mixed arable and pasture in temperate forest brown earths.

Use Table 3 to help you to investigate your chosen pair. Present your results in an appropriate form.

Soil type and biome type	E.g. oxisol of tropical rainforest	Contrasting agricultural choices	
		E.g. ranching	E.g. cereal growing
Inputs	E.g. fertilizer, water, pest-control, labour, seed (GM or not), breeding stock, livestock growth promoters, machines		
Outputs	E.g. food quality, food storage, yield, pollutants, transportation, processing, packaging		
System characteristics	Diversity		
	Sustainability		
Environmental impacts	Pollution		
	Habitat loss		
	Biodiversity loss		
	Soil erosion or degradation		
	Desertification		
	Disease epidemics		
Socio-economic factors	Subsistence or cash crop		
	Traditional or commercial		
	For export or local consumption		
	For quality or quantity		
	Employment of workforce		

▲ Table 3

Palm oil—rainforest in your shopping

The oil palm is a tropical palm tree indigenous to West Africa and Central America but imported to Southeast Asia in the early 1900s. Here it is grown for its oil. Half of the large plantations are in Malaysia and the rest are in Indonesia and other Southeast Asian countries. In Indonesia, the area of land occupied by palm oil plantations has doubled in the last 10 years and is still increasing. According to Friends of the Earth, an NGO, demand for palm oil is the most significant cause of rainforest loss in Malaysia and Indonesia.

Palm oil is high in saturated fats and semi-solid at room temperature. It is extremely versatile and useful and is found in close to 50% of packaged products sold in supermarkets, including:

- cooking oil and margarine

- processed foods: chocolate, bread, crisps, pizza

- cosmetics (lipsticks), detergents and soaps (Sunlight Soap and Palmolive), shampoo.

It is also used in lubricants, animal feeds and biofuel.

What do you use that contains palm oil?

Oil palm plantations provide employment and exports. Growing a few oil palms can bring an income for a subsistence farmer and large oil plantations and processing plants provide much-needed employment.

However, oil palm plantations often replace tropical rainforest and, in Malaysia and Indonesia, primary rainforest has been cleared for oil palm. Habitat for orang-utans, pygmy elephant and Sumatran rhino is lost. Often this forest is on peat bogs which are then drained and habitats lost.

▲ Figure 16 Oil palm

To maintain the monoculture of oil palms, herbicides and pesticides are used on the plantations and these poison other animal species. Animals that were in the rainforest, such as elephants, move into the plantations seeking food and are killed as pests.

Are there solutions? Yes.

Palm oil can be produced more sustainably and there are certification schemes to try to ensure that deforestation and exploitation are not occurring when oil palm is the crop. But palm oil supplies 40% of the world's vegetable oil demand, and it is so efficient at converting sunlight to oil that it needs less than 6% of the land used to produce all other vegetable oils. To get the same amount of alternative oils like soybean, coconut or sunflower oil, you would need between 4 and 10 times more land. This would just shift the problem to other parts of the world and threaten other habitats, species and communities. Also, there are millions of smallholder farmers who depend on producing palm oil for their livelihoods. Boycotting palm oil is not the answer. More action is needed to tackle the issues and problems associated with palm oil.

Soil degradation

Degraded land is land that has lost some degree of its natural productivity due to human-caused processes but there is no generally agreed definition. On a global scale, around 25% of total land area is degraded.

Soil can be considered "black gold", and we are running out of it. The UN has declared soil finite and predicted catastrophic loss within 50 years.

Table 4 shows how much land had been degraded by 2019. "Degraded" here means loss of productivity due to land management practices.

Proportion of degraded land over total land area	Countries
0–10%	Australia, Chad, Finland, Mali, Nepal, Peru, Poland, Venezuela
10–20%	Botswana, Cambodia, Canada, Ecuador, Italy, Spain, Turkey
20–30%	China, Nigeria, Paraguay, South Africa, Vietnam
30–40%	Benin, Eritrea, Panama, Ukraine
40–50%	Philippines
Over 50%	Mexico
No data	Argentina, Brazil, Egypt, Japan, New Zealand, Norway, Thailand, USA

▲ Table 4 Examples of countries with varying proportions of degraded land over total land area (in 2019)
Data from multiple sources compiled by the UN. Published online at OurWorldInData.org. Retrieved from: https://ourworldindata.org/grapher/share-degraded-land (CC BY 4.0)

Environmental issues include climate change, loss of biodiversity, lack of drinking water, poor sanitation and the depletion of fuel wood supplies due to unsustainable rates of use. All of these are significant, but it could be argued that land degradation is the most pressing environmental and social problem facing society today, particularly affecting the world's poor.

Connections: This links to material covered in subtopic 1.3.

It is estimated that an area equal to the size of China and India combined is now classified as having impaired biotic function (damaged ecosystem structure) as a result of poor land management resulting in soil loss. As populations expand, and as social and cultural changes occur, greater and greater demands are being made on larger areas of landscape and soil.

In HICs where there has been a relatively long tradition of agriculture on an industrial scale, the agricultural culture includes a knowledge of land management that aims for sustained soil fertility and strives to avoid soil erosion. However, even here there are occasions when climate and intensive agriculture conspire to bring about unprecedented levels of soil erosion.

Where in the world is soil degradation happening?

Degradation and erosion are highest in Asia, Africa and South America: a loss of 30–40 tonnes of soil per hectare per year. The rate of soil loss is about 17 tonnes per year in Europe and the USA.

Solutions to soil degradation

Numerous alternative farming approaches have been developed in relation to the current ecological crisis, to promote soil regeneration. These approaches attempt to address food sustainability, water quality and local economic stability, as well as restoring and conserving soils.

Soil regeneration is the process of improving the quality of soil by adding organic matter, which helps to improve drainage, water retention, and nutrition for plants. There are many ways to regenerate soils but, as you have read, this is a long-term process as soils develop slowly.

In **rewilding** (see subtopic 2.5), nature is allowed to restore itself. There is still human management but intensive cropping is not part of this and soil fertility builds up.

Permaculture ("permanent agriculture") is tackling the issue of how to grow food, build houses and create communities, while minimizing environmental impact. It promotes multi-cropping and integrated farming systems, using land areas of any size from a balcony to a backyard to a large farm. It is highly ethical farming that aims to care for the Earth, care for people and share any surplus.

Non-commercial cropping means growing crops not for sale but for personal consumption or for soil improvement.

Zero tillage or no-till farming on arable land is when the soil is not disturbed (tilled) by ploughing, or by harrowing or cultivating between harvest and sowing the next crop. Seed is sown directly into the remains of the previous crops. Soil erosion is decreased as plant roots of the previous crop anchor it and there is less compaction of soil so it retains more water. However, the cost of zero tillage equipment is high and there may be an increase in herbicide use as sometimes the previous crop remains are killed off (glyphosate is often used).

Cover cropping is where a crop (such as alfalfa, rye, clover or buckwheat) is grown to cover the soil, not for harvest. The cover crop is then either killed off or cultivated into the soil, increasing organic matter content.

Mulching also involves covering the soil and mulches can be anything from cardboard, sawdust, compost, dead plants to plastic sheeting. All mulches reduce evaporation or erosion and those with organic material improve fertility.

▲ Figure 17 Digging a cover crop into the soil

Crop rotation has been used since agriculture started, once people realized that growing the same crop year after year on the same land degraded the soil. Ancient Near Eastern farmers alternated legume and cereal growing to maintain fertility. Sometimes the fields are left fallow (not cropped) in a two-field rotation. The three-field rotation was practised in Europe from the Middle Ages to the 20th century. This was a year of cereals, then one of legumes, then a fallow year. But it meant that one-third of the land each year did not produce. The Norfolk four-course rotation—wheat, turnips, barley, clover, in that order—gave a crop on all land and the turnips fed livestock which could then be kept through winters. Clover adds nitrogen to the soil as it is a legume with root nodules, while turnips are a root crop used as animal feed and have deep roots which mix the soil layers. Wheat and barley can be cash crops or animal feed.

Regenerative farming

These systems use animals to clear vegetation and plough the land, or use mob grazing to improve soil.

Until the last 70 years, most farms had livestock and grew crops—mixed farming. The animal manure added nutrients and organic matter to the soil in which the crops grew. Some crops were grown as animal feed (e.g. turnips) and some for cash cropping. Rotations of animals and grass fields with crops such as legumes

▲ Figure 18 Mulch of organic matter around a young fruit tree

and cereals kept the soil in good condition. With increased mechanization of farms and bigger machines, hedges have been removed to make bigger fields (which are easier to work) and modern farms tend to specialize in crops or animals. This means farmers have to buy fertilizer for their crops or buy animal feed for their animals.

Now, there is a move back to mixed farming called regenerative farming. This aims to regenerate the soil through practices such as:

- **mob grazing**, which is short-duration, high-density grazing with a longer than usual grass recovery period. Move a herd of cattle on average once a day and leave the grass to recover for between 40 and 100 days. Electric fences make this possible. If chickens are allowed into the areas after the cattle are moved, they scratch in the soil and dung to eat worms or insects. This disperses the dung, which is then incorporated into the soil more quickly by worms

- allowing **pigs** to root with their snouts in an area of ground. This mimics the actions of wild boar, a keystone species. Allowing pigs to live in temperate woodland means they clear the ground layer, removing plants such as nettles, brambles, bracken, willow and couch grass. They eat all tree saplings which are small enough, reducing the number of non-native or unwanted tree species. They can also be used to clear fallen seeds such as acorns which can be toxic to horses and cattle

- **herbal leys**, which are mixes of grasses, legumes and herbs that farmers plant for their animals to graze. The roots go to different depths in the soil and some herbs go so deep that they bring nutrients up to shallower-rooted plants. Legumes add nitrogen; carbon sequestration and biodiversity are increased; and animals benefit as some species have medicinal roles.

You may wonder why farmers moved away from these seemingly commonsense approaches in the first place. Look back at the information about the Green Revolution and consider how humans have been able to feed a growing population through intensive agriculture on finite land resources. The economics of specialized farms make business sense when inputs (chemicals, fertilizers, fuel) are relatively cheap but there has been a lack of understanding of sustainability in agriculture and now this is recognized.

Technological improvements

These can lead to very high levels of productivity, as seen in modern high-tech greenhouse and vertical farming techniques that are increasingly important for supplying food to urban areas.

One approach to reducing soil degradation is to not use soil as a growing medium at all. Hydroponics involves growing plants in water with mineral nutrients, with no soil.

In vertical farming, plants are stacked in layers in a controlled environment with artificial lighting. Vertical farms may be put in disused underground tunnels, shipping containers, mine shafts or buildings. There is increased crop yield per unit area of land but start-up and input costs are high and crops tend to be salads, strawberries or medicinal plants.

▲ Figure 19 A robot assistant detecting weeds and spraying chemicals in a vertical farming system

Organic agriculture

On 18 January 2016, Sikkim, India became the first 100% organic state in the world; duly acknowledged by Prime Minister Narendra Modi.

Organic farming uses no synthetic fertilizers or chemicals and no genetically modified plants or animals. But it does use organic fertilizers, crop rotation and biological pest control. This is how everyone farmed before the advent of synthetic chemicals, but now, high synthetic input farming is called "conventional".

Organic agriculture is practised in 187 countries, and 72.3 million hectares of agricultural land are managed organically by at least 3.1 million farmers. At the time of writing, however, certified organic farming comprises only 1.5% of farming globally.

The question is: can organic farming feed the world? Data on yields is not yet clear but some crop yields from organic farming are the same as or greater than those from conventional farming. Organic foods have a premium value so the economies may make sense for farmers.

According to the FAO, global sales of organic food and drink reached more than USD 112 billion in 2019. But global food market revenue in 2021 was about USD 8.3 trillion, so most food is not produced organically yet.

Genetically modified crops

Genetically modified (GM) crops have DNA of one species inserted into the crop species to form a transgenic plant. GM crops of soybean, cotton and maize are the most common but the issue of GM crops is surrounded by controversy in some countries over ethics, food security and environmental conservation.

Proponents of GM crops say they are part of the solution to increasing food production. By making a crop disease or pest resistant, fewer chemicals need to be used and less is lost in spoilage. We are only doing what selective breeding has done since farming began—albeit on a molecular scale. Golden rice has been made to synthesize beta-carotene, the precursor to vitamin A; as a humanitarian tool, this could prevent the vitamin A deficiency suffered by 124 million people. Golden rice has not yet been grown commercially.

Opponents of GM crops say we do not know what we are releasing into the environment. Could the GM plants cross-pollinate with other varieties so the introduced DNA escapes into wild populations? Could the DNA cross the species barrier? If a GM crop can kill a pest species that feeds on it, will that species die out? What will the effect on food chains be? Labelling of GM foods is now demanded and some countries have rejected them totally.

What you eat matters

Different diets have different environmental impacts, due to variations in supply chain efficiency, the distance that food travels and specific details regarding production of particular crops. Plant-based diets are usually more sustainable but it depends on factors such as:

- how and where the plants are grown

- how they are transported and how far

> **TOK**
>
> Eating plants is more sustainable than eating animals. How can we reconcile mixed farming, in which raising livestock and sending animals to market to enter the human food chain is part of sustainable agriculture? What is your perspective on this? Do different groups (your peers, teachers, other adults) have different views? Ask them.

> Connections: This links to material covered in subtopic 1.1.

- how much energy and what type of energy is needed to grow them

- how much is wasted in storage or leftover foods.

Consider the diets of two people living in Sweden. A diet including avocados, imported soft fruits, tuna and beef will have a very different sustainability value from a diet with local seasonal fruits, locally caught mackerel and local chicken.

In general, seasonal foods, grown locally with short supply chains and low fossil fuel inputs, and low in animal protein are more sustainable in terms of carrying capacity.

The planetary health diet

In 2019 an independent group of scientific experts (the EAT-Lancet Commission) produced a report which addressed the question: "Can we feed a future population of 10 billion people a healthy diet within planetary boundaries?"

They concluded that it was possible, but only if there is "widespread multi-sector, multi-level action" including:

- a substantial global shift towards healthy dietary patterns

- large reductions in food loss and waste

- major improvements in food production practices.

A sustainable global food system by 2050 means sufficiently healthy food for all with: no additional land-use conversion for food; protection of biodiversity; reduced water use; decreased nitrogen and phosphorus loss to waterways; net zero carbon dioxide emissions; and significantly lower levels of methane and nitrous oxide emissions.

The planetary health diet recognizes the following:

- Agriculture occupies nearly 40% of global land, making agroecosystems the largest terrestrial ecosystems on the planet.

- Food production is responsible for up to 30% of global GHG emissions and 70% of freshwater use. Land conversion for food production is the single most important driver of biodiversity loss.

- Foods sourced from animals, especially red meat, have relatively high environmental footprints per serving compared with other food groups. This has an impact on GHG emissions, land use and biodiversity loss. This is particularly the case for meat from grain-fed livestock.

- What is or is not consumed are major drivers of malnutrition in various forms. Globally, over 820 million people go hungry every day, 150 million children suffer from long-term hunger that impairs their growth and development, and 50 million children are acutely hungry due to insufficient access to food.

- The world is also experiencing a rise in obesity. Today, over two billion adults are overweight or obese, and diet-related non-communicable diseases including diabetes, cancer and heart disease are among the leading causes of global deaths.

- Urban food environments often pose particular challenges to health and sustainability given the concentrated availability of junk foods and related advertising.

The report suggests our diet should be as in Figure 20—based on whole grains, fruits, vegetables and nuts with a small amount of dairy and meat or fish. An average adult should eat no more than 2,500 kcal per day.

It also matters where your food comes from, how it is grown and how sustainably it is produced or harvested.

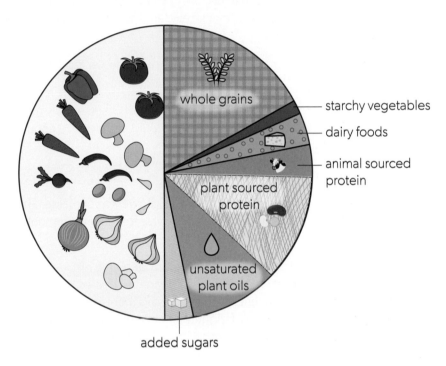

▲ Figure 20 The planetary health diet suggestion
Adapted from: EAT Foundation

(ATL) Activity 14

ATL skills used: thinking, communication, social

Is it better to eat organic, locally grown produce, or produce that has been grown in Kenya and air-freighted?

In small groups, evaluate the advantages and disadvantages of each and prepare a two-minute presentation for the class.

▲ Figure 21 Picking green beans for export

▲ Figure 22 Locally grown seasonal vegetables

If all humans alive today adopted the diet of Europeans, we would need 2.5 more planets like the Earth to support us. If we all ate organic food, we could not produce enough.

Commercial intensive agriculture has done a remarkable job in managing to feed the rapidly increasing human population. But it may not in the future.

What has to change?

In HICs, there is an increasing supply of organically produced foodstuffs, which are sold at a premium. The idea is that they are better for our health and free of pesticide residues. But there is little scientific evidence to support some claims for organic foods.

Locally grown foods are also marketed as "better" due to lower food miles and artisanal local producers.

However, large commercial farms and large supermarket chains supply most of our food—efficiently and at the lowest prices. We could not produce enough food if we all ate organic, local produce.

The tricky truth about food miles

Food miles measure how far food has moved from grower to your table. The premise is that foods with higher food miles are worse for the environment, so we should buy locally produced foods instead.

But this can be a fallacy: we should instead be considering the efficiency of food production and transport. Growing green beans in a heated greenhouse in northern Europe uses far more energy than growing them in Kenya, even including the carbon emissions of air-freighting them to Europe. In fact, Kenya's bean industry is very environmentally cost-efficient. Beans are grown using manual labour (no fossil fuels), using organic fertilizer from cows and with low-tech irrigation schemes.

It costs more in carbon emissions to drive 10 km in a fossil-fuelled car to a supermarket than to fly a packet of beans from Kenya to that supermarket.

According to the WWF, these are the best ways to eat sustainably and healthily.

- Eat more plants—enjoy vegetables and wholegrains.

- Eat a variety of foods—have a colourful plate.

- Waste less food—one-third of food produced for human consumption is lost or wasted.

- Moderate your meat consumption, red and white—enjoy other sources of protein such as peas, beans and nuts.

- Buy food that meets a credible certified standard. Consider Marine Stewardship Council (MSC), Aquaculture Stewardship Council (ASC), free range and Fairtrade food labels.

- Eat fewer foods high in fat, salt and sugar. Keep food such as cakes, sweets and chocolate, as well as cured meat, fries and crisps, to an occasional treat.

- Choose water, avoid sugary drinks and remember that juices only count as one of your five a day, however much you drink.

- Avoid ultra-processed foods (UPFs) which include ready meals, ice cream, ham, sausages, crisps, mass-produced bread, breakfast cereals, carbonated drinks, fruit-flavoured yogurts, instant soups and noodles.

Is it a sustainable diet?

Terrestrial versus aquatic food production systems

In terrestrial food production systems, food is usually harvested at the first (crops) or second trophic level (meat usually originates from primary consumers like cows, pigs and chickens). This means that these production systems are making a rather efficient use of solar energy. However, terrestrial systems do have higher losses when it comes to skeletal waste—land-based animals have more energy tied up in their skeletons as they have to support themselves on land.

In aquatic food production systems, most food comes from higher trophic levels. Typical food fish tend to be carnivorous and are quite often at trophic level 4 or higher. Because of the energy losses at each trophic level, the energy efficiency of aquatic food production systems is lower than that of terrestrial systems. Although the energy conversions along the food chain tend to be more efficient in aquatic ecosystems (because fish are cold-blooded), the initial intake of solar energy is less efficient than in terrestrial ecosystems because of the absorption and reflection of sunlight by water. Also, energy losses in the form of heat are higher in water than on land.

Harvesting from natural ecosystems

Harvesting wild species from ecosystems by traditional methods may be more sustainable than land conversion and cultivation. But it is mostly for subsistence and not suitable for large-scale production. A variety of secondary forest products are harvested, such as Brazil nuts, honey, wild animals, fruits (e.g. mango, guava), coconuts, cola nuts, cacao, vanilla, cinnamon, coffee, truffles (subterranean fungi), bamboo shoots, insects.

Here are three examples of foods harvested from natural ecosystems.

1. **Brazil nuts**, *Bertholletia excelsa*

   Brazil nut trees are some of the largest in the Amazon rainforest. They can live for 500 years and are found individually in pristine forest or sometimes in backyards. It is not yet possible to cultivate Brazil nut trees together at scale on farms so the nuts are harvested in the wild by migrant workers or indigenous peoples. Nuts in groups of 8–24 are found inside a heavy fruit, protected by hard shells. As long as some nuts are left for rodents such as agoutis to bury and disperse, harvesting can be sustainable with minimal environmental impact.

   However, there are problems, including forest clearance and illegal logging of the trees for their timber. Also, the collectors are not paid much for the nuts as they sell to buyers who sell them on for processing and eventual sale in shops.

   Is this fairtrade? What can be done to ensure this harvesting is sustainable?

▲ Figure 23 Why don't we eat more insects?

▲ Figure 24 The cane rat is often hunted for bushmeat

▲ Figure 25 A pangolin—said to be the most smuggled animal in the world

2. **Entomophagy**—humans eating insects

Humans eat most things that are not toxic—mammals (but rarely carnivores), birds, fish, plants, fungi—but not many insects. This is odd as other animals get much of their food from insects and in some cultures, insect-eating is common. Insects are high in protein and often low in fats, there are many species and they are abundant, e.g. locusts, mealworms. Why don't most humans eat insects?

3. **Bushmeat** is any wild animal killed for food. In some countries, this is called game; in others, hunting wild animals. The term bushmeat is highly politicized as many see it as illegal hunting, particularly in Central and West Africa and tropical rainforests. Although bushmeat can be many species, the emotive element focuses on the killing of the Great apes for meat, often orphaning their young.

Trade in bushmeat is increasing, because logging roads built into the forests make access easier for hunters and the meat has a high price in markets.

The cane rat is a large rodent pest of crops in West Africa, which is hunted for bushmeat. Some farmers in Benin and Togo are successfully starting to farm cane rats for meat.

Pangolins live in Asia and Africa. Their scales are a delicacy in some Asian countries. They are used in traditional medicine, rituals and as food. All eight species of pangolin are on the IUCN Red List as threatened with extinction.

(ATL) Activity 15

ATL skills used: thinking, communication, social, research

1. For what reasons do humans hunt and eat wild animals? Discuss this in class and make a list of your reasons. Can you put these reasons into different categories?

2. a. When, in your view, can eating wild animals be ethical?

   b. What, if any, do you think are the differences between eating wild animals and eating those raised for meat?

3. Subsistence agriculture was practised for thousands of years by people all around the world, fed the population and was sustainable with smaller global populations. What do you think is the place of low input and low productivity systems in feeding the population today?

Malnutrition

Malnutrition is an umbrella term for "bad" nutrition and it is the result of a diet that is unbalanced. Nutrients may be:

- lacking—undernourishment, usually a lack of calories

- excessive—overnourishment, usually too many calories leading to obesity

- unbalanced—the wrong proportion of micronutrients.

The FAO estimates that in 2020 between 720 and 811 million people in the world did not get enough energy from their food—they suffered from undernourishment. About 418 million were in Asia, 282 million in Africa and 60 million in Latin America and the Caribbean. Of these, about 200 million were children and infants. Chronic undernourishment during childhood years leads to permanent damage: stunted growth, intellectual disability and social and developmental disorders.

Many people are also suffering from an unbalanced diet—their food contains enough energy, but lacks essential nutrients like proteins, vitamins and certain minerals. The FAO estimates that three billion of us do not have a healthy diet.

Food is one of the most important resource issues facing global society today, alongside drinking water. As populations increase, as global trade expands and as market choice develops, greater and greater demands are being made on food supplies and food production systems. In the last 50 years, technology and science have made huge advances in agricultural practice and agricultural production, but the human population has increased too.

In many HICs, food is relatively cheap. Most people purchase foods out of choice and preference rather than basic nutritional need. Seasonality of produce has disappeared due to imports. Exotic foods are freely available all year round. Modern technology and transport systems mean that New Zealand lamb, beans from Kenya, dates from Morocco and bananas from the tropics can be bought in almost any HIC supermarket anywhere in the world, all year round.

In LICs, many populations struggle to produce enough food to sustain their population. There may be political and economic agendas as well as simple environmental limitations on food production. Much of their food may be exported to gain foreign currency.

Cereals may be grown for export and revenue generation (cash cropping), rather than to feed indigenous populations. Non-food crops may also be grown as cash crops: coffee, tobacco, hemp, flax and biofuels. These crops occupy land that could be used for food production and arable land is in finite supply.

Increasing sustainability of food supplies—a summary

Feeding the world's population in the future is a challenge, as the population is growing and the area available to agriculture is decreasing. The FAO has stated that by 2050, there will be another two billion humans to feed and this needs an increase in food production of 70% of current levels.

Factors that contribute to the decrease in agricultural land are:

- soil erosion

- salinization

- desertification

- urbanization.

We could improve sustainability of food supplies in the following five ways.

1. **Maximize the yield** of food production systems, without unsustainable practices. For example:

   a. **improve technology** of agriculture

      • Mixed cropping and interplanting conserve water and soil.

      • No-plough tillage (drilling seeds into the stubble of the previous crop) also conserves water and soil.

      • Buffer zones planted around agricultural land absorb nutrient run-off and provide a habitat for wildlife.

      • Biological control of pests and integrated pest management reduce losses.

      • Trickle irrigation wastes less water.

   b. **alter what we grow** and how we grow it

      • GM foods—inserting into cereals the gene from legumes that allows them to fix nitrogen would reduce the need for nitrogenous fertilizers.

      • Aquaculture and hydroponics.

      • Soil conservation measures (discussed earlier in this subtopic)

   c. **a new Green Revolution**

      • Agroecology—recycling nutrients and energy on farms within closed systems, with crops and animals balancing inputs and outputs.

      • Breeding plants more adapted to drought, increasing shade and keeping bare soil covered—drought-proofing farms.

2. **Reduce food waste** by improving storage and distribution

   a. In LICs, food waste is mostly in **production and storage**, e.g. loss through pest infestations, severe weather, lack of good storage such as no refrigeration, no canning factory nearby.

   b. In HICs, food waste is mostly in **consumption**, e.g. consumers buy more food than they need and let it go off; supermarkets have too strict standards (round apples, red strawberries) so reject edible if misshapen food; packaging preserves food but contaminates waste food that could otherwise be recycled for livestock.

3. **Increase monitoring and control**

   a. By governmental and intergovernmental bodies—to regulate imports and exports to reduce unsustainable agricultural practices.

   b. By multinational and national food corporations to raise standards and practices on supplier farms.

   c. by individuals and in NGO pressure groups.

4. **Change our attitudes** towards food and our diets

   a. Eating different crops.

   b. Eating less meat.

   c. Improving education about food.

   d. Increasing consumption of insects, a protein source that reproduces rapidly and in large number.

   Changing people's diet can improve the efficiency of food production. If we obtain more food from lower trophic levels (plants), we will greatly increase the amount of food available. People in HICs eat more meat than they actually need, so they could simply replace some of their meat with food taken from the first trophic level. However, the trend is the other way. More people in LICs and LMICs are eating more meat. We cannot all eat more meat. Even though animals graze on some land that would not support crops, it is energetically inefficient to feed our grain to animals which we then eat. In HICs, we mostly eat more than we need to sustain ourselves. Obesity has become a problem in some countries, famine in others. We do not seem able to get the balance right.

5. **Reduce food processing, packaging and transport** and increase consumer awareness of food production efficiency, e.g. by improving labelling of foodstuffs. While local foods have lower food miles, they may cost more in energy used to produce them. For example, growing tomatoes in a heated greenhouse in a temperate country such as the UK may use more total energy than growing them in the tropics and flying them into the UK.

Predictions for future food supplies

The FAO has made the following predictions on food supply for 2030.

1. Human population will grow to about 8.6 billion by 2030; in 2023 it was 8 billion. Demand for agricultural products will increase in LICs where consumption will increase.

2. The human population will be increasingly well-fed with per capita calorie consumption increasing.

3. More people will eat more meat.

4. Most increased production will be from higher yields and more irrigation, not more land.

5. GM crops, no-tillage planting, soil conservation measures and improved pest control will all increase productivity.

6. Aquaculture will increase.

TOK

- Consumer behaviour plays an important role in food production systems. Are there general laws that can describe human behaviour?

- Our understanding of soil conservation has progressed in recent years. What constitutes progress in different areas of knowledge?

- How are food choices influenced by culture and religion?

▲ Figure 26 Sustainable Development Goal 2 (SDG2)

According to the UN, SDG 2 is about creating a world free of hunger by 2030.

In 2020, between 720 million and 811 million people worldwide were suffering from hunger, roughly 161 million more than in 2019.

Also in 2020, a staggering 2.4 billion people—above 30% of the world's population—were moderately or severely food insecure, lacking regular access to adequate food. The figure increased by nearly 320 million people in just one year. Globally, 149.2 million children under five years of age (22%), were suffering from stunting (low height for their age) in 2020, a decrease from 24.4% in 2015.

ATL Activity 16

ATL skills used: thinking

Ecocentric or technocentric view of soil fertility management?

As agricultural systems, even those on the relatively small scale, become more and more intensive and more and more commercial, greater and greater demands are made on soil per unit area. The soil is required to "work harder", produce greater volumes per unit area, and cope with greater demands from genetically modified, nutrient-hungry plants. The soil in its natural state and operating under ambient environmental conditions can no longer fulfil the demands being made upon it. Therefore the soil environment must be modified artificially with the addition of fertilizers and additional irrigation water. This is the technocentrist's response.

Very quickly, farmers face spiralling costs as they attempt to maintain or increase productivity per unit area while at the same time maintaining soil health and thus guaranteeing long-term sustainability of the industry. It costs relatively large amounts to buy synthetic fertilizers and the hardware for irrigation. Therefore, the farmer must produce more to get the revenue to pay for this investment in chemicals and hardware. Increased production puts more strain on the health of the soil system, so more external inputs are required.

What could an ecocentric farmer do to avoid this vicious cycle?

Check your understanding

In this subtopic, you have covered:

- land is a finite natural resource

- yields or efficiencies need to increase to feed the human population

- food distribution and consumption are not equitable

- there are many different types of agricultural system

- the increased production of food has been due to fertilizers

- soils are degrading

- soils can be conserved and regenerated by conservation measures

- dietary shifts may be more sustainable

→

AHL

- choice of agricultural system depends on soil type

- there are farming systems that promote soil health, biodiversity and productivity

- different diets and food production systems have different impacts on the environment and different efficiencies

- malnutrition is common.

To what extent can the production of food be considered sustainable?

1. Describe the difference between land availability and human population growth.

2. Explain why food supply is unequally distributed.

3. Explain why agricultural systems vary.

4. List four pairs of different agricultural systems.

5. Discuss the role of traditional agricultural systems in sustainability.

6. Explain how food supply has increased.

AHL

7. Comment on reasons for soil degradation.

8. Evaluate ways to regenerate soils.

9. Discuss ways to increase global food supply.

10. Compare two farming systems on the same soil type.

11. Describe three alternative farming approaches.

12. Compare meat-based and plant-based diets in terms of their sustainability.

13. State reasons for malnutrition.

14. Predict if food supply will meet human needs in 2300.

Taking it further

- Investigate the impacts of your own diet

- Organize a school menu aligned with the planetary health diet or "Meat Free Mondays".

- Study interactive maps of world hunger (e.g. on the Visual Capitalist website) and research global trends of displaced people (e.g. on the UNHCR website). Use the information to inform your community of the facts in a way you think will have an impact.

- Engage with an organization like the World Food Programme and look for opportunities to support access to food. This could include volunteering at a local "soup kitchen".

- Learn to cook vegetarian or vegan meals and reflect on the carbon footprints of these meals as compared with other meals.

Exam-style questions

1. a. Explain the link between soil fertility, primary productivity and human activity. [7]

    b. Using examples, discuss how social, cultural, political and economic factors influence societies in their choice of food production systems. [9]

2. With reference to named examples, discuss the significance of diversity in the sustainability of food production systems. [9]

3. a. Outline the processes involved in the formation of fertile soils from bare rock. [4]

    b. Explain how negative and positive feedback mechanisms may influence the growth of decomposer populations in the soil. [7]

4. To what extent can the different environmental value systems improve the sustainability of food production? [9]

5. **Figure 1** refers to a typical western European diet. This example shows recommended consumption of food types (on the basis of health) and the environmental impact of their production.

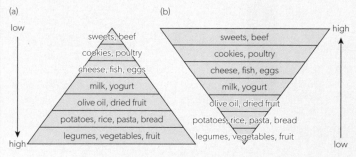

▲ Figure 1 (a) Recommended consumption pyramid and (b) Environmental impact pyramid

    a. With reference to **Figures 1(a)** and **1(b)** state the food that has the highest environmental impact. [1]

    b. With reference to **Figures 1(a)** and **1(b)** state the food that has the highest recommended consumption. [1]

    c. Describe the relationship between the pyramids in **Figures 1(a)** and **1(b)**. [2]

    d. Identify **two** environmental impacts associated with producing the foods near the base of the recommended consumption pyramid (**Figure 1(a)**). [2]

    e. Describe how foods high on the environmental impact pyramid, shown in **Figure 1(b)**, are likely to affect the ecological footprint of global food production. [2]

    f. Outline **two** reasons why the composition of a typical diet in other regions of the world may differ from the Western European diet shown in **Figure 1(a)**. [2]

6. **Figure 2** shows a typical soil profile.

◄ Figure 2 A typical soil profile

Adapted from: WilsonBiggs / Hridith Sudev Nambiar / US Department of Agriculture / Wikimedia (CC BY-SA 4.0)

    a. State **one** transfer of matter occurring within the soil profile. [1]

    b. State **one** transformation process occurring within the soil profile. [1]

    c. Identify **one** example of an output to the atmosphere from the soil system. [1]

    d. Describe **two** characteristics of soil with high primary productivity. [2]

    e. Outline **two** conservation methods that could be used to reduce soil erosion. [2]

7. With reference to **four** different properties of a soil, outline how each can contribute to high primary productivity. [4]

8. Discuss how human activities impact the flows and stores in the nitrogen cycle. [9]

9. Compare and contrast the impact of **two** named food production systems on climate change. [7]

10. To what extent is pollution impacting human food production systems? [9]

11. Climate can both influence, and be influenced by, terrestrial food production systems.

    To what extent can terrestrial food production strategies contribute to a sustainable equilibrium in this relationship? [9]

12. Soil quality is important for global food production systems.

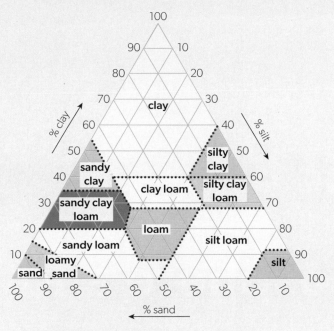

▲ Figure 3 Soil texture triangle

   a. With reference to **Figure 3**, state the soil texture that has the following composition: 20% clay; 55% silt; 25% sand. [1]

   b. Describe how the addition of sand to a silty clay loam could alter its characteristics for healthy plant growth. [2]

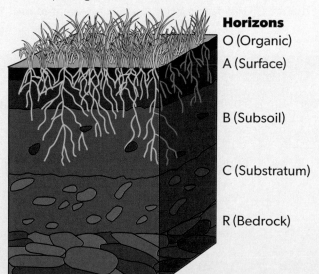

Horizons
O (Organic)
A (Surface)
B (Subsoil)
C (Substratum)
R (Bedrock)

   c. Draw a flow diagram to show the flows of leaching and decomposition associated with the mineral storage in the "A" horizon in **Figure 4**. [2]

   d. Identify **one** other input to the mineral storage in the "A" horizon in **Figure 4**. [1]

   e. Identify **one** other output from the mineral storage in the "A" horizon in **Figure 4**. [1]

   f. Outline why leaving arable farmland fallow (unused) between growing seasons could lead to soil degradation. [2]

13. a. Outline how soil can be viewed as an ecosystem. [4]

    b. Compare and contrast the impact of humans on the carbon and nitrogen cycles. [7]

14. a. The soil system includes storages of inorganic nutrients. Identify **two** inputs to these storages. [2]

    b. Identify **two** outputs from these storages or inorganic nutrients. [2]

    c. Solid domestic waste may contain non-biodegradable material and toxins that have the potential to reduce the fertility of soils.

       Explain how strategies for the management of this waste may help to preserve soil fertility. [7]

    d. The provision of food resources and assimilation of wastes are two key factors of the environment that determine its carrying capacity for a given species.

       To what extent does the human production of food and waste influence the carrying capacity for human populations? [9]

15. a. Outline **one** climatic and **one** edaphic (soil) factor which affect the final climax community in an ecosystem. [4]

    b. Explain **two** examples of soil degradation and the appropriate soil management strategies from a named farming system. [6]

◀ Figure 4 Horizons (layers) in a typical soil profile
Adapted from: WilsonBiggs / Hridith Sudev Nambiar / US Department of Agriculture / Wikimedia (CC BY-SA 4.0)

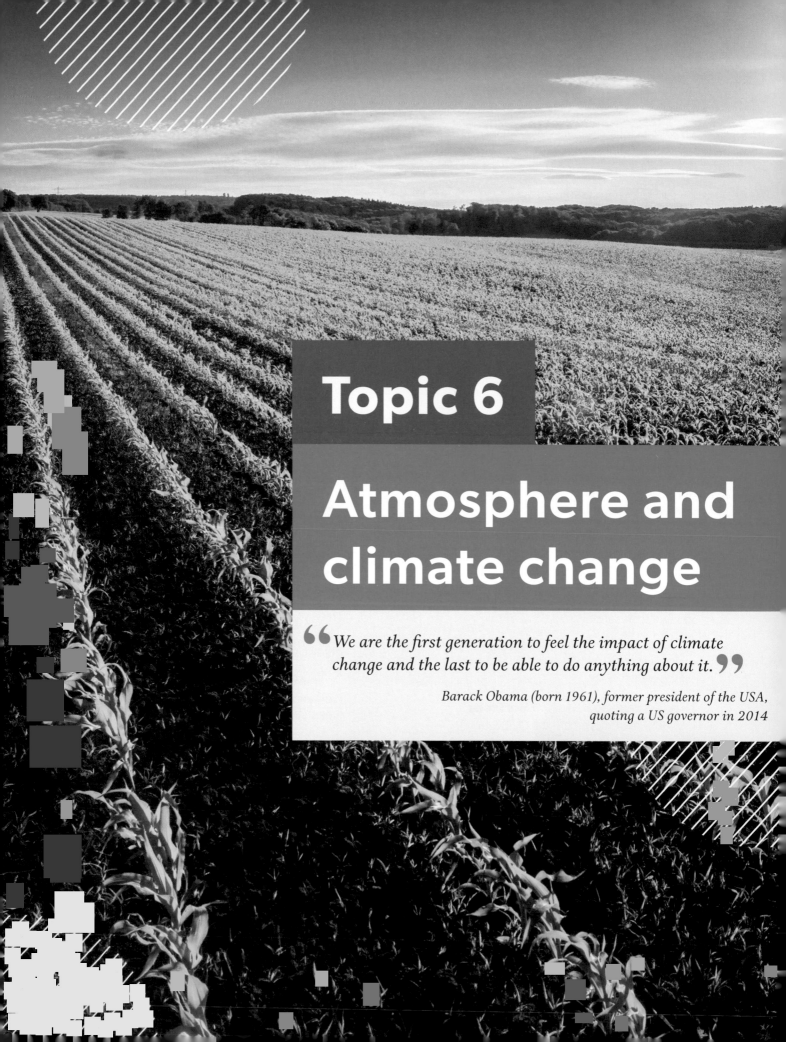

Topic 6

Atmosphere and climate change

"We are the first generation to feel the impact of climate change and the last to be able to do anything about it."

Barack Obama (born 1961), former president of the USA, quoting a US governor in 2014

Guiding question

- How do atmospheric systems contribute to the stability of life on Earth?

Understandings

1. The atmosphere forms the boundary between Earth and space. It is the outer limit of the biosphere and its composition and processes support life on Earth.
2. Differential heating of the atmosphere creates the tricellular model of atmospheric circulation that redistributes the heat from the equator to the poles.
3. Greenhouse gases (GHGs) and aerosols in the atmosphere absorb and re-emit some of the infrared (long-wave) radiation emitted from the Earth's surface, preventing it from being radiated out into space. They include water vapour, carbon dioxide, methane and nitrous oxides (GHGs) and black carbon (aerosol).
4. The greenhouse effect keeps the Earth warmer than it otherwise would be due to the broad spectrum of the Sun's radiation reaching the Earth's surface and infrared radiation emitted by the warmed surface then being trapped and re-radiated by GHGs.
5. The atmosphere is a dynamic system, and the components and layers are the result of continuous physical and chemical processes.
6. Molecules in the atmosphere are pulled towards the Earth's surface by gravity. Because gravitational force is inversely proportional to distance, the atmosphere thins as altitude increases.
7. Milankovitch cycles affect how much solar radiation reaches the Earth and lead to cycles in the Earth's climate over tens to hundreds of thousands of years.
8. Global warming is moving the Earth away from the glacial–interglacial cycle that has characterized the Quaternary period, toward new, hotter climatic conditions.
9. The evolution of life on Earth changed the composition of the atmosphere, which in turn influences the evolution of life on Earth.

AHL

What makes up the atmosphere?

The atmosphere forms the boundary between the Earth and space and has layers. It is the outer limit of the biosphere and its composition and processes support life on Earth. It is made up of various gases: 78% nitrogen, 21% oxygen and 1% other gases including carbon dioxide, argon, neon, hydrogen, ozone and water vapour (Figure 1).

The atmosphere is a dynamic system with inputs, outputs, storages and flows. Heat and pollutants are carried across the Earth by air currents in the atmosphere.

- The amount of water vapour in the atmosphere varies. It is measured by relative humidity and weather forecasts usually show this.

- The tricellular model of circulation (subtopic 2.4) disperses heat over the Earth.

- Wind redistributes the atmosphere and its gases.

- Circulation of the atmosphere and circulation of the oceans are how heat is redistributed around the Earth. Energy from the Sun drives this circulation due to uneven distribution of solar energy—greatest at the equator, least at the poles. If the circulation did not happen, the poles would be too cold for life and the equator too hot.

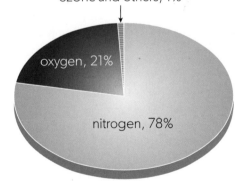

carbon dioxide, argon, water vapour, ozone and others, 1%

oxygen, 21%

nitrogen, 78%

▲ Figure 1 **Atmospheric composition**

There is very little carbon dioxide in the atmosphere (0.04% of the total gases) but it and other greenhouse gases (GHGs) are increasing through anthropogenic activities (activities of humans).

The equator gets the most direct sunlight and energy because the Sun's rays reach the equator at a higher angle—almost 90 degrees.

Much less solar energy reaches the poles because the angle is lower and because snow and ice reflect more of the energy than water and land. The difference in the amount of solar energy drives atmospheric circulation.

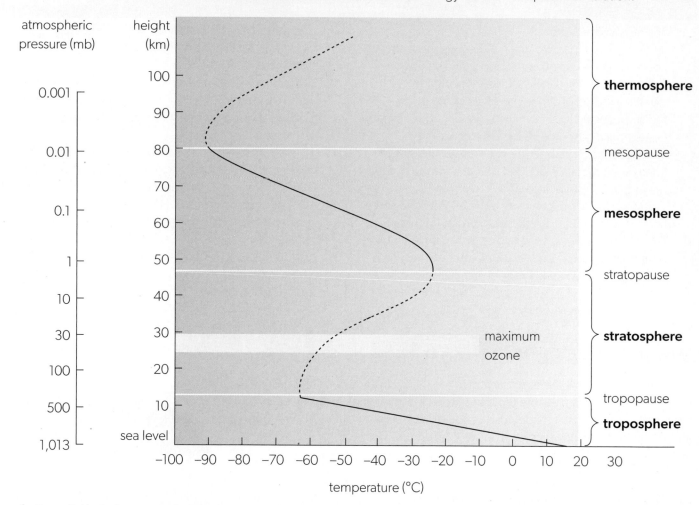

▲ Figure 2 Vertical structure of atmosphere

Connections: This links to material covered in subtopics 1.2 and 2.4.

Although the atmosphere is approximately 1,100 km in depth above the surface of the Earth, the **stratosphere** (10–50 km) and the **troposphere** (less than 10 km) are where most reactions affecting life occur, e.g. ozone formation and cloud formation.

Human activities and activities of other organisms impact atmospheric composition by altering inputs and outputs of the system. And the atmospheric composition affects activities of humans and other organisms.

Over geological time, the composition of the atmosphere has changed greatly. Changes in the concentrations of atmospheric gases such as ozone, carbon dioxide, methane and water vapour have significant effects on ecosystems and living organisms.

The Earth's energy budget

The Earth maintains a balance between the overall amount of incoming and outgoing energy at the top of the atmosphere. This is called the Earth's energy budget or the Earth's radiation budget.

Incoming energy from the Sun is short-wavelength energy. The Earth emits long-wavelength energy back to space. For the Earth's temperature to be stable over long periods of time (for the energy budget to be in balance), the amounts of incoming and outgoing energy must be equal. If incoming energy is more than outgoing energy, the Earth will warm. If outgoing energy is greater than incoming energy, the Earth will cool.

The energy reaching the Earth from our Sun is mostly visible light but also some ultraviolet and some infrared wavelengths. Averaged over the Earth, a square metre receives about 340 W of energy, which is slightly less than the output of a 60 W incandescent light bulb or a 9 W LED bulb.

About 30% of the Sun's energy is reflected back into space and 70% is absorbed by the Earth's surfaces. This absorbed solar energy is used in several processes including photosynthesis, heating the land and oceans, and evaporation. As the Earth's surface warms up, the solar energy is converted to heat energy— infrared radiation which has a longer wavelength than visible light. Most of this is radiated back to space.

Some of the radiated energy is reflected back to the Earth's surface (back radiation) by gases in the atmosphere—the greenhouse gases (GHGs). The surface of the Earth then gains more heat and warms up, and more heat energy is reflected back to space.

The greenhouse effect

This trapping of heat close to the Earth's surface is the **greenhouse effect**. It is essential for life as we know it to exist on Earth. The greenhouse effect is a natural and necessary process that maintains suitable temperatures for living systems—a good thing for life on Earth.

The nearest planets to the Earth are Mars (with a surface temperature of −53°C) and Venus (with a surface temperature of +450°C). The Earth's average surface temperature is a comfortable 15°C—just right for life and 33°C warmer than it would be without the greenhouse effect. A key feature of this temperature is the fact that water is liquid. The Earth is in the so-called "Goldilocks Zone"—not too hot nor too cold for biological life.

The greenhouse effect is caused by gases in the atmosphere reducing heat losses by radiation back into space. These GHGs trap heat energy that is reflected from the Earth's surface and re-radiate it—some back to space and some back to the Earth. A common misconception is that this works in the same way as a glass greenhouse. But greenhouses trap heat by reducing convection, while the atmosphere reduces loss of heat by radiation.

If there were no GHGs in the atmosphere, this heat would go straight back into space and the temperature on Earth would fall drastically every night. Instead, some heat is absorbed by gases in the atmosphere which re-emit it as heat energy back to the Earth. It is a bit like having a blanket around the Earth.

Key term

The **greenhouse effect** is a process that occurs when gases in the Earth's atmosphere trap the Sun's heat energy.

TOK

Words such as "greenhouse" and "Goldilocks" are culturally bound and the concepts they refer to may not be familiar to everyone.

To what extent are the words "greenhouse" and "Goldilocks" valid and useful only within one culture?

Key terms

The **enhanced greenhouse effect** refers to the accumulation of greenhouse gases by human (anthropogenic) activity leading to global warming (increasing mean global temperature).

Global warming means increasing mean global temperatures.

Climate change includes global warming but refers to the broader range of changes that are happening on Earth as a result.

The enhanced greenhouse effect

As humans increase emissions of some GHGs, the greenhouse effect is exaggerated or **enhanced**. This is causing **global warming** and **climate change**.

There are three points that may be confusing when reading information on climate change. So be careful that you understand:

1. the role of ozone and chlorofluorocarbons (CFCs)

2. the role of water vapour

3. whether figures refer to total GHG effects or the enhanced (anthropogenic) greenhouse effect.

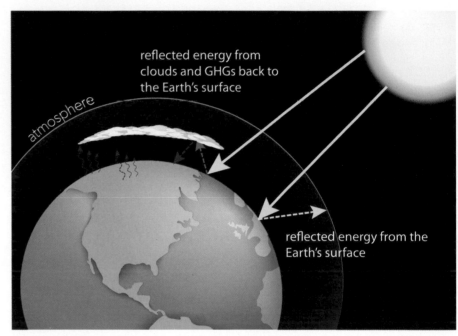

▲ Figure 3 **The greenhouse effect on Earth**

ATL Activity 1

ATL skills used: thinking, communication

1. Review the tricellular circulation pattern (subtopic 2.4) and check that you understand how it works.

2. Examine Figure 4 and draw a systems diagram of the atmospheric system. Label the inputs, stores, flows and outputs.

Greenhouse gases

Greenhouse gases (GHGs) include: carbon dioxide; water vapour; methane; chlorofluorocarbons (CFCs) and hydrochlorofluorocarbons (HCFCs); nitrous oxide; and ozone.

However, the main GHGs are **water vapour**, **carbon dioxide** and **methane**. They cause our atmosphere to be warmer than it would be without them. Only molecules with two or more bonds joining the atoms can absorb and re-emit energy in this way. Nitrogen and oxygen make up most of the atmosphere but are not greenhouse gases.

Although water vapour is a significant GHG it is usually excluded from climate models as the percentage in the atmosphere changes. Relative humidity measures this. Also it is essential for life and so cannot be mitigated against.

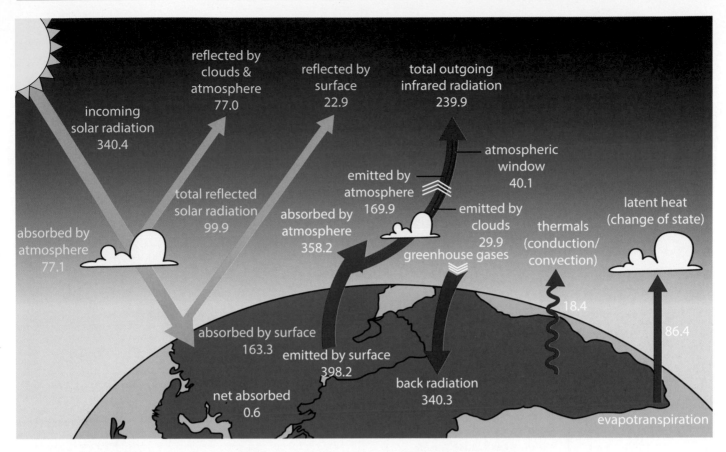

▲ Figure 4 Diagram of the Earth's energy budget. Units are W m⁻² (watts per square metre of surface)
Adapted from: NASA

Aerosols

Aerosols are very small solid particles suspended in the atmosphere. These are particulate matter and are less than 2.5 μm in diameter. They are called $PM_{2.5}$. 1 μm is 10^{-6} m.

Most aerosols (90%) come from natural processes such as volcanic eruptions and natural forest fires. Dust from sandstorms and sea salt are aerosols if the particles are small enough. The other 10% of aerosols are anthropogenic.

Black carbon or soot is pure carbon and is an aerosol. It is produced by incomplete combustion of fossil fuels, biomass and biofuel. This means it comes from human activities such as use in industry and power stations, and transport including ships, cars and aircraft. It also comes from forest fires and incomplete burning of organic matter. Since the start of the industrial era (usually stated as 1750), black carbon has caused air pollution and damage to human health. In the 1950s, it was found in Arctic haze when a reddish-brown haze was seen over the Arctic. Black carbon settling on snow and ice reduces albedo (ability to reflect solar radiation) and is perhaps why the Arctic is warming faster than other regions.

Within houses and dwellings there are many sources of aerosols: cigarettes, cooking stoves, burning wood, candles and spray cans (e.g. of paint, perfumes, cream). Aerosols have an impact on the climate because, just like greenhouse gases, they are able to change the Earth's **radiative forcing** or energy balance.

Key term

Radiative forcing (RF) is the difference between incoming and outgoing radiation of the Earth.

Aerosols change the amount of energy that is absorbed in the atmosphere and the amount that is scattered back out to space. Most aerosols are cooling—they reflect the Sun's energy back out into space. Only one aerosol—black carbon—contributes to global warming by boosting the warming effects of GHGs in the atmosphere. Climate models estimate that aerosols have masked about 50% of the warming that would otherwise have been caused by GHGs trapping heat near the surface of the Earth. Without the presence of these aerosols in the air, the Earth could be about 1°C hotter.

Radiative forcing

In accordance with the laws of thermodynamics, as the Earth absorbs energy from the Sun, it eventually emits an equal amount of energy to space. The difference between incoming and outgoing radiation is known as a planet's **radiative forcing (RF)**: radiative forcing = incoming energy − outgoing energy.

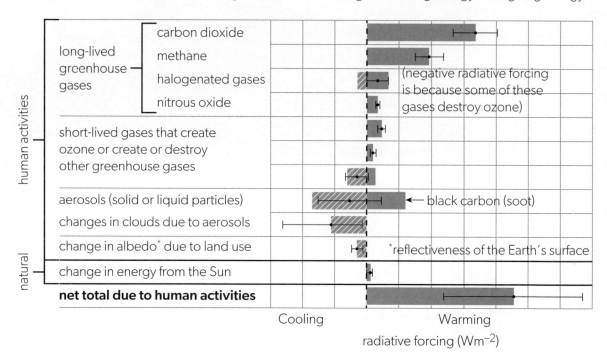

▲ Figure 5 Bar graph showing the total amount of radiative forcing caused by human activities—including indirect effects—between 1750 and 2011
Adapted from: IPCC (Intergovernmental Panel on Climate Change). 2013. Climate change 2013: The physical science basis. Working Group I contribution to the IPCC Fifth Assessment Report. Cambridge, United Kingdom: Cambridge University Press

AHL

(ATL) Activity 2

ATL skills used: thinking, communication, social

Work in small groups. Study Figure 5 and think about what the data is showing. Try to explain what it means to other members of your group.

Dynamic nature of the atmosphere

Space has no atmosphere. The Earth has an atmosphere with layers (Figure 2) but the boundaries are not clearly defined and move slightly depending on latitude and season. Almost all weather happens in the troposphere where terrestrial life exists.

The layers are mixed by physical and chemical processes. The physical processes are global warming and air movements due to temperature and pressure differences. Chemical processes causing mixing include production of ozone from oxygen.

Gravity is the force pulling molecules in the atmosphere towards the Earth. The atmosphere has weight and gravity acts on it keeping it around the Earth. Gravitational force is inversely proportional to distance, so it weakens as you go further from the surface of the Earth. Therefore, the atmosphere is denser nearer the surface and it thins with increasing altitude. This is why mountaineers on high mountains take oxygen tanks with them and why airplanes are pressurized.

Temperature also decreases with altitude—it gets colder and can be measured by standard lapse rate (about 1°C for every 100 m altitude). This is because less dense air means fewer molecules and total heat is related to the amount of matter present. Less matter = less heat.

Contrails

Above the troposphere is the stratosphere. The stratosphere is ideal for plane flight because it has strong horizontal winds but little turbulence. The jet stream is also there—blowing either west to east or north to south. Flying with the jet stream can speed up planes while flying against it can slow them down.

Aircraft contrails—the plumes that form behind planes as they fly—have a significant impact on the climate. When jet fuel burns, it creates water vapour (the primary exhaust emission), along with carbon dioxide, small amounts of unburnt hydrocarbons, oxides of nitrogen, soot particles and carbon monoxide. As the water droplets leave the engine, they quickly freeze to ice. If the conditions are right for clouds to form, the ice crystals forming behind the plane trigger the formation of a larger cloud.

Often, aircraft appear to be at the same level with one causing a contrail and the other not. However, the regions of humid air that cause contrails are known to be wide but shallow. A difference in flight level of 300 m is enough for one aircraft to cause a contrail and the other not. Contrails of more efficient engines with cooler exhaust gases can form at lower altitudes than those of less efficient engines.

Persistent contrail cirrus, formed by aircraft, can affect the reflection of solar radiation and have an impact on the Earth's climate. These clouds trap outgoing long-wave radiation and their radiative forcing **may** be larger than that of the carbon dioxide emitted by all aircraft ever.

Clouds and climate change

Essentially, all clouds are reflective and create a cooling effect on the climate, but some are also very good at trapping heat, acting as a blanket across the planet and helping to warm it further. So some clouds decrease global warming, while others increase it.

Clouds are made from lots of droplets of water and maybe a few crystals of ice; ice crystals form around aerosol particles such as sea salt, desert dust, soot from burning fossil fuels, and sulfuric acid. If you increase particles, you get more droplets or ice crystals in that cloud, and that changes its properties.

▲ Figure 6 The air on Mt Everest is one-third as dense as the air at sea level. This means there is only one-third as much oxygen available. Climbing high mountains therefore often requires breathing apparatus

▲ Figure 7 Contrails from airplanes

Connections: This links to material covered in subtopics 1.2 and 3.1.

Key term

Global warming potential (GWP) is a relative measure of how much heat a known mass of a GHG traps over a number of years compared with the same mass of carbon dioxide.

Global warming potential

Different gases have different **global warming potential (GWP)**. Carbon dioxide has a GWP of 1. Methane has a GWP of 27–30 over 100 years. So methane traps 27–30 times as much heat as the same mass of carbon dioxide. But it lasts for about 10 years in the atmosphere which is far less than carbon dioxide. Methane is also a precursor to ozone so more methane leads to more ozone formation.

Ozone forms a layer in the stratosphere that absorbs much of the ultraviolet radiation from the Sun. Ozone is a GHG in the troposphere but it acts in the stratosphere to cool the Earth. There is no direct link between global warming and ozone depletion but the climate is complex and there are indirect links. Thinning of the ozone layer does allow more ultraviolet radiation to reach the Earth's surface but this amounts to less than 1% of solar radiation reaching the Earth and is not significant in causing warming.

CFCs are chemicals made by humans—they are not present in the atmosphere as a result of any natural processes. CFCs break down ozone when they reach the stratosphere and act as GHGs in the troposphere. There are many types, e.g. CFC11 and CFC12 as well as HCFCs. Although their concentration in the atmosphere is measured in parts per trillion (10^{-9}), they have a large contribution to the enhanced greenhouse effect because each molecule has a high GWP and a long lifetime in the atmosphere. Their GWPs may be thousands of times that of carbon dioxide. That means a molecule of a CFC is up to 10,000 times more effective at trapping long-wave radiation than a molecule of carbon dioxide (GWP = 1).

When you look at data, consider whether the contribution of water vapour is included or excluded. Water vapour has the largest effect on trapping heat energy so it is the most potent GHG but it is not usually listed. This is because the concentration of water vapour is variable and it is constantly condensing to liquid water, snow and ice, which stops it acting as a GHG. Around 36–66% of the greenhouse effect is due to water vapour. The IPCC and most scientists omit water vapour from their calculations but the IPCC work on the figure of a 50% contribution to the greenhouse effect by water vapour. Clouds may contribute up to 25% (depending on the type of cloud and its altitude) and other GHGs cause the rest, with carbon dioxide having the largest effect.

Remember that most GHGs in the atmosphere are there through natural processes. The exceptions are CFCs and HCFCs, which are human-made. It is the **increase in GHGs due to anthropogenic activities** that is of concern. Carbon dioxide concentration may be higher now than at any time during the last 160,000 years. The recent rapid rate of increase is unprecedented and is due to human activities.

The amount of carbon added to the atmosphere each year due to human activities may not seem much when measured in parts per million (417 ppm in February 2023) but this equates to an increase of 3.2 to 4.1 GtC in the form of carbon dioxide each year over the last 25 years, according to the IPCC. A Gt is a gigatonne or one billion tonnes (10^9). So an increase of 3.2 to 4.1 GtC is up to 4,100,000,000 tonnes above the natural carbon cycle and does not include the carbon in methane. Natural sinks (oceans, plants) absorb about half of this carbon each year (see subtopic 2.3).

Greenhouse gas	Pre-industrial concentration (ppm)	Present concentration (ppm)	100 year GWP	% contribution to enhanced greenhouse effect	atmospheric lifetime (years)
Carbon dioxide, CO_2	270	400	1	75	50–200
Methane, CH_4	0.7	1.774	27–30	18	12
Nitrous oxide, N_2O	0.27	0.31	273	4	140
Fluorinated gases	0	0.00025	varies	4	45
Tropospheric ozone	not known	variable	2,000	variable	few weeks

▲ Table 1 Greenhouse gas (GHG) data

Milankovitch cycles

Milankovitch cycles affect the solar radiation reaching the Earth and lead to cycles in the Earth's climate over tens to hundreds of thousands of years. These very long-term cycles are due to variations in the orbit of the Earth round the Sun. Milutin Milankovitch was a Serbian astrophysicist who, in the early 20th century, began investigating the cause of the Earth's ancient ice ages.

There are three Milankovitch cycles (Figure 8) due to:

- the shape of the Earth's orbit, which is an ellipse (oval shape)—eccentricity

- the angle of tilt of the Earth—obliquity

- the direction in which the Earth's axis of rotation is pointing (how much it wobbles)—precession.

The changes in these cycles do not explain the current rapid warming of the Earth but do explain other changes such as the seasons and Ice Ages.

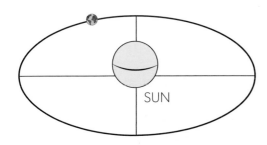

100,000-year cycles
changes in eccentricity* (orbit shape)

*exaggerated so the effect can be seen: the Earth's orbit shape varies between 0.0034 (almost a perfect circle) to 0.058 (slightly elliptical)

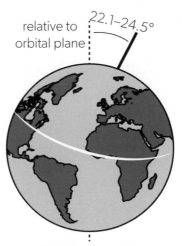

relative to orbital plane
22.1–24.5°

41,000-year cycles
changes in obliquity
(tilt)

26,000-year cycles
axial precession
(wobble)

▲ Figure 8 Milankovitch cycles
Adapted from: NASA / JPL-Caltech

ATL Activity 3

ATL skills used: thinking, communication

Watch video clips on the NASA website that explain Milankovitch cycles.

Eccentricity (orbit shape)

1. Explain why this changes slowly.

2. Describe the changes that eccentricity causes.

3. The current ellipse is the most circular. Explain the effect this has on length of seasons.

Obliquity (tilt of angle of rotation on axis)—cycle of about 41,000 years

This explains why we have seasons. The nearer you are on Earth to the Sun, the warmer it is. The greater the tilt, the more extreme the seasons.

Larger tilt leads to more extreme seasons and warmer summers when glaciers retreat and ice melts. Currently we are midway in this cycle of about 41,000 years.

4. Explain the effect that obliquity has on seasons.

5. Describe the effect of obliquity on ice.

6. State if this effect is positive or negative feedback.

7. Explain the reasons for your choice in 6.

Axial precession (wobble)—cycle of 25,771.5 years

The Earth wobbles on its axis as it rotates because the Moon and Sun pull the Earth through gravity.

8. Describe the effect of this wobble on seasons.

Currently, southern hemisphere summers are hotter and northern seasons more moderate. In 13,000 years this will flip.

Milankovitch thought that obliquity was the most important in changing climate but that is still debated.

9. To what extent is anthropogenic climate change affected by Milankovitch cycles?

Carbon dioxide levels

By analysing air trapped in ice core samples during the last ice ages, we know that carbon dioxide levels ranged from 180 to 280 ppm. This is due to Milankovitch cycle-driven changes to the Earth's climate. These fluctuations provided an important feedback to the total change in the Earth's climate that took place during those cycles. Through positive feedback loops, this causes either decreasing concentrations of atmospheric carbon dioxide (with cooling and glaciation) or increasing carbon dioxide concentrations (with warming and interglacial conditions). These cycles occur over many thousands of years but do not explain current warming.

Climate change caused by human activity is far faster. Burning fossil fuels in quantity only started in about 1850 in the Industrial Revolution. In that time the Milankovitch cycles have not changed much. In fact, over the last 40 years, solar radiation reaching the Earth has decreased a little. Check how atmospheric carbon dioxide levels have changed during your lifetime.

Geological timescale changes

Over the Earth's history there have been at least five major ice ages when glaciers expanded and sea levels fell. Each ice age is hundreds of millions of years long and we are still in one called the Quaternary period. This started 2.5 million years ago—but this is less than 0.1% of geological time.

Within each major ice age, there are interglacial periods when temperatures increase but ice sheets still exist—like the Greenland and Antarctic ice sheets now. And there are glaciation periods (sometimes also called "ice ages") when glaciers increase. The last was some 18,000 years ago and followed by the interglacial Holocene epoch which started about 11,700 years ago. We are currently in the fairly warm interglacial Holocene epoch of the Quaternary period. Climate has changed over geological time without human influence but current changes are unprecedentedly rapid. Some scientists now call our epoch the Anthropocene epoch.

The Anthropocene is considered to have started in the 1850s as the Industrial Revolution began. This was the time when human activity started to have a significant impact on the Earth's climate and ecosystems. The word "Anthropocene" is derived from the Greek words *anthropo*, for "man", and *cene* for "new".

Global warming is moving the Earth away from the glacial–interglacial cycle that has characterized the Quaternary period, towards new, hotter climatic conditions.

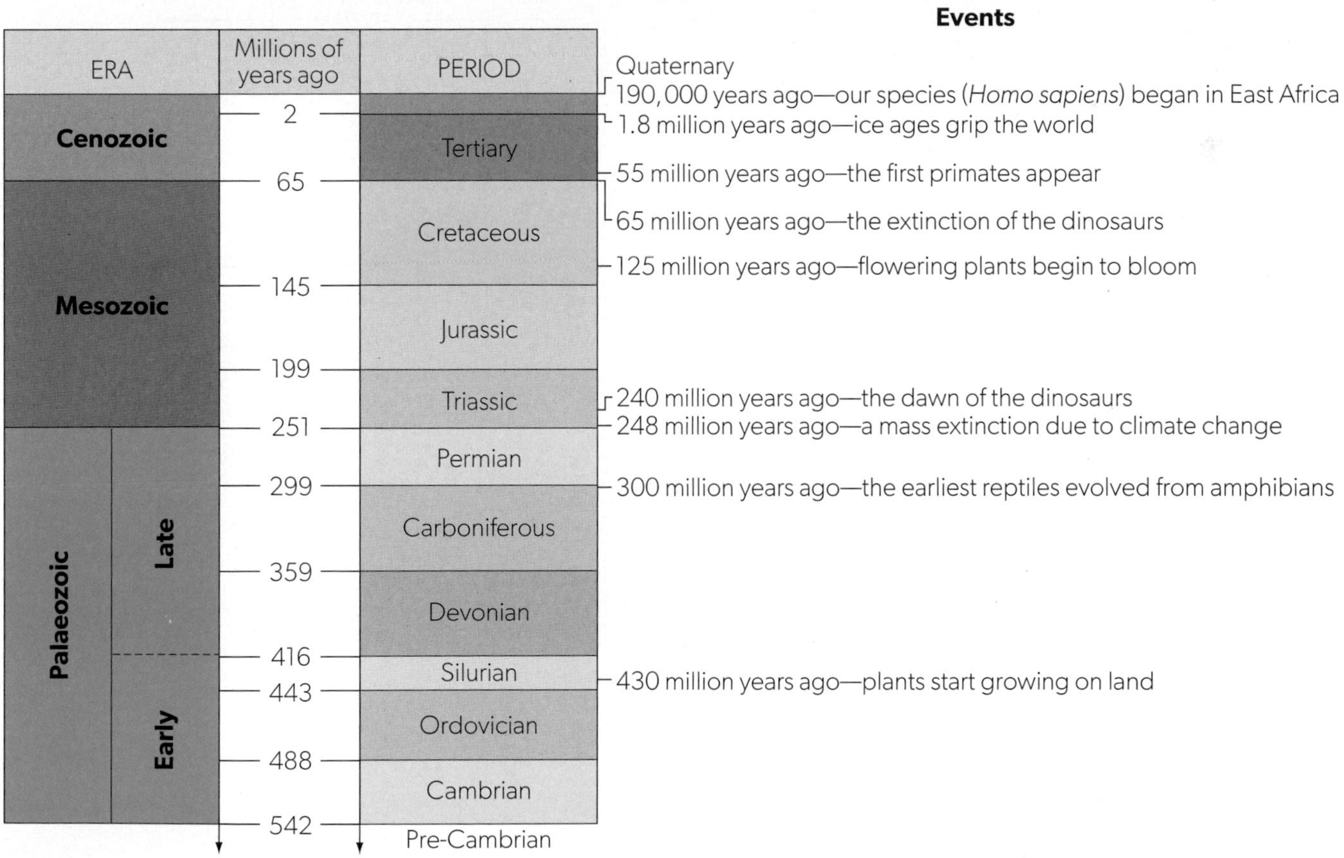

Events

ERA	Millions of years ago	PERIOD	
Cenozoic	2		Quaternary
		Tertiary	190,000 years ago—our species (*Homo sapiens*) began in East Africa
	65		1.8 million years ago—ice ages grip the world
			55 million years ago—the first primates appear
Mesozoic		Cretaceous	65 million years ago—the extinction of the dinosaurs
	145		125 million years ago—flowering plants begin to bloom
		Jurassic	
	199		
		Triassic	240 million years ago—the dawn of the dinosaurs
	251		248 million years ago—a mass extinction due to climate change
Palaeozoic (Late)		Permian	
	299		300 million years ago—the earliest reptiles evolved from amphibians
		Carboniferous	
	359		
		Devonian	
	416	Silurian	430 million years ago—plants start growing on land
	443		
Palaeozoic (Early)		Ordovician	
	488		
		Cambrian	
	542	Pre-Cambrian	

▲ Figure 9 **The geological time scale (see also subtopic 3.1, Figure 15)**

Past atmospheric changes

The atmosphere and climate of the Earth have always changed. The evolution of life on Earth changed the composition of the atmosphere, which in turn influences the evolution of life on Earth. Climate is unstable and has fluctuated greatly in the past. Climate is influenced by abiotic factors (mainly temperature and precipitation) and biotic factors (plants and animals).

We cannot measure precipitation in the distant past but we can measure temperature both directly and indirectly by proxy. We can also measure atmospheric gas concentration in bubbles trapped in ice. Various direct and indirect measurements are taken on sediments or fossilized animal shells from the period, but it is hard to say how accurate they are.

The percentage composition of the pre-biotic Earth's atmosphere was very different from its current composition. Before plants evolved to photosynthesize, there was no free oxygen in the atmosphere on Earth. Photosynthesis decreased the carbon dioxide concentration and increased the oxygen concentration. This enabled stratospheric ozone to form and the oxidation of metals to occur, e.g. the formation of iron ore. Oxygen in the atmosphere gradually increased to about 35% at the end of the Carboniferous period about 300 MYA (million years ago).

(ATL) Activity 4

ATL skills used: thinking, research

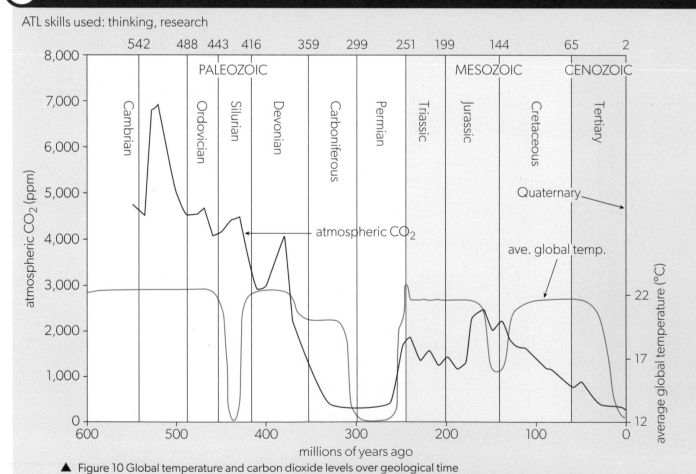

▲ Figure 10 Global temperature and carbon dioxide levels over geological time

1. Study Figure 10. What is the relationship between carbon dioxide and temperature in the graph?
2. What other factors besides atmospheric carbon may influence Earth temperatures?
3. Can we rely on the data collection methods used?

In the geological timescale, the average temperature on Earth was 20°C in the Early Carboniferous period 350 MYA and this cooled to 12°C later in the Carboniferous, slightly lower than our average of 15°C today. When the temperature was 20°C, the carbon dioxide concentration in the atmosphere was probably about 1,500 ppm (parts per million) and this decreased to about 350 ppm when the average temperature was 12°C. Today the concentration of carbon dioxide in the atmosphere is about 417 ppm (0.04%), less than in the previous 600 million years except for the Late Carboniferous period.

Practical

Skills:

- Inquiry 1: Exploring and designing

1. Develop a research question to investigate the impact of

  - albedo

  - any GHG

  on the temperature of a closed system.

Hints:

- A closed system could be a sealed bottle of liquid with a temperature probe or thermometer inside.

- On a hot day, is dark or light clothing more cooling?

- The independent variable will be the colour of the gas contained in the sealed container.

- The dependent variable will be temperature.

2. Develop your hypothesis and consider the method you would use.

Check your understanding

In this subtopic, you have covered:

- heat is dispersed over the Earth by the tricellular circulation pattern

- greenhouse gases (GHGs) prevent heat from radiating back to space

- what the greenhouse gases are

- the atmosphere is dynamic and thins with altitude

- Milankovitch cycles

- global warming is occurring

- the atmosphere changes over time influenced by life evolving.

How do atmospheric systems contribute to the stability of life on Earth?

1. Describe the structure of the atmosphere.

2. Explain how heat is dispersed over the Earth's surface.

3. Explain the greenhouse effect.

4. Explain the enhanced greenhouse effect.

5. Describe the changes in atmosphere with increased altitude.

6. Briefly describe Milankovitch cycles and explain why they cannot account for current global warming.

7. Over geological time, explain what has influenced changes in atmospheric composition.

>> Taking it further

- Watch for contrails in the air above your school if you are on a flight path. When do they form? How long do they last?

- Hold a class discussion on the future in the Anthropocene epoch.

- Look for citizen science and voluntary agencies that offer opportunities to participate in gathering knowledge of the atmosphere and the impacts of air pollution.

- Set up a school weather station to record temperature and other variables over time.

6.2 | Climate change—causes and impacts

Guiding questions

- To what extent has climate change occurred due to anthropogenic causes?
- How do differing perspectives play a role in responding to the challenges of climate change?

Understandings

1. Climate describes the typical conditions that result from physical processes in the atmosphere.
2. Anthropogenic carbon dioxide emissions have caused atmospheric concentrations to rise significantly. The global rate of emission has accelerated, particularly since 1950.
3. Analysis of ice cores, tree rings and deposited sediments provide data that indicates a positive correlation between the concentration of carbon dioxide in the atmosphere and global temperatures.
4. The greenhouse effect has been enhanced by anthropogenic emissions of GHGs. This has led to global warming and, therefore, climate change.
5. Climate change impacts ecosystems at a variety of scales, from local to global and affects the resilience of ecosystems and leads to biome shifting.
6. Climate change has an impact on (human) societies at a variety of scales and socio-economic conditions. This impacts the resilience of societies.
7. Systems diagrams and models can be used to represent cause and effect of climate change with feedback loops, either positive or negative, and changes in the global energy balance.
8. Evidence suggests that the Earth has already passed the planetary boundary for climate change.
9. Perspectives on climate change for both individuals and for societies are influenced by many factors.
10. Data collected over time by weather stations, observatories, radar and satellites provides opportunity for the study of climate change and land-use change. Long-term data sets include the recording of temperature and GHG concentrations. Measurements can be both indirect (proxies) and direct. Indirect measurements include isotope measurements taken from ice cores, dendrochronology and pollen taken from peat cores.
11. Global climate models manipulate inputs to climate systems to predict possible outputs or outcomes, using equations to represent the processes and interactions that drive the Earth's climate. The validity of the models can be tested via a process known as hindcasting.
12. Climate models use different scenarios to predict possible impacts of climate change.
13. Climate models show the Earth may approach a critical threshold with changes to a new equilibrium. Local systems also have thresholds or tipping points.
14. Individual tipping points of the climate system may interact to create tipping cascades.
15. Countries vary in their responsibility for climate change and also in vulnerability, with the least responsible often being the most vulnerable. There are political and economic implications and issues of equity.

Causes of climate change

The cause of climate change is us. But the most important thing to remember when you read this subtopic is that there is still time. Time to reduce emissions, time to reduce the effects of climate change, time to act. As Greta Thunberg says:

> If we do not act soon, anthropogenic environmental changes will bring serious harms to the future. We have a moral obligation to avert harms to the future, so as to leave a world as rich in life and possibility as the world we inherited. Therefore, we have a moral obligation to act, and act now.

This is true for every generation. You have read that in the 1970s human use of natural resources tipped over the line from sustainable to unsustainable. Since then we have been trying, with some small successes, to get back over that line.

Your actions matter. Humans have been slow to recognize the threat of global warming but, at last, it cannot be denied and governments are acting. So can individuals and, if enough of us act, then the worst scenarios of the Intergovernmental Panel on Climate Change (IPCC) predictions can be avoided. The following have all happened because enough individuals got together and acted:

- reducing plastic bags and one-use plastics

- increases in energy efficiency and in use of renewables

- conservation projects

- clean-up activities

- willingness to hold transnational companies and industry to account for pollution.

Don't think that your actions are pointless. Don't continue to buy fast fashion. Think about your use of energy in heating and cooling, driving or flying unnecessarily, using one-use plastics, and in your diet. Recycle, reuse, repurpose, reduce. If enough of us live lives with a smaller carbon footprint, it can have an effect.

▲ Figure 1 A comic illustration of global warming

Climate and weather

Weather is the daily result of changes in temperature, air pressure, wind speed, and precipitation and humidity in our atmosphere. Weather varies from place to place, sometimes over very short distances. We try to predict weather but can only do so with some accuracy for about five days ahead. There are many variables that interact in complex ways. Weather can fluctuate wildly—a very hot day or a very cold one does not mean that average temperatures are changing nor does one scorching summer or one exceptionally cold winter.

Climate is the average weather pattern over about 30 years for a location on Earth. Climate may show long-term trends and changes if records are kept for long enough.

Weather and climate operate on very different timescales, but they are both affected by ocean and atmospheric circulatory systems.

Both are also affected by the following.

- Clouds—may trap heat underneath them or reflect sunlight away from the Earth above them.

- Forest fires—release carbon dioxide, a GHG, but regrowth traps it again in carbon stores.

- Volcanic eruptions—release huge quantities of ash which circulate in the atmosphere, cooling the Earth. For example, Mt Pinatubo in the Philippines erupted in 1990 lowering global temperatures for a few years.

- Human activities—burning fossil fuels and keeping livestock release GHGs.

Connections: This links to material covered in subtopic 2.4.

Climate change is long-term change and has always happened (see subtopic 6.1). Factors that influence climate change include:

- fluctuations in solar insolation affecting temperature

- Milankovitch cycles

- changing proportions of gases in the atmosphere released by organisms.

Temperature and carbon dioxide levels

For climate to change on a global scale, inputs and outputs must change, e.g. heat input increases, heat output decreases, or both. (See radiative forcing, subtopic 6.1)

GHGs reduce heat loss from the atmosphere. If there are more GHGs, less heat is lost. The system changes in a dynamic equilibrium which may stabilize or reach a new equilibrium if a **tipping point** is passed (see subtopic 1.2).

Long-term records show that the global average surface temperature of the Earth is increasing (Figure 2), although there are fluctuations from year to year. Global average temperature data sets from NASA, NOAA, Berkeley Earth, and meteorological offices of the UK and Japan, show substantial agreement concerning the progress and extent of global warming.

Global average temperature has increased by about 0.85°C since 1880 (Figure 2). Most of the increase has been since 1980.

Connections: This links to material covered in subtopics 1.1, 1.2 and 2.4.

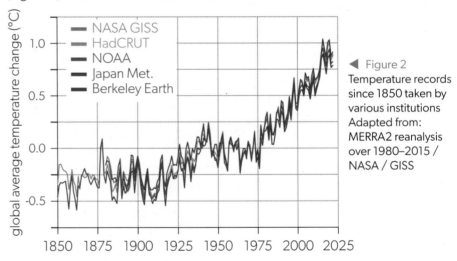

◀ Figure 2
Temperature records since 1850 taken by various institutions
Adapted from: MERRA2 reanalysis over 1980–2015 / NASA / GISS

Anthropogenic carbon dioxide (CO_2) emissions have caused atmospheric concentrations to rise significantly. The global rate of emission has accelerated, particularly since 1950 (Figure 3).

This increase started in the Industrial Revolution in late 19th century Europe. But it accelerated from the 1960s onwards as industrialization spread and more fossil fuels were discovered and burned. Within each year, there are fluctuations in CO_2 levels (Figure 4) but the general trend is increasing.

There is a very strong correlation between carbon dioxide levels in the atmosphere and temperature (Figures 5, 6 and 7). Data comes from direct measurements since about 1850 and, before that, from indirect measurements based on the analysis of ice cores, tree ring data and deposited sediments. All indicate a positive correlation between the concentration of carbon dioxide in the atmosphere and global temperatures.

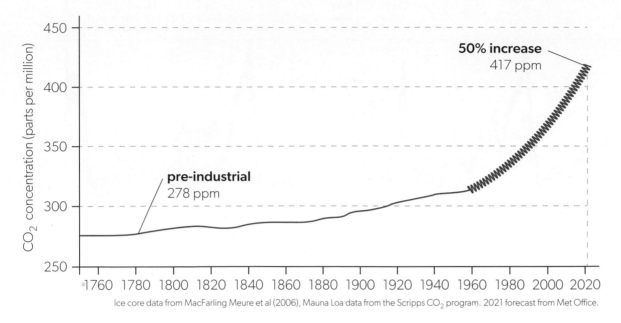

Ice core data from MacFarling Meure et al (2006), Mauna Loa data from the Scripps CO₂ program. 2021 forecast from Met Office.

▲ Figure 3 Global atmospheric CO_2 concentrations from 1700 to 2021

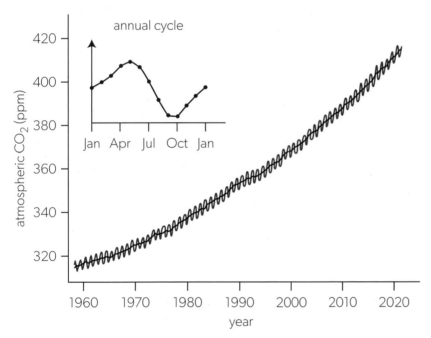

▲ Figure 4 Graph of carbon dioxide in the atmosphere as measured at the Mauna Loa Atmospheric Baseline Observatory by NOAA and the Scripps Institution of Oceanography. This graph is known as the Keeling Curve
Adapted from: Scripps Institution of Oceanography / UC San Diego / NOAA

Levels of CO^2 in the atmosphere have corresponded closely with temperature over the past 800,000 years. The temperature changes were started by Milankovitch cycle variations in the Earth's orbit which led to cycles of ice ages and warm periods. Increased global temperatures released CO_2 into the atmosphere, which in turn warmed the Earth. Antarctic ice core data shows the long-term correlation until about 1900 when human activity started moving carbon from the slow to the fast carbon cycle.

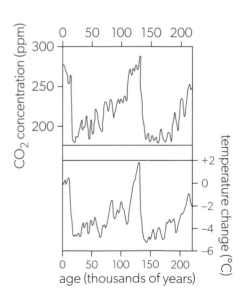

▲ Figure 5 Graph showing comparison of changes in atmospheric carbon dioxide and temperature

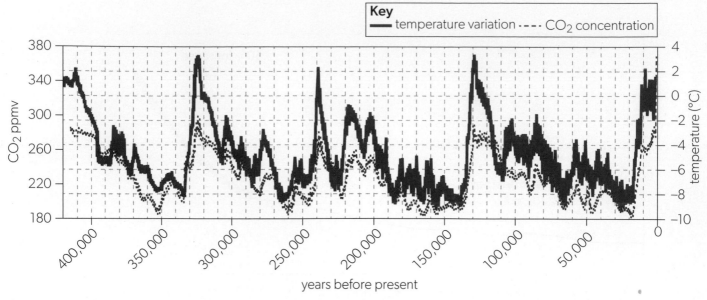

▲ Figure 6 Graph showing the temperature and CO_2 concentrations in the Antarctic atmosphere over the last 400,000 years from ice core data

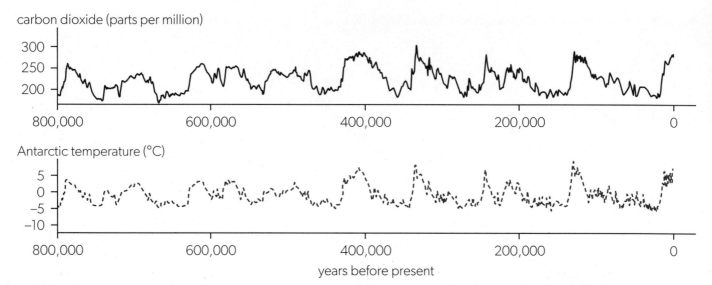

▲ Figure 7 Graphs showing the temperature and CO_2 concentrations in the Antarctic atmosphere over the last 800,000 years from ice core data
Data from: Lüthi et al., 2008, and Jouzel et al., 2007

TOK

The data that supports climate change as an actual phenomenon is extensive.

To what extent does this support the idea that science is the supreme form of all knowledge?

ATL Activity 5

ATL skills used: thinking, communication, research

Use Figures 3 to 7 to answer these questions.

1. Explain why data in Figure 3 is provided by five institutions.

2. When and why did atmospheric CO_2 concentrations suddenly start to rise?

3. Calculate how much greater the CO_2 concentration is now compared with pre-industrial levels.

4. Explain why there is an annual cycle of CO_2 in the atmosphere.

5. Describe the correlation between CO_2 levels and temperature over:

    a. the last 250,000 years   b. the last 800,000 years.

Climate change and the carbon cycle

Anthropogenic emissions of GHGs have enhanced the greenhouse effect. Human activity has caused emissions into the atmosphere of large quantities of carbon dioxide, methane and nitrous oxide and smaller quantities of other GHGs. This has led to global warming and climate change.

Without human interference, the carbon in fossil fuels leaks slowly into the atmosphere through volcanic activity over millions of years in the slow carbon cycle. By burning coal, oil and natural gas, we accelerate this process. Every year, humans release into the atmosphere huge amounts of carbon that took millions of years to accumulate (Figure 8). In 2009, humans released about 8.4 billion tonnes of carbon into the atmosphere by burning fossil fuels and by cement production which is responsible for up to 8% of emissions of CO_2. About half of these emissions are removed by the fast carbon cycle each year, the rest remain in the atmosphere.

> Connections: This links to material covered in subtopics 1.1, 2.3 and 2.5.

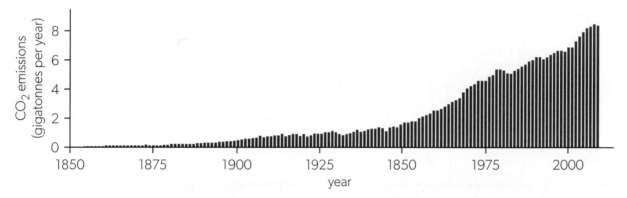

▲ Figure 8 Graph of global CO_2 emissions, 1850–2020
Data from: Carbon Dioxide Information Analysis Center and Global Carbon Project, NASA Earth Observatory

In every million molecules in the atmosphere, about 417 are now carbon dioxide. This is the highest concentration in two million years. Methane concentrations have risen from 715 parts per billion in 1750 to 1,895 parts per billion in 2021, the highest concentration in at least 650,000 years. Methane has a global warming potential (GWP) far greater than CO_2 (see subtopic 6.1). In the late 1960s, CO_2 levels were rising at a rate of about 1 ppm per year. By the early years of the 21st century, this rise had accelerated to 2 ppm per year, and it has accelerated further to nearly 2.5 ppm annually in the last 10 years. These may sound like small amounts but it is clear that the effect of the increase is huge.

ATL Activity 6

ATL skills used: thinking, research

The Keeling Curve is a daily record of global atmospheric carbon dioxide concentration maintained by the Scripps Institution of Oceanography at UC San Diego. It has become iconic because it is an accurate measure that is long-term and independent.

Find the Keeling Curve website at UC San Diego.

1. In March 2023 the reading of CO_2 levels was 421.25 ppm. What is it now?

2. Why does atmospheric CO_2 concentration vary throughout one year?

3. View the curves for each time period from one week to 800K years. In which curve did CO_2 levels decrease? When was this? Why?

Climate change impacts

Climate change impacts the resilience of ecosystems and leads to biome shifting. The effects are global and local, as shown in Figure 9.

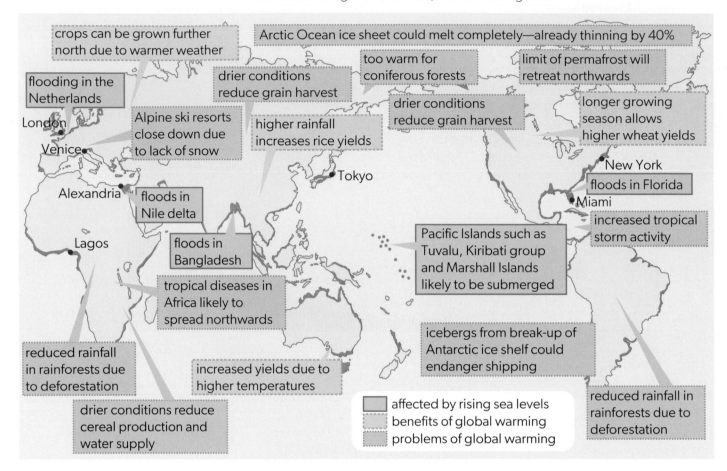

▲ Figure 9 World map showing some effects of climate change

- Global changes include changes to ocean circulation and sea-level rise.

- Local changes include coral bleaching, desertification and changes in productivity. These changes impact the health, food and water supplies, and infrastructure of human societies. But the impacts are not equitable across the world.

ATL Activity 7

ATL skills used: thinking, communication, research

In Figure 9, changes are identified as benefits, problems or affected by rising sea levels.

1. Draw a table with three columns labeled: benefits of global warming, problems of global warming and areas affected by rising sea levels. Use the examples in Figure 9 and add others that you think of to each column.

2. List the effects of climate change on an ecosystem you have studied.

3. Research the impact of temperature change in the oceans on coral reefs.

4. For the region where you live or go to school, investigate the potential and current impacts of climate change.

On oceans and sea levels

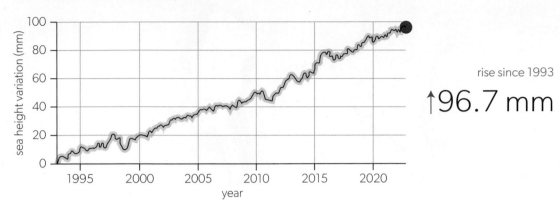

rise since 1993

↑96.7 mm

▲ Figure 10 Graph of sea level height 1993–2023, based on satellite sea level observations
Data from: NASA's Goddard Space Flight Center

Sea levels are rising (Figure 10) because:

- water expands as it heats up—thermal expansion

- ice melting on land slips off the land and into the sea increasing the volume of seawater.

The Greenland and the Antarctic ice sheets are on land and are thinning. This and the thermal expansion of the seas will mean that sea levels rise more. By how much is not clear but predictions are becoming more accurate as climate modelling improves with more variables entered into the programmes. An increase of between 1.5 and 4.5°C could mean a sea level rise of 15–95 cm (IPCC data). But this assumes a proportional relationship. If a tipping point threshold is exceeded, sea levels could rise by many metres. Up to 40 nations will be affected. Low–lying states such as Bangladesh, the Maldives and the Netherlands will lose land area and some, such as Tuvalu, will disappear completely.

The oceans absorb carbon dioxide and this makes them slightly acidic. Ocean acidity has increased by about 25% from pre-industrial times to the early 21st century, as they have absorbed about half the carbon produced by anthropogenic activities. This affects marine organisms, particularly corals. But also as they warm, they absorb less carbon dioxide.

On polar ice caps

Melting of land ice in Antarctica and Greenland (Figure 12) will cause sea levels to rise as water flows into the oceans. Melting of the floating ice cap of the Arctic will not increase the volume of water because ice has the same displacement as liquid water. But glaciers melting into the seas will increase the volume of water. The Greenland ice sheet could melt completely and slow down or stop the North Atlantic Drift current (NAC) by diluting the saltwater. If the NAC and the Gulf Stream slow or even shut down, the climate of the UK and Scandinavia will be much colder (see oceanic conveyor belt, subtopic 2.4).

Melting in the Arctic could open up trade routes, make travel in the region easier and allow exploitation of undersea minerals and fossil fuel reserves.

Methane clathrate is a form of ice under the Arctic Ocean floor that traps methane. If this were to melt and reach the surface, the release of methane could trigger a rapid increase in temperature.

▲ Figure 11 Ice breaking off from a glacier

Connections: This links to material covered in subtopics 1.2 and 4.2

▶ Figure 12 **Maps showing the shrinking of the ice sheets in Greenland in just 10 years**

▲ Figure 13 **Location and date marker for glacier in Jasper National Park, Canada**

On glaciers

In the Little Ice Age between about 1550 and 1850, glaciers increased in size. They then decreased (except for the period 1950–80 when global dimming due to air pollution possibly masked some global warming) and have continued to decrease in size. Some have melted completely. Loss of glacier ice leads to flooding and landslides. Glacier summer melt provides a fresh water supply to people living below the glacier. This has provided water to many major Asian rivers (Ganges, Brahmaputra, Indus, Yellow, Yangtze) which are fed by the Himalayan glaciers. Glacier melt is causing significant drought problems in Tanzania where the Kilimanjaro glacier has lost over 80% of its volume.

The Thwaites glacier is nicknamed the Doomsday Glacier. It is in west Antarctica and is one of the widest glaciers on Earth. It holds enough water to raise sea levels by up to 5 metres. It has a high risk of collapse as it is eroding along its underwater base. It is grounded on the seabed not land and so warmer oceans could cause it to destabilize.

On weather patterns

More heat means more energy in the climate. The weather will become more violent and unpredictable with bigger storms and more severe droughts. Global precipitation may increase by up to 15%, causing more soil erosion. Lack of water in some regions will mean more irrigation and consequent salinization. There is evidence that severe weather and more extreme rainfall or droughts are occurring. Cyclones are more intense and slower moving which increases destruction where they make landfall. Monsoon rains fail more often than they used to.

Examples of extreme weather events:

• Hurricane Ian, Florida 2022

• Storm Filomena snowfall, Madrid 2021—heaviest snowfall in 50 years

• Cyclone Ana, Fiji and Cyclone Seroja, Indonesia 2021

• longest drought in 40 years in the Horn of Africa

• flooding in Kimberley region of western Australia—a 1 in 100 year event

• over 7 million hectares burned in bushfires, eastern Australia 2019–20

• ENSO (El Niño Southern Oscillation, see subtopic 2.4) events are more extreme and last longer.

On food production

Warmer temperatures should increase the rate of biochemical reactions. This means photosynthesis should increase. But respiration will increase too so there may be no overall increase in net primary productivity (NPP). In Europe, the crop-growing season has expanded with warmer climate.

If biomes shift away from the equator, there will be winners and losers. It very much depends on the fertility of the soils as well. If production shifts northwards from Ukraine with its rich black earth soils to Siberia with its thinner, less fertile soils, NPP will fall. Various predictions state huge ranges of changes from −70% to +11%. There are just too many variables to be certain. What is certain is that some crop pests will spread to higher latitudes because they will not be killed by cold winters.

In the seas, a small increase in temperature can kill plankton, the basis of many marine food webs. Heatwaves and drought kill livestock on land.

▲ Figure 14 **Crop killed by drought**

On drylands

Desertification is land degradation when a previously fertile area becomes desert and no longer supports a population, whether animal or human. Drylands where precipitation equals evaporation cover 41% of the land surface and two billion people live on them. Examples include the Sahel, Gobi desert, Mongolian grasslands, East African drylands and South American drylands.

Case study 1

The Sahel

The Sahel is a belt of land across Africa south of the Sahara Desert. It spans 10 countries. It is grassland and savanna with more rainfall (up to 600 mm) in the south and less (100–200 mm) in the north where it meets the Sahara. About 75 million people live in the Sahel, which has been called "the Hunger Belt". There are alternating periods of abundant and scarce rainfall but the length and severity of droughts are increasing. In 1984–5, a massive drought caused famine and most livestock died. Before this, fairly good rainfall had led to overstocking—the number of animals exceeded the biocapacity of the region to provide water and pasture.

It is estimated that 200 million people may live in the Sahel by 2050. Many are expected to migrate due to a combination of climate issues, conflict, and the threat from the militant group Boko Haram.

In 2007, a project to fight desertification was started by a combined effort of the governments affected. It was called the Great Green Wall. This was an initiative to plant millions of trees in a strip 15 km wide and 8,000 km long across the Sahel. The area covered 100 million hectares. This ambitious project had critics and has been hampered by lack of funds and political inertia. So far, 18% has been completed on 20 million hectares but progress on the rest is slow.

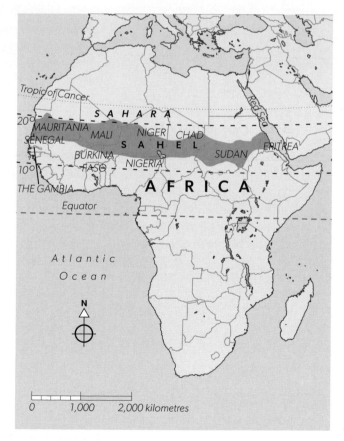

▲ Figure 15 The Sahel—a belt of grassland and savanna spanning ten countries

▲ Figure 16 **Koala, indigenous to Australia**

On biodiversity and ecosystems

Melting of the tundra permafrost releases methane which is trapped in the frozen soils (see subtopic 5.1). In Alaska, Canada and Russia, permafrost is melting. Houses built on it are shifting as it thaws.

Animals can move to cooler regions but plants cannot. The distribution of plants can shift as they disperse seeds which germinate and grow in more favourable habitats. But this happens slowly at about 1 km per year—perhaps too slowly to stop them becoming extinct. Species in alpine or tundra regions have nowhere to go, neither towards higher altitudes nor towards higher latitudes.

Polar bears could become extinct in the wild or mate more often with grizzlies (see Case study 2). Birds and butterflies have already shifted their ranges to higher latitudes. Plants are breaking their winter dormancy earlier. Loss of glaciers, decreased salinity of marine waters and changes to ocean currents alter habitats.

If droughts increase, then wildfires are more likely to wipe out species and their habitats. In the Australian bushfires of 2019–20, there were 143 million native mammals including 61,000 koalas in the path of the flames. It is still not clear how many died.

Indonesian forest fires have set fire to the peat bogs which have burned continually for years. The amount of carbon released by these adds significantly to carbon in the atmosphere. Pine forests in British Columbia are being devastated by pine beetles which are not being killed off in milder winters.

Increase in temperature of fresh and saltwater may kill sensitive aquatic species. Corals are very sensitive to increased sea temperature. An increase of one degree can cause coral bleaching as the mutualistic zooxanthellae algae in the corals are expelled and the coral dies. Corals are the basis for many food webs. If the corals die, the ecosystem dies.

Case study 2

Grolar or pizzly bears

Both in the wild and in captivity, there are rare records of bear hybrids produced when a polar bear and a grizzly bear mate. Climate change may result in more of these hybrids as grizzlies move northwards in milder winters and, due to a reduction in the number of salmon, eat seals which are a polar bear's prey. These changes will mean that the two species are in closer proximity.

▶ Figure 17 Pizzly bear

On human health

Heat waves killed many in Europe in 2006. These heatwaves may increase. Insect vectors of disease will spread to more regions as they survive in warmer winters. Malaria, yellow fever and dengue fever are spreading to higher latitudes.

Algal blooms may be more common as seas and lakes warm. Some are toxic (red tides) and can kill humans and other species.

In a wetter climate, fungal disease will increase. In a drier climate, dust increases leading to asthma and chest infections.

Warmer temperatures in higher latitudes would reduce the number of people dying from the cold each year and reduce heating bills for households. Fewer snowstorms and icy roads could mean lower death tolls on roads.

On water supplies

Increased evaporation rates may cause some rivers and lakes to dry up. Without a water supply, populations would have to move away. According to the UN, 2.4 billion people live in the river basins fed by the Himalayas and their water supply is reducing. In Europe and North America, glaciers are also in retreat. "Water wars" and conflict over water shortages and inequitable sharing are increasing.

On human migration

If people cannot grow food or find water locally, they will move to regions where they can. Global migration of millions of environmental refugees is quite possible. This would have implications for nation states, services and economic and security policies. The IPCC estimates there will be 150 million refugees from climate change by 2050.

On national economies

Some will suffer if water supplies decrease or drought, fires, flood or cyclones occur. Others will gain if it became easier to exploit mineral reserves (e.g. tar sands of Canada and Siberia) that have been frozen in permafrost or under ice sheets. If rivers do not freeze, hydroelectric power generation is possible at higher latitudes.

The Northwest Passage is a sea route for shipping from the Atlantic to the Pacific via the Arctic Ocean north of Canada. Many explorers tried to find this route in the northern summers but were stopped by sea ice. In 2007, the passage was navigable for the first time in recorded history.

On indigenous peoples

Nomadic pastoralists in dryland regions have indigenous knowledge and have been able to maintain a livelihood by recognizing the limits of their environment. The Mbororo of the Sahel can migrate long distances with their herds over a year using pasture intensively then allowing it to recover. Traditional knowledge of weather, climate, animal breeding cycles, trees, fruiting, insect presence, and wind direction tells them what to expect and to move accordingly. But pressures from increasing population and droughts make transhumance (seasonal movement of livestock) harder for pastoralists.

On equity in societies

Climate change is more than an environmental crisis—it is a social crisis too. According to the World Bank report on the social dimensions of climate change, the poorest and most vulnerable often bear the brunt of climate change even though they contribute the least to the issue. We must address issues of inequality on many levels: between richer and poorer countries; between rich and poor within countries; and between generations.

▲ Figure 18 **A migrant seasonal farm worker picking and packaging strawberries**

Examples of climate change increasing societal inequality

Certain social groups are particularly vulnerable to crises, e.g. children, landless tenants, migrant workers, female-headed households, persons with disabilities, indigenous people and ethnic minorities, displaced persons, sexual and gender minorities, older people, and other socially marginalized groups.

The causes of the inequity are a combination of their geographical locations; their financial, socio-economic, cultural and gender status; and their access to resources, services, decision-making power, and justice.

The IPCC has highlighted the need for climate solutions that conform to principles of distributive justice for more effective development outcomes.

The most vulnerable are often disproportionately impacted by measures to address climate change. In the absence of well-designed and inclusive policies, climate change mitigation measures can place a higher financial burden on poor households. For example, policies that expand public transport or carbon pricing may lead to higher public transport fares which can disproportionately impact poorer households.

The energy crisis of 2022–3, when Russian oil flow was stopped to Europe, resulted in huge energy price rises and then food price rises. These drove inflation to high levels. The effect was felt more by the poor who spend a larger proportion of their income on energy and food.

If not designed in collaboration with affected communities, approaches such as limiting forestry activities to certain times of the year could adversely impact indigenous communities that depend on forests year-round for their livelihoods.

A challenge facing many countries is engaging citizens who may not understand climate change. Support has to be gathered from those who are concerned that they will be unfairly impacted by climate policies. This requires transparency, access to information and citizen engagement on climate risk and green growth. This should create coalitions of support or public demand to reduce climate impacts and to overcome behavioural and political barriers to decarbonization.

Gains and losses from climate change

Overall, there will be gains and losses for national economies. Agricultural production may rise in higher latitudes but fall in the tropics. Africa will probably lose food production and rainfall. Northern Darfur has seen desertification on a massive scale already and many more millions of hectares may undergo desertification. Extracting fossil fuels or minerals in higher latitudes may get easier.

To put a monetary value on this is difficult but the Stern Report (from the former chief economist of the World Bank) suggested in 2006 that 1% of global GDP should then have gone to mitigating the effects of climate change to save up to 20% of global GDP in a recession later.

Estimates of the cost of global climate change by 2050 range from USD 14 trillion to USD 23 trillion. No one really knows though.

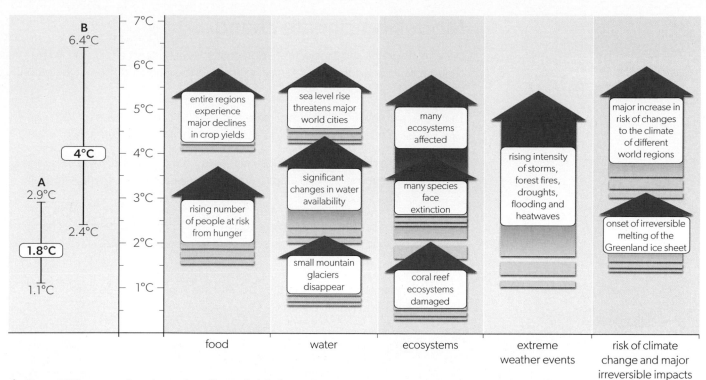

▲ Figure 19 **Some predicted negative effects of global warming**

(ATL) Activity 8

ATL skills used: thinking, communication, social, research, self-management

1. Divide the class into two groups.

   - Group 1: Collate the benefits of climate change and do any further research.

   - Group 2: Collate the disadvantages of climate change (Figure 19) and do any further research.

   As a class, debate whether the benefits of climate change exceed the disadvantages, or vice versa.

2. Copy and complete Table 1, referring to the data in this section.

Benefits of climate change	Disadvantages of climate change
The Northwest Passage will improve shipping	Africa will lose food production

▲ Table 1

Hold a debate between the two groups.

3. Consider these lists.

Who suffers the most?

- coastal cities

- the Maldives and low-lying islands

- the most vulnerable

- LICs.

Who benefits the most?

- Scandinavian countries

- people in higher latitudes

- people in higher altitudes

- Siberia, northern Canada or Greenland

- those profiting from oil extraction in the Arctic

- carbon-offset companies

- renewable energy companies.

The impacts of climate change are not equitably distributed across the Earth. Make your own lists of "winners and losers".

Modelling climate change

Systems diagrams and models can represent the causes and effects of climate change. Feedback loops, both positive and negative, and changes in the global energy balance can be shown.

There are five ways in which the climate can change over time due to a change in GHG levels in the atmosphere. Figure 20 shows the following scenarios.

There may be a **direct relationship**—increase in solar radiation, lowered albedo, increased methane release all lead to (force) increased climate change through positive feedback loops (a).

There may be a **buffering action** in which forcing increases but climate change does not follow in a linear way. It is insensitive to change (b) or negative feedback loops operate.

It may respond slowly at first but then **accelerate** until it reaches a new equilibrium (c) with positive feedback.

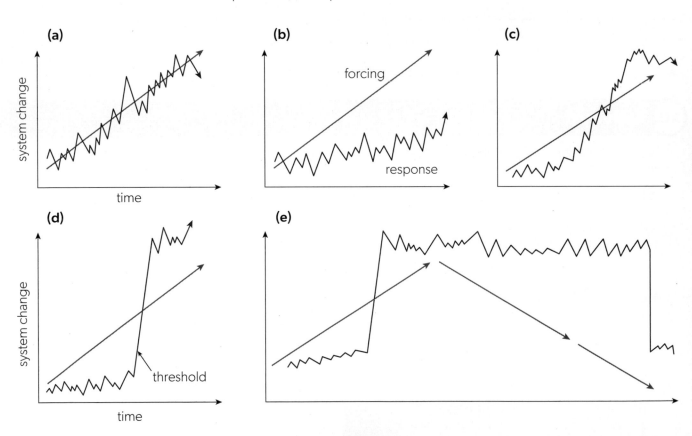

▲ Figure 20 **Possible climate change responses to a forcing mechanism**

It may reach a **tipping point**—the climate does not respond to changes until it reaches a **threshold**, at which point it changes rapidly until a new, much higher equilibrium is reached (d).

In addition to the threshold change, climate change may then get **stuck at the new equilibrium**, even when the forcing decreases, until it then tips over a new threshold and falls rapidly. These threshold changes could occur in just a few decades (e).

A way of visualizing this is to imagine the climate as a car that you (the forcing mechanism) are pushing uphill. In (a) you push steadily uphill. In (b) you push with the same force but the car moves much more slowly—it has more resistance. In (c) you reach a part of the hill with a shallower gradient and the car moves more easily before the hill gets steeper again. In (d) you push until the car reaches the edge of a cliff and then falls over it. Can you explain (e) in this analogy?

Our problem is that we do not know which of these situations we are living in. So how do we decide what to do? See subtopic 6.3 for more information.

TOK

Is it possible for members of the public to judge whether or not to accept scientific findings when there are so many possibilities?

Planetary boundary for climate change

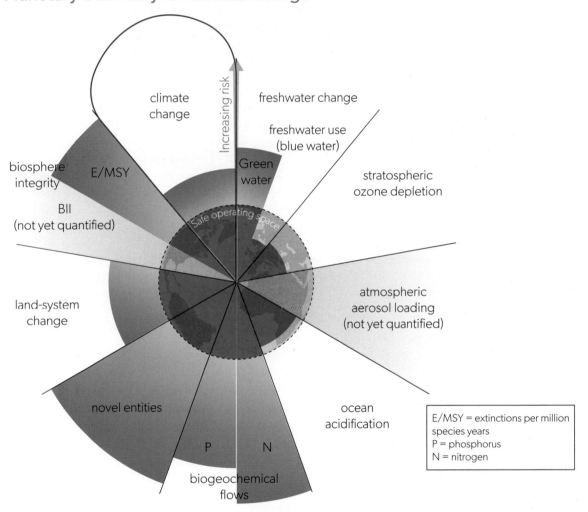

▲ Figure 21 **Planetary boundaries model with climate change boundary crossed**
Adapted from: Azote for Stockholm Resilience Centre, based on analysis in Wang-Erlandsson et al 2022 / Stockholm Resilience Centre (CC BY-NC-ND 3.0)

According to the Paris Agreement, it is necessary to keep the global temperature increase below 2°C—ideally no more than 1.5°C—above pre-industrial levels. This global limit is calculated based on the concentration of CO_2 in the atmosphere. The threshold must be below 350 ppm. We crossed that some time ago so have crossed this boundary.

Connections: This links to material covered in subtopics 1.1 and 1.3.

Perspectives on climate change

Psychological research finds people use denial of climate change as a defence mechanism. This is because, for much of the time, the idea of climate change is too large and scary to understand.

Some reject the idea that human-caused climate change exists; others have argued that human-made climate change is occurring but that the extent to which climate is changing and the precise impact of human activity is uncertain.

One survey in the UK, Ireland, Norway, Poland, Italy and Germany found that:

> Large majorities of people are worried about the impact of climate change. 81% of people on average say they are worried about the impact of climate change for future generations, and 80% say this for humanity in general. Most people also think climate change is harmful now, or will be harmful within the next 10 years.[1]

There are widespread misconceptions of scientists' views on climate change. Across all six countries, the average estimate for the proportion of scientists who have concluded that human-caused climate change is happening is 68%—far lower than the reality, of 99.9%. There is also some doubt over the causes of climate change. Three-quarters of people (74%) on average say that climate change is mainly caused by human activities. This figure is 82% in Italy, but only 61% in Norway. Large numbers of people also believe that oil companies are hiding technology that could enable cars to run without petrol or diesel.[1]

Perspectives on climate change for individuals and for societies are influenced by many factors. This results in debate on how best to respond to climate change.

Why, ultimately, should nations act to control greenhouse gases, rather than just letting climate turmoil happen and seeing who profits? One reason is that the cost of controls is likely to be much lower than the cost of rebuilding the world. Coastal cities could be abandoned and rebuilt inland, for instance, but improving energy efficiency and reducing GHG emissions in order to reduce rising sea levels should be far more cost-effective. Reforms that prevent major economic and social disruption from climate change are likely to be less expensive than reacting to the change. The history of anti-pollution programmes shows that it is always cheaper to prevent emissions than to reverse any damage they cause.

More on evidence for climate change

Weather forecasts and climate predictions are improving all the time because of information from satellites, weather stations, observatories, radar and computerized data collection. Such long-term data enables the study of climate change and changes in land use. These long-term data sets include temperature and greenhouse gas concentrations.

Models help us to understand complex systems by testing hypotheses. Models are based on collected data. The more data there is and the more accurate it is, the better the modelling.

---

[1] Public perceptions on climate change
https://www.kcl.ac.uk › assets › peritia-climate-change

TOK

To what extent is our perspective determined by belonging to a particular culture?

Practical

Skills:

- Tool 1: Experimental techniques

- Tool 2: Technology

Create a survey to investigate perspectives on climate change in your class or school or community.

AHL

Connections: This links to material covered in subtopic 1.2.

Measurements are both direct and indirect (or proxy) measurements. Direct measurements are in the present; indirect measurements can help us to understand past climate data. Both direct and indirect measurements contribute to creating climate models. Using past data informs modelling and makes the model more accurate.

Direct measurements include measuring and recording temperature, humidity, precipitation, air pressure, wind speeds and direction, gases in the atmosphere, cloud formations and pollution levels. Direct measurements of carbon dioxide in the atmosphere only go back to the 1950s but indirect measurements extend much further back so we get a longer-term view of how climate has changed.

Indirect (proxy) measurements include ice core sampling, tree rings, sediment cores, fossil organisms and pollen from peat cores.

What do ice cores show?

- Carbon dioxide levels were stable over the last 10,000 years at 280 ppm.

- In the early 1800s, they started to rise when humans started burning coal during the Industrial Revolution.

- Carbon dioxide levels were at 417 ppm in February 2023. What are they now?

- Methane has also increased to over double its pre-industrial level.

- Antarctic ice shows a cycle of 100,000-year glacial periods—ice ages—interspersed with warmer interglacial periods.

- Carbon dioxide concentrations were low in cold times and higher in warmer.

▲ Figure 22 Ice core collecting—a proxy method to measure changes in climate over time

What do tree rings show?

- Some trees can live for 1,000 years.

- Trees need water and sunlight to photosynthesize and grow.

- Higher temperatures speed up growth.

- Each year, trees lay down a layer of stored material on top of the previous year like a coat.

- The cells laid down are larger in the summer and smaller in winter, forming clear rings of growth.

- In warmer, wetter years the rings are wider than in cooler years.

- By felling a tree or taking a core across the trunk, we can count the tree rings.

- These rings tell both how old the tree is and how the climate changed each year (Figure 23).

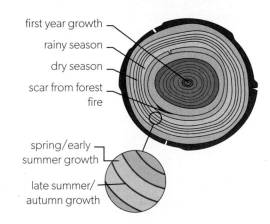

first year growth
rainy season
dry season
scar from forest fire

spring/early summer growth
late summer/autumn growth

▲ Figure 23 Tree rings provide data on past climate
Adapted from: NASA

Pollen and fossil analysis

- Like ice cores, peat bogs and lake sediments are laid down over long periods.

- Pollen falls into bogs and lakes and onto ice. It is preserved.

- Sampling the cores and identifying the pollen tells us what was living at the time.

- Fossils of organisms and fossil pollen found in rock strata can be identified and dated. This tells us what was living at the time.

More on climate models

Climate models are based on collecting long-term data sets and are also called general circulation models (GCMs). Global climate models manipulate inputs to climate systems to predict possible outputs or outcomes, using equations to represent the processes and interactions that drive the Earth's climate. The validity of the models can be tested via a process known as hindcasting.

Hindcasting is a process of testing computer models (e.g. climate change models), by comparing them to the actual historical observations to see how well they match, e.g. starting a model at 1950 conditions and running it for 75 years then comparing the results with actual measurements. Hindcasting runs a model backwards from the present time to check its validity.

A model that fails to match will not produce realistic projections, so it requires altering. If the model matches the observed conditions, it is more reliable and its predictions of the future may be more reliable too.

Although very accurate now, no model can be 100% certain. Today's models are based on well-understood physical processes of Earth systems. Even 50-year-old models of climate accurately predicted a temperature rise of 0.9°C since 1970.

Climate change scenarios

The future of the Earth's climate is inherently uncertain but not totally unknowable. Climate models use different scenarios to predict possible impacts of climate change.

You may have seen graphs in which projections of future temperature or population or other variables show different possibilities. These are models of scenarios showing projections of what may happen given the data we have. We do not know what will change in the future, hence the projections.

Temperature change scenarios

Projections can show "what can happen if nothing changes" and "what should happen with various changes". Climate change projections have a baseline scenario with no policy intervention and then various pathways towards desired goals. Figures 24–26 show scenarios for temperature and sea level changes. The same projections are done for population change in subtopic 8.1.

TOK

Certainty about the past is more difficult to attain than certainty about the present or the future.

Discuss this statement with reference to evidence for climate change.

Connections: This links to material covered in subtopics 1.2, 2.4, HL.a and HL.b.

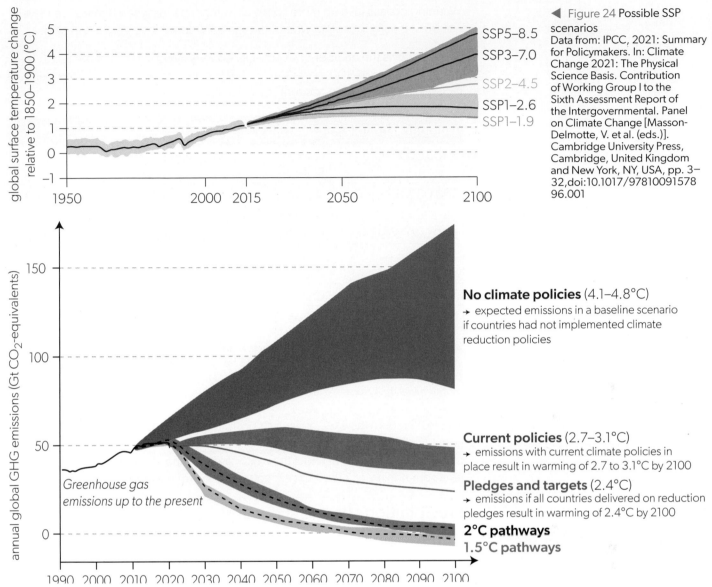

◀ Figure 24 Possible SSP scenarios
Data from: IPCC, 2021: Summary for Policymakers. In: Climate Change 2021: The Physical Science Basis. Contribution of Working Group I to the Sixth Assessment Report of the Intergovernmental. Panel on Climate Change [Masson-Delmotte, V. et al. (eds.)]. Cambridge University Press, Cambridge, United Kingdom and New York, NY, USA, pp. 3–32,doi:10.1017/97810091578 96.001

No climate policies (4.1–4.8°C)
→ expected emissions in a baseline scenario if countries had not implemented climate reduction policies

Current policies (2.7–3.1°C)
→ emissions with current climate policies in place result in warming of 2.7 to 3.1°C by 2100

Pledges and targets (2.4°C)
→ emissions if all countries delivered on reduction pledges result in warming of 2.4°C by 2100

2°C pathways

1.5°C pathways

▲ Figure 25 Annual GHG emission scenarios. The width of each shaded "strip" shows the uncertainty in that scenario. Warming refers to the expected global temperature rise by 2100, relative to pre-industrial temperatures
Adapted from: Hannah Ritchie & Max Roser / Our World in Data (CC BY 4.0)

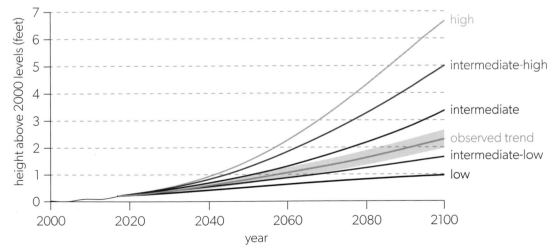

▲ Figure 26 Observed sea level from 2000 to 2018, with future sea level through 2100 for six future pathways. The pathways differ based on future rates of GHG emissions and global warming and differences in the plausible rates of glacier and ice sheet loss.
Data from: NOAA Climate.gov / Sweet et al. 2022

Warming oceans take up more space as water expands when heated. Sea levels rise but it is hard to predict where, though confidence levels are increasing.

Table 2 shows the number of people per country living on land expected to be under sea level by 2100, based on a 50–70 cm rise in sea levels (2°C temperature increase, not taking into account ice sheet instability).

Number of people	Countries
<100,000	Chile, Kenya, Namibia, Papua New Guinea, Portugal, Sweden
100,000–499,000	Australia, Canada, Colombia, Madagascar, Mexico, Tanzania
500,000–999,000	Ecuador, France, Italy, Mozambique, Pakistan, Turkey
1–9 million	Brazil, Egypt, Germany, Japan, Myanmar, Nigeria, Philippines, UK, USA
10–50 million	China, India, Indonesia, Thailand, Vietnam
No data	Bolivia, Botswana, Greenland, Kazakhstan, Mongolia

▲ Table 2 Global predictions of the effects of sea level rise—the number of people living on land in various countries who are expected to be under sea level by 2100
Data from: Scott A. Kulp and Benjamin H. Strauss: New elevation data triple estimates of global vulnerability to sea-level rise and coastal flooding, Nature Communications

In its 2019 report, the IPCC projected 0.6–1.1 m of global sea level rise by 2100 (or about 15 mm per year) if GHG emissions remain at high rates. By 2300, seas could be as much as 5 m higher under the worst-case scenario. If countries do cut their emissions significantly, the IPCC expects 0.3–0.6 m of sea level rise by 2100.

Precipitation change scenarios

Climate change is not limited to temperature changes or sea level rise. Changes in precipitation—rain and snow—will have impacts. Because there are so many complex variables, there is less agreement on how precipitation will change but there will be more evaporation and surface drying with increased heat. It may be that dry areas become drier and wet areas become wetter. There is some agreement that:

- both the tropical Pacific and high-latitude areas will have more precipitation in the future. India, Bangladesh and Myanmar will all become wetter, as will much of northern China

- the Mediterranean region and southern Africa will have less precipitation

- there will be reduced precipitation in southwest Australia around Perth, in southern Chile, the west coast of Mexico and over much of the tropical and subtropical Atlantic Ocean.

Critical threshold on climate change

Figure 27 shows the drastic increase in global temperature in the last 150 years. Is it possible that this would not have an impact on global systems, both abiotic and biotic?

(ATL) **Activity 9**

ATL skills used: thinking, communication, social, research, self-management

Look at all the information on changes in temperature, sea level and precipitation.

In small groups, produce a poster, graphic or animation to show the absolute worst and the absolute best scenarios.

The Earth may be approaching a critical climate threshold (tipping point) beyond which unanticipated, rapid and potentially catastrophic changes may occur, with an irreversible change away from the current equilibrium due to positive feedback loops.

There is local evidence that some regions have reached this threshold, e.g.

- the Antarctic ice sheet

- the slowing of the Atlantic thermohaline circulation

- the Amazon Rainforest–Cerrado transition.

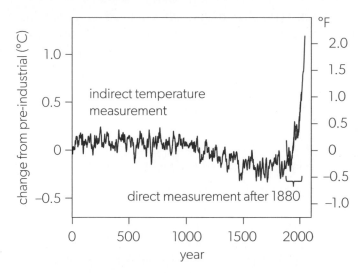

▲ Figure 27 Global temperature since 0 CE
Adapted from: Efbrazil (CC BY-SA 4.0)

Tipping cascades

It is possible in complex systems for one tipping point to increase the likelihood of another and another—a tipping cascade. A domino effect is when one element triggers another and that triggers another as in a row of bricks on end. Reaching a tipping point in one part of the system may then trigger another tipping point and so on. This is a tipping cascade. As much is unknown, this makes climate change predictions very uncertain. Non-linear relationships such as these make it difficult to predict the scale and pace of climate change.

Examples of potential tipping cascades:

- permafrost melting leading to boreal forest dieback leading to more GHGs in the atmosphere leading to higher temperatures, leading to sea level rises …

- Greenland ice sheet is lost causing the West Antarctic ice sheet to go as warm seawater reaches it, leading to huge sea level rise, leading to failure of the Atlantic thermohaline circulation, leading to …

Do not confuse these cascades with trophic cascades (see subtopic 2.5) where predators limit the density or behaviour of their prey in a trophic level and enhance survival of the next lower trophic level.

TOK

The statement "with the least responsible often being the most vulnerable" is political, ethical and economic.

In what ways is factual evidence sometimes used, abused, dismissed and ignored in politics, economics and ethics?

More on inequity and vulnerability to climate change

Countries vary in their responsibility for climate change and also in vulnerability, with the least responsible often being the most vulnerable. There are political and economic implications and issues of equity.

At the 2022 COP 27 climate talks in Egypt, government negotiators debated which countries should receive funds to address the loss and damage caused by climate change. The G77 (a block of 134 LICs) and China wanted all LICs to be eligible for the funds. The EU, which has caused a lot of climate change and so will be expected to pay into the fund, wanted the money to go only to "particularly vulnerable" countries. Eventually, it was decided to set up a Loss and Damage Fund but as of 2023, the details are still to be worked out.

Vulnerability has three aspects:

- risk of floods, storms, heatwaves, droughts, sea level rise

- peoples at risk

- financing, expertise and government capability to act.

Australia, for example, is at risk of flood and fire but also has funds and expertise to deal with them. Other countries may not yet face climate threats but could they deal with them if they happen? Measuring vulnerability is difficult. The IPCC stated in 2022 that there are global hotspots of high human vulnerability, particularly in west, central and east Africa, south Asia, central and south America, small island developing states and the Arctic.

ATL Activity 10

ATL skills used: thinking, research

Measurement of GHG emissions varies. It may consider total emissions in a country per year, per capita emissions in each country, or cumulative emissions over time by country.

Visit the Our World in Data website and review the data in the section "Who emits the most CO_2?"

1. Find the data showing country emissions of CO_2 in 2017.

   a. List the five largest emitting countries in 2017.

   b. Find data on the five largest emitters now. What has changed and why?

2. Find the most recent data relating to annual CO_2 emissions by country.

   a. List five countries where total CO_2 emissions have decreased. Suggest reasons for this.

   b. What do these countries have in common?

   c. List five countries where emissions continue to increase.

   d. What do these countries have in common?

   e. What are the other variables in this data?

3. On the same Our World in Data page, find per capita emissions by country over time.

   a. List the 10 largest emitting countries per capita in 2021.

   b. Where in the world are most of these?

   c. Why do they have such large per capita emissions?

   d. Why is their contribution to global emissions fairly small?

   e. Find the country where you live or are now. What are the per capita emissions in 2021 for that country?

   f. Is that country listed as most vulnerable to climate change?

Check your understanding

In this subtopic, you have covered:

- climate is due mainly to temperature and precipitation

- human activities have increased carbon dioxide levels in the atmosphere

- evidence is conclusive that temperature and carbon dioxide levels are linked

- the enhanced greenhouse effect is causing climate change

- there are many impacts of climate change

- feedback loops enhance or reduce global warming

- there is evidence that the planetary boundary for climate change has been passed

- responses vary depending on perspectives

- direct and indirect measurements give evidence of changes

- models predict future changes with various scenarios

- a climate tipping point may result in a new equilibrium and tipping cascades

- vulnerable countries are often those that emit fewer GHGs

- risk to climate change effects is inequitable.

AHL

To what extent has climate change occurred because of anthropogenic causes?

How do differing perspectives play a role in responding to the challenges of climate change?

1. Describe how CO_2 levels in the atmosphere have changed over the last 200 years.

2. Explain what has caused this change.

3. Evaluate the evidence for the correlation between CO_2 levels and global temperature.

4. Briefly describe negative and positive impacts of climate change.

5. Evaluate the accuracy of climate change modelling.

6. Discuss the factors influencing perspectives on climate change.

7. Explain how climate models are developed including hindcasting.

8. Discuss the value of using different scenarios to predict climate changes.

9. Explain why the climate threshold is difficult to predict.

10. Discuss the implications of climate justice for all.

AHL

>> Taking it further

- Create a presentation or display for the school to raise awareness about the issue of climate change.

- Participate in climate action events.

- Join a citizen action group or youth parliament to create policies around climate change.

- Knit the climate stripes developed by Professor Ed Hawkins at the University of Reading in 2018 or create your own garment.

6.3 Climate change—mitigation and adaptation

Guiding question

- How can human societies address the causes and consequences of climate change?

Understandings

1. To avoid the risk of catastrophic climate change, global action is required, rather than measures adopted only by certain states.
2. Decarbonization of the economy means reducing or ending the use of energy sources that result in CO_2 emissions and their replacement with renewable energy sources.
3. A variety of **mitigation** strategies aim to address climate change.
4. Adaptation strategies aim to reduce adverse effects of climate change and maximize any positive consequences.
5. Individuals and societies on a range of scales are developing **adaptation** plans, such as National Adaptation Programmes of Action (NAPAs), and resilience and adaptation plans.

6. Responses to climate change may be led by governments or a range of non-governmental stakeholders. Responses may include economic measures, legislation, goal setting commitments and personal life changes.
7. The UN has played a key role in formulating global strategies to address climate change.
8. The IPCC has proposed a range of emissions scenarios with targets to reduce the risk of catastrophic climate change.
9. Technology is being developed and implemented to aid in the mitigation of climate change.
10. There are challenges to overcome in implementing climate management and intervention strategies.
11. Geoengineering is a mitigation strategy for climate change, treating the symptom not the cause.
12. A range of stakeholders play an important role in changing perspectives on climate change.
13. Perspectives on the necessity, practicality and urgency of action on climate change will vary between individuals and between societies.
14. The concept of the tragedy of the commons suggests that catastrophic climate change is likely unless there is international cooperation on an unprecedented scale.

Key terms

Mitigation involves reduction and/or stabilization of GHG emissions and their removal from the atmosphere. It is anthropogenic intervention to reduce the anthropogenic forcing of the climate system; it includes strategies to reduce GHG sources and emissions and enhance GHG sinks. (IPCC, 2007)

Adaptation is the adjustment in natural or human systems in response to actual or expected climatic stimuli or their effects, which moderates harm or exploits beneficial opportunities. (IPCC, 2007)

Global action on climate change

To avoid the risk of catastrophic climate change, global action is required and is happening. For over 30 years, the UN has persuaded national governments to act to reduce emissions of GHGs, particularly carbon. Individual nations are acting to reduce their GHG emissions but some are acting faster than others. Reductions in countries with the largest carbon footprints will have the biggest effect globally but this does not mean you should not "do your bit"—everything we do matters.

(ATL) Activity 11

ATL skills used: thinking, communication, research

Five countries emit the most GHGs: USA, China, Russia, Japan and India.

Figures 1–3 show emissions from these five countries plus Sweden. Sweden has had a carbon tax for 30 years and aims to be carbon neutral by 2045; it is reducing its carbon emissions annually.

Per capita CO$_2$ emissions

Carbon dioxide (CO$_2$) emissions from fossil fuels and industry[1]. Land-use change is not included.

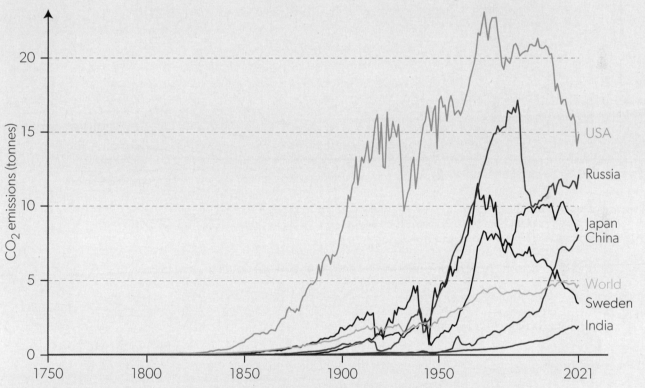

1. Fossil emissions measure the quantity of CO$_2$ emitted from the burning of fossil fuels, and directly from industrial processes such as cement and steel production. Fossil CO$_2$ includes emissions from coal, oil, gas, flaring, cement, steel, and other industrial processes. Fossil emissions do not include land-use change, deforestation, soils, or vegetation.

▲ Figure 1 Per capita carbon dioxide emissions by the world and six named countries
Adapted from: Our World in Data. Data from: Global Carbon Budget (2022); Gapminder (2022); UN (2022); HYDE (2017); Gapminder (Systema Globalis) (CC BY 4.0)

1. Consider the data in Figure 1 for the periods 1850–1975 and 1975–2021.

   a. Describe the change in per capita CO$_2$ emissions of the countries in Figure 1 during these two periods.

   b. Describe the change in world per capita emissions during these two periods of time.

   c. Explain why world per capita emissions are not greater than those of the US, Russia, Japan and China.

2. Explain why some countries have managed to reduce their total emissions but others have not.

3. Describe total world per capita emissions changes since about 1960.

Annual CO$_2$ emissions

CO$_2$ emissions from fossil fuels and industry. Land-use change is not included.

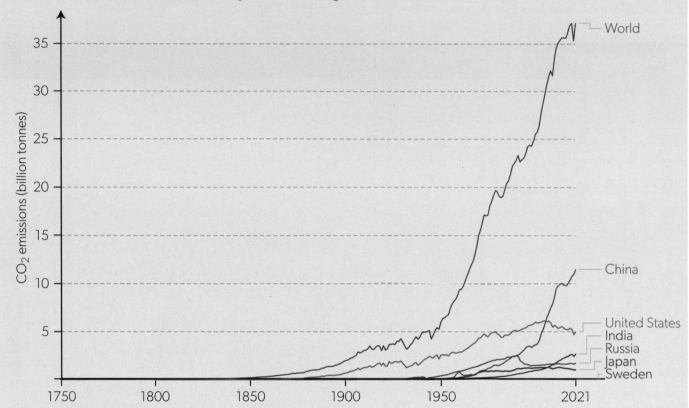

▲ Figure 2 Annual carbon dioxide emissions by the world and six named countries
Adapted from: Our World in Data. Data from: Global Carbon Budget (2022); Gapminder (2022); UN (2022);
HYDE (2017); Gapminder (Systema Globalis) (CC BY 4.0)

4. Now consider Figure 2, which shows the total annual emissions of these countries and the world.

   a. Explain why these country emissions graphs are different from the per capita emissions graphs.

   b. Explain the shape of the world annual emissions graph.

5. Figure 3 shows total cumulative emissions over time by the same countries and the world. Explain why the country order from biggest emitter to least has changed on the cumulative emissions graph.

6. On the Our World in Data website, find the cumulative CO$_2$ emissions graph.

   a. Add two other countries of your choice.

   b. Describe how these countries compare with the countries on the graph in Figure 3.

   c. Explain why there may be differences.

Cumulative CO$_2$ emissions

Cumulative emissions are the running sum of CO$_2$ emissions produced from fossil fuels and industry since 1750. Land-use change is not included.

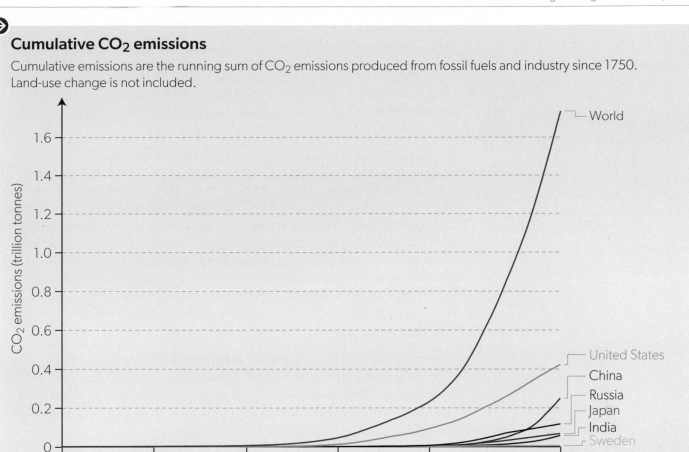

▲ Figure 3 Cumulative carbon dioxide emissions by the world and six named countries
Adapted from: Our World in Data. Data from: Global Carbon Budget (2022); Gapminder (2022); UN (2022); HYDE (2017); Gapminder (Systema Globalis) (CC BY 4.0)

UN conventions

At the 1992 Earth Summit in Rio de Janeiro, Brazil, governments founded three sister "Rio **Conventions**" in response to challenges of climate change, desertification and biodiversity loss:

- the United Nations Framework Convention on Climate Change (UNFCCC, also known as UN Climate Change)

- the Convention on Biological Diversity (CBD, also known as UN Biodiversity)

- the United Nations Convention to Combat Desertification (UNCCD).

UN Framework Convention on Climate Change

The ultimate objective of all agreements under the UNFCCC is to stabilize GHG concentrations in the atmosphere, at a level that will prevent dangerous human interference with the climate system and within a time frame which allows ecosystems to adapt naturally and enables sustainable development. Today, the UNFCCC has near-universal membership. The 199 countries that have ratified the Convention are called Parties to the Convention.

Key term

A UN **convention** or treaty is an agreement between different countries that is legally binding to the contracting States.

Key term

A UN **protocol** creates legally binding obligations as international law. "Protocol" is used for agreements less formal than a "treaty" or "convention", often a subsidiary to a treaty.

Conference of the Parties (COP) summits are annual meetings of the UNFCCC parties (states). In 2023, COP 28 was held in Dubai. Where is the latest one held?

The UNFCCC entered into force on 21 March 1994 and is the parent treaty of the 1997 Kyoto Protocol and the 2015 Paris Agreement. The main aim of the **Paris Agreement** is to keep the global average temperature rise this century as close as possible to 1.5°C above pre-industrial levels.

The **Kyoto Protocol** was adopted on 11 December 1997. Owing to a complex ratification process, it entered into force only on 16 February 2005. Currently, there are 192 Parties to the Kyoto Protocol. It only binds developed countries (HICs) because it recognizes that they are largely responsible for the current high levels of GHG emissions in the atmosphere. In its Annex B, the Kyoto Protocol set binding emission reduction targets for 37 industrialized countries and economies in transition and for the European Union. Overall, these targets added up to an average 5% emission reduction compared with 1990 levels over the five-year period, 2008–2012 (the first commitment period). In the Kyoto Protocol, international emissions trading was introduced to help states meet their targets.

In Doha, Qatar, in 2012, the **Doha Amendment** to the Kyoto Protocol was adopted and extended to 2020. It set a goal of reducing GHG emissions by 18% compared with 1990 levels for the 194 industrial countries which agreed to it. It was a legally binding emission reduction target. It entered into force in 2020 after the mandated minimum of 144 states agreed to it.

Intergovernmental Panel on Climate Change (IPCC)

The IPCC was established by the UN and the World Meteorological Organization in 1988 to provide policymakers with regular scientific assessments on the current state of knowledge about climate change. It has 195 member states. It has produced six assessment reports so far.

The overwhelming view of member states, says Christiana Figueres, head of the UN Climate Change Secretariat, is that any agreement "has to be much more collaborative than punitive"—if it is to happen at all. "Even if you do have a punitive system, that doesn't guarantee that it is going to be imposed or would lead to any better action."

Connections: This links to material covered in subtopic 1.3.

To critics, the absence of a legal "stick" to enforce compliance is a deep—if not fatal—flaw in the Paris process, especially after all countries agreed in 2011 that an agreement would have some form of "legal force". They warned that a deal already built on sometimes vague promises from member states could end up as a toothless addition to the stack of more than 500 global and regional environmental treaties, while the rise in global temperatures mounts past a UN ceiling of 2°C with the prospect of more floods, droughts and heatwaves.

Even the European Union, which has long argued for a strong, legally binding deal, is increasingly talking about a "pledge and review" system under which national commitments would be re-assessed every five years against a goal of halving world emissions by 2050.

Strategies to alleviate climate change

We can try to reduce the impact of (mitigate) climate change or adapt to it or do both.

There are three routes we can take on this issue: do nothing; wait and see; or take precautions now.

Science cannot give us 100% certainty on the issue of global warming or predict with total accuracy what will happen. What it can do is collect data and provide evidence. How that evidence is interpreted and extrapolated will depend on individual viewpoints, scientific consensus, economics and politics.

The consensus has finally swung to accepting that the evidence is overwhelming and climate change is happening fast. Science has been saying this for 50 years or more but it has taken a long time for policymakers to listen and act.

The precautionary principle

The **precautionary principle** is the majority choice: act now, in case.

The danger in the "wait and see" strategy is that it takes a long time for actions to have results. To move the global economy away from a fossil fuel base is a long, slow process. The possible disruption of national economies by the process was thought by some not to be necessary. But it is possible that we will reach the **tipping point** when our actions will have little effect because positive feedback mechanisms change the climate to a new equilibrium which could be 8°C warmer than it is now. It is better to be safe than sorry.

These precautions can be divided into three categories: international commitments; national actions; and personal lifestyle changes.

▲ Figure 4 Lifestyle choices make a difference in commuters cycling not driving

---

(ATL) Activity 12

ATL skills used: thinking, research

Carbon neutrality by 2050 is the goal of the UN. In 2020, over 110 countries had pledged to do this and the European Union had committed to do so.

Some countries—e.g. the USA, UK, France, Spain, Italy, and many others—have managed to reduce emissions (even when we correct for trade) while increasing GDP. We can make progress on reducing emissions, but it is currently slow, perhaps too slow.

1. Visit the Our World in Data website and find the map showing the status of net-zero commitments of countries around the world.
2. List which countries have committed to a zero carbon emissions target and whether this is in (a) law or (b) a pledge or policy. Are there any similarities between the countries that have net zero in law?
3. List countries which have achieved net zero. Do these countries have similarities?

Key term

Decarbonization of the economy means reducing or ending the use of energy sources that result in CO_2 emissions and replacing them with renewable energy sources.

Decarbonization

Decarbonization means finding alternative ways of living and working that (a) reduce carbon emissions and (b) capture and store carbon in our soil and vegetation. It requires a radical change in the current economic model which is focused on growth. This means transforming how energy is generated and changing energy sources, changing how we build and move around, and how land resources are managed. We must either drastically reduce our consumption or switch to low emission technologies and renewable alternatives.

Actions include the following.

1. **Move towards low-carbon electricity by reducing carbon intensity (carbon produced per unit of energy) by:**
   - using renewables
   - having more nuclear energy
   - burning gas not coal (gas emits less CO_2 per unit of energy).
2. **Use more electric transport, less fossil-fuelled transport.** Some energy sectors (e.g. transport) are harder than others to decarbonize but it can be done.
3. **Develop low-cost low-carbon energy and battery technologies.** To do this quickly, and allow LICs to avoid high-carbon development pathways, low-carbon energy needs to be cost-effective and the default choice.
4. **Improve energy efficiency.**

Reducing emissions from food production

1. **Reduce meat and dairy consumption, especially in HICs.** Shift dietary patterns towards lower-carbon food products. This includes eating less meat and dairy and also substituting high-impact meats (e.g. beef and lamb) with chicken, fish or eggs. Meat substitutes, vegetarian or vegan diets are all part of this.
2. **Promote lower-carbon meat and dairy production.** Grass-fed meat and milk production, not intensive feedlots.
3. **Improve crop yields.** Sustainable intensification of agriculture produces more food on less land. This helps prevent deforestation from agricultural expansion. Vertical farming and hydroponics are part of this.
4. **Reduce food waste.** Around one-third of food emissions come from food that is lost in supply chains or wasted by consumers. Improve harvesting, storing and transport and reduce consumer waste.

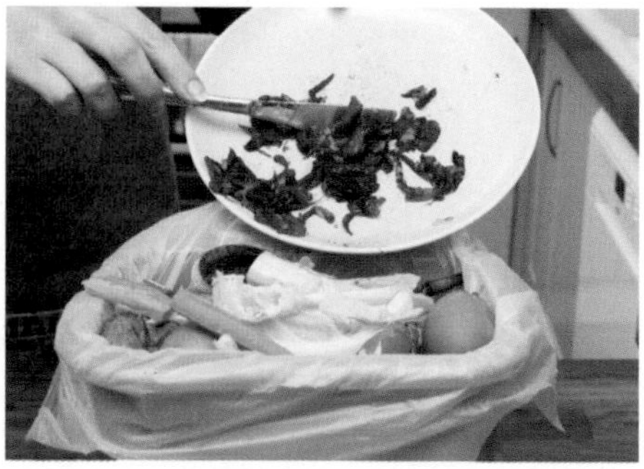

▲ Figure 5 Reduce food waste

Mitigation strategies

Mitigation means reducing the severity of something. There are two ways to do this regarding climate change:

- Reduce or stop production of what is causing climate change (GHGs).
- Remove the cause once it is produced.

Since emissions of GHGs are causing climate change, the strategies are as follows.

Reduce GHG emissions

1. Reduce CO_2 emissions.

2. Reduce emissions of nitrogen oxides

3. Reduce methane emissions from agriculture, by:

    *   using alternatives to fossil fuels and ending fossil fuel subsidies

    *   increasing energy efficiency

    *   increasing use of renewable energy

    *   building low-carbon, resilient cities and homes

    *   putting a price on carbon—a carbon tax

    *   reducing deforestation

    *   decreasing the use of fossil-fuelled transport

    *   decreasing methane emissions from ruminants or decreasing global livestock numbers.

Remove carbon dioxide from the atmosphere

This may be achieved through:

1. afforestation

2. rewilding and land management practices (see subtopic 3.3)

3. carbon capture and storage (CCS) and bioenergy with carbon capture and storage (BECCS)

4. geoengineering.

hydroelectric dam

nuclear power station

offshore wind turbines

solar farm

solar panels on house

electric vehicles

▲ Figure 6 **Mitigation strategies for climate change**

But even if mitigation strategies drastically reduce future emissions of GHGs, past emissions will continue to have an effect for some time as the gases are circulating in the atmosphere.

There is a difference between stabilizing GHG emissions and stabilizing concentrations of GHGs in the atmosphere (Figure 7). If tomorrow we could stabilize GHG emissions at today's levels, their concentrations in the atmosphere would continue to rise. This is because human activities are adding GHGs to the atmosphere faster than natural processes can remove them (see carbon cycle, subtopic 2.3).

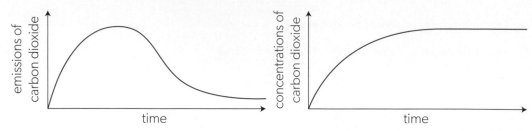

▲ Figure 7 Graphs of emissions and concentration of carbon dioxide over time

To stabilize GHG concentrations in the next 100 years, we would need to reduce emissions by 80% of peak emission levels. This is extremely unlikely to happen.

Stabilize or reduce GHGs

Reduce energy consumption

- Reduce energy waste by using it more efficiently, e.g. improve fuel economy in motor car engines, hybrid or electric vehicles; insulate and cool buildings more efficiently; use energy efficient light bulbs and appliances; educate people about the impacts of their actions.

- Reduce overall demand for energy and electricity by being more efficient and using less, e.g. change lifestyles and business practices; use less private transport; cycle or walk rather than driving; eat less meat; adopt a circular economy; stop flying or fly less often.

- Adopt carbon taxes and remove fossil fuel subsidies.

- Set national limits on GHG production and a carbon credit system.

- Have personal carbon credits which can be traded and encourage people to reduce their carbon footprint.

- Change development pathways and socio-economic choices—change priorities in government and educate to change social attitudes, e.g. London has a toll charge for cars driving into the city and low emission zones for trucks.

- Improve efficiency of energy production.

Reduce emissions of nitrous oxides and methane from agriculture

- Reduce methane production, e.g. methane from cows can be reduced by changing their diets.

- Capture more methane produced from landfill sites.

- Develop sustainable agriculture.

Use alternatives to fossil fuels

- Replace high GHG emission energy sources with low GHG emission ones, e.g. hydroelectric and other renewables and nuclear power generation instead of burning fossil fuels.

Remove carbon dioxide from the atmosphere

This is termed carbon dioxide removal (CDR).

1. **Increase amounts of photosynthesis** to increase the rate at which atmospheric carbon dioxide is converted into a biomass carbon sink (see subtopic 2.3). This could be done by reforesting, decreasing deforestation rates and restoring grasslands, e.g. the UN-REDD programme (UN Reducing Emissions from Deforestation and forest Degradation) in LICs.

2. **Carbon capture and storage (CCS) and bioenergy with CCS (BECCS).** This means capturing carbon dioxide in emissions from power stations, oil refineries and other industries which emit large amounts of carbon dioxide. However, this increases the cost of energy and products. BECCS involves producing energy from biomass—which still releases carbon dioxide—and then capturing the emissions from this.

   To store the carbon dioxide, it is transported to areas with suitable rocks and then pumped into the rocks under pressure. An alternative is to store it in mineral carbonates by reacting carbon dioxide with metal oxides at high temperatures. Limestone is calcium carbonate so this is like making limestone, but a huge amount of energy is needed for this.

   A few pilot plants have carried out CCS but there are no large-scale CCS power stations yet (2023).

3. **Use more biomass as a source of fuel.** If the same crop is planted in the following year, an amount of carbon dioxide equal to that released by burning the fuel is then captured by photosynthesis. This should mean the fuel is effectively carbon-neutral.

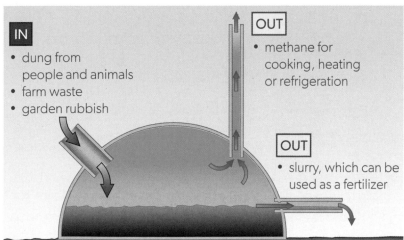

IN
- dung from people and animals
- farm waste
- garden rubbish

OUT
- methane for cooking, heating or refrigeration

OUT
- slurry, which can be used as a fertilizer

▲ Figure 8 **Diagram of a biogas anaerobic digester**

Biomass can be used as a fuel:

- directly, by burning it to generate heat or electricity

- indirectly to produce biofuels, e.g. biogas from animal waste or waste food on a small or larger scale in biogas anaerobic digesters (Figure 8); or biodiesel and ethanol from waste organic matter or waste vegetable oils or from planting crops such as sugar cane.

Geoengineering

Geoengineering or climate engineering involves large-scale intervention projects. This is different from the other mitigation strategies because so far these proposals are hypothetical or computer models. They have not been tried and they raise ethical questions.

Examples of geoengineering projects

1. Scatter iron, nitrates or phosphates on oceans to increase algal blooms which take up more carbon and act as a carbon sink.

2. Release sulfur dioxide from airplanes to increase global dimming.

3. Place mirrors on satellites between the Earth and the Sun to deflect solar radiation.

4. Build with light-coloured roofs to increase albedo and reflect more sunlight.

Adaptation strategies

Adaptation aims to reduce the adverse effects and maximize any positive effects of climate change. These initiatives and measures aim to reduce the vulnerability of natural and human systems against actual or expected climate change effects.

But who pays? This depends on the technological and economic resources available and on the will of a country, industry, company or individual. This is called the **adaptive capacity**.

Having adaptive capacity is a necessary condition for the design and implementation of effective adaptation strategies to reduce the damaging outcomes resulting from climate change. HIC nations can provide support to LIC nations.

Possible actions include the following.

1. Change land use through planning legislation:

   a. Do not allow building on flood plains or low-lying areas.

2. Build to resist flooding:

   a. Plan water catchment and run-off to minimize flooding.

   b. Build houses on stilts or with garages underneath which can be flooded with little damage.

Key term

Adaptive capacity is the ability or potential of a system to respond successfully to climate variability and change, and includes adjustments in both behaviour and in resources and technologies. (IPCC 2007)

 Practical

Skills:

- Tool 1: Experimental techniques

- Inquiry 1: Exploring and designing

- Inquiry 2: Collecting and processing data

Create a survey to investigate attitudes in your school or community to a proposal to mitigate climate change.

3. Change agricultural production:

   a. Irrigate more effectively in drought areas.

   b. Store rainwater for times of water shortage.

   c. Breed drought-tolerant crops.

   d. Grow different crops.

4. Manage the weather:

   a. Seed clouds to encourage rainfall.

   b. Plant trees to encourage more rainfall.

5. Migrate to other areas.

6. Vaccinate against water-borne diseases, e.g. typhoid.

7. Manage water supplies:

   a. Use desalination plants.   c. Harvest run-off more effectively.

   b. Increase reservoirs.   d. Use water harvesting from clouds.

Some strategies are both mitigating and adapting. A green roof mitigates climate change by absorbing carbon dioxide and adapts the building by naturally cooling it.

ATL Activity 13

ATL skills used: thinking, communication, research

1. Research and discuss strategies (a) already adopted and (b) which could be adopted to alleviate climate change by:

   - you
   - your school
   - your community
   - your country.

2. Management strategies may focus on altering activity, regulating and reducing emissions, or clean up and restoration. Copy and complete Table 1.

Strategy for reducing global emissions	Example of action
Altering the human activity producing pollution	
Regulating and reducing the pollutants at the point of emission	
Clean up and restoration	

▲ Table 1

Carbon offset and carbon emissions trading

Carbon offsets occur when a polluting company buys a carbon credit to make up for carbon it has emitted (Figure 9). The money should be used to fund action somewhere in the world that removes the same amount of carbon from the air or prevents other carbon emissions.

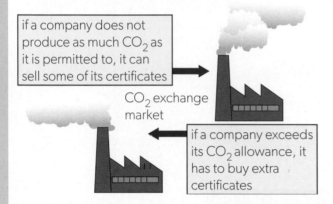

industrial companies receive emissions certificates from their governments, allowing them to emit a certain amount of CO_2

if a company does not produce as much CO_2 as it is permitted to, it can sell some of its certificates

CO_2 exchange market

if a company exceeds its CO_2 allowance, it has to buy extra certificates

▲ Figure 9 How carbon offsetting and trading works

The concept of **carbon emissions trading** evolved as part of the Kyoto Protocol. In this scheme, countries that go over their CO_2 emissions quota (set by international agencies) can buy carbon credits from countries that do not meet their quotas. In this way, global emissions limits are still met. The system is very complex to operate. It is difficult enough for an industry to monitor carbon emissions; it is even more difficult for a country. Consider these issues:

Questions

1. Which country owns the emissions from an international flight or container ship? Is it where the journey starts or ends? Or where the airline or shipping company is based or registered?

2. Who sets the quotas?

3. Are there penalties for going over a limit on emissions?

The market that has grown up for trading carbon emission permits is volatile. The EU emissions trading scheme, for example, saw the value of carbon credits fall due to an overestimate of the allocation required when it started. The scheme does not encourage

industries or countries to reduce their emissions: they can simply buy permission to continue emitting. Carbon emissions trading is an alternative to a carbon tax. Such a tax would be paid by organizations releasing carbon dioxide.

There are now voluntary schemes that allow individuals and companies to **offset carbon emissions**. If you book a plane ticket today, you may be asked if you want to pay to offset the carbon emissions produced by your flight. If you agree to this, the money should go to a company that invests it in a scheme that reduces carbon emissions (e.g. a renewable energy scheme such as wind turbines, tree planting or hydroelectric power generation). The market is increasing as environmentally aware individuals invest. But the money must be invested in a scheme that would not otherwise have happened. Just taking the money and planting trees that would have been planted anyway is not recapturing any more carbon dioxide.

Clearly, it would be better not to create the emissions in the first place—but sometimes it is unavoidable.

Some people have suggested **personal carbon allowances** (PCAs)—individual allowances for carbon emissions. A person who travels a lot or lives in an energy inefficient home would either need to change their lifestyle or buy "credits" on the open market from people who produce less carbon dioxide.

4. Would you support PCAs?

Becoming **carbon neutral** is also a goal. All carbon you release is balanced by an equivalent amount that is taken up or offset, e.g. by planting trees, buying carbon credits or taking part in projects to reduce future GHG emissions.

5. A country aims to become carbon neutral by 2050. This will affect different communities in different ways. Consider and discuss the effect on:

    • indigenous groups

    • disadvantaged groups

    • rural communities

    • urban centres.

NAPAs

A NAPA is a National Adaptation Programme for Action on climate change. These are submitted to the UNFCCC by LICs. In a NAPA, the vulnerable country identifies its urgent needs for adaptation. For example, Bangladesh is at high risk of flooding due to sea level rise and to drought if monsoon rains fail. In its NAPA, it identified building of coastal embankments, flood shelters, improved irrigation and drought-resistant crops.

A UN fund provides financing for these actions but it is currently small for the amount of work that is required.

ATL Activity 14

ATL skills used: thinking, communication, social, research, self-management

Working as a class, allocate tasks to find the information requested below. Then determine how to share the information with the whole class.

1. Find out what the latest COP summits have decided.

2. a. What is your own government's policy on climate change and carbon emissions?

   b. What alternative energy sources to fossil fuels are available in the country in which you live?

   c. What are the advantages and disadvantages of these alternatives?

3. List the possible ways that countries could reduce their carbon emissions.

4. a. Use www.carbonfootprint.com to calculate your own carbon footprint. This measures your carbon use in tonnes of CO_2 not hectares as in ecological footprints.

   b. List as many ways as you can of reducing your own carbon footprint.

   c. How many of these will you actually do?

   d. Which of the strategies you have listed would be most effective in reducing CO_2 emissions and why? Consider whether the strategies need others to act as well, if they reduce your quality of life, if the technology is available, and how easy they are to follow. Are the strategies ecocentric or technocentric?

5. What do you think are the ethical issues surrounding geoengineering strategies?

More on responses to climate change

Responses to climate change are led by the UN, groups of nations, national governments, NGOs, activists, the media and others. They include:

- economic measures—e.g. pricing, carbon emissions trading, subsidies for renewables or tariffs for fossil fuels

- legislation—international, national, community specific

- goal setting commitments—e.g. corporate social responsibility (CSR), B-Corp branding, company commitments to the public or shareholders

- personal lifestyle changes—adding solar panels or wind turbines to a house, reducing heating or AC use, changing diet, travelling less.

ATL Activity 15

ATL skills used: thinking, research

Research one example of a response to climate change and its impact in each of these areas:

- economic

- legislative

- industry

- individual.

Evaluate the effectiveness of the response.

AHL

AHL

Connections: This links to material covered in subtopics 1.1, 1.3, HL.a and HL.b.

Key term

Carbon neutrality means having a balance between emitting carbon and absorbing carbon from the atmosphere in carbon sinks.

Impact of the UNFCCC

The role of the UNFCCC is to stabilize GHG concentrations at a level that prevents interference with the climate system. The UNFCCC works through the IPCC and COP summits.

The IPCC has proposed a range of emission scenarios with targets to reduce the risk of catastrophic climate change.

The scenarios are plausible representations of unknown future GHG emissions based on current evidence and data. Some call them best guesses.

There are five families of scenario, from **carbon neutrality** by 2050 to twice current carbon dioxide levels in 2050 and three times these in 2100.

The aim of these scenarios is not to predict the future—no probability is associated with the different scenarios. It is to take into account the uncertainty linked to future human activities and to inform the decisions of states and their societies.

The scenarios are called Shared Socioeconomic Pathways (SSPs) and are projected possibilities up to 2100. The scenarios from the 6th IPCC report of 2021 are:

- SSP1: Sustainability

- SSP2: Middle of the road

- SSP3: Regional rivalry

- SSP4: Inequality

- SSP5: Fossil-fuelled development

Scenario 1—Most optimistic: 1.5°C increase by 2050

- Global CO_2 emissions are cut to net zero around 2050.

- Societies switch to more sustainable practices.

- Focus shifts from economic growth to overall well-being.

- Investments in education and health go up.

- Inequality falls.

- Extreme weather is more common, but the world avoids the worst impacts of climate change.

Scenario 2—Next best: 1.8°C increase by 2100

- Global CO_2 emissions are cut severely, but not as fast as in Scenario 1.

- Net zero is reached after 2050.

- Same socio-economic shifts towards sustainability as SSP1 but temperatures stabilize around 1.8°C higher by 2100.

Scenario 3—Middle of the road: 2.7°C increase by 2100

- CO_2 emissions stay around current levels before starting to fall by 2050.

- Do not reach net zero by 2100.

- Socio-economic factors continue as they are now with development and income growing inequitably.

- Progress toward sustainability is slow.

- Temperatures rise 2.7°C by 2100.

Scenario 4—Dangerous: 3.6°C increase by 2100

- Emissions and temperatures rise steadily and CO_2 emissions roughly double from current levels by 2100.

- Countries become more competitive with one another.

- Countries shift towards national security and ensuring their own food supplies.

- By 2100, average temperatures have risen by 3.6°C.

Scenario 5—Avoid at all costs: 4.4°C increase by 2100

- Current CO_2 emissions levels roughly double by 2050.

- The global economy grows quickly.

- Growth is fuelled by exploiting fossil fuels and energy-intensive lifestyles.

- By 2100, the average global temperature is 4.4°C higher.

Figures 10–12 show possible CO_2 emissions, temperature changes and sea level changes in the five scenarios.

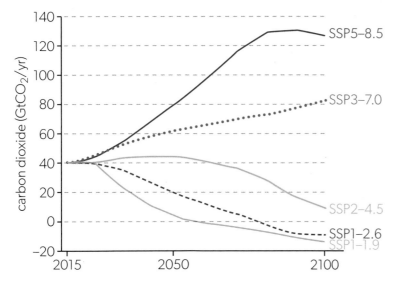

▲ Figure 10 Future CO_2 emissions in the five illustrative scenarios. The numbers after the SSP1–5 labels are the expected levels of radiative forcing in 2100: 1.9 to 8.5 W/m²
Data from: Sixth Assessment Report of IPCC Working Group I, 2021

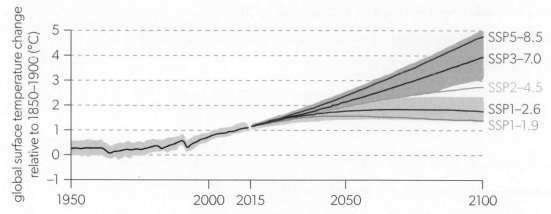

▲ Figure 11 Future global surface temperature change relative to 1850–1900 in the five illustrative scenarios
Data from: Sixth Assessment Report of IPCC Working Group I, 2021

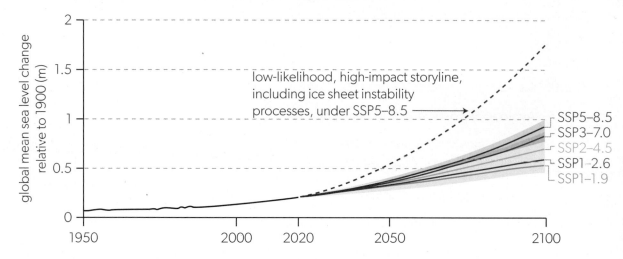

▲ Figure 12 Future global mean sea level rise in the five illustrative scenarios
Data from: Sixth Assessment Report of IPCC Working Group I, 2021

Practical

Skills:

- Tool 2: Technology
- Inquiry 2: Collecting and processing data
- Inquiry 3: Concluding and evaluating

1. Research the graphs of the IPCCC scenarios in Figures 10–12 and find others in the IPCC *Summary for Policymakers*.

2. Evaluate the implications of each scenario for (a) the world and (b) the country where you live.

3. Investigate mitigation and adaptation policies of the national government where you are living.

The IPCC cannot say which scenario is most likely because that will depend on future factors including government policies. But it does show how choices today will affect the future. In every scenario, warming will continue for at least a few decades. Sea levels will continue rising for hundreds or thousands of years, and the Arctic will be practically free of sea ice in at least one summer in the next 30 years. But how quickly sea levels rise and how dangerous the weather becomes will depend on the pathway chosen by us.

The results of the five scenarios call for immediate action on adaptation and mitigation. In the 2021 report, the IPCC also notes that the objective of limiting warming to 1.5°C in the long term is not lost, via the SSP1–1.9 scenario in which warming is limited to 1.4°C at the end of the century after temporarily exceeding 1.5°C. However, this requires drastic and immediate decisions for mitigation.

Impact of technology

Technology helps mitigate climate change. This technology may include apps on smartphones and in smart cities giving information on EV charging points, public transport routes, recycling points and peer-to-peer markets to reduce waste (subtopic 8.3).

Industrial research is developing new technology that is more efficient at reducing or removing carbon emissions.

More on geoengineering

Geoengineering is a mitigation strategy for climate change, treating the symptom not the cause. It is a deliberate, ambitious and large-scale intervention in the Earth's climate system.

But it has potential high costs, uncertainty of impacts, political hesitancy, lack of convincing trials and the potential for geopolitical conflict.

It can be as straightforward as afforestation but more radical approaches are:

- space reflectors to stop some solar radiation reaching the Earth

- increasing cloud reflectiveness

- adding aerosols to the stratosphere to reflect sunlight

- GGR (greenhouse gas removal) by charring biomass and burying it so carbon is locked in soil

- BECCS (see under Mitigation strategies earlier in this subtopic)

- ambient air capture, which removes CO_2 from the atmosphere and stores it

- ocean fertilization to add nutrients to increase NPP and store carbon

- adding alkaline minerals (ground-up limestone) to oceans to increase its pH.

Challenges in combating climate change

There are challenges to overcome in implementing climate management and intervention strategies:

- lack of belief that climate change is a serious problem

- lack of financial resources or planning strategies from national governments

- lack of leadership from a range of stakeholders, e.g. individuals, NGOs, political leaders and transnational companies

- international inequalities, e.g. economies that profit from fossil fuels and those that do not, differences in LIC and HIC societies

- differences in perspective between younger and older, coastal or low-lying and inland or upland communities.

More on perspectives on climate change

Perspectives on the necessity, practicality and urgency of action on climate change will vary between individuals and between societies.

ATL Activity 16

ATL skills used: thinking, research

Consider the implementation of one technology to mitigate climate change in a named society.

▲ Figure 13 Geoengineering with satellites to launch mirrors into space to reflect sunlight

ATL Activity 17

ATL skills used: thinking, communication, social research

Research three examples of geoengineering and consider arguments for and against their potential.

Share your findings in a three-minute presentation to the class.

Connections: This links to material covered in subtopics 1.1, 1.3, HL.a, HL.b and HL.c.

(ATL) Activity 18

ATL skills used: thinking, communication, social, research, self-management

1. As a class, list the stakeholders involved in changing perspectives on climate change action. This may include your friends, teachers, celebrities, the media, scientists, local government, NGOs, national governments, and more.

2. As a class, sort the stakeholders into three groups—most effective, least effective, neutral. Explain your choices.

3. As an individual, which group influences your perspective on climate change the most? Why?

4. Now look at these pairs of groups:

    *   younger and older

    *   advantaged and disadvantaged groups within a society

    *   HIC and LIC societies

    *   coastal and inland communities

    *   countries that export fossil fuels and countries that import them.

    What do you think are the perspectives of most people in each of these groups?

5. Debate in class why different groups do not share the same perspectives.

The tragedy of the commons with respect to climate change

The atmosphere is common to all. But when one nation benefits from an action that harms the atmosphere (e.g. burning of fossil fuels), the costs are shared by all nations. The reverse scenario is also true: the costs of restoring the atmosphere (e.g. by CCS) might be borne by a single nation, but the benefits are gained by all nations.

International agreement is beginning to happen. COP 27, the 27th conference of the parties to the UNFCCC (IPCC) was held in 2022 in Sharm el-Sheikh, Egypt. It agreed, after much debate, to establish and operate a loss and damage fund, particularly for nations most vulnerable to the climate crisis. This will see LICs that are particularly vulnerable to the adverse effects of the climate crisis supported for losses attributed to climate change. Such losses may arise from droughts, floods, rising seas and other disasters. Support could take the form of building sea walls or creating drought-resistant crops. It could cost HICs anywhere from USD 160 to 340 billion annually by 2030.

However, the wording "phase down use of fossil fuels" instead of "phase out" was prevalent and there was a focus on low emission energy production (natural gas) rather than renewable energy sources. Delegates from almost 200 countries attended COP 27. The final agreement did mention "the urgent need for deep, rapid and sustained reductions in global greenhouse gas emissions" to limit global warming to 1.5°C above pre-industrial levels, the most ambitious goal of the Paris Agreement.

Check what the latest COP decided. Is this progress?

TOK

Discuss how environmental ethics influence approaches to achieving a sustainable future.

Check your understanding

In this subtopic, you have covered:

- global action is needed now and is happening

- mitigating climate change can be by decarbonizing economies, using energy more efficiently, using renewable energy and capturing carbon emissions

- adapting to climate change can be by flood defence, drought-resistant crops and vaccination

- there are various responses to climate change

- UNFCCC, IPCC reports and COP summits have an effect

- IPCC scenarios predict possible futures

- there are technological solutions

- there are challenges to implementation of change

- geoengineering can offer solutions

- many stakeholders influence individuals and attitudes vary

- international cooperation is happening but slowly.

AHL

How can human societies address the causes and consequences of climate change?

1. Explain why action on climate change needs to be global as well as national.

2. Describe how decarbonizing the economy can be achieved.

3. Evaluate three mitigation strategies to address climate change.

4. Evaluate three adaptation strategies to address climate change.

5. Describe the different levels at which measures on climate change actions are taken.

6. List three examples of government measures.

7. Evaluate the role of the UN in climate change mitigation and adaptation.

8. Describe the five IPCC scenarios.

9. To what extent is geoengineering feasible today?

10. Comment on why perspectives on climate change actions vary.

AHL

≫ Taking it further

- Create information posters for the school about personal behaviours that can be taken to mitigate climate change.

- Form a student council on climate change actions.

- Engage with doughnut economics groups around the world and implement a plan for the school.

- Create a social media channel to inform others about behaviour change to mitigate climate change.

- Visit a local power production site, energy from waste site, recycling centre or carbon offsetting project.

6.4 Stratospheric ozone

Guiding questions

- How does the ozone layer maintain equilibrium?

- How does human activity change this equilibrium?

Understandings

1. The Sun emits electromagnetic radiation in a range of wavelengths, from low frequency radio waves to high frequency gamma radiation.
2. Shorter wavelengths of radiation (namely, UV radiation) have higher frequencies and, therefore, more energy, so pose an increased danger to life.
3. Stratospheric ozone absorbs UV radiation from the Sun, reducing the amount that reaches the Earth's surface and, therefore, protecting living organisms from its harmful effects.
4. UV radiation reduces photosynthesis in phytoplankton and damages DNA by causing mutations and cancer. In humans, it causes sunburn, premature ageing of the skin and cataracts.
5. The relative concentration of ozone molecules has stayed constant over long periods of time due to a steady-state of equilibrium between the concurrent processes of ozone formation and destruction.
6. Ozone-depleting substances (ODSs) destroy ozone molecules, augmenting the natural ozone breakdown process.
7. Ozone depletion allows increasing amounts of UVB radiation to reach the Earth's surface, which impacts ecosystems and human health.
8. The Montreal Protocol is an international treaty that regulates the production, trade and use of chlorofluorocarbons (CFCs) and other ODSs. It is regarded as the most successful example yet of international cooperation in management and intervention to resolve a significant environmental issue.
9. Actions taken in response to the Montreal Protocol have prevented the planetary boundary for stratospheric ozone depletion being crossed.
10. ODSs release halogens, such as chlorine and fluorine, into the stratosphere, which break down ozone.
11. Polar stratospheric ozone depletion occurs in the spring due to the unique chemical and atmospheric conditions in the polar stratosphere.
12. Hydrofluorocarbons (HFCs) were developed to replace CFCs as they can be used in similar ways and cause much less ozone depletion, but they are potent GHGs. They have since been controlled by the Kigali Amendment to the Montreal Protocol.
13. Air conditioning units are energy-intensive, contribute to GHG emissions and traditionally have contained ODSs.

Electromagnetic radiation

The Sun emits radiation across the **electromagnetic spectrum** (Figure 1). The peak of this emission occurs in the visible portion of the spectrum. The Earth's surface is protected by the atmosphere from most higher energy waves that are damaging to cells. Important gases protecting the Earth in this way are water vapour, carbon dioxide and ozone.

Sunlight is visible light. This light plus infrared and ultraviolet reach the Earth through the "optical window" in the atmosphere. Infrared is radiant heat which humans cannot see but can feel as heat.

The ecosphere absorbs infrared heat and this energy drives processes in the atmosphere, hydrosphere and biosphere.

Visible light wavelengths are ones that human eyes detect. Other organisms detect slightly different wavelengths. For example, bees see ultraviolet tram lines on flowers which guide them to the nectar source.

Humans see visible light as white but it is made up of multiple colours. Think of a rainbow or when white light is shone through a prism to split up the wavelengths. Red light has the longest wavelength and violet the shortest.

Photosynthesis in green plants occurs because leaf pigments (mostly chlorophylls and carotenoids) absorb certain wavelengths of light. Chlorophylls absorb blue and red light but reflect green light which is why we see leaves as green. Carotenoids absorb violet and blue-green light and so we see fruits as red (tomato) or yellow (corn).

Ultraviolet light is too short a wavelength for human eyes to see.

Key terms

Electromagnetic (EM) radiation is the complete range of radiation that has magnetic and electric fields and travels in waves.

The **electromagnetic spectrum** is the range of all frequencies of EM radiation from very long-wave radio waves to very short-wave gamma rays.

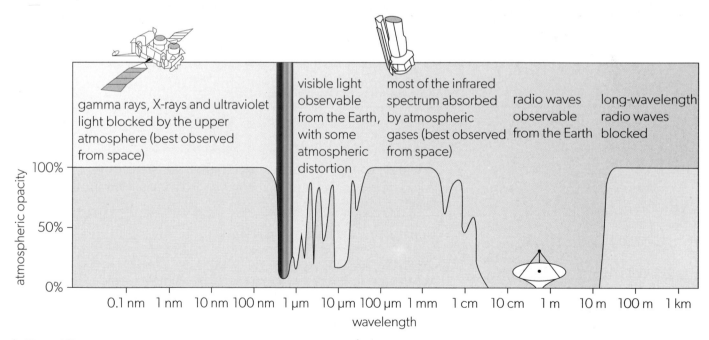

▲ Figure 1 The electromagnetic spectrum and the Earth's atmosphere showing "windows" in the atmosphere
Adapted from: National Aeronautics and Space Administration, Science Mission Directorate. (2010). Introduction to the Electromagnetic Spectrum

TOK

"...bees see ultraviolet..."

What is the role of inductive and deductive reasoning in scientific inquiry, prediction and explanation?

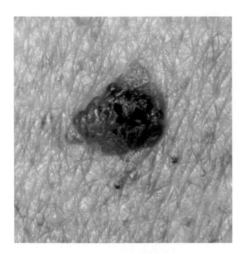

▲ Figure 2 Melanoma caused by UV exposure

Connections: This links to material covered in subtopic 2.2.

Effects of ultraviolet radiation (UV)

All categories of UV can damage organisms.

UV-A:

- comprises 95% of terrestrial UV radiation and has the longest wavelengths

- causes wrinkles and premature ageing of skin and sometimes skin cancer as it can penetrate skin down to the middle layer

- penetrates glass and clouds.

UV-B:

- comprises 5% of terrestrial UV radiation and has slightly shorter wavelengths than UV-A

- only penetrates the top layer of skin and can be stopped by glass

- causes skin to redden and burn; strongly linked to skin cancer as it damages skin's DNA, causing melanoma cancers

- sunscreen of SPF30 or more will deflect UV-B.

UV-C:

- has the shortest wavelength (highest frequencies)

- absorbed by stratospheric ozone, which protects the Earth; this ozone also absorbs most incident UV-B rays

- can be used intentionally in tertiary water treatment as a disinfectant

- is also emitted by lasers, welding torches or old tanning beds

- is particularly harmful to eyes.

Damaging effects of UV radiation

Increased exposure to UV radiation has a variety of damaging effects.

- Aquatic ecosystems produce more than 50% of the biomass on our planet, and are key components of the Earth's biosphere. UV damage could have serious consequences for the productivity of phytoplankton and ecosystems, particularly near the poles.

- Suppression of the human immune system.

- Cataract formation in eyes as protein of the lens denatures and turns cloudy instead of clear. This causes blindness if untreated. Worldwide, it is estimated that 15 million people are blind due to cataracts. Of these cases, some 10% may be due to exposure to UV radiation (according to WHO data).

- Skin cancers caused primarily by exposure to UV, either from the Sun or from artificial sources such as sunbeds. Globally in 2020, over 1.5 million cases of skin cancer were diagnosed and over 120,000 skin cancer-associated deaths were reported. Excessive sun exposure in children and adolescents contributes to skin cancer in later life. This is especially true in Australia and

New Zealand where the number of cases of skin cancer in humans has increased strongly. In New Zealand, the daily weather report in summer includes isolines to show burn times.

- There is a "sun safety" slogan in Australia: Slip on a shirt; Slop on the sunscreen (factor 30+); Slap on a hat; Seek shelter; Slide on the shades (sunglasses).

- As our knowledge and understanding of the impacts of UV radiation has progressed, so has this slogan. The original slogan (1981) was just "slip, slop, slap" but it was updated in 2007 to include "seek" and "slide".

Beneficial effects of UV radiation

- In animals, UV radiation stimulates the production of vitamin D. In our bodies, vitamin D deficiency causes rickets. This is a condition in which a child's bones are short of calcium and so are too soft to support the body.

- Used to treat psoriasis and vitiligo, both skin diseases.

- Used as a sterilizer because it kills pathogenic bacteria; also an air and water purifier.

- Industrial uses in lasers, viewing old scripts, forensic analysis, lighting.

Introduction to ozone

Oxygen gas is made up of molecules containing two oxygen atoms (diatomic), O_2. Ozone is a molecule made up of three oxygen atoms, O_3. Ozone is found in two layers of the atmosphere.

In the stratosphere, it:

- is "good"

- blocks incoming UV radiation from the Sun

- protects life from damaging UV

- is not a cause of global warming.

In the troposphere (see subtopic 6.1), it:

- is considered "bad" because it acts as a GHG

- acts as a GHG but concentration varies enormously—it is not everywhere, and it is short-lived.

Stratospheric ozone depletion is not a cause of global warming. But in the troposphere, ozone is a GHG. Do not get confused over its different environmental effects.

ATL Activity 19

ATL skills used: thinking, social

List the ways you protect your skin against the Sun's radiation.

Discuss in class the benefits and damaging effects of solar radiation on humans.

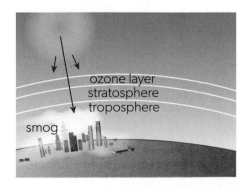

▲ Figure 3 **Ozone in the atmosphere**

The ozone layer

Ozone is a reactive gas mostly found in the so-called ozone layer in the lower stratosphere. The highest ozone concentrations are usually seen at altitudes between 20 and 40 km (or between 15 and 20 km at the poles). This layer is very thin—about 1–10 ppm ozone.

The ozone layer absorbs 99% of incoming UV-C radiation, most incoming UV-B radiation and 5% of UV-A. What reaches the surface of the Earth does not cause much damage because UV-C has the shortest wavelength and is the most penetrating and damaging.

The ozone layer is an example of a steady state of dynamic equilibrium because ozone is continuously made from oxygen atoms and is continuously converted back to oxygen. But if the rates of ozone formation and depletion become unequal, the equilibrium will tip. In both the formation and the destruction of ozone, UV radiation is absorbed. Under the influence of UV radiation, oxygen molecules split into two oxygen atoms. Oxygen atoms are extremely reactive, so these two atoms can combine with two other oxygen molecules to form two ozone molecules. Ozone molecules can also absorb UV radiation and split into an oxygen molecule and an oxygen atom. The oxygen atom can react with another oxygen molecule, making another ozone molecule (Figure 4).

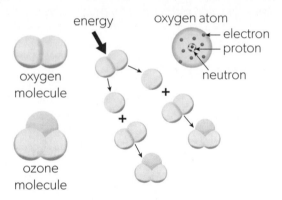

▲ Figure 4 **Formation of ozone**

The absorption of UV radiation by the ozone layer is crucial. Without it, life on land would be impossible.

Depletion of stratospheric ozone—the ozone hole

Since the 1950s, scientists have been measuring the amount of ozone in the stratosphere above Antarctica. Before 1979, scientists had not observed atmospheric ozone concentrations below 220 Dobson units. But in the early 1980s, through a combination of ground-based and satellite measurements, they began to realize that the Earth's natural sunscreen was thinning dramatically over the South Pole each spring. This thinning of the ozone layer over Antarctica came to be known as the ozone hole.

The amount of ozone decreased strongly during the spring (September and October in the southern hemisphere) and increased again in November. Apart from this annual ozone cycle, the scientists discovered that the ozone hole was growing (Figure 5). The minimum thickness of the ozone layer had reduced drastically and recovery had been taking longer. These results were later confirmed by NASA satellite data. Reductions in the amount of stratospheric ozone had been observed in other areas as well, including the Arctic region.

Ozone depletion allows increasing amounts of UV-B radiation to reach the Earth's surface, which impacts ecosystems and human health. Ozone depletion has affected the stratosphere over the whole Earth, but more so at the poles.

◀ Figure 5 **Maximum yearly extent of the ozone hole over Antarctica since 1979** Credit: CAMS

Ozone-depleting substances

Ozone depletion is the result of air pollution by chemicals that are mostly human-made (Tables 1 and 2). Ozone-depleting substances (ODSs) destroy ozone molecules, augmenting the natural ozone breakdown process and altering the steady state of equilibrium. The most important ODSs are halogenated organic gases (e.g. chlorofluorocarbons (CFCs) and others).

ODSs include CFCs, hydrochlorofluorocarbons (HCFCs) and hydrofluorocarbons (HFCs). Do not be confused by these substances.

- CFCs and HCFCs are ODSs because they contain chlorine atoms.

- HCFCs and CFCs were destroying the ozone layer high in the Earth's atmosphere. This layer is essential for protecting life from harmful UV radiation.

- HCFCs are ODSs and GHGs.

- HFCs are not ODSs but are GHGs.

- HCFCs and HFCs are GHGs because they have hydrogen atoms.

CFCs	HCFCs	HFCs
phased out	phased out	replace CFCs and HCFCs
ODSs	ODSs	not ODSs
not GHGs	GHGs	GHGs

▲ Table 1 **Comparing ozone-depleting substances**

Substance	Use or source	Remarks
CFCs (or freons)	Propellants in spray cans; plastic foam expanders; refrigerants	Release chlorine atoms
HCFCs	As replacements for CFCs	Release chlorine atoms but have a shorter lifetime in the atmosphere Stronger GHG
Halons	Fire extinguishers	Release bromine atoms
Methyl bromide	Pesticide	Releases bromine atoms
Nitrogen oxides (NO, NO_2, N_2O, often summarized as NO_x)	Bacterial breakdown of nitrites and nitrates in the soil (intensive farming) High-flying supersonic aircraft	Converted to NO, which reacts with ozone

▲ Table 2 The most important ozone-depleting substances
Adapted from: Six Red Marbles and OUP

The action of ODSs

All ODSs are human-made. When CFCs (or freons) were developed during the 1930s, they seemed to be the answer to many technological problems, largely because they are inert (non-reactive) at ground level.

They were used as:

- propellants in aerosols

- expanders of gas-blown plastics (plastic foam)

- pesticides

- flame retardants

- refrigerants (replacing earlier refrigerants which were toxic and flammable).

CFCs are extremely stable at ground level and it was a long time before it was discovered that they were not so stable when exposed to UV radiation in the stratosphere. UV radiation releases chlorine atoms from CFCs. These chlorine atoms can react with ozone, which results in ozone destruction. They can also react with oxygen atoms, thereby preventing ozone formation. In both processes, the chlorine atoms are re-formed so they can react with ozone or oxygen atoms again. One chlorine atom can thus destroy many molecules of ozone in a chain reaction with positive feedback.

It has been relatively easy to replace CFCs in spray cans and as blowing agents for plastic foam. However, it is much more difficult to find a suitable refrigerant. The refrigerants used before the introduction of CFCs are not an option because of their dangerous properties. The most suitable CFC replacements are HCFCs. These substances are nearly as good as refrigerants as CFCs and are also non-toxic and non-flammable. However, HCFCs also destroy ozone and contribute to the greenhouse effect. Some HCFCs have a GWP 2,000 times that of carbon dioxide. Only their shorter lifetime in the atmosphere makes them less harmful to the ozone layer than CFCs.

CFCs are extremely stable and therefore persist in the atmosphere for up to 100 years after their release. Measures taken to prevent release of CFCs into the atmosphere will take a long time to result in a thicker ozone layer.

Strategies for reducing ozone depletion

The removal of CFCs as aerosol propellant is as good example of how the actions of individuals make a difference. When it became known that CFCs were causing the depletion of the ozone layer, various bodies released informational campaigns about it. Many people then checked the "ingredients" list on spray cans and stopped purchasing aerosols that contained CFCs. This was an effective tool that forced manufacturers to make changes. In 1975, Johnson and Sons banned the use of CFCs in their products globally. We can make a difference.

Strategy for reducing pollution	Example of action
Altering the human activity producing pollution	Replace gas-blown plastics
	Replace CFCs and HCFCs with carbon dioxide, propane or air as a propellant
	Replace aerosols with pump action sprays
	Replace methyl bromide pesticides (but most gases that can be used to replace CFCs are GHGs)
Regulating and reducing the pollutants at the point of emission	Recover and recycle ODSs from refrigerators and AC units
	Legislate to have fridges returned to the manufacturer and coolants removed and stored
	Capture ODSs from scrap car AC units
Clean up and restoration	Add ozone to or remove chlorine from stratosphere—not practical but it was once suggested that ozone-filled balloons were released

▲ Table 3 Strategies for reducing ozone depletion
Adapted from: Six Red Marbles and OUP

The Montreal Protocol (1987)

The United Nations organization involved in protecting the environment is the UNEP (United Nations Environmental Programme). The UNEP:

- forges international agreements, including the Montreal Protocol

- studies the effectiveness of these agreements and the difficulties in implementing and enforcing them

- gives information to states, organizations and the public.

The Montreal Protocol is an international treaty that regulates the production, trade and use of CFCs and other ODSs. All ODSs controlled by the Montreal Protocol contain either chlorine or bromine. Substances containing only fluorine (e.g. HFCs) are not ODSs. The Montreal Protocol is regarded as the most successful example of international cooperation in management and intervention to resolve a significant environmental issue.

The discovery of the ozone hole led to a fast response on national and international levels. But even before governments and international organizations

Connections: This links to material covered in subtopics 1.1 and HL.a.

reacted, the public in many HICs started to boycott products containing CFCs and other ODSs (mainly spray cans). The aerosol industry reacted quickly by changing to ozone-friendly spray cans. Before CFCs were forbidden by law, hardly any CFC-containing spray cans were produced anymore.

A total of 197 countries signed up to the protocol. They agreed to freeze consumption and production of many ODSs to 1986 levels by 1990 and to strongly reduce consumption and production of these substances by 2000. After 1987, the original Montreal Protocol was strengthened in a series of seven amendments. In the protocol, a distinction was made between HICs and LICs; LICs were given more time for implementation with staggered phase-out schedules.

Timeline in CFC reduction

1970s	Ozone-depleting properties of CFCs are recognized. In 1974, the USA and Sweden banned them from non-essential aerosol uses. Concerns continue to mount through 1980s.
1985	British Antarctic Survey reports the ozone hole.
1987	The Montreal Protocol is organized by the UNEP. Over 30 countries agree to cut CFC emissions by half by 2000.
2006	NASA and NOAA record the Antarctic ozone hole as the largest ever measured.
2007	The phasing out of HCFCs by 2030 is agreed.
2012	The 25th anniversary of signing of the protocol.
2019	Kigali Amendment ratified to phase down HFC production by late 2040s.

Significance and success of the Montreal Protocol

Had the world not banned ODSs, what would have happened? By 2050, there would have been ozone hole-like conditions over the whole Earth, and the planet would probably have become uninhabitable. Three factors contributed to the relatively fast action:

- the clear and present danger to human health made it personal

- vivid satellite imagery made it perceptible

- there were practical solutions—ODSs could be replaced fairly quickly and easily.

A model for other agreements?

The Montreal Protocol has been largely successful (Figure 7). This protocol is:

- the best example of international cooperation on an environmental issue

- an example of the precautionary principle in science-based decision making

- an example of many experts in their different fields coming together to research a problem and find solutions

- written so that different countries could phase out ODS chemicals at different times depending on their economic status

- the first protocol with regulations that were carefully monitored.

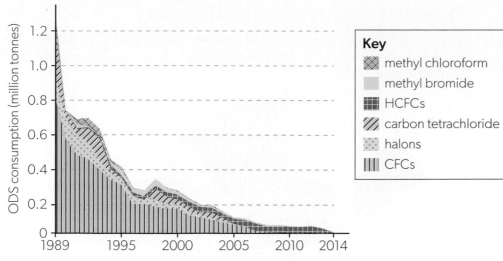

Note: In some years, gases can have negative consumption values. This occurs when countries destroy or export gases that were produced in previous years (i.e. stockpiles).

◀ Figure 6 ODS global consumption, 1989–2014. ODS consumption is measured in ODS tonnes—the amount of ODS consumed, multiplied by its ozone depleting potential value Adapted from: Our World in Data. Data from: UN Environment Programme (CC BY 4.0)

However, this is not the end of the story. Due to the long life of CFCs in the atmosphere, chlorine did not reach its peak in the stratosphere until 2005 nor will it return to pre-ODS levels much before 2050. LICs are still allowed to make and use some HFCs. There is also an illegal market in ODS chemicals.

Today, the ozone hole still exists, forming every year over Antarctica in the southern hemisphere spring and closing in the summer. This is because some ODSs persist in the atmosphere for 50 to 150 years before decaying.

Major natural volcanic eruptions also result in short-term ozone losses. Nitrous oxide, a powerful GHG emitted from fertilizer applications in agriculture, is also a potent ODS. It is not controlled by the Montreal Protocol, and emissions are growing.

Look back at Figure 5 and you will see a small blip in the data—the hole does not seem smaller in 2020–2022. This is probably due to the following.

1.   An unexpected increase in emissions of trichlorofluoromethane (CFC-11).

2.   Global warming having a cooling effect in the middle and upper stratosphere, as the greenhouse effect reduces the exchange between the different layers of the Earth's atmosphere. This stratospheric cooling trend could be affecting the fairly large and long-lasting ozone holes of recent years.

3.   For the 2022 ozone hole in particular, the massive eruption of the Hunga Tonga-Hunga Ha'apai volcano under the South Pacific on 15 January could have had an effect.

Monitoring of the ozone hole and of ODS production is still needed.

While the Montreal Protocol shows that we can agree to tackle large environmental problems such as climate change, there are differences. ODSs were a defined issue and replaceable components of a few products. The size of climate change makes it considerably more difficult to address. Fossil fuel use has driven the global economy and stopping it is hard—but not impossible.

Planetary boundary for ozone depletion

The actions taken in response to the Montreal Protocol have succeeded in preventing the planetary boundary for stratospheric ozone depletion being crossed (Figure 7).

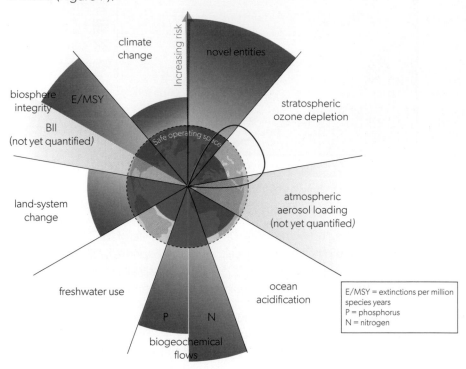

▲ Figure 7 Ozone depletion boundary not crossed
Adapted from: Azote for Stockholm Resilience Centre, based on analysis in Persson et al 2022 and Steffen et al 2015 (CC BY-NC-ND 3.0)

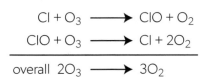

▲ Figure 8 Chlorine breaks down ozone in the stratosphere

More on ozone reactions with CFCs and HCFCs

Ozone-depleting substances release halogens such as chlorine and fluorine into the stratosphere, where they break down ozone (Figure 8).

Ozone is necessary for life on Earth. It forms a thin layer of protection over the whole planet through atmospheric circulation. Without ozone in the stratosphere, UV radiation would bombard living cells causing high mutation rates and cell death.

Evolution of photosynthesis by blue-green algae in oceans around two billion years ago released oxygen into the atmosphere. This led to the formation of ozone. By around 600 million years ago, a thin layer of ozone had formed, protecting life on land from harmful UV radiation. This facilitated the Cambrian Explosion when the diversity of life on land "exploded".

Why ozone holes at the poles?

The ozone holes at the poles are areas of lower ozone concentration due to ozone destruction by human-made chemicals.

Severe depletion of stratospheric ozone in the Antarctic ("the ozone hole") occurs in the southern hemisphere springtime (September–October) due to the unique chemical and atmospheric conditions there. The very low winter temperatures in the Antarctic stratosphere cause polar stratospheric clouds (PSCs) to form. Reactions that occur on the surfaces of PSCs, combined with the isolation of polar stratospheric air in the polar vortex, allow chlorine and bromine reactions to break down ozone and produce an ozone hole.

Volcanic eruptions release volcanic aerosols or particles and these also accelerate ozone destruction by providing a surface on which the reactions occur.

A similar hole is found over the Arctic pole at the northern hemisphere spring equinox but weather is less cold here than in the Antarctic and the hole is less severe.

The issue of HFCs

Hydrofluorocarbons (HFCs) were developed to replace ODSs as they can be used in similar ways. However, they are very potent greenhouse gases.

HFCs:

- are not ODSs

- are potent GHGs with GWP 12–14,800

- have a short lifespan in the atmosphere of 10–20 years

- if eliminated, could avoid a 0.5°C global temperature rise by 2100.

HFCs are now widespread in air conditioners, refrigerators, aerosols, foams and other products (Figure 9).

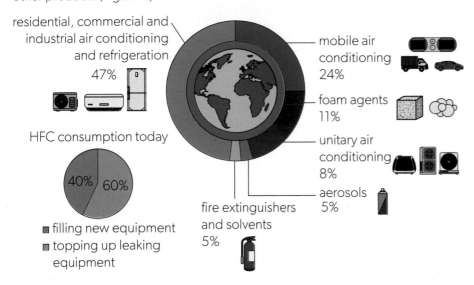

▲ Figure 9 Uses of HFCs
Adapted from: Climate & Clean Air Coalition

Overall HFC emissions are growing at a rate of 8% per year (Figure 10) and annual emissions are projected to rise to 7–19% of global CO_2 emissions by 2050.

China has produced 70% of global HFCs but in 2022 barred companies from expanding their production as it ratified the Kigali Amendment. This is part of China's phase-out programme for HFCs.

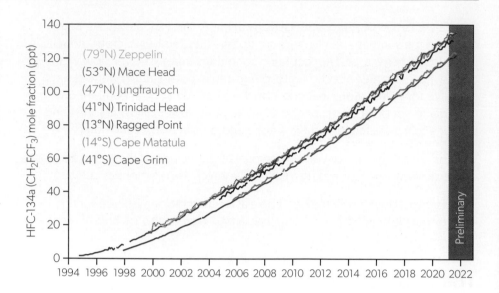

▶ Figure 10 HFC-134a atmospheric concentrations
Adapted from: Jens Mühle. Data from: Agage / NASA (CC BY-SA 4)

Connections: This links to material covered in subtopic 8.3.

▲ Figure 11 **An old AC unit**

Fridges and air conditioning units

Facts:

- There are some 1.9 billion air conditioning units (ACs) in the world.

- AC units are energy-intensive, contribute to GHG emissions, and traditionally have contained ODSs.

- Over 200 million fridges are sold each year and there are probably about 1.4 billion working worldwide.

- Fridges and AC units are the cooling industry, which is useful but also incredibly polluting. It accounts for around 10% of global CO_2 emissions. That is three times the amount produced by aviation and shipping together.

How we dispose of old AC units and fridges matters because there are so many of them and the older ones contain CFCs, HCFCs and HFCs.

Each fridge and AC unit uses electricity and many run continually. While helpful in the home, these appliances contribute to carbon emissions in the electricity they use. They could also contribute to ozone depletion if the refrigerant is not captured after disposal.

Both AC units and fridges rely on the same basic process—evaporation. Each unit contains a fluid called a refrigerant that is used for heat transfer. The refrigerant absorbs heat at low temperature and releases heat at a higher temperature and pressure. In the process, the refrigerant changes from liquid to gas and back again. With evaporation to the gaseous state, the substance cools—just as your body cools when sweat evaporates.

You may have noticed many pipes inside AC units and fridges. These pipes contain the refrigerant but what happens when an AC unit or fridge is no longer wanted or goes wrong? Where does the refrigerant go? ODSs in old units can be captured and sent to a waste management facility, where, under high temperature incineration, at least 99.99% of the chemicals are destroyed. However, not all ODSs are destroyed in this way.

The use of air conditioning around the world is rapidly rising and is expected to continue doing so as global temperatures increase.

The most common HFC found in domestic fridges is HFC-134a, which has a GWP 3,400 times that of carbon dioxide. A typical fridge can contain 0.05–0.25 kg of refrigerant. If this leaks into the environment, the resulting emissions would be equivalent to driving 675–3,427 km (420–2,130 miles) in an average family-sized car.

One of the most damaging HFCs is HFC-23. This is a by-product in the production of HCFC-22, which is a common propellant and refrigerant used in AC units. HFC-23 has the highest GWP of all HFCs and, according to a recent study, global emissions of HFC-23 reached an all-time high in 2018—despite international efforts to reduce them. The rise suggests that not enough is being done to collect and destroy HFC-23 during manufacturing processes. Ineffective enforcement is allowing large-scale illegal HFC trade and some of these HFCs may end up in devices sold to consumers.

Alternative refrigerants

Manufacturers have started turning to climate-friendly chemicals, known as natural refrigerants, which have comparatively low or zero GWP. Major global brands such as Coca Cola, PepsiCo and Unilever have set goals to phase out HFCs and are already starting to use alternatives. Ammonia, certain hydrocarbons and CO_2 are the most popular options. Most supermarkets in Europe now use CO_2 in their fridges and freezers after EU regulations to phase out HFCs were introduced in 2015. But ammonia, for example, is highly toxic. It would present a health risk should it escape through a leak. The flammable gas propane is also used.

Alternatives to AC units

Not having a fridge or AC used to be the norm for everyone. Structures such as wind houses in the Middle East, north-facing cold rooms in old houses in Europe, and underground ice houses where ice was stored through the summer to chill food could all be adapted in modern designs.

▲ Figure 12 **An old fridge**

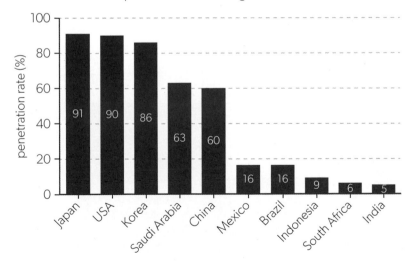

◀ Figure 13 Graph of percentage of households with AC in selected countries Data from: Statista

Practical

Skills:

- Tool 3: Mathematics

- Inquiry 3: Concluding and evaluating

1. Consider the data in Figure 13. What factors do the countries with the (a) highest and (b) lowest percentage of households with AC have in common?

2. Present the data in a different graphical form.

Building design and planning make a difference.

- Greening environments cools them down. Greening of cities with green walls or roofs, hedges, grass, street trees, and parks reduces the heat island effect. Shading of streets also helps (subtopic 8.3).

- Buildings made of stone, brick or concrete have a high thermal mass so they heat and cool slowly, making days cooler and nights warmer.

- Eco-houses can be built underground or half-underground with thick earth walls.

- Compact houses with a lower surface area, shaded porches and roofs that overhang all cool a building.

- In offices, increasing air flow with fans or through flow, window blinds, and reducing heating from lights all help.

Wind towers or windcatchers

A windcatcher, wind tower or wind scoop is a traditional architecture from Iran (Figure 14). It is used to create cross-ventilation and passive cooling in buildings.

Night air is colder than daytime air—much cooler in arid climates. Wind towers (or courtyards) in hot climates fill with cold air at night; the cold night air flows in because it is more dense than the rising warm air it is displacing. The air becomes stratified, with hot air floating on top of cooler air and little mixing; the cool air is "trapped" because the openings in the wind tower are at the top. During the day, the cold air then flows from the tower or courtyard into adjacent rooms, cooling them down.

Wind towers can be even more effective if linked to a qanat or underground tunnel in which water flows from an aquifer or well to where it is needed (Figure 15). Being underground, there is little evaporation and the system is less vulnerable to earthquakes or floods. Cool air is pulled up from the qanat by the wind tower due to the Bernoulli effect. This can cool the basement room by 15°C. This system has been used for 1,000 years.

▲ Figure 14 Wind towers are made of masonry, like this eight-sectioned example in Souq Waqif, Doha, Qatar

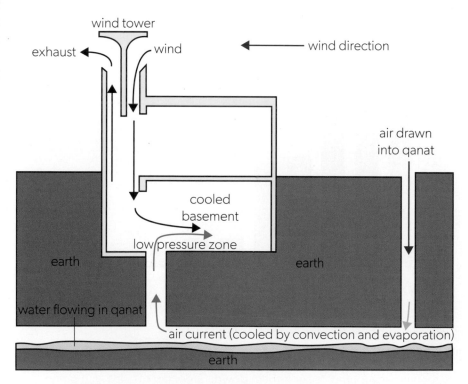

▲ Figure 15 Diagram of qanat and wind tower cooling system
Adapted from: Williamborg / Wikimedia Commons

Check your understanding

In this subtopic, you have covered:

- the Sun's radiation includes shorter wavelength UV radiation, which damages living cells by burning, mutations, skin cancer and cataracts

- stratospheric ozone protects life by absorbing this radiation

- this ozone was in a steady state of equilibrium until human-made ODSs destroyed it

- ozone depletion is greater at the poles

- the Montreal Protocol, an international treaty, limited production of ODSs and stopped the crossing of the planetary boundary

- more about ODSs

- why the poles have more ozone depletion

- the use of HCFCs to replace CFCs, although they are GHGs

- the Kigali Amendment controls HFCs

- AC units contained ODSs

- possible replacements for AC units.

AHL

How does the ozone layer maintain equilibrium?

How does human activity change this equilibrium?

1. State the wavelengths of solar radiation that reach the Earth's surface and describe their effects.

2. Describe the role of stratospheric ozone in the atmosphere.

3. Describe the ways in which UV radiation can damage life.

4. Identify the type of equilibrium of stratospheric ozone.

5. Explain why this equilibrium has altered.

6. Explain the effect of the change on ecosystems and on human health.

7. Describe the impact of the Montreal Protocol.

8. Explain why this protocol was successful and give evidence for this success.

9. Describe the chemical equations for ozone formation and destruction.

10. Describe the effect of CFCs on ozone.

11. Explain why the ozone hole is over the poles at the equinoxes.

12. Discuss the role of HFCs in the atmosphere.

13. Describe how HFCs have been controlled.

14. Evaluate alternatives to air conditioning units.

AHL

⟩⟩ Taking it further

- Discuss the extent to which the Montreal Protocol sets a precedent for how environmental issues can be addressed at a global scale.

- Present findings on alternatives to air conditioning to the school leadership.

- Produce information about protection against UV light during the highest risk periods of the year.

Exam-style questions

1. a. Outline the role of the greenhouse effect in regulating the temperature on Earth. [4]

    b. Using examples, discuss the potential impacts of climate change on ecosystem services. [9]

2. Compare and contrast the adaptation strategies to climate change for **two** societies. [7]

3. a. State where the ozone hole referred to in **Figure 1** is located. [1]

    b. Describe the changes in mean ozone hole area between 1979 and 2016. [2]

    c. Identify **one** possible reason for the changes shown during the 1980s. [1]

    d. Explain how the data in **Figure 1** can be used in judging the success of the Montreal Protocol in addressing ozone depletion. [4]

4. a. Explain how the atmosphere plays a role in maintaining life-supporting temperatures over the Earth's surface. [7]

    b. In addressing environmental issues, mitigation strategies may be seen as primarily ecocentric and adaptation strategies as primarily technocentric.

    To what extent is this view valid in the context of named strategies for addressing the issue of global warming? [9]

5. To what extent does the development of different societies impact their choice of mitigation and adaptation strategies for climate change? [9]

6. To what extent does sustainability play a role in making decisions about energy and climate change policies at national and international levels? [9]

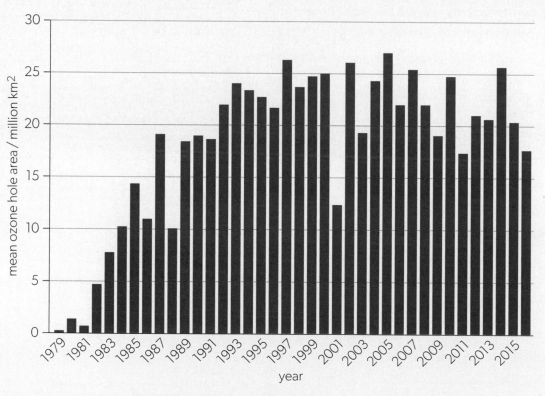

▲ Figure 1 Mean ozone hole area between 1979 and 2016

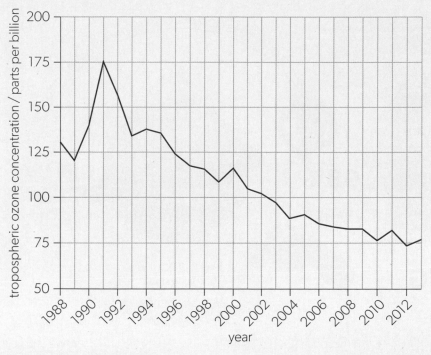

▲ Figure 2 Tropospheric ozone levels in Mexico City

7. a. With reference to **Figure 2**, calculate the difference between the highest concentration and lowest concentration of tropospheric ozone. [1]

    b. State **two** factors necessary for the chemical formation of ozone in the troposphere. [2]

    c. Outline why a high concentration of ozone in the troposphere is a direct problem for humans, while in the stratosphere it is a benefit to humans. [2]

    d. Suggest possible reasons for the overall trends of tropospheric ozone levels in **Figure 2**. [4]

8. The hole in the ozone layer over Antarctica, discovered in the 1980s, was caused by chlorofluorocarbons (CFCs). The Montreal Protocol requires the use of hydrochlorofluorocarbons (HCFCs) or hydrofluorocarbons (HFCs) instead of CFCs (**Figure 3**). However, these two gases are also linked to environmental problems (**Figure 4**).

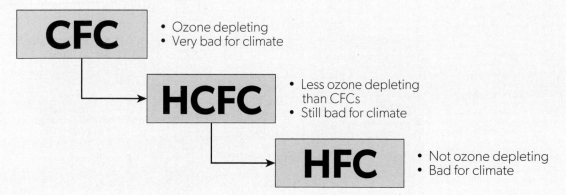

▲ Figure 3 Comparison of the effects of CFCs, HCFCs and HFCs
Data from: Avipsa Mahapatra, Climate Lead, Environmental Investigation Agency, Washington D.C.

BY 2050 HFCs COULD CAUSE

12% OF ALL THE
WARMING
IN THE WORLD

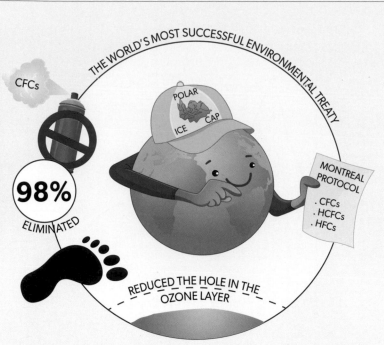

◀ Figure 4 HCFCs and HFCs cause less damage than CFCs but still affect the environment
Data from: © 2016 Cognitive www.wearecognitive.com / Children's Investment Fund Foundation (CIFF) www.ciff.org

a. Identify **two** possible consequences for life on Earth resulting from the depletion of stratospheric ozone. [2]

b. Outline why the Montreal Protocol may be considered the world's most successful environmental treaty. [2]

c. Outline why governments agreed to phase out the use of HFCs from 2019 in the Kigali Amendment to the Montreal Protocol. [2]

d. Identify **one** advantage of staggered dates for the phasing out of HFCs for countries at different levels of economic development. [1]

e. Identify **one** disadvantage of staggered dates for the phasing out of HFCs for countries at different levels of economic development. [1]

9. In 2016, the Earth's atmospheric levels of carbon dioxide reached 400 ppm. Suggest the potential impacts of high levels of greenhouse gases on human societies in different locations. [7]

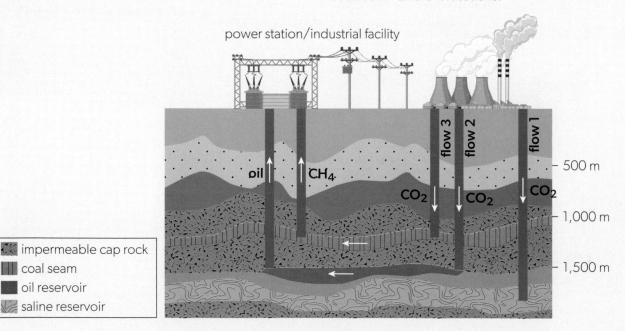

▲ Figure 5 Carbon capture and storage flow chart
Data from: http://www.wri.org/resources/charts-graphs/carbon-capture-sequestration-flow-chart (CC BY 4.0)

10. **Figure 5** shows the process of carbon capture and storage (CCS) that can be used to manage climate change. Carbon dioxide (CO_2) is pumped into three different underground locations, where it is stored.

Flow 1 pumps CO_2 into an underground saline reservoir.
Flow 2 pumps CO_2 into an oil reservoir; CO_2 replaces oil; oil is produced.
Flow 3 pumps CO_2 into a coal seam; CO_2 replaces methane (CH_4); methane is produced.

   a.   Outline the evidence that CO_2 acts as a greenhouse gas.   [1]

   b.   State a greenhouse gas other than CO_2.   [1]

   c.   Outline how the mitigation strategy shown in **Figure 5** is different to an adaptation strategy for managing climate change.   [2]

   d.   Identify **two** mitigation strategies to manage climate change, other than carbon capture and storage.   [2]

   e.   Outline how Flows 1 and 2 shown in **Figure 5** may contribute to the capture and storage of atmospheric carbon.   [2]

11. To what extent do anthropocentric value systems dominate the international efforts to address climate change?   [9]

12. There are concerns that increased carbon dioxide (CO_2) emissions are leading to changes in the global climate.

   a.   Use the data in **Figure 6** to calculate the projected percentage increase from 2007 to 2030 in CO_2 emissions for Russia.   [1]

   b.   Outline how CO_2 emissions may cause a change in the global climate.   [2]

   c.   Identify **two** possible reasons for the projected change in CO_2 emissions for China.   [2]

   d.   Identify **one** reduction strategy that the USA might use to achieve its projected change in CO_2 emissions.   [1]

   e.   Identify **one** adaptation strategy that could be used to reduce the impacts of climate change.   [1]

   f.   Explain how the ability to implement mitigation and adaptation strategies may vary from one country to another.   [4]

13. a.   Distinguish between the causes of recent global warming and those of ozone depletion.   [4]

   b.   Explain the impact of global warming and ozone depletion on coastal ecosystems.   [6]

   c.   Environmental value systems may lead to different approaches to addressing the issue of global warming. Discuss which environmental value system(s) you consider to be most appropriate in the management of global warming.   [8]

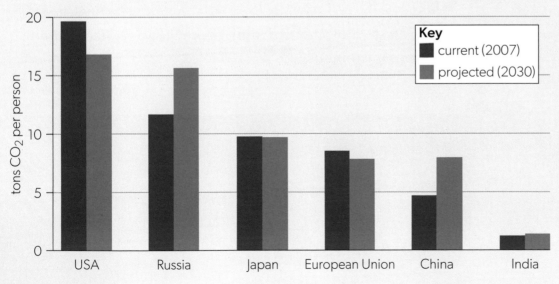

◄ Figure 6 CO_2 emissions for selected countries in 2007 and 2030 (projected) Data from: World Resources Institute, http://www.wri.org/resources/charts-graphs/capita-co2-emissions-selectmajor-emitters-2007-and-2030-projected.

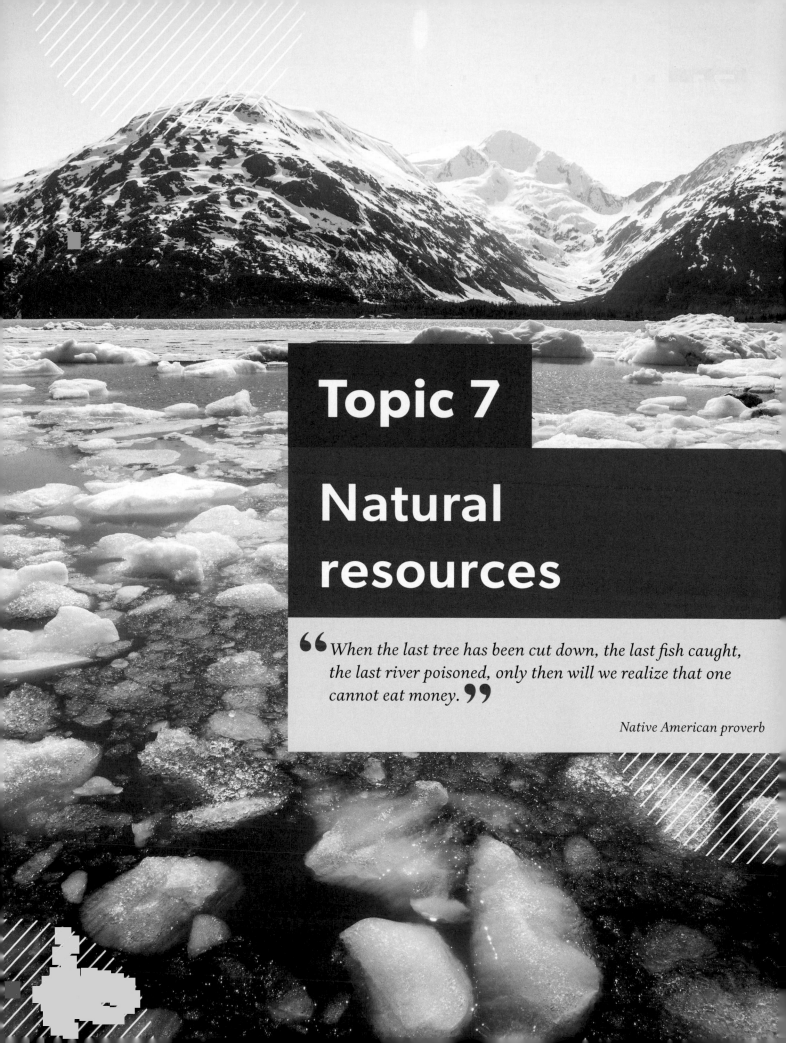

Topic 7
Natural resources

When the last tree has been cut down, the last fish caught, the last river poisoned, only then will we realize that one cannot eat money.

Native American proverb

Guiding questions

- How does the renewability of natural capital have implications for its sustainable use?

- How might societies reconcile competing perspectives on natural resource use?

- To what extent can human societies use natural resources sustainably?

Understandings

1. Natural resources are the raw materials and sources of energy used and consumed by society.
2. Natural capital is the stock of natural resources available on Earth.
3. Natural capital provides natural income in terms of goods and services.
4. The terms "natural capital" and "natural income" imply a particular perspective on nature.
5. Ecosystems provide life-supporting ecosystem services.
6. All resources are finite. Resources can be classified as either renewable or non-renewable.
7. Natural capital has aesthetic, cultural, economic, environmental, health, intrinsic, social, spiritual and technological value. The value of natural capital is influenced by these factors.
8. The value of natural capital is dynamic in that it can change over time.
9. The use of natural capital needs to be managed in order to ensure sustainability.
10. Resource security depends on the ability of societies to ensure the long-term availability of sufficient natural resources to meet demand.
11. The choices a society makes in using given natural resources are affected by many factors and reflect diverse perspectives.
12. A range of different management and intervention strategies can be used to directly influence society's use of natural capital.
13. The SDGs provide a framework for action by all countries in global partnership for natural resources use and management.
14. Sustainable resource management in development projects is addressed in an environmental impact assessment (EIA).
15. Countries and regions have different guidance on the use of EIAs.
16. Making EIAs public allows local citizens to have a role as stakeholders in decision-making.
17. While a given resource may be renewable, the associated means of extracting, harvesting, transporting and processing it may be unsustainable.
18. Economic interests often favour short-term responses in production and consumption which undermine long-term sustainability.
19. Natural resource insecurity hinders socio-economic development and can lead to environmental degradation and geopolitical tensions and conflicts.
20. Resource security can be brought about by reductions in demand, increases in supply or changing technologies.
21. Economic globalization can increase supply, making countries increasingly interdependent, but it may reduce national resource security.

Connections: This links to material covered in subtopics 1.1, 1.2, 1.3, HL.a, HL.b and HL.c.

Natural resources

Natural resources support life and meet human needs through providing goods or services. They include any natural substance which may be used for many different things and valued in different ways—see Figure 1.

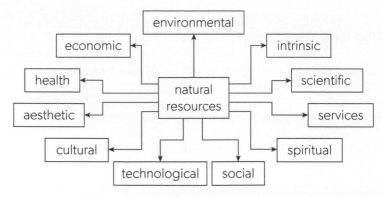

▲ Figure 1 Different ways to view or value natural resources

ATL Activity 1

ATL skills used: thinking, communication

1. Copy Figure 1.

2. Can you think of any other ways to value natural resources? Add them to your figure.

3. Add a single visual representation to each of the categories in your figure.

4. In which category would you put the following natural resources? Oil, metals, stone, air, sunlight, soil, water.

5. Make sure you have an example for each category.

Natural capital and natural income

See also subtopic 1.3.

- Natural capital is the stock of natural resources available on Earth; it is a resource which has some value to humans.

- The terms natural resource and natural capital are used interchangeably.

- Natural income is the yield or harvest from natural resources.

In the past, economists spoke of capital as the products of manufacturing human-made goods, and separated these from land and labour. We now recognize that capital includes:

- **goods**: natural resources with a perceptible value to us, e.g. trees, soil, water, living organisms and mineral ores

- **services**: natural resources that support life, e.g. flood and erosion protection provided by forests; photosynthesis which provides oxygen for life forms to respire; processes that maintain healthy ecosystems.

The way we see the world shapes the way we treat it. If a mountain is a deity, not a pile of ore; if a river is one of the veins of the land, not potential irrigation water; if a forest is a sacred grove, not timber; if other species are biological kin, not resources; or if the planet is our mother, not an opportunity—then we will treat each other with greater respect. Thus is the challenge, to look at the world from a different perspective.

David Suzuki (born 1936), scientist and environmental activist

The term **natural capital** is now used to describe goods and services that are not manufactured but do have value to humans. They are the environmental benefits from the physical, chemical and biological functions of ecosystems. They can be improved or degraded and given a value—we can begin (with difficulty) to give monetary values to ecosystems. We may be able to process the goods and services of natural capital to add value to them. For example, we may mine for metals or turn trees into timber, but the metal and trees are still natural capital.

Just as economic capital yields economic income, natural capital yields **natural income**. This is the yield, harvest or services that can be obtained without depleting the original capital. For example:

- cherry trees produce cherries

- the water cycle provides us with fresh water

- forests provide timber.

The measure of the true wealth of a country must include its natural capital, e.g. how many mineral resources, forests or rivers it has. In general, high-income countries (HICs) add value to natural income by manufacturing goods from it whereas low-income countries (LICs) may have greater unprocessed natural capital. The World Bank now calculates the wealth of a country by including the rate of extraction of natural resources and the ecological damage caused by this, including carbon dioxide emissions.

Natural capital is classified on a sliding scale based on how long it takes to restock compared with how fast we use it. Ultimately, all resources are finite.

Non-renewable natural capital:

- exists in finite amounts on Earth

- is not renewed or replaced after it has been used or depleted

- is re-made eventually but only on a geological timescale

- includes minerals, soil, water in aquifers and fossil fuels.

As these resources are used, the stocks are depleted and new sources of the stock or alternatives need to be found.

TOK

To what extent can economic principles be applied to natural capital?

Key term

Non-renewable natural capital is either irreplaceable or only replaced over geological timescales, e.g. fossil fuels, soil and minerals.

▲ Figure 2 Coal is a non-renewable resource; a forest is a renewable resource

Renewable natural capital can be generated and/or replaced as fast as it is being used.

Renewable natural capital:

- has the capacity to be replaced by the natural growth rate or through recurring processes

- can run out if the standing stock (how much is there) is harvested unsustainably (i.e. more is taken than should be)

- includes living species and ecosystems that use solar energy and photosynthesis

- includes non-living items such as groundwater and the ozone layer.

Natural capital can be used sustainably or unsustainably (see subtopic 1.3). If renewable natural capital is used beyond its natural income, this use is unsustainable.

The depletion of natural resources at unsustainable levels and efforts to conserve these resources are often a source of conflict within and between political parties and countries. The impacts of extraction, transport and processing of renewable natural capital may cause damage and make this natural capital unsustainable.

For example, the natural capital of water may be renewable or non-renewable:

- In regions of high rainfall where most rain is collected and used for drinking, water is renewable natural capital.

- In drier regions where underground aquifers refill slowly at rates longer than an average human lifetime, water is non-renewable natural capital.

Recyclable resources

Iron ore is a non-renewable resource. Once the ore has been mined and processed it is not replaced in our lifetime. However, from iron ore we produce iron which can be cast into numerous forms and represents a significant commodity within modern societies. About 65% of a car is made from iron or iron-derived products like steel. However, steel and iron can be recycled. Old or damaged cars can be broken down. Their parts can be used to replace parts in other cars, or remanufactured into new metal objects. Although iron ore is non-renewable, the iron extracted from the ore becomes a renewable resource. The same is true for aluminium.

Forests as natural capital and natural income

A forest is a stock of natural resources (natural capital) which provides the natural income of **timber**. To continue to provide timber, the forest must be managed sustainably. This means it must be harvested at a rate that:

- allows replacement through natural growth

- does not damage ecosystems, watersheds, or wildlife

- leaves enough trees in place to reproduce the next generation.

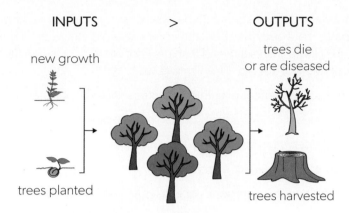

INPUTS   >   OUTPUTS

new growth

trees die
or are diseased

trees planted

trees harvested

forest = natural capital

▲ Figure 3 Sustainable use of the forest stock

In Figure 3, the inputs must be equal to or greater than the outputs (MSY—see subtopic 4.3) and the forest stock must remain at the same level. If the forest is not managed in this way, it is unsustainable and will not provide future generations with timber or other forest products. Many logging operations (both legal and illegal) use clear fell techniques, which are highly destructive and can lead to:

- increased release of GHGs into the atmosphere

- changes in temperature

- changes in moisture content in soil and atmosphere

- flooding and soil erosion

- loss of hatitat and extinction of some wildlife

- problems for indigenous people and social conflict.

Fish as natural capital and natural income

As discussed in subtopic 4.3, fish as natural capital have been comprehensively overexploited. Unsustainable fishing has led to some significant issues and there is a danger that some species will go extinct. An example of this is the bluefin tuna, one of the largest and fastest fish in the oceans. Progression of technology and high demand for bluefin tuna means that the spawning population (2023) is down to approximately 30% of the 1970 population.

Life-support ecosystem services

Many of these life-support services are not tangible; you cannot see them or touch them. This means they do not have capital and income in the same way that goods do.

▲ Figure 4 Damage of clear felling

Practical

Skills:

- Tool 3: Mathematics

- Tool 4: Systems and models

A stock of cod in an area is 100,500. Each year, the birth rate is 600 and the death rate is 500. The number of fish migrating into the school of cod is 1,000, and only 100 are leaving.

1. Draw a systems diagram, to scale, to show this information.

2. Calculate the maximum number of cod that could be sustainably harvested.

With good sustainable management, these services are renewable. However, if they are badly managed, they become non-renewable. Between 2001 and 2005, the UN sponsored the Millennium Ecosystem Assessment (MEA) (see also subtopic 1.3), the aim of which was to:

- assess the consequences of anthropocentric-initiated ecosystem changes

- assess the impact of such actions on human well-being

- enhance conservation efforts based on scientific research

- enhance sustainable use of ecosystems.

The MEA established four categories of ecosystem services (Figure 5).

Supporting services are the services on which the other three categories rely—the underlying processes that maintain consistency and allow the Earth to sustain life. These essential services include nutrient cycling, soil creation, photosynthesis and the water cycle.

Regulating services are the basic services that make life possible; they are processes that "cooperate" with each other to keep ecosystems clean, functional, resilient and sustainable. Through these processes carbon is stored, climate is regulated, organic matter is decomposed, erosion and flooding are controlled, plants are pollinated and water is purified.

Cultural services result from our interactions with the natural world, from our ancestral past to modern humans. They are non-material benefits that have contributed to human development and cultural advancement. Our interactions and reliance on nature have built knowledge and spread ideas thus contributing to local, national and global cultures. From this has come art, architecture, music and recreation.

Provisioning services are any type of benefit humans can extract from nature. They include plants for food, medicine, making cloth, oil for cooking and biogas; trees for fuel or timber; and drinking water.

▲ Figure 5 Ecosystem services according to the MEA

ATL Activity 2

ATL skills used: thinking, communication, social

In groups of two or three, copy and complete Table 1, choosing one specific service for each category.

- In the column headed "Sustainable", explain all the actions we should take to keep the service sustainable.
- In the column headed "Unsustainable", explain all the actions we may take that makes the service unsustainable.

Life-supporting service	Sustainable	Unsustainable
Supporting		
Regulating		
Cultural		
Provisioning		

▲ Table 1

Services provided by vegetation

We all know that vegetation is important to humans; it provides obvious goods such as food, clothing and medicine. However, it also provides significant life-support services.

Water replenishment is where surface water moves from land surfaces into groundwater stores. Artificial replenishment was discussed in subtopic 4.2. Vegetation can aid replenishment of aquifers and help to regulate water flows in the hydrological cycle. All vegetation acts as a temporary store for precipitation: water lands on the vegetation (on leaves, stems, stalks and branches) then slowly trickles down to the ground surface. This slows delivery of water to the soil store, allowing more time for surface water to soak into the soil and rocks. Thus, water flows into the stores rather than over the surface to streams and rivers.

▲ Figure 6 Vegetation provides a range of services

These effects reduce the volume of floodwater and slow its passage through the water cycle, thus reducing flooding and soil erosion (the roots of the vegetation hold the soil in place). In contrast, the lack of vegetation in urban areas increases flood risk, because there is nowhere for the precipitation to go except over the land.

Wetlands are particularly good at flood protection and soil preservation, acting as a natural sponge which traps the water and releases it slowly. The Mississippi River has hardwood **riparian** wetlands (banks next to the river) along it. In the past, these areas stored floodwater for up to 60 days but now, due to clearance and draining, they hold the waters for only 12 days.

Plants also mitigate air and water pollution—another life-supporting service. We looked at eutrophication in subtopic 4.4. Reed beds can be used as a riparian buffer zone between agricultural fields and nearby rivers. The reed beds remove excess nutrients from the soil water that flows beneath them. This reduces the risk of eutrophication in the river.

Trees are often planted in urban areas because they are known to have air purification functions. Vegetation is known for its **carbon sequestration** processes but it also removes ammonia, calcium, carbonyl, nitrates, ozone, organic compounds, particulate matter, sulfates and volatile organic compounds (VOC) from the air that we breathe.

Practical

Skills:

- Tool 1: Experimental techniques
- Tool 2: Technology
- Inquiry 1: Exploring and designing
- Inquiry 2: Collecting and processing data
- Inquiry 3: Concluding and evaluating

Design a questionnaire to investigate the value the school community places on the different ecosystem services.

You could:

- select 10 different life-supporting services and find an image for each one
- ask respondents to rate how important that thing is to them.

For example: On a scale of 1 to 4, how important is it for you to have clean air to breathe?

1. Not at all important
2. Not important
3. Important
4. Essential

Valuing natural capital

There are many ways to value natural capital, but we must consider the breadth of meaning of the word "value". Value may relate to:

- what something is worth in financial terms

- how important or useful something is

- an ethical standpoint, when we have to decide what is right or wrong.

There are different ways in which we can categorize natural capital. One method uses "use-valuation" vs "non-use valuation".

- Use-valuation: natural capital that we can put a price on, such as marketable goods and services.

- Non-use valuation: natural capital that it is almost impossible to put a price on, such as intrinsic rights, future knowledge sources, value for future generations.

Many people feel that the only way to make others realize the importance of non-use valuation natural capital is to find some way to put a price tag on it. Others feel this may just encourage exploitation of such resources.

It is essential for us to know whether or not a resource can be used sustainably. We may think that agriculture is sustainable because crops are eaten and then more are planted. But agriculture is only sustainable if soil fertility and structure are maintained, and the environment is not degraded. If biodiversity is lost due to agriculture, can the agriculture be sustainable?

Slash and burn agriculture (shifting cultivation) and sporadic logging in virgin forest are both sustainable as long as the environment has time to recover. This is dependent on low human population densities. Are we currently giving the environment enough time to recover?

Since the early 1980s, the UN Environmental Programme (UNEP) has been using a system of integrated environmental and economic accounting (Socio-economic Environmental Assessment—SEEA) to try to value the environment and track resource depletion. If countries would include the cost of degrading their natural resources within their GNP (gross national product), it would be easier to see the real cost and health of the nation.

From the UN Earth Summit in Rio de Janeiro in 1992 came Agenda 21 (see subtopic 1.1). In consultation with their communities, local councils were supposed to produce their own plans—a local Agenda 21.

What do we value and how?

The value of natural capital is viewed in different ways. Value is not static—it changes through space and time—see Table 2.

Value of natural capital	What is it?	Examples	Uniqueness
Aesthetic	Something in nature that is beautiful, our appreciation of that beauty An interaction between an individual and the beauty that is there	a forest a beach the Taj Mahal	This can be completely personal. Does one person have the exact same tastes as another?
Cultural	Tangible buildings, books and artworks. Intangible folklore, language, traditions and knowledge The natural heritage of the landscape and biodiversity	Uluru the Washington Monument the *Mona Lisa*	This is specific to a particular culture—local, regional or global.
Economic	Something that is tangible and that has monetary value	gold fossil fuels gemstones	This value is generally accepted on a global level.
Environmental	Natural assets and the goods and services they provide to support human life; this could be seen as the whole planet	fresh air water beautiful views	Valuation of this will vary enormously. Ecocentrics, anthropocentrics and technocentrics have very different viewpoints.
Health	The natural environment keeps us healthy in many ways	fresh water fresh air beautiful scenery pets	Depends on how you view health and what makes you healthy; this varies according to perspective.
Intrinsic	The inherent existence of things: nature has a value in and of itself, nothing to do with human usage	Siberian tiger huntsman spider giant panda	Some organisms are more likely to be valued for their intrinsic value than others. Top predators and insect pests are not valued by most people.
Social	The relationships we have with each other and communities	family football team	This is very personal and can change very quickly.
Spiritual	Many people say this is the religions of the world and the buildings that represent them It is also humans' independent spirituality	Angkor Wat Sacre Coeur Stonehenge	This may be destroyed by groups of people who have different beliefs.
Technological	The know-how in research, development, organization and knowledge	computers mobile phones the internet	Changes status extremely quickly—it has built in obsolescence and is prone to changes in fashion. Advances in technology change the status of other natural capital.

▲ Table 2 How we value natural capital

(ATL) Activity 4

ATL skills used: thinking, communication, social, research, self-management

Putting a value on the environment

Work in groups.

1. a. Look at Table 2. Sort the values of natural capital into three sets and give each set a name. Make sure you can justify the sets you have chosen and the names you have given them.

    b. Sort the following items into your three sets:

your school	Polio virus	Antarctica
your city	the Amazon rainforest	Great Barrier Reef
your home	the Sahara desert	Shanghai
your local park or protected area	Lake Superior	Tokyo
tigers	San Francisco	
mosquitoes	tundra in Siberia	

    c. Share your three sets with another group and justify your choices.

    d. Repeat (b) using the other group's sets.

    e. Compare your sets and discuss any differences.

2. What are the difficulties in categorizing natural resources and human environments?

3. How do you decide what characteristics need to be taken into consideration when trying to do this?

4. Do you think that environments can have their own intrinsic value?

Dynamic nature of natural capital

The importance of types of natural capital varies over time and through space for many reasons.

- A resource available today may not be available in the future (e.g. fossil fuels).

- A resource valued at one time may be useless in the future (e.g. arrowheads made from flint).

- A resource valued in the past may no longer be useful or may be considered unethical (e.g. whale oil for lamps, cheetah skins for coats).

Our use of natural capital also depends on cultural, social, economic, environmental, technological and political factors.

- Technocentrists believe that new discoveries will provide new solutions to old problems: for example, replacing hydrocarbon-based fuel with hydrogen fuel cells, or harvesting algae as a food source.

- Uranium is in demand as raw material for nuclear power by fission but will lose value if we can harness the energy of nuclear fusion—the hydrogen economy.

Examples of changing value of natural capital

1. Cork forests

Cork from the bark of the cork oak tree has been essential for centuries to seal glass bottles. Now plastic corks, screw-top bottles and plastic lids are replacing cork. Many of these are not biodegradable like cork, and some are made from fossil fuels.

Cork oak forests in the Mediterranean region have high biodiversity, second only to that of the Amazon rainforest. The bark is harvested by hand every nine years, without killing the trees. Unfortunately, as cork forests lose their value as natural capital to humans, they are cut down and the land used for other purposes.

It may be that cork will regain value as a resource as many counties are banning plastics more and more.

2. Lithium

▲ Figure 7 Harvesting cork in a cork oak forest, Andalucia, Spain

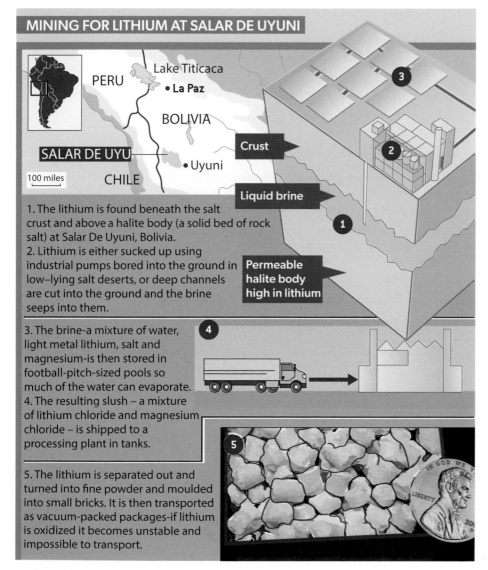

MINING FOR LITHIUM AT SALAR DE UYUNI

PERU
Lake Titicaca
• La Paz
BOLIVIA
SALAR DE UYU
• Uyuni
CHILE
100 miles

Crust
Liquid brine
Permeable halite body high in lithium

1. The lithium is found beneath the salt crust and above a halite body (a solid bed of rock salt) at Salar De Uyuni, Bolivia.
2. Lithium is either sucked up using industrial pumps bored into the ground in low–lying salt deserts, or deep channels are cut into the ground and the brine seeps into them.

3. The brine-a mixture of water, light metal lithium, salt and magnesium-is then stored in football-pitch-sized pools so much of the water can evaporate.
4. The resulting slush – a mixture of lithium chloride and magnesium chloride – is shipped to a processing plant in tanks.

5. The lithium is separated out and turned into fine powder and moulded into small bricks. It is then transported as vacuum-packed packages-if lithium is oxidized it becomes unstable and impossible to transport.

▲ Figure 8 Mining for lithium in Bolivia

We use lithium carbonate batteries if we have a mobile phone, tablet or electric car. Thirty years ago, we had little idea where lithium-containing ores were in the world because we did not use much of it as a resource. Now we cannot get enough of them.

More than half the world's known reserves of lithium are underneath desert salt plain in Bolivia and under the Chilean Atacama desert. China, Australia and Argentina have big reserves too. But the annual production of lithium is not nearly enough to power electric cars if they were to replace cars with petrol engines.

(ATL) Activity 5

ATL skills used: thinking, communication

1. Using Figure 8, list the advantages and disadvantages of lithium mining in Bolivia.

2. Lithium is a finite resource but could be recycled once extracted.

    a. State whether lithium is a renewable or non-renewable resource.

    b. Discuss the sustainability of human use of lithium.

3. All mining activity disrupts ecosystems. Explain reasons for the increase in human demand for lithium.

"Capital" and "income"?

Most people associate the terms "capital" and "income" with the economy, and with the idea of gain from some venture. Some people argue that the terms "natural capital" and "natural income" encourage a philosophical perspective that nature is there for human exploitation (an anthropocentric stance).

On the other hand, any student of economics knows that such gains can only be achieved through careful management. From an environmental perspective, that means careful management for the sustainable use of resources.

TOK

To what extent do the terms "natural capital" and "natural income" imply values, assumptions and perspectives on our relationship with nature?

(ATL) Activity 6

ATL skills used: thinking, research

Read Case study 1 and do further research if you want to. Then copy and complete Table 3.

Environment	Evidence that the terms "natural capital" and "natural income" support the idea that nature is for human exploitation	Evidence that the terms "natural capital" and "natural income" encourage sustainable use of resources
The Arctic		
The Antarctic		

▲ Table 3

Exploiting the poles

The Arctic and Antarctic are perhaps the last wildernesses on Earth and are beautiful. Their ecosystems are fragile and contain much biodiversity found nowhere else. Any disturbance has a long recovery time; growth is slow because temperature is limiting. On land, water is also limiting because it is frozen for much of the year and unavailable to plants.

▲ Figure 9 Map of the Arctic

The Arctic

Until recently, humans could not exploit the resources of the Arctic on a large scale because the seas are frozen for all but a few months of the year and conditions are harsh. But there are mineral riches—especially hydrocarbons—locked under the Arctic Ocean and surrounding land masses.

The world's oil supply comes from many countries. Many countries would like a national source of oil, as they would then not be dependent on importing oil. Some 40% of oil comes from and is exported by OPEC (Organization of Petroleum Exporting Countries).

There are 12 oil-exporting countries whose economies rely on oil exports, and they control oil prices and supply. The USA produces more oil than any other country, then Saudi Arabia and then Russia. Remaining oil is dispersed across a number of other countries. The price of a barrel of crude oil varies greatly:

- for much of the 1990s it was around USD 30

- in 2008 and 2011 it was around USD 100

- in early 2023 it was just over USD 75.

With climate change causing the Arctic to warm up, there are more ice-free days. High oil prices mean that reserves that were once uneconomic to extract are no longer so. Thus, the Arctic could be the next "goldmine"—or environmental disaster, depending on your environmental worldview. At 2008 prices, the estimated value of the Arctic's minerals is USD 1.5–2 trillion. There are crude oil reserves under northwestern Siberia and Alberta, Canada. There is also oil right under the North Pole. Humans have the technology to extract this oil. So why not?

Who owns the Arctic?

There is no land at the North Pole: it is ice floating on water. Under the UN Convention on the Law of the Sea (UNCLOS), a state can claim a 200-nautical-mile (370 km) zone and beyond that up to 150 nautical miles (278 km) of rights on the seabed. A state with such a claim may fish or exploit the minerals exclusively in this zone and other countries may not. This distance is not measured from the border or edge of a country but from the edge of the continental shelf, which may be some distance away from the border of the country under the sea.

In August 2007, a Russian submarine expedition planted a Russian flag on the seabed at the North Pole, two miles under the Arctic ice cap. They claimed that the seabed under the pole, called the Lomonosov Ridge, is an extension of Russia's continental shelf and thus Russian territory.

Six countries—Canada, Denmark, Iceland, Norway, Russia and the USA—have Arctic Ocean coastlines and Denmark has sent its own scientific expeditions to study the opposite end of the Lomonosov Ridge to see if they can prove it is part of Greenland—a Danish territory.

The Antarctic

Antarctica is a continent of which 98% is covered by ice and snow. No large mineral or oil reserves have been found but humans exploit the continent through tourism, fishing, sealing and whaling. Nearly 100,000 tourists are expected to visit Antarctica in the 2022–23 season. No one country owns Antarctica but seven have staked territorial claims via the Antarctic Treaty. This treaty was perhaps the first step in recognition of international responsibility for the environment. It was signed in 1959 by 12 countries including the US, UK and USSR who signed it in the middle of the Cold War. The treaty was strengthened in 1991 and covers all land south of latitude 60°S.

The agreement is that:

- the area will be free of nuclear tests and nuclear waste; for peaceful purposes; a preserved environment; undisputed as a territory

- there will be prevention of marine pollution; cleaned-up sites; no commercial mineral extraction; sealing with annual limits

- commercial whaling is now tightly regulated.

Fishing is less of a success story. There is overfishing of many species because fishing is hard to regulate in the seas around Antarctica. This is causing a collapse of many penguin and seal populations.

The ice on Antarctica represents approximately 61% of all freshwater on Earth. If this melted, global sea level would rise by 70 m.

It appears that the ice is melting and some large ice sheets are calving (breaking up and slipping away from the land). Over three weeks in 2002, Larsen B—a huge ice shelf, over 3,000 km^2 and 220 m deep—broke up and floated out to sea. On 22 January 2023, a new iceberg broke away with an area of 1,550 km^2. But in other areas, the ice is getting thicker.

Question

Compare and contrast the (a) sociopolitical and (b) environmental pressures on the Arctic and Antarctic.

▲ Figure 10 Map of the Antarctic

Globalization

Did you know?

- 51 of the world's top 100 economies are corporations.

- Transnational corporations (TNCs) control 66% of world trade, control 80% of foreign investment, and employ just 3% of the world's labour force of 2.5 billion.

- Walmart may be bringing 38,000 people out of poverty per month in China.

Globalization is the concept that every society on Earth is connected and unified into a single functioning entity. The connections are mostly economic but also allow the easy exchange of goods, services, information and knowledge.

Globalization has been facilitated by new technologies, air travel and the communications revolution. The World Trade Organization (WTO) controls the rules of this global trade. Information is only an email, text, website or phone call away. Everyone can access the global market—as long as they are connected. Ebay, for example, allows someone in Europe to purchase goods from another individual in the USA.

Global trade is not new. The Ancient Greeks and Romans traded across their world. The Han dynasty in China traded across the Pacific Basin and India. European empires and the Islamic world traded via trade routes around the world. What is new is the speed and scale of the trade and the communication. Since the end of the Second World War, protectionism of markets has decreased and free trade has increased. The World Bank and the International Monetary Fund (IMF) were set up in 1944. Since then, they have influenced development and world finance, including third world debt. Some think that globalization only leads to higher profits for TNCs. But there is evidence that poverty has decreased in countries with increased global contacts and economies, e.g. China. Ecologically, international agreements on global issues such as climate change or ozone depletion have tended to be easier to conclude with increased globalization. There is a tendency for globalization to westernize some countries.

Globalization is not internationalism.

- Internationalism recognizes and celebrates different cultures, languages, societies and traditions. It promotes the nation state as a unit.

- Globalization sees the world as a single unit or system. It does not recognize the differences of internationalism. Globalization is making individuals more aware of the global community, its similarities and its differences. Globalization is both a positive and negative force. On the one hand, it can make us aware of the plight of others on the other side of the globe; on the other, it can make us aware of what one society has and we do not have.

Many, if not most, products are now traded on a global scale. They are part of what is referred to as the "global market" and minerals mined in South Africa or Australia are traded and shipped globally.

Managing natural capital

If we deplete renewable natural resources faster than they can be regenerated, that is not sustainable in the longer term. If we release waste into the environment at a rate faster than it can be removed, degraded or cleaned up, we are not living sustainably.

(ATL) Activity 7

ATL skills used: thinking, communication

1. Look back at subtopic 1.3 and review the definitions of sustainability.

2. Reread the sections on:
    *   Minamata (subtopic 1.1)
    *   Chernobyl or Fukushima Daiichi (subtopic 1.1)
    *   Grand Banks cod fisheries (subtopic 4.3)
    *   grey wolves in Yellowstone (subtopic 2.5)
    *   the carbon cycle and climate change (subtopics 2.3 and 6.2)
    *   the nitrogen cycle and eutrophication (subtopics 2.3 and 4.4)
    *   biodiversity loss (subtopic 3.2)
    *   water pollution (subtopic 4.4)
    *   soil degradation (subtopic 5.2).

3. Write short paragraphs answering these questions.

    a. What are the common causes of resource loss and polluting waste?

    b. What do you think can be done to limit destruction of natural capital?

Resource security and choices

Resource security relates to the ability of a nation to provide enough natural resources for its people.

The capacity of an ecosystem to renew biomass is the biocapacity of that ecosystem. An increasing number of people live in countries where there is a biocapacity deficit—they are living beyond the means of the environment to support them. Short-term solutions are:

*   reducing demand or increasing supply

*   importing goods and natural services (e.g. food and water)

*   changing technologies

*   in extreme cases, emigration to other countries.

But none are long-term solutions.

Food security

While being fully dependent on imported foods makes a country vulnerable, being fully self-sufficient has the same effect. If there is a local disaster or disease (e.g. potato blight in Ireland), global trade in food is a safety net. Countries hit by famine or food shortages can import food to sustain the population.

A study of food security in different countries found that most could not become self-sufficient. There is not enough suitable agricultural land.

A country that is self-sufficient in food can reduce its carbon footprint as food transport is reduced but even large countries would not be able to become food independent.

A study at Leiden University in the Netherlands identified countries that are most dependent on food imports. These include Egypt (which would need more than 7 times its land area to be self-sustainable); Papua New Guinea (5 times); the Democratic Republic of Congo (4.3 times); Pakistan (4.2 times); the Philippines (3.6 times); India (nearly 3 times); Indonesia (2 times); and Ethiopia (1.8 times).

At the other end of the food security scale, Argentina could feed itself with just 5.5% of its land; the USA and Canada need only 12–13%; Norway would need 123% of its land area to be self-sufficient; the Netherlands would need 120%; Switzerland 119%; Belgium 113%; Sweden 13%; Ireland 14.56%; Finland 18%; Germany 54%; the Czech Republic, the UK, Spain, and Portugal each need 60% of their land to be self-sufficient.

However, self-sufficiency assumes a diet much reduced in variety—with only foods that grow locally—and it would require far less meat eating. Cattle do eat grass but most are fed with soy and grains grown in Brazil, Russia and Ukraine. Growing feed for animals locally means less land for food directly fed to humans. Eating less meat is one option but even eating no meat would not create food security for a population with no imported foods.

Water security

Brazil has the largest freshwater resource in the world, mostly held in the Amazon region. In Europe, Croatia, Sweden and Finland have high levels of freshwater resource. Review subtopic 4.2 on water access, water stress and water security.

The concept of **virtual water** can be applied to the production of goods and it is the volume of water needed to make a particular product. For example, a can of Coca-Cola holds 0.33 l of liquid but about 200 l of water is required to grow and process the sugar in the can.

The water footprint of a country considers use of both local and global water resources—that is, it accounts for actual water used and for the virtual water content of products. On this basis, the UK is only 38% self-sufficient in water because it imports so many goods and foodstuffs.

Most of the freshwater in the world is used in agriculture and the biggest users are China, India and the USA. But Japan is the biggest importer of virtual water as it is a big importer of goods, followed by Mexico, Italy and Germany.

It is estimated that 20% of the world's total water footprint is exported in traded goods. While water stress is an increasingly important issue, the dependence of countries on virtual water is often ignored.

A self-sustaining society?

A self-sustaining or self-sufficient society is one which can grow and maintain enough to fulfill its needs indefinitely. Such societies are rare. Many forms have been tried, e.g. "back to the land" movements, agrarianism (which promotes subsistence agriculture), various "preppers", deep ecologists. But self-sufficiency in everything is hard. Many small societies seek energy self-sufficiency before attempting food or water security.

Here are some examples of societies which attempt to live sustainably.

▲ Figure 11 Production of a can of Coca-Cola requires 200 litres of water

▲ Figure 12 The first house in Findhorn ecovillage

ATL Activity 8

ATL skills used: thinking, communication, social, research

1. Select a named food resource or supply of fresh water used in your local community.

2. Investigate how this resource is produced, transported, used and disposed of. For example, if you select bananas: Where do they come from? How are they transported? Where and why do you buy them? What happens to the banana skins?

3. Repeat your investigation for a named natural resource other than food or water, e.g. a fuel, a metal or something made from wood.

4. Discuss your findings in class.

- Findhorn ecovillage, Scotland—about 200 families of different nationalities live here with energy from wind turbines and biomass, biological sewage treatment and water heating by solar energy. But they only grow some of their food although their carbon footprint is half the UK average.

- Masdar City in the UAE aims to be 100% environmentally friendly. Solar panels provide power and urban transport but it is in the middle of the desert and cannot grow its own food or supply enough water.

Can you find an example of a society or community that lives in a self-sustaining way?

Making choices in resource use

Governments, societies, companies and individuals have to make choices all the time about which natural resources to use. These choices change over time and in response to global events.

The impacts of climate change have forced change in technologies in order to reduce carbon emissions. That leads to change in demand for natural resources (e.g. lithium not coal), which leads to shifts in power and wealth between countries. That may lead to social unrest, population shifts and changing political relationships between countries. Political systems may rise and fall over natural resource use or lack of resources. The natural environment is impacted by our choices.

Management strategies and natural capital

Natural capital is everything that nature provides for "free". It is the most fundamental capital there is as it provides the basic conditions for human societies to exist, and it sets the limits for our socio-economic systems. It is limited and it is vulnerable. Think about that. In order to ensure it continues to support humanity, we must have management and intervention strategies in place to provide guidance for the way society uses it.

Management and intervention strategies can be at an individual, local, regional, national or international scale. At a governmental level, actions may include:

- implementation of national action plans for the SDGs
- introduction of pollution taxes, fines or legislation
- demands for compensation for damage caused to the environment
- subsidies for green energy or heavy taxation of fossil fuels
- investment in research to improve sustainability
- campaigns on resource saving, e.g. use less energy, insulate, don't waste food
- education programmes relating to sustainability.

At a more local level, individuals, NGOs and businesses or industry can:

- individually take steps to reduce waste, recycle more, conserve water, improve energy efficiency
- transition as businesses to the circular model or doughnut economy by removing waste from production processes and using renewable materials, (e.g. manufacture concrete that can store carbon dioxide)
- launch campaigns and use social media to encourage community action.

Case study 3

Australian government and plastics

The Australian government has a mission which is: "an 80% reduction in plastic waste entering the Australian environment by 2030".

Here is the problem, as identified by the Department of Agriculture, Water and the Environment in the National Plastics Plan 2021.

- In 2018–19, Australians used 3.5 million tonnes of plastics, including one million tonnes of single-use plastic and 70 billion pieces of "scrunchable" plastic as food wrappers.

- 60% of the plastic was imported and 85% ends up in landfill.

- Only 13% of the plastic is recycled.

- 130,000 tonnes of plastic leak into the oceans each year and plastic in the oceans is expected to outweigh fish by 2050.

- Plastic use is set to double by 2040.

▲ Figure 13 Single-use plastic bottles

Over-use of plastic must be addressed at all levels of production—design, use, recovery and reuse. It will require action and commitment from government, industry and consumers. To this end, in 2019 the Australian government set out the National Waste Policy Action Plan.

Year	
2019	Government established a timetable to ban export of plastic waste
	National Waste Policy Action Plan agreed
2020	First National Plastics Summit
	Recycling and Waste Reduction Act 2020
	Microbeads phased out in all products
2021	January: Circular economy road map for plastics released
	July: Regulation of unsorted mixed plastic exports
	Review of used packaging material
	National Plastics Design Summit
2022	Regulation of unprocessed waste plastic exports
	Phase-out of non-compostable plastic packaging, expanded polymers, and PVC packaging labels
2023	80% of supermarket products must display Australasian Recycling Label
2025	National Packaging Targets • 100% of packaging must be reusable, recyclable or compostable • 70% of packaging must be recycled or composted • 50% of packing must be from recycling • Single-use plastic packaging phased out

▲ Table 4 Australian National Waste Policy Action Plan

As of 2023, the following items are banned in the majority of Australian states if they are made from plastic:

- lightweight plastic bags

- drinking straws and drink stirrers

- cutlery, plates and bowls

- cotton bud sticks

- polystyrene food and drink containers

- microbeads.

AHL

By 2026, some Australian states will ban:

- heavyweight plastic bags

- plastic bags for fruit and veg

- plastic cups and lids

- coffee cups containing plastic

- plastic straws attached to juice boxes

- plastic takeout containers.

The Australian government is using legislation to deal with the over-use of plastic in Australia, but this is a worldwide problem.

Questions

1. List all natural resources that use of plastic is exploiting.

2. What can you do to reduce your use of plastics?

3. What has Australia done to reduce plastic use?

Local movements

There are hundreds of different management or intervention strategies that can be put into place at a local scale, and they can be targeted at almost any natural resource. Think about your school or your local community and the measures being taken to increase sustainable approaches to resource use. These schemes are successful because we all take ownership in them and the educational aspect helps people to understand the benefits they bring.

Possible actions taken by schools include:

- banning the use of single-use drink bottles, and only allowing reusable bottles for water or other drinks

- installing water fountains that can be used to refill reusable bottles

- avoiding printed handouts and newsletters—sending electronic files instead

- always photocopying or printing on both sides of the paper

- ensuring every classroom has a waste bin for rubbish and a paper recycle bin to ease recycling

- reusing paper that has only been used on one side

- setting up a second-hand uniform shop.

TOK

To what extent could indigenous knowledge systems support a move to sustainable lifestyles?

Practical

Skills:

- Tool 1: Experimental techniques

- Tool 2: Technology

- Inquiry 1: Exploring and designing

Design an investigation to study the measures your school or local community is taking to move towards sustainability. Refer to subtopic 1.4 on how to design questionnaires and surveys.

Environmental impact assessments

An environmental impact assessment (EIA) is a process used to identify the likely impacts of a proposed development that will alter land use of an area—for example, to plant a forest or convert fields to a golf course. The EIA was developed to help decision makers consider whether or not the project should go ahead by weighing the relative advantages and disadvantages. The report that is produced must give appropriate information about environmental, social and economic impacts connected with the proposal. The EIA is made public so citizens can have a stake in decision-making.

EIAs look at what the environment is like now and forecast what may happen if the development occurs. Both negative and positive impacts are considered as well as other options to the proposed development. While EIAs mainly deal with questions about the effect on the natural environment, they must also consider the likely effects on human populations. This is especially important where a development might impact human health or have an economic effect for a community.

What does an EIA need to include?

There is no set way of conducting an EIA, but various countries have minimum expectations for what should be included. It is possible to break down an assessment into three main tasks.

> Connections: This links to material covered in subtopic HL.a.

1. **Identifying impacts (scoping)**

   - Identify and assess the possible impacts and effects on the environment, economy and society.

   - Canvas all stakeholders about the project.

   - Conduct a baseline study to investigate the current abiotic environment and biotic community, to enable assessment of what might change if the development goes ahead.

2. **Predicting the scale of potential impacts**

   - Try to quantify potential changes to microclimate, biodiversity, scenic and amenity value resulting from the proposed development.

3. **Limiting the effect of impacts to acceptable limits (mitigation)**

   - Consider reasonable alternatives.

   - Assess the efficacy of proposed measures to avoid or minimize the impacts.

   - Assess any monitoring, management and planning measures that are to be used to minimize or mitigate the negative impacts.

What are EIAs used for?

EIAs are often, though not always, part of the planning process that governments set out in law when large developments are considered. They provide a documented way of examining environmental, economic and social impacts that can be used as evidence in the decision-making process for any new development. The developments that need EIAs differ from country to country, but certain types of development tend to be included in the EIA process in most parts of the world.

These include: major new road networks; airport and port developments; building power stations; dams and reservoirs; proposed measures quarrying; and large-scale housing projects.

Where did EIAs come from?

In 1969, the US Government passed the National Environmental Policy Act (NEPA). The NEPA made it a priority for federal agencies to consider the natural environment in any land use planning. This gave the natural environment the same status as economic priorities. In the USA, environmental assessments (EAs) are carried out to determine if an EIA (called EIS—environmental impact statement) needs to be undertaken and filed with the federal agencies. Within 20 years of NEPA becoming law in the USA, many other countries also included EIAs as part of their planning policy.

Weaknesses of EIAs

- Different countries have different standards for EIAs, which makes it difficult to compare them.

- There is no requirement for the environment to take a high priority—decisions makers just have to justify their decisions.

- It is difficult to determine where the boundaries of the investigation should be. How large an area and how many variables should it consider? How far will the impacts spread from the proposed development site?

- It is very difficult to consider all indirect impacts of a development so some may be missed.

Summary

EIAs are models of the system under study and allow us to predict the effects of the proposed change. A model is only as good as its parameters and asking the right questions is crucial. A change of land use will always have an effect but whether this is a net positive or negative one depends on the criteria used to measure it. Simplistically, if a factory blocks your view of the mountains that may be a loss to you but the factory may bring employment to the area, produce goods that would otherwise be imported and reduce the country's ecological footprint.

The precautionary approach or principle was stated in the 1992 Rio Declaration: where there are threats of serious or irreversible damage, lack of full scientific certainty shall not be used as a reason for postponing cost-effective measures to prevent environmental degradation. Where there is reasonable concern of harm but scientific information is incomplete, then prevention rather than remediation should operate.

Cost-benefit analysis measures impacts of a development or change of land use by assigning monetary values. In theory, this puts all costs into the same units of measure—money—so they can be assessed. Of course, how the assessment is made is critical to the values assigned and there are several ways to do this. For example, valuations may be based on the cost of restoring the environment to its previous state (e.g. after an open cast mine operation), or made by asking people which of several options they would select or be prepared to pay for.

Strategic environmental assessment tries to measure the social and environmental costs of a development but this can be subjective or inaccurate. Does it also depend on the environmental worldview of those planning the assessment?

ATL Activity 9

ATL skills used: thinking, communication, social, research, self-management

1. As a class, identify or imagine a development or change of land use in an area near to your school or home. Discuss the following questions.

    • What criteria would you use to identify the factors you think will change (e.g. number of jobs provided, net profit, land degradation, habitat and biodiversity loss or gain, pollution)?

    • How would you value these factors? (Is there another way of measuring them apart from financially?)

    • How would you weigh up the evidence to decide whether the project should proceed or proceed in a modified state?

2. Now hold a discussion, with different students taking the role of different stakeholders—for example:

    • environmentalist

    • local politician

    • unemployed young adult

    • older retired person living nearby

    • local government official

    • business owner proposing the project

    • other stakeholder.

TOK

Angkor Wat is a temple complex dedicated to Vishnu, the second god of the Hindu triumvirate. When it was built a natural ecosystem was destroyed.

▲ Figure 14 Angkor Wat, Cambodia

To what extent do history and religion play a role in decisions to go ahead with developments that will alter land use?

SDGs and sustainable resources

The SDGs recognize that economic and social development are linked to tackling environmental issues and sustainable resource management. The only resources that can be managed sustainably are renewable ones. Non-renewable resources such as fossil fuels can never be used sustainably. Even though renewable resources can be managed sustainably, there are certain aspects of their use—such as extraction, harvest, transportation and processing—that may not be sustainable.

Examples of resource management

SDG 6 Water resource management: northern Clarendon, Jamaica

There is an issue of water stress in northern Clarendon. Lack of water causes problems with health and hygiene, food production and even school attendance. The area suffers from severe droughts, so people have had to pay large amounts of money to have water brought into the area (also creating carbon dioxide emissions from transport).

The UNEP worked with Clarendon Parish to introduce sustainable solutions to address water scarcity, sanitation issues and pollution issues, and to raise community awareness about environmental protection.

Measures taken included:

- installation of rainwater harvesting systems, including installation of water tanks (to increase storage capacity) and gutters (to catch rainwater and divert it to the tanks)

- installation of handwashing stations

- wastewater reuse and greywater recycling

- installation of water tanks to increase storage capacity

- appropriate training about water safety

- setting up a solar power system to pump water from the tanks (reducing carbon emissions).

▲ Figure 15 Rainfall harvesting in an open area

After these measures, the people of Northern Clarendon are catching their own water through rainfall harvesting and this will help to solve the water stress issues.

SDG 12 Food production and sustainability: The Philippines

Food production can be sustainable and if it is, then soil fertility is maintained or improved, there is less soil erosion, water, land and air pollution, and biodiversity improves. These are all crucial for the health of the environment and the continuation of the services it provides for humanity.

Sustainability can be severely challenged by a single catastrophic event such as Typhoon Nock-ten, which hit the Philippines on Christmas Eve 2016. It killed 11 people and stranded 11,000 without electricity or food. A state of emergency was declared in some areas as 80% of the agricultural land was destroyed, crops were strewn across the landscape and farmers lost everything.

A young reality TV chef, Louise, went into action mode and rallied the villagers. The only option for them to survive was to grow their own food and they started asking for vegetable seeds which would grow quickly. However, this typhoon was not a one-off occurrence: they are common in the Philippines so there was a need for longer-term solutions to build resilience. There is also a strong community respect for the environment and an ethos to pull together.

Louise noticed that the cocoa trees were not affected by the storm. They are a high-income potential crop that is climate-resilient, grows quickly and is ready for harvest within two or three years. With this knowledge she initiated the Cacao Project with the aim of creating a holistic, sustainable food production system that would empower the farmers to build resilience into the system. It is stated that over 200 farmers have been trained in agroforestry techniques and they have planted cacao trees on 70 hectares of land.

▲ Figure 16 Typhoon damage

ATL Activity 10

ATL skills used: thinking, social

Review the UN SDGs (subtopic 1.3). These goals are an urgent call for action by all countries to protect the Earth, end poverty and achieve peace and prosperity for all.

Select two more SDGs that are about sustainable resource management. Discuss briefly how your chosen goals may be achieved.

Still unsustainable?

Food production can be sustainable, but there are still problems with extraction, harvesting, transport and marketing, not least because commercial farming is driven by a desire for economic gain. In total, nearly one-third of all the food grown is wasted; this is quite apart from the burning of fossil fuels to run machinery at all stages.

Agriculture in one form or another provides us with the food we eat. This means there is a wide range of crops and they all have different harvesting, transportation, storage, processing and marketing strategies and techniques.

Harvesting losses of fruit and vegetables

- Crops may be damaged before they are even harvested, due to disease, pests, bad weather and other natural factors.

- There may be a lack of technology to harvest the crop, or poor harvesting technology may damage the crop.

- Inadequate storage may mean that the harvested produce spoils before it reaches the market.

- A glut in the market may mean that prices are so low that it is not worth harvesting the crop, so it is left to rot.

- Ugly fruit and vegetables may be discarded during harvest and left to rot.

- Many farmers overproduce to ensure they can meet demand. If they cannot meet the demand, they lose the contract to supply. Buyers have the upper hand in this situation, and they will change or cancel orders if they find a better deal with another supplier.

ATL Activity 11

ATL skills used: thinking, communication, social, research

1. Research waste of fruit and vegetables locally at harvesting and identify any other causes. Then draw a systems diagram to show stores, flows, inputs and outputs.

2. Attempt to find the percentages lost at each stage—harvesting, transporting, storage and processing.

3. Explain how these losses impact the sustainability of food production.

4. Discuss any possible solutions to these issues.

5. Research harvesting losses for a grain crop of your choice.

▲ Figure 17 Getting vegetables to market in Malawi

Transport and storage losses

Losses during transport and storage of fruit and vegetables can be due to:

- unrefrigerated closed vehicles or storage facilities, which allow deterioration of the produce

- open trucks, which allow produce to fall off during transport

- careless handling of the produce during loading and unloading of vehicles, which causes damage

- poorly designed packaging or poor packing, which causes damage

- overloading of vehicles or storage facilities, leading to collapse of the stacks and damage to produce

- failure to protect crops from the weather during loading and unloading

- poor storage of crops at the farm or once they have been delivered to be processed or sold, allowing insect infestation, fungal moulds and bacterial rot, over-ripening, or damage by rodents or birds.

Processing losses

Fruit and vegetables are processed after harvest and each stage of processing involves losses.

- Crops are cleaned by being soaked or sprayed with water. In some cases the water is stagnant or polluted in some way. Fruit and some vegetables can be damaged at this stage.

- Produce is sorted and graded by size, quality and colour and items that do not meet the "marketable standard" are discarded.

- Excessive peeling and trimming may take place when fruit and vegetables are prepared and processed for canning or freezing.

- Juicing creates heavy losses because the bulk of the fruit or vegetable is discarded.

Consumer level losses

This is an area in which individual actions really can make a difference. Food waste is often made up of safe, high-quality food that is left uneaten and thrown away.

According to the Feeding America website, in the USA:

- just under 54 billion kg of food, worth USD 408 billion, is wasted every year

- that waste could provide 130 billion meals

- 40% of all food is wasted

- 39% of all food waste—19 billion kg—is household food waste

- 61% of all food waste—just under 30 billion kg—is commercial food waste.

There are many organizations like Feeding America that work to recover or rescue food that would otherwise be wasted, in order to distribute it to people facing hunger. They work with farmers, retailers and manufacturers to collect and distribute unwanted and discarded food. Such organizations also work with restaurants and hotels to collection food left over at the end of each day that would otherwise go to landfill or composting.

There are various issues at the root of this problem, especially as regards fruit and vegetables.

- Perfectly good produce is discarded because it does not meet retailers' standards for colour and appearance.

- Large, pre-packaged containers may mean that not all the produce is eaten.

- Portion sizes in restaurants are often too large, so much goes to waste. A solution here is take-out containers.

- Supermarkets often encourage over-buying, e.g. "buy one, get one free" offers.

- People do not understand what a "best before" date really means so they throw away food that passes that date. It actually means what it says—the food it at its best (or optimum quality) before that date.

- People do not know how to store food properly, so they throw away food because it "doesn't look right".

▲ Figure 18 Fruit and vegetable being sold at a supermarket

 Practical

Skills:

- Tool 1: Experimental techniques

- Inquiry 3: Concluding and evaluating

Investigate how much food you waste on a daily basis for a week—store it temporarily, take pictures, and record the weight, type and other relevant factors.

Make a list of ways you can reduce your food waste.

Case study 5

Lithium extraction: the answer to clean energy?

Review Figure 8 earlier in this topic on lithium extraction in Bolivia.

Natural resource security and insecurity have profound, wide ranging impacts on socio-economic development, environmental degradation and geopolitical tensions and conflicts. In fact, geopolitics and natural resources are powerful allies in a quest for power and prosperity. The demand for lithium is likely to have some interesting consequences.

What is lithium?

Look at the properties of lithium shown in Figure 19. Some of them seem to be pretty undesirable for a natural resource. Yet lithium is in very high demand because it is used in rechargeable batteries for phones, laptops, digital cameras and electric vehicles plus non-rechargeable batteries in heart pacemakers.

Many governments are trying to reduce their carbon footprint by increasing the use of electric vehicles, which are powered by lithium-ion batteries.

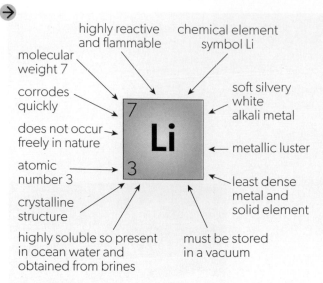

molecular weight 7

highly reactive and flammable

chemical element symbol Li

corrodes quickly

soft silvery white alkali metal

does not occur freely in nature

metallic luster

atomic number 3

least dense metal and solid element

crystalline structure

highly soluble so present in ocean water and obtained from brines

must be stored in a vacuum

▲ Figure 19 Properties of lithium

The majority of the world's lithium, also known as white gold, comes from the Lithium Triangle (Figure 20). According to the US Geological Survey, this is an area of about 400,000 km² located in Argentina, Bolivia, and Chile. Investors are keen to collaborate with these countries to exploit the lithium. However, the financial gains are high so it can be difficult to gain rights to extract.

Economist Patricia Vásquez, in her book *The Lithium Triangle: The Case for Post-Pandemic Optimism,* outlines the opportunities this natural resource presents to Argentina, Bolivia and Chile. But it is not straightforward: the countries must decide whether to become leading producers of lithium (adding value by processing it themselves), or to simply export the raw material and allow other countries to make more money.

In Argentina, private mining companies now control lithium production. In 2022, Rio Tinto paid USD 825 million for the Rincon lithium project while Livent extended extraction in the Fenix lithium mine. This is not likely to favour the Argentine government's plans to expand lithium mining and enter battery production.

Local governments have gone for short-term profit, not long-term improvement of the country's economy.

Most of the lithium reserves in Chile are in Salar Flat of the Atacama and production levels are matching those of Australia. In a government referendum, the people of Chile were opposed to nationalization of the mining industry. Internal politics aside, Chile will play some part in the global lithium battery market.

Bolivia is unlikely to feature highly in this resource exploitation because there is a lack of internal collaboration.

▲ Figure 20 The Lithium Triangle
Adapted from: The Economist Newspaper Limited

So what's the problem?

To reduce carbon emissions, we are heavily dependent on lithium batteries to store wind and solar power and to power electric cars. When we had a similar reliance on fossil fuel combustion, we were ignorant of the long-term negative impacts of such reliance. Humanity must not make the same mistake with lithium extraction as we are well aware of the issues. We are therefore responsible to future generations and the planet to proceed with care.

The negative impacts of lithium extraction are as follows.

- It takes 500,000 litres of water to extract one tonne of lithium.

- Extraction causes changes to the hydrological cycle.

- Water released back into the environment is often polluted, poisoning reservoirs and causing health problems such as respiratory issues.

- The Lithium Triangle is located partially in the Atacama Desert so extraction depletes freshwater supplies in an area already lacking water.

- There are increased carbon emissions during battery production. It has been estimated that the carbon emissions from producing a car battery are about 70% more than those from producing a whole car in Germany.

- As with all mineral extraction, there are a lot of waste products.

ATL Activity 12

ATL skills used: thinking, communication

1. Define environmental sustainability.

2. List advantages and disadvantages of lithium use in batteries.

3. Compare and contrast the use of lithium and the use of fossil fuels in energy production for transport.

More on globalization and resources

Review the short section on globalization earlier in this subtopic. The term globalization is used frequently, covers a broad range of ideas and has many definitions. The globalization processes shown in Figure 21 are an interconnecting web of threads that spread around the world, linking individuals and societies.

TOK

There is a paradox here between the "clean" energy revolution and the "dirty" lithium mines. Yes, electric cars reduce carbon emissions, but what are the costs of lithium extraction?

Explain how developments in scientific knowledge like these may trigger political and humanitarian controversies.

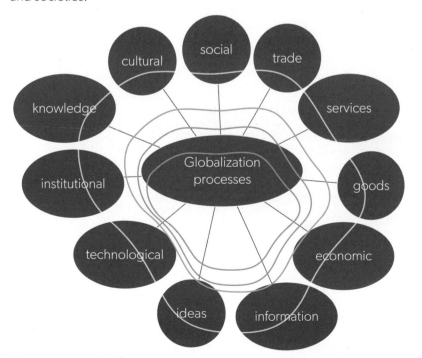

▲ Figure 21 The global web

Globalization is by no means a modern phenomenon. The "Silk Road" is actually many ancient trade routes used between 50 BCE and 250 CE. These trade routes linked China to the Mediterranean and all the places in between. As technology improved, coins were created, road building improved and the links spread further and further. People have always travelled and in doing so exchange the ideas, knowledge and goods shown in Figure 21. However, they also exchange diseases and devastation (e.g. COVID-19), fake news and a globalized culture—for better or worse. As technology improves and processes speed up, the spread is farther and faster.

Globalization and food systems and cultures

Globalization is having a significant impact on many aspects of the food industry. On a global scale, food systems and food culture are changing; the changes may be viewed as good or bad depending on your perspective. Coffee is now a global drink, available in most places in a wide range of styles. On the other hand, it is highly unlikely that most of us would welcome the idea of eating spiders and scorpions.

Food cultures are being uprooted and moved to new, sometimes distant locations. In the new location they may maintain a traditional flavour, or they may merge with the indigenous food culture to create a new hybrid cuisine. For example, the Japanese restaurant Wawa has something called a Katsu Ramen burger—a burger infused with katsu curry spices and held in a ramen noodle bun. Similarly, chicken tikka masala may have origins in northern India but a Bengali chef working in Scotland also laid claim to inventing it.

Globalization is also having an impact in food systems. For many people, there is greater availability and diversity of food but for others, food access is not good. One of the results of globalization is the entry into the marketplace of multinational fast food and supermarket chains.

As these multinationals boom, small local suppliers "bust". Traditional food markets, "street food" vendors and local grocery stores cannot compete. Supermarkets do bring improvements, such as:

- higher standards of food quality and better safety standards

- competitive prices

- convenience—get all you need in one place

- more variety.

However, they also bring:

- ultra-processed foods (UPF), often high in empty calories, salt, sugar and various additives

- a culture of easy food preparation resulting in a generation that has not learned to cook from raw ingredients

- often higher cost for the consumer

- the potential for cravings and addiction to unhealthy foods

- increased obesity, particularly in childhood.

Connections: This links to material covered in subtopic 5.2.

▲ Figure 22 Ultra-processed foods from fast-food chains

These changes in food systems impact availability and access throughout the supply chain (production to consumer). This has bought about a shift in food culture on a global scale, with a shift in consumption patterns and nutritional standards that is not equitable across the socio-economic strata. The lower socio-economic groups tend to have poorer quality diets that involve energy-dense, cheap foods—e.g. fast foods.

Globalization and food security

According to the 1996 World Food Summit, food security is "when all people, at all times, have physical and economic access to sufficient safe and nutritious food that meets their dietary needs and food preferences for an active and healthy life".

Food security is an issue in many LICs and some of this can be attributed to globalization. As protectionism and international tariffs have reduced, food products can move rapidly around the world. LICs are often forced to become export centres for a small number of high demand foods and this causes home food insecurity. For a farmer, growing a cash crop (sold to the market) instead of self-sufficiency crops means there may be more income but often this is spent on buying food. The subsistence-based economy is transformed into an export-based one. The local markets are flooded with cheap, subsidized food from HICs and local farmers wishing to make more money change to growing crops for export. This means staple foods have to be imported, usually at a higher price.

> **Connections:** This links to material covered in subtopic 5.2.

> ### TOK
>
> To what extent do wealthy inhabitants of HICs have a moral responsibility towards those facing food insecurity?

ATL Activity 13

ATL skills used: social, research

The food you eat may have been processed locally but if you are eating mangoes, avocados, oranges or bananas in the middle of winter in a cold climate, they are not locally produced.

1. Check the labels on some of the food products in your home. Where did the foods come from originally?

2. Does the country of origin have food security issues?

3. Choose an exotic fruit and trace the origin.

4. Discuss the issues surrounding such purchases.

Globalization and water security

Globalization is not the only factor that causes water insecurity; climate change also plays a large part. However, it could be said that globalization is also a major cause of climate change. A large proportion of the world population is affected by climate change, globalization and water insecurity.

Water access is now impacted by market forces and cross-basin trading of water. Cross-basin trading is the movement of water from one river basin to another using human-made means, e.g. pipes, canals, aqueducts. Globalization has increased rates of urbanization and industrialization—both of which put heavy demands on the water supply.

> **Connections:** This links to material covered in subtopic 4.2.

Water supply has been marketized (exposed to market forces) so it is seen as a commodity that is subject to economic and policy decisions, as are other goods. This has meant that water resources are managed in a way that changes access for some people. In the past, people generally had open access to water but as it is now seen as an economic resource it follows market rules and is no longer supplied for a nominal price by the government. Consumers are charged for using water, either at a flat rate or at variable rates using monitored water meters. When drought hits and water supply falls or even fails, then water prices can rise. A water supplier must maintain a surplus to ensure supply and they need to make a profit to cover infrastructure improvements, mending leaks in pipes and delivery costs. In some cases, the water companies will cut supplies to ensure payment. Impacts are felt by people on lower incomes in particular.

 Practical

Skills:

- Tool 1: Experimental techniques

- Tool 2: Technology

- Tool 3: Mathematics

- Inquiry 1: Exploring and designing

- Inquiry 2: Collecting and processing data

- Inquiry 3: Concluding and evaluating

Design an investigation to study renewable freshwater withdrawals per capita.

The data to investigate this is available on the Our World in Data website. Search for water use stress and find the data for water withdrawals per capita. (You can download the raw data in table format, and look at changes over time or differences between countries.)

Identify a suitable research question.

Construct a number of graphs to find the best one to answer your research question.

Remember: one data source for secondary data is not adequate for reliable results. Consider data from other reliable sources, e.g. World Bank, UN, Gapminder.

Globalization and energy security

Connections: This links to material covered in subtopics 7.2 and HL.b.

Globalization of energy has been advantageous because the advances in technology have meant energy can flow between nations so creating an international energy market.

- Oil and LPG (liquified petroleum gas) tankers carry these fuels from supplier to consumer country.

- Pipelines carry natural gas from producer to consumer country.

- Wood pellets used to fuel some power stations are transported in shipping from where the wood is grown to where it is burned.

- Electricity—whether generated by coal, wood, gas, oil, solar, wind or nuclear—is bought and sold internationally and transported via interconnectors (very high voltage cables, sometimes laid on the sea bed) from countries with surplus generation to those which need it. For example, the UK has five interconnectors with France, the Netherlands, Belgium and Norway. One to Denmark was under construction in 2023. Electrical energy can flow in either direction. Sharing surplus electricity reduces carbon emissions.

Energy production and consumption can become more efficient as technology is shared, leading to reduced exploitation costs. It is also possible to conserve energy with less wastage from, for example, wind turbines or solar panels.

The development of renewable energy has been boosted and will continue to be boosted through globalization. Renewable energy technologies offer a trade opportunity in which all participants may benefit from the progress made by one trading partner. Through globalization, renewable energy is attracting foreign investment into LICs, e.g. those with high solar radiation inputs for solar farms or high wind speeds for wind farms.

 Practical

Skills:

- Tool 1: Experimental techniques
- Tool 2: Technology
- Inquiry 1: Exploring and designing
- Inquiry 2: Collecting and processing data
- Inquiry 3: Concluding and evaluating

1. Design an investigation into the ways two contrasting countries use a named resource and how that has changed over time. You might investigate use of steel, concrete, inorganic fertilizer or fossil fuels, or a resource of your choice. The countries might be HIC and LIC, larger and smaller, with much natural resource and with little natural resource, highly industrialized and less industrialized, or a pair of your choice.

2. Use secondary data sources to select relevant data.

3. Present your data using appropriate tables and graphs, including appropriate axis labels, titles and error bars.

4. Analyse your data using appropriate statistical tests.

Remember: You should use at least two, preferably three, data sources for each country. Reliable secondary data sources include Gapminder, Our World in Data and World Bank. Always acknowledge your sources properly.

Check your understanding

In this subtopic, you have covered:

- what natural resources are and how humans use them
- the difference between natural capital and natural income
- the differences between goods and services
- renewable and non-renewable natural resources
- sustainable use of resources
- the value of natural capital can change over time
- how we value natural capital and how that value changes with time
- how a particular perspective on nature influences how we view it
- resource security depends on sustainable use

- management strategies on resources

- SDGs as a framework for actions

- EIA use and content

- a renewable resource may be processed unsustainably

- geopolitical tensions are the result of competition for resources

- national resource security is a priority for governments.

How does the renewability of natural capital have implications for its sustainable use?

How might societies reconcile competing perspectives on natural resource use?

To what extent can human societies use natural resources sustainably?

1. Outline natural resources, natural capital and natural income.

2. Distinguish goods and services.

3. Compare and contrast renewable and non-renewable resources.

4. Describe the various ways in which natural capital has value.

5. Explain the dynamic nature of natural capital.

6. Demonstrate how the use of natural capital can be managed to ensure sustainability.

7. Explain how the value of natural resources varies depending on the perspective of the society.

8. Discuss the range of management and intervention strategies that can be used to influence use of natural capital.

9. With the use of relevant examples, examine how the UN SDGs can be used to address environmental issues and sustainable resource management.

10. Describe how EIAs impact the sustainability of future development projects.

11. Evaluate the inclusion of all stakeholders in the role of an EIA.

12. Explain how a renewable resource can be used unsustainably due to the extraction, harvesting, transportation and processing of it.

13. Outline how short-term economic interests undermine long-term sustainability.

14. Explain how resource insecurity is linked to socio-economic development, environmental degradation, geopolitical tensions and conflicts.

15. Discuss how resource security can be achieved.

16. Explain the links between economic globalization and interdependence of countries and national resource supply and security.

≫ Taking it further

- Produce infographics about how a particular resource's value has changed over time.

- Investigate and promote the ecological services provided by local or regional ecosystems, such as water provision and soil stability.

- Investigate the sustainability of local food production.

- Compare the carbon and/or water footprint of local food and imported food.

- Organize a fundraising fair that promotes items that are sustainable.

- Produce a short video to show how the use and value of resources has changed over time.

- Investigate and evaluate an example of an EIA (e.g. the story of Nauru and resource depletion).

- Monitor local news regarding the long-term availability of natural resources.

- Gather knowledge of the sustainable use of natural resources for a local citizen science group.

- Engage with an SDG and design a project which raises awareness of possible solutions related to this SDG.

- Participate in a local food production project that increases food security.

- Promote renewable energy production for the school to increase energy security.

AHL

7.2 Energy sources—uses and management

Guiding questions

- To what extent can energy consumption be equitable across the world?

- How can energy production be sustainable?

Understandings

1. Energy resources are both renewable and non-renewable.
2. Global energy consumption is rising with increasing population and with per capita demand.
3. The sustainability of energy sources varies significantly.
4. A variety of factors will affect the energy choices that a country makes.
5. Intermittent energy production from some renewable sources creates the need for energy storage systems.
6. Energy conservation and energy efficiency may allow a country to be less dependent on importing a resource.

7. Energy security for a country means access to affordable and reliable sources of energy.
8. The global economy mostly depends on finite reserves of fossil fuels as energy sources; these include coal, oil and natural gas.
9. Nuclear power is a non-renewable, low-carbon means of electricity production.
10. Battery storage is required on a large scale to meet global requirements for reduction of carbon emissions, but it requires mining, transporting, processing and construction, all of which produce emissions and pollution, and cause sociopolitical tensions.

Energy and energy resources

All energy on Earth comes from our Sun. Without it, the planet would be at absolute zero, which is −273°C, and there would be no life forms. The Sun's energy drives the climate, geochemical cycles, photosynthesis, animal life—everything. Humans rely on energy from the Sun via plants and also supplement it by using other sources of energy to power our civilization.

Fossil fuels are simply stored solar energy. These non-renewable energy sources are the compressed, decomposed remains of organic life from millions of years ago. Now humans extract and burn them to release that energy. The combustion of fossil fuels releases carbon dioxide that was locked up by photosynthesis when the organisms were living. This emission of carbon dioxide is having a large effect on atmospheric carbon dioxide levels.

Most of the energy we need to power the world economy and supply our own requirements comes from fossil fuels. Theoretically, we could get that energy from renewable resources. At the moment, on a worldwide scale we obtain a small percentage from renewable resources, but it is rising. Norway, Iceland and New Zealand have high levels of renewable energy and much is from hydropower. China, the USA and Brazil are leading countries in installing renewable energy capacity.

Connections: This links to material covered in subtopics 1.1, 1.2 and 1.3, and in topic 6.

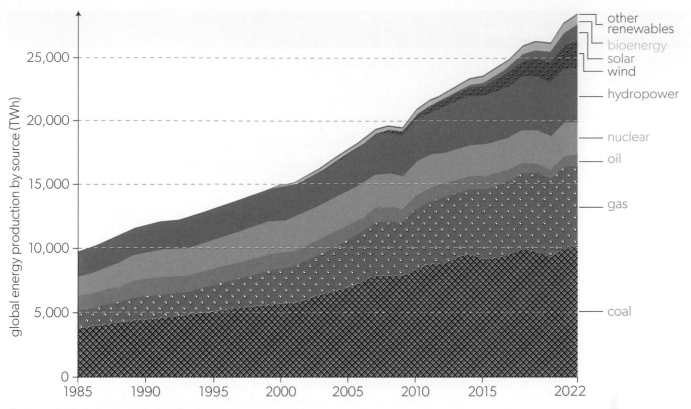

▲ Figure 1 Global energy production by source
Adapted from: Our World In Data (CC BY 4.0)

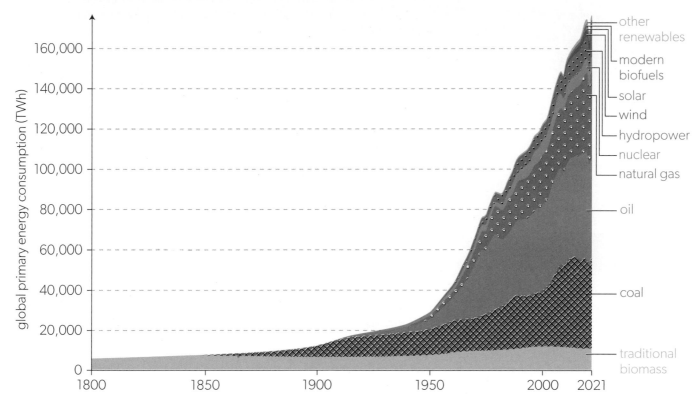

▲ Figure 2 Global energy consumption by source. Primary energy is calculated based on the substitution method which takes account of the inefficiencies in fossil fuel production by converting non-fossil energy into the energy inputs required if they had the same conversion losses as fossil fuels
Data from: Energy Institute Statistical Review of World Energy (2023); Vaclav Smil (2017)
1800 / Our World In Data (CC BY 4.0)

Data-based questions

Compare the graphs in Figures 1 and 2. TWh is a terawatt hour, which is 10^{15} watts per hour.

Study Figure 1.

1. Describe briefly what it shows.

2. State which three fuels provided the most energy for electricity production in 2022.

3. Explain briefly why sources of electricity production changed between 1985 and 2022.

Study Figure 2.

4. Describe briefly what it shows.

5. State which three fuels were the largest sources of global primary energy consumption in 2021.

6. Explain reasons for the changes in energy source between 1900 and 2021.

Considering data in both graphs:

7. Explain why global energy consumption is so much greater than energy use to produce electricity.

We know that we must increase the percentage of our energy produced from renewable resources and progress is being made. In 2021, the EU and the USA got about 22% of their energy needs from renewables. Investment in research on making renewable energy sources more efficient, for example in wind, wave and tidal power or solar cells, is increasing. But seeking new fossil fuel resources continues for the following reasons.

- TNCs (transnational corporations) and heavy industry are committed to the carbon economy—machines are made to run on fossil fuels—and the scale of change required is difficult to imagine (although it is happening).

- The cost of energy from fossil fuels used to be cheapest but now renewables are the cheapest form of energy (according to a UN report in 2021).

- Countries are locked into the resource that they currently use—by trade agreements or convenience.

All renewable sources are location-dependent:

- Hydropower and pumped storage hydroelectric power (HEP) need suitable terrain.

- Wave or tidal power are not possible for land-locked countries.

- Solar energy requires a sunny climate for maximum efficiency.

- Wind power has a range of wind speeds within which it can operate effectively.

Table 1 shows advantages and disadvantages of various energy sources.

TOK

The world economy continues to run on fossil fuels despite the knowledge that it is detrimental to do so.

What role do reason and emotion play in these decisions?

Energy source and facts	Advantages	Disadvantages
Non-renewable		
Coal • Fossilized plants laid down in the Carboniferous period • Mined from seams of coal which are in strata between other types of rock; mines may be opencast (large pits) or tunnels underground • Burned to provide heat directly, or to create steam that drives turbines in power stations	• Plentiful supply • Easy to transport as a solid • Needs no processing • Relatively cheap to mine and convert to energy by burning • Up to 250 years of coal left	• Non-renewable energy source—cannot be replaced once used (same for oil and gas) • Burning releases carbon dioxide • Some coals contain up to 10% sulfur—burning sulfur forms sulfur dioxide which causes acid deposition • Soot from burning coal produces smog, leading to lung disease • Coal mines leave degraded land and pollution • Lower heat of combustion than other fossil fuels, i.e. less energy released per unit mass
Oil (fossil fuel) • Fossilized plants and microorganisms that are compressed to a liquid and found in porous rocks • Crude oil is refined by fractional distillation to give a variety of products from lighter jet fuels and petrol to heavier diesel and bitumen • Many oil fields are under oceans so extraction is dangerous • Extracted by oil wells: pipes are drilled down to the oil-bearing rocks to pump the oil out • Most of the world economy runs on oil either burned directly in transport and industry or to generate electricity	• High heat of combustion • Many uses • Once found, relatively cheap to mine and to convert into energy	• Only a limited supply—may run out in 20–50 years' time • Gives off carbon dioxide when burned • Oil spill danger from tanker accidents • Risk of damage to oil pipelines, either accidental or through terrorism • Extraction of oil from tar sands and oil shales is expensive and particularly polluting
Natural gas (fossil fuel) • Methane gas and other hydrocarbons trapped between seams of rock • Extracted by drilling like crude oil • Often found with crude oil • Used directly in homes for domestic heating and cooking	• Highest heat of combustion • Lots of energy gained from it • Ready-made fuel • Relatively cheap form of energy • Cleaner fuel than coal and oil	• Only limited supply of gas but more than oil • About 70 years' worth at current usage rates • Gives off less carbon dioxide than coal and oil

▲ Table 1 Advantages and disadvantages of energy sources

Energy source and facts	Advantages	Disadvantages
Nuclear fission • Uranium is the raw material; this is radioactive and is split in nuclear reactors by bombarding it with neutrons. As it splits into other elements, massive amounts of energy are released • Uranium is mined. Australia has the most known reserves; Canada exports the most; other countries have smaller amounts • About 80 years' worth left to mine at current rates but could be extracted from seawater	• Raw materials are relatively cheap once the reactor is built, and can last quite a long time • Small mass of radioactive material produces a huge amount of energy • No carbon dioxide or other pollutants released (unless there are accidents)	• Extraction costs are high • Nuclear reactors are expensive to build and run • Nuclear waste is radioactive and highly toxic for a long time—needs storage for thousands of years, in mine shafts or under the sea • Accidental leakage of radiation can be devastating • Accidents are rare (worst nuclear reactor accident at Chernobyl, Ukraine in 1986) • Risk of uranium and plutonium being used to make nuclear weapons
Renewable		
Hydroelectric power (HEP) • Energy harnessed from movement of water through rivers, lakes and dams to power turbines to generate electricity 	• High-quality energy output compared with low-quality energy input • Creates water reserves as well as energy supplies • Reservoirs used for recreation • Safety record good	• Costly to build • Can cause flooding of surrounding communities and landscapes • Dams have major ecological impacts on local hydrology • May cause problems with deltas—no sediment means they are lost • Silting of dams • Downstream lack of water (e.g. Nile) and risk of flooding if dam bursts
Pumped-storage HEP • Pumped-storage reservoirs also power turbines • Two water reservoirs at different elevations that can generate power as water moves down from one to the other (discharge), passing through a turbine	• Stores and releases large amounts of energy • Can balance energy use—load-balancing as low-cost off-peak electricity pumps water uphill to the top reservoir • Also stores energy from intermittent sources • Energy efficiency about 70–80%	• High capital cost • Few sites for two large reservoirs and dams • Needs hilly terrain • Mostly small scale

▲ Table 1 (*cont.*)

Energy source and facts	Advantages	Disadvantages
Biomass • Decaying organic plant or animal waste is used to produce methane in biogas generators or burned directly as dung or plant material • More processing can give oils (e.g. oilseed rape, oil palms, sugar cane) which can be used as fuel in vehicles instead of diesel fuel = biofuels	• Cheap and readily available energy source • If the crops are replanted, biomass can be a long-term, sustainable energy source	• May replace food crops on a finite amount of crop land, leading to starvation • When burned, it still gives off atmospheric pollutants, including GHGs • If crops are not replanted, biomass is a non-renewable resource
Wood • From felling or coppicing trees • Burned to generate heat and light	• Cheap and readily available source of energy • If trees are replaced, wood can be a long-term, sustainable energy source	• Low heat of combustion—not much energy released for its mass • When burned, it gives off atmospheric pollutants, including GHGs • If trees are not replanted wood is a non-renewable resource • High cost of transportation as high volume
Solar—photovoltaic cells • Conversion of solar radiation into electricity via chemical energy	• Potentially infinite energy supply • Single dwellings can have own electricity supply • Safe to use • Low-quality energy converted to high	• Manufacture and installation of solar panels can be costly • Intermittent supply—need sunshine, do not work in the dark • Need maintenance—must be cleaned regularly
Concentrated solar power (CSP) • Mirrors arranged to focus solar energy on one point where heat energy generated drives a steam turbine to make electricity	• Solar energy is a renewable source • Cost of power stations equivalent to fossil fuel power stations	• Requires area of high insolation—so usually in tropics • Relatively new technology but improving all the time
Solar—passive • Using buildings or panels to capture and store heat	• Minimal cost if properly designed	• Requires architects who can design for solar passive technology

▲ Table 1 (*cont.*)

Energy source and facts	Advantages	Disadvantages
Wind • Wind turbines (modern windmills) turn wind energy into electricity • Can be found singularly, but usually many together in wind farms (onshore or offshore) 	• Clean energy supply once turbines made • Little maintenance required	• Need the wind to blow • Often windy sites not near highly populated areas • Manufacture and installation of wind farms can be costly • Noise pollution—though this is decreasing with new technologies • Some local people object to onshore wind farms, arguing that visual pollution spoils countryside • Question of whether birds are killed or migration routes disturbed by turbines
Tidal turbines • Movement of sea water in and out drives turbines • A tidal barrage (a kind of dam) is built across estuaries, forcing water through gaps • In future, underwater turbines may be possible out at sea and without dam	• Should be ideal for an island country such as the UK • Potential to generate a lot of energy this way • Tidal barrage can double as a bridge, and help to prevent flooding	• Construction of barrage is very costly • Only a few estuaries are suitable • Opposed by some environmental groups as having a negative impact on wildlife • May reduce tidal flow and impede flow of sewage out to sea • May disrupt shipping
Wave • Movement of sea water into and out of a cavity on the shore compresses trapped air, driving a turbine	• Should be ideal for an island country • More likely to be small local operations, rather than done on a national scale	• Construction can be costly • May be opposed by local or environmental groups • Storms may damage them
Geothermal • Uses heat under the Earth in volcanic regions • Cold water is pumped into the ground and comes out as steam • Steam can be used for heating or to power turbines creating electricity	• Potentially infinite energy supply • Used successfully in some countries, such as New Zealand	• Can be expensive to set up • Only works in areas of volcanic activity • Geothermal activity might decrease, leaving power station redundant • Dangerous underground gases have to be disposed of carefully

▲ Table 1 (*cont.*)

ATL Activity 14

ATL skills used: thinking, research

Use Table 1 to answer these questions.

1. Which energy sources do not release carbon dioxide when used to produce energy?

2. Which energy resources do not release carbon dioxide:

   a. during extraction or transport?

   b. during construction (e.g. of wind turbines or power stations)?

   c. during end of life disposal?

3. Are any energy sources truly "clean"?

4. Find out what percentage of energy is from renewable sources (a) globally and (b) in your home country.

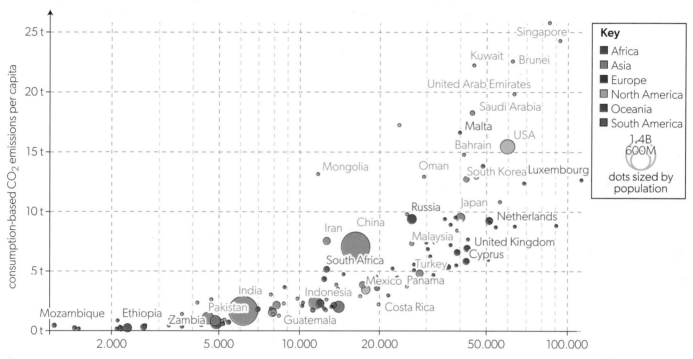

▲ Figure 3 Emissions of carbon dioxide from energy consumption per capita vs gross domestic product (GDP) per capita in 2020
Data from: Our World In Data (CC BY 4.0)

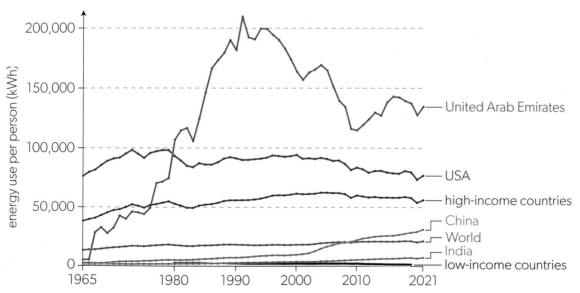

◄ Figure 4 Energy use per person for the world and for selected countries. Energy use not only includes electricity, but also other areas of consumption such as transport, heating and cooking
Data from: Our World In Data (CC BY 4.0)

ATL **Activity 15**

ATL skills used: thinking, social, research

In groups of two or three, consider the data in Figures 3 and 4. On average, an individual in Singapore produced nearly 25 tonnes of carbon dioxide in 2020 from using energy. An individual in most African countries produced very little carbon dioxide from energy consumption.

1. What is the relationship between GDP per capita of a country and carbon dioxide emissions per capita ?

2. Which three countries have the largest populations?

3. Research which of these countries produces most total carbon dioxide emissions and explain why.

4. List reasons for high energy use in the United Arab Emirates and give reasons why this has fallen since 1990.

5. List reasons for high energy use in the USA.

6. Explain why China's energy use per capita has increased more than that of India.

7. Go to the Our World in Data website, find these graphs and add the country in which you are living. How does it compare?

8. Research how much of the energy your country uses is from renewable sources and how this has changed over the last 20 years.

Case study 6

Energy use in construction

The five construction materials that dominate energy use in material production are steel, cement, paper, plastics and aluminium.

Concrete

After water, humans use concrete more than any other material. It is incredibly useful and versatile and is made from cement plus small stones and water. The Romans made concrete and their constructions (e.g. the Colosseum in Rome, Hadrian's Wall in Britain) still exist. Portland cement, the most common type of cement, is made from limestone. It is a fine powder produced by heating limestone and clay minerals in a kiln to form clinker nodules which are ground up and a little gypsum (calcium sulfate) is added.

Cement is easy to transport as it is a dry powder. This can be mixed with water and small stones on site to make concrete, which then sets hard quickly in an exothermic (giving out heat) reaction. We add steel frameworks to concrete to make it rigid for homes and offices, skyscrapers and roads, dams, bridges and hospitals. It degrades very slowly, is waterproof and a brilliant building material. Look around and it is everywhere. It is said that the weight of concrete is more than that of all plants on Earth.

But there are issues with concrete.

*   Concrete production is the most energy-intensive of all manufacturing industries. The production of 1 m³ of concrete requires 2,775 MJ of energy.

*   In all stages of production, concrete is said to be responsible for 4–8% of the world's carbon dioxide emissions and for 10% of industrial water use. Some 75% of this consumption of water is in water-stressed regions.

*   Dust from concrete construction sites contains $PM_{2.5}$ and this leads to silicosis and other lung diseases.

*   Sand used in concrete manufacture has to come from somewhere, e.g. beaches and river beds.

*   Limestone for cement is quarried leaving landscapes with large holes and having lost biodiversity.

*   Concreted areas do not absorb rainfall and run-off causes flooding.

In 2022 these countries produced the most concrete (in million metric tonnes):

China	2,100	Vietnam	120
India	370	USA	95

During the post-Second World War construction boom, Japan embraced concrete and now, according to one estimate, there is 30 times as much concrete per m² in Japan than in the USA.

In the USA after 1930, a lot of new infrastructure was built using concrete. For example, the Hoover Dam contains 3.3 million m³ of concrete.

China produces the most concrete now and uses half the world's supply.

▲ Figure 5 Construction site using concrete

▲ Figure 6 Rescue workers in Turkey after the 2023 earthquake

According to Transparency International, a watchdog group, the construction industry worldwide can be corrupt and there is a lot of bribery. The earthquakes in Syria and Turkey in 2023 resulted in many housing blocks collapsing, killing the occupants. It appears that poor building standards and lack of oversight meant buildings were constructed in violation of regulations.

What can be done?

- Use alternative building materials where possible, but be aware of their problems. For example, steel, asphalt and plasterboard use more energy than concrete in their construction. Wood is renewable but we are harvesting at unsustainable rates. Clay is used for building (see subtopic 8.2) and more could be used in 3D printed houses.

- Maintain or reuse existing concrete structures to reduce the need for new concrete. When concrete buildings are taken down, most concrete goes to landfill or is crushed to make aggregate. But concrete slabs could be reused in new builds.

- Use alternative mixes for making cement and heat to a lower temperature to reduce the carbon footprint of production.

- Use alternative materials that are by-products of other industries instead of clinker; this can mean stronger and lighter concrete.

- Research ways to make lower carbon concrete, such as:

  - carbonation—adding carbon dioxide to concrete as it is made or cured helps to reduce carbon emissions

  - embedding dye-sensitive solar cells in concrete allows for energy generation in the day and any excess can be stored in batteries

  - storing potential energy in concrete blocks— energy is used to lift the blocks which are allowed to fall and rotate a motor when there is demand for electricity

  - waterless concrete production using sulfur, not water.

TOK

To what degree do political leaders and officials have ethical obligations and responsibilities to act on the knowledge that lives were lost through disregard for regulations?

→ Steel

▲ Figure 7 Steel structure of a new commercial building

Here are some key facts, taken from the GreenSpec website.

- Steel is "iron with most of the carbon removed".

- Iron constitutes about 5% of the Earth's crust and is the fourth most abundant element in the crust.

- Iron ore is mined in opencast mines; most of these are in China (23%), Australia (18%) and Brazil (18%). 98% of the iron ore mined is used to make steel.

- Converting iron ore to iron needs limestone, coke (from coal) and a lot of heat in a blast furnace.

- Steel represents around 95% of all metal produced, and 51% of global steel is used for construction.

- China is the biggest producer of steel by far, followed by the EU, Japan, USA and India.

- Steel use per capita increases annually.

- 6.5% of global carbon dioxide emissions derive from iron and steel production (IEA 2010).

- Iron and steel are the most recycled materials—globally, about 85% of steel in construction is recycled and 42% of crude steel is recycled material.

Steel, concrete or something else?

Steel is stronger than concrete but reinforced concrete—with steel rods within the structure—is stronger still.

But steel can rust, it conducts electricity and heat and it takes longer to make than concrete.

Alternative building materials are straw bales, clay, bags of earth, plastic, wood, bricks, bamboo, cork and stone.

Think about the advantages and disadvantages of some of these alternative building materials. Remember, everything humans create has a cost for the environment.

Sustainability of energy resources

Fossil fuel use is not sustainable. Fossil fuels are a finite, non-renewable resource. Even today, oil, coal and gas provide 80% of global energy needs. There are two main methods for removing fossil fuels from the ground: mining and drilling.

Oil

Oil is in underground reservoirs; in the cracks and pores of sedimentary rock or in tar sands near the Earth's surface. It is recovered by drilling (sedimentary rock sources) and strip mining (tar sands). Once extracted, oil is transported to refineries via supertanker, train, truck or pipeline. Fractional distillation separates it into usable fuels such as gasoline, propane, kerosene, and raw materials for products such as plastics and paint.

Coal

Coal is a very dirty fuel and produces significant carbon emissions.

It is extracted via two methods: underground mining uses heavy machinery to cut coal from deep underground deposits, whereas surface mining (also known as strip mining) removes entire layers of soil and rock to access coal deposits below. Opencast mines cover huge areas and are destructive.

Natural gas

Natural gas is located in porous and permeable rock beds or mixed into oil reservoirs and can be recovered via standard drilling. To transport it easily, it is liquified (LNG). The liquefaction process removes water, oxygen, carbon dioxide and sulfur compounds contained in the natural gas. As a liquid, natural gas is reduced to 1/600th of its original volume. LNG produces 40% less carbon dioxide than coal and 30% less than oil, which makes it the cleanest of the fossil fuels.

Hydraulic fracturing, also known as fracking, is a technique used to get oil and gas out of the ground. Water, sand and a cocktail of chemicals are pumped deep underground at high pressure to open up cracks (or fractures) in the rock, releasing the oil or gas trapped inside. Fracking is controversial as it uses large amounts of water, can cause earth tremors and leaks methane into the atmosphere.

Uranium

Nuclear fuels are mined and are non-renewable. Uranium is a naturally occurring element with an average concentration of 2.8 parts per million in the Earth's crust. Traces of it occur almost everywhere. It is more abundant than gold, silver or mercury, about the same as tin and slightly less abundant than cobalt, lead or molybdenum. Vast amounts of uranium also occur in the world's oceans, but at very low concentrations.

▲ Figure 8 A uranium mine

The current global demand for uranium is about 67,000 tU/yr (tonnes uranium per year). The vast majority is consumed by the power sector with a small amount used for medical and research purposes, and some for naval propulsion.

Current uranium reserves are expected to be depleted by 2100, and new sources are hard to find. As a result, uranium prices have been steadily rising, with some estimates predicting a doubling of prices by 2030. Canada, Kazakhstan, Namibia and Australia all have mines with uranium concentrations greater than 1,000 parts per million. Australia has the world's largest uranium reserves but ranks second in global uranium production, behind Kazakhstan.

Pollutants from the mining of uranium can contaminate aquatic ecosystems for hundreds of years, threatening downstream human communities and fish and wildlife. Even small amounts of some pollutants can poison fish, accumulate in the food chain, and cause deformities and reproductive problems for aquatic species.

Uranium itself is only slightly radioactive. However, radon—a radioactive inert gas—is released to the atmosphere in very small quantities when the ore is mined and crushed. Radon decays into other toxic elements known collectively as "radon daughters". These give off radiation in the lungs when they are breathed in, causing lung cancer. The longer miners work or the greater their exposure to radon daughters, the greater their risk of lung cancer.

Rare earth elements

Rare earth elements and other metals are mined and extracted from their ores. They are also a finite non-renewable resource but theoretically could be recovered and recycled in a circular economy.

Renewable energy

The only renewable energy resource is our Sun, which drives heating of the surface and creates winds and currents. Even that will cool and become a red giant eventually, but not for about five billion years from now.

Environmental cost

All energy production systems that humans use have environmental costs. Once we find a resource, we move it, process it, and then convert it using a power station, wind turbine or solar panel. These processes require other materials (concrete, steel, glass, alloys, wood), producing waste products and taking up land. When an energy production unit reaches the end of its useful life, it is dismantled and moved. Sometimes parts are recycled if it is economic to do so. Often they are not. We must get better at recycling these materials.

Energy security

Worldwide demand for energy is high and rising by about 2% per year. The need for energy to power the global economy is rising and we keep burning non-renewable fossil fuels. But fossil fuels will run out soon.

Energy price fluctuations and political changes can be risky for governments so they need to spread the risk. This can be achieved by using several energy sources from several places including their own (if they have one).

Energy sources are not equally spread over the world. Fossil fuel deposits are clumped in some areas—there are large deposits of coal in China, oil in the Middle East, gas in Russia and Qatar—so some countries have to buy fuels from others. Depending on other countries for a nation's energy can work well if there is peace and it is economic to do so. But it increases **energy security** risks if something goes wrong.

Even with improvements in energy efficiencies and energy conservation measures, it is inevitable that we will run out of some energy resources. Energy conservation can limit growth in energy demand and contribute to energy security but it only has a small impact on total use at the moment. Our need for more and more energy is insatiable.

Key term

Energy security for a country is access to affordable and reliable sources of energy.

ATL Activity 16

ATL skills used: thinking, research

There are always wars somewhere in the world. Consider recent armed conflicts and identify whether any were caused by demand for energy or energy sources.

▶ Figure 9 One estimate of years of global coal, oil and natural gas left, reported as the reserves-to-product (R/P) ratio which measures the number of years of production left based on known reserves and present annual production levels. These values can change with time based on the discovery of new reserves, and changes in annual production

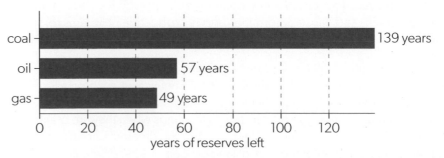

About half of the world's fossil fuel assets will be worthless by 2036 under a net zero transition to 2050. Countries that are fossil fuel producers will lose income, yet others with large investment in producing renewables will gain. Some nations with so called "stranded assets" will perhaps want to slow down decarbonization but those which import fossil fuels at high cost will want faster action.

Norway, for example, is rich in oil yet high in renewable energy production. It exports most of its oil and could find that its GDP declines unless it diversifies away from an oil-exporting economy.

Energy choices

The energy choices made by a society depend on many factors.

- Availability of supply—within national borders or imported

- Technological developments—finding new sources of energy, e.g. shale oils and harnessing wave power

- Politics—can lead to conflict over energy supplies or choices to use more expensive domestic supplies for increased security; this may also impact the decision to use nuclear energy or not

- Economics—globalization of economies can make it uneconomic to produce your own power and cheaper to import it

- Cultural attitudes—our love of private vehicles powered by fossil fuels means we are very reluctant to give them up or change to electric cars

- Sustainability—only renewable energy sources are sustainable, yet they account for a small percentage of world energy supply

- Environmental considerations—pollution, GHG emissions, backlash against nuclear power generation (being phased out in Germany when people felt it was too dangerous after the Fukushima accident (subtopic 1.1)).

Case study 7

Energy generation around the world

In the **USA**, in 2022, about 60% of electricity generation was from fossil fuels—coal, natural gas, petrol and other gases. About 18% was from nuclear energy, and about 22% was from renewable energy sources. There is a proposed target to run on 80% renewable energy by 2030. Although the USA imports crude oil, 40% of its oil requirement is from within its borders. It is a net exporter of coal and gas.

Roughly 60% of **Russia's** electricity is generated by fossil fuels, 20% by hydroelectricity, 20% by nuclear reactors. Renewable energy sources other than HEP generate less than 1% of Russia's energy. Russia exports electricity to Latvia, Lithuania, China, Poland, Turkey and Finland.

Some 76% of **India's** electricity generation needs are met by three fuels: coal, oil and solid biomass.

Coal has underpinned the expansion of electricity generation and industry, and remains the largest single fuel in the energy mix. Net energy import was 40% of total in 2021–22. There is investment in renewables but it is small.

Coal is currently the foundation of **China's** energy system. It has huge reserves of coal but has to import oil and gas. In 2021 China consumed 54% of total world consumption of coal covering close to 70% of the country's primary energy needs. China produces and consumes more coal than any other country. It is the largest emitter of carbon dioxide with nearly 20% of the world's population. However, coal is a short-term filler for China's energy supply and it has pledged that carbon emissions will peak before 2030, then decline. There are long-term climate commitments to renewable energy production. Installed wind and solar capacity in China is now 35–40% of the global total.

Denmark pioneered wind turbine power 50 years ago. Currently, about 47% of its electricity is from wind power and it aims to reach 84% by 2035. Both onshore and offshore wind farms exist. The Danish government drove this change. In the 1970s, much of Denmark's energy was from coal-fired power stations but the government wanted to reduce carbon emissions. There was a ban on nuclear power plants and wind power was seen as the solution. Although wind speeds in Denmark are not particularly high, there are shallow waters offshore where the turbines can be sited and then linked to the national grid onshore. Denmark is linked to the electricity grids of neighbouring countries and can buy electricity from them if the wind drops, and sell it if their own demand is less than that generated. This is, of course, one disadvantage of wind power—you need the wind to blow. There is little evidence that wind turbines kill many birds that fly into them. Most birds fly over or round the turbines and adapt their migration routes. Power lines kill more birds than do wind turbines.

In **Iceland**, 100% of the population has access to electricity. Of the energy produced in Iceland (excluding transport), about 70% is hydroelectric and 30% is geothermal. Wind power is increasing but small at 0.03%. Fossil fuels accounted for 0.01% of all energy produced in Iceland in 2021. However, Iceland participates in the international carbon credit market which means foreign companies can "buy" Iceland's green energy to provide green certificates of origin for their own energy. In this way electricity from a coal-fired power station can be classed as "green".

(ATL) Activity 17

ATL skills used: thinking, research

1. From Figure 10, calculate the total percentage of energy consumption in 2019 from fossil fuels.

2. List the types of renewable energy.

3. Explain why coal, which is the dirtiest fossil fuel, is used for energy production.

4. Figure 10 shows data from 2019. Research the latest figures.

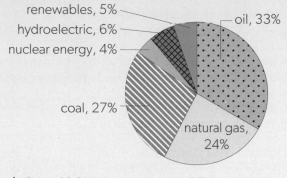

▲ Figure 10 Global energy sources 2019
Data from: BP Statistical Review 2020

Energy storage solutions

Energy is stored in our food, our bodies, fossil fuels, biomass and biofuels. Humans store energy by hydropower, rechargeable batteries, green hydrogen, thermal storage (tanks of hot water or other liquid, or hot solids), and fuel cells.

A battery storage power station uses a bank of batteries to "peak-shave" and level out supply and demand.

A battery is a device that stores chemical energy and converts it to electrical energy. There are many types but the basis is the same: electrons flow between two metal electrodes—the anode and cathode—creating an electric current. In rechargeable batteries, electricity is converted into chemical energy and back to electricity when needed. Types include sodium–sulfur, metal–air, lithium-ion, and lead–acid batteries. In electric vehicles (EV), there are lead–acid batteries, nickel metal hydride (NiMH) batteries, and lithium-ion (Li-ion) batteries. Because of increasing demand for battery storage, much research goes into new and more efficient types. A sodium-ion battery is promising and sodium is abundant on Earth. A solid-state lithium-metal battery could also be efficient.

However, batteries can be very expensive to produce—in terms of both raw materials and energy use (Figure 11).

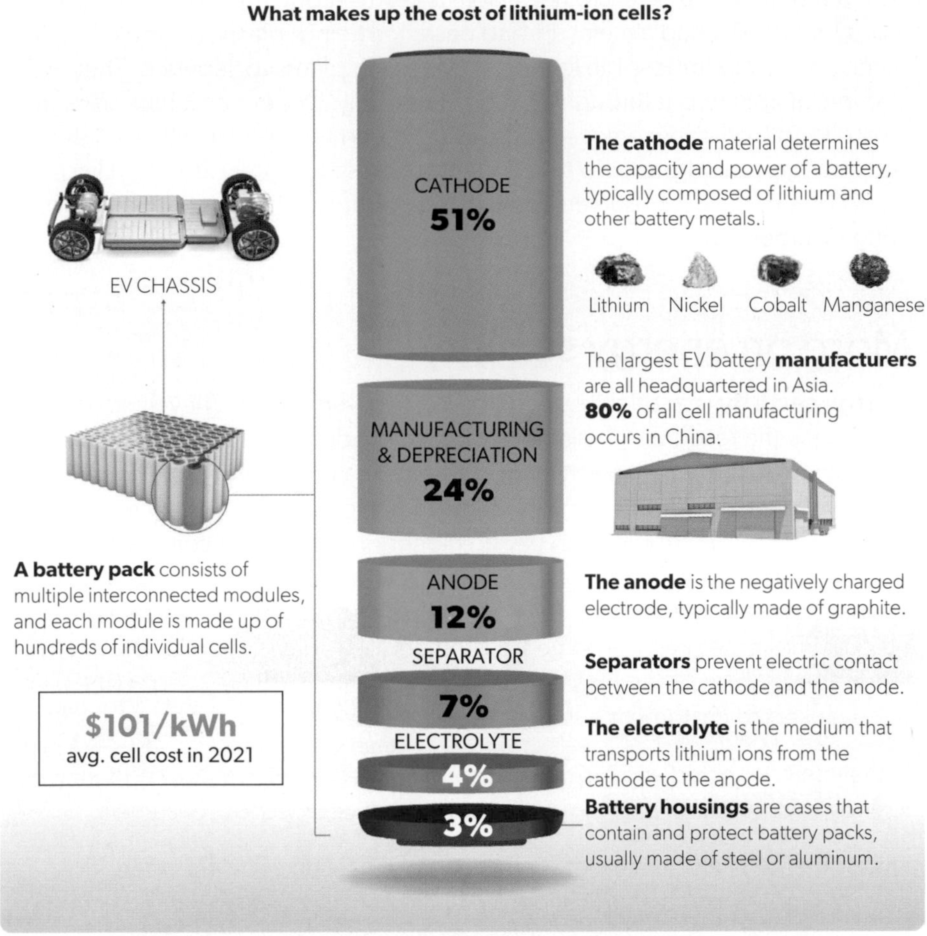

What makes up the cost of lithium-ion cells?

CATHODE
51%

The cathode material determines the capacity and power of a battery, typically composed of lithium and other battery metals.

Lithium   Nickel   Cobalt   Manganese

MANUFACTURING & DEPRECIATION
24%

The largest EV battery **manufacturers** are all headquartered in Asia. **80%** of all cell manufacturing occurs in China.

ANODE
12%

The anode is the negatively charged electrode, typically made of graphite.

SEPARATOR
7%

Separators prevent electric contact between the cathode and the anode.

ELECTROLYTE
4%

The electrolyte is the medium that transports lithium ions from the cathode to the anode.

3%

Battery housings are cases that contain and protect battery packs, usually made of steel or aluminum.

EV CHASSIS

A battery pack consists of multiple interconnected modules, and each module is made up of hundreds of individual cells.

$101/kWh
avg. cell cost in 2021

▲ Figure 11 Cost of an EV (electric vehicle) battery cell (note that percentages do not add to 100 due to rounding)

Energy conservation and energy efficiency

Energy conservation is changing our behaviour to reduce consumption of energy.

Examples include:

- turning off lights

- reducing use of heating or air-conditioning

- travelling less by fuel-driven vehicles.

It is also about using increased energy-efficient technologies—for example:

- designing housing to conserve or remove heat

- using low energy intelligent lighting

- designing methods of shipping with sails

- designing goods to be easily recycled (circular economy).

LEDs (light emitting diodes) for lighting homes and businesses are very efficient. Lighting accounts for nearly 5% of global carbon dioxide emissions. LED lighting achieves energy savings of 50–70% (or more) compared with old technologies. Changing to LED lights is an obvious way to save energy and money.

ATL Activity 18

ATL skills used: research, communication

Select one example of energy conservation and one of energy efficiency. Carry out some research and produce a two-minute talk on each example.

▲ Figure 12 Pringles tubes: recyclable or not?

Connections: This links to material covered in subtopics 1.2, 1.3 and HL.b.

Designing goods to be easily recyclable is key to reducing waste. The original tubes for Pringles had a metal rim and base, foil lining, plastic lid and cardboard outer. They were impossible for normal recycling plants to separate. They were also impossible to put through a circular economy. Under consumer pressure, the Kellogg Company removed the plastic lid and replaced it with a foil seal and arranged a complex way to recycle the tubes separately. But recycling has to be simple for consumers to do it easily. Check what has happened to Pringles tubes now.

More on energy security

Energy security is a serious issue for national governments. They must decide how to use the resources available to them to produce useful energy, while also maintaining national security. It may be cheaper to buy energy from another country but what happens if you fall out with that country? If fuel prices increase, how long will your people keep paying higher prices without complaint? The 2022 energy crisis in Europe and halting of Nord Stream pipeline gas from Russia to Europe forced European countries to find new sources of energy quickly. Russia also had to find new customers for its gas. Qatar has the world's third-biggest proven natural gas reserves. It ships liquified natural gas (LNG) in huge tankers to customer countries around the world.

In the interconnected and globalized world market for energy, the sudden need to find new sources of energy was a shock and prices rose for consumers.

Through energy efficiency measures, decreasing reliance on imported energy supplies and diversification, a country can improve its energy security. Increasing renewable energy sources and storage is one strategy that many countries are adopting.

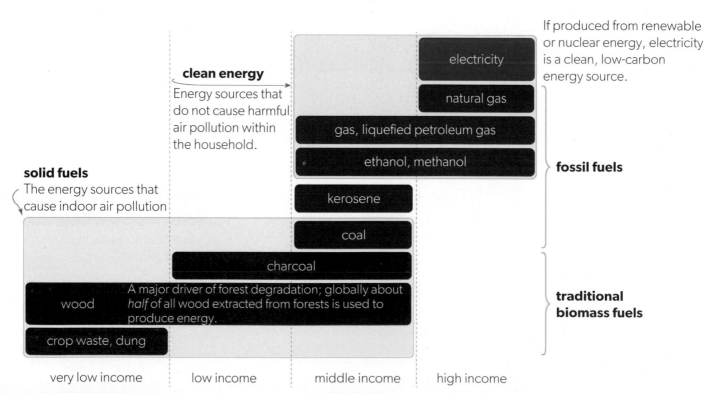

▲ Figure 13 The energy ladder: the dominant energy source for cooking and heating, by level of income

The richer you are, the more likely you are to use gas or electricity to cook and heat your home (Figure 13). Of the world's population, 40% burn biomass or fossil fuels directly for energy to cook. Often this is on open fires or stoves and indoor air pollution from these is far higher than in the most polluted cities.

Data-based questions

The graphs in Figures 14–16 show sources of energy from 1965 to 2021 for the world (Figure 14), the USA (Figure 15) and India (Figure 16).

In 2023, the population of the world was 8 billion, USA 333 million, India 1.419 billion.

In 1965, the population of the world was 3.3 billion, USA 194 million, India 500 million.

Study the graphs and note the scales on the *y*-axes.

1. Describe how energy use has changed from 1965 to 2021 in each graph.

2. Compare energy use and population sizes in the USA and India. Account for the differences.

3. Describe and explain the trends in selection of energy sources over the period of the graphs.

4. Suggest reasons for the high dependency on coal for energy production in India.

5. Calculate if India or the USA used more coal for energy consumption in 2021.

6. Look up the population sizes of India and the USA in 2021 and calculate energy consumed per capita for each country.

7. Visit the Our World in Data website and find the page on energy production and consumption.

   a. Select the country in which you are living now. Describe energy source changes in this graph and give possible reasons for the changes from 1965 to 2021.

   b. Select Norway and estimate from the graph what percentage of energy came from renewable sources in 2021. Explain how Norway can have such a high percentage of renewable sources.

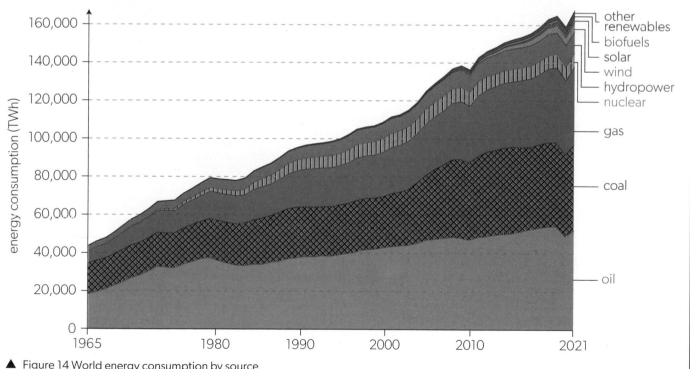

▲ Figure 14 World energy consumption by source

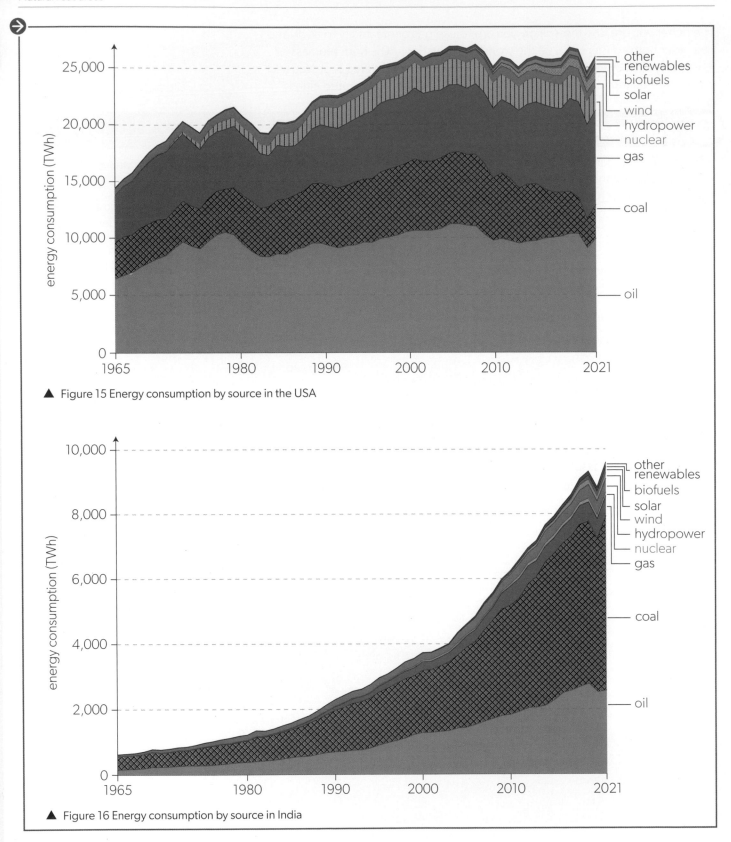

▲ Figure 15 Energy consumption by source in the USA

▲ Figure 16 Energy consumption by source in India

How much longer for fossil fuels?

The global economy mostly depends on finite reserves of fossil fuels as energy resources. The intention and promise of most governments worldwide is to reach net zero carbon emissions by 2050, and humans do have the technology to make use of alternative—renewable—energy sources. Yet energy demand continues to rise and nations are turning back to fossil fuels to meet this demand. In 2018, over 70% of the growth in global energy demand was met using oil, natural gas and coal, resulting in energy related carbon emissions rising by 1.7%.

If we continue to burn fossil fuels at the current rate and only have the known stocks (i.e. do not find new ones), these resources will run out. One estimate for the time until current fossil fuel reserves are gone is shown in Figure 9. Another is shown in Figure 17; this predicts that we will run out of oil by 2052, gas by 2060 and coal by 2090 (this is by far the dirtiest fossil fuel to burn).

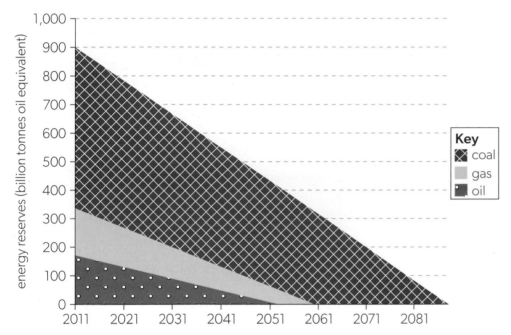

▲ Figure 17 Energy reserves in billions tonnes of oil equivalent

Climate change, campaign groups and energy security needs are focusing governments' policies and forcing them to act. So too is depletion of fossil fuel reserves—no country wants to run out of energy.

Timelines for final depletion of fossil fuels depend on:

* rate of consumption—increased efficiency and energy conservation measures slow down the rate

* discovery of new deposits—and these are still being discovered

* developments in technology for extraction—for example, from tar or oil sands of Canada and Venezuela

* increased use of renewables or nuclear power—which replace fossil fuel use.

Case study 8

Oil sands and tar sands of Alberta

In Alberta, Canada, the tar sands are a site of oil extraction which is larger than England and one of the biggest industrial projects on Earth. Growing on the land are boreal forest and muskeg ecosystems.

- The amount of oil here is huge. It is mined in open pits, destroying the vegetation.

- The oil is held in sand or sandstone and is a heavy oil (bitumen) which is thick and hard to extract.

- The energy cost and water requirements of extraction are high. Toxic waste water is kept in tailing dams which take up more land.

Because the cost of extraction is higher than for other oils, the break-even point is reached only when oil prices are over USD 100 per barrel. This leads to boom and bust cycles in extraction.

Environmental issues include:

- loss of ecosystems, soil erosion and fragmented wildlife habitats

- deformed aquatic life

- air and water pollution by heavy metals extracted with the oil

- water extraction from the Athabasca River reducing flow rates

- GHG emissions greater than those of the conventional oil industry

- harm to the indigenous peoples of the region—their lands are ruined and they suffer illnesses including higher rates of cancers and heavy metal poisoning.

▲ Figure 18 Map of the Alberta oil sands

Nuclear power

Most nuclear power stations obtain energy by fission reactions of uranium or plutonium. The energy released from the reaction is used to heat a liquid (usually water) which boils and drives turbines to generate electricity.

Construction costs of a nuclear power station are high. But once constructed, nuclear power stations are very cost-effective because nuclear fission is 8,000 times more efficient at producing energy than fossil fuels.

Advantages of nuclear power:

- low-cost, zero-carbon energy

- constant supply, not intermittent

- 90% of the fuel can be recycled

- low maintenance once built.

Disadvantages of nuclear power:

- high cost of construction

- safety risks of a nuclear accident and radiation sickness

- risk of fuels being used in nuclear weapons

- disposal of spent fuel which can be toxic and radioactive for 10,000 years so must be stored safely

- mining for uranium has negative effects

- thermal pollution from release of cooling water into local seas from power stations—this changes water chemistry.

Battery storage

Many renewable energy sources are intermittent in nature (solar, wind, tidal). This means it is necessary to store the energy on a large scale to meet our requirements. Batteries are the main solution at present.

Batteries require resources to be mined, transported and processed. Manufacture of batteries takes energy and there is the question of how to recycle them and reclaim materials. All these processes produce emissions and pollution and cause sociopolitical tensions.

The main elements and oxides required to produce effective batteries are lithium, cobalt and rare earth elements. Mining and processing of these minerals creates toxins and results in pollution of land and oceans. Failures of mine tailing dams have also occurred. The necessary elements for producing batteries are found only in certain countries, but demand is global. This has unintended consequences such as geopolitical conflicts.

Case study 9

Rare earth elements, lithium, cobalt and graphite

Review the section on lithium mining in subtopic 7.1. Then consider the rare earth elements (REEs) and other elements required for new technologies to decarbonize the world economy.

For a sustainable future, we must decouple (unlink) economic growth from carbon emissions and link it with international mineral extraction. This is already happening.

When domestic energy sources change, global power shifts too. Extracting oil from tar sands has allowed the USA to produce enough oil for its own needs (although it does still export and import oil). This has weakened the OPEC control of oil prices.

The demand for REEs and other metals is high because they are needed for high-tech consumer products—including cell phones, computer hard drives, electric and hybrid vehicles, flat-screen monitors and televisions, high-performance magnets, high-tech catalysts, electronics, wind turbines, glass, ceramics and alloys. National defence systems also require REEs.

Lithium is needed for lithium-ion batteries, particularly in electric vehicles (EVs) which need about 8 kg each. Bolivia has the highest identified lithium resources in the world, with 20 million tonnes. Chile and Australia also have large reserves.

Nickel is essential as the cathode in lithium-ion batteries. A Tesla EV battery contains 50 kg of nickel. Indonesia is the biggest producer of nickel, followed by the Philippines then Russia.

Cobalt is used in alloys in planes and gas turbines. It is also in EV lithium-ion batteries (up to 14 kg per battery). The Democratic Republic of Congo (DRC) accounted for more than two-thirds of global cobalt production in 2022, making it the world's largest cobalt producer by a large margin. Cobalt is the DRC's largest source of export income.

Graphite is the largest component of lithium-ion batteries, comprising over 50% of every lithium-ion battery and over 95% of a battery's anode. Currently it is the only material that can be used at the anode. Up to 70 kg of graphite is used in a single EV. China is the leading global producer, accounting for 79% of production in 2021.

The **rare earth elements** (REEs) are a set of 17 metallic elements with similar properties—the 15 lanthanides, plus scandium and yttrium. They are not particularly rare and are often found together but only certain rocks have enough of them to make mining profitable. The main deposits are in China (particularly in Inner Mongolia), Russia, Kyrgyzstan, Kazakhstan, the USA and Australia. There are also resources in India, Vietnam, Malaysia, Thailand, Indonesia, South Africa, Namibia, Mauritania, Burundi, Malawi, Greenland, Canada and Brazil.

The minerals containing REEs are mined and the elements separated from the ore. This is energy intensive and there is pollution from the chemicals used.

In 1993, 38% of world production of REEs was in China, 33% in the USA, 12% in Australia, and 5% each in Malaysia and India. Several other countries, including Brazil, Canada, South Africa, Sri Lanka and Thailand, made up the remainder. In 2008, China accounted for more than 90% of world production of REEs, and by 2011, China accounted for 97% of world production and most of the processing. Since then, other countries have increased production and China was responsible for 60% of REEs in 2021 but nearly all processing.

If all cars in the world were EVs requiring batteries, the amount of minerals mined would need to increase from 400 kilotonnes in 2021 to 11,800 kilotonnes by 2040 according to the International Energy Authority (IEA).

▲ Figure 19 A Nissan Leaf EV chassis showing part of the battery

Question

Discuss the range of arguments for mining REEs and other elements for use in consumer products.

Check your understanding

In this subtopic, you have covered:

- there are renewable and non-renewable energy resources with different levels of sustainability

- intermittent energy supply is overcome with storage

- global energy demand is constantly increasing

- efficiency and energy conservation reduce demand but not by enough

- energy choice of a country depends on many factors

- choices are often about energy security

- fossil fuels are finite and will run out

- nuclear energy is one option but controversial

- resources needed for battery storage create environmental and political tensions.

AHL

→

To what extent can energy consumption be equitable across the world?

How can energy production be sustainable?

1. List types of renewable and non-renewable energy resources.

2. Explain the reasons why global energy demand is increasing.

3. Describe the environmental costs of a named renewable and a named non-renewable energy source.

4. Describe the factors that influence a country in its energy choices.

5. Describe solutions to intermittent energy production.

6. Explain how a named country can become less dependent on an imported energy source.

7. Describe, with examples, how a country can improve its energy security. **AHL**

8. Explain why the time of final depletion of fossil fuels is hard to predict.

9. Tabulate the advantages and disadvantages of using nuclear energy for electricity production.

10. Predict the future for use of batteries for energy storage.

▶▶ Taking it further

- Evaluate use of energy and consider whether it should be reduced individually, in the school and in the community. Consider how this reduction could occur.

- Carry out a school survey or questionnaire to see if energy use can be reduced in school.

- Consider and adopt ways your own household could reduce energy use.

- Consider your personal perspective on using nuclear power relating to your own environmental value system.

TOK

The choice of energy source is controversial and complex.

To what extent does one country's choice of energy source have impacts on other countries?

How can we distinguish between a scientific claim and a pseudoscience claim when making choices?

7.3 Solid waste

Guiding question

- How can societies sustainably manage waste?

Understandings

1. Use of natural resources generates waste that can be classified by source or type.
2. Solid domestic waste (SDW) typically has diverse content.
3. The volume and composition of waste varies over time and between societies due to socio-economic, political, environmental and technological factors.
4. The production, treatment and management of waste has environmental and social impacts, which may be experienced in a different location from where the waste was generated.
5. Ecosystems can absorb some waste, but pollution occurs when harmful substances are added to an environment at a rate faster than they are transformed into harmless substances.

6. Preventative strategies for waste management are more sustainable than restorative strategies.
7. Different waste disposal options have different advantages and disadvantages in terms of their impact on societies and ecosystems.
8. Sustainable options for management of SDW can be promoted in societies.
9. The principles of a circular economy provide a holistic perspective on sustainable waste management.

Natural resources and waste

According to the UN Sustainable Development Goal (SDG) 12:

- if the global population reaches 9.8 billion by 2050, the equivalent of almost three planets will be required to provide the natural resources needed to sustain current lifestyles

- in 2015, almost 12 tonnes of resources were extracted per person, with electronic waste as the fastest-growing sector.

Connections: This links to material covered in subtopics 1.1, 1.2, 1.3 and HL.b.

Since the beginning of the 20th century industrialized countries have been wasting approximately 30% of the natural resources they extract or harvest. This is a problem because most industrial societies rely on non-renewable resources to function (e.g. fossil fuels). Natural resources are being overexploited and depleted, the environment is being polluted and changed irreversibly, there is loss of biodiversity and social conflicts arise. None of this is sustainable.

There are many ways to classify waste. For example, it can be classified by source or type.

The major types of waste are liquid, solid, organic, recyclable and hazardous. These originate from:

- households (domestic waste): from the day-to-day use of a domestic home

- industry: material that is no longer of any use in the manufacture of products

- agriculture: unwanted or unsaleable produce

- E-waste: this is very varied, ranging from plugs and electronic components to TVs, computers and mobile phones

- food: any food that is discarded between harvesting and consumption (see subtopic 8.3)

- biohazardous waste: potentially infectious pathogens and other materials.

(ATL) Activity 19

ATL skills used: thinking, research

Copy and complete Table 1, giving an example of each type of waste.

	Liquid	Solid	Organic	Recyclable	Hazardous
Households					
Industry					
Agriculture					
E-waste					
Food					
Biohazardous					

▲ Table 1

Can you think of alternative ways to categorize waste?

Human societies generate unmanageable amounts of waste. The "take, make, waste" concept is discussed in subtopic 8.2; Figure 1 is an expansion of that idea. What humanity once thought possible is not realistic with human population growth.

▼ Figure 1 Linear thinking

W	A	S	T	E
there is an infinite supply of natural resources	we can extract those resources for everything we want	we can process those resources to make whatever we want	we can distribute those products all over the world	we can buy and use all the products we want	when we are bored with the products we throw them away and get new ones	the Earth has infinite capacity to absorb our waste

W   A   S   T   E

This type of linear thinking is not sustainable: we must switch to a circular concept if we are to make a difference.

Solid domestic waste

Solid domestic waste (SDW) is our rubbish (trash or garbage) from residential and urban areas. It is a mixture of paper, packaging, organic materials (waste food), glass, dust, metals, plastic, textiles, paint, old batteries, electronic waste (e-waste) and more (see Figures 3 and 4). It is collected from homes and shops and, although it only makes up about 5% of total waste, which includes agricultural and industrial waste; it is waste that we can control.

▲ Figure 2 Solid domestic waste

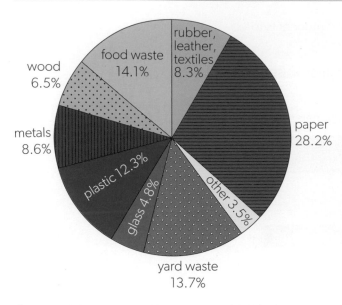

▲ Figure 3 Composition of SDW in USA in 2013 (before recycling)

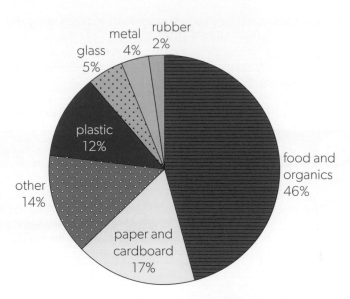

▲ Figure 4 Composition of global SDW
Data from: World Bank Data (CC BY 4.0)

(ATL) Activity 20

ATL skills used: thinking, communication, social, research, self-management

1. Make a list of all the waste you produce in a day and in a week.

2. Now do the same for your household.

3. Categorize the items you have listed as:

   - recyclable

   - biodegradable

   - hazardous or toxic

   - waste electrical and electronic equipment (WEEE).

4. Create a pie chart for your household waste, showing the percentage in each category. Share this in class and discuss differences with others.

5. How much do you recycle?

6. How do you manage recycling in your home? Who does the recycling?

7. Where do the recycled materials go?

▲ Figure 5 Landfill in the Philippines where people desperate for money make a living sorting through SDW

When is something waste?

A resource has value to humans. One human's waste may be another human's resource—it depends on how we value it. That is why in many LICs there are whole industries set up to collect SDW. People travel round residential areas looking for useful items in communal bins. In many LICs, families live on and around landfill sites so they can trawl through the waste that arrives from the city.

Waste is material which has no value to its producer. If it is not recycled it becomes a problem and needs to be disposed of. We create waste in most of the processes we carry out—energy production, transport, industrial processes, construction, selling of goods and services, and domestic activities.

Volume and composition of waste

The following data is from a 2018 World Bank report.

- On average, the world generates 0.74 kg of waste per capita per day. However, average values vary significantly between countries, from 0.11 to 4.54 kg per capita per day.

- Generally, the higher the level of urbanization and income, the more waste.

- An estimated 2.01 billion tonnes of municipal (domestic) solid waste were generated in 2016. This number is expected to grow to 3.40 billion tonnes by 2050, at current rates.

- The total quantity of waste generated in LICs is expected to increase by more than three times by 2050.

- Currently, the East Asia and Pacific region is generating most of the world's waste (23%). The Middle East and North Africa region is producing the least in absolute terms (6%).

- Recyclables make up a substantial fraction of waste streams, ranging from 16% paper, cardboard, plastic, metal and glass in LICs to about 50% in HICs.

- As countries rise in income level, the quantity of recyclables in the waste stream increases, with paper increasing most significantly.

Volume

Of the 2.01 billion tonnes of SDW each year, 33% is not managed in an environmentally safe manner. This is not evenly distributed between countries.

Table 2 lists the HICs which produced the most and least waste per year in 2018–20.

HICs producing most waste (kg per capita per year)		HICs producing least waste (kg per capita per year)	
• Denmark	850.6	• Japan	336.9
• USA	810.9	• South Korea	413
• Luxembourg	795.2	• Turkey	419.7
• New Zealand	781.1	• Canada	423.6
• Iceland	735.1	• UK	456.7

▲ Table 2 HICs producing most and least waste per year, 2018–2020

These figures include waste from households, commerce, small businesses and garden waste—this is called municipal solid waste (MSW). Consider reasons for the differences between countries.

Data-based questions

The data in Table 3 can be shown in a number of ways—see Figures 6–8.

When you are doing your internal assessment, you need to think about which graph best suits your purpose and which comparison links most clearly to your research question.

	2016	2030	2050
Middle East & North Africa	129	177	255
Sub-Saharan Africa	174	269	516
Latin America & Caribbean	231	290	369
North America	289	342	396
South Asia	334	466	661
Europe & Central Asia	392	440	490
East Asia & Pacific	468	602	714

▲ Table 3 Waste generation (actual and projected), by region (millions of tonnes/year)
Data from: World Bank Group

As these graphs show, the amount of SDW produced in all regions of the world is projected to increase significantly.

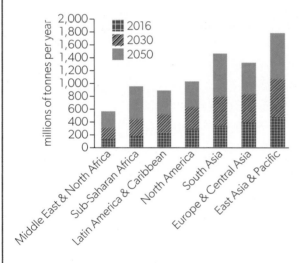

▲ Figure 6 Waste generation by region
Data from: World Bank Data (CC BY 4.0)

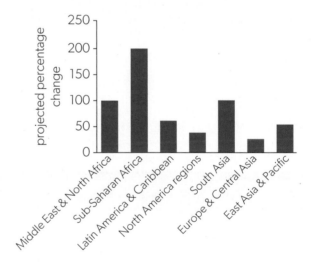

▲ Figure 7 Projected percentage change in waste generation between 2016 and 2050
Data from: World Bank Data (CC BY 4.0)

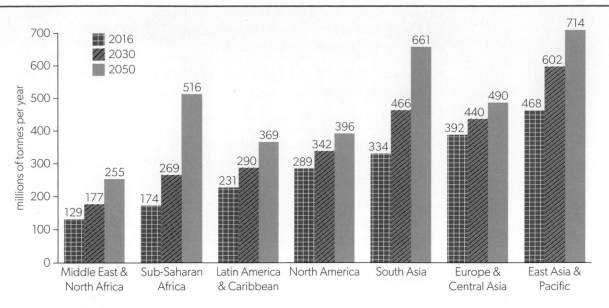

▲ Figure 8 Waste generation by region
Data from: World Bank Data (CC BY 4.0)

1. Which graph do you find easier to understand, Figure 6 or Figure 8? Justify your choice.

2. Calculate the percentage change in waste generation between 2016 and 2050 for South Asia.

3. Describe and explain the changes in waste generation between 2016 and 2050.

4. Describe and explain the changes in waste generation for the region that you live in.

5. Figure 6 suggests that the regions of North America and Europe & Central Asia are becoming more sustainable as regards the treatment of waste.

   To what extent is this statement true?

What is the cause of this universal increase in waste generation?

- Growing population—more waste being generated.

- Urbanization—a much higher density of population and easy access to consumer goods. People can buy more items and that means more waste.

- Technological progress—changes in technology mean changes in packaging. In the past milk and soda came in glass jars that were returned to the manufacturer to be cleaned and reused. Now most drinks come in plastic bottles, many of which are not biodegradable.

- Improvements in the standard of living—as we earn more money, we can buy more goods.

- Changes in consumer shopping habits—online shopping is now big business. We no longer have to go out to buy the goods that we want, we can order almost anything online to be delivered to us. Generally, that means more packaging and so more waste.

- Appearances—for many people, the way they look, the technology they carry and the luxury goods they have all say something about their status. This prompts a throwaway societal attitude, in which everything is replaceable and must be replaced if it is old or goes out of fashion.

Figures 6–8 show that there is a difference in the speed of increase of waste generation; North America and Europe & Central Asia appear to be showing a slower rate of increase. This could be due to:

- the introduction of laws that require recyclable items to be recycled, not thrown out as waste

- the availability of technology necessary to recycle or reprocess various materials

- social pressure from the population for action to be taken to reduce the waste of resources, to lessen the environmental damage caused by waste disposal methods

- the action of certain companies to reduce the amount of packaging used for their products—for example, Coca-Cola bottling partners produce bottles made from 100% recycled polyethylene terephthalate (rPET).

Composition

Just as the volume of waste varies both spatially and temporally, so does the composition of waste. Figure 4 shows the mean composition of SDW globally. There are also differences between countries, as shown by Figures 9–12.

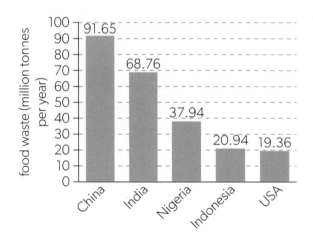

▲ Figure 9 Food waste (million tonnes per year)
Data from: UNEP

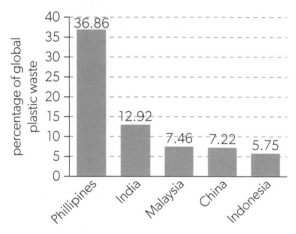

▲ Figure 10 Percentage of global plastic waste
Data from: Our World in Data (CC BY 4.0)

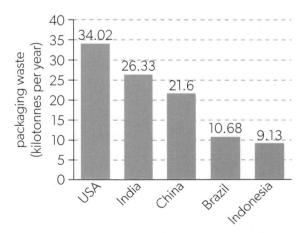

▲ Figure 11 Packaging waste (kilotonnes per year)
Data from: World Population Review

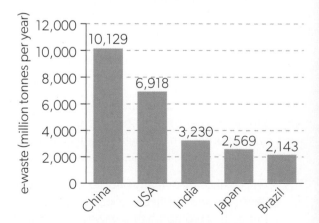

▲ Figure 12 E-waste (million tonnes per year)
Data from: The Global E-waste Statistics Partnership (CC BY-NC-SA 3.0 IGO)

You must be very careful when interpreting charts like these. Watch out for:

Population size and per capita numbers. India produces nearly 50 million metric tonnes more food waste than the USA but if you look at the per capita numbers the situation changes; in the USA, food waste is 59 kg per capita per year whereas in India it is 50 kg per capita per year.

It is a similar situation with e-waste. China as a country produces 10,129 million tonnes per year whereas the USA produces 6,918 but, looking at the per capita numbers the situation is reversed. China produces only 7.2 kg per capita whereas the USA produces 20 kg per capita.

- Where waste is occurring may vary. In many LICs a large proportion of food is lost due to poor agricultural harvesting, storage and transportation facilities. The FAO estimates that nearly 40% of the food wasted in India is for these reasons, before the food is processed or eaten.

- Some waste—especially food production waste—may occur due to natural events such as flooding, drought and storms.

- There is an international trade in waste. Waste moves between countries for treatment, disposal, or recycling. Often hazardous and toxic waste is shipped from HICs to LICs for disposal.

- E-wastes are moved legally and illegally between countries so some of the waste shown in Figure 12 may not be from the country where it is "counted".

These factors influence the differences in the composition of waste between countries. The factors affecting the volume of waste produced (discussed earlier) will also impact the composition of waste produced.

(ATL) Activity 21

ATL skills used: communication, social, research

Choose one country from each of Figures 9–12.

Investigate the reasons for the production of that waste in that country (e.g. food waste in India).

Complete Table 4, then discuss your findings with the rest of the class.

Type of waste and country	Reasons for the amount of waste produced

▲ Table 4

Managing SDW

We can minimize waste, or we can dispose of it somewhere, but we cannot throw it away—it has to go somewhere. We have a choice as to how we deal with waste. The options are:

- recycling (or the 3Rs)
- landfill sites
- incineration
- anaerobic digestion
- composting organic waste
- waste to energy schemes
- export the waste to other countries.

Strategies to minimize waste

In the past, minimizing was based around the "3Rs"—Reduce, Reuse, Recycle. However, many more Rs have been added (subtopic 8.2), including Refuse, Rethink, Repurpose, Repair, Remanufacture, Recover and Refurbish. All of these can minimize the amount of waste to be managed.

Reduce

The best action we can take is to produce less waste in the first place. This requires us to use fewer resources. We do not have to stop consuming, we just need to cut back and adjust our lifestyles, thinking about what we need rather than what we want.

- Learn how to maintain your possessions so they last longer.

- Change your shopping habits: buy things that will last; look for items with less packaging; buy products made from recycled materials; choose products that are energy efficient; be aware of the resources you are using in the home, such as water and electricity; shop smart by buying only the food you need; buy second-hand or vintage clothes.

- Make sure you know what can and cannot be recycled and follow the rules of sorting in your area.

Reuse

This is where products are used for something other than their original purpose or they are returned to the manufacturer and used repeatedly.

- Where possible, return packaging (such as bottles) to the manufacturer for reuse.

- Compost food waste.

- Hire, don't buy.

- Read e-books or second-hand books.

- Donate unused food to food banks or shelters.

- Donate old clothes to charity shops, or use them as cleaning rags.

Recycle

This is probably the best-known R. Many towns and cities now have kerbside recycling, where people sort their waste into separate containers for recycling before it leaves the home.

- In Germany, for example, each household has four recycling bins.

- In the UK, there is discussion about charging households more if they produce more than the standard amount of waste.

▲ Figure 13 **Recycling bins on Orchard Road, Singapore**

Recycling involves collecting and separating waste materials and processing them for reuse. (If materials are separated from the waste stream and washed and reused without processing in some way, this is reuse.) The economics of recycling determine whether it is commercial or not and this can vary with the market cost of the raw materials or cost of recycling. Some materials have a high cost of production from the raw material, so recycling is particularly worthwhile commercially. Aluminium cans are probably the best example of this.

ⒶⓉⓁ Activity 22

ATL skills used: thinking, communication, social, research, self-management

The US Environmental Protection Agency (EPA) produces many resources that illustrate the different ways to minimize our waste production. Visit the EPA website and do some research.

Create a media product of your choice that will encourage young children to be active and enthusiastic about minimizing waste.

Case study 10

Germany: what belongs where?

Many, but not all, German households are allocated three or four waste bins of different colours, for different types of waste.

Brown bin (biological waste)

- Kitchen waste: old bread, egg shells, coffee powder and filters, food leftovers, tea leaves and tea filters.

- Fruit and vegetables: peels, apple cores, leaves, nutshells, fruit stones and pips, lettuce leaves.

- Garden waste: soil, hedge trimmings, leaves, grass clippings, weeds, dead flowers and twigs.

- Other: feathers, hair, kitchen towels, tissues, sawdust and straw.

Blue bin (paper)

- Envelopes, books, catalogues, illustrations, cartons, writing pads, brochures, writing paper, school books, washing detergent cartons without plastic, newspapers, paper boxes.

Yellow bin or yellow plastic bags (plastic, etc.)

- Plastic bags, margarine tubs, milk sachets, plastic packaging trays for fruit and vegetables, bottle tops, detergent bottles, carrier bags, vacuum packaging, dishwashing liquid bottles.

Grey bin (household waste)

- Ash, wire, carbon paper, electrical appliances, bicycle tubes, photos, broken glass, bulbs, chewing gum, personal hygiene articles, nails, porcelain, rubber, plastic ties, broken mirrors, vacuum cleaner bags, street sweeping dirt, carpeting pieces, diapers, cigarette butts, miscellaneous waste.

Households that do not have a brown bin put their biological waste in a grey bin.

Questions

1. What do you think happens to the contents of collected brown bins?

2. Explain why the contents of blue and yellow bins can be recycled.

3. List the possible options for disposal of the contents of grey bins.

4. Describe ways in which the contents of grey bins could be reduced.

Recycling plastics

Plastics are made from oil and the world's annual consumption of plastic materials has increased from around 5 million tonnes in the 1950s to nearly 100 million tonnes today. Of this, 10% ends up in the oceans. As much as 8% of the world's oil production may be used to make plastics and we throw most of this away as it is used mainly in single-use packaging.

Plastic is a difficult material to recycle as there are many different types of plastic (Figure 14) and it is bulky and light. Some types of plastic are worth more than others to recyclers but these have to be sorted from the rest. However, plastic recycling is carried out to some extent.

A report on the production of carrier bags made from recycled rather than virgin polythene concluded that the use of recycled plastic resulted in the following environmental benefits:

- reduction of energy consumption by two-thirds

- production of only a third of the sulfur dioxide and half of the nitrous oxide

- reduction of water usage by nearly 90%

- reduction of carbon dioxide generation by two-and-a-half times.

A different study concluded that 1.8 tonnes of oil are saved for every tonne of recycled polythene produced.

Recycled plastic can be made into fleeces and anoraks, cassette cases, window frames, bin bags, seed trays and a range of other products. It takes 25 two-litre plastic drinks bottles to make one fleece garment. Sadly, most plastic is used once and then put in holes in the ground.

Many LICs have an informal recycling sector.

Symbol	Code	Description
1	PET	**Polyethylene terephthalate**—fizzy drink bottles and oven-ready meal trays
2	HDPE	**High-density polyethylene**—bottles for milk and washing-up liquids
3	PVC	**Polyvinyl chloride**—food trays, cling film, bottles for squash, mineral water and shampoo
4	LDPE	**Low density polyethylene**—carrier bags and bin liners
5	PP	**Polypropylene**—margarine tubs, microwaveable meal trays
6	PS	**Polystyrene**—yoghurt pots, foam meat or fish trays, hamburger boxes and egg cartons, vending cups, plastic cutlery, protective packaging for electronic goods and toys
7	OTHER	**Any other plastics** that do not fall into any of the above categories. An example is melamine, which is often used in plastic plates and cups

▲ Figure 14 Types of plastic and their uses

Why is the plastic bag so successful?

Plastic bags are everywhere—but only since the late 1980s when we traded reusable shopping bags for plastic ones.

Fast facts

- Plastic bags are clean, cheap to produce, waterproof and convenient. It costs 1 US cent to produce a single plastic bag, compared with 4 cents to make a paper bag. They are so cheap that stores have given them away!

- An estimated 500 billion–1 trillion plastic bags are made each year.

- Most are used once and then thrown away because they are so thin: the average "working life" of a plastic bag is 15 minutes.

- Most are made from oil and take 200–1,000 years to break down. A few degrade in sunlight if they are made of biodegradable starch polymer materials.

- Discarded plastic bags end up in landfills, oceans, turtle stomachs, on trees on deserted islands—anywhere and everywhere.

- Burning plastic bags releases toxins.

There are many possible solutions to the problems of plastic bag use. One of them is plastax—a plastic bag tax. In South Africa, Ireland, Australia, Taiwan and Bangladesh, governments have acted to ban or tax plastic bags.

- Ireland has a tax on plastic bags which has led to a 95% decrease in their use.

- South Africa has banned thin bags; the thicker ones can be reused, must be paid for and do not float around the country.

- China started a ban on free plastic bags from supermarkets in 2008. In 2020 they banned non-biodegradable plastic bags and in January 2021 banned non-degradable plastic straws.

▲ Figure 15 Plastic in the oceans

Connections: This links to material covered in subtopic 7.1.

Plastic cups and alternatives

| Plastic cup | • Use once and throw away
 • Made from non-renewable oil
 • Made using petroleum and natural gas (same amount as a paper cup) | Paper cup  | • Use once and throw away
 • Made from renewable wood
 Compared with a plastic cup, it:
 • uses the same amount of petroleum and natural gas to make
 • consumes 12 times as much steam
 • consumes 36 times as much electricity
 • uses twice as much water for cooling
 • costs 2.5 times more.
 It does not degrade due to the design of landfills that stop decay! |
| Ceramic cup | • Use repeatedly—wash and reuse
 • Made from non-renewable clay
 • Requires hot water and detergent for cleaning but will not catch up to the plastic cup until it has been washed 1,000 times | | |

▲ Figure 16 Which cup would you use?

(ATL) Activity 23

ATL skills used: thinking, communication, social, research, self-management

Just about every form of plastic has the same advantages and disadvantages as plastic bags.

1. Make a list of anything around you that is plastic.

2. Discuss and create a single list for the whole class.

3. How many of the items on the list do you use each day?

4. How many of the items do you recycle or dispose of properly?

5. Why do you use these plastic items? Think carefully.

6. Are there alternatives available? If so, what are they?

7. The facts in Figure 16 are from a study by Canadian scientist Martin Hocking. Which cup would you use? Justify your decision.

Strategies for waste disposal

If waste materials are not recycled or reused, we must manage them.

Landfill

▲ Figure 17 A waste truck unloading in a landfill site, Wales

Landfill is the main method of disposal. Waste is taken to a suitable site and buried there. Hazardous waste can be buried along with everything else and the initial cost is relatively cheap. Landfill sites are not just holes in the ground (Figure 18). They are carefully selected to be not too close to areas of high population density, water courses and aquifers. They are lined with a special plastic liner to prevent leachate (liquid waste) seeping out. The leachate is collected in pipes. Methane produced as a result of fermenting organic material in the waste is either collected and used to generate electricity or vented to the atmosphere. Soil is pushed over the waste each day to reduce smells and pests. New landfill sites are getting harder to find as we fill up the ones we have at a faster and faster rate.

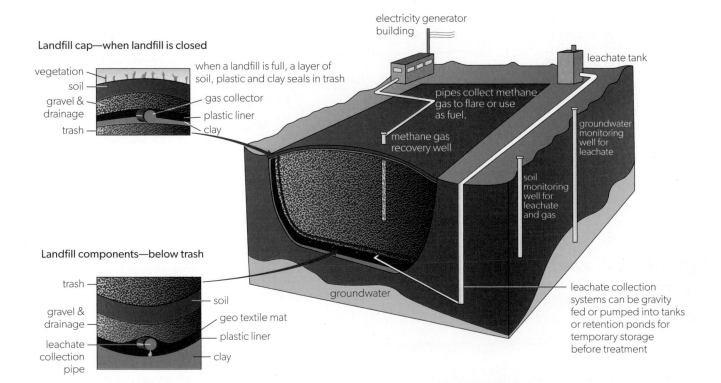

▲ Figure 18 The technology of a landfill site

Table 5 summarizes the advantages and disadvantages of landfills.

Advantages	Disadvantages
Modern landfills are eco-friendly as there are strict laws and regulations about standards.	They are at least partially responsible for climate change—every 1 kg of non-recycled waste generates 700 g of CO_2. One tonne of biodegradable waste can produce 400–500 m^3 of gas; some of it is methane.
Removes the waste from urban and suburban areas.	Negative impacts on wildlife as animals ingest plastic.
Segregates hazardous waste—modern sites have a separate area to place hazardous materials away from the public.	Dioxin emissions are also present and they are hazardous to the environment.
They are cheap because waste travels a short distance and that reduces pollution.	Landfill sites can collapse due to heavy rain or over-dumping of material. Sri Lanka 2017: Meethotamulla site collapsed killing 30 people and destroying 140 houses.
Supports local jobs and businesses—SDW must be handled and managed effectively.	Methane is highly flammable at very low concentrations—5–15% of air volume.
They may be the site of a waste-to-energy scheme where the gases (carbon dioxide and methane) that accumulate due to decomposition are filtered out and used for energy production.	Water and soil contamination are a problem if the protective membranes rupture leaking hazardous chemicals into the local environment and polluting ground water.
Land can be reused after the site is closed but there may be settling issues.	Studies have linked long-term proximity to landfill to cancer, respiratory distress and developmental defects in children.

▲ Table 5 Advantages and disadvantages of landfill

Incineration

Incinerators burn waste at high temperatures, up to 2,000°C. In some cases, waste is pre-sorted to remove incombustible or recyclable materials. The heat produced is often used to generate steam to drive a turbine or heat buildings directly. This is called waste-to-energy incineration.

In other cases, all the waste is burned but this can cause air pollution, particularly release of dioxins (from burning plastics, heavy metals (lead and cadmium, from burning batteries) and nitrogen oxides. The ash from incinerators can be used in road building and the space taken up by incinerated waste is far smaller than that in landfills. Plants are expensive to build though, and need a constant stream of waste to burn, so do not necessarily encourage people to reduce their waste output.

There are many advantages to incineration.

▲ Figure 19 Waste incinerator in Vienna, designed by the architect Friedensreich Hundertwasser as something in which the city can take pride

- Reduces waste quantity—decrease volume by up to 95% so far less land is required to accommodate the waste. This is especially beneficial for countries with a shortage of land, e.g. Japan.

- Reduces pollution compared with landfill sites. Landfill sites release more GHGs and contaminate water supplies—incinerators do not.

- Can be used for waste-to-energy schemes. Sweden generates nearly 10% of its heating needs from incinerators.

- Well-installed incinerators filter the smoke before it is released.

- Reduces waste transportation costs because incinerators can be placed near the source of the waste.

- Allows more control over noise, smell and visual pollution because the waste does not lie around and decay to produce a smell.

- Prevents methane gas production because there is no decay process.

- High temperatures destroy any harmful germs that are present, so it is very good for clinical waste.

- Effective way to recycle metals as the temperatures are not high enough to melt the metals.

- Operated by computer, thus reducing human error.

- Ash from the incineration can be used in construction or sent to landfills.

There are far fewer disadvantages.

- Set-up and maintenance costs are high.

- Smoke does cause some pollution as there is the release of carcinogen dioxins, particulates and nitrogen oxide.

- Does not promote recycling and so feeds into the linear economy.

Anaerobic digestion

Anaerobic digestion (AD) is when biodegradable matter is broken down by microorganisms in the absence of oxygen.

The organic waste includes food waste, crop residues and manure. It is sealed in a reactor where the microorganisms break it down to release biogas (mostly methane but also carbon dioxide, hydrogen sulfide and water vapour).

The following are advantages of AD.

- Biogas can be purified into renewable natural gas.

- Liquid and solid digestate is left after the process. Liquids are a good fertilizer. Solids are used for fertilizer, animal bedding, compost (Figure 21).

Disadvantages of AD include the following.

- Carbon dioxide is released, along with pollutants such as carbon monoxide, sulfur dioxide and nitrogen oxides.

- There is risk of toxic spills of liquid digestate.

- There may be odour issues for neighbours.

- Food waste must be collected and transported to the AD site.

- It requires a constant input of waste.

- It is expensive.

▲ Figure 20 A modern biodigester plant for organic waste

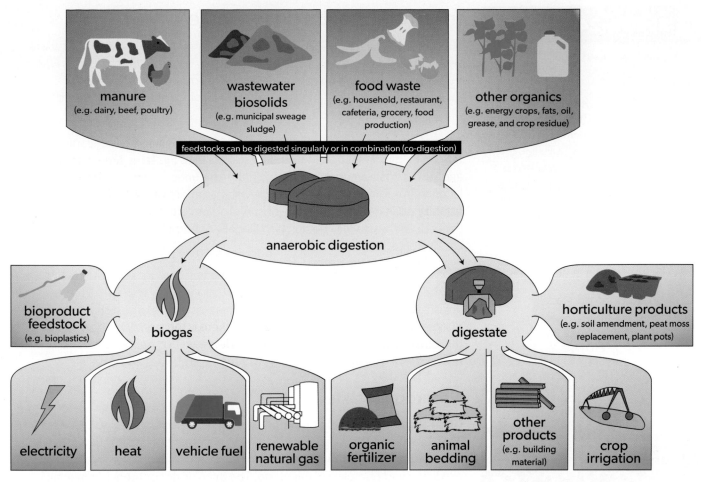

▲ Figure 21 Inputs and outputs of anaerobic digestion

Composting

Domestic organic waste can be composted into fertilizer. Composting can be done at home on a small scale or local government authorities can collect home organic waste and compost it on a larger scale, before selling the composted materials (fertilizers) back to the public.

Advantages of composting include the following.

- It reduces the need for expensive inorganic fertilizer.

- Compost is non-toxic and improves soil quality in an environmentally friendly way—delivers trace minerals and nutrients; breaks up and aerates soil; improves drainage; adds beneficial bacteria.

- It can be done indoors and easy to do at home.

- It reduces carbon emissions.

- There are a few disadvantages of composting.

- There are set-up costs for households wishing to compost organic waste.

- It is time-consuming and hard work, takes up space and can produce an unpleasant odour.

- Diseased plants can spread infection.

ATL Activity 24

ATL skills used: research

The WEEE Directive is from the European Community. It set the target of a minimum recycling rate of 4 kg per capita of electrical and electronic goods by 2009. It failed to meet this target, but awareness was increased and more recycling of WEEE goods did happen.

1. Find out what happens to unwanted or broken electrical and electronic goods in your country.

2. What laws are in place to control or track the export or import of e-waste?

▲ Figure 22 Biohazard storage

ATL Activity 25

ATL skills used: thinking, communication, research

Research and find out more about one of the waste trades discussed here, or another one that interests you.

1. How much of this waste is moved around the world?

2. Who does the exporting?

3. Who does the importing?

4. What laws, treaties or agreements are in place to try to control this trade?

International trade in waste

Strange though it may seem, countries import and export a vast array of waste products, most commonly plastics, recyclables, e-waste and hazardous waste. EU shipments of recyclable items have increased by over 70% since 2000. Exporting plastic is thought to be one of the driving forces behind unsustainable production and consumption of plastic.

Most of the movement is from HICs to LICs. This is because it is cheaper to export the waste than to set up recycling infrastructure. At the same time, both energy and labour costs are reduced and so is the amount of waste going to national landfills. In addition, it encourages the establishment of new businesses to be established in the LICs, creating jobs. A problem is that this movement is not well tracked, and waste may be dumped illegally or burnt instead of recycled. There are also other issues.

E-waste is discarded electronics that have a battery or plug. These items may be unwanted because they are too slow, don't work anymore or are just unfashionable. Small electronic equipment makes up the largest share of e-waste—phones, microwaves, vacuum cleaners and kettles. Around 50 million metric tonnes of e-waste are generated every year—about 7 kg per person. Most of this waste goes to LICs in Asia and Africa where it is processed and recycled.

Hazardous waste

Once again, HICs benefit by exporting their hazardous waste and LICs accept it for treatment, disposal or recycling. The movement of hazardous waste has drawn international attention due to the environmental problems caused by exporting and dumping of this waste. The threat is not only environmental; it is also a health hazard to the millions of people who live in the LICs. However, many people in HICs turn a blind eye because the problems do not directly affect them. The environmental and health issues have been moved elsewhere.

The EU alone exported nearly 8.2 million tonnes of hazardous waste in 2020.

Impacts of waste trading

This trade has impacts on environment health, social well-being, economic development and human health. The impacts are worst in the receiving countries. Illegal dumpsites are now widespread, therefore so are the impacts.

Effects on humans

The lack of safe recycling processes means that people come into contact with toxic waste with no safety equipment or protective clothing. The results of this include:

*   burning of the skin

*   toxins absorbed through the skin, inhaled from the air, and taken in through eating and drinking contaminated food and water

*   thousands of lives cut short due to sickness and health problems including cancer, diabetes, hormone disruption, skin lesions, emphysema and reproductive damage.

Effects on the environment

Lack of proper disposal of the waste leads to:

- pollution of water sources and soil by heavy metals, toxins and chemicals from waste

- poisoning of the environment, leading to death of wildlife

- concentrations of persistent organic pollutants (POPs) which bioaccumulate and biomagnify putting birds, fish and other wildlife at risk.

Biodegradability and half-life

The length of time for which these waste materials present a problem depends on their biodegradability and half-lives.

Half-life (applies to non-biodegradable materials) is the time it takes for the material to reduce by half of its original value. It can take a significant amount of time for harmful pollutants to become harmless.

Biodegradable material breaks down through biological degradation of organic materials by living organisms. Such materials break down into the base elements of water, carbon dioxide and methane. They break down quickly and so do not accumulate in the environment.

Non-biodegradable material cannot be broken down or decomposed by biological processes. Most inorganic material is non-biodegradable and therefore causes problems because it accumulates and remains in the environment for a long time.

(ATL) Activity 26

ATL skills used: thinking, communication, research

Table 6 shows the half-life of a selection of non-biodegradable pollutants that we regularly release into the environment.

1. Add some other biodegradable materials and find out how long they take to decay or decompose.

2. Outline the impacts these different decomposition rates will have on the environment and ecosystems.

Non-biodegradables	Approximate half-life	Biodegradable material	Decomposition time
Methane in the soil	9.6 years	Carrots	1–2 weeks
Lead in soil	700 years		
Cadmium	30 years		
Chromium	3.74 million years		
Arsenic	6.5 years		
Mercury	3 months		
Plastics	up to 250 years or more		
Dioxins	11 years		
Barium	6 years		

▲ Table 6

Preventative waste management strategies

It is not straightforward to select the best management strategy for waste. Not polluting may seem the obvious strategy, but by being alive we all emit carbon dioxide and other greenhouse gases and produce waste. It would be impossible not to. Economies depend on production of goods and these need raw materials. Politicians have to make difficult choices which sometimes come down to a balance between employment for people or protecting the environment. All SDW management strategies have advantages and disadvantages but choices now should focus on increasing the background sustainability of everything we do.

Culturally, we may not be willing to change or we may not have the finance to invest to do so. However, Table 7 outlines some options for waste management.

Process of pollution	Level of pollution management
Human activity producing pollutant	Alter human activity, to reduce consumption: • reduce packaging • recycle goods • reuse clothes, goods, containers • compost organic matter.
Release of pollutant into environment	Control release of pollutants, often with legislation on waste disposal: • separate waste into different types • legislate about waste separation • educate for waste separation • tax disposable items.
Impact of pollutant on ecosystems	Clean-up and restore damaged systems: • reclaim landfills • incinerate SDW for energy • collect plastics, e.g. from the Great Pacific Garbage Patch.

▲ Table 7 Three-level model of waste management
Adapted from: Six Red Marbles and OUP

The sustainable circular economy

Most goods are produced in a linear model—"take, make, dump" (see subtopic 7.3, Figure 1 and subtopic 8.2). We find the raw materials or natural capital (take) and use energy to produce goods (make). When the goods become redundant or break down, we then discard and replace them with others (dump).

Our global economy has been built on this unsustainable premise, but the Earth and its resources are finite. We cannot really throw things away because there is no "away". Even reducing fossil fuel use and becoming more efficient at obtaining resources only delays the inevitable dwindling of natural capital available to humans.

The circular economy is a model that is sustainable (see subtopic 1.1).

It aims to:

* restore the environment

* use renewable energy sources

* eliminate or reduce toxic waste

* eradicate waste through careful design.

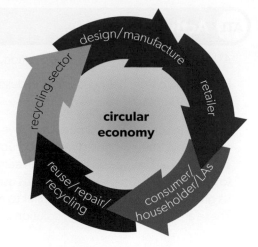

▲ Figure 23 The circular economy

To do these things, the model relies on manufacturers and producers retaining ownership of their products and taking responsibility for recycling them or disposing of them when the consumer has finished using them. The producers act as service providers, selling use of their products, not the products themselves. This means that they take back products when they are no longer needed, disassemble or refurbish them and return them to the market.

This model has similarities with agricultural practices in which good husbandry and soil conservation lead to sustainable growth of foodstuffs (Table 8).

Key principles of the circular economy model are summarized in Table 8. The Ellen McArthur Foundation website has more information on this model.

Principle	Agricultural sustainable practices	Circular economy practices
Design out waste	Reduce or eliminate food waste	Recycle plastics, metals
Build resilience through diversity	Manage for complex ecosystems	Build for connections and reuse of components
Use renewable energy sources	Use more solar energy, human labour, fewer chemicals and fossil fuels	Shift taxation from labour to non-renewable energy
Think in systems	Systems are non-linear, feedback-rich and interdependent; emphasize storages and flows	Increase effectiveness and interconnectedness in manufacturing
Think in cascades	Use all stages of a process. Decomposition recycles all nutrients. Burning wood shortcuts this and breaks down nutrients	Do not produce waste. Use it to produce more products

▲ Table 8 Similarities between sustainable agriculture and the circular economy

TOK

The circular economy is a paradigm shift.

To what extent can we be sure that a paradigm shift is for the better?

ATL Activity 27

ATL skills used: thinking, communication, social, research

Refer back to the Shell ad campaign described in Case study 2 of subtopic 1.1.

The text in some forms of the ad conveyed that if we had a magic bin, then we could make our rubbish disappear. The advert also suggests that we can actively find ways to recycle. For example, greenhouses use CO_2 to help us grow flowers, and our waste sulfur to make concrete. These are examples of energy solutions.

Consider your responses to these questions.

1. What do you think about the ethics of the advert?

2. Were Friends of the Earth right to complain or should we take advertising with a "pinch of salt"?

3. Find another advert (any format) about the environment and society that you think may be misleading. Discuss your findings in class.

The circular economy of a plastic product

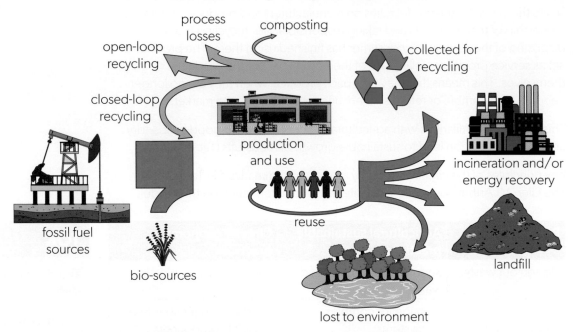

▲ Figure 24 Possible life cycles for plastic products
Adapted from: Philosophical Transactions of the Royal Society of London. Series A: Mathematical, Physical and Engineering Sciences

The circular economy model can be applied to everything we use, but it does take a monumental shift in thinking. Putting this model into action may require taxes, incentives, social policies, legislation, education campaigns and improved access to disposal facilities.

For example, a plastic product may need the following.

- Taxes on the production companies if they use fossil fuels instead of renewable energy sources.

- Incentives so that people return old, worn out or broken products to the manufacturer for reuse or recycling. This is done in western Australia where collection depots pay 10c for every plastic bottle given to them.

- Social policies that make it usual to return plastic products to the manufacturer.

- Legislation against the use of landfill.

- Education campaigns that teach people about the importance of sustainability and the circular economy.

- Improved access to disposal facilities so plastic products are placed in the right place to return to the manufacturers or to be recycled properly.

(ATL) Activity 28

ATL skills used: thinking, research

1. Draw a simplified version of Figure 24 for a product of your choice.

2. List the taxes, incentives, social policies, legislation, education campaigns and so on that could be needed to move your product to the circular economy model.

Check your understanding

In this subtopic, you have covered:

- the different types of waste that are generated and how production, treatment and management of it varies spatially and temporally

- the impacts of the pollution caused by waste and how to prevent the damage and restore the environments

- the advantages and disadvantages of different waste disposal options

- sustainable management of SDW

- circular economy solutions and a holistic and sustainable approach to waste management.

How can societies sustainably manage waste?

1. State the sources and types of waste generated from resource exploitation.

2. List the contents of SDW.

3. Discuss how the volume, composition and management of waste varies with time and space.

4. Outline the factors that impact the speed at which waste in the environment decays.

5. Discuss the sustainable strategies for waste management and how they can be promoted.

6. Explain how the circular economy is a holistic way to manage waste sustainably.

➤➤ Taking it further

- Investigate the many websites that have data on waste production and management. Look at changes through time, and spatially.

- Investigate the working conditions for waste disposal in LICs and raise awareness through a campaign.

- Visit your local waste management facility.

- Investigate how your society manages all types of waste.

- Ask the school administration what happens to waste generated by the school and whether this can be improved to decrease waste or involve the circular economy.

- Become involved in a repair cafe or a library of things.

Exam-style questions

1. With reference to named societies, to what extent do environmental value systems influence the use of resources? [9]

2. a. Using examples, evaluate **two** solid domestic waste disposal strategies as methods to mitigate climate change. [7]

   b. Using examples, discuss the potential impacts of climate change on ecosystem services. [9]

3. a. With reference to **Figure 1**, identify the recycling rate in England in 2018. [1]

   b. Outline **one** reason for the shape of the recycling rate curve from 2013 to 2018. [1]

   c. Estimate the reduction in solid domestic waste (in million tonnes) going to landfill from 2001 to 2018. [1]

   d. Describe **three** reasons why the proportions of solid domestic waste being recycled or composted and incinerated have changed. [3]

   e. Outline **one** reason why there has been an overall change in recorded total solid domestic waste between 2001 and 2018. [1]

4. Hydropower is a resource that can be exploited from rivers. Explain how the value of this resource to a society may vary over time. [7]

5. To what extent is the use of solid domestic waste as an energy source beneficial to a society? [9]

6. a. Outline how the concept of sustainability can be applied to managing natural capital. [4]

   b. To what extent does sustainability play a role in making decisions about energy and climate change policies at national and international levels? [9]

7. a. Identify **four** strategies for limiting the impact of burning fossil fuels without reducing their use. [4]

   b. Even though there is growing global support for ecocentric values, the global consumption of fossil fuels continues to rise each year.

      With reference to energy choices in named countries, discuss possible reasons for this situation occurring. [9]

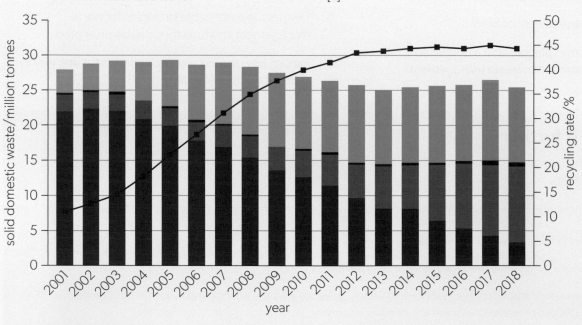

▲ Figure 1 Management of solid domestic waste in England, 2001–2018

8. a. With reference to **Figure 2**, state the country that has the highest level of recycling or composting. [1]

    b. Outline **two** possible reasons for greater use of landfills in the USA compared with the European countries shown in **Figure 2**. [2]

    c. Outline **two** strategies for reducing the environmental impact of landfill sites. [2]

    d. Identify **two** problems associated with **one** of the waste disposal choices of Germany. [2]

9. Examine the driving factors behind the changing energy choices of different countries, using named examples. [9]

10. a. Outline **four** different ways in which the value of named resources has changed over time. [4]

    b. The use of renewable resources is not always sustainable due to the activities involved in their production.

        Justify this statement for a named source of renewable energy. [7]

    c. Increasing concern for energy security is likely to lead to more sustainable energy choices.

        Discuss the validity of this statement, with reference to named countries. [9]

11. a. Outline the reasons why natural capital has a dynamic nature. [4]

    b. Explain how the inequitable distribution of natural resources can lead to conflict. [7]

    c. The management of a resource can impact the production of solid domestic waste.

        To what extent have the three levels of the pollution management model been successfully applied to the management of solid domestic waste? [9]

12. Solid domestic waste may contain non-biodegradable material and toxins that have the potential to reduce the fertility of soils.

    Explain how strategies for the management of this waste may help to preserve soil fertility. [7]

13. Evaluate **one** possible pollution management strategy for solid domestic waste. [8]

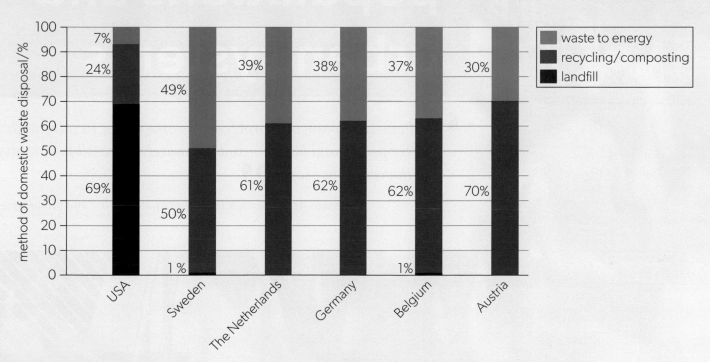

▲ Figure 2 Methods of domestic waste disposal for selected countries

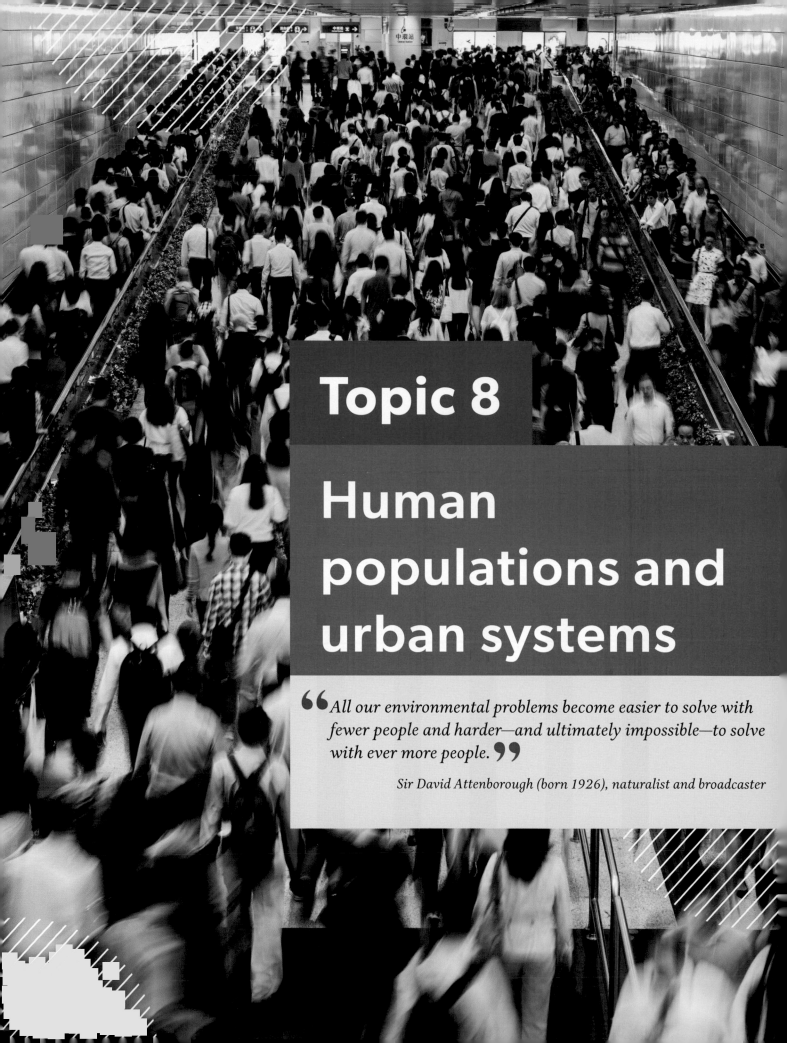

Topic 8

Human populations and urban systems

"All our environmental problems become easier to solve with fewer people and harder—and ultimately impossible—to solve with ever more people."

Sir David Attenborough (born 1926), naturalist and broadcaster

Guiding questions

- How can the dynamics of human populations be measured and compared?

- To what extent can the future growth of the human population be accurately predicted?

Understandings

1. Births and immigration are inputs to a human population.
2. Deaths and emigration are outputs from a human population.
3. Population dynamics can be quantified and analysed by calculating total fertility rate, life expectancy, doubling time and natural increase.
4. The global human population has followed a rapid growth curve. Models are used to predict the growth of the future global human population.
5. Population and migration policies can be employed to directly manage growth rates of human populations.
6. Human population growth can also be managed indirectly through economic, social, health, development and other policies that have an impact on births, deaths or migration.
7. The composition of human populations can be modelled and compared using age–sex pyramids.
8. The demographic transition model (DTM) describes the changing levels of births and deaths in a human population through different stages of development over time.
9. Rapid human population growth has increased stress on the Earth's systems.
10. Age–sex pyramids can be used to determine the dependency ratio and population momentum.
11. The reasons for patterns and trends in population structure and growth can be understood using examples of two countries in different stages of the DTM.
12. Environmental issues such as climate change, drought and land degradation are causing environmental migration.

AHL

Human population

- Are there too many humans alive on Earth today?

- Have we exceeded the carrying capacity of the Earth?

- Are we heading for a population crash?

The difficulty we have in trying to answer these questions is that humans are able to manipulate the environment. We can increase carrying capacity locally, live in large cities, live in regions that cannot grow enough food for the population, and use technology. Here we look at demographics—the study of the dynamics of population change.

Type "world population clock" into a search engine and you will find a number of websites that give an estimate of the human population on Earth.

Practical

Skills:

- Tool 2: Technology
- Tool 3: Mathematics

1. Use the numbers in Figure 1 to calculate the:

    a. number of births per year

    b. number of deaths per year

    c. population growth rate per year.

2. Find more recent values for the data in Figure 1.

3. Calculate the differences between February 2023 and now.

Figure 1 shows the world's population statistics at the time of writing.

Births per day	World population	Deaths per day
118,688	8,017,163,487	59,588

▲ Figure 1 Global population statistics, February 2023
Data from: Worldometer

Demographic indicators

Various quantitative measures are used to assess human populations. The human population is a system and, like all systems, it has inputs and outputs (the processes are not considered here). Figure 2 is not for the global population—at the time of writing, there is no immigration to or emigration from planet Earth!

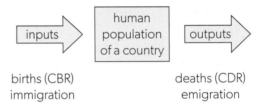

inputs → human population of a country → outputs

births (CBR)
immigration

deaths (CDR)
emigration

▲ Figure 2 Human population as a system
Data from: Worldometer

Inputs to the system

- Fertility rate is the number of births per thousand women of child-bearing age (15–44). The global average fertility rate is around 2.3 children per woman today. Over the last 50 years the global fertility rate has halved. Fertility rates in a country do not include migration data. Fertility and birth rates measure slightly different things.

- **Crude birth rate (CBR)** is the number of births per 1,000 or per hundred (%) of the total population, not per woman of child-bearing age. CBR is the number of births per thousand individuals—male and female, young and old.

- Fertility rate and CBR can be calculated at a global, regional or local scale.

- **Immigration rate** is the number of people entering a region or country to take up residence on a permanent or semi-permanent basis (not for a holiday).

Outputs from the system

- **Crude death rate (CDR)**, like CBR, is expressed as deaths per 1,000 or per hundred (%) of the total population, regardless of age or gender and the differences they make. This can be calculated globally, regionally or locally.

- **Emigration rate** is the number of people leaving a region or country on a permanent or semi-permanent basis.

Key terms

Crude birth rate (CBR) is the number of births per thousand individuals in a population per year.

Immigration rate is the number of immigrants per 1,000 population per year.

Crude death rate (CDR) is the number of deaths per thousand individuals in a population per year.

Emigration rate is the number of emigrants per 1,000 population per year.

Population dynamics

Key terms

Natural increase rate (NIR) is the rate of human growth expressed as a percentage change per year.

Doubling time (DT) is the time in years that it takes for a population to double in size.

Total fertility rate (TFR) is the average number of children each woman has over her lifetime.

Life expectancy (LE) is the average number of years that a person can be expected to live.

The basic measures that are used to assess human populations are also used to measure population size and how it is changing. The basic measures are used to calculate **natural increase rate (NIR)** and **doubling time (DT)**. More specific measures such as **total fertility rate (TFR)** and **life expectancy (LE)** are also indicators of population dynamics.

The TFR has a crucial number: the replacement fertility. This value—2.1—is the number of children each woman should have to maintain the population at zero growth rate. If the TFR is higher than 2.1, the population increases; if it is lower than 2.1, the population decreases. The reason it is 2.1 is because it is an average; some couples do not have any children, some have many. In reality, replacement fertility is around 1.75 in most HICs and many LICs. However, in Africa many countries have a rate in excess of 3.5 due to infant and childhood mortality and cultural and societal choices.

Natural increase rate (NIR) is calculated as follows.

$$NIR = \frac{\text{crude birth rate} - \text{crude death rate}}{10}$$

Dividing by 10 is essential to get a percentage. You must remember to calculate the NIR correctly (dividing by 10) to be able to know the doubling time.

Doubling time (DT) of a population is calculated as follows:

$$DT = \frac{70}{NIR}$$

Now do the maths:

- A population with NIR of 1% will double in size in 70 years.

- A population with NIR of just 2% will double in about 35 years. Such NIRs are pretty common, especially in Africa.

Data-based questions

Region	Australia	Brazil	Canada	India	Switzerland	Tanzania	World
Land area (km²)	7,741,220	8,515,767.049	9,984,670	3,287,263	41,285	947,303	
Population	26,141,369	217,240,060	39,292,355	1,408,000,000	8,636,896	61,741,120	
Pop. density	3.38	25.51	3.94	428.32	209.20	65.18	
NIR%							
CBR/1,000/year	12.3	13.96	10.17	16.42	10.5	33.3	
CDR/1,000/year	6.7	6.81	8.12	9.42	8.13	5.1	
TFR	1.73	1.8	1.57	2.0	1.46	4.4	
LE (years)	83.1	75.9	83.8	67.7	83.8	70.9	
DT (years)							

▲ Table 1

1. Do some research and complete the World column of Table 1.

2. Using the data provided, calculate for each country, and for the world:

   a. the NIR (natural increase rate)

   b. the DT (doubling time).

3. Describe the differences in the CBR between Tanzania and Switzerland.

4. Explain the differences in the CDR between India and Canada.

5. Discuss the impact of the DT for Tanzania.

6. Which country has the "best" DT? Outline why you picked this country.

Human population growth

The rate of global human population growth has, until now, followed an exponential curve. This is when population follows an accelerating rate of growth which is proportional to the population size. For example, a population increases in each generation from 2 to 4, 4 to 8, 8 to 16, etc.

Connections: This links to material covered in subtopics 1.2, 1.3 and 2.1.

Our current population growth rate is phenomenal—each year about 90 million people are born. Predictions are that, even with slowing growth rates, it will double again within another 100 years.

ATL Activity 1

ATL skills used: thinking, communication

Copy and complete Table 2 using the data in Figure 3.

Calculate the doubling time in years for the human population to increase:

a. from 2 to 4 billion

b. from 4 to 8 billion.

Years	Population	Doubling time
1804–1927	1–2 billion	123 years

▶ Table 2

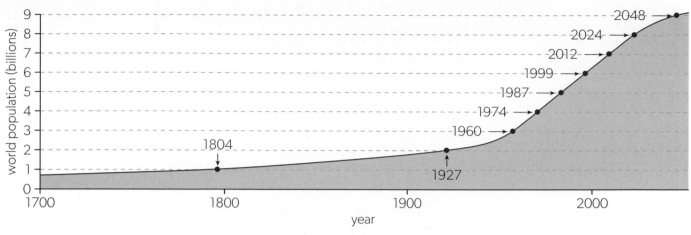

▲ Figure 3 World human population growth since 1700

TOK

Inequalities of life

If we could reduce the world's population to a village of precisely 100 people, with all existing human ratios remaining the same, the demographics would look something like this:

- The village would have 60 Asians, 14 Africans, 12 Europeans, 8 Latin Americans, 5 from the USA and Canada, and 1 from the South Pacific

- 51 would be male; 49 would be female

- 82 would be non-white; 18 white

- 80 would live in substandard housing

- 67 would be unable to read

- 50 would be malnourished and 1 dying of starvation

- 33 would be without access to a safe water supply

- 39 would lack access to improved sanitation

- 24 would not have any electricity (and of the 76 with electricity, most would only use it for light at night)

- 7 people would have access to the internet

- 1 would have a college education

- 1 would have HIV

- 2 would be near birth; 1 near death

- The 5 people from the USA would control 32% of the entire village's wealth

- 33 would be receiving—and attempting to live on—only 3% of the income of the village.

These figures are taken from the *State of the Village Report* by Donella Meadows, first published in 1990 and updated in 2005. They are controversial because some people think Meadows was biased in her use of statistics.

At the time of writing:

- the male:female ratio is 1.05:1

- almost 80% of the world's population is literate

- less than one-sixth of the world's population is malnourished

- about 3% of the world's population have a college education

- about 9% own a computer

- the USA controls no more than 30% of the world's wealth.

However, the "global village" values are still a stark demonstration of the inequalities of life.

Choose one of the following questions and discuss in class.

- How does the use of language influence how we perceive this knowledge?

- How do our values and assumptions influence the way we perceive this knowledge?

- To what extent is this an example of propaganda in favour of a particular viewpoint?

Of the eight billion humans alive today, about half live in poverty.

We know we are living unsustainably but we can only estimate what the future human population growth rate will be, and when and at what number the exponential curve will start to level out and even decrease (see Figure 4).

If you look at any statistics or any graph of projected human population growth, estimates vary enormously. This is because they are based on past and current trends. We can apply mathematical formulas to current figures but they assume human behaviour is predictable. It is also very hard to build in the impact of the demographic structure of the population. The two extremes assuming a CBR of 10 per 1,000 are as follows.

- An ageing population has fewer people of child-bearing age and more in the older age groups. That implies a lower birth rate and a higher death rate.

- A youthful population has a high proportion of the population in the reproductive age range (15–49 years old). This is called a demographic dividend—it causes population momentum, thus having a significant impact on population growth but a beneficial impact on economic growth potential.

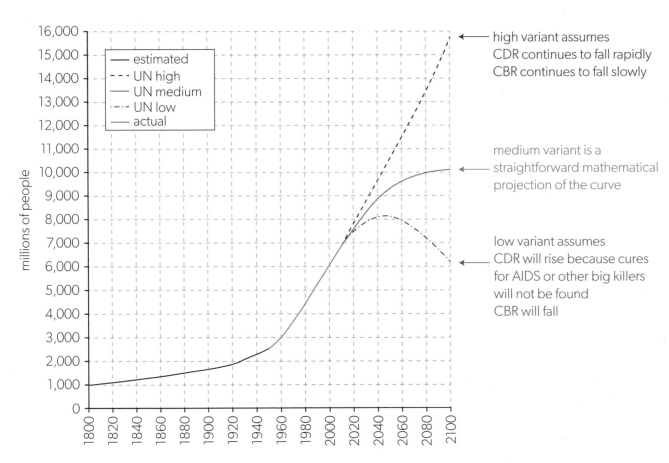

▲ Figure 4 Global human population growth rate over time and three predictions for the future
Adapted from: Six Red Marbles and OUP

Practical

Skills:

- Tool 2: Technology

- Tool 3: Mathematics

1. Find a graph showing the change in human population size from about 1,500 CE to now. Sketch this graph.

2. Exponential growth is characterized by increasingly short doubling times.

   Using Figure 3, copy and complete Table 3 with doubling times for the global population (in billions).

3. Use the data in your completed table to draw a line graph to show the global population and the doubling time.

Date	Population	Doubling time (yrs)	Date	Population	Doubling time (yrs)
1500	0.5	1,500		5.0	
1800	1.0	300		6.0	
1927	2.0			7.0	
	3.0			8.0	
	4.0			9.0	

▲ Table 3

What will be the future world population?

Figure 5 shows the changes in the global population from 1700 and projected to 2100. Projections are problematic because they involve predicting human behaviour, which is notoriously unpredictable. We can look at past trends but we cannot be sure they will continue.

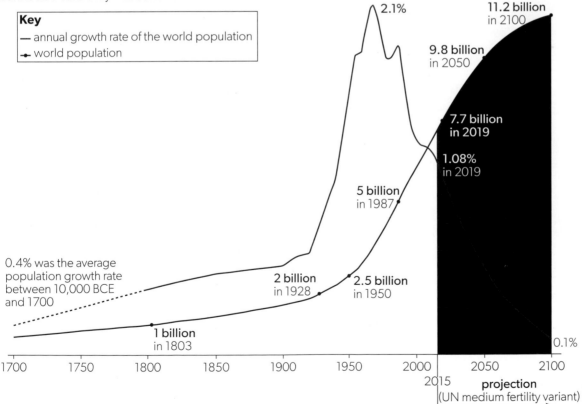

▲ Figure 5 World population growth, 1700–2100
Adapted from: Max Roser. Data from: Our World in Data based on HYDE, UN, and UN Population Division (2017 revision) (CC BY 4.0)

Why are fertility rates high?

It appears that the decision to have children is not correlated with GNP of a country nor with personal wealth. Some possible reasons for high fertility rates are outlined here.

1. High infant and childhood mortality: according to UNICEF, 5 million children under 5 die each year due to malnutrition and disease. Having lots of children increases the likelihood that at least some of them will reach adulthood.

2. Security in old age: traditionally, children take care of their parents in their old age. The more children the more secure the parents, and the less the burden for each child. If there is no social welfare network, children look after their parents.

3. Children are an economic asset in agricultural societies. They work on the land as soon as they are able. More children means more help—but also more children to feed. In HICs, children are dependent on their parents during their education and take longer to contribute to society.

4. Status of women: traditionally, women are regarded as subordinate to men. In many countries, women are deprived of many rights, such as owning property, having their own career, getting an education. Instead, they do most of the agricultural work and are considered worthy only to make children; their social status depends on the number of children they produce, particularly boys. Breaking down some of these barriers of discrimination (social or religious), and allowing girls to have an education and gain status outside the context of bearing children have contributed to low fertility rates in HICs.

5. Unavailability of contraceptives: in HICs, contraception is the prime way of reducing fertility. In LICs, an effective family-planning programme can aid socio-economic development. Women may not obtain contraception for many reasons—poverty, availability, religious beliefs, cultural practices.

6. Inheritance: in some countries girls cannot legally inherit property or wealth and therefore many families want a male heir to carry on the family business, inherit the wealth and continue the genetic line.

Ways to reduce fertility rates

1. Provide education in the form of basic literacy to children and adults.

2. Improve health by preventing the spread of diseases through simple measures of hygiene (boiling water), by improving nutrition, and by providing some simple medication and vaccines.

3. Make contraceptives and family counselling available.

4. Enhance income through small-scale projects focusing on the family level. Microlending, as in the Gramin Bank, is a practice that has had high success. Small loans are given, e.g. for a farmer to buy some seed and fertilizer to grow tomatoes; for a woman to buy pans to bake bread; for a weaver to buy yarn; for an auto mechanic to get some tools. Thus, small enterprises may start that will feed a whole family. Return of the loan is guaranteed through credit associations formed by members of the community.

5. Improve resource management. Local people may grow tree seedlings for transplanting in reforestation projects, or prevent erosion through soil conservation measures. (Note that large projects in LICs often do not work. Major projects like building dams for HEP leave LICs in debt (Third World debt) and force the population into cash cropping, e.g. tobacco, oil palm).

ATL Activity 2

ATL skills used: thinking, communication

The status of women

According to the UN, women's rights are key to reducing population growth rate. During this century, less than 20% of countries will account for nearly all of the world's population growth. Those countries (e.g. the least developed nations in sub-Saharan Africa and south Asia) are also where girls are less likely to attend school, child marriage is common, and women often lack basic rights.

1. Outline the issues which may maintain the low status of women.

2. Suggest proposals which might lead to a lowering of fertility rates.

3. List four reasons why educating women will reduce fertility or birth rates.

Mortality

Obviously a high CBR is not the only thing that increases population growth rates: the NIR involves both the CBR and the CDR. A major cause of global population increase is the falling death rate, which historically has fallen before the CBR falls, giving a wider gap between the two and an increasing population. Factors that impact the CDR are shown in Figure 6.

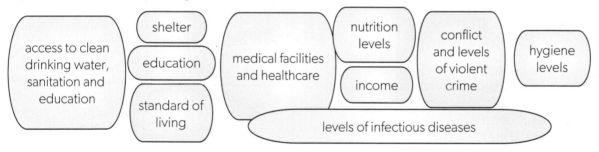

▲ Figure 6 Factors that impact CDR

ATL Activity 3

ATL skills used: thinking, communication, social

Consider the factors shown in Figure 6. Discuss how each factor may increase or decrease the CDR.

TOK

To what extent might the information in Figure 6 influence government policies in various countries?

Can this image change the way we interpret the world?

When is a country overpopulated?

CBR and CDR determine the global human population size. But optimum population sizes of different countries vary depending on economic factors and resource availability.

If the optimum population is when the population produces the highest economic return per capita, using all available resources, then some countries may have a higher optimum population density than others. The UK and the Netherlands have high population densities but can support this population with a high living standard. Brazil, with two people per km² in the north, is overpopulated as resources are much scarcer. The problem is that the richer countries have to import goods and services from elsewhere.

> Overconsumption and overpopulation underlie every environmental problem we face today.
>
> Jacques-Yves Cousteau

Population and migration policies

We may be able to count how many people are alive, what age they are and where they live. We can even predict changes in the future. But we do not really know how many people and other species the Earth can support. All the evidence we have at the moment is that we are using the Earth's resources unsustainably. But are we inventive enough to either live within our means or find ways to increase productivity?

There are direct and indirect policies and actions that can change population growth rates. They may involve cultural, religious, economic, social and political factors.

ATL Activity 4

ATL skills used: thinking, communication

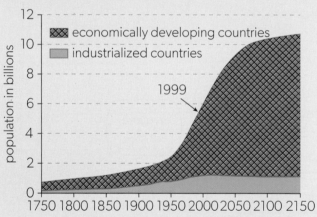

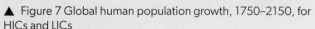

▲ Figure 7 Global human population growth, 1750–2150, for HICs and LICs

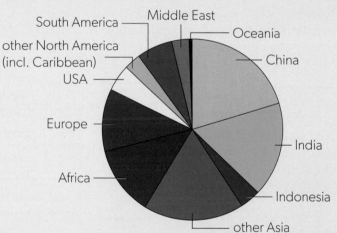

▲ Figure 8 Distribution of world population in 2022

1. With reference to Figure 7, calculate the percentage of the world's population living in LICs and HICs in 1950 and predicted in 2150.

2. Estimates of human population size vary greatly. Describe three reasons why estimates might be greater or smaller than the true figures.

3. Identify the limiting factors on human population growth.

4. Using Figure 8, estimate the percentage of the world population living in Asia in 2022.

5. The LICs comprise a large percentage of the global population. To what extent will this be the same in the future?

6. a. Using Figure 8, state the region in the world with the highest percentage of population in 2022.

   b. Predict how this may change by 2050.

Direct policies are implemented by governments and include the following.

- Anti-natalist policies—introduced to reduce the birth rate, e.g. the one-child policy in China, which operated between 1980 and 2016.

- Pro-natalist polices—aim to increase the birth rate, e.g. Romania, China's policy in 2023 (which allows three or more children per family).

- Immigration policies—control who can enter a country.

Indirect policies are less direct and often involve increasing levels of development. These include improvements in factors such as pensions schemes, gender equality, public healthcare, education and welfare.

Table 4 shows how these factors can increase or decrease population growth.

> Connections: This links to material covered in subtopics 1.3, HL.a and HL.b.

Policies that may reduce population growth rates	Policies that may increase population growth rates
• Parents in subsistence communities may be dependent on their children for support in their later years and this may create an incentive to have many children. If the government introduces pension schemes, the CBR comes down. • If you pay more tax to have more children or even lose your job, you may decide to have a smaller family. • Policies that stimulate economic growth may reduce birth rates as a result of increased access to education about methods of birth control. • Urbanization may be a factor in reducing CBR as fewer people can live in the smaller urban accommodation. • Policies directed towards the education of women, and enabling women to have greater personal and economic independence, may be most effective in reducing fertility and population pressures.	• Agricultural development, improved public health and sanitation may lower death rates and stimulate rapid population growth without significantly affecting fertility. • Lowering income tax or giving incentives and free education and healthcare may increase birth rates, e.g. the baby bonus in Australia. • Encouraging immigration, particularly of workers—for example, Russia allows migrants to work who do not have qualifications to fill the gap in manual labour.

▲ Table 4 Policies that may reduce or increase population growth rates

ATL Activity 5

ATL skills used: thinking, communication

1. Create your own table of policies that may reduce or increase population growth rates of countries. Add an example of a country that operates each policy.

2. Compare and contrast policies that reward the population with those that penalize or punish people for not following the policy.

10 REDUCED INEQUALITIES

▲ Figure 9 SDG 10

Migration policies

Migration is common in a globalized world. It is so common that it is in the 2030 SDGs (Goal 10) as part of the efforts to reduce inequalities. The aim is to encourage the implementation of planned and well-managed migration policies.

Basic facts

- The number of international migrants was 280 million in 2020. That is close to 4% of the world's human population and is nearly a 50% increase on 2000.

- Since the Second World War, 26 million migrants are refugees and asylum seekers.

- Reasons for migration include economic, social and environmental factors, political instability and conflict.

The SDG targets are to:

- retain health workers in LICs

- provide scholarships to study abroad

- end human trafficking

- respect the rights of migrant workers, especially women

- reduce the costs of remittance transfers

- provide legal identity for migrants.

Migration is beneficial for countries at origin and destination.

- It fills the labour gap in countries with an ageing population—15% of governments have policies to encourage migration for this reason.

- In 2016, migrants sent USD 429 billion home in remittances. China, India, Mexico and the Philippines benefit most from these remittances.

According to the UN data booklet *International Migration Policies* (2017):

- 61% of countries have policies to maintain current levels of immigration, e.g. member states of the European Union (EU)

- 13% of countries have policies to reduce the level of immigration, e.g. many Asian countries

- 14% of countries have no official policies about immigration

- 44% of countries have policies to encourage immigration of highly skilled workers, e.g. the Netherlands

- 41% of countries aim to maintain current levels of immigration of highly skilled workers, e.g. Chile

- 99% of governments use fines, detention or deportation of irregular or illegal migrants, e.g. Argentina

- 77% use penalties for employers of irregular or illegal migrants, e.g. Argentina.

ATL **Activity 6**

ATL skills used: thinking, communication, social, research

1. Research the migration policies of a country of your choice. State if the country is pro- or anti-natalist.

2. Produce a poster about the policies and explain them to your classmates.

3. To what extent does a country's development depend on its economy and its demographics to influence its development policies?

4. Discuss the cultural, historical, religious, social, environmental, political and economic factors that influence population and migration policies.

TOK

What kinds of knowledge should inform policy decisions regarding migration and population control?

To what extent are political decisions made on ethical grounds?

To what extent do political leaders and officials have ethical obligations and responsibilities?

Population pyramids

Population pyramids are also referred to as "age–sex pyramids" or "age–gender pyramids". They are a graphical illustration used to show the age and gender distribution of a population; they are very informative but rarely actually pyramid shaped. Figure 10 highlights some of the features of a population pyramid.

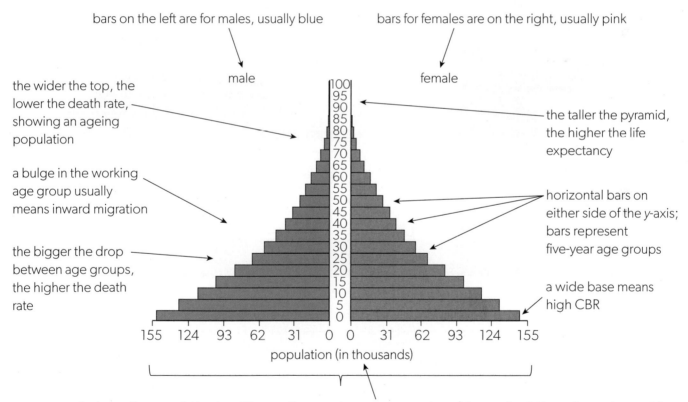

bars on the left are for males, usually blue

bars for females are on the right, usually pink

the wider the top, the lower the death rate, showing an ageing population

the taller the pyramid, the higher the life expectancy

a bulge in the working age group usually means inward migration

horizontal bars on either side of the y-axis; bars represent five-year age groups

the bigger the drop between age groups, the higher the death rate

a wide base means high CBR

population (in thousands)

x-axis shows the population in millions or thousands or as a percentage (always check the units on the x-axis)

▲ Figure 10 Annotated diagram of a generic population pyramid
Adapted from: Barking Dog Art / Q2A Media

There is a range of shapes of population pyramids and several different ways they can be displayed. Be very careful to check the unit for the population on the x-axis.

Connections: This links to material covered in subtopics 1.2 and 1.4.

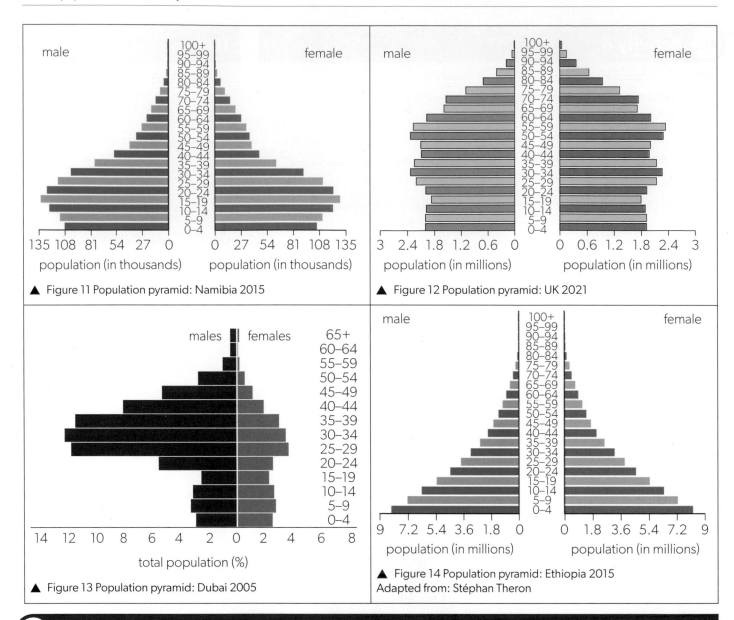

▲ Figure 11 Population pyramid: Namibia 2015

▲ Figure 12 Population pyramid: UK 2021

▲ Figure 13 Population pyramid: Dubai 2005

▲ Figure 14 Population pyramid: Ethiopia 2015
Adapted from: Stéphan Theron

ATL Activity 7

ATL skills used: thinking, research

1. Look at Figures 11–14. Match each description below to the correct pyramid and explain your choice.

a. • high birth rate • high under 5 mortality rate • fairly high death rate • low life expectancy	b. • falling birth rate • very high death rate 40–50 years ago	c. • steady birth rate • birth rate that fell consistently between 20 and 35 years ago • low death rate • high life expectancy

2. For the pyramid that did not match any of the descriptions, describe its shape.
   Explain the shape and the imbalance of males and females of working age.

3. Search online and find the Population Pyramids of the World website.
   a. Find the pyramid for your own country and comment on the change over time.
   b. Find the current world population pyramid and discuss the impacts of the shape for older people, working age people and those under working age.

Pro-natalist and anti-natalist policies: Singapore

Singapore has had both pro- and anti-natalist population policies at different times which reflect the changing dynamics of the city state.

Anti-natalist policies, 1972–87: "Stop at Two"

In 1972, Singapore's government started the anti-natalist policy with the slogan "Stop at Two". It introduced a series of policies to encourage lower birth rates—these included:

- establishing the Family Planning and Population Board (FPPB)

- access to low-cost contraception

- creating family planning clinics and enabling easy access to these clinics

- using media to promote smaller families

- free education for smaller families

- access to low-cost healthcare for smaller families

- promoting sterilization.

As the population pyramids for Singapore in 1970 (Figure 15) and 1980 (Figure 16) show, the anti-natalist policy had some success. This is shown by the "tucking in" of the base, showing falling birth rates.

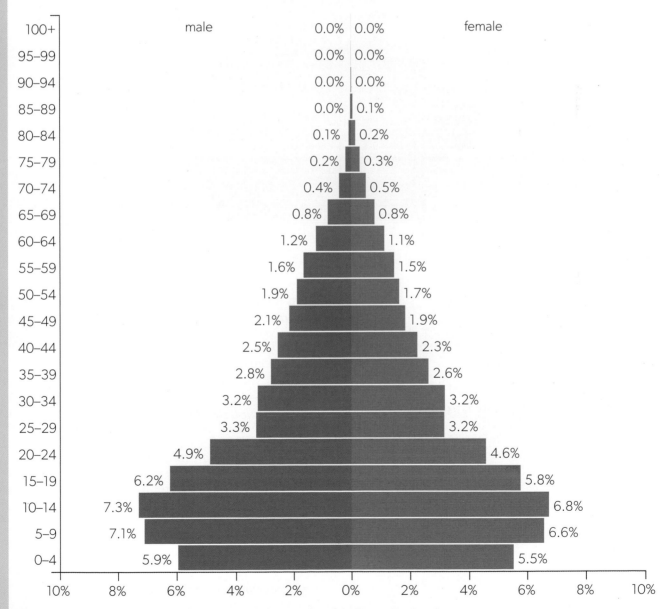

▲ Figure 15 Population pyramid for Singapore in 1970, before the "Stop at Two" policy

Case study 1

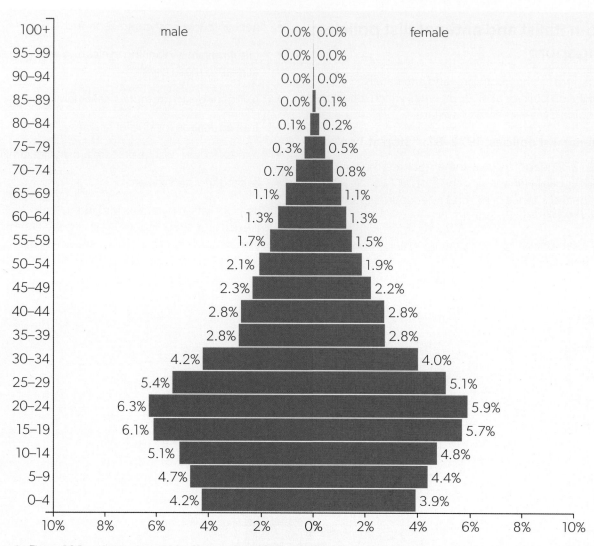

▲ Figure 16 Population pyramid for Singapore in 1980, after the "Stop at Two" policy

Success of the anti-natalist policy?

For most countries, including Singapore, anti-natalist policies tend to be successful. The reasons may be as follows.

- The ability of women to control their fertility and focus on education or jobs is a popular change.

- A change in attitudes to family size often means that financial incentives are not effective.

- Businesses are very supportive of such measures as it lessens their costs in maternity leave.

- Urbanization means smaller living spaces so a reduced family size is popular.

Pro-natalist policies, 1987 onwards: "Have three or more if you can afford it"

The success of the "Stop at Two" policy meant that Singapore faced a shrinking population, and all the issues associated with an ageing population. So, in 1987, Singaporeans were actively encouraged to "Have three or more if you can afford it". The strategies for this policy included:

- an increase in maternity leave to 12 weeks

- maternity leave covered for the first four children

- increase in child benefits to reduce concern about the costs of having children

- government-sponsored dating agencies

- government-subsidized childcare

- discouraging sterilization and abortion
- Family Planning and Population Board eliminated
- parents with good education (usually degree level) get enhanced child benefits.

Pro-natalist policies tend to be far less successful because (Figure 17) of the following.

- Once women have rights and understand that they can control their fertility they rarely want to return to having multiple children.
- In the broadly democratic HICs, government rarely has a strong influence over personal decisions.
- Governments are seen as being "controlling" and this is not popular.

- Businesses cannot always afford to support the initiative for extended maternity leave and this may lead to discrimination against women in the workplace.

Question

Copy Table 5 and list as many reasons as you can—just keep adding rows until you run out of ideas.

You can do some additional research to boost your thinking.

Anti-natalist policies		Pro-natalist policies	
Reasons for success	Reasons for failure	Reasons for success	Reasons for failure

▲ Table 5

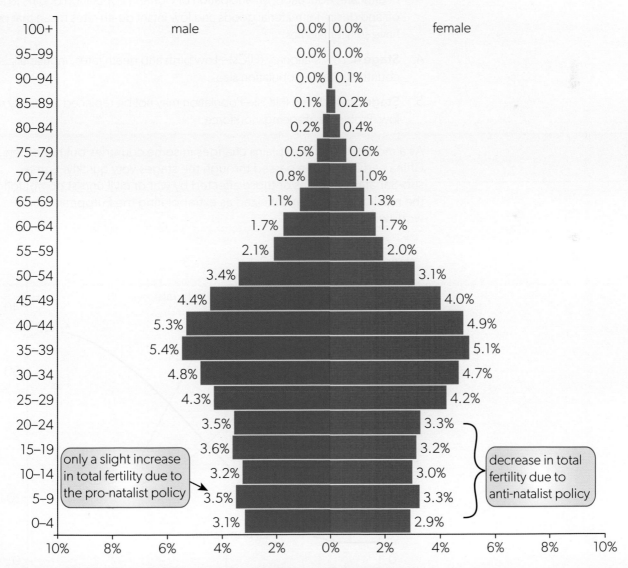

▲ Figure 17 Population pyramid for Singapore in 2000, after the "Have three or more…" policy

Demographic transition model

The **demographic transition model** (**DTM**) is the pattern of decline in mortality and fertility (natality) of a country as a result of social and economic development. Demographic transition can be described as a five-stage population model, which can be linked to the stages of the sigmoid growth curve.

1. **Stage 1**: High stationary (pre-industrial societies)—High birth rate due to no birth control, high infant mortality rates, cultural factors encouraging large families. High death rate due to disease, famine, poor hygiene and little medicine.

2. **Stage 2**: Early expanding (LICs)—Death rate drops as sanitation and food improve, disease is reduced so lifespan increases. Birth rate is still high so population expands rapidly and child mortality falls due to improved medicine.

3. **Stage 3**: Late expanding (wealthier LICs)—As a country becomes more developed, birth rates also fall due to access to contraception, improved healthcare, education, emancipation of women. Population begins to level off and desire for material goods and low infant death rates mean that people have smaller families.

4. **Stage 4**: Low stationary (HICs)—Low birth and death rates, industrialized countries. Stable population sizes.

5. **Stage 5:** Declining (HICs)—Population may not be replaced as fertility rate is low. Problems of ageing workforce.

As a model, the DTM explains changes in some countries but not others. China and Brazil have passed through the stages very quickly. Some sub-Saharan countries or those affected by war or civil unrest do not follow the model. It has been criticized as extrapolating the European model worldwide.

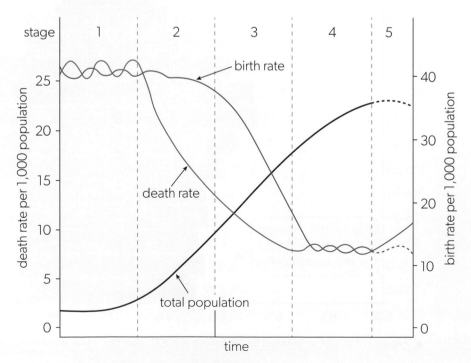

▲ Figure 18 The demographic transition model

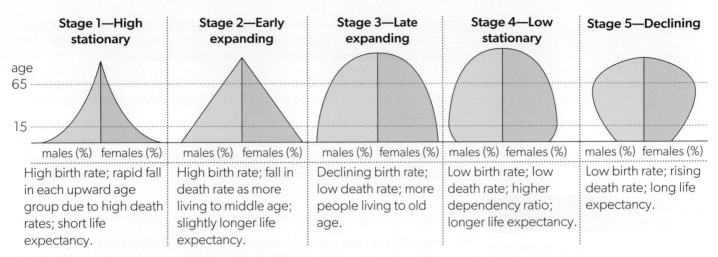

Stage 1—High stationary	Stage 2—Early expanding	Stage 3—Late expanding	Stage 4—Low stationary	Stage 5—Declining
High birth rate; rapid fall in each upward age group due to high death rates; short life expectancy.	High birth rate; fall in death rate as more living to middle age; slightly longer life expectancy.	Declining birth rate; low death rate; more people living to old age.	Low birth rate; low death rate; higher dependency ratio; longer life expectancy.	Low birth rate; rising death rate; long life expectancy.

▲ Figure 19 The five stages of the demographic transition model
Adapted from: Six Red Marbles and OUP

ATL Activity 8

ATL skills used: thinking, communication, research

1. Copy and complete Table 6 to summarize the characteristics of each pyramid.

Stage of DTM	High stationary	Early expanding	Late expanding	Low stationary	Declining
Birth rate					
Death rate					
Life expectancy					
Population growth rate					
Example					

▲ Table 6

2. For each pyramid below, identify the stage. Comment on the birth rate, death rate, life expectancy, gender differences and stage of development of the country.

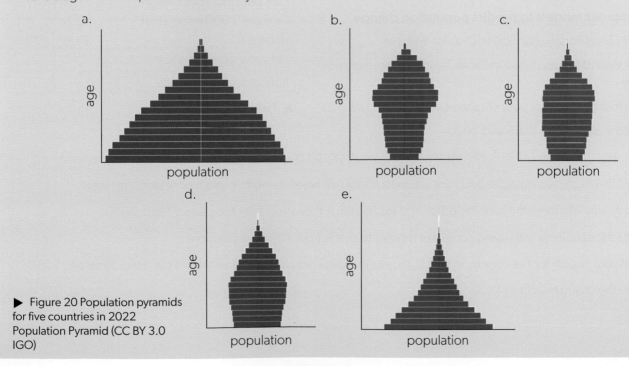

▶ Figure 20 Population pyramids for five countries in 2022
Population Pyramid (CC BY 3.0 IGO)

The DTM is controversial because it is based on change in several industrialized countries yet it suggests that all countries go through these stages. Like all models, it has limitations and strengths. The limitations are as follows.

- The initial model did not include the fifth stage, which has only become clear in recent years as countries such as Germany and Sweden have fallen into population decline.

- The fall in the death rate is not always as steep as the DTM suggests, because movement from the countryside to cities may create large urban slum areas which have poor or no sanitation and consequent high death rates of the young and infirm.

- Deaths from AIDS-related diseases and other pandemics may also affect this.

- The fall in the birth rate assumes availability of contraception and that religious practices allow for this. It also assumes increasing education of and increased literacy rates for women. This is not always the case.

- Some countries and regions have compressed the timescale of these changes. The Asian "Tiger economies" of Singapore and Hong Kong, for example, have leapt to industrialized status far more quickly than others.

- This is a Eurocentric model and assumes that all countries will become industrialized. This may not be the case in some "failed states", for example.

What are the strengths of the DTM?

TOK

To what extent are the methods of the human sciences "scientific"?

Practical

Skills:

- Tool 2: Technology

- Inquiry 2: Collecting and processing data

- Inquiry 3: Concluding and evaluating

Using computer models to predict population change

Find the UN Data Portal—Population Division website.

1. Select the indicators shown in Table 7 and two more of your choice.

2. Select four countries of your choice—two HICs and two LICs.

3. Select the date range 2015 and 2020.

4. Search and complete Table 7, adding columns for your chosen countries.

5. a. Explain which two variables you think are most important when trying to explain population growth.

   b. Justify why you think they are important and explain what they show over the period 1950–2050.

6. Discuss the problems with using computer models to predict population expansion.

7. Repeat your search for the country in which you live (or where you hold nationality) and the dates 1950 and 2050.

   Explain the changes in the various indicators.

Indicators	World	
	2015	2020
Median age of population (years)	28.5	29.7
Rate of population change (%)	1.2	0.9
CBR per 1,000 population	19.2	17.2
CDR per 1,000 population	7.5	8.1
LE (years)	71.8	72.0
Indicator of your choice		
Indicator of your choice		

▲ Table 7

Practical

Skills:

- Tool 1: Experimental techniques
- Tool 2: Technology
- Tool 3: Mathematics
- Tool 4: Systems and models

- Inquiry 1: Exploring and designing
- Inquiry 2: Collecting and processing data
- Inquiry 3: Concluding and evaluating

You may wish to collect secondary data for use in your individual investigation (II).

Useful websites include Gapminder, World Bank, Our World in Data, Nationmaster and IndexMundi.

Explore these websites and collect data to investigate the relationship between countries' level of development (as measured by GDP USD per capita) and a demographic factor of your choice.

You may choose to pair any of the following:

1. Level of development (independent variable):

    - human development index (HDI)

    - education (years in school)

    - literacy level (%).

2. Demographic factor (dependent variable):

    - infant mortality rate
    - CBD
    - CDR

    - NIR
    - ageing dependency
    - youthful dependency.

For a valid study, you should refer to at least three websites per variable and check the validity of the sites. The exception is the HDI, for which there is only one website—the UN Data Portal.

Human population growth: the implications

Remind yourself about the global human population growth rate and the UN predictions shown in Figure 4 in this subtopic. The different projections are based on the CBR, CDR and mathematical projections. The flaw is that the projections assume that past trends will continue.

> Connections: This links to material covered in subtopics 1.1, 1.2 and 1.3.

As predicted, the CBR has fallen steadily. Between 1950 and 2023:

- globally it fell by 19.9 per 1,000 per year

- in HICs, the fall was less at 13.5 per 1,000 per year

- in LICs, the fall was greater at 25.4 per 1,000 per year.

Also, as predicted, the CDR has fallen rapidly. Between 1950 and 2023:

- globally it fell by 11.9 per 1,000 per year

- in HICs, the fall was minimal at 0.8 per 1,000 per year

- in LICs, the fall was again greater at 16.5 per 1,000 per year.

AHL

As can be seen, the prediction for the high variant is relatively in line with past trends. The CBR is unlikely to increase by much in HICs as pro-natalist policies tend to be unsuccessful in the long run. The CDR is likely to increase in HICs simply because death is inevitable. There is also the unpredictable impact of pandemic viruses such as COVID-19. According to the WHO (17 February 2023) there were 756,581,850 confirmed cases of COVID-19 with 6,844,267 deaths globally. What impact this will have in the future is unknown.

Practical

Skills:

- Tool 2: Technology
- Tool 3: Mathematics

- Inquiry 2: Collecting and processing data
- Inquiry 3: Concluding and evaluating

Table 8 shows global population data, actual and predicted, from 2000 to 2100 (with 1950 for comparison).

Year	Global population	Year	Global population	Year	Global population
1950	2,499,322,156	2035	8,879,397,400	2075	10,370,994,198
2000	6,067,758,458	2040	9,188,250,492	2080	10,414,637,188
2005	6,558,176,119	2045	9,467,543,574	2085	10,430,679,101
2010	6,985,603,104	2050	9,664,516,145	2090	10,423,541,036
2015	7,426,597,536	2055	9,908,304,869	2095	10,396,305,844
2020	7,840,952,880	2060	10,067,733,606	2100	10,396,305,844
2025	8,191,988,453	2065	10,195,964,900		
2030	8,546,141,326	2070	10,297,166,711		

▲ Table 8

1. Transfer the data in Table 8 to a computer program that will allow you to draw graphs.

2. Experiment with different ways to show actual and projected global population growth from 2000 to 2100. Be imaginative: use at least three different types of graph and do not stick to the "standard" graphs.

3. In groups of two or three, analyse each other's graphs and comment on any patterns emerging.

Humanity's 21st century challenge is to meet the needs of all within the means of the planet.

Kate Raworth (born 1970), economist

TOK

How has the use of technology transformed different modes of human communication?

The doughnut economy

Meeting the needs of humanity means everyone has the essentials in life, such as food, housing, healthcare and political freedom of expression. The doughnut hole—the one in the middle—shows the proportion of people who are not getting these essentials. Part of the challenge is to get everyone out of the hole; the SDGs aim to do exactly that. However, going outside the doughnut's ecological ceiling puts pressure on life-supporting systems.

Humans today are pushing the limits of the Earth's ability to support us. In 2017, we overshot land use change, biodiversity loss, climate change and nitrogen and phosphorus loading. This means that four of the fundamental life-support systems are under threat. Check the model today to see if other ceilings have been crossed.

At the same time, however, we cannot afford to be overshooting the doughnut's outer crust if we are to safeguard the Earth's systems.

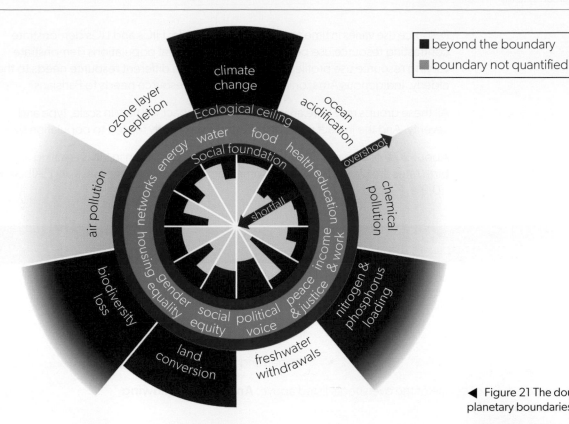

◀ Figure 21 The doughnut of social and planetary boundaries (2017)

A large proportion of people do not have the essentials.

Boundaries are overshot to provide both essentials and non-essentials, e.g. excess food or leisure activities.

▲ Figure 22 Overshooting the boundaries of the Earth's ability to support humans results in more inequity
Adapted from: Leremy / Shutterstock and norph / Shutterstock

Human population growth and resource use

A set of simple facts underlies population growth: more people require more resources and produce more waste; people usually want to improve their standard of living; the more people there are, the greater the impact they have.

If we can control population increase and resource demand, levels of sustainability should increase and there is a chance that we can come back within the ecological ceiling.

Size of population is not the only factor responsible for our impact on our resource base and on the environment in which we live. We also need to consider the wealth of a population, resource desire and resource need (or use). Many population impact models assume that all individuals (or all populations of a similar size) have the same resource needs and thus have the same impact environmentally (based on resource use and waste associated with exploiting a resource). However, individual resource use (and population resource use) is a dynamic principle.

Resource use varies in time and space. For example: HICs and LICs demonstrate contrasting resource use per capita. Urban and rural populations demonstrate varying resource use profiles. Young people have different resource needs to the elderly. Indigenous Amazonians have different resource needs to Parisians.

All these groups may have an impact, but the impact will vary in scale, type and severity. And the impact may not necessarily be linearly related to population size.

About 20% of us live in HICs, 80% in LICs. The proportion in HICs is falling as birth rates are higher in LICs and sometimes negative in some HICs (e.g. Albania −0.9%, Finland −0.2%).

ATL Activity 9

ATL skills used: thinking communication, social, research, self-management

The doughnut model (Figure 21) shows that we have overshot four of the planetary boundaries:

- land conversion
- climate change
- biodiversity loss
- nitrogen and phosphorus loading.

In groups of four, each student chooses one of the overshoots listed above. **Answer the following questions for your overshoot only**.

1. Outline all the human activities that you think are contributing to the overshoot.

2. Consider the factors involved in social foundation, shown in Figure 21.

    - food
    - housing
    - income and work
    - water
    - gender equality
    - education
    - energy
    - social equity or political voice
    - health
    - networks
    - peace and justice

    Copy and complete Table 9, considering each factor in turn.

Factor involved in social foundation	Does impact ✓ Does not impact ✗	Outline how the overshoot has an impact	Give a real-life example

▲ Table 9

3. Discuss your ideas with your group. Discuss whether you agree with each other and add or ideas.

Dependency ratios and population momentum

Age–sex pyramids provide a wealth of information. In addition to what was discussed earlier in this section we can also calculate the dependency ratio and population momentum.

Dependency ratio

$$\text{dependency ratio} = 100 \times \frac{\text{number of dependents (0–14) + (65+)}}{\text{working population (15–64)}}$$

This can be further broken down into

$$\text{youth dependency} = 100 \times \frac{\text{number of dependent children (0–14)}}{\text{working population (15–64)}}$$

$$\text{old-age dependency} = 100 \times \frac{\text{number of old-age dependents (65+)}}{\text{working population (15–64)}}$$

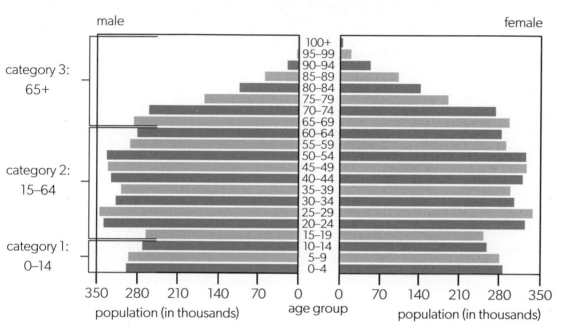

▲ Figure 23 Population pyramid for Sweden in 2015, showing the age groups for the dependency ratio

These ratios indicate changes in population age structure over time. They are a good indicator of the effects these changes may have on social and economic development. They allow governments to plan ahead and calculate how much of their expenditure will be needed to provide healthcare, social security and education for future populations.

Important features of the dependency ratio are as follows.

- A ratio of 100 means that every worker is paying taxes so the government can provide services for themselves plus one other person.

- It allows governments to predict future economic, social and health needs.

- As the ratio increases, it means more people are dependent on the working population; this has tax implications.

- As the percentage of dependents rises, there are proportionally fewer workers so there is less money going to the government in taxes. This often leads to an increase in taxes to boost government funds.

- At the top end (old-age dependency) there will be more people in need of pensions, health benefits and social support systems.

- Age determines the dependency ratio of the indigenous or local population, but this can be changed by government policies. For example, China's previous one-child policy decreased the youthful dependency significantly; policies to attract foreign workers to compensate for an ageing population will lower the dependency ratio.

- Breaking down the ratio into youthful and ageing allows for better planning—a large youthful population bodes well for the future as it will provide a large working population, but it will need to be provided with schooling and appropriate healthcare; a high level of ageing dependency is another matter. Older dependents will need age care facilities and very different healthcare provision, and this is generally very costly. They will also require pensions. The upside is that with improved health they often increase the incomes of many leisure facilities and businesses.

- Dependency ratio tends to be high in populations with a very high or very low fertility rate.

- A low dependency ratio is better than a high one as that means less burden on the workforce to support dependents. Table 10 shows highest and lowest dependency ratios globally.

	Dependency ratio	Youthful dependency	Ageing dependency
Niger	108.92	103.50	5.42
Qatar	18.38	16.17	2.22

▲ Table 10 Highest and lowest dependency ratios globally

In Niger, every 100 workers have to work to provide taxes for the government to provide services for themselves plus 109 other people because there are so many dependent young.

Qatar has the lowest dependency ratio in the world, with a youth unemployment rate of 0.5% of the total population and a reliance on migrant workers for large-scale infrastructure projects.

Problems with the dependency ratio

The delineation of the age groups does not really apply to all countries:

- In many LICs, children under 14 years old work.

- In most HICs, many children do not finish their education and work to pay taxes until they are at least 21.

- The age of retirement is now very fluid and varies between countries. In Turkey it is 52 years old for men and 49 for women, whereas in Norway and Iceland it is 67 years old for both genders.

There are other reasons people may not be economically active. They may be students, have an illness or disability, be stay-at-home parents or unemployed.

▲ Figure 24 Children working in a cotton mill in Macon, Georgia, in January 1909

Population momentum

It is obvious that fertility and mortality rates will indicate the natural increase rate; however, these are not the only factors that are important. **Population momentum** plays a key role, and it explains why populations continue to grow even when the TFR has fallen to or below replacement level. Population momentum and population dividend are closely linked and both can be "good" or "bad" for a country.

The reasons for population momentum relate to the population structure and the percentage of the population in the reproductive age group. The higher the proportion of young people, the higher the population growth. Even if this generation maintains the fertility rate at replacement level, the sheer number of women who can have children will mean that there are more births than deaths.

Table 11 shows statistics for two theoretical populations.

- Both countries have the same total population of 100 at the start.

- Both countries have the same TFR of 2 (below replacement rate).

- The LIC has a higher proportion of the population in the reproductive age range, so the same TFR does not mean the same number of births!

- The LIC has 80 more births than the HIC, leading to a faster population growth rate population momentum.

- The HIC has a higher proportion of the population in the older age groups so in addition to a lower CBR there is a higher CDR.

> ## Key term
>
> **Population momentum** is when a population continues to grow even if the fertility rate declines.

Data	15–49	50+	Total population	TFR	Number of births	Final population
LIC	90	10	100	2	$90 \times 2 = 180$	280
HIC	50	50	100	2	$50 \times 2 = 100$	200

▲ Table 11 Demographic statistics for two theoretical populations for a year

Country fact file: Niger

Demographics

- Total population: 25.25 million.

- Population growth rate is one of the highest in the world.

- Population is predominately youthful with approximately 49% under 15 years of age.

- Only 2.7% of the population over 65 years of age.

- Population is mainly rural—only 21% live in urban areas.

- Fertility rate is the highest in the world—6.9 per woman in her lifetime.

- A wide range of ethnicities but just over 55% of the population are Hausa.

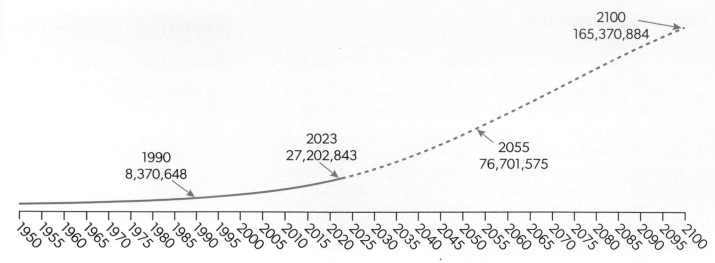

▲ Figure 25 Population of Niger: 1950 and projection to 2100
Data from: Population Pyramid (CC BY 3.0 IGO)

Population momentum

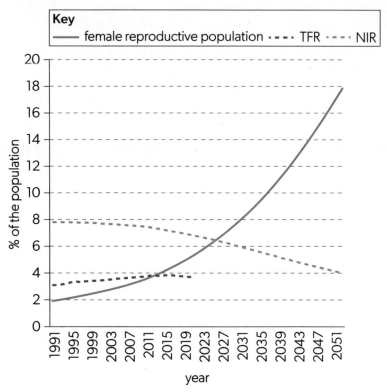

▲ Figure 26 Niger population momentum
Data from: The World Bank (CC BY 4.0) / Population Pyramid (CC BY 3.0 IGO)

ATL **Activity 10a**

ATL skills used: thinking, communication, research

1. Draw a sketch map of Niger to show where it is in Africa. Mark on the map the Sahara the Sahel, and the areas where most of the population of Niger live.

2. Explain why most of Niger's population is rural.

3. Outline the impact of high fertility rates on future population growth in Niger.

Education

- Literacy rate is one of the lowest in the world with only 37% of the population being able to read and write. Male literacy is 46%, female literacy is 29%.

- Primary education is compulsory for six years but the enrolment rate is only 46%—children have to help with the harvest rather than attend school.

- 50% of the children who finish education are males.

Health and well-being

- Health and well-being conditions are poor.

- Clean drinking water access is only 47%—and 39% in rural areas.

- Only 16% of the people have access to safely managed sanitation.

- Infant mortality rate is one of the highest in the world: 133 per 1,000 live births. Under-5 mortality rate is 330 per 1,000 live births.

- Maternal mortality is 820 deaths per 100,000 live births.

- Nurses are in short supply at 0.2 per 100,000 persons.

TOK

To what extent does the government have a moral or ethical obligation to improve healthcare and education in Niger?

ATL Activity 10b

ATL skills used: thinking, communication, research

4. Discuss how poor education and low literacy rates in females may impact fertility and mortality rates.

5. Draw an annotated diagram to show the interrelationship of health, education and demographics in Niger.

6. Using Figure 26, explain how the percentage of reproductive women impacts the TFR and the NIR.

Population structure and the demographic transition model

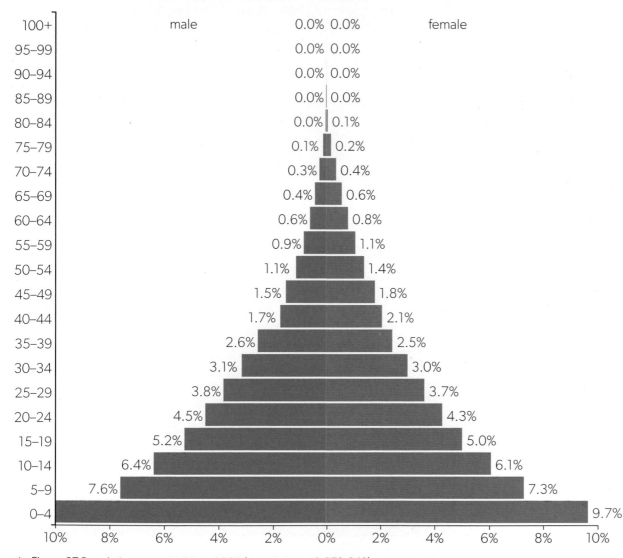

▲ Figure 27 Population pyramid: Niger 1990 (population = 8,370,648)

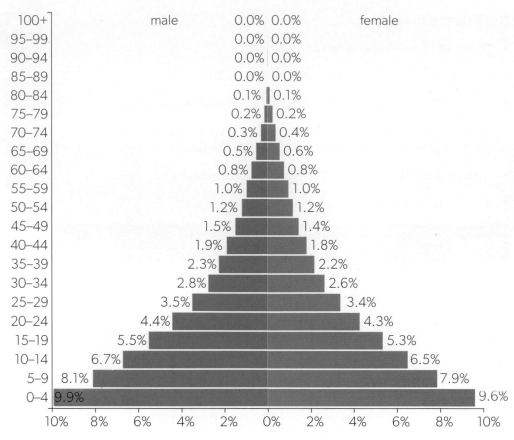

▲ Figure 28 Population pyramid: Niger 2023 (population = 27,202,843)

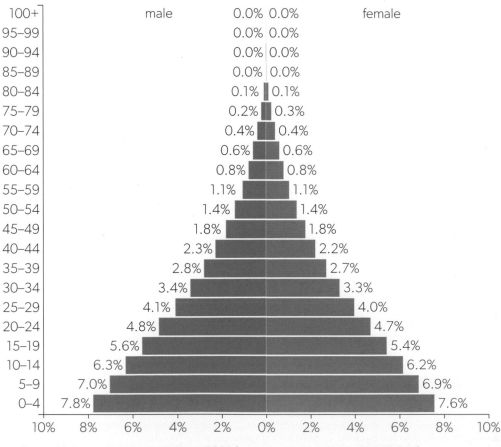

▲ Figure 29 Population pyramid: Niger 2050 (projected population = 67,043,296)

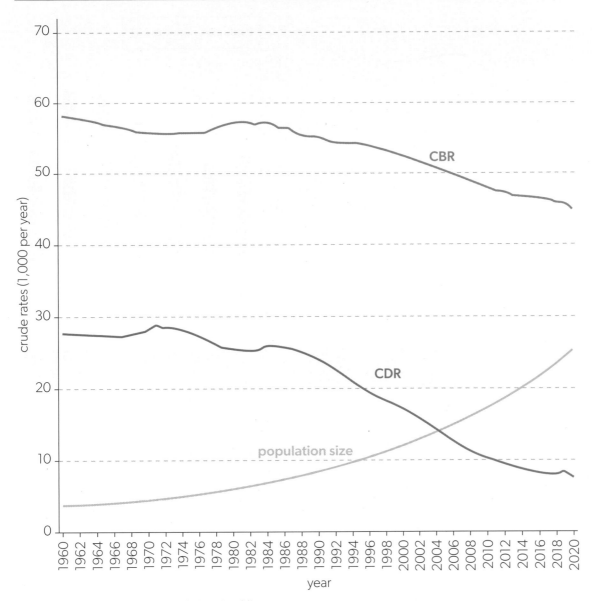

▲ Figure 30 CBR, CDR and population size: Niger
Data from: The World Bank (CC BY 4.0)

The dependency ratio

ATL Activity 10c

ATL skills used: thinking, communication, research

7. State what stage of the DTM Niger was at in 1990, and what stage it is predicted to be at in 2050.

8. Explain what progress Niger has made in the DTM between 1950 and 2023.

9. Analyse the impact of population momentum on the progress Niger is making in the DTM.

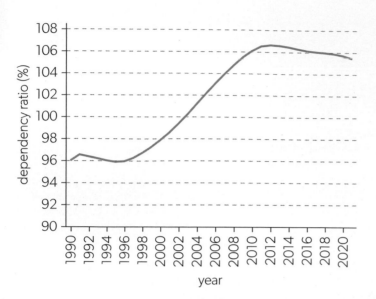

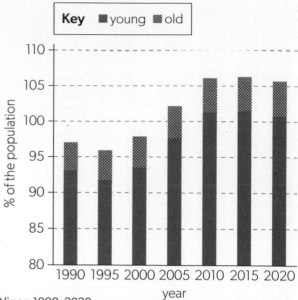

▲ Figure 31 Changes in the dependency ratio and young and old population in Niger: 1990–2020
Data from: The World Bank (CC BY 4.0)

(ATL) Activity 10d

ATL skills used: thinking, communication, social, research

10. Describe the changes in the dependency ratio of Niger between 1990 and 2020.

11. Using the data in Figure 31, predict the dependency ratio in Niger in 2050.

12. Justify your prediction.

13. Discuss the benefits of a high youthful dependency ratio.

Country fact file: Japan

Demographics

- Total population is 125.4 million, 98% of whom are Japanese.

- 92% of the population lives in urban areas.

- Population growth rate is −0.5%.

- Total fertility rate is 1.34 per woman in her lifetime, well below replacement rate.

- Population structure is 38% above 60 years of age; 28.7% above 65; median age 48.4.

- With an increase in the number of people remaining childless, the population is expected to drop to 88 million by 2065.

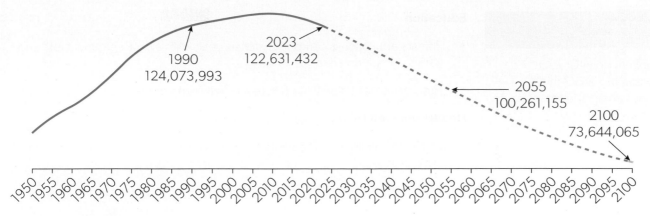

▲ Figure 32 Population of Japan since 1950, with projection to 2100
Data from: Population Pyramid (CC BY 3.0 IGO)

Population momentum

- In Japan, the TFR dropped below replacement rate in 1985.

- The population hovered around zero increase in 2009 when the population peaked at 129 million.

- The population started to shrink in 2010 and has been shrinking ever since.

- The stable population size between 1985 and 2009 was due to the inertia of Japan's youth entering reproductive years.

- The rate of decline is currently accelerating.

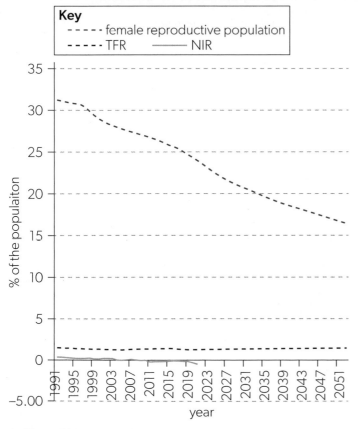

▲ Figure 33 Japan population momentum
Data from: Population Pyramid (CC BY 3.0 IGO)

ATL **Activity 11a**

ATL skills used: thinking, communication, research

1. Draw a sketch map of Japan to show where it is in the world. Annotate your map to show the biomes of Japan and where most of the population live.

2. Explain why most of Japan's population is urban.

3. Outline the impact of such low fertility rates on future population growth.

TOK

To what extent do HICs such as Japan have moral or ethical obligations to help LICs improve education and healthcare?

ATL Activity 11b

ATL skills used: thinking, communication, research

4. Discuss how education and literacy have impacts on fertility and mortality.

5. Compare and contrast Japan and Niger in terms of education and health.

6. Using Figure 33, explain how the percentage of reproductive women impacts the TFR and the NIR.

Education

- Primary education is compulsory for six years with a 99% enrolment rate and a 99% completion rate.

- Literacy rate is 99% with no difference between genders.

Health and well-being

- Healthcare is provided by the government and universal insurance schemes are available.

- Elderly people have been covered by government-sponsored insurance since 1973.

- Life expectancy is 84.62 years (81.64 years for males; 87.74 years for females).

- 99% of the population has access to clean drinking water.

- 81% of the people have access to safely managed sanitation.

- Infant mortality rate is 5 per 1,000 live births.

- Under 5 mortality rate is 6 per 1,000 live births.

- Maternal mortality is 4 deaths per 100,000 live births.

- Nurses are in adequate supply at 12.7 per 100,000 persons.

Population structure and the demographic transition model

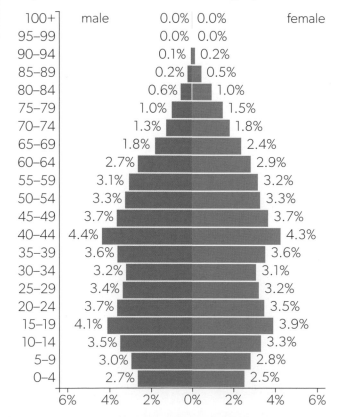

▲ Figure 34 Population pyramid: Japan 1990 (population = 123,686,321)

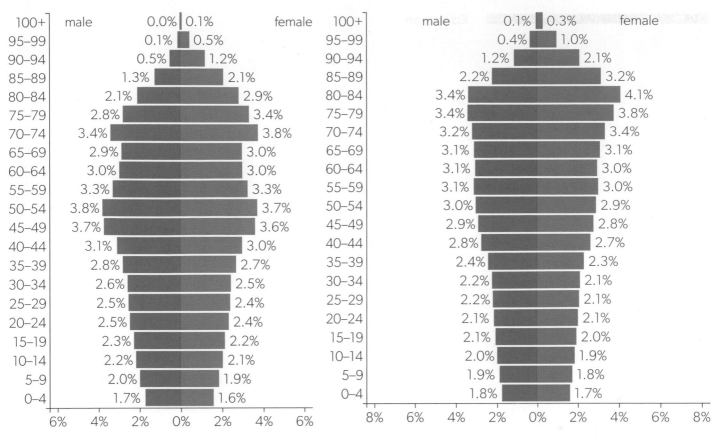

▲ Figure 35 Population pyramid: Japan 2023
(population = 123,294,513)

▲ Figure 36 Population pyramid: Japan 2055
(projected population = 100,261,155)

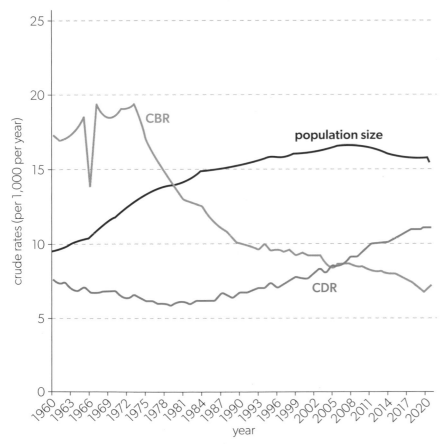

▲ Figure 37 CBR, CDR and population size: Japan
Data from: The World Bank (CC BY 4.0)

ATL Activity 11c

ATL skills used: thinking, communication, research

7.   State what stage of the DTM Japan is at in:

   • 1990

   • 2023

   • 2055.

8.   Research what happened in Japan when the CBR dropped from 18.7 in 1965 to 13.8 in 1966.

9.   Explain what problems Japan will face in 2055 if the projected population pyramid becomes reality.

10. Analyse the impact of population momentum on Japan's population structure.

The dependency ratio

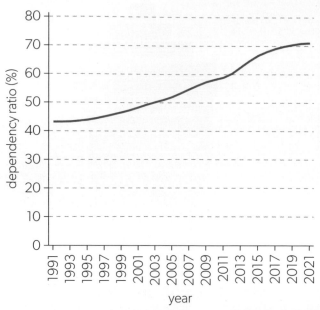

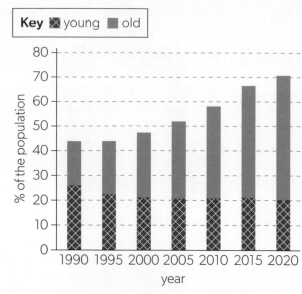

▲ Figure 38 Changes in the dependency ratio and young and old population in Japan: 1990–2020
Data from: World Bank Group (CC BY 4.0)

ATL **Activity 11d**

ATL skills used: thinking, communication, social, research

11. Describe the changes in the dependency ratio of Japan between 1990 and 2020.

12. Explain the differences between the dependency ratios of Japan and Niger.

13. Discuss the costs and benefits of high old age dependency.

14. Suggest strategies to deal with high old age dependency.

Migration of populations as a result of failing ecosystems, vulnerability to natural hazards and gradual climate-driven environmental changes causing poverty and hunger.

UNEP

Connections: This links to material covered in subtopics 1.2, 1.3 and 6.2.

Environmental issues and migration

People leave their homes because of issues caused by climate change.

Changes can be:

- short term—extreme weather, droughts, disrupted seasonal weather
- gradual—sea-level rise, desertification, salinization.

Changes compromise well-being and livelihoods by:

- disrupting economic well-being
- impacting health and safety
- compromising food supply as rising temperatures cause crop failures
- destruction of housing.

In 2021 extreme weather events displaced 23.7 million people. Some could return home, for others the move was permanent.

LICs suffer particularly badly as they already have insufficient resources.

Categories of environmental migration

Environmental emergency migrants flee temporarily due to an environmental disaster or sudden event such as a hurricane or tsunami.

Environmental forced migrants leave their home areas because the environment is deteriorating or degrading (for example, rising sea levels lead to coastal erosion).

The people of Kiribati, Nauru and Tuvalu are forced to migrate due to climate change which is causing sea level rise, saltwater intrusion and drought.

In the last 10 years, 94% of households in Kiribati, 74% of households in Nauru, and 74% of households in Tuvalu have been impacted by climate change. People have had to find new homes to ensure some sort of livelihood—income and a place to live. And 23% of the migrants in Kiribati and 8% in Tuvalu named climate change as the reason for migrating.

Environmental motivated migrants move to avoid expected climate-induced threats in the future. For example, as crop productivity falls due to rising temperatures, decreased precipitation and encroaching desertification, people leave the area to find more suitable conditions.

▲ Figure 39 Aerial photo of the island of Tuvalu

The Pacific Islands have many examples of environmental motivated migrants. These people have contributed little to the causes of climate change, but they suffer the negative consequences:

- extreme weather events are becoming more severe

- rising temperatures are changing patterns of precipitation

- rising sea levels

- ocean acidification.

For example, Narikoso in Fiji is threatened by rising sea-levels: the coastline has moved 15 metres inland in 30 years and many houses are now standing in the water. New homes are now being built 150 metres away from the original village so they will not be flooded, but remain close enough to the old village to maintain community. People are also receiving training to adjust farming practices to suit the new climate conditions.

 Practical

Skills:

- Tool 2: Technology

- Tool 3: Mathematics

- Inquiry 1: Exploring and designing

- Inquiry 2: Collecting and processing data

Demographic factors provide information on demographic processes and their outcomes (for example, CBR, CDR, TFR, migration levels, mortality rates, population growth rates).

Socio-economic indicators track economic progress and social change, and generally portray a people's state of well-being and quality of life (for example, employment levels, housing quality, family incomes).

1. Devise a hypothesis on the relationship between any socio-economic indicator and a demographic factor.

2. Use secondary data from sources such as Gapminder, World Bank and Our World in Data to test your hypothesis.

3. Use a suitable statistical tool such as the Social Science Statistics calculator.

Check your understanding

In this subtopic, you have covered:

- population dynamics, the vital rates and how they change

- how the global population is growing and how that growth can be managed

- population composition and how it changes through time using population pyramids and the demographic transition model (DTM)

- how human population growth is stressing the Earth's systems

- dependency ratios and population momentum and the impact they have on population structure

- environmental migration.

AHL

How can the dynamics of human populations be measured and compared?

To what extent can the future growth of the human population be accurately predicted?

1. Define: demographic inputs and outputs, CBR, CDR, TFR, LE, NIR, doubling time, immigration and emigration.

2. Discuss the reasons for rapid growth of the human population and the uncertainties in predicting future trends.

3. Compare and contrast population and migration policies and how they impact growth rates of the human population.

4. Discuss the factors that impact success of population and migration policies.

5. Analyse the use of population pyramids.

6. Describe the DTM and the links between it and population pyramids.

7. Analyse how human population growth has put a strain on Earth's systems and crossed planetary boundaries in the doughnut economy model.

8. Determine the dependency ratio and population momentum from sampled population pyramids.

9. Discuss reasons why two named countries are in different stages of the DTM.

10. Examine two examples of environmental migration.

AHL

▶▶ Taking it further

- Assess and discuss issues regarding population change in your local or regional area.

- Investigate traditional migration routes that still exist.

- Investigate how the traditional migration routes are being threatened and what the solutions may be.

- Work with an appropriate agency to volunteer in a local refugee centre.

- Engage with local NGOs supporting seasonal or indigenous communities.

- Support the UNHCR.

Guiding questions

- To what extent are urban systems similar to natural ecosystems?

- How can reimagining urban systems create a more sustainable future?

Understandings

1. Urban areas contain urban ecosystems.
2. An urban area is a built-up area with a high population density, buildings and infrastructure.
3. An urban area works as a system.
4. Urbanization is the population shift from rural to urban areas.
5. Due to rural–urban migration, a greater proportion of the human population now live in urban rather than rural systems, and this proportion is increasing.
6. Suburbanization is due to the movement of people from dense central urban areas to lower-density peripheral areas.
7. The expansion of urban and suburban systems results in changes to the environment.
8. Urban planning helps to decide on the best way to use land and buildings.
9. Modern urban planning may involve considering the sustainability of the urban system.
10. Ecological urban planning is a more holistic approach that treats the urban system as an ecosystem, understanding the complex relationships between its biotic and abiotic components.
11. Ecological urban planning will follow principles of urban compactness, mixed land use and social mix practice.
12. Societies are developing systems that address urban sustainability by using models such as a circular economy or doughnut economics to promote sustainability within the urban system.
13. Green architecture minimizes harmful effects of construction projects on human health and the environment, and aims to safeguard air, water and earth by choosing environmentally friendly building materials and construction practices.

The urban area

Many of us live in urban areas but what is an urban area? Whether we call it an urban area, city, suburb or town, all refer to the same concept in this section—a human settlement with high population density and built infrastructure.

An urban area is:

- a city or town and the region that surrounds it (the suburbs)

- where the population is linked together by social and economic interactions—these may be regional, national or global in scale

- where most inhabitants are in non-agricultural jobs

- a built-up agglomeration of human settlement with grey infrastructure (roads, metros, railways, buildings and utilities that determine the layout of a city); generally high human population density; and environmental pressures

Ultimately we need to recognize that while humans continue to build urban landscapes, we share these spaces with other species.

David Suzuki

▲ Figure 1 The urban race—expansion of urban space into suburbs and the countryside

Connections: This links to material covered in subtopics 1.2 and 5.3.

- created through urbanization

- efficient if well planned, designed and managed as green infrastructure brings nature into cities to provide social, ecological and economic benefits

- problematic if poorly planned due to: fragmentation of natural ecosystems; pollution; higher temperatures; sealing of the soil stopping infiltration; increased consumption of natural resources.

Urbanization and the urban area

Key term

Urbanization is the population shift from rural to urban areas.

In 1800 less than 10% of the global population lived in an urban area. This is no longer the case. In 2007, the urban population was 3.35 billion—slightly more than the rural population of 3.33 billion. This shift from rural to urban areas is called **urbanization** and is due to a variety of factors.

Figure 2 shows global urbanization from 1960 to 2017. You can see that the increase in urban populations has been consistent and rapid whereas the increase in rural populations levelled off around 2004.

Urbanization is caused by industrialization. In the past, secondary industries such as manufacturing attracted people to cities. Now, tertiary (service sector) and quaternary (knowledge and information based) industries are the attraction, especially in HICs. People think that cities will offer higher paid jobs, better education and healthcare options along with entertainment, communications and better lifestyles.

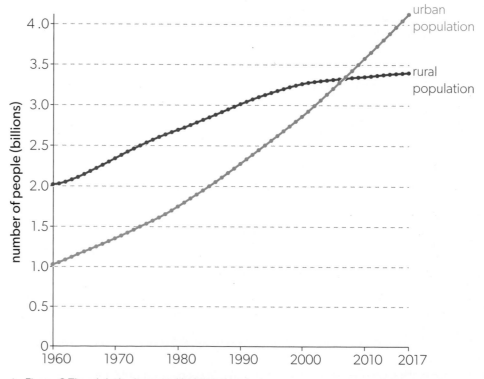

▲ Figure 2 The global urban-rural balance

Practical

Skills:

*   Tool 3: Mathematics

Table 1 shows data from Our World in Data: urban and rural world human population numbers in 1960 and 2017.

	Urban	Rural
1960	1,019,029,723.00	2,012,175,449.00
2017	4,121,601,045.00	3,395,754,782.00
% change		

◀ Table 1

1. Calculate the percentage change between 1960 and 2017 for the:

    a. urban population
    b. rural population.

2. Draw a bar graph to show the data in Table 1.

3. Explain if it is better to have years or the rural or urban population on the *x*-axis.

The relatively recent move to urban areas means that people now live in increasingly dense concentrations, have different work habits and move and communicate in different ways. So what does all this mean?

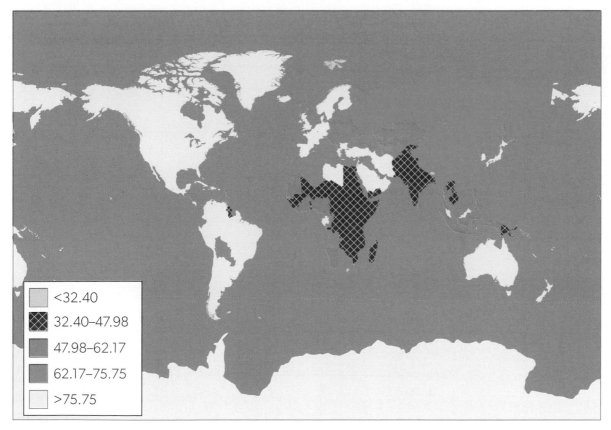

	<32.40
	32.40–47.98
	47.98–62.17
	62.17–75.75
	>75.75

▲ Figure 3 World map showing the urban population as a percentage of total population in 2021
Adapted from: The World Bank. Data from: United Nations Population Division. World Urbanization Prospects: 2018 Revision (CC BY-SA 4.0)

If you search for "Urbanization" on the Our World in Data website, you will find lots of interesting information about this topic.

The highest levels of urbanization are in HICs where over 80% of the population is urbanized. In LICs, most of the population is still rural. Despite the recent expansion of urban areas, only around 1% of global land is built-up. To maintain this percentage, urban areas must be operated sustainably.

Urbanization is likely to continue as the mechanization of agriculture means less employment in that sector and a move to the cities for employment. This increase in the number of urban dwellers is relevant because there will be a larger number of people in a small space. All these people need resources for education, employment, housing, healthcare and transport—all of which need to be well planned to ensure sustainability. This is in line with the UN's SDG11: "make cities inclusive, safe, resilient and sustainable". To do that, governments must know how many people they have to provide for.

11 SUSTAINABLE CITIES AND COMMUNITIES

▲ Figure 4 SDG 11 Sustainable cities and communities

ATL Activity 12

ATL skills used: thinking, communication, social, research, self-management

Working in pairs, complete Table 2 to show the differences and similarities between urban and rural areas.

- In class, identify any additional activities you can think of and add them to Table 2.

- Individually, find a single picture for each box that you feel shows the differences or similarities between urban and rural areas.

Activities	Urban areas	Rural areas
Population density		
Buildings density		
Energy		
Cultural		
Productivity		
Trade		
Social		
Microclimate		
Transport		
Water supply		
Sanitation		
Plants		
Animals		
Services		
Pollution		

▲ Table 2

The urban system

Urban areas can be considered as systems, with inputs, processes and outputs. Any urban area is a complex interconnected system of buildings, microclimate, transport, goods and services, energy supply, water supply, sewage facilities, humans, plants and other animals.

To identify the inputs, processes and outputs, it is important to understand what happens in an urban area and what impacts the activities have on the environment and the inhabitants of the area. Many cities have been functioning as cities for a very long time so their functions and activities have changed.

Activities and functions of a city

- Defence: many of the cities that were established a long time ago had defences built to protect them from attack. They may still have defensive walls and buildings.

- Capital: this is usually the government administrative centre and may change through time. The reasons for the changes are many and varied. For example, the capital of Brazil was first Salvador (Portuguese colonial administration centre), then Rio de Janeiro from 1763 to 1960 and now, after 1960, it is Brasilia. Similarly, in 2022, Indonesia announced its capital was to move from Jakarta to a new city named Nusantara.

- Law and order: all cities have a law enforcement.

- Trading hubs: all urban areas are centres of trade to varying degrees and at varying scales, from local to regional to international. Trade often requires a vast amount of infrastructure to support it.

- Manufacturing areas: in most towns the heavy industries (such as engineering) have disappeared or moved to the outskirts as they require large amounts of land.

- Centres of tertiary and quaternary industry—tertiary industry is often referred to as the service sector (e.g. finance, medical services, education); quaternary industry is knowledge- and information-based (e.g. financial planning, consultancy, IT, research and design).

- Homes to headquarters or regional offices for many corporations.

- Sources of jobs, goods and services: the type of jobs and services will vary depending on the size of the settlement and will vary through time.

- Transportation hubs: if you study any map, you will see that various transport routes focus on major towns and cities, largely due to the other functions that they serve.

- Religious centres: nearly all urban areas have a wide range of religious buildings or sites such as temples, churches and community centres.

- National monuments and museums are found in most cities and often show the historic element of a settlement; they can also be a symbol of independence.

- Centres of diversity and the arts because they attract so many people from different backgrounds and cultures.

- Media centres for news and entertainment: the central positions of urban areas make them advantageous for these organizations.

- Sources of basic essentials for living such as shelter, water, sanitation and energy sources (usually electricity).

Fairtrade: Bonn, Germany

Greenpeace: Amsterdam, the Netherlands

▲ Figure 5 Headquarters for two major companies and organizations

These activities have both beneficial and harmful impacts:

- Urban areas shape cultures, human mobility, social structures, the natural environment and political and economic systems.

- Traffic congestion contributes significantly to global climate change.

- Over-crowding causes stress for many people and disease may spread easily.

- Most cities suffer from segregation issues based on wealth, race and religion.

- It is not easy to provide all the necessary services—education, health facilities, water and sanitation and waste disposal.

- There are significant air, water and noise pollution problems.

- Cities have their own microclimate due to the density of buildings, thermal response of the buildings, increased emissions and lack of natural surfaces. This is the heat island effect.

- Loss of agricultural land and other natural environments.

ATL Activity 13

ATL skills used: thinking, communication, research

Urban ecosystems, like all ecosystems, have biotic and abiotic components, stores and flows, inputs and outputs.

Stores include biotic components (plants, animals, other forms of life) and abiotic components (soil, water, air, buildings, roads, other materials).

Flows into and out of the system are water and food, fuels, energy, materials, waste, labour and services.

Draw a systems diagram of an urban area, including as many inputs, outputs and stores as you can. Here is one example.

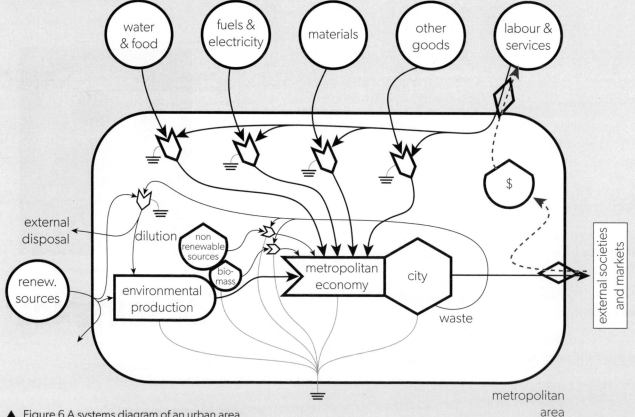

▲ Figure 6 A systems diagram of an urban area
Adapted from: Frontiers Media S.A (CC-BY 4.0)

Are urban areas sustainable?

With such a high proportion of the world population living in urban areas, there are large amounts of waste, high resource consumption and high energy use—this is neither sustainable nor resilient. One suggested solution is "circular cities". Figure 7 shows what it means to have a circular city.

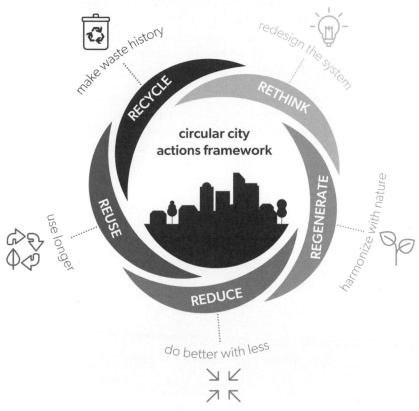

▲ Figure 7 The "circular city" model
Adapted from: ICLEI – Local Governments for Sustainability, Circle Economy, Metabolic, and Ellen MacArthur Foundation, 2021. Circular City Actions Framework: Bringing the circular economy to every city. Bonn, Germany.

Circular models are seen to be advantageous because they use nature as a template. A circular city has a thriving local economy that incorporates the Rs—reuse, reduce, regenerate, rethink, and recover products, materials and human potential within the city. Different sectors collaborate and materials are not wasted but cycled back into the system. A system that achieves this is sustainable, resilient and environmentally friendly.

Green infrastructure

Green infrastructure integrates natural and semi-natural areas into cities. Trees along streets, green facades, green roofs, parks, lakes and wetlands can all help to manage storm water, improve air quality, reduce heat stress and increase biodiversity.

- Green infrastructure extends the longevity of exterior surfaces as it protects against weathering and sunlight.

- Green roofs act as extra insulation, providing protection against extreme temperatures. This reduces the need for heating and air-conditioning.

ATL **Activity 14**

ATL skills used: thinking, communication, research

Draw a new systems diagram of an urban area but base it on a circular economy, similar to the one in Figure 7.

TOK

To what extent has technology extended and modified our ability to solve the problems presented by urban living?

▲ Figure 8 A green roof offers many environmental benefits

- Green roofs sequester greenhouse gases and help to remove air pollutants such as nitrous oxide, sulfur dioxide and particulate matter. They also provide a food source as many people use the space to grow vegetables and some keep honey bees.

- Storm water can be a resource not a threat. Heavy rain in cities is a problem because the sealed surfaces deliver the water to storm drains faster than it can be assimilated. This means that pollutants from the streets are washed directly into the rivers. Green infrastructure absorbs rainfall and thus mitigates flooding. In Toronto, green roofs have been mandatory for certain new developments since 2009.

Amsterdam has a rich culture and is taking the lead in developing urban circularity to transform from a linear economy to a circular one. It is invested in food and biomass, consumer goods, and construction to try to take a balanced approach to providing for people's basic needs. Basic needs include the aesthetics of the city.

The urban ecosystem

As you learned in subtopic 2.1, ecosystems are systems with living and non-living components which interact with each other.

▲ Figure 9 Central Park, New York

The urban ecosystem is made up of natural and constructed systems. The interactions are complex because there is a concentration of human population and this means socio-economic factors and biophysical processes are involved. The ecological processes happen in islands within cities, connected by corridors (streets) and buildings. These corridors may enable or restrict migration and dispersal between the islands.

(ATL) Activity 15

ATL skills used: thinking, communication, research

Consider Central Park (Figure 9) as an example of an urban ecosystem.

1. List all the:

    a. biotic components

    b. abiotic components

    c. inputs and outputs

    d. flows and stores.

2. Compare and contrast the Central Park ecosystem with one in a rural nature reserve.

Urban areas have certain impacts on the ecosystem, some of them unique to the urban situation.

- Transport networks facilitate entry for alien species—cats, dogs, other pets, urban foxes, insects, microorganisms.

- There are different physical and chemical properties such as pollution and tall buildings causing wind tunnels.

- The natural ecosystems are highly fragmented in most cities, e.g. Central Park in New York, or your local urban green space.

- A microclimate is created by the unique qualities of the urban area and thus the ecosystems in urban settings have different biotic conditions compared with the surrounding areas.

Green infrastructure—parks, trees, road verges, gardens, green roofs and urban forests—is designed to make urban ecosystems more contiguous and to deliver a wide range of ecological services. Services provided include:

- regulation of air quality, flooding, noise, temperature and pollination

- provision of food and water, recreation, education and cultural heritage.

Urban planning and sustainability

Urban planning aims to develop and design urban areas that are sustainable and meet the needs of all stakeholders in the community. This is a large task as it must address physical, domestic, environmental, commercial, industrial, financial and health needs.

The aims of urban planning are:

- to work on a large citywide scale and involve many different experts such as architects, economists, sociologists, public health experts and more

- to ensure urban areas are functional now and in the future

- to produce a cohesive plan that considers the numerous elements of city life, including new and pre-existing land and land use, buildings, roads and so on

- to locate specific functions in specific zones. Land is often reserved for a particular land use and legislation prevents other uses being located there

- to consider growing vulnerability to climate change, pressures of migration, and spatial inequalities

- to have an overarching strategic plan to guide the sustainable development of the city. This plan should have a clear goal such as to ease congestion, create more community spaces or improve natural spaces.

Practical

Skills:

- Tool 4: Systems and models

- Inquiry 1: Exploring and designing

Design an investigation to study the differences between an urban ecosystem and a natural ecosystem.

Connections: This links to material covered in subtopics 1.3 and 6.3.

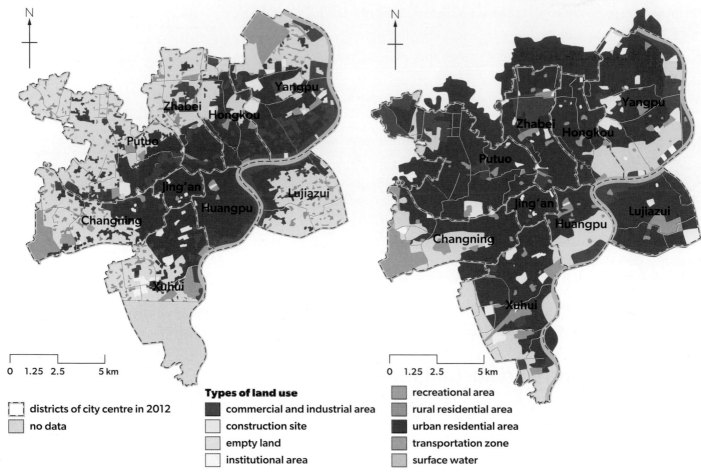

Types of land use

[ ] districts of city centre in 2012
[ ] no data

■ commercial and industrial area
□ construction site
□ empty land
□ institutional area

■ recreational area
■ rural residential area
■ urban residential area
■ transportation zone
■ surface water

▲ Figure 10 Urban land use zones, Shanghai, in 1973 (left) and 2012 (right)

City zones can be clearly identified in land use maps of Shanghai (see Figure 10).

- The commercial and industrial areas are mainly along the river.

- There are a few recreational areas spread throughout the city.

- Most of the city is for urban residential land use.

The map of Shanghai shows empty land which may be greenfield sites—these are areas that have never been developed before. Careful planning is necessary for these sites and there will be no infrastructure in place to support proposed land use developments. All stakeholders, professional consultants and experts should be involved in the planning process. This ensures that different perspectives are considered.

ATL Activity 16

ATL skills used: thinking, communication, research

1. Investigate maps of the city or urban area closest to you and see how:

   a. the city has developed and expanded over time

   b. urban land use has change through time.

2. Using Figure 10, describe the major changes in land use areas in Shanghai between 1973 and 2012.

3. Suggest reasons for the changes.

Deurbanization

Some urban areas may be in decline, often due to deurbanization. Decline may be identified by failing businesses, decreasing population growth and deterioration of the buildings and infrastructure. These areas are often the focus of revitalization through good planning. Depending on the way in which decline is happening, revitalization may include road repairs, pollution clean-up (often involving citizens), the addition of parks and other recreational facilities. Some areas experiencing urban renewal have old structures cleared and rebuilt more fit for purpose. The land use zoning may change, and old industrial buildings may be re-purposed.

The London Docklands are an example. This riverfront land used to be the docks for the Port of London. The docks were closed in the 1980s as container shipping took over; they soon slipped into serious decline. They were then the focus of redevelopment for commercial and residential use. This created wealth and established thriving communities as people and service industries moved into the area.

Cities must also plan with economic development in mind and to do that planners must identify areas of growth where they can boost financial success. This is done by making the area more attractive to businesses, that then hire local workers and create economic growth. This leads to a "snowball" effect as the workers use local facilities such as restaurants and shops.

None of this planning will help if the infrastructure cannot support it. Fundamental facilities must be well planned for an area to function effectively. This means that the following facilities must be in place or planned for:

- public works: electricity supply, communications, water, sewage

- community infrastructure: schools, parks, hospitals, shops

- transport and well-being: roads, police, fire service, ambulance service, refuse collection.

▲ Figure 11 London: River Thames looking towards regenerated docklands and Canary Wharf in 2018

Sustainability in the urban area

Efforts are being made in many countries to integrate sustainability and resilience into the planning of urban processes. This is essential to maintaining quality of life in cities, to meet SDG 11. According to the UN: "Cities are drivers of economic growth and contribute more than 80 per cent of global GDP". To achieve SDG 11, planners need to address issues of:

- affordability of housing
- integrated public transport systems
- green spaces and green buildings
- security
- education and employment

- use of renewable resources
- reuse and recycling of waste
- energy efficiency
- involvement of the community
- pollution—air, water, noise.

SDG 11 is about making cities and human settlements inclusive, safe, resilient and sustainable.

The Sustainable Development
Goals Report 2022

This means that environmental planners must look carefully at the relationship between natural and human systems. Sustainable cities are designed with consideration of social, environmental and economic impacts, and integrate eco-friendly initiatives into the design. Human comfort and mental health are also considered by reducing costs and creating cultural opportunities.

Activity 17

ATL skills used: thinking, communication, research

Consider the issues that need to be addressed to achieve sustainable cities.

Research one of the following examples and analyse to what extent the city is sustainable.

- Brasilia in 1950

- Forest City, Malaysia

- Copenhagen (reduction or removal of car use)

- San Francisco (installation of electric vehicle (EV) charging stations)

- Dubai (water conservation).

▲ Figure 12 Copenhagen is one of the top five most sustainable cities in the world

Characteristics of green cities

Ideas about what constitutes a green city vary between countries, cities and experts. Here are some common themes.

Transport improvements need to be made in most cities. Internal combustion engines are responsible for 75% of carbon monoxide pollution and 27% of total greenhouse gas emissions in the USA today. Solutions may involve improving public transport or introducing measures that reduce the use of cars, e.g.:

- Improve public transport to make it cost-effective, flexible and accessible.

- Introduce electric trolleybuses, metro systems, underground railways and maglev trains.

- Incentivize alternative travel options such as bus travel (build bus lanes to reduce travel times), use of EVs (install charging points), or carpooling (for people who work or go to school in the same area).

- Introduce congestion charges. In place in London, such charges have reduced city traffic by 33% since 2003. It currently costs GBP 15 per day to drive in the city centre. The revenues generated go to improving public transport.

- Introduce parking and traffic controls—remove parking spaces and replace some traffic routes with car-free zones, bike routes and walkways. In Oslo , such measures have reduced car usage by nearly 20%. In Copenhagen there are 675,000 bicycles compared with 120,000 cars and 42% of residents walk or bike to work. This is due to the existence of cycle and walking highways.

- Create limited traffic zones to limit the number of cars that can enter the city and the times at which they can enter. Rome restricts cars from entering the city at certain times and only allows residents to enter. There is also an annual fee to enter even at restricted times. This has reduced car traffic by 20% during restricted times.

▲ Figure 13 Cycleways can help reduce car usage

- Encourage use of public transport. The local government in Utrecht has worked with private companies to provide free public transport passes for employees, leading to a 37% reduction in car usage.

- Create apps that allow people to track their mobility. They gain points for walking, biking, or using public transport and thus get rewards—as individuals, teams or participating companies. In Bologna, 73% of people have reported a decreased use of cars because of this.

These changes can make a significant contribution to making a city greener by reducing noise, water and air pollution, increasing the use of renewable energy in some cases, and improving public transport systems.

For a city to become sustainable, there must be a switch to clean energy—a popular choice is solar energy. All forms of renewable energy will improve air quality, but solar panels are compact and versatile. Solar panels are readily available, and many governments offer subsidies or tax incentives to encourage people to install solar panels. This incentive lowers utility bills, supports local economies, and promotes energy independence. In the UK the government (EC04) scheme will give up to GBP 14,000 towards solar panels and other home improvements. In addition, no tax is levied on solar panels.

A large percentage of global GHG emissions comes from buildings. In 2022, 35% of GHG emissions in the EU were from the building sector. This can be addressed by reconditioning old buildings or constructing new ones that require less maintenance and have lower utility costs. The LEED (Leadership in Energy and Environmental Design) certification is widely used to rate sustainability of buildings. Buildings can gain this certification through use of:

- low cost, low-impact eco-friendly building material—cellulose insulation is made from repurposed newspapers and this reduces water use too

- cool roofs, which are made from light-coloured or reflective material which redirects sunlight and keeps the buildings cooler. Residential heating, cooling and air conditioning releases 441 million tonnes of GHG emissions per year

- green roofs

- solar panels or other renewable energy options

- natural building materials or even recycled materials

- good insulation, heating and ventilation

- rainwater harvesting.

▲ Figure 14 Solar panels in residential areas

Urban farming

Urban farming is a relatively new phenomenon which is helping to feed citizens in an eco-friendly way. It involves agricultural practices such as horticulture, aquaculture and hydroponics that are within the city or surrounding areas. People can grow their own food, save money and reduce transport costs and emissions of agricultural produce. Farmers' markets were established in the UK in 1997 and there are now over 550 of them nationwide.

▲ Figure 15 Urban farming in open spaces

Community-supported agriculture is seen in farmers' markets, indoor farming, vertical farming, beekeeping, roof-top farming and many other ways to support locally produced and sold foods. This:

- increases food security as people are growing their own food

- creates a sense of community

- provides healthier food options which people value more

- uses land efficiently

- improves food safety and quality.

Water conservation is essential in cities. Many of the ways to address water conservation are covered in subtopic 4.2. Good tactics in the city include rainwater harvesting and use of waterless hardware such as waterless toilets. In addition, green infrastructure uses the natural water cycle to clean supplies instead of water treatment plants. Trees and wetlands do this task and can be restored in cities.

To be sustainable, the population must have access to public resources, such as:

- green spaces: these act as conservation areas, control pollution, improve mental and physical well-being and can become an edible landscape if they are used to grow food

- public buildings such as hospitals, recreational facilities and cultural or community centres.

Waste management needs to be a circular process not a linear one.

- The 3Rs are part of this: reduce, reuse, recycle. In San Francisco, recycling and composting is mandated and this has diverted 77% of waste out of landfills.

- Paperless approaches are helped by technology and many institutions are now paperless.

The good news

Sustainable cities are another tool to mitigate climate change. Sustainable infrastructure will decrease flooding, heatwaves, the spread of disease, property damage, insurance losses and casualties. As they reduce their ecological footprint to zero, sustainable cities can become an inspiration for all.

Ecological urban planning

Ecological planning should be part of the overall urban planning process as it takes a holistic approach to the urban system by viewing it as an ecosystem. It aims to:

- integrate natural and human social environments and maintain ecological services

- make urban areas more compact by reducing urban sprawl; reduce car dependency; reduce energy consumption; improve public transport; and increase environmental justice through better accessibility and social equality.

These aims are achieved by merging all the elements discussed in the previous section through resilience planning, urban ecology or biophilic design.

Biophilic design

Biophilic design is a concept that aims to increase the connectivity between humans and the natural environment, by using nature directly within urban spaces. This will improve physiological and psychological health, performance, wealth and the environment for the people within the urban environment. It is a recent innovation but biophilic design has been around since 562 BCE in the Hanging Gardens of Babylon.

Stephen Kellert, a Yale professor of social ecology, created a framework for biophilic design around the basic principle that design should celebrate and show respect to nature and enrich the urban environment. Study Figure 16 carefully—it shows the biophilic elements that should be found in all buildings to make them inviting and relaxing.

Biophilia is the passionate love of life and of all that is alive; it is the wish to further growth, whether in a person, a plant, an idea, or a social group.

Erich Fromm (1900–1980), psychoanalyst

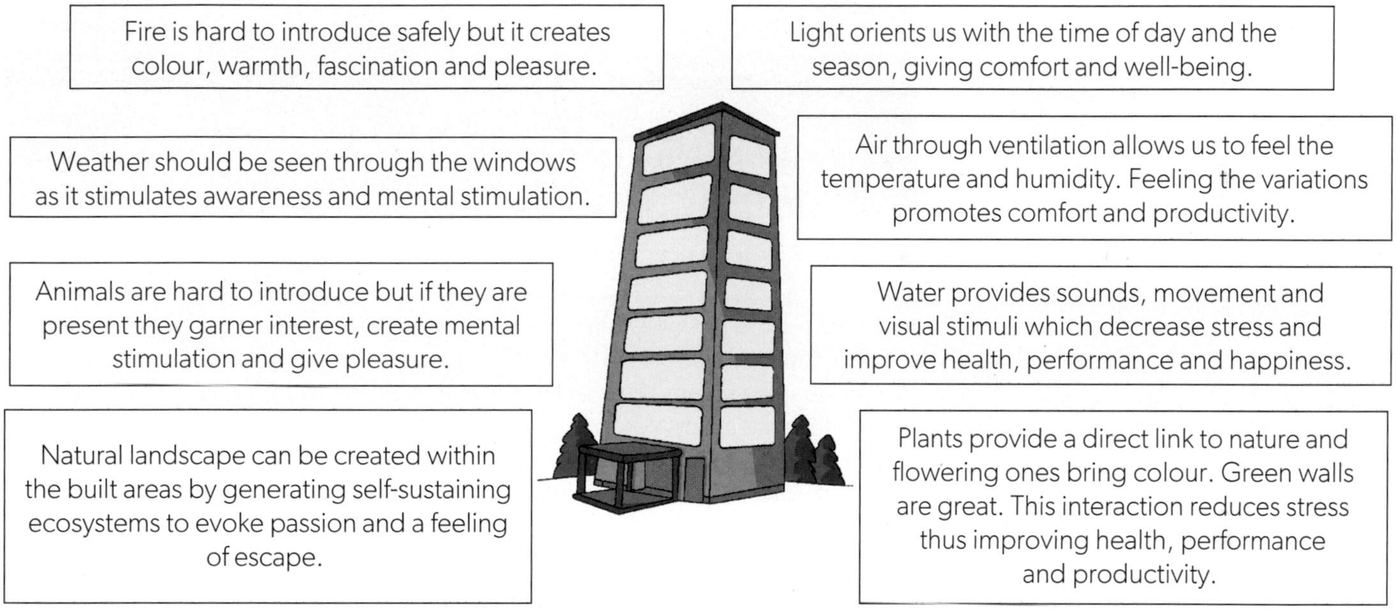

Fire is hard to introduce safely but it creates colour, warmth, fascination and pleasure.

Light orients us with the time of day and the season, giving comfort and well-being.

Weather should be seen through the windows as it stimulates awareness and mental stimulation.

Air through ventilation allows us to feel the temperature and humidity. Feeling the variations promotes comfort and productivity.

Animals are hard to introduce but if they are present they garner interest, create mental stimulation and give pleasure.

Water provides sounds, movement and visual stimuli which decrease stress and improve health, performance and happiness.

Natural landscape can be created within the built areas by generating self-sustaining ecosystems to evoke passion and a feeling of escape.

Plants provide a direct link to nature and flowering ones bring colour. Green walls are great. This interaction reduces stress thus improving health, performance and productivity.

▲ Figure 16 The elements needed in a biophilic building

ATL Activity 19

ATL skills used: thinking, communication, social, research

1. Study Figure 16 and list the benefits that biophilic design brings.

2. Look around your school or home and list the biophilic elements that are present.

3. Investigate how the missing features could be introduced into your school or home.

4. State which ecosystem services are maintained by biophilic design.

5. Research a local example of biophilic design.

6. Discuss your findings in class.

TOK

In 1983, E.O. Wilson defined biophilia as "the innate tendency to focus on life and lifelike processes".

Fromm's definition suggests our link with nature is physiological; Wilson claims it has a genetic basis.

To what extent is the interpretation of these concepts linked to our cultural or indigenous heritage?

Benefits of biophilic design

1. For health

Plants and water features provide sound and visual stimuli which can help to reduce stress and improve health and happiness. Plants can also improve air quality.

Changing patterns of lighting and temperature through the day, to mimic natural changes, helps to maintain our natural 24-hour circadian cycle.

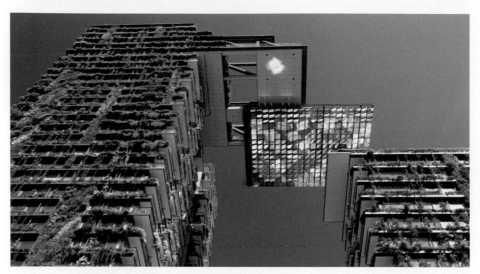

▲ Figure 17 Biophilic design in a residential building

In the UK, "green gymnasiums" are developing. People help to clear areas where the vegetation is overgrown to establish places where people can exercise. This is good for physical and mental fitness as well as quality of life. In such areas, health disparities between rich and poor neighbourhoods are lower. These areas also maintain ecosystem services such as carbon storage, flood management, pollution control, erosion control, and nutrient cycling.

2. Economic

These natural environments come at a cost—both in their establishment and in their maintenance. However, the payback is also high. Research in New York suggested that biophilic design increased worker productivity to the value of USD 470 million; decreased crime-related expenses by USD 1.7 billion; increased foot traffic (because streets with vegetation are more pleasant to be in); and so increased customer spending by 25%. Biophilic designed buildings also fetch a higher price on the markets.

3. Environmental

Including plants provides the following benefits.

- Flood management caused by the decrease in impermeable surfaces and thus better infiltration.

- Use of greywater for watering the plants reduces the amount of wastewater that enters the sewerage system.

- Pollution control for both air and water as plants are biofilters.

- Reduction in the heat island effect as natural surfaces are cooler than paved surfaces.

- Increased biodiversity as there are more habitats for organisms. The Khoo Teck Puat Hospital, Singapore has over 100 species of butterflies onsite as they are attracted to the vegetation in the building exteriors.

- Carbon sequestration by plants helps to reduce the carbon footprint of the urban areas.

- Cool roof designs are reflective thus increasing the albedo and reducing the amount of heat absorbed in the city. This reduces temperature fluctuations and decreases heating and cooling needs.

- Trees provide shade and cool the air through evaporation.

- Additional urban green trails encourage people to walk more, thus reducing carbon emissions.

4. Sustainability and resilience

Biophilic design incorporates natural protection as the vegetation provides ecosystem services such as flood management to protect urban areas. If the inhabitants of cities are happy and value the natural environments they are more likely to take care of the area which helps sustainability. As people make the most of opportunities to be outside and exercise, they become healthier and physically fitter and that makes them more resilient.

Case study 2

Singapore

▲ Figure 18 Central Singapore with green spaces

The city of Singapore was planned to be a biophilic city from the start. This has been achieved by:

- creating extensive interconnected green spaces

- using linear parks to connect the large green spaces, allowing citizens to walk easily around the city in a pleasant green environment

- educating the citizens on how they can maintain the green aspects

- creating vertical gardens and sky-gardens (for example, on the Oasia Hotel Downtown), to increase the proportion of natural environments by "building" upwards.

Rural–urban migration

This is the movement of people from rural to urban areas and it is partially responsible for urbanization. This is the most probable direction of migration as urban areas are more developed and are perceived to be the solution to problems of unemployment, poor education and healthcare services, and lack of entertainment.

LICs tend to have higher rates of rural–urban migration because there is a higher proportion of their population in rural areas, and a higher proportion of younger people.

Rural–urban migration can be:

- national: within the country (e.g. to Tunis, Tunisia—Figure 19)

- international: between countries (e.g. to Dubai, UAE—Figure 19)

- voluntary: the migrants choose to relocate

- forced: environmental or other factors force people to leave.

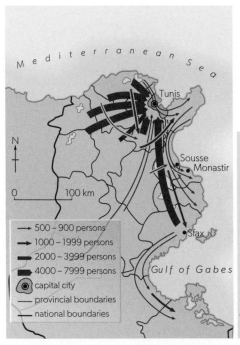

▲ Figure 19 Comparison of internal (left) and international (right) migration from North Africa, the Middle East, South Asia, and Southeast Asia
Adapted from: Stefan Chabluk, Hart McLeod

Causes of rural–urban migration

Urban areas are growing and transforming through economic expansion, population growth and migration. Rural areas lack the same level of development or the same access to services.

Rural life

- Life is integrated with and dependent upon the natural environment.

- Natural disasters may be common, e.g. floods, droughts, wildfires, storms.

- Environmental degradation such as desertification, pollution, water scarcity and soil erosion are all problems in rural areas.

- Agriculture is often at a subsistence level, which does not provide an income—food insecurity is a constant threat.

- Provision of education and healthcare facilities is often limited or non-existent.

- Government spending on services is limited.

- There are very few recreational or entertainment opportunities.

- Shopping malls are absent.

- Employment opportunities are mostly restricted to agriculture and related activities.

- Mechanization of farming further reduces job opportunities in agriculture.

This amounts to quite a lot of **push factors** in rural areas. Most urban **pull factors** are the opposite of these push factors—urban areas are perceived to have all the solutions, particularly for younger, mobile people (Table 3).

Perception	Reality
• Increased access to education and healthcare facilities • More public services available such as recreation and entertainment • Large variety of shops so more choice • Plenty and varied employment opportunities • Salaries are higher • Better housing and basic utilities such as water and electricity	• The influx of migrants puts great pressure on all services and migrants living in peripheral areas do not have access to or the money to afford any of these "luxuries" • Unemployment rates are high and there are no jobs or migrants are paid less than is legally acceptable • Lack of housing means migrants live in squatter settlements or poor quality housing in peripheral areas; standard of living is very low

▲ Table 3 Perceptions and realities of rural–urban migration

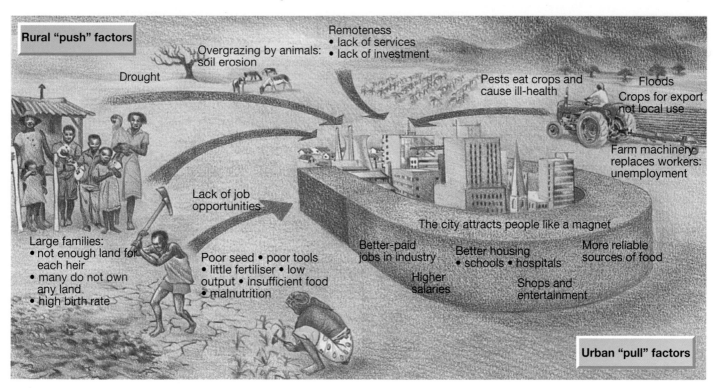

▲ Figure 20 Push-pull factors of rural–urban migration
Adapted from: Hardlines, Richard Morris, Angela Knowles, Dave Russell

Cities struggle to keep pace with the natural growth rate in urban populations; add to that the influx of migrants and the quality of life for many falls dramatically. However, migrants do bring advantages to the city, providing a ready supply of cheap labour which promotes economic development and industrial growth.

Rural–urban migration in Brazil

Between 2000 and 2010, 2.8 million people of working age migrated from rural areas into Brazilian cities. This had a positive impact in the urban areas, increasing productivity and development. However, it also resulted in poor socio-economic conditions for the migrants, including:

- underemployment or informal employment
- social conflict and inequality between migrants and local inhabitants
- limited access to public services
- poor healthcare
- high crime rates.

TOK

To what extent do logic and imagination shape peoples' perspective of the place to which they will migrate?

The urban–rural shift

Although urban areas have many attractions, they also have some negative elements such as traffic congestion, pollution, overcrowding, lack of green spaces and high crime rates. Even though there are a lot of people in cities, there is a sense of "loneliness" and a lack of personal connection with others because people are just "too busy". Many people are away from their cultural roots and family. These factors lead to deurbanization—people move back to the rural areas.

Urban–rural migration has a number of pull factors:

- lower living expenses in rural areas as housing is cheaper
- larger houses with gardens
- lower population densities
- returning to family; emotional attachment to where they grew up
- a less polluted environment
- dislike of the urban lifestyle
- new industrial and business developments
- out of town shopping centres
- quieter and safer environment
- good travel links to the city for work.

Rural living has become far more attractive in recent years as infrastructure has improved, basic services are in place and transport networks have become more extensive.

Suburbanization

The end result of urban–rural migration is the growth and spatial reorganization of cities. The population shifts from the compact central urban areas into the low density peripheral areas or suburbs. This is referred to as suburbanization or urban sprawl. The initial movement may be to the villages and rural communities that surround the city. However, as time passes, the intervening rural spaces are developed and become subsumed into the urban sprawl. This adds to the environmental harm of cities as they spread over a wider area. Suburbanization is often seen as one of the causes of urban decay because some areas of the central city become run down and abandoned. This is where the migrants may end up.

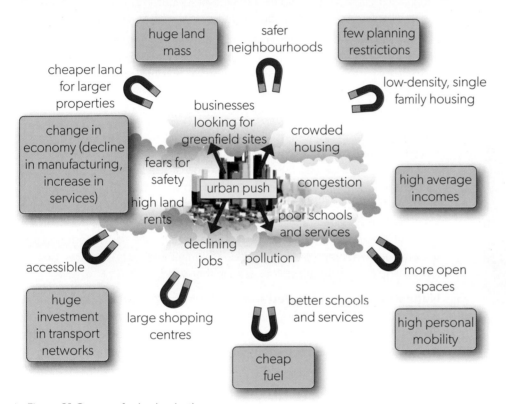

▲ Figure 21 Causes of suburbanization

Many of the residents of the suburbs still work in the central urban area and commute into the city by car or public transport. Some people can work from home due to technological advances and the COVID-19 pandemic which showed that it was not necessary to "go to work" because the same output can be achieved working online.

The following are consequences of this shift to the suburbs.

- As more people want to live in the suburbs, house prices increase.

- The original inhabitants of the rural or suburban areas cannot afford the increased prices so they leave.

- There is development of commuter settlements which are almost empty during the day: local services (such as local shops and bus services) are not utilized so they close down, and the small local community loses out.

ATL Activity 21

ATL skills used: thinking, communication, research

Research the area where you live and find some examples of urban–rural migration.

1. List the pull factors of the rural areas.

2. List the push factors of the urban areas.

3. Draw a flow diagram to show these changing circumstances.

- There is increased demand for restaurants so these businesses capitalize on the opportunity.

- Agricultural land is sold for housing developments and the areas become more built-up.

- Commutes to work increase noise, congestion and pollution.

Urban expansion and the environment

The expansion of urban areas and the creation of suburban zones results in changes to the natural environment. Some are negative changes, but not all of them.

1. **Habitat loss** and deforestation are necessary to facilitate the expansion of urban areas. This decreases the population of some species and changes the range of their habitat. This in turn changes the interactions between organisms and species. There is a decrease in biodiversity around urban areas due to the spread of pollutants beyond their boundaries.

2. The interconnected nature of cities means there is an increase in the movement of **invasive species**. This can be by accident—in shipments and imports—or it can be through the escape of species from homes and gardens. Invasive plant species are spread along transport routes. For example, *Buddleia* is an invasive plant that has spread through the UK via railway lines as the trains pass at high speed and spread the seeds. It outcompetes native plant species. Invasive species are also introduced into the environment when pet owners release unwanted pets. Big cats, dogs, snakes and other reptiles are released in this way.

3. **Natural cycles are disrupted** and that impacts reproduction and distribution of species. Many birds live in urban areas and they have adjusted feeding habits to the new environments. Beak shape of some has altered to more easily access the seeds from human-made bird feeders. Birdsong is at higher frequency or louder in urban areas due to noise pollution and some birds sing at night if streetlights are on.

4. The **abiotic** environment is changed due to the urban heat island effect as sunlight is more easily absorbed by dark-coloured materials. This in turn alters precipitation patterns and decreases the likelihood of snowfall.

5. The prevalence of **impermeable** surfaces in cities (pavements, buildings, tarmac) increases surface run-off and decreases soil quality. Water moves through the area quickly as surface run-off, picking up pollutants and sediment and thus causing degradation of water quality and flooding. Many urban planners now factor this in and ensure there are trees and other plants to counter the issue.

6. **Pollution** in general spreads out from the city into surrounding areas; great reliance on vehicles increases air pollution. Disposal of solid domestic waste in landfills causes a range of pollution issues. In the USA, the increase in the number of motor vehicles has outpaced population growth and people are commuting longer distances.

▲ Figure 22 *Buddleia* is an invasive species

7. Agricultural land and natural environments are replaced by **urban sprawl**. Between 1950 and 1995, Chicago's population increased by 48% while the land coverage of the city increased by 165%. Large areas of land are taken up by roads, leading to fragmentation of natural areas.

8. One of the unexpected successes of expanding urban areas is that some **wildlife** has adapted to the new conditions and is thriving. Some species take advantage of what cities can offer, such as good food, shelter and protection from predators. They live longer and have more offspring in the urban setting than they do in wild habitats. Animals that do well in urban environments include red-tailed hawks, sparrows, starlings, rock doves, skylarks, common pipistrelle bats, house mice, rats, European badgers, foxes, coyotes and racoons, along with a very wide variety of arthropods, gastropods and reptiles. The species vary depending on the country and climate.

▲ Figure 23 Urban fox

Successful urban animals have common characteristics. They:

*   are omnivorous generalists and not fussy about food or shelter—they make the most of bird feeders, garbage and pet food

*   are tolerant of human disturbances

*   are adaptive and change their behaviour to fit the new environment.

ATL Activity 22

ATL skills used: thinking, communication, social, research, self-management

In small groups, research the urban area where you live and produce a podcast about the successful urban wildlife there.

ATL Activity 23

ATL skills used: thinking, communication, social, research, self-management

An infographic is a great way to portray the benefits of sustainable development. Search online for some examples.

Work on your own or with a partner to create your own infographic that shows the factors that you think are most important to sustainable urban areas.

Ecological urban planning: the next step

A lot has been discussed in the previous sections about urban planning and how it can be sustainable. Urban areas need to be more compact. Look back to Figure 10 to see how far Shanghai spread between 1973 and 2012. Urban areas also need to be designed to ensure social mixing. These are both considered to be achievable by having more integration of different land uses—then people do not have to travel so far for work, entertainment and socializing. This would then improve the chances of achieving the aims set out in the previous section on ecological urban planning.

AHL

(ATL) Activity 24

ATL skills used: thinking, communication, social, research, self-management

Consider these quotes about urban planning. You may agree or disagree with them—both are fine. However, you have to decide how to apply the three HL lenses to the matter of urban planning.

> 66
> Urban planning is the marriage of technical expertise and social values, where the goal is to create liveable, just, and sustainable communities for all. 99
>
> Pierre Clavel

> 66
> By far the greatest and most admirable form of wisdom is that needed to plan and beautify cities and human communities. 99
>
> Socrates

> 66
> The recovery of sprawl to vibrant places is literally our generation's greatest challenge. 99
>
> Steve Mouzon

> 66
> Urban planning is not just about buildings and streets; it is about creating places that are liveable and sustainable, places that enhance the well-being of their residents. 99
>
> Michael Pyatok

> 66
> Urban planning is the art of creating liveable, sustainable, and resilient cities, where people can live, work, and play in harmony with their environment. 99
>
> Andy Merrifield

> 66
> Urban planning is not just about designing buildings and spaces, it's about creating a vision for the future and shaping the liveable environments of tomorrow. 99
>
> Mike Lipske

> 66
> Urban planning is a critical tool for shaping the future of cities, ensuring that they remain dynamic, vibrant, and sustainable over time. 99
>
> Richard Florida

TOK

To what extent does the presentation of these quotes impact your perception of the information?

Work in teams to design a city that addresses SDG 11 (safe, resilient and sustainable). In your design, make sure you consider environmental economics, environmental ethics and environmental law.

Present your final design, using any media you choose (podcast, poster, animation, 3D model).

Models for sustainable urban systems

Models, covered in subtopic 1.2, are "a simplified representation of reality that can be used to understand how a system works and to predict how it will respond to change". Keep this in mind when reading this section. Remember that although the simplification can make a model less accurate, models are still useful.

The following models were developed in the context of economics but they can easily be adapted to urban areas. The original way of looking at the economy was in a linear fashion (Figure 24). If you are interested in the linear model and the problems of living this way, find the video "The Story of Stuff" on YouTube.

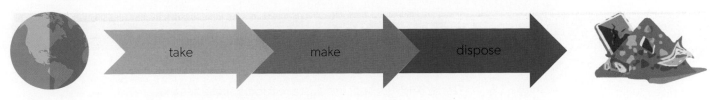

▲ Figure 24 Take, make, dispose: the linear model
Adapted from: Q2A Media Services Pvt. Ltd & OUP

There are clear issues with this linear model approach because it leads to overexploitation of natural resources and unmanageable accumulations of waste in the environment. It is not sustainable and it is being replaced by the circular economy and doughnut economy.

Circular models for urban areas

These models are based on natural systems and the interdependence between society, the economy and the environment; they are more likely to be sustainable. This acceptance of interdependence is missing from the linear models, which shows an indifference to the negative consequences of urban growth.

In the urban context, circular models emphasize circulating resources rather than scrapping them and extracting new resources. Such models view waste as an error in the system—the resources used should be either organic (able to be regenerated) or inorganic (able to be recovered and reused).

Because waste is an error, the models suggest strategies that can be used to eliminate or decrease waste significantly. Many of these strategies have grown from the traditional 3Rs: reduce, reuse, recycle (Table 4).

Connections: This links to material covered in subtopics 1.2, 1.3, 5.2, 6.3, 7.1 and HL.a.

Strategy	Meaning
Reduce	Decrease the number and quantity of raw materials used.
Reuse	Use the product again and again to get full benefit from it before it is broken down to component parts for recycling.
Refuse	This is about personal choices—don't buy something you don't need and don't buy products that can't be reused.
Rethink	Redesign products and components and the packaging they come in to make everything safe with a longer lifespan.
Repurpose	Use products or materials for functions that they were not originally made for.
Repair	Maintain the usefulness of products: don't throw them away when they break, replace worn out parts.
Remanufacture	Remake the product to make it as good as new or better than it was. It also has a guarantee that it will work well.
Recover	Aim to recover all the resources from all products to reduce waste.
Refurbish	Collect any rejected or unwanted items, clean them and repair them to make them look nicer.
Recycle	This is to be avoided because it means the product was poorly designed in the first place. Recycling breaks the product down into its component parts so it can no longer be used.

◀ Table 4 Waste reduction strategies

ATL **Activity 25**

ATL skills used: thinking, communication, research

Table 4 lists most of the waste-reducing strategies that are used in circular models.

Copy Table 4 and add a third column, giving an example of each strategy in the urban system. These examples can be from anywhere within the urban system. Think about your own city or your own habits. Remember, clothes, scrap material, old electronics, paper and card, glass, metals, and some plastics can all be kept out of the waste system.

Hope for the future

Perhaps the most drastic strategy is rethinking. The Ellen MacArthur Foundation has some very interesting videos about rethinking. Rethink the way we view ownership, rethink the whole operating system.

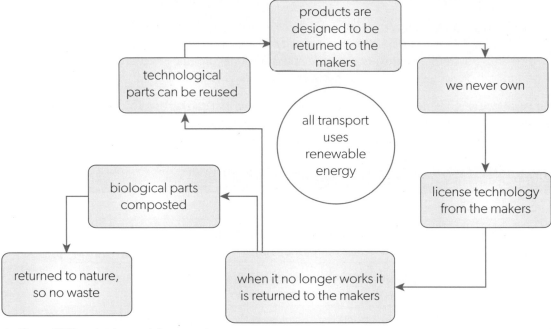

▲ Figure 25 The circular model

One in 58 Londoners were homeless in 2022. **99**

Evening Standard, *11 January 2023*

TOK

To what extent do politicians have a moral or ethical obligation to act upon this knowledge to address such deprivation?

Doughnut model in the urban area

The doughnut economics model (devised by Kate Raworth; Figure 26) is based on economics, but the model is broadly applicable in many other areas as well. We look at ecosystems through lenses and one of them is an environmental economics lens. Take this further and consider this model's applicability to the urban environment.

If you look at a doughnut, there is nothing in the middle, just a hole. Think about what this means for the people living in that hole. Some or all of their basic needs are missing or in short supply; they live in a state of deprivation. This is a global problem and London (the capital of the UK) has the most acute homelessness problem in the UK. This is not a problem that is restricted to poor countries. There is nothing extravagant about the 12 basic human needs of the social foundation—they are all addressed in the SDGs.

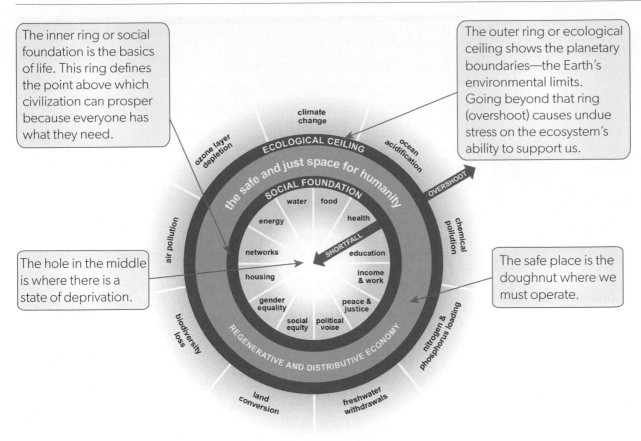

The inner ring or social foundation is the basics of life. This ring defines the point above which civilization can prosper because everyone has what they need.

The outer ring or ecological ceiling shows the planetary boundaries—the Earth's environmental limits. Going beyond that ring (overshoot) causes undue stress on the ecosystem's ability to support us.

The hole in the middle is where there is a state of deprivation.

The safe place is the doughnut where we must operate.

▲ Figure 26 The doughnut model
Adapted from: Q2A Media Services Pvt. Ltd & OUP

ATL Activity 26

ATL skills used: thinking, communication, social, research, self-management

Work in groups.

1. List the 12 basic human needs within the social foundation.

2. Go to the UN Sustainable Development website and find out which SDG addresses each of the 12 needs.

3. Investigate a city to determine what percentage of people are without these basic needs. Each person in the group investigates a different city.

4. Compare and contrast your cities to see which have high percentages of the population without basic needs and which are more equitable.

5. Explain how the inequity could be addressed by the law, ethics or economics.

The generous city

The name "generous city" was coined by Janine Benyus and it refers to cities that nestle within the natural ecosystem. The way to make a city generous is for the architects to be inspired by the natural world around them. This is a process which comes in steps that are regenerative not degenerative.

1. Observe the indigenous ecosystem the city is in. For example, Dubai is in the desert so it must be planned to harmonize with that surrounding desert environment.

2. Investigate the baseline parameters of that environment. How much solar energy is there? How does rainwater flow and how is it stored? What fertilizes the soil? How is carbon sequestered?

3. Use this data as the baseline for city planning so that the city takes and gives in the same way as the surrounding natural environment.

4. Use the strategies discussed earlier in this subtopic to ensure the city blends with the natural environment. The addition that makes a generous city is connections—everything must be linked through an infrastructure web (instead of a food web) that winds through wildlife corridors and urban farming areas.

5. To complete the model, ensure the city has an inherently distributive design.

   a. Every household has renewable energy to provide its own needs.

   b. Housing is affordable for everyone.

   c. Public transport takes priority over all other transport so that it is the cheapest, quickest way to travel.

   d. Neighbourhoods are self-supporting and work, education and homes are close together.

   e. There is a communal feel so that everyone maintains and cares for the environment.

Is this realistic? Some say it must be; otherwise we will overshoot the planetary boundaries and the resultant collapse of the natural systems will mean the planet is no longer able to support us. There are many examples of cites that are moving towards the regenerative, distributive model.

▲ Figure 27 Wildlife corridors

ATL Activity 27

ATL skills used: thinking, research

Sustainable cities of the future could be built from scratch or created by transforming existing cities.

Park 20/20 in Hoofddorp, Amsterdam is a business park that makes human experience central. The design is called "Cradle to Cradle" as opposed to the original adage "Cradle to Grave". It has high end, innovative building design in a healthy sustainable environment. The cycle is continuous using recycled materials, integrated energy supply, solar energy, water storage, recycling, with filtration and wildlife habitats throughout.

Neom in Saudi Arabia is planned as a 50-year building programme to make a sustainable city—a smart city with vertical living, walkable communities and much more. Search online to find out more.

According to the Arcadis Sustainable Cities Index, the top 10 sustainable cities in 2022 were:

1. Oslo
2. Stockholm
3. Tokyo
4. Copenhagen
5. Berlin
6. London
7. Seattle
8. Paris
9. San Francisco
10. Amsterdam

Investigate one or more of these cities and identify the strategies that are being used to make a positive difference to city living.

Green architecture

Green architecture and civil engineering combine new and indigenous knowledge systems to provide essential services.

- Architects design and plan the structures.

- Civil engineers manage the design to completion process.

When architects and civil engineers collaborate, they design a system that is greater than the sum of its parts. The aim is to choose environmentally friendly building materials and construction practices that:

- minimize the harmful effects of construction projects on human health and the environment

- safeguard air, water and soil.

Within this partnership, the civil engineers use circular environmental models by switching to renewable materials such as timber and hemp. There is also the emergence of innovative technologies like improved prefabrication of building components. This is popular because it is more efficient and produces a better product by reducing cost, waste and carbon. In some cases, the materials used are carbon negative: during their lifetime of use they remove more carbon from the atmosphere than they release.

Green architecture is generally based on a circular model. The idea of cradle to cradle is once again central. A common way to make this model a reality is to use biobased materials. Biobased materials, also called bioproducts, have certain characteristics—they:

- are modern materials that have had extensive processing

- include chemicals and energy derived from renewable biological resources

- are nearly all biodegradable.

There are a lot of different biobased materials, such as:

- cellulose fibres (reconstituted cellulose)

- casein (the "fats" extracted from milk to make low-fat milk), which can be processed to make plastic, glue, paints, fibres

- bioplastics based on soy, corn, sugar cane, wheat or potatoes

- corn starch made from maize, which is used to make packing pellets

- grease made from vegetable oils that can be used to replace petroleum-based lubricants.

TOK

To what extent do models have a role in the acquisition of knowledge in urban planning?

▲ Figure 28 Corn-starch cup

Bale construction

This type of construction uses straw as a part of a building. Straw houses have been built in Africa since the Palaeolithic era (2.58 million years ago) so this method is not new. Straw bales were used in construction in Germany 400 years ago and loose straw has been an insulator (on the floor or as a thatch roof) in North America, northern Europe and Asia for centuries. Today bale construction uses straw bales as structural elements or as an insulator. The bales must be 45 cm thick.

▲ Figure 29 Constructing a straw bale house

▲ Figure 30 A church made from straw bales in Nebraska, USA

Advantages

- It fits well with circular models, because straw bales are made from the waste left after grain is harvested.

- The farmer makes money from the waste product.

- Straw is an excellent insulator; the thicker the layers the better the insulation. Thicker walls help reflect more sunlight and keep buildings cooler in summer.

- Building with straw is simple and the skills are easily learned and passed on.

- Wheat and other grains use solar energy to grow so that reduces energy costs of raw materials.

- Straw is totally biodegradable so the straw bales can be ploughed into the soil when the building is no longer serviceable.

- Well-constructed straw walls are flame retardant as there is very little air in the compressed bales.

- The bales are versatile and can be easily cut into different shapes for aesthetics.

Disadvantages

- Straw bale walls need to be kept dry to avoid rot. This can be done through proper sealing.

- The bales are expensive to move over long distances so there must be a raw material source nearby.

- This is an "unusual" method of construction so building codes and expertise may be absent.

Check your understanding

In this subtopic you have covered:

- urban areas are a system

- the pull and push factors of urban areas and of rural areas

- rural–urban and urban–rural migration

- urbanization, deurbanization and suburbanization

- the urban ecosystem

- impacts urban areas have on the environment

- a brief introduction to urban planning

- ecological urban planning

- modelling sustainable urban systems

- examples of green architecture.

AHL

To what extent are urban systems similar to natural ecosystems?

How can reimagining urban systems create a more sustainable future?

1. List the biotic and abiotic components of the urban ecosystem.

2. Define urbanization and describe an urban area.

3. Draw a diagram of the urban areas system.

4. Explain why urbanization is happening.

5. Describe the process of suburbanization.

6. Compare and contrast rural–urban migration and urban–rural migration.

7. Outline the impacts of urban and suburban systems and their expansion.

8. Describe how urban planning can improve sustainability in an urban area.

9. Discuss the holistic approach of ecological urban planning.

10. To what extent does ecological urban planning follow the principles of compactness, land use and social mixing?

11. Explain how circular or doughnut economics promote sustainability in urban systems.

12. Evaluate the role of green architecture in reducing the harmful effects of construction projects.

AHL

❯❯ Taking it further

- Discuss impacts and management options for the problems created by urbanization.

- Evaluate the extent to which a local urban environment is sustainable, referring to SDG 11.

- Volunteer with an organization that works to support people who may have suffered from social and environmental inequity locally.

- Propose smart city functionality for your school community.

8.3 Urban air pollution

Guiding question

- How can urban air pollution be effectively managed?

Understandings

1. Urban air pollution is caused by inputs from human activities to atmospheric systems, including nitrogen oxides (NOx), sulfur dioxide, carbon monoxide and particulate matter.

2. Sources of primary pollutants are both natural and anthropogenic.

3. Most common air pollutants in the urban environment are derived either directly or indirectly from combustion of fossil fuels.

4. A range of different management and intervention strategies can be used to reduce urban air pollution.

5. Nitrogen oxides (NOx) and sulfur dioxide react with water and oxygen in the air to produce nitric and sulfuric acid, resulting in acid rain.

6. Acid rain has impacts on ecology, humans and buildings.

7. Management and intervention strategies are used to reduce the impact of sulfur dioxide and NOx on ecosystems and to minimize their effects.

8. Photochemical smog is formed when sunlight acts on primary pollutants causing their chemical transformation into secondary pollutants.

9. Meteorological and topographical factors can intensify processes that cause photochemical smog formation.

10. Direct impacts of tropospheric ozone are both biological and physical.

11. Indirect impacts of tropospheric ozone include societal costs and lost economic output.

Urban air pollution

Air pollution is contamination of the indoor or outdoor environment by any chemical, physical or biological agent that modifies the natural characteristics of the atmosphere.

Particulate matter (PM) is made up of very small particles of solids or liquids in air.

Air pollution is caused by solid and liquid particles and gases that are suspended in the air. These particles and gases can come from natural and anthropogenic sources.

- Natural sources include volcanoes, dust storms, wildfires, pollen and mould spores.

- Anthropogenic sources include vehicle exhausts, industry, forest fires and indoor spray cleaning products.

These may be primary or secondary pollutants.

1. **Primary pollutants** are emitted directly from a process. Most common anthropogenic air pollutants are from the combustion of fossil fuels. Primary pollutants include:

    *   carbon monoxide—from incomplete combustion of fossil fuels

    *   carbon dioxide

    *   unburned hydrocarbons

    *   nitrogen oxides (NOx)—especially nitrogen dioxide, a brown gas, but also nitrous oxide and nitric oxide

    *   sulfur dioxide—from coal with high sulfur content

    *   particulate matter (PM)—small particles that are inhaled

    *   mercury—from coal-fired power stations

    *   sulfur—from coal with high sulfur content.

2. **Secondary pollutants** are formed when primary pollutants undergo a variety of reactions with other chemicals already present in the atmosphere. Sometimes these are **photochemical reactions** in the presence of sunlight Secondary pollutants include:

    *   tropospheric ozone, which forms when hydrocarbons combine with NOx in the presence of sunlight

    *   particulates produced from gaseous primary pollutants

    *   peroxyacetyl nitrate (the most common PAN), which is a stable oxidant of ozone

    *   nitrogen dioxide, which forms when nitrogen oxide combines with oxygen

    *   sulfuric acid and nitric acid, formed when sulfur dioxide or nitrogen oxides react with water; these lead to acid deposition.

Facts about air pollution

*   Air pollution—the combination of outdoor and indoor particulate matter, and ozone—is a risk factor for many of the leading causes of death including heart disease, stroke, lower respiratory infections, lung cancer, diabetes and chronic obstructive pulmonary disease (COPD).

*   Globally, air pollution contributed to 11.65% of all deaths in 2019.

*   Air pollution tends to be greater in LICs (because of indoor pollution due to a reliance on solid fuels for cooking) and in MICs that are industrializing (due to outdoor air pollution).

*   According to the WHO, 2.5 billion people are exposed to air pollution seven times higher than the WHO guidelines.

*   Over 90% of urban air pollution in LICs comes from old motor vehicles that are poorly maintained.

▲ Figure 1 Combustion of fossil fuels produces air pollution

Connections: This links to material covered in subtopics 1.2, 6.1 and 6.2.

TOK

Some unpleasant facts about air pollution are given here.

Are intuition, evidence, reasoning, consensus and authority all equally convincing methods of justifiably accepting such facts?

Other causes of air pollution

Air pollution is caused by chemicals, particulates and biological materials.

Wildfires

Wildfires are unplanned fires that occur in natural areas such as forests. Nearly 90% of wildfires are caused by anthropogenic actions such as stubble burning on farms, discarded cigarette butts, untended campfires and, sadly, arson. They are also started naturally by lightning. Apart from the obvious immediate danger to natural and human systems, these fires produce pollutants such as soot and dust particles ($PM_{2.5}$) and smoke containing toxic chemicals which can stay in the air for days. They also increase the average temperature (thermal pollution) and that increases the risk of more fires.

Chemical industries

Many industries such as chemical and textiles release many compounds both chemical and organic. The natural process of microbial decay by fungi and bacteria breaks down these compounds, releasing toxic methane gas.

Construction and demolition

As urbanization and population growth continue, construction and demolition are becoming a major source of pollution. These activities are a source of particulate matter and of pollutants released from the heavy machinery required, which uses fossil fuels. There are also other impacts such as noise and visual pollution.

Indoor air pollution

This type of pollution is caused by the following.

▲ Figure 2 Demolition of an old building

- Inefficient use of fuels in homes that are poorly ventilated. Smoke from these fuels can generate 100 times more fine particles than is considered safe by the WHO.

- Poorly maintained and older wood-burning stoves or open fires.

- Smoke from tobacco products.

- Soft furnishings, synthetic carpets, paints and adhesives can give off VOCs (volatile organic compounds), some of which are carcinogenic.

- Some cleaning products.

To put this into perspective, here are some statistics from the WHO factsheet on air pollution.

- One-third of the global population cooks using unsafe fuels or technologies.

- In 2020 household air pollution was responsible for approximately 3.2 million deaths.

- In 2020, 650 children under five died per day from household air pollution.

- Ambient and household air pollution combined cause 6.7 million premature deaths every year.

- Exposure to household air pollution causes strokes, heart disease, pulmonary diseases and lung cancer.

- Women and children are at greatest risk.

Practical

Skills:

* Inquiry 1: Exploring and designing

An indicator species measures pollution by looking at its relative abundance in an environment.

Design an experimental method using an indicator species to measure the levels of pollution in a local environment.

Particulate matter (PM)

Particulate matter has become an increasingly problematic component of air pollution. PM can be microscopic liquid droplets or solid particles that are inhaled. Exposure is continuous to a greater or lesser extent, but it tends to be worse in urban or industrial areas where there are more sources.

* PM_{10} is 10 micrometres (μm) or less in diameter. These particles can damage the lungs physically or they may be absorbed into the blood. $1 \, \mu m = 10^{-6} \, m$

* $PM_{2.5}$ is 2.5 micrometres or less in diameter. There is a greater chance that these will damage people's and animals' health as they invade the lungs and bloodstream more easily.

How much damage is done depends on the chemical and physical properties of the particles, how concentrated they are, their size and the length of time the organisms are exposed to the pollution. The impacts are surprisingly diverse.

* If the PM contains toxic material such as cadmium (cigarette smoke) or lead (combustion of fossil fuels), these substances can be absorbed into the bloodstream. High lead levels cause nerve or kidney damage.

* Some people are allergic or sensitive to the particles causing sneezing and runny eyes.

* Living organisms such as bacteria and fungi are included in PM and they can cause infections.

* Although asbestos is used less in buildings today, when construction or demolition of old buildings takes place asbestos may be released. That can cause asbestosis, an incurable lung disease that develops when small asbestos fibres stay in the lungs.

* PM can cause irritation of mucus membranes.

* PM can cause increased respiratory stress and asthma.

▲ Figure 3 An indicator species: lichen

TOK

This knowledge about particulate matter may be new to you.

As knowers, to what degree do we have a responsibility to share such knowledge?

Air pollution management strategies

Various management and intervention strategies can be used to reduce urban air pollution at different levels. They can be applied to different pollution sources and they have advantages and disadvantages.

- Alter the human activity that is producing the pollutant. This is the best place to start as it stops the problem before it is a problem. This is usually the least expensive way to approach air pollution and involves alternative technologies and changing lifestyles through campaigns, education, government legislation and economic incentives or deterrents.

- Control the release of the pollutant into the environment through regulatory measures imposed by legislation. This may be achieved by using technologies that extract the pollutant from emissions.

Here are a few possible urban air pollution strategies. As you read this section, think about whether each strategy is altering human activity or controlling release of the pollutant.

1. Improve public transportation such as buses, subways, trains, ferries and trams. Public transport systems are designed to move people efficiently and cheaply. They are often funded and operated by local governments. Improved public transport may decrease air pollution by:

   - using vehicles powered by alternative fuel sources such as electricity or renewable fuels

   - transporting more people using fewer vehicles, so there are fewer private cars on the road. By not using the car, a person can reduce their carbon emissions by up to 10 kg per day

   - reducing road congestion—because there are fewer cars on the roads, there is less congestion and thus lower emissions.

2. Improve the infrastructure for cycling. Many people choose not to cycle in cities because it is dangerous due to cars and bicycles using the same space. This can be solved by installing dedicated cycleways where cars are excluded. There are many benefits to this other than the obvious improvements in mental and physical health.

   - Human power is used instead of fuels so there are no harmful emissions. A moderate increase in bicycle use could save between 6 and 14 million tonnes of carbon dioxide emissions per year.

   - Noise pollution is reduced as bikes do not make much noise.

3. Plant more trees and other vegetation in urban areas. Vegetation improves the aesthetics and air quality of the city. PM, odours and gases such as ammonia, nitrogen oxides and sulfur dioxide settle on the leaves of trees, where they are absorbed via stomata. This cleans the air. Vegetation also reduces tropospheric ozone and releases oxygen.

▲ Figure 4 Dedicated cycleways reduce air pollution

4. Legislate for the installation of catalytic converters in all diesel or petrol fuelled vehicles. The exhaust fumes go through the catalyst-coated metal housing and up to 98% of the pollutants are removed. The catalyst—usually platinum or palladium (expensive and rare)—speeds up the chemical reactions between oxygen and the pollutants to convert them to less toxic by-products such as carbon dioxide, nitrogen gas and water vapour. However, catalytic converters are expensive, use non-renewable metals and are susceptible to theft.

5. Pedestrianize town centres, by banning all non-essential vehicles. This leads to a significant reduction in noise pollution and pollution from combustion of fossil fuels. It encourages walking as the environment is safer. It is also more pleasant, and it encourages social and cultural activities.

▲ Figure 5 Pedestrianized High Street in Newbury, Berkshire, UK

ATL Activity 29

ATL skills used: thinking, communication, social, research, self-management

Here are some more strategies that may help to reduce urban air pollution.

- Buy efficient, EV low-polluting vehicles, or cycle or walk.

- Carpool or limit car use.

- If using a car with an internal combustion engine, check tyre pressures, do not idle the engine, maintain the engine, and drive economically.

- Introduce tight fuel and emissions standards.

- Increase use of green walls, green barriers and street trees.

Work in groups.

1. Research how the strategies listed here help reduce urban air pollution.

2. Add any other strategies that you think will reduce urban air pollution.

3. Complete Table 1 by adding all the strategies mentioned.

Alter human activity	Control release of the pollutant
Plant more vegetation	Catalytic converter legislation

▲ Table 1

4. Each person in the group selects one strategy from each side of the table and lists the advantages and disadvantages of that strategy.

5. Share your work with the class.

Acid deposition

Acid rain, or acid deposition, includes any form of precipitation with acidic components, such as sulfuric or nitric acid, that falls to the ground from the atmosphere.

There are two forms of acid deposition:

- wet deposition: the acid comes down in the form of rain, snow, fog or hail

- dry deposition: the acid comes down as ash or dry particles.

Acidity

Connections: This links to material covered in subtopics 1.3 and 2.2.

Acids are chemicals that are able to release hydrogen ions (H^+). The acidity of solutions is measured using the pH scale (Figure 6). On this scale, a pH value of 7 is neutral (pure water). Values below 7 indicate acidic solutions; values above 7 indicate basic (alkaline) solutions. The pH scale is not a linear scale, it is logarithmic. A solution with pH 2 is 10 times more acidic than a solution with pH 3.

Unpolluted rain is slightly acidic and has a pH of about 5.6. This is caused by the presence of carbon dioxide in the atmosphere. Precipitation is called acidic when its pH is below pH 5.6. Certain pollutants increase the acidification of rain, which can sometimes fall lower than pH 2.

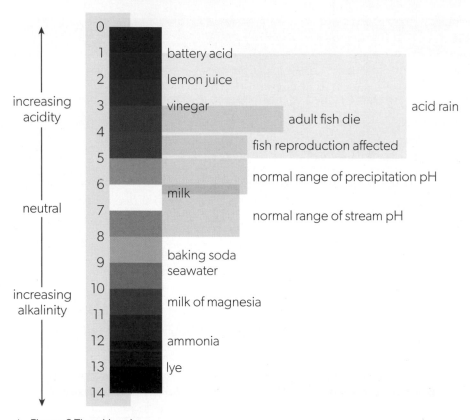

increasing acidity

neutral

increasing alkalinity

pH	
0	
1	battery acid
2	lemon juice
3	vinegar
4	adult fish die
5	fish reproduction affected
6	normal range of precipitation pH
	milk
7	normal range of stream pH
8	
9	baking soda / seawater
10	milk of magnesia
11	
12	ammonia
13	lye
14	

acid rain

▲ Figure 6 The pH scale

Main acid deposition pollutants and sources

Pollutants

The pollutants causing acid deposition may be primary or secondary.

- Primary pollutants are ones that leave the chimney of a factory or the exhaust pipe of a car, e.g. sulfur dioxide (SO_2) and nitrogen oxides (NOx).

- Secondary pollutants are made when primary pollutants react with water to form strong acids—sulfuric and nitric acids.

Carbon dioxide (CO_2) is also acidic but forms a weak acid—carbonic acid.

▲ Figure 7 Industrial air pollution

Sources

Naturally, sulfur dioxide is produced by volcanic eruptions and nitrogen oxides are produced by lightning.

The most important human activities that lead to the emission of these pollutants are the combustion of fossil fuels in motor cars, industry and power stations which use fossil fuels (coal, oil, gas) to produce steam to drive the turbines.

Sulfur dioxide is formed when sulfur-containing fuels are burned. Sulfur is common in coal and oil, but is usually absent in natural gas.

Nitrogen oxides are produced in combustion processes, partly from nitrogen compounds in the fuel, but mostly by direct combination of atmospheric oxygen and nitrogen at high temperatures.

If primary air pollutants remain in the atmosphere for a sufficiently long time, a variety of secondary air pollutants can be formed. Sulfur dioxide can react with oxygen from the atmosphere to form sulfur trioxide (SO_3). Both sulfur dioxide and sulfur trioxide can react with water to form sulfurous acid (H_2SO_3) and sulfuric acid (H_2SO_4) respectively. The nitrogen oxides can also react with water and form nitric acid (HNO_3). These secondary pollutants are very soluble in water, and are removed from the air by precipitation in the form of rain, hail and snow (wet deposition).

nitrogen dioxide + oxygen + water → nitric acid

$$4NO_2\,(g) + O_2\,(g) + 2H_2O\,(l) \rightarrow 4HNO_3\,(aq)$$

sulfur dioxide + oxygen + water → sulfuric acid

$$2SO_2\,(g) + O_2\,(g) + 2H_2O\,(l) \rightarrow 2H_2SO_4\,(aq)$$

carbon dioxide + water → carbonic acid

$$CO_2\,(g) + H_2O\,(l) \Leftrightarrow H_2CO_3\,(aq)$$

Acid deposition damage became a focus of environmental attention in the early 1970s when Germany's Black Forest showed a dieback or "waldsterben". In this event, trees of all ages, both coniferous and deciduous, showed signs of physical damage. But before that, in the Industrial Revolution (which started in 1750), people noticed acidic rain and acid air in cities; the term "acid rain" was first used in 1872.

Effects of acid deposition

Acid deposition can have direct and indirect effects on soil, plants and water:

- direct effects—for example, acid falling on forests weakens tree growth; acid falling on lakes and ponds decreases the pH of the water and affects aquatic organisms

- indirect effects—for example, acid increases solubility of metal ions such as aluminium which are toxic to fish and plant roots; acid deposition causes leaching of nutrients.

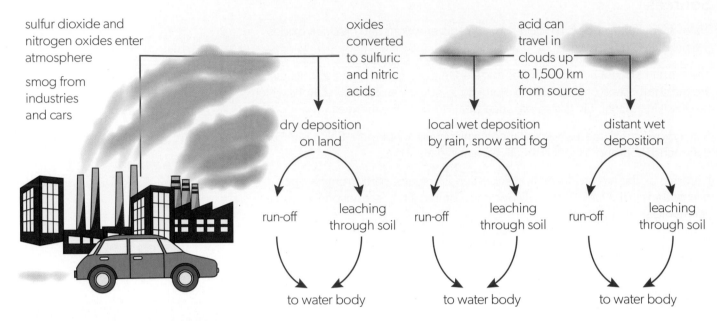

▲ Figure 8 The effects of acid deposition on soil, water and living organisms
Adapted from: Six Red Marbles and OUP

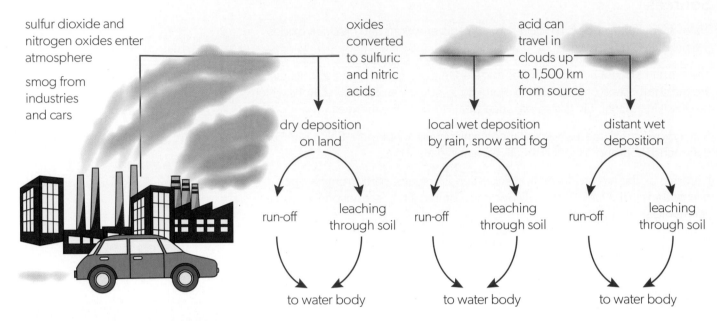

▲ Figure 9 Forest death from acid deposition

Effects of acid deposition on coniferous forests

Acid deposition affects coniferous forests in several ways.

1. Leaves and buds show yellowing (loss of chlorophyll) and damage in the form of lesions and thinning of wax cuticles.

2. These and other changes reduce growth, and allow nutrients to be leached out and washed away and pathogens and insects to gain entry.

3. Symbiotic root microbes are killed and this greatly reduces the availability of nutrients, further reducing tree growth.

4. It reduces the ability of soil particles to hold on to nutrients such as calcium, magnesium and potassium ions, which are then leached out.

5. It releases toxic aluminium ions from soil particles and these ions damage root hairs.

As a result of these effects, trees are weakened and may die.

Toxic effects of acid deposition

1. Aluminium ions affect fish and other aquatic organisms.

Aluminium is a common element in the soil. Acid precipitation decreases the pH of the soil, making aluminium more soluble. The aluminium released from the soil eventually ends up in streams and rivers. Fish are particularly sensitive to aluminium in water. At low concentrations, aluminium disturbs the fish's ability to regulate the amount of salt and water in its body. This inhibits the normal intake of oxygen and salt. Fish gasp for breath and the salt content of their bodies is slowly lost, leading to death. At higher concentrations, a solid is formed on the fish's gills, leading to death by suffocation. Apart from aluminium, other toxic metals can dissolve because of increased acidity.

2. Lichens

Lichens, which are a symbiotic pairing of an alga and a fungus (subtopic 2.1), are found growing on trees and buildings. They are particularly sensitive to gaseous pollutants like sulfur dioxide and are used as indirect measures of pollution (see Figure 3). Immediately downwind of a heavily polluting industrial region only a few tolerant species are found. These are indicator species of high levels of air pollution. As the distance from the source of pollutants increases, more and more species are able to survive. Tables of lichen indicator species are used to estimate pollution levels.

3. Nutrient removal effect on soil fertility

As described above, acid rain affects the soil by reducing the ability of soil particles to hold on to nutrients, such as calcium, magnesium and potassium ions. These nutrients are then leached out. Acid rain also inhibits nitrogen-fixing bacteria and so reduces their ability to add nitrate ions to the soil.

4. Buildings

Acid deposition affects human constructions. The acids dissolve marble, limestone (used in buildings and statues) and corrode steel, paintwork and other constructions.

Limestone buildings and statues (including many with great archaeological and historical value) react with acid and dissolve (Figure 10).

5. Peat bogs

Recent research has found that peat bogs affected by acid rain produce up to 40% less methane than before. This is because the bacteria that use the sulfates as a food source outcompete the ones that produce methane. This reduction in methane production reduces methane, a GHG, in the atmosphere.

6. Human health

Dry deposition is in the form of $PM_{2.5}$ of sulfates and nitrates and these penetrate into houses and into our lungs. Premature deaths from lung disease such as asthma and bronchitis can result from this.

 ## Practical

Skills:

- Tool 1: Experimental techniques
- Tool 2: Technology
- Inquiry 1: Exploring and designing
- Inquiry 2: Collecting and processing data
- Inquiry 3: Concluding and evaluating

The impacts of acid deposition can be a good internal assessment topic. You can investigate the impact of acid deposition on plants by measuring factors such as growth rates, germination rates, or rate of photosynthesis.

Design and conduct an experiment to measure the impact of acid deposition on plants.

▲ Figure 10 Effect of acid deposition on a limestone sculpture—1908 (top), 1969 (bottom)

Regional effects of acid deposition

The effects of acid deposition are regional, in contrast to climate change or ozone depletion, which are global. This is because, before the pollutants can spread over long distances, they return to the surface as dry or wet precipitation. The acids seldom travel further than a few thousand kilometres.

- Dry deposition usually occurs quite close to the source of the acidic substances. It consists of sulfur dioxide, sulfur trioxide and nitrogen oxides.

- Wet deposition occurs at slightly longer distances from the sources of the primary pollutants. It consists of sulfurous acid, sulfuric acid and nitric acid.

In general, the areas downwind of major industrial regions are strongly affected (Figure 11). For example, Scandinavian forests and lakes are mainly affected by acid rain originating in the UK, Poland and Germany, brought over by prevailing south-westerly winds. Industrial pollution from the USA is blown by prevailing winds towards Canadian forests. Industrialization in China affects Southeast Asia.

Soils and bodies of water are most often affected by acid rain. However, the impact of acid rain depends very much on the geology of the area on which it falls. Acid rain does little harm to soils derived from calcium carbonate rocks (limestone and chalk) because these soils are alkaline and neutralize (or buffer) the acids. However, acid (non-alkaline) rocks produce soils that are very sensitive to acid rain. Acid rain leaches out minerals from these soils. This reduces biodiversity and run-off affects nearby lakes.

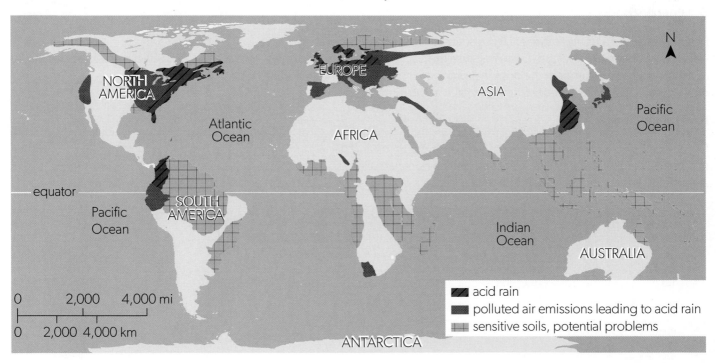

▲ Figure 11 Regions of the world with most acid deposition impacts

Pollution management strategies for acid deposition

As discussed under water pollution management strategies, pollution can be approached at different levels—shown in Table 2.

(ATL) Activity 30

ATL skills used: thinking, communication, research, self-management

Choose one of the following:

a. Canada affected by acid deposition from the USA

b. Sweden and Norway affected by acid deposition from Poland, Germany, and the UK

c. China.

1. Research the effects of acid deposition in your chosen area.

2. Research the effectiveness of intergovernmental agreements or legislation to reduce the effects of acid deposition.

3. Draw a poster to show the impact and pollution management strategies.

Reducing the impacts of acid deposition

1. Liming lakes to neutralize acidity

By 1990, over 400 Scandinavian lakes were virtually lifeless; the loss of many fish and invertebrate species was linked to high levels of lake acidity. In the 1980s, Sweden experimented with adding powdered limestone to lakes and rivers. However, results were mixed. The pH of treated lakes rises quickly, but this is short-lived because incoming water is still acidic—liming treats the symptoms and not the cause.

Biodiversity was not immediately restored; the lime seemed to affect nutrient balance as nutrients other than calcium were absent.

2. Reducing emissions

One way to reduce the emission of sulfur dioxide and nitrogen oxides is to reduce the combustion of fossil fuels by:

• reducing the need for electricity

• reducing car use and developing more efficient or electric-powered cars

• switching to alternative energy sources, such as wind or solar energy

• switching to biofuels, though this may add to the loss of food crops and increase the malnutrition problems discussed in subtopic 5.2

• switching to nuclear-powered electricity generation, to reduce SO_2 and NOx emissions.

Strategy for reducing pollution	Example of action	Evaluation
Alter the human activity producing pollution	Reduce fossil fuel use by using alternatives: • bioethanol to run engines in vehicles or planes • renewable energy sources for electricity.	Such measures also reduce CO_2 emissions but we live in a fossil fuel reliant economy. Demand for power is ever increasing, particularly in India and China as they industrialize.
	Reduce overall demand for electricity—education campaigns to turn lights off, insulate houses. Use less private transport—carpool, public transport, walk, cycle.	
	Use low sulfur fuels, remove sulfur before burning, or burn mixed with limestone.	
Regulate and reduce pollutants at the point of emission	Clean-up technologies at "end of pipe" locations (points of emission), e.g. scrubbing in chimneys to remove sulfur dioxide.	Expensive and costs passed on to consumer. Catalysers are cost-effective if well maintained but expensive to buy.
	Catalytic converters convert nitrous oxides back to nitrogen gas.	
Clean up and restore damaged sites	Lime fields, acidified lakes and rivers. Recolonize damaged areas.	Effective in restoring pH but has to be repeated regularly. Costly. Affects biodiversity in other ways. Treats symptoms and not the cause. Agreements are difficult to establish and to monitor.
	Lime forestry plantations. Trees acidify soils as they remove nutrients.	
	International agreements.	

▲ Table 2 Pollution management strategies for acid deposition

ATL Activity 31

ATL skills used: thinking, communication

Table 2 shows both positive and negative impacts of the different strategies.

Write a short essay entitled: Evaluation of strategies for reducing impacts of acid deposition.

3. Precombustion techniques

These techniques aim to reduce SO_2 emissions by removing sulfur from fuel before combustion. The sulfur removed from fuel can be obtained in several useful forms:

• as the element sulfur which can be used in the chemical industry

• as gypsum, which can be used in construction

• as sulfur dioxide which can be used in the production of sulfuric acid, one of the most commonly used chemicals.

4. End of pipe measures

End of pipe measures remove sulfur dioxide and nitrogen oxides from waste gases. Examples include waste gas scrubbers in electricity plants which remove sulfur dioxide, or catalytic converters in motor cars which remove nitrogen oxides, together with other pollutants.

(ATL) **Activity 32**

ATL skills used: thinking, communication, research

Figure 12 shows the progress made in reducing wet deposition in the USA between 1983 and 1997.

1.   Research the most recent situation in terms of acid deposition in the USA.

2.   Select another country and compare their progress with that of the USA.

3.   Evaluate the success (and/or failure) of one strategy to reduce the effects of acid deposition.

sulfate wet deposition

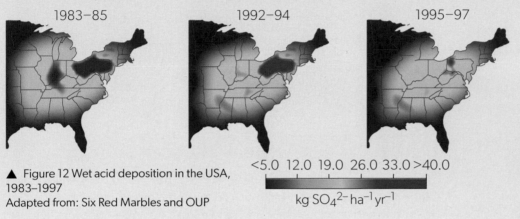

▲ Figure 12 Wet acid deposition in the USA, 1983–1997
Adapted from: Six Red Marbles and OUP

<5.0 12.0 19.0 26.0 33.0 >40.0

kg SO_4^{2-} ha^{-1} yr^{-1}

Photochemical smog

Photochemical smog is a brown haze common in some cities. It forms when ozone, nitrogen oxides and volatile organic compounds (VOCs) from the combustion of fossil fuels react in sunlight to produce a toxic mixture of ozone, nitric acid, aldehydes and peroxyacyl nitrates (PANs). Photochemical smog is mainly nitrogen dioxide and ozone but is a complex mixture of about 100 different primary and secondary air pollutants. The biggest contribution to photochemical smog is from motor vehicle exhausts in cities.

Connections: This links to material covered in subtopic 6.4.

The formation of photochemical smog

On warm, sunny days with lots of traffic, photochemical smog can be formed over cities. Although usually associated with the combustion of fossil fuels, forest burning can also contribute to photochemical smog.

In Kalimantan, Indonesia, forest fires cause smog over much of Southeast Asia. The 1997 and 2019 fires were particularly bad but there are now fires every year, especially in El Niño years.

Complex reactions create many chemicals in photochemical smog including VOCs, PANs, ozone, aldehydes, carbon monoxide and nitrous oxides. Highly reactive VOCs oxidize nitrogen oxide into nitrogen dioxide but it doesn't break down any ozone molecules. This process causes a build-up of ozone near ground level and smog formation.

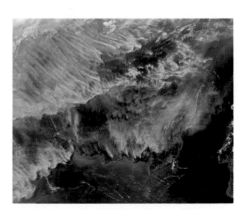

▲ Figure 13 Satellite photos of smoke from forest fires over Kalimantan, September 2019

AHL

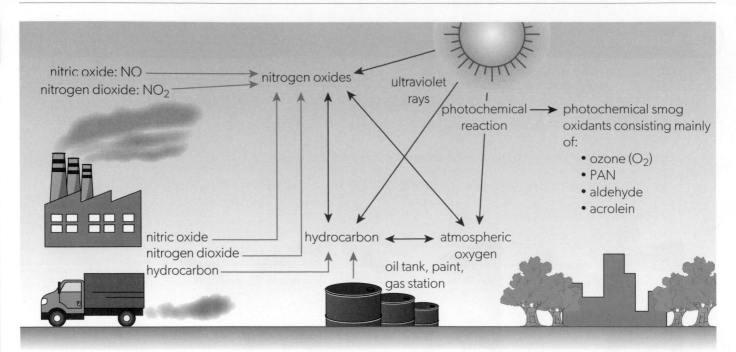

▲ Figure 14 Formation of photochemical smog

Because nitrogen dioxide is an important component of smog, smog can be seen as a brown hue above the city. All the chemicals in smog are strongly oxidizing and affect materials and living things. At higher concentrations, smog can cause coughs and decreased ability to concentrate.

Even though the main primary pollutants—nitrous oxides and hydrocarbons—reach a maximum concentration during the morning and evening rush hours, photochemical smog is at its maximum in the early afternoon. This is because the important smog-causing reaction is a photochemical reaction, so it reaches its peak in the afternoon sun (Figure 15).

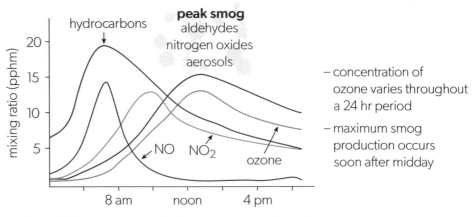

▲ Figure 15 Concentrations of main components of photochemical smog in a day

▲ Figure 16 Photochemical smog over Los Angeles

Because photochemical smog first caused problems in Los Angeles, it is often called Los Angeles-type smog. Other cities that frequently suffer from this type of smog are Santiago, Mexico City, Rio de Janeiro, Sao Paulo, Beijing and Athens.

The occurrence of photochemical smog is governed by a large number of factors, including local topography, climate, population density, and fossil fuel use.

Smog is most often formed over large cities that are low-lying or in valleys. The hills or mountains surrounding these cities take away most of the wind and on warm, calm days severe smog can occur.

Thermal inversion makes things worse. Normally, air over cities is relatively warm and has a tendency to rise. On warm days, however, an even warmer layer of air on top of the warm polluted air can prevent this air rising, trapping the pollution at ground level (Figure 17). This occurs most often in warm, dry climates. Weather plays an important role in the disappearance of smog: rain cleans the air of pollutants while winds can disperse the smog.

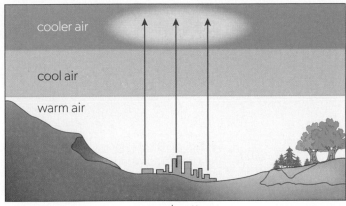

▲ Figure 17 Thermal inversion trapping smog over the city

Under the above conditions the concentration of pollutants can reach harmful or even lethal levels. Smog does not only affect life in the city itself. Often, smog is blown out of the city by the wind and causes damage in the countryside, sometimes up to 150 km away from the city where the smog was formed.

(ATL) Activity 33

ATL skills used: thinking, communication, research, self-management

Research photochemical smog in Santiago, Mexico City, Rio de Janeiro, Sao Paulo, Beijing, Athens, or a city close to you. Investigate the following.

1. The causes of photochemical smog specific to that location.

2. The extent of the problem in terms of:

   a. the area impacted

   b. changes over time—has the problem worsened

   c. diurnal changes in urban air pollutants

   d. measures being taken to reduce the impacts.

3. Use a statistical tool to test the significance level of change over a period of time.

Tropospheric ozone

Most people think of ozone as a "good" gas because it protects us from the Sun. However, the ozone in the troposphere—about 10% of the ozone in the atmosphere—is not good for life. According to the EPA, ozone levels above 0.7 ppm for over 8 hours are unhealthy. Ozone is also a GHG with a global warming potential 2,000 times that of carbon dioxide.

Connections: This links to material covered in subtopics 1.3, 2.4 and 6.1.

Practical

Skills:

- Tool 2: Technology
- Tool 3: Mathematics

- Tool 4: Systems and models
- Inquiry 1: Exploring and designing

Figure 18 $PM_{2.5}$ for HICs, MICs, LICs
Data from: The World Bank (CC BY 4.0)

1. Study Figure 18. Explain why MICs and LICs have higher $PM_{2.5}$ emissions than HICs.

2. Predict what the future trends will be for LICs, MICs and HICs.

This variation can form the basis of practical work or an individual investigation. There are plenty of apps that can be used to check the air quality index (AQI) of your own area and many other places in the world. You can check the AQI three times a day for a week or longer, or you can measure AQI in other ways.

3. Design an experiment to investigate changes or variation in the AQI.

Formation of tropospheric ozone

The pollutants emitted during the combustion of fossil fuels include hydrocarbons and nitrogen oxides. Hydrocarbons are emitted because not all the fuel is combusted. Nitrogen oxides are formed when oxygen and nitrogen (both originating from air) react as a result of high temperatures during combustion reactions.

Nitrous oxide contributes to the formation of tropospheric ozone (O_3). First, nitric oxide (NO) reacts with oxygen to form nitrogen dioxide (NO_2)— a brown gas that contributes to urban haze. Other pollutants like hydrocarbons and carbon monoxide accelerate the formation of nitrogen dioxide. When NO_2 absorbs sunlight, it breaks up into nitric oxide and oxygen atoms. These oxygen atoms subsequently react with oxygen molecules, forming ozone.

Under normal conditions, most ozone molecules oxidize nitric oxide back into nitrogen dioxide, creating a virtual cycle that leads to only a very slight build-up of ozone near ground level.

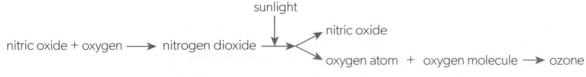

▲ Figure 19 Formation of tropospheric ozone

Impacts of tropospheric ozone

Tropospheric ozone is a major component of photochemical smog. It is a toxic gas and a powerful oxidant. Ozone and particulate matter have direct and indirect impacts on Earth—biological, societal and economic.

Impacts on biological elements of the environment include:

- irritation of the respiratory system, causing coughing and wheezing, sore throat, asthma, pulmonary disease and possibly lung cancer

- reduced lung function caused by the muscles in the respiratory system constricting—this makes breathing difficult and reduces vigorous activity

- increased susceptibility to respiratory infections such as bronchitis and emphysema

- eye irritation

- weakened immune system

- inhibited plant growth—ozone, even in small quantities, degrades chlorophyll and can reduce or stop photosynthesis in plants

- damage to plant cuticles and membranes—tobacco, tomato and spinach plants are particularly sensitive to ozone

- necrosis on the upper surfaces of leaves on trees, affecting their productivity.

Ozone also has negative impacts on inorganic materials such as rubber, cellulose, paint, plastics, fabrics and metals:

- damages the materials and decreases their aesthetic appearance

- reduces elasticity in rubber and fabrics

- reduces the lifetime of car tyres

- shortens the lifespan of all the listed materials

- bleaches fabrics.

This has economic implications as it increases the costs of maintenance, upkeep and replacement.

Reducing the impacts of tropospheric ozone

As with other forms of pollution, strategies for reducing tropospheric ozone are to:

- alter human activity producing it

- regulate and reduce at point of emission

- clean up and restore.

ATL Activity 34

ATL skills used: thinking, communication, social, research, self-management

Listed here are some direct impacts of tropospheric ozone.

Research:

- additional direct impacts

- indirect impacts—consider the healthcare system, the workforce, outdoor activities, food production.

As a class produce an infographic about the impacts of tropospheric ozone.

You might wish to check the AQI for your area and take steps to reduce your exposure. This could be by:

- staying indoors

- doing no sporting activity when the AQI indicates that tropospheric ozone levels are high

- using air filters in your home or a suitable respirator if you have to go outside

- avoiding burning wood fires, candles, incense—all increase $PM_{2.5}$

- not smoking.

Check your understanding

In this subtopic, you have covered:

- the causes, sources, impacts and management of urban air pollution

- causes, impacts and management of acid deposition

- photochemical smog formation

- the anthropogenic, meteorological and topographical factors that cause or intensify photochemical smog formation

- direct and indirect biological, physical, societal and economic impacts of tropospheric ozone.

AHL

How can urban air pollution be effectively managed?

1. Draw a systems diagram of urban air pollution.

2. State the sources of natural and anthropogenic air pollutants.

3. Identify the most common activity that causes urban air pollution.

4. Discuss the different management and intervention strategies that can be used to reduce urban air pollution.

5. Outline how acid rain is produced.

6. Describe the impacts of acid deposition.

7. Explain the management and intervention strategies used to minimize the impacts of acid deposition.

8. Describe the formation of photochemical smog.

9. Explain the meteorological and topographical factors that intensify the formation of photochemical smog.

10. Outline the direct biological and physical impacts of tropospheric ozone.

11. Discuss the extent to which tropospheric ozone impacts society and the economy.

AHL

▶▶ Taking it further

- Investigate and debate the causes and consequences of urban air pollution in the local environment, and strategies that could be used to reduce pollution.

- Participate in citizen science air quality projects by installing a networked weather station in your school.

- Advocate for improved walking and cycling options for your school.

Exam-style questions

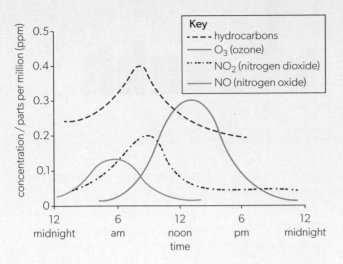

▲ Figure 1 Concentration of atmospheric pollutants associated with photochemical smog

1. a. Identify **one** primary pollutant from the pollutants shown in **Figure 1**. [1]

    b. Outline why the pollutant named in (a) is referred to as a primary pollutant. [1]

    c. Outline **one** reason why there is an increase in nitrogen oxides and hydrocarbons early in the day. [1]

    d. Explain the changes in ozone concentration over the period shown in **Figure 1**. [3]

    e. State **one** environmental impact of the accumulation of ozone shown in **Figure 1**. [1]

    f. Outline **two** local conditions that may increase the severity of photochemical smog. [2]

    g. Outline the role of catalytic converters in reducing photochemical smog. [1]

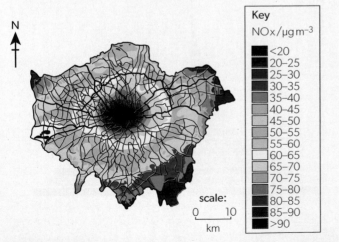

▲ Figure 2 Annual mean oxides of nitrogen (NOx) concentrations measured in London air, 2016
Data from: Greater London Authority

2. a. With reference to **Figure 2**, explain why the highest levels of NOx are found in the centre of London. [3]

    b. Evaluate **one** strategy to reduce NOx emissions from transport. [3]

    c. Identify **two** potential impacts of improved air quality on London and its population. [2]

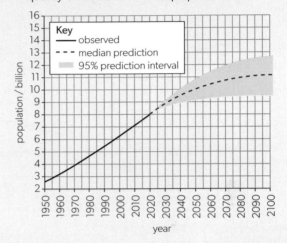

▲ Figure 3 World population figures, 1950–2019, and predictions, 2020–2100
Data from: World Population Prospects, United Nations, DESA, Population Division (CC BY 3.0 IGO)

3. a. Using **Figure 3**, identify the year in which the median prediction of the world population will reach 10 billion. [1]

    b. Outline **one** reason for the uncertainty in predicting the world's population in **Figure 3**. [1]

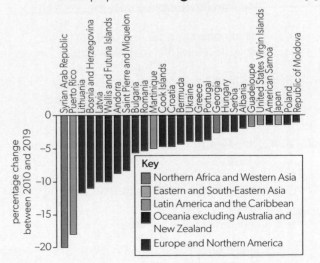

▲ Figure 4 Countries and regions where population decreased by at least 1% between 2010 and 2019
Adapted from: United Nations, Department of Economic and Social Affairs, Population Division (2019). World Population Prospects 2019. (CC BY 3.0 IGO)

c. Using **Figure 4**, identify the region that has the most countries with a decrease in the percentage change in population between 2010 and 2019. [1]

d. Outline **two** factors that could contribute to a reduction in population in the countries in **Figure 4**. [2]

e. Discuss how a country's stage in the demographic transition model (DTM) might influence its national population policy. [4]

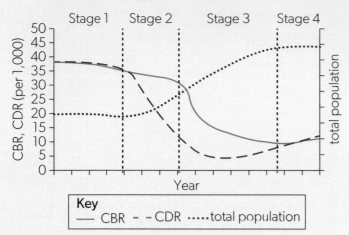

▲ Figure 5 Demographic transition model

4. Costa Rica has a crude birth rate (CBR) of 15.3 and a crude death rate (CDR) of 4.8.

a. Identify the stage in which Costa Rica would be placed on the demographic transition model shown in **Figure 5**. [1]

b. Calculate the natural increase rate (NIR) for Costa Rica. [1]

c. Calculate the doubling time for Costa Rica. [1]

d. Outline **one** strength and **one** limitation of the demographic transition model. [2]

e. Outline the socio-economic factors that may cause a society to move from Stage 2 to Stage 3 on the demographic transition model. [3]

5. a. Using **Figure 6** identify the region with the highest fertility rate in the period 2005–2010. [1]

b. Outline **two** possible reasons for the projected change in total fertility rate in Sub-Saharan Africa in the period 2045–2050. [2]

c. Identify **two** reasons for the projected increase in total fertility rate in Europe by the period 2045–2050. [2]

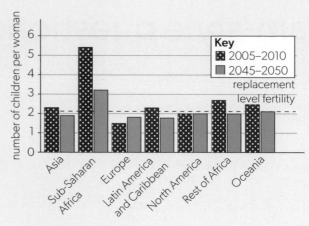

▲ Figure 6 Current and projected total fertility rates by region
Data from: World Resources Institute

6. Explain the causes and effects of acid deposition on natural ecosystems. [7]

7. With reference to **Figure 7**, outline **two** reasons for differences between the age–gender pyramids for Indonesia and Timor-Leste. [2]

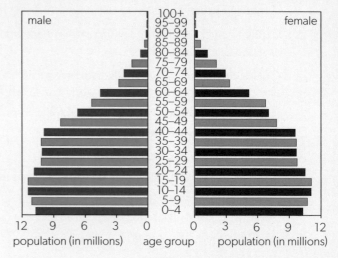

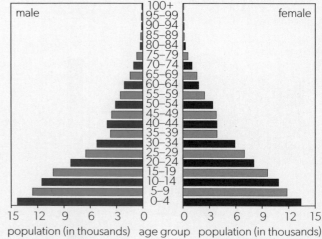

▲ Figure 7 Age–gender pyramids for Indonesia (top) and Timor-Leste (bottom), 2016
Data from: Central Intelligence Agency

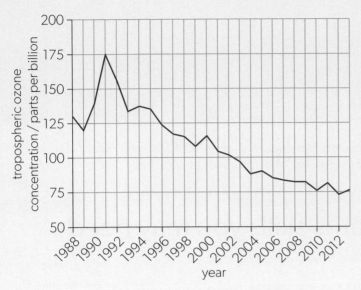

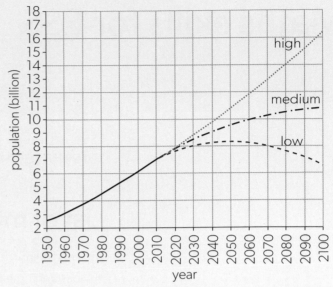

▲ Figure 8 Tropospheric ozone levels in Mexico City
Adapted from: Gobierno de la Ciudad de México /
Calidad del Aire

▲ Figure 10 Three projections for world population from the
present day to 2100. The three lines indicate the high, medium
and low projections for population size
Adapted from: Population Matters. Data from: United Nations

8. a. With reference to **Figure 8**, calculate the difference
        between the highest concentration and lowest
        concentration of tropospheric ozone.   [1]

    b. State **two** factors necessary for the chemical
        formation of ozone in the troposphere.   [2]

    c. Outline why a high concentration of ozone in the
        troposphere is a direct problem for humans, while
        in the stratosphere it is a benefit to humans.   [2]

    d. Suggest possible reasons for the overall trends of
        tropospheric ozone levels in **Figure 8**.   [4]

9. From the shape of the age–gender pyramid in **Figure 9**,
    suggest how the population in Brazil is likely to change
    in the next 30 years.   [3]

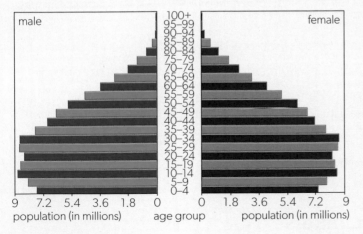

▲ Figure 9 Age–gender pyramid for Brazil, 2014
Data from: Central Intelligence Agency

10. a. Using **Figure 10**, calculate the range between
        the highest and lowest projected population
        size for 2100.   [1]

    b. Identify **two** factors that could explain the variation
        in the projected population growth for the world.   [2]

    c. Outline **one** economic implication of the highest
        projection for world population being realized.   [1]

    d. Outline **one** environmental implication of the
        highest projection for world population being
        realized.   [1]

    e. Outline **one** advantage of modelling future
        human population sizes.   [1]

    f. Outline **one** disadvantage of modelling future
        human population sizes.   [1]

11. Outline **four** ways in which urbanization may
    influence processes in the hydrological cycle.   [4]

12. Explain how acid deposition falling on a forest
    may impact a nearby aquatic ecosystem.   [7]

13. a. Outline how demographic tools can be used
        to study a human population.   [4]

    b. Urban air pollution can become a problem as
        human populations develop. Evaluate urban air
        pollution management strategies at the three
        levels of intervention.   [7]

Maths skills for ESS

ESS is not a maths course. However, you may need to do a few calculations as you go through the course and when you work on your individual investigation.

List of maths skills you may need

You should know:

- basic arithmetic functions—adding, subtracting, multiplying and dividing

- about approximations, reciprocals and frequencies

- how to calculate fractions and decimals, percentages and ratios

- how to calculate the mean, median and mode

- relevant SI units for time, length, area and volume

- standard notation (e.g. 2×10^6 m = 2 million metres)

- about direct and inverse proportion: direct proportion is when one variable increases, so does the other; inverse proportion is when one variable increases, the other decreases

- how to draw graphs of various types and interpret the data, including the significance of gradients and changes in gradients, intercepts and areas

- how to interpret data in various forms: scatter plot, point-to-point line, line of best fit, box-and-whisker plot, bar chart, stacked histogram, pie chart, kite diagram

- how to evaluate data through statistical tests such as standard deviation, chi-squared test, t-test, analysis of variance (ANOVA), correlation coefficient and Spearman's rank

- how to calculate Simpson's reciprocal diversity index

- how to calculate the Lincoln index for estimating population size of motile organisms

- how to calculate natural increase rates of populations and population doubling times.

You do not have to memorize equations.

Some basic maths

It is assumed that you are able to add, subtract, multiply and divide numbers in your head, on paper, or using a calculator or smartphone.

An **approximation** is anything that is similar to but not exactly equal to something else. You can approximate a number by rounding it up or down.

Example:

Round 95,637 to the nearest 10, 100 and 1,000.

95,637 to the nearest 10 is 95,640 (if last digit is 0–4, round down, if 5–9, round up); to the nearest 100 is 95,600; to the nearest 1,000 is 96,000.

Fractions, decimals and percentages are ways of showing numbers that are parts of a whole. You will have learned to convert between fractions, decimals and percentages in your maths lessons.

Example:

$$\frac{1}{4} = 0.25 = 25\%$$

A **ratio** shows how much there is of one thing compared with another, usually written as $a:b$. For example, if you mix oil and water with 1 part oil to 4 parts water, the ratio is 1:4.

Frequency is the number of times a value occurs in a set of data. If 5 of your class gained a grade 7, the frequency of grade 7 is 5.

The **mean** (average) of a data set is found by adding all the values in the data set and then dividing by the number of values in the set.

The **median** is the middle value when a data set is ordered from least to greatest.

The **mode** is the value that occurs most often in a data set—the most common number.

Example:

data set: 3, 4, 5, 5, 5, 8, 8, 11, 15, 16

mean: $\dfrac{3 + 4 + 5 + 5 + 5 + 8 + 8 + 11 + 15 + 16}{10} = \dfrac{80}{10} = 8$

median: middle values are 5 and 8 so median is $\dfrac{5 + 8}{2} = 6.5$

mode: most common value is 5

Units and notation

Units

The International System of Units (SI)—also known as the metric system—sets international standards for measurement.

Table 1 shows some units used in ESS.

Unit	Examples	Symbol(s)
time	second, minutes, hour, year	s, min, h, yr
length	micrometre, millimetre, metre	μm, mm, m
area	metre squared, hectare	m^2, ha (= 10,000 m^2)
volume	cubic metre	m^3
mass	milligram, gram, kilogram, tonne	mg, g, kg, t

◀ Table 1 Units used in ESS

Prefixes

Table 2 shows prefixes, used to indicate multiples of the original unit (e.g. 1 kilometre = 1,000 metres).

Prefix	Base 10	Symbol
peta	10^{15}	P
giga	10^{9}	G
kilo	10^{3}	k
milli	10^{-3}	m
micro	10^{-6}	μ

▲ Table 2 Unit prefixes

Combinations of units

You will come across various combinations of units.

Common examples are:

- in productivity—gram per metre squared per year = $g/m^2/yr$ or $g\ m^{-2}\ yr^{-1}$

- in the carbon cycle—gigatonnes of carbon per year = GtC/yr or GtC yr^{-1}

Other abbreviations

Particulate matter is pollution by particles in the air.

PM_{10} are particles with diameters generally 10 μm and smaller.

$PM_{2.5}$ are fine inhalable particles 2.5 μm in diameter and smaller.

(A human hair is about 70 μm in diameter.)

1 Gt (gigatonne or gigaton) = 1 billion metric tonnes = 10^{9} tonnes.

Direct and inverse proportion

In **direct proportion**, the ratio between two quantities remains the same.

In **inverse** or indirect proportion, as one quantity increases, the other decreases. For example, as the speed of a vehicle increases, the time taken to complete a journey decreases. Speed and time are inversely proportional.

Graphs

Throughout this book, there are many examples of graphs. Scan through the book now to remind yourself of the different types and the information they show.

Steps in graph plotting

You may draw graphs using any spreadsheet software or by hand. Whatever method you choose, follow these steps.

Step 1: Identify the variables. A variable is something that can change in an investigation.

- The independent variable is the one the investigator changes (e.g. temperature, light intensity, distance from the sea).

- The dependent variable is the one that changes because the independent variable has changed.

- A controlled variable is one that is kept constant.

    Plot the independent variable (the one controlled by the investigator) on the x-axis—the horizontal axis.

- Plot the dependent variable on the y-axis—the vertical axis.

Step 2: Determine the variable range and scale of the graph.

Step 3: Label each axis, including units, and give the graph a title

Step 4: Determine the data points and plot on the graph.

Step 5: Draw the graph. It may be a straight line (direct proportion), nearly a straight line (draw a line of best fit) or a curve (draw line freehand).

The **gradient of a line** is a measure of the steepness of a straight line.

The **gradient** of a graph indicates the rate of change of one measure against another. The steeper the gradient, the higher the value of the rate of change (increasing). If a gradient has a negative value, then the rate of change is negative (decreasing).

Figure 1 shows a straight line with a positive gradient. To find the gradient of a curved graph, you would need to draw a tangent to the curve.

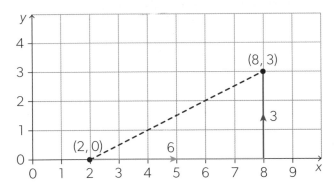

▲ Figure 1 This line has a positive (uphill) gradient

Linear and non-linear relationships

A linear relationship creates a straight line when plotted on a graph. A non-linear relationship creates a curve when plotted on a graph: changes in output are not proportional to changes in any of the inputs.

Intercepts

The x-intercept is the point where a line crosses the x-axis, and the y-intercept is the point where a line crosses the y-axis.

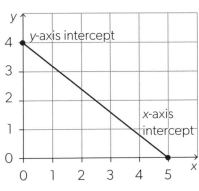

▲ Figure 2 Intercepts

Scatter plots

To see if there is a relationship between two variables—for example, air temperature and ice cream sales—you can use a scatter plot. Each pair of values (e.g. air temperature and number of ice creams sold on a particular day) is plotted, allowing you to see if there is a pattern.

As Figure 3 shows, correlation may be:

- positive—as one variable increases, the other variable increases
- negative—as one variable increases, the other decreases
- not present—if there is no correlation, there is no clear relationship between the variables.

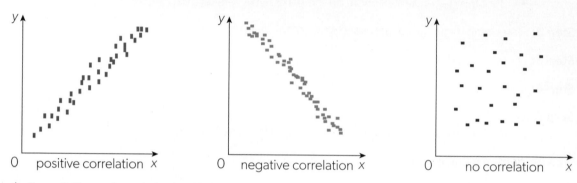

▲ Figure 3 Types of correlation in scatter plots

Point-to-point line

A **point-to-point line** joins adjacent points with straight lines (Figure 4). Graphs like this are used to show changes over time.

In a scatter graph, it is not appropriate to join the points like this—you should draw a **line of best fit** instead.

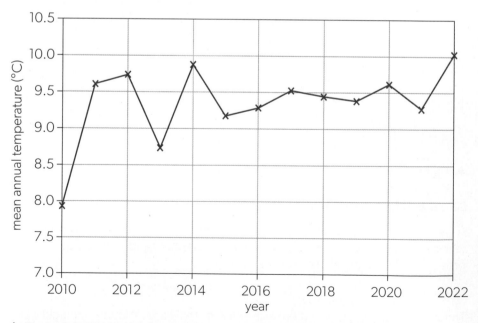

▲ Figure 4 Point-to-point showing UK mean annual temperature, 2010–2022

Box-and-whisker plot

A **box-and-whisker plot** is useful when there are outliers in the data. The line in the centre indicates the median; the edges of the box show the upper and lower quartiles; and the lines show the range of the data, to the maximum and minimum values. Any outliers are marked with small crosses or dots above or below the main plot. (In Figure 5, there is an outlier for site 2.)

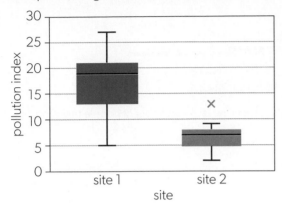

◀ Figure 5 Box-and-whisker plot to compare pollution levels at two sites
Adapted from: American Society for Quality

Categorical and qualitative data

The distinction between categorical and qualitative data is important when drawing graphs and using statistical tests.

Categorical data can be divided into groups. Examples of categorical variables are nationality, age groupings and educational level.

Quantitative data can be counted or measured in numerical values. Height, age in years, and distance are examples of quantitative data. Discrete quantitative data can take particular values only (e.g. shoe size); continuous quantitative data can take any value (e.g. wingspan).

Bar charts and histograms

A bar chart or bar graph is used to show categorical data. The bars may be horizontal or vertical. Bar length is proportional to the quantity of data. All bars are the same width and there are spaces between the bars.

Stacked bar charts (Figure 6) are more complex as each categorical group has two or more colour-coded stacks representing different groupings.

A histogram (Figure 7) is used to show quantitative data. The bars can be different widths and there is no space between them.

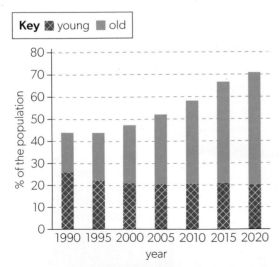

◀ Figure 6 Stacked bar chart showing changes in the dependency ratio in Japan, 1990–2020

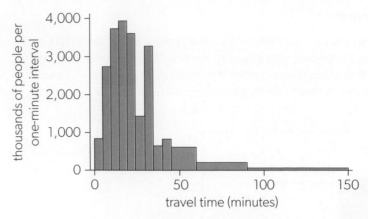

▲ Figure 7 Histogram of travel time to work in the USA vs number of people travelling

Pie charts

Pie charts are circular charts used to show categorical data. The segments show the relative sizes of the categories. Look through the book for examples of these. Pie charts may be hollow, in which case they are referred to as "doughnut" pie charts (Figure 8).

▶ Figure 8 Example of a doughnut pie chart
Adapted from Boltor / Wikimedia Commons
(CC BY-SA 3.0)

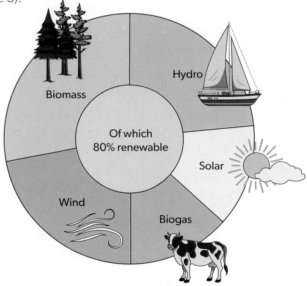

Kite diagrams

Kite diagrams (see subtopic 2.6) show observations at various points along a transect. They contain a great deal of information, including species presence or absence, species abundance, and distance along transect. Relationships between biotic and abiotic factors can be hypothesized from data in kite diagrams. However, they are difficult to draw accurately by hand and few software packages include them.

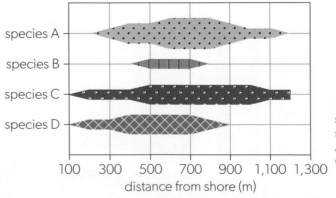

◀ Figure 9 A kite diagram showing abundance of four species along a transect
Adapted from: OriginLab Corporation

Statistical tests

There are many statistical tests and it can be difficult to know which to use. The main ones that you may need in ESS are:

- standard deviation
- chi-squared test
- *t*-test

- analysis of variance (ANOVA)
- correlation coefficient
- Spearman's rank.

The following guidance may help you to decide which test to use, but you can also ask your ESS teacher or your maths teacher. It is best to ask **before** you start collecting data, at the stage of designing your research question and methodology.

You can also find some useful information online—for example, on the websites Statistics How To, Social Science Statistics or Scribbr.

Standard deviation

Standard deviation (SD) is a measure of dispersement in statistics. "Dispersement" tells you how much your data is spread out. Specifically, SD shows you how much your data is spread out around the mean. The standard deviation equation is shown in Figure 10. You do **not** need to learn this.

$$s = \frac{\sum(x - \bar{x})^2}{n - 1}$$

▲ Figure 10 The standard deviation equation

The **normal distribution** curve (Figure 11) is a bell shape.

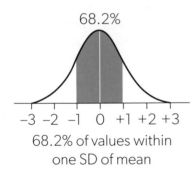
68.2%

−3 −2 −1 0 +1 +2 +3
68.2% of values within one SD of mean

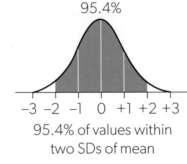
95.4%

−3 −2 −1 0 +1 +2 +3
95.4% of values within two SDs of mean

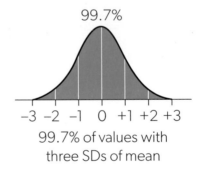
99.7%

−3 −2 −1 0 +1 +2 +3
99.7% of values with three SDs of mean

▲ Figure 11 Normal distribution and standard deviations
Adapted from: Investopedia / Sabrina Jiang

The mean or average is in the centre. Each segment represents one standard deviation away from the mean. For all normal distributions, 68.2% of the observations fall within plus or minus (+/−) one standard deviation of the mean; 95.4% of the observations fall within +/− two standard deviations; and 99.7% fall within +/− three standard deviations.

Data sets relating to height of an adult human population, blood pressure levels, or income level may show a normal distribution.

Normal distribution curves can be skewed to one end or the other.

Chi-squared test, χ^2

$$\chi^2 = \sum_i \frac{(O_i - E_i)^2}{E_i}$$

▲ Figure 12 Chi-squared equation

A chi-squared test is used to determine if there is a statistically significant association between, for example, the distribution of two species. It indicates whether a difference between two categorical variables is due to chance or to a relationship between the variables.

1. Start with two hypotheses: the null hypothesis is that there is no difference in the distribution of the two species; the alternative hypothesis is that there is a difference.

   Observed values (O) are those you obtain. Expected values (E) are the frequencies expected, based on the null hypothesis.

2. Produce a frequency table of observed results (O) and expected results (E). Then apply the chi-squared formula: find the sum of observed minus expected results squared $(O - E)^2$ and divide by the expected value (E).

3. Find an appropriate published **probability table**—a chart showing all possible outcomes of a situation under specified conditions and the likelihoods of occurrence.

4. Work out the "degrees of freedom" (D_f)—the number of independent values that can vary in the sample. To calculate D_f, subtract 1 from the number of items in the data sample.

$$D_f = N - 1 \text{ where } N \text{ is the sample size}$$

5. In the probability table, look up the p-value. The p value is a proportion. A statistically significant test result ($p \leq 0.05$) means that the test hypothesis is false or should be rejected. A p-value greater than 0.05 means that no effect was observed.

   If the calculated p-value is less than 0.05, the null hypothesis is considered to be false (nullified—hence the name "null hypothesis"). If the value is greater than 0.05, the null hypothesis is considered to be true.

The *t*-test

A ***t*-test** is a statistical test that compares the means of two samples. It is used in hypothesis testing, with a null hypothesis that the difference in group means is zero and an alternative hypothesis that the difference in group means is not zero.

The *t*-test is used to reduce subjective influence when testing a null hypothesis.

Use the *t*-test when you have collected a small, random sample from some statistical "population" and want to compare your sample mean with another value. The value for comparison could be a fixed value or the mean of a second sample.

To conduct a *t*-test, you calculate a p-value (similar to chi-squared testing). If the p-value is less than 0.05, then the result is said to be statistically significant. If the p-value is greater than 0.05, then the result is insignificant.

The *t*-value measures the size of the difference relative to the variation in your sample data. The greater the magnitude of *t*, the greater the evidence against the null hypothesis.

What is the difference between the chi-squared test and the *t*-test?

Both a chi-squared test and a *t*-test can test for differences between two groups.

A chi-squared test of independence is used when you have two categorical variables.

A *t*-test is used when you have a dependent quantitative variable and an independent categorical variable (with two groups).

ANOVA

Analysis of variance (ANOVA) is a statistical formula used to compare variances across the means (or average values) of different groups. ANOVA is a generalization of the *t*-test. The *t*-test is used to compare the means of two groups, whereas ANOVA is used to compare the means of three or more groups. It is testing groups to see if there is a difference between them.

A one-way ANOVA uses one independent variable, whereas a two-way ANOVA uses two independent variables.

Like the other tests, **ANOVA** is a way to find out if survey or experimental results are statistically significant—to identify whether to reject or accept the null hypothesis.

Correlation coefficients

Correlation coefficients are used to measure the strength of the relationship between two variables. Remember, however, that correlation does not necessarily indicate causation. For example, there is a strong correlation between ice cream sales and shark attacks, but this does not mean eating ice cream causes shark attacks! Instead, it is because, in warmer weather, more people buy ice cream and more people go into the sea.

There are several types of correlation coefficient, but a popular one is **Spearman's rank correlation coefficient**. This is commonly used in linear regression.

Linear regression is a type of statistical analysis used to predict the relationship between two variables. It assumes a linear relationship (straight line on a graph) between the independent variable and the dependent variable and aims to find the best-fitting line that describes the relationship.

Formulae used to calculate correlation coefficients give a value between −1 and +1 (Figure 13).

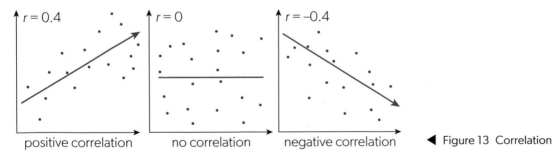

◀ Figure 13 Correlation

- A correlation coefficient of +1 indicates a strong positive relationship. For every positive increase in one variable, there is a proportional positive increase in the other. For example, shoe sizes go up in (almost) perfect correlation with foot length.

- A correlation coefficient of −1 indicates a strong negative relationship. For every positive increase in one variable, there is a proportional negative decrease in the other.

- A correlation coefficient of 0 means that there is no relationship between the two variables.

The **Spearman's rank correlation coefficient (ρ)** is used to discover the strength of a link between two sets of data.

While a scatter graph of two data sets may give a hint about whether the two have a correlation, Spearman's rank gives a numerical value on the degree of correlation (or non-correlation).

The Spearman's rank correlation has a value from +1 to −1 where:

- a value of +1 means a perfect association of rank

- a value of 0 means that there is no association between ranks

- a value of −1 means a perfect negative association of rank.

Simpson's reciprocal index

Species diversity is a function of the number of species and their relative abundance; it is a measure of species richness and species evenness.

Ecologists try to express diversity using a number. The higher the number, the greater the species diversity. This makes it possible to compare ecosystems or to see whether ecosystems are changing over time. The most common way to turn diversity into a number is by using **Simpson's reciprocal index** (see also subtopic 2.6).

$$D = \frac{N(N-1)}{\sum n(n-1)}$$

where:

D = Simpson's reciprocal index

N = total number of organisms of all species found

n = number of individuals of a particular species

Be careful. The term "Simpson's *diversity* index" (which you may read elsewhere) actually describes three related indices (Simpson's index, Simpson's index of diversity and Simpson's reciprocal index). Here we are using **Simpson's reciprocal index** in which 1 is the lowest value (when there is just one species) and a higher value means more diversity. The highest value is equal to the number of species in the sample. Values for the other indices range from 0 to 1.

Lincoln index

The **Lincoln index** (see also subtopic 2.6) is a statistical measure used to estimate the population size of a motile population. It is also called the capture–mark–release–recapture method.

Follow these steps to calculate the Lincoln index:

- In a defined and marked area, capture a sample of individuals. The number of individuals captured is n_1.

- Mark the captured individuals. Marking must not be easily removable and must not adversely affect the organism's survival or behaviour.

- Release the marked individuals and allow them to rejoin the population.

- Capture a second sample of individuals; the number captured is n_2.

- Count the number of marked individuals in the second sample; this is m_2.

Apply this formula:

$$N = \frac{n_1 \times n_2}{m_2}$$

Population calculations

See also subtopic 8.1.

Natural increase rate (NIR) is the birth rate minus the death rate of a population over time:

$$NIR = \frac{\text{number of births} - \text{number of deaths}}{\text{population}}$$

Population growth rate is the change in the number of individuals over a specific period of time.

$$GR = \frac{N}{t}$$

- GR = growth rate (measured as number of individuals)

- N = change in population

- t = time

The **doubling time** is the time it takes for a population to double in size.

$$\text{doubling time} = \frac{70}{\text{annual growth rate}} = \frac{70}{GR}$$

Example:

For a population with an annual growth rate of 3.5%, doubling time $= \frac{70}{3.5}$ = 20 years for the population size to double.

Index

p448: ridjin / Getty Images; p450: Simon McGill / Getty Images; p456: Luciano Queiroz / Shutterstock; p458: marcovarro / Shutterstock; p465: Adrian Reynolds / Shutterstock; p475: MintArt/Shutterstock; p476: guentermanaus / Shuttersock; p478: Shchipkova Elena / Shutterstock; p484 (t): Caleb Holder / Shutterstock; p484(b): Paula Bailey / Getty Images; p485: hxdbzxy / Shutterstock; p489(t): Frontpage / Shutterstock; p489(b): Toa55 / Shutterstock; p491: Wasu Watcharadachaphong / Shutterstock; p493(t): Denis Pogostin / Shutterstock; p493(b): photozi / 123RF; p494: Zapp2Photo / Shutterstock; p497(l): mikeledray / Shutterstock; p497(r): Alison Hancock / Shutterstock; p500(t): D. Kucharski K. Kucharska / Shutterstock; p500(m): Aurélia Zizo; p500(b): Eugene Troskie / Shutterstock; p504: MintArt / Shutterstock; p508: Chris McLoughlin / Getty Images; p515(t): Galyna Andrushko / Shutterstock; p515(b): Mimadeo / Shutterstock; p529: Netta Arobas / Shutterstock; p530: Matty Symons / Shutterstock; p531: Sawat Banyenngam / Shutterstock; p532(t): Julian W / Shutterstock; p532(b): Corradox / Wikimedia Commons (CC BY-SA 3.0); p534: Ververidis Vasilis / Shutterstock; p539: Avalon.red / Alamy Stock Photo; p551: connel / Shutterstock; p552: photographyfirm / Shutterstock; p553(tl): zulazhar / Shutterstock; p553(tm): SpaceKris / Shutterstock; p553(tr): Teun van den Dries / Shutterstock; p553(bl): ES_SO / Shutterstock; p553(bm): manfredxy / Shutterstock; p553(br): Scharfsinn / Shutterstock; p563: abidal / 123RF; p568: Nathalie Speliers Ufermann / Shutterstock; p571: Copernicus Atmosphere Monitoring Service / ECMWF; p578: ND700 / Shutterstock; p579: liubomir / Shutterstock; p580: Diego Delso, delso.photo (CC BY-SA 4.0); p586: timandtim / Getty Images; p589(l): Patty Chan/Shutterstock; p589(r): Lillac/Shutterstock; p591: Photodisc / Getty Images; p592(t): Daniel Prudek / Shutterstock; p592(m): Arindam Banerjee / 123RF.com; p592(b): Photodisc / Getty Images; p593: wirestock / Freepik Company; p597: joserpizarro / 123RF; p603(t): M. Unal Ozmen / Shutterstock; p603(b): W. L. Tarbert (CC BY-SA 3.0); p605: Frankvr / Shutterstock; p609: Bule Sky Studio / Shutterstock; p610(l): Brad Ingram / Shutterstock; p610(r): Richard Whitcombe / Shutterstock; p612: derejeb / Depositphotos; p613: I. Pilon / Shutterstock; p616: Nitr / Shutterstock; p626: Maxim Burkovskiy / Shutterstock; p628: majeczka / Shutterstock; p631(t): Rob Wilson / Shutterstock; p631(b): ZUMA Press Inc / Alamy Stock Photo; p632: Jim Parkin / Shutterstock; p633: John Carnemolla / Shutterstock; p634: Our World In Data / (CC BY 4.0); p637: BloombergNEF; p638(t): Bored Photography/Shutterstock; p638(b): Max Roser / Our World In Data (CC BY 4.0); p639: Energy Institute Statistical Review of World Energy (2023) / Our World In Data (CC BY 4.0); p640: Energy Institute Statistical Review of World Energy (2023) / Our World In Data (CC BY 4.0); p644: David Bleeker - London / Alamy Stock Photo; p647: Ulrich Mueller / Shutterstock; p648: Riza Azhari/Getty Images; p654: sunsetman / Shutterstock; p656: Shane Gross / Shutterstock; p657(tl): Picsfive / Shutterstock; p657(bl): diplomedia / Shutterstock; p657(r): Judith Collins / Alamy Stock Photo; p658: Huguette Roe / Shutterstock; p659: Muellek / Shutterstock; p660: Sander van der Werf / Shutterstock; p662: momente / Shutterstock; p670: Martin Puddy / Getty Images; p682: checy / Shutterstock; p685: Population Pyramid (CC BY 3.0 IGO); p686: Population Pyramid (CC BY 3.0 IGO); p687: Population Pyramid (CC BY 3.0 IGO); p688: Population Pyramid (CC BY 3.0 IGO); p689: Population Pyramid (CC BY 3.0 IGO); p696: Pictorial Press Ltd / Alamy Stock Photo; p699: Population Pyramid (CC BY 3.0 IGO); p700: Population Pyramid (CC BY 3.0 IGO); p704: Population Pyramid (CC BY 3.0 IGO); p705: Population Pyramid (CC BY 3.0 IGO); p707: Romaine W / Shutterstock; p713(t): Ralf Liebhold / Shutterstock; p713(b): Greenpeace; p716(t): Mike Harrington / Iconica / Getty; p716(b): Photo Spirit / Shutterstock; p720 (t): Frank Bach / 123RF; p720(b): SpeedKingz / Shutterstock; p721: Martin Shields / Alamy Stock Photo; p722: littleny / Shutterstock; p724: SAKARET / Shutterstock; p725: oksana.perkins / Shutterstock; p730: David Chapman / Alamy Stock Photo; p731: Jamie Hall / Shutterstock; p736: Andrew F. Kazmierski / Shutterstock; p737: Alpha and Omega Collection / Alamy Stock Photo; p738(l): Alicia Fox / Alamy Stock Photo; p738(r): Adrian Sherratt / Alamy Stock Photo; p741: zhuda / Shutterstock; p742: Luke Schmidt / Shutterstock; p743: Teresa Azevedo / Shutterstock; p744: Kevin George / Shutterstock; p745: Peter Titmuss / Shutterstock; p746: chuyuss/Shutterstock; p748: Aquila / Shutterstock; p749(t): TopFoto; p749(b): TopFoto; p753: NASA; p754: Photodisc / Getty Images; p760: Our World in Data (CC BY 4.0).

Artwork by Q2A Media, Aptara Inc., Six Red Marbles, David Mackin, Hardlines, Dave Russell, Barking Dog Art, GreenGate Publishing, Mark Walker, Wearset Ltd, Peter Bull Art Studio, HL Studios, James Stayte, Tech-Set Ltd, IFA Design, Clive Goodyer, Angela Knowles, Richard Morris, Martin Sanders, Mark Ruffle, Stéphan Theron, F-0876, Archivo Oxford, Steve Evans, Hart McLeod, Alan Rowe, Stefan Chabluk, Tim Jay, Peters & Zabransky, and Oxford University Press.

Extracts from IPCC Fourth Assessment Report: Climate Change 2007, published by Intergovernmental Panel on Climate Change. Reproduced with permission by Intergovernmental Panel on Climate Change.

Quotes by David Suzuki. Reproduced with permission by The David Suzuki Foundation.

Quote by Dr. Jane Goodall, from 5 Biodiversity Lessons from Dr Jane Goodall by Amy Richardson, published in The Future Forest Company. Reproduced with permission by Jane Goodall Institute.